Engineering Mechanics, Strength of Materials and Elements of Structural Analysis

Engineering Mechanics, Strength of Materials and Elements of Structural Analysis

Prof. C. Venkatramaiah
M.Sc.(Engg.), Ph.D.

Ex-Professor, Department of Civil Engineering,
Sri Venkateswara University College of Engineering, Tirupati

Formerly at Engineering College, Kakinada; and
College of Engineering, Guindy, Chennai

Prof. A.V. Narasimha Rao
B.E., M.Tech., Ph.D.

Professor, Department of Civil Engineering,
Sri Venkateswara University College of Engineering, Tirupati

CBS

CBS Publishers & Distributors Pvt. Ltd.

New Delhi • Bengaluru • Chennai • Kochi • Kolkata • Mumbai
Hyderabad • Nagpur • Patna • Pune • Vijayawada

ISBN: 81-239-1025-8

First Edition: 2003
Reprint: 2007, 2009, 2011, 2012, 2014, 2018

Published by **Satish Kumar Jain** and produced by **Varun Jain** for
CBS Publishers & Distributors Pvt. Ltd.,
4819/XI Prahlad Street, 24 Ansari Road, Daryaganj, New Delhi - 110002
delhi@cbspd.com, cbspubs@airtelmail.in • www.cbspd.com
Ph.: 23289259, 23266861, 23266867 • Fax: 011-23243014

Corporate Office: 204 FIE, Industrial Area, Patparganj, Delhi - 110 092
Ph: 49344934 • Fax: 011-49344935
E-mail: publishing@cbspd.com • publicity@cbspd.com

Branches:
• *Bengaluru:* 2975, 17th Cross, K.R. Road, Bansankari 2nd Stage,
  Bengaluru - 70 • Ph: +91-80-26771678/79 • Fax: +91-80-26771680
  E-mail: cbsbng@gmail.com, bangalore@cbspd.com
• *Chennai:* No. 7, Subbaraya Street, Shenoy Nagar, Chennai - 600030
  Ph: +91-44-26681266, 26680620 • Fax: +91-44-42032115
  E-mail: chennai@cbspd.com
• *Kochi:* Ashana House, 39/1904, A.M. Thomas Road, Valanjambalam,
  Ernakulum, Kochi • Ph: +91-484-4059061-65
  Fax: +91-484-4059065 • E-mail: cochin@cbspd.com
• *Kolkata:* 6-B, Ground Floor, Rameshwar Shaw Road, Kolkata - 700014
  Ph: +91-33-22891126/7/8 • E-mail: kolkata@cbspd.com
• *Mumbai:* 83-C, Dr. E. Moses Road, Worli, Mumbai - 400018
  Ph: +91-9833017933, 022-24902340/41 • E-mail: mumbai@cbspd.com

Representatives:

• Hyderabad: 0-9885175004	• Nagpur: 0-9021734563
• Patna: 0-9334159340	• Pune: 0-9623451994
• Jharkhand: 0-9811541605	• Uttarakhand: 0-9716462459

Printed at:
J.S. Offset Printers, Delhi (India)

Foreword

Professor C. Venkatramaiah is an excellent teacher. He is excellent in English with superb communicative skills. He had the uncanny understanding of his students of different intellectual capacities and made the whole group comfortable with his teaching. He taught many courses for students of other departments besides his own.

Professor A. V. Narasimha Rao has been known to be a sincere, systematic and hard working teacher.

Very few good books are there in this area from Indian authors, although there are a few excellent books from foreign authors.

The subject matter of this book is exhaustive and covers all that is required by students of many engineering disciplines in their first few basic courses. The exposition of the various concepts has been very lucid and the book has been written wholly in S.I. Units. The subject matter has been amply illustrated by incorporating a good number of well graded examples of wide variety. At the end of each chapter, a few practice problems have been added for students to solve them independently.

The large number of problems and objective questions included herein will help the student prepare for competitive examinations too.

The authors' long experience and maturity in teaching this subject have gone into the preparation of this book, which should prove very useful to young engineering students and teachers of engineering institutions. The volume can be advantageously used by practicing professional engineers to refresh and update their knowledge.

The references listed at the end of the book may be used by the readers for in-depth review.

I wish the authors great success in all their academic endeavours.

Tirupati
14 January 2003

Prof. G. Ramakrishnan
First Principal
S.V. University College of Engineering

Preface

The authors do not feel apologetic for adding yet another title in the field of Engineering Mechanics, which is one of the most basic and important part of any Engineering Curriculum. There are a few standard titles from foreign authors, but those available from Indian Authors are by no means exhaustive. Hence the authors have given due weightage to presentation of the concepts in lucid language, and tried to maintain a logical sequence of the topics. Coverage of a wide variety and adequate number of numerical problems, including practice problems, is another feature of this book.

This book is divided into three parts–Part-I STATICS, Part-II DYNAMICS, and Part-III ELEMENTS OF SOLID MECHANICS AND STRUCTURAL ANALYSIS. Here the authors would like to give the reason for including Part-III which does not fall wholly under the purview of Engineering Mechanics. Some Universities include the portions in Part-III in the syllabus of Engineering Mechanics for the convenience of students of disciplines other than Civil Engineering (this subject is usually taught to all engineering students in the first year itself), so that they need not study these topics later in their second year. In the absence of this, they study Part-III in a later semester.

There are ten chapters each in Parts I, II & III, thus making the treatment very exhaustive. The number of illustrative problems as well as practice problems is more than adequate, which includes a wide variety. Another special feature of the book is the inclusion of over 500 objective questions, with a significant number from competitive examinations like GATE and IES. The answers for practice problems as well as objective questions are given separately at the end of the book. References are also given at the end of the book. Thus the book will be useful for two/three full courses of the engineering curriculum in the annual/semester pattern.

The treatment is entirely in the S.I. Units. Mechanical properties of certain common engineering materials are given in Appendix-A. The notation, which is made as consistent and less confusing as possible, is set out in Appendix-B. The objective questions will cater to the needs of students preparing for competitive examinations and recruitment tests such as GATE and IES.

The authors acknowledge the help of the works of stalwarts in the field as they do not claim any originality for the material: however, they do claim their own stamp and style of presentation, logicality of sequence of topics, sub-topics, and notation.

The authors are grateful to several of their past students for their enthusiastic response to their teaching of this subject, which has been the primary motivation in undertaking this venture.

The authors specially acknowledge with thanks the support and encouragement received from their wives-Ms Lakshmi Suseela and Ms Lalitha- during the preparation of the manuscript. The senior author places on record the appreciation received from his sons-in-law, Mr. Marutheswar and Mr Kishore Kumar, and the inspiration received from his grand children, Masters Sriharsha and Yasaswi, and Miss Snigdha, and, likewise, the junior author acknowledges the encouragement and appreciation received from his children Ms Padmaja and Master Ramakrishna, and his son-in-law, Mr. Ajiesh, in respect of this venture.

Suggestions for improvement are welcome from the academic community.

TIRUPATI   Prof. C. VENKATRAMAIAH
JUNE 2002   Prof. A. V.NARASIMHA RAO

Contents

PART-II DYNAMICS

CHAPTER-11 RECTILINEAR MOTION 237

CHAPTER-12 PROJECTILE MOTION 253

PART-III ELEMENTS OF SOLID MECHANICS AND STRUCTURAL ANALYSIS

Introduction to Engieering Mechanics

'Mechanics' is that branch of Physics, relating to the study of the action of forces on material bodies, with regard to their state of rest, motion, or deformation. It is the oldest of all Physical Sciences.

Modern knowledge of gravity and motion is due to the work of Isaac Newton (1642-1727); his laws laid the foundation of Newtonian Mechanics. Einstein's theory of relativity (1905) placed certain limitations on Newton's work, which led to the development of 'relativistic mechanics'. The differences are limited to the situations when the speed of the body approaches that of light ($\approx 3 \times 10^8$ m/s), as in the case of large-scale phenomena of astronomy and small-scale phenomena involving subatomic particles.

'Engineering mechanics' is the application of the principles of Mechanics to engineering problems. For the bulk of such problems the principles of Newtonian mechanics apply.

Depending upon the nature of the body involved, Engineering Mechanics can be further subdivided into Mechanics of rigid bodies, Mechanics of deformable bodies (or Strength of Materials) and Mechanics of Fluids.

Broadly speaking, Mechanics of rigid bodies is subdivided into 'Statics' and 'Dynamics'. Statics is the study of bodies at rest or in equilibrium under the action of a system of forces. Dynamics is the study of bodies in motion; it is further subdivided into 'Kinematics' and 'Kinetics'. Kinematics is the study of motion of a body without reference to the causes of motion, while Kinetics deals with the motion as well as the forces responsible for it.

Forces have certain other effects on bodies such as generation of heat when rubbing of two bodies is involved (thermal), production of electric current (piezoelectric), or change in optical properties (photo-elastic), and so on. These are not of interest to us in this text.

STATICS

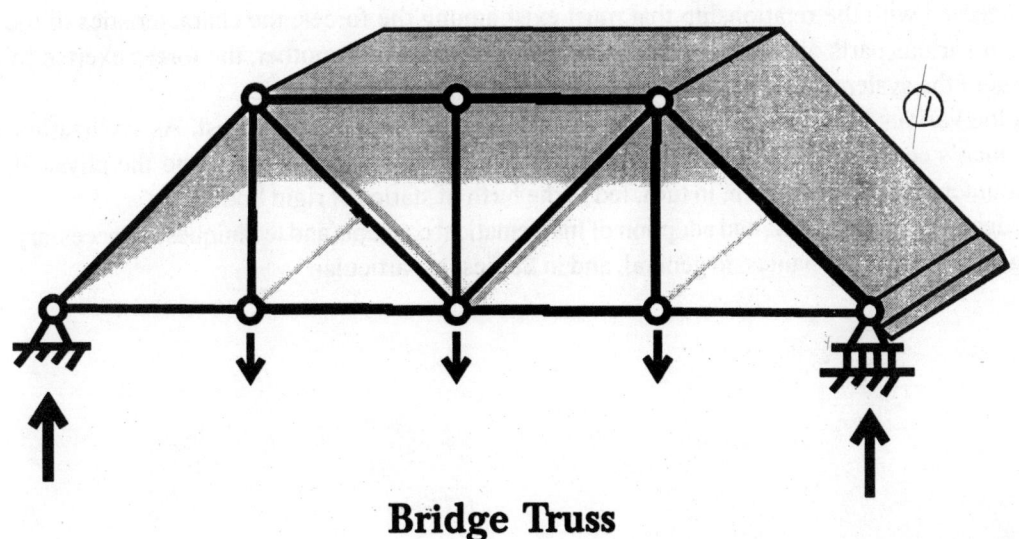

Bridge Truss

PART-I

Introduction to Statics

In Statics, a body or bodies in equilibrium, or at rest, under the action of a set of forces, are dealt with. We are concerned with the relationship that must exist among the forces, the characteristics of the forces which various parts of a body or different bodies exert on one another, the forces exerted by the supports of the system, the configuration of the body or bodies and so on.

Among the various branches of mechanics, statics of rigid bodies is the oldest. As civilization advanced, man's construction activity progressed: this led to the need to understand the physical forces encountered in nature, which, in turn, led to the birth of statics of rigid bodies.

Knowledge of geometry, logic, and adoption of mathematical concepts and techniques are necessary for solving problems in Mechanics, in general, and in Statics, in particular.

Chapter 1

Force and Force Systems

1.1 INTRODUCTION

'Force' may be visualised as the agent or cause which changes or tends to change the position of rest or uniform motion of a body. The action of one body on another may also be described as a force. Force may be produced by direct contact of the bodies or without contact; gravitational force and magnetic force are examples of the latter type.

When more than one force act on a body, they form a force system. Systems of forces may be classified according to their arrangement in a number of ways, which will be seen in detail in Section 1.6.

1.2 DEFINITIONS

Certain definitions are very fundamental to our study and are given below.

Matter: Matter is actually made up of atoms and molecules. Since this is too complex, matter is taken to be continuously distributed—a continuum; it may be rigid or deformable.

Particle: A material with negligibly small dimensions is called a 'particle', the mass being concentrated at a point. The concept is analogous to that of a point in geometry, which has position but no dimensions. However, there is a difference in that a particle, being a material, has mass while a point, being only a geometrical concept, has none.

Body: A conglomeration of particles is called a 'body'. It may be a rigid body or an elastic or deformable body.

Rigid Body: The particles in a rigid body are so firmly connected together that their relative positions do not change irrespective of the forces acting on it. Thus the size and shape of a rigid body are always maintained constant.

Elastic Body: A body whose size and shape can change under forces is a deformable body. When the size and shape can be regained on removal of forces, the body is called an elastic body.

Scalar Quantity: A quantity which is fully described by its magnitude only is a scalar. Arithmetical operations apply to scalars. Examples are: Time, mass, area and speed.

Vector Quantity: A quantity which is described by its magnitude and also its direction is a vector. Operations of vector algebra are applicable to vectors. Examples are: Force, velocity, moment of a force and displacement.

A 'Tensor' quantity needs two or more directional aspects besides its magnitude for its complete description. The order of a tensor is the number of directional aspects needed. Examples are: stress, inertia and strain.

Scalars and vectors are special cases of tensors of order zero and one, respectively.

1.3 MASS

The quantum of matter in a body is characterised by its 'mass', which is a measure of its inertia, or its resistance to change of velocity. It is the property of a body by virtue of which it can experience attraction to other bodies. The symbol used is 'm'.

1.4 FORCE

According to Newton's first law of motion, 'force' produces acceleration, change of state of rest or motion of a body (Rest and motion are relative terms, requiring a reference).

According to Newton's second law of motion, the acceleration produced is proportional to the force applied, and the constant of proportionality is the 'inertial mass' or simply the 'mass' of the body. This provides a way to define the unit of force. The force needed to produce unit acceleration (1 m/s^2) for a body of unit mass (a kilogram—a certain platinum cylinder kept at the Bureau of Weights and Measures at Sevres, France) is designated as a unit force, one Newton, named after Newton (N).

The force with which any body of a certain mass is attracted towards the centre of the earth by its gravitational force is known as 'Weight' of that body (the acceleration produced is called the accelertion due to gravity—$g = 9.81$ m/s^2, nearly). The weight of a body of mass 1 kg is therefore 9.81 N. (The weight of 100 gm will be 0.981 N or 1 N, nearly). The symbol used is $W (= m \cdot g)$.

The effect of a force on a body depends on its magnitude, its direction, and its point of application. Force is thus a vector quantity. By 'direction' is meant the inclination of the line of action of the force with a fixed reference line in space. By 'sense' is meant the arrow-head showing the action or the way it acts—forwards and backwards along the line of action. Thus force is special vector—a 'localised' vector.

A force is thus completely described by specifying its magnitude, direction and sense, and point of action.

Graphically, a force is represented by the segment of a straight line along its line of action, the length of the line being proportional to the force. The direction or sense of the force is indicated by placing an arrow head on the straight line. Either the head or the tail of the vector may be used to indicate the point of application of a force; however, the convention used must be consistent.

A point force is an idealisation commonly used, although a force can be transmitted only through a finite area, however small that be. The loss of accuracy is insignificant.

1.4.1 Kinds of Forces

There are three basic kinds of forces as mentioned below:

1. Tensile force or pull
2. Compressive force or push
3. Shear force

1.4.1.1 Tensile Force or Pull

When equal and opposite forces are applied at the ends of a rod or a bar away from the ends, along its axis, they tend to pull the rod or bar. This kind of a force is called a tensile force or tension (Fig. 1.1).

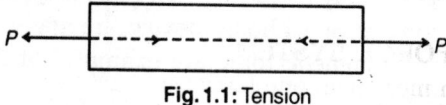

Fig. 1.1: Tension

1.4.1.2 Compressive Force or Push

When equal and opposite forces act at the ends of a rod or a bar towards the ends along its axis, they tend to push the rod or the bar. This kind of force is called a compressive force or compression (Fig. 1.2).

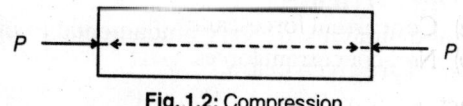

Fig. 1.2: Compression

1.4.1.3 Shear Force

When equal and opposite forces act on the parallel faces of a body (Fig. 1.3), shear occurs on

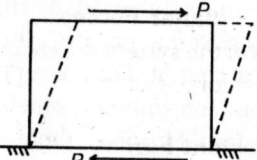

Fig. 1.3: Shear Force

these planes. This tends to cause an angular deformation as shown.

1.5 PRINCIPLE OF TRANSMISSIBILITY OF FORCES

This principle states that the condition of equilibrium or of motion of a rigid body will remain unchanged if the point of application of a force acting on it is transmitted to act at any other point along its line of action (Fig. 1.4).

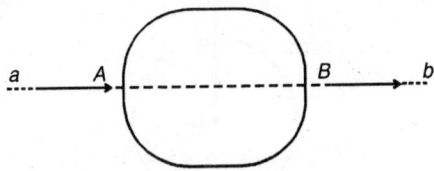

Fig. 1.4: Transmissibility of a Force

However, this principle cannot be used to determine the internal forces and deformations of the body; this is because tension may become compression and vice-versa, if this principle is applied.

1.6 FORCE SYSTEMS

When more than one force act on a body, they constitute a 'force system', this system may be classified according to the line of action and the arrangement of the forces.

(1) Coplanar forces, and
(2) Non-coplanar forces

Another way to classify them is as

(*a*) Concurrent forces, and
(*b*) Non-concurrent forces.

In fact, each of (1) and (2) may fall into either (*a*) or (*b*).

1.6.1 Coplanar Forces

All the forces of the system lie in the same plane.

1.6.2 Non-Coplanar Forces

All the forces of the system do not lie in the same plane (a spatial force system).

1.6.3 Concurrent Forces

All the forces of the system pass through a common point, or have a common point of application.

1.6.4 Non-Concurrent Forces

All the forces of the system do not pass through a common point; some or all of the forces may be parallel to each other.

1.6.5 Coplanar-Concurrent Forces

All the forces of the system lie in the same plane and also are concurrent.

1.6.6 Coplanar-Non-Concurrent Forces

All the forces of the system lie in the same plane but are not concurrent. If all the forces are parallel to each other; it is a coplanar parallel force system.

1.6.7 Non-Coplanar-Concurrent Forces

All the forces of the system do not lie in a single plane but are concurrent. This is a concurrent force system in space.

1.6.8 Non-Coplanar-Non-Concurrent Forces

All the forces of the system do not lie in a single plane and also are not concurrent. If the forces are parallel to each other, it is a non-coplanar parallel force system; otherwise, it is a general force system.

A parallel force system may be 'like' or 'unlike', depending upon whether all the forces act in the same direction or otherwise.

1.7 EQUILIBRIUM

When a body is stationary, or at rest, it is said to be in equilibrium. In this case, either the body is not acted on by any force or force system, or the effect of the force system is null and void.

1.8 RESULTANT

If a force system acting on a body can be replaced by a single force, with exactly the same effect on the body, this single force is said to be the 'resultant' of the force system.

Further, each one of the forces of the system is called a 'component' of the resultant.

1.9 EQUILIBRANT

A force, equal and opposite to the resultant of a force system, is called the 'equilibrant' of the system; this is because this force along with the force system will keep the body in equilibrium.

1.10 ACTION AND REACTION

When two bodies are in contact with each other, both tend to exert forces on each other. If the force exerted by the first body on the second is called 'action', that exerted by the second body on the first is called 'reaction', and vice-versa. Action and reaction are equal and opposite according to Newton's third law of motion.

1.11 TYPES OF FORCES ON A BODY

Broadly speaking, the forces on a body may be externally applied or non-applied or inherent.

1.11.1 Applied Forces

Forces exerted on a body by any other body, or by any external agency, are called 'applied forces' on the body. Generally, contact is required in this case.

1.11.2 Non-applied Forces

Forces inherent in the body without the effect of any external agency are 'non-applied forces'; the weight of a body, for example, is a force due to the gravitational attraction of the mass of the body towards the centre of the earth. No contact with any external agency is needed. Often, forces induced inside a body owing to externally applied forces are called internal forces.

1.12 FREE BODY DIAGRAM

For the consideration of the equilibrium of body, all the forces acting on it must be considered. A diagram showing the body concerned under isolation with all the forces applied on it, including those exerted on it by other bodies, is very useful in this regard. Such a diagram is known as the "free body diagram".

If a body or a structure is in equilibrium, any part of it is also in equilibrium; this is so if the forces, often internal, at the surface of separation of the particular part of the body from the rest are also considered along with the forces acting on that part.

In a free body diagram, all the physical supports and the actions are removed and are replaced by the forces exerted by these on the body.

The free body concept is extremely important because it is only through it that the equilibrium or the motion of any chosen portion of a body or a structure can be understood.

ILLUSTRATIVE PROBLEMS

PROBLEM 1.1: A roller is placed on a smooth horizontal surface as shown in Fig. 1.5 (*a*). Draw the free body diagram for the roller.

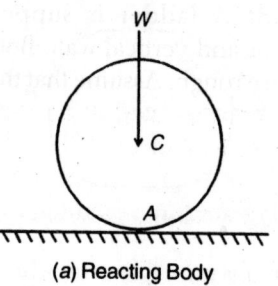

(*a*) Reacting Body

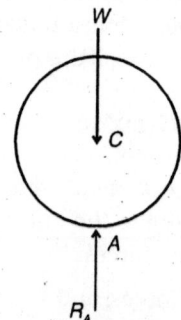

(*b*) Free Body Diagram for the Roller

Fig. 1.5 (Illustrative Problem 1.1)

Problem 1.2: A short cylinder weighing W rests on a symmetrical horizontal V-notch of angle 2α as shown in Fig. 1.6 (*a*). Draw the free body diagram for the cylinder.

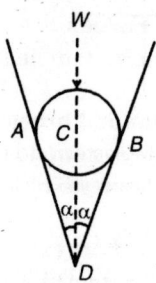

(*a*) Reacting Body

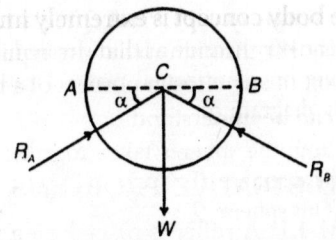

(b) Free Body Diagram for the Short Cylinder

Fig. 1.6 (Illustrative Problem 1.2)

Problem 1.3: A ladder is supported by a horizontal floor and vertical wall. Both the floor and the wall are rough. Assume that the weight of the ladder is concentrated at its centroid. The ladder supports a man in addition to its own weight as shown in Fig. 1.7 (*a*). Draw the free body body diagram for the ladder.

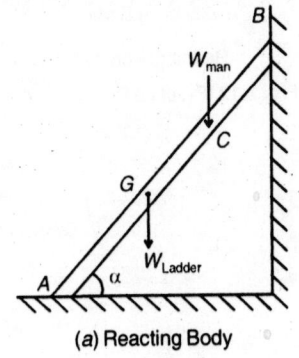

(a) Reacting Body

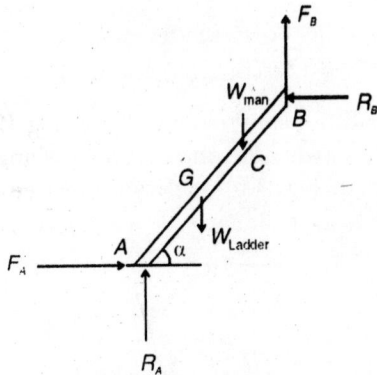

(b) Free Body Diagram for the Ladder

Fig. 1.7 (Illustrative Problem 1.3)

Problem 1.4: Two blocks of weights W_1 and W_2 $(W_1 > W_2)$ are connected by inextensible flexible wire running around a smooth pulley as shown

in Fig. 1.8 (*a*). The heavier block is resting on a smooth floor. Draw the free body diagram for the block weighing W_1.

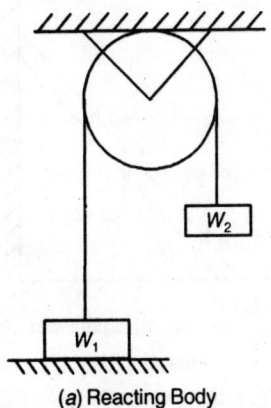

(a) Reacting Body

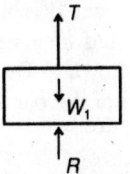

(b) Free Body Diagram for the Heavier Block

Fig. 1.8 (Illustrative Problem 1.4)

PRACTICE PROBLEMS

1.1 A ball is supported in a vertical plane by a string *BC* and a smooth wall *AB* as shown in Fig. 1.9. Draw the free body diagram for the ball.

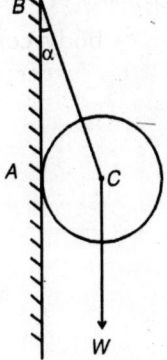

Reacting Body

Fig. 1.9 (Practice Problem 1.1)

1.2 A uniform ladder of weight W rests against a smooth vertical wall and a rough horizontal floor making an angle of 45° with the horizontal as shown in Fig. 1.10. Draw the free body diagram for the ladder.

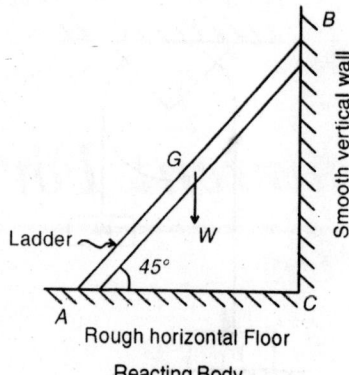

Fig. 1.10: (Practice Problem 1.2)

1.3 A bar *AB* of weight *W* is hinged at *A* to the wall and is supported in a vertical plane by the string *BD* as shown in Fig. 1.11. Draw the free body diagram of the bar.

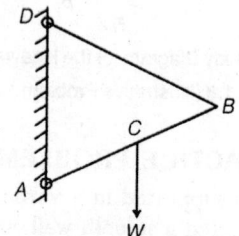

Fig. 1.11: (Practice Problem 1.3)

1.4 Two similar spheres *P* and *Q* (Fig. 1.12) each of weight *W* rest inside a hollow cylinder which is resting on a horizontal plane. Draw the free body diagram of:

(*a*) Both the spheres taken together
(*b*) The sphere *P*
(*c*) The sphere *Q*

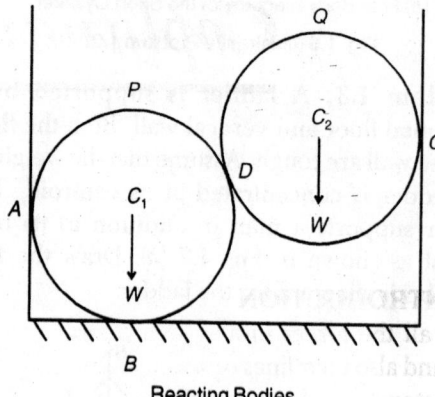

Fig. 1.12 (Practice Problem 1.4)

Chapter 2

Coplanar Concurrent Forces

2.1 INTRODUCTION

When all the forces in a system lie in a single plane and also their lines of action are concurrent, the system is said to constitute a coplanar concurrent system of forces.

In this chapter, composition, resolution, and equilibrium of such force systems are considered. There are several engineering applications, which will be included in the illustrative problems.

2.2 LAW OF PARALLELOGRAM OF FORCES

The law of parallelogram of forces, which gives the resultant of two forces acting at a point, was enunciated by VARIGNON and NEWTON (1687), based on experimental work. Although it cannot be proved analytically, it is taken to be valid since it can be verified experimentally.

This law can be stated thus:

"If two forces acting at a point are represented in magnitude and direction by the adjacent sides of a parallelogram, then their resultant is represented by the diagonal passing through the point of intersection of the two sides representing the forces" (Fig. 2.1). This affords a graphical approach to the problem.

If the two forces act at an angle θ between them, the resultant R can be obtained analytically from the geometry of the figure as follows:

If X is the foot of the perpendicular from C on AB produced [Fig. 2.1 (*b*)].

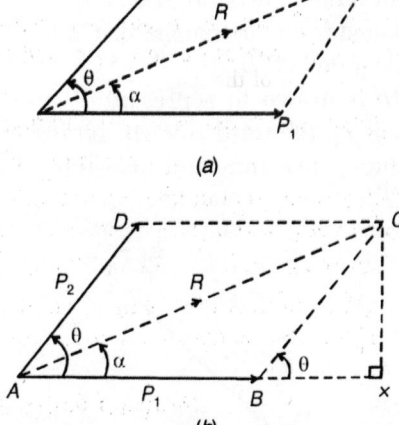

Fig. 2.1: Parallelogram Law of Forces

$R^2 = AX^2 + CX^2 = (AB + BX)^2 + CX^2 = AB^2 + BX^2 + 2AB \cdot BX + CX^2$

But $BX^2 + CX^2 = BC^2$ from $\triangle BXC$, and also $BX = BC \cdot \cos\theta$

$\therefore \quad R^2 = AB^2 + BC^2 + 2AB \cdot BX$

$\qquad = AB^2 + BC^2 + 2AB \cdot BC \cdot \cos\theta$

$\therefore \quad R = \sqrt{P_1^2 + P_2^2 + 2P_1P_2\cos\theta} \qquad$ ...(Eq. 2.1)

If α is the angle made by R with respect to P_1,

$$\left.\tan\alpha = \frac{CX}{AB + BX} = \frac{P_2\sin\theta}{(P_1 + P_2\cos\theta)}\right\}$$

or $\qquad \alpha = \tan^{-1}\left[\dfrac{P_2\sin\theta}{P_1 + P_2\cos\theta}\right]$ $\qquad$ ...(Eq. 2.2)

Special Cases

When $\theta = 0°$, $R = P_1 + P_2$

When $\theta = 90°$, $R = \sqrt{P_1^2 + P_2^2}$

When $\theta = 180°$, $R = P_1 - P_2$

$\theta = 0°$ and $180°$ mean that the forces are collinear; in these cases, the resultant is the algebraic sum of the two forces.

The process of finding the resultant of a force system is called 'Composition of forces'.

Note: It can be observed that the process is commutative—that is, the resultant will be the same irrespective of the sequence or order in which they are taken.

2.3 TRIANGLE LAW OF FORCES

It may be observed from Fig. 2.1(*b*) that the resultant *R* of the two forces, P_1 and P_2 may also be obtained from the triangle *ABC* instead of the parallelogram *ABCD*.

If *AB* is drawn to represent P_1 and *BC* to represent P_2, the third side *AC* represents *R* in magnitude, direction and also line of action (Fig. 2.2).

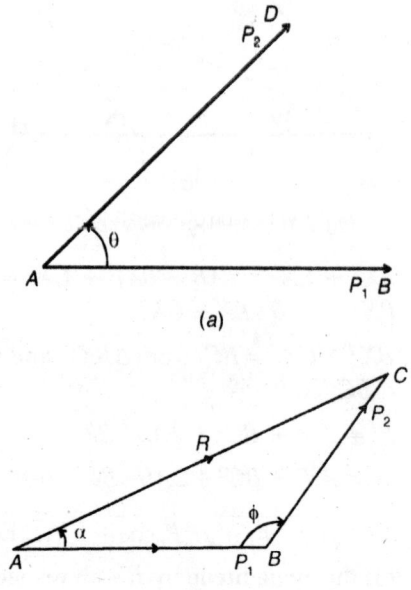

(a)

(b)

Fig. 2.2: Triangle Law of Forces

The law may be stated thus:

"If two forces acting at a point are represented by the two sides of a triangle taken in a certain order, their resultant is represented by the third-side taken in the opposite order". This represents a graphical approach.

The resultant may also be got analytically from the geometry of the triangle *ABC* as follows:

From the cosine rule,

$$R = \sqrt{P_1^2 + P_2^2 - 2P_1P_2\cos\phi}$$

θ being the angle between P_1 and P_2 in the triangle $(\theta + \phi = 180°)$ ...(Eq. 2.3)

Also, the angle α may be got by the sine rule

$$\frac{P_2}{\sin\alpha} = \frac{R}{\sin\phi}$$ ...(Eq. 2.4)

Note: The resultant is commutative.

In this context, it is necessary to differentiate between 'equality' and 'equivalence' of forces.

Two forces may be said to be *equal* if they have the same magnitude and direction with different points of application [*BC* is thus equal to *AD* in Fig. 2.2 (*b*)].

However, they are said to be *equivalent* if they have the same effect on a rigid body. The resultant of a force system is thus the equivalent force (*R* is equivalent to P_1 and P_2 taken together).

Note: Equivalence has to be necessarily based on some specific kind of effect. For example, two fifty paise-coins have the same buying capacity as one rupee coin, but the former will not be useful to operate a public telephone or a weighing machine which is designed to work when a one-rupee coin is dropped in the appropriate slot.

2.4 LAMI'S THEOREM

The statement of this theorem is as follows:

If a body is in equilibrium under the action of three coplanar concurrent forces, each force is proportional to the sine of the angle between the other two.

This may be proved as given below:

Let three concurrent forces P_1, P_2 and P_3 keep a body in equilibrium, as shown in Fig. 2.3.

Let the three forces be represented by *ab*, *bc* and *ca*, respectively, in the triangle *abc*, making these sides parallel to the respective forces—P_1,

P_2 and P_3. The vector diagram has to necessarilly close by the triangle law of forces as these forces keep the body in equilibrium.

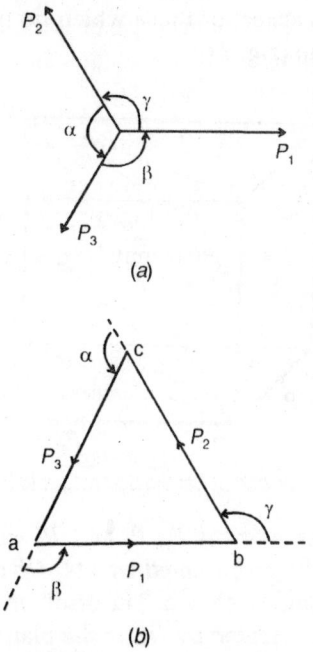

(a)

(b)

Fig. 2.3: Lami's Theorem

Applying the sine rule for the triangle *abc,*

$$\frac{P_1}{\sin(180°-\alpha)} = \frac{P_2}{\sin(180°-\beta)} = \frac{P_3}{\sin(180°-\gamma)}$$

or $\quad\dfrac{P_1}{\sin\alpha} = \dfrac{P_2}{\sin\beta} = \dfrac{P_3}{\sin\gamma}$　　...(Eq. 2.5)

which proves the theorem.

Note:

1. All the forces shall act either towards or away from the point of concurrency. If not, the unlike force may be transmitted through the body using the principle of transmissibility.

2. The converse statement that—if only three forces keep a body in equilibrium, they shall be concurrent, is also true and can be proved. The proof is given in the next chapter.

2.5 RESOLUTION OF FORCES

This is the *reverse* process of composition of forces or determination of the resultant. It involves the determination of a number of component forces which will have the same effect as the given force. The procedure of resolution is just opposite to that of composition.

A force may be resolved into its components in *any two* given directions, or, for that matter, in *any number* of given directions. Theoretically, therefore, a given force may be resolved into an infinite number (of sets) of components.

These are illustrated in Fig. 2.4.

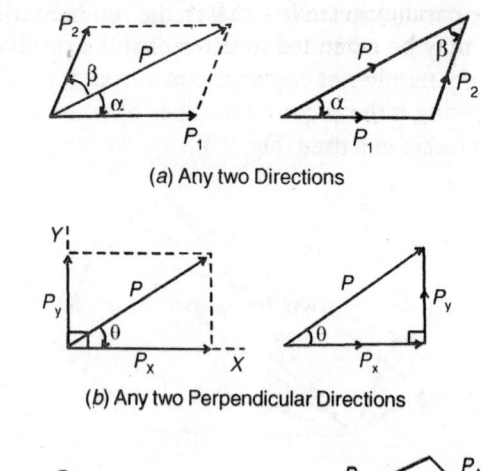

(a) Any two Directions

(b) Any two Perpendicular Directions

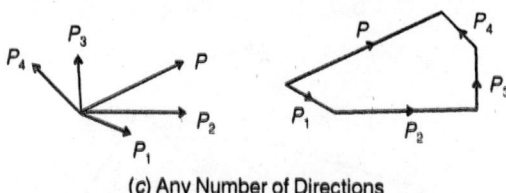

(c) Any Number of Directions

Fig. 2.4: Resolution of Forces into Components

In Fig. 2.4 (*a*), the component of the given force *P* in the two directions making angles α and β with *P* as shown are determined using either the parallelogram law or the triangle law of forces.

In Fig. 2.4 (*b*), the components of *P* in any two perpendicular directions *X* and *Y* as shown are determined similarly.

In Fig. 2.4 (*c*), the components of *P* in any number of given directions are shown; however, the procedure for this is not as simple as in (*a*) & (*b*).

Of these, the rectangular components in (*b*), as they are called, are the most popular, owing to the relative simplicity of the trigonometric relations involved:

$$P_X = P\cos\theta$$
$$P_Y = P\sin\theta$$
$$P = \sqrt{P_X^2 + P_Y^2}$$
$$\tan\theta = P_Y / P_X$$
...(Eq. 2.6)

2.6 EXTENSION OF LAW OF PARALLELO-GRAM OF FORCES

The parallelogram law of forces given in Section 2.2 may be extended to determine the resultant of any number of coplanar concurrent forces by applying it the required number of times taking two forces at a time (Fig. 2.5).

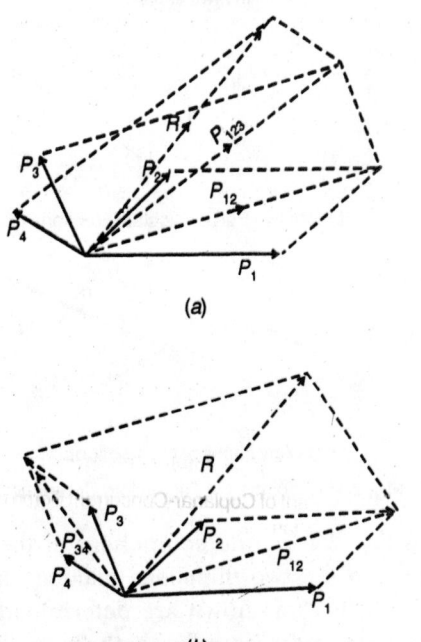

(a)

(b)

Fig. 2.5: Extension of Parallelogram Law of Forces

Obviously, there can be more than one sequence in such cases; however, the resultant of the force system will be the same finally, irrespective of the sequence employed.

In Fig. 2.5 (*a*), P_1 and P_2 are combined to get P_{12}; P_{12} and P_3 to get P_{123}; and finally, P_{123} and P_4 are combined to get the resultant R. In Fig. 2.5 (*b*), P_1 and P_2 are combined to get P_{12}; P_3 and P_4

to get P_{34}; and finally, P_{12} and P_{34} are combined to get the resultant R, which is same as in (*a*).

In addition to the foregoing, it may be noted that the parallelogram law can be extended to forces in space, or those which do not all act in one plane (Fig. 2.6).

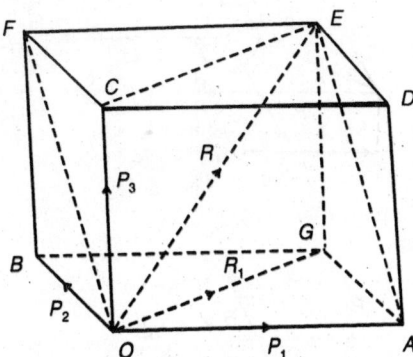

Fig. 2.6: Extension of Parallelogram Law to Forces in Space

Let three forces P_1, P_2 and P_3 which act at a point O be represented by OA, OB and OC, not in one plane as shown. The resultant of P_1 and P_2 is R_1, represented by OG in the plane $OAGB$, the three forces being in this plane. The resultant of R_1 and P_3 is R, represented by OE in the plane $OGEC$.

R is thus the required resultant of P_1, P_2 and P_3, obtained by applying the parallelogram law twice in different planes. In fact, R is represented by the diagonal of the rhomboid from the point of concurrency, the sides of the rhomboid representing, P_1, P_2 and P_3 (This may got in more than one route, in fact).

The principle may be extended to more than three such concurrent forces in space; however, the graphical approach is laborious and confusing in this case. An analytical approach is preferred here.

This has application in the analysis of forces in what are known as "Space Frames".

2.7 THEOREM OF RESOLVED PARTS
The theorem of resolved parts states:

"The component of the resultant of a coplanar concurrent force system in a given direction is equal to the algebraic sum of the components of all the forces of the system in that direction".

This can be easily understood for two forces as shown in Fig. 2.7.

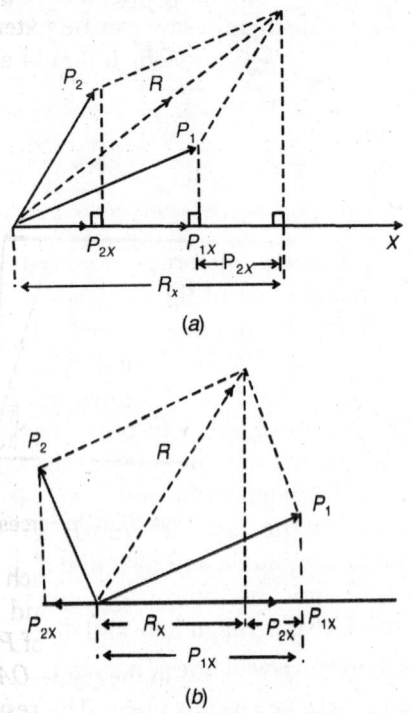

(a)

(b)

Fig. 2.7: Theorem of Resolved Parts (for two forces)

R is the resultant of forces, P_1 and P_2 (by the parallelogram law). The components of P_1 and P_2 parallel to the x-direction are, say, P_{1x} and P_{2x}. From the geometry of the figure $P_{1x} + P_{2x} = R_x$, the component of R parallel to the x-direction.

In (a) P_{1x} and P_{2x} add to each other. However, if the directions of P_{1x} and P_{2x} are opposite as in (b), their algebraic sum will be $P_{1x} \sim P_{2x}$, which will be equal to R_x.

This can be easily extended for any number of concurrent forces ..., $P_1, P_2, P_3, ...$

If their resultant is R, with its x-component as R_x,

$$R_x = P_{1x} + P_{2x} + P_{3x} + ...,\text{ algebraically,}$$

or $R_x = \Sigma P_x$

This idea can be used to determine the resultant of any number of coplanar concurrent forces analytically, since this is applicable to resolved parts in any direction. What is

conveniently done is to consider the resolved parts or components of all the forces in any two perpendicular directions and apply the theorem of resolved parts for the components of the resultant of the system (Fig. 2.8).

$$R_x = \Sigma P_x$$
$$R_y = \Sigma P_y$$
$$\left. \begin{array}{l} R = \sqrt{R_x^2 + R_y^2} = \sqrt{(\Sigma P_x)^2 + (\Sigma P_y)^2} \\ \tan \alpha = \Sigma P_y / \Sigma P_x \end{array} \right\} \quad ...(\text{Eq. 2.7})$$

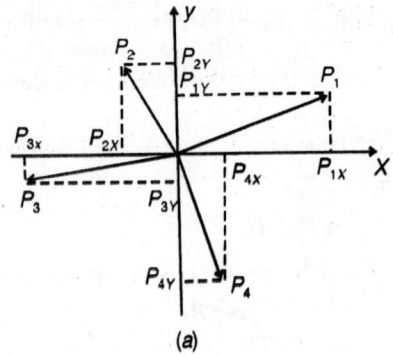

(a)

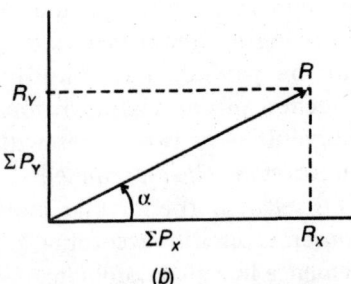

(b)

Fig. 2.8: Resultant of Coplanar-Concurrent Force System by Resolution

It is better to show the resultant and its rectangular components separately as in (b) for clarity.

In this case, the rectangular components of any force which makes an angle θ with respect to the x-direction are given by;

$$\left. \begin{array}{l} P_x = P\cos\theta \\ P_y = P\sin\theta \end{array} \right\} \quad ...(\text{Eq. 2.8})$$

Similarly, $R_x = R\cos\alpha$

and $R_y = R\sin\alpha$

It is convenient to use the acute angle θ with respect to the axis, and consider the component directed towards the +ve direction of the axis as

+ve. Thus this procedure amounts to composition of forces by resolution.

2.8 EQUILIBRIUM OF SEVERAL CO-PLANAR CONCURRENT FORCES

We have seen how to determine the resultant of a system of coplanar concurrent forces using rectangular components with respect to a set of perpendicular axes.

If the resultant of the system reduces to zero, the particle or the body on which such a system acts will be in equilibrium. This enables us to determine the conditions necessary for the equilibrium of the system of coplanar concurrent forces.

For R to be zero, R_x and R_y shall be zero

$$\left. \begin{array}{l} \text{or} \quad R_x = \Sigma P_x = 0 \\ \text{and} \quad R_y = \Sigma P_y = 0 \end{array} \right\} \qquad \ldots(\text{Eq. 2.9})$$

This affords an analytical approach to the problem.

Note:

(1) Since there are two equations, problems involving not more than two unknowns can be solved. The magnitude and direction of one unknown force or the magnitudes of two forces with known direction may be determined.

(2) The set of perpendicular axes can be chosen arbitrarily according to the convenience in a given problem.

2.9 LAW OF POLYGON OF FORCES

The law of triangle of forces may be extended to cover the case of more than two forces. Taking two forces at a time, the triangle law can be applied the required number of times to determine the resultant of the concurrent force system graphically (Fig. 2.9).

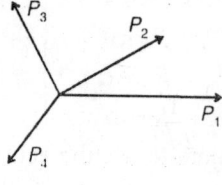

(a)

Fig. 2.9: Polygon of Forces

In (*a*) the concurrent force system is shown. In (*b*), the triangle law of forces is applied three times—P_1 and P_2 combined to yield R_1, R_1 and P_3 to give R_2, and finally R_2 and P_4 to result in the required resultant R of the system. Instead, the polygon *ABCDEA* may be viewed as the graphical construction to represent the forces of the system in sequence; the vector *AE*—from the first point to the last—obviously gives the resultant in magnitude and direction if the sides of the polygon are made to represent the respective forces in magnitude and direction.

The 'Polygon law of forces' may thus be stated:

If a number of coplanar concurrent forces are represented in magnitude and direction by the sides of a polygon, taken in a particular order, then the resultant of the system is represented by the closing side in magnitude and direction, taken in the reverse order.

Note:

(1) Since the 'equilibrant' of a system has been defined as being equal and opposite to the resultant, the closing side of the polygon, taken in the same order, represents the equilibrant.

(2) Since the resultant of a system has to be zero for equilibrium of the system, if the force polygon of the system is a closed one, it follows that the force system keeps a particle or a body in equilibrium. In such case, any one force may be treated as the equilibrant for all the rest of the forces.

The polygon law of forces also has several practical applications in mechanics.

ILLUSTRATIVE PROBLEMS

Problem 2.1: Find the magnitude of the two forces, such that they act at right angles, their resultant is 30 N. But if they act at 60° their resultant is 35 N.

Solution:

$$R = \sqrt{P^2 + Q^2 + 2PQ\cos\theta}$$

$$R = 30 \text{ N}, \text{ when } \theta = 90°$$

$$\therefore \quad 30 = \sqrt{P^2 + Q^2 + 2PQ\cos 90°}$$

$$30 = \sqrt{P^2 + Q^2} \qquad \text{(Since } \cos 90° = 0\text{)}$$

or $\quad P^2 + Q^2 = 30^2 = 900$

$$P^2 + Q^2 = 900 \qquad \qquad ...(i)$$

$$R = 35 \text{ N}, \qquad \text{when } \theta = 60°$$

$$35 = \sqrt{P^2 + Q^2 + 2PQ\cos 60°}$$

$$35 = \sqrt{P^2 + Q^2 + PQ} \text{ (Since } \cos 60° = 1/2\text{)}$$

$$P^2 + Q^2 + PQ = 35^2 = 1225$$

$$P^2 + Q^2 + PQ = 1225 \qquad \qquad ...(ii)$$

Substituting (i) in (ii)

$$900 + PQ = 1225$$

$$PQ = 1225 - 900 = 325$$

$$PQ = 325 \qquad \qquad ...(iii)$$

$$(P+Q)^2 = P^2 + Q^2 + 2PQ = 900 + 2 \times 325$$

$$(P+Q)^2 = 1550$$

or $P + Q = 39.37 \qquad \qquad ...(iv)$

$$(P-Q)^2 = P^2 + Q^2 - 2PQ$$

$$= 900 - 2 \times 325$$

$$(P-Q)^2 = 250$$

$$P - Q = 15.81 \qquad \qquad ...(v)$$

$$P + Q = 39.37$$

$$P - Q = 15.81$$

Adding $2P = 55.18$

$$\therefore \quad P = \frac{55.18}{2} = 27.59 \text{ N} \qquad \qquad ...(vi)$$

Substituting (vi) in (iv)

$$27.59 + Q = 39.37$$

$$Q = 39.37 - 27.59$$

$$Q = 11.78 \text{ N}$$

$$P = 27.59 \text{ N and } Q = 11.78 \text{ N}$$

PROBLEM 2.2: Find the angle between two equal forces P, when their resultant is (i) equal to P, and (ii) equal to P/2.

Solution:

(i) $\qquad P = P$

$$Q = P$$

$$R = P$$

$$R = \sqrt{P^2 + Q^2 + 2PQ\cos\theta}$$

$$P = \sqrt{P^2 + P^2 + 2 \cdot P \cdot P\cos\theta}$$

$$= \sqrt{2P^2(1 + \cos\theta)}$$

$$P = P\sqrt{2(1 + \cos\theta)}$$

$$1 = 2(1 + \cos\theta)$$

$$1 + \cos\theta = \frac{1}{2}$$

$$\cos\theta = -\frac{1}{2}$$

$$\theta = \cos^{-1}(-1/2) = 120°$$

$\theta = 120°$ (The angle between the two forces)

(ii) $\qquad P = P$

$$Q = P$$

$$R = P/2$$

$$R = \sqrt{P^2 + Q^2 + PQ\cos\theta}$$

$$\frac{P}{2} = \sqrt{P^2 + P^2 + 2 \times P \times P\cos\theta}$$

$$\frac{P}{2} = \sqrt{2} \cdot \sqrt{P^2} \cdot \left(\sqrt{1 + \cos\theta}\right) = \sqrt{2} \cdot P \cdot \sqrt{(1 + \cos\theta)}$$

$$\sqrt{1 + \cos\theta} = \frac{1}{2\sqrt{2}}$$

$$1 + \cos\theta = \frac{1}{8}$$

$$\cos\theta = \frac{1}{8} - 1 = -\frac{7}{8}$$

$$\theta = \cos^{-1}(-7/8) = 151°02'42''$$

The angle between the two forces $= 151°02'42''$

PROBLEM 2.3: Two equal forces are acting at a point with an angle of 45° between them. If the resultant force is 40 N, find the magnitude of each force.

Solution:

$$P = P$$

$$Q = P$$

$$\theta = 45°$$

$$R = 40 \text{ N}$$

$$R = \sqrt{P^2 + Q^2 + 2PQ\cos\theta}$$

$$40 = \sqrt{P^2 + P^2 + 2\cdot P\cdot P\cos 45°}$$

$$40 = \sqrt{2P^2 + 2P^2\left(1/\sqrt{2}\right)}$$

$$40 = \sqrt{3.4142P^2}$$

$$3.4142P^2 = 40^2 = 1600$$

$$P^2 = \frac{1600}{3.4142} = 468.63$$

$$P = \sqrt{468.63} = 21.65 \text{ N}$$

The magnitude of each force is 21.65 N

PROBLEM 2.4: Two forces of magnitude 250 N and 150 N are acting at point O as shown in Fig. 2.10 (a). If the angle between the forces is 60°, determine the magnitude of the resultant force. Also determine the angle β & γ as shown in the Fig. 2.10 (a).

Solution:

$$P = 250 \text{ N}$$

$$Q = 150 \text{ N}$$

$$\theta = 60°$$

$$R = \sqrt{P^2 + Q^2 + 2PQ\cos 60°}$$

$$= \sqrt{250^2 + 150^2 + 2(250)(150)(1/2)}$$

$$= \sqrt{122500}$$

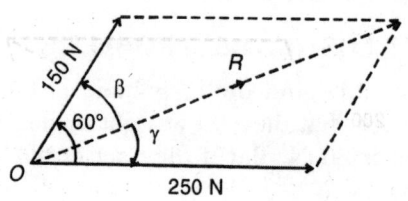

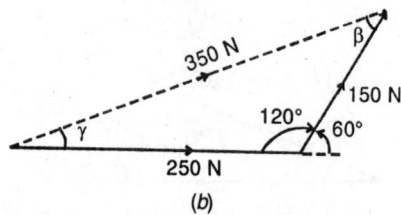

(b)

Fig. 2.10 (Illustrative Problem 2.4)

$$= 350 \text{ N}$$

Applying sine rule

$$\frac{150}{\sin\gamma} = \frac{350}{\sin 120°} = \frac{250}{\sin\beta}$$

$$\therefore \quad \sin\gamma = \frac{150}{350}\sin 120° = 0.3711$$

$$\gamma = \sin^{-1}(0.3711) = 21.79°$$

$$\sin\beta = \frac{250}{350}\sin 120° = 0.6186$$

or $$\beta = \sin^{-1}(0.6186) = 38.21°$$

$$R = 350 \text{ N}; \gamma = 21.79° \And \beta = 38.21°$$

PROBLEM 2.5: Two forces P and Q are acting at a point O as shown in Fig. 2.11 (a). The force $P = 300$ N and force $Q = 200$ N. If the resultant of the forces is equal to 400 N, then find the values of angles β, γ and α.

Solution:

$$R = \sqrt{P^2 + Q^2 + 2PQ\cos\alpha}$$

$$R = 400 \text{ N}; P = 300 \text{ N}; Q = 200 \text{ N}$$

$$\therefore \quad 400 = \sqrt{300^2 + 200^2 + 2\times 300\times 200\cos\alpha}$$

$$160000 = 90000 + 400000 + 1200000\cos\alpha$$

$$120000\cos\alpha = 30000$$

$$\cos\alpha = \frac{30000}{120000} = \frac{1}{4} = 0.25$$

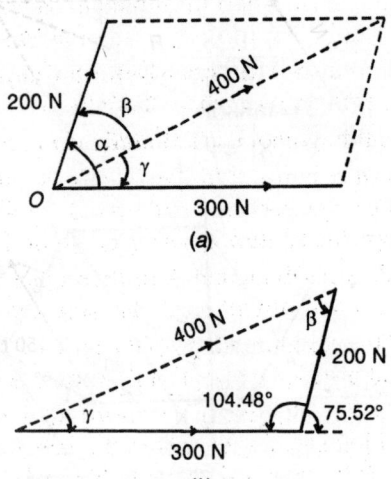

Fig. 2.11 (Illustrative Problem 2.5)

$$\alpha = \cos^{-1}(0.25) = 75.52°$$

$$\frac{300}{\sin\beta} = \frac{400}{\sin 104.48°} = \frac{200}{\sin\gamma}$$

$$\sin\beta = \sin 104.48° \times \frac{300}{400}$$

$$= 0.7262$$

$$\beta = \sin^{-1}(0.7262) = 46.57°$$

$$\sin\gamma = \sin 104.48° \times \frac{200}{400}$$

$$= 0.4841$$

$$\gamma = \sin^{-1}(0.4841)$$

$$= 28.95°$$

$$\beta = 46.57°; \gamma = 28.95° \text{ and } \alpha = 75.52°$$

PROBLEM 2.6: A vector of magnitude 10 units is directed 60° North of East. Determine its components due East and North both graphically and analytically.

Solution:

Graphically,

Required vector = '*OP*'

Its length *OP* = 10 units to a chosen scale

Its direction = 60° with the East

Its components due east and north are found by measuring its projections on the *X* and *Y* axes

respectively to the same scale, are 5.00 units and 8.70 units.

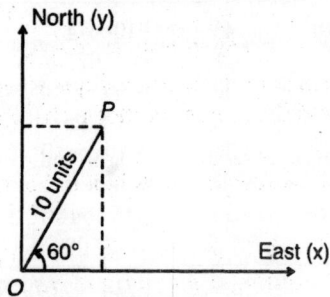

Fig. 2.12 (Illustrative Problem 2.6)

Analytically,

Its component due east is *OP* cos 60°

$$= 10 \times 0.5$$

$$= 5 \text{ Units}$$

Its component due north is *OP* sin 60°

$$= 10 \times \frac{\sqrt{3}}{2} = 8.66 \text{ units}$$

PROBLEM 2.7: Fig. 2.13 (*a*) shows four forces 10 kN, 15 kN, 20 kN and 5 kN acting at a point *O*. Determine the direction and magnitude of the resultant force with reference to the reference axes.

Solution:

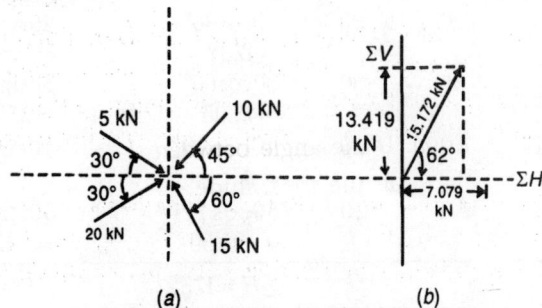

Fig. 2.13 (Illustrative Problem 2.7)

$$R = \sqrt{(\Sigma H)^2 + (\Sigma V)^2} \text{ (see table on the next page)}$$

$$= \sqrt{(7.079)^2 + (13.419)^2}$$

$$= 15.172 \text{ N}$$

$$\tan\theta = \frac{\Sigma V}{\Sigma H} = \frac{13.419}{7.079} = 1.8956$$

$$\theta = 62°11'$$

S.No.	Force (kN)	θ (in deg.)	Horizontal Component (kN)	Vertical Component (kN)
1	10	45	$-10\cos45°$ $=-7.071$	$-10\sin45°$ $=-7.071$
2	5	30	$+5\cos30°$ $=+4.330$	$-5\sin30°$ $=-2.500$
3	20	30	$+20\cos30°$ $=17.320$	$+20\sin30°$ $=+10.000$
4	15	60	$-15\cos60°$ $=-7.500$	$+15\sin60°$ $=+12.990$
			$\Sigma H = +7.079$ N	$\Sigma V = +13.419$

The magnitude of the resultant $= 15.172$ N and its direction is $62°11'$ with the horizontal.

PROBLEM 2.8: The forces 10 N, 20 N, 30 N, 40 N and 50 N are acting on one of the angular points of regular hexagon, towards the other five angular points, taken in order.

Find the direction and magnitude of the resultant force.

Solution:

S.No.	Force (N)	θ (in deg.)	Horizontal Component (N)	Vertical Component (N)
1	10	0	$10\cos0°$ $=10.00$	$10\sin0°$ $=00.00$
2	20	30	$20\cos30°$ $=17.32$	$20\sin30°$ $=10.00$
3	30	60	$30\cos60°$ $=15.00$	$30\sin60°$ $=25.98$
4	40	90	$40\cos90°$ $=0$	$40\sin90°$ $=40$
5	50	120	$50\cos120°$ $=-25.00$	$50\sin120°$ $=+43.30$
			$\Sigma H = 17.32$	$\Sigma V = 119.28$

$$R = \sqrt{(\Sigma H)^2 + (\Sigma V)^2}$$

$$= \sqrt{(17.32)^2 + (119.28)^2}$$

$$= 120.53 \text{ N}$$

Let $\theta =$ Angle, which the resultant makes with the horizontal (i.e., AB)

$$\tan\theta = \frac{\Sigma V}{\Sigma H}$$

$$= \frac{119.28}{17.32}$$

$$= 6.8868$$

$$\theta = 81°44'$$

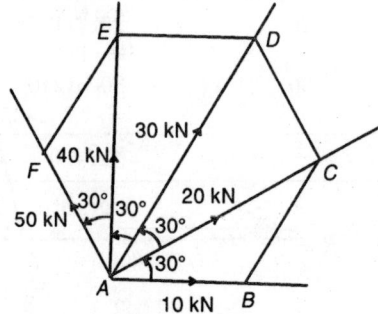

Fig. 2.14 (Illustrative Problem 2.8)

PROBLEM 2.9: Three forces of magnitude 50 kN, 20 kN and 30 kN are acting at a point O as shown in Fig. 2.15 (*a*). The angles made by 50 kN, 20 kN and 30 kN forces with x-axis are 60°, 120° and 240° respectively. Determine the magnitude and direction of the resultant.

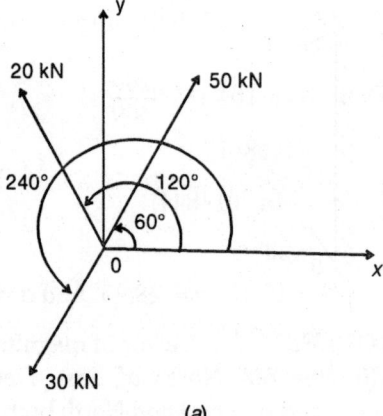

(a)

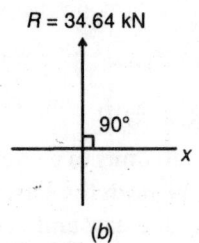

(b)

Fig. 2.15 (Illustrative Problem 2.9)

Solution:

S.No.	Force (N)	θ (in deg.)	Horizontal Component (kN)	Vertical Component (kN)
1	50	60	$50\cos 60°$ $= +25$	$50\sin 60°$ $= +43.30$
2	20	120	$20\cos 120°$ $= -10$	$20\sin 120°$ $= +17.32$
3	30	220	$30\cos 240°$ $= -15$	$30\sin 240°$ $= -25.98$
			$\Sigma H = 0$	$\Sigma V = +34.64$ N

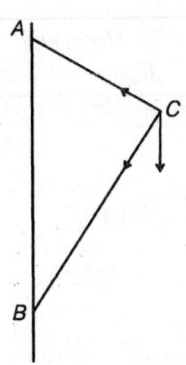

(d)

$$R = \sqrt{(\Sigma H)^2 + (\Sigma V)^2}$$

$$= \sqrt{(0)^2 + (+34.64)^2} = 34.64 \text{ N}$$

$$\theta = \tan^{-1} = \frac{\Sigma V}{\Sigma H}$$

$$= \tan^{-1}\left(\frac{34.64}{0}\right) = 90°$$

PROBLEM 2.10: Fig. 2.16 (*a*) shows a simple derrick crane which carries a load of 10 kN. If *AB*, *BC*, *CA* are 4 m, 4 m and 2 m long, find the forces in *AC* and *BC* using the law of triangle of forces.

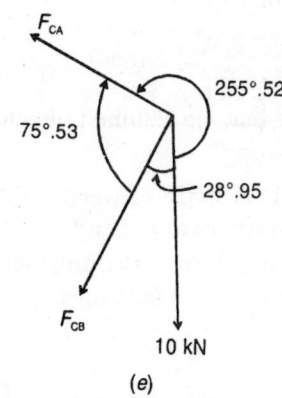

(e)

Fig. 2.16 (Illustrative Problem 2.10)

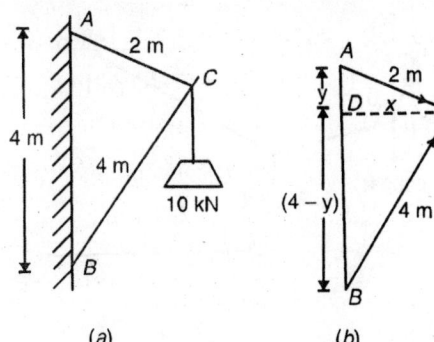

(a)   (b)

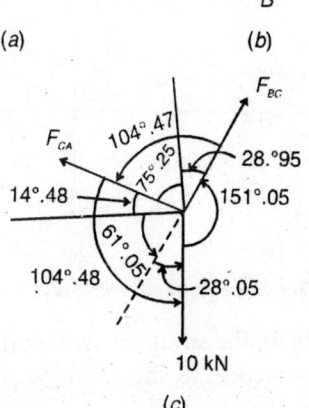

(c)

Solution:

$$x^2 + y^2 = 2^2 = 4$$

or $$x^2 = 4 - y^2 \qquad \ldots(i)$$

$$x^2 + (4 - y)^2 = 4^2 = 16$$

or $$x^2 = 16 - (4 - y)^2 \qquad \ldots(ii)$$

Equating (*i*) and (*ii*)

$$4 - y^2 = 16 - (4 - y)^2$$
$$4 - y^2 = 16 - 16 + 8y - y^2 4$$

∴ $$y = \frac{4}{8} = 0.5$$

$$x^2 - 4 - y^2 = 4 - (0.5)^2$$

$$x = \sqrt{3.75} = 1.9365 \text{ m}$$

$$\angle ACD = \tan^{-1}\left(\frac{y}{x}\right) = \tan^{-1}\left(\frac{3.5}{1.9365}\right)$$

$$= 14.48°$$

$$\angle BCD = \tan^{-1}\left(\frac{4-y}{x}\right) = \tan^{-1}\left(\frac{3.5}{1.9365}\right)$$

$$\frac{10}{\sin 104.67°} = \frac{F_{BC}}{\sin 104.48°} = \frac{F_{CA}}{\sin 151.05°}$$

$$F_{BC} = 10 \text{ kN}$$

$$F_{CA} = \left(\frac{\sin 151.05°}{\sin 104.67°}\right) \times 10$$

$$= 5 \text{ kN}$$

or $\quad \dfrac{F_{CA}}{\sin 28.95°} = \dfrac{10}{\sin 75.53°} = \dfrac{F_{CB}}{\sin 255.52°}$

$$F_{CA} = 10 \times \left(\frac{\sin 28.95°}{\sin 75.53°}\right) = 4.999 \text{ say 5 kN}$$

$$F_{CB} = 10 \times \left(\frac{\sin 255.52°}{\sin 75.53°}\right) = -9.9995 \text{ say } -10 \text{ kN}$$

-ve sign indicates that the assumed direction for the force F_{CB} is wrong.

PROBLEM 2.11: A body of weight 500 N is suspended by two strings of 4 m and 3 m lengths, attached at the same horizontal level 5 m apart. Find the tension in strings.

Solution:

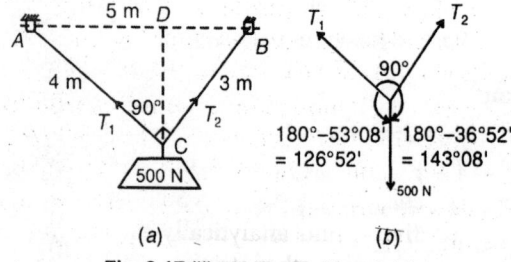

(a)   (b)

Fig. 2.17 (Illustrative Problem 2.11)

Given

Load at $C = 500$ N

$\qquad AC = 4$ m

$\qquad BC = 3$ m

$\qquad AB = 5$ m

Since in triangle ABC

$$AB^2 = AC^2 + CB^2$$

$$5^2 = 4^2 + 3^2$$

∴ $\qquad \angle ACB = 90°$

Since $\angle CAB = 3/5 = 0.6$

∴ $\qquad \angle CAB = 36.87°$ or $36°52'$

and $\angle CBA = 90° - 36°52' = 53°08'$

$$\angle ACD = 90 - \angle CAD = 90 - \angle CAB$$

$$= 90 - 36°52' = 53°08'$$

$$\angle BCD = 90° - \angle CBD = 90 - \angle CBA$$

$$= 90 - 53°08' = 36°52'$$

Applying Lami's theorem at joint 'C'

$$\frac{T_1}{\sin 143.08°} = \frac{500}{\sin 90°} = \frac{T_2}{\sin 126.87°}$$

$$T_1 = \frac{\sin 143.08°}{\sin 90°} \times 500$$

$$= 300.35 \text{ N}$$

$$T_2 = \frac{500}{\sin 90°} \sin 126.87° = 400 \text{ N}$$

$$T_1 = 300.35 \text{ N and } T_2 = 400 \text{ N}$$

PROBLEM 2.12: A string *ABCD*, attached to two fixed points *A* and *D* has two equal weights 6000 N attached to it at *B* and *C*. The weights rest with the portions *AB* and *CD* inclined at angles of 30° and 60° respectively, to the vertical as shown in Fig. 2.18 (*a*). Find the tension in the strings.

Solution:

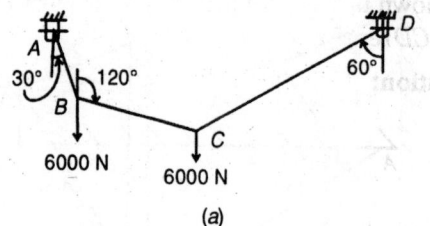

(a)

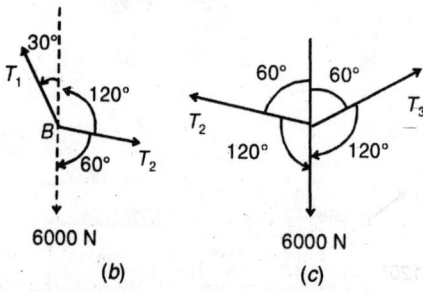

(b)   (c)

Fig. 2.18 (Illustrative Problem 2.12)

Let us split up the string into two parts, for the sake of convenience, as shown in Fig. 2.18 (*b*) & (*c*)

Let T_1 = Tension in the portion AB of the string

T_2 = Tension in the portion BC of the string

T_3 = Tension in the portion CD of the string

Applying Lami's theorem at joint B,

$$\frac{T_1}{\sin 60°} = \frac{6000}{\sin 150°} = \frac{T_2}{\sin 150°}$$

$$T_1 = \frac{6000}{\sin 150°}\sin 60°$$

$$= 10392.3 \text{ N}$$

$$T_2 = \frac{6000}{\sin 150°} \times \sin 150° = 6000 \text{ N}$$

Again applying Lami's equation at joint C

$$\frac{T_1}{\sin 120°} = \frac{T_3}{\sin 120°} = \frac{T_2}{\sin 120°}$$

$$T_2 = 6000\frac{\sin 120°}{\sin 120°} = 6000 \text{ N}$$

$$T_3 = \frac{6000}{\sin 120°}\sin 120° = 6000 \text{ N}$$

$$T_1 = 10392.3 \text{ N}; \; T_2 = 6000 \text{ N}$$

and $\;\; T_3 = 6000$ N

PROBLEM 2.13: A rope supported at A and B carries a load of 25 kN at C and a load of W at D as shown in Fig. 2.19 (*a*). Find the value of W so that CD remains horizontal.

Solution:

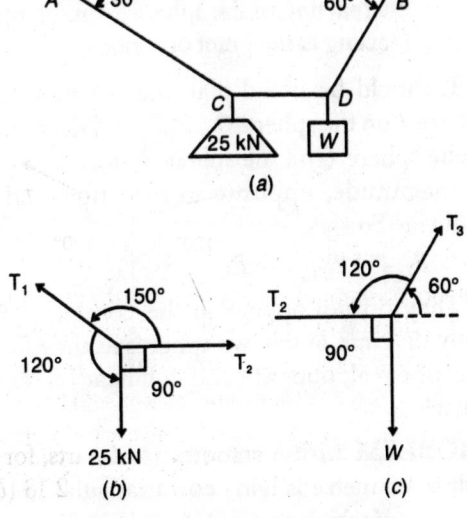

(a)

(b)

(c)

Fig. 2.19 (Illustrative Problem 2.13)

Free body diagram of the points C and D are shown in Fig. 2.19 (*b*) & (*c*).

Applying Lami's theorem for the system of forces acting at 'C'.

$$\frac{T_1}{\sin 90°} = \frac{25}{\sin 150°} = \frac{T_2}{\sin 120°}$$

$$T_2 = 25\frac{\sin 120°}{\sin 150°} = \frac{25 \times 0.8660254}{0.5}$$

$$= 43.30127 \text{ kN}$$

Applying Lami's theorem for the system of forces acting at 'D'

$$\frac{T_2}{\sin 150°} = \frac{W}{\sin 120°} = \frac{T_3}{\sin 90°}$$

$$\therefore \quad W = T_2\left(\frac{\sin 120°}{\sin 150°}\right)$$

$$= 43.30127\left(\frac{\sin 120°}{\sin 150°}\right)$$

$$= \frac{43.30127 \times 0.8660254}{0.5000}$$

$$W = 75 \text{ kN}$$

PROBLEM 2.14: Five strings are tied at a point and are pulled in all directions, equally spaced, from one another. If the magnitude of the pulls on three consecutive strings is 50 N, 80 N and 60 N respectively, find analytically the magnitude of the pulls on two other strings.

Solution:

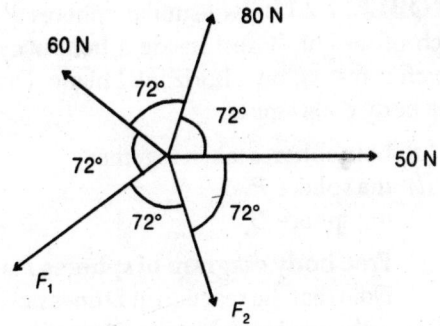

Fig. 2.20 (Illustrative Problem 2.14)

S.No.	Force (N)	θ (in deg.)	Horizontal Component (N)	Vertical Component (N)
1	50	0	$50\cos 0°$ $=50$	$50\sin 0°$ $=0$
2	80	72	$80\cos 72°$ $=24.72$	$80\sin 72°$ $=76.08$
3	60	144	$60\cos 144°$ $=-48.54$	$60\sin 144°$ $=35.27$
4	F_1	216	$F_1\cos 216°$ $=-0.809F_1$	$F_1\sin 216°$ $=-0.588F_1$
5	F_2	288	$F_2\cos 288°$ $=0.309F_2$	$F_2\sin 288°$ $=0.951F_2$
			$\Sigma H=$	$\Sigma V=$

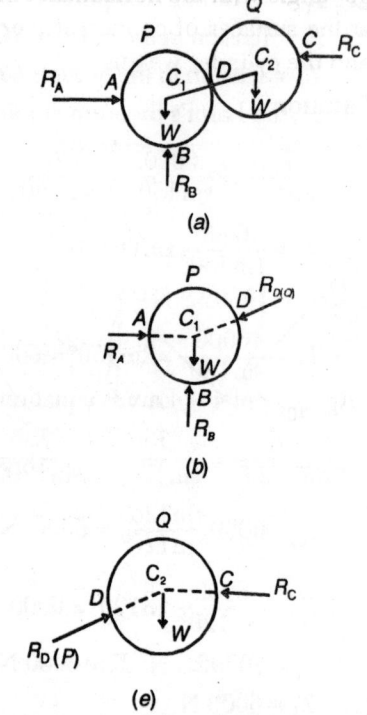

$$\Sigma H= -0.809F_1 + 0.309F_2 + 26.18$$
$$\Sigma V= -0.588F_1 - 0.951F_2 + 111.35$$

For the equilibrium of the point
$$\Sigma H= 0 \ \& \ \Sigma V= 0$$

i.e., $-0.809F_1 + 0.309F_2 + 26.18 = 0$

or $F_1 - 0.382F_2 = 32.36$ ...(i)

$-5.88F_1 - 0.951F_2 + 111.35 = 0$

$F_1 + 1.617F_2 = 189.37$ ...(ii)

Subtracting $\overline{F_1 - 0.382F_2 = 32.36}$

$1.999F_2 = 157.01$

$$F_2 = \frac{157.01}{1.999}$$

$F_2 = 78.54$ N

$F_1 = 32.36 + 0.382 (78.54)$

$= 32.36 + 30.00$

$= 62.36$ N

$F_1 = 62.36$ N and $F_2 = 78.54$ N

PROBLEM 2.15: Two similar spheres P and Q each of weight W rest inside a hollow cylinder which is resting on a horizontal plane. Draw the free body diagrams of

(a) both spheres taken together
(b) the sphere P
(c) the sphere Q

(a) **Free body diagram of spheres P and Q:** Note that the reaction at D does not appear in the free body diagram, it being an internal force between the two spheres.

Fig. 2.21 (Illustrative Problem 2.15)

(b) **Free body diagram of sphere P:** $R_{D(Q)}$ is the reaction of the sphere Q on the sphere P at the point of contact D, acting in the direction normal to the surface. That is along $C_1 C_2$.

(c) **Free body diagram of sphere Q:** $R_{D(P)}$ is the reaction of the sphere P on the sphere Q acting at the point of contact D.

It should be noted that, the reaction of the sphere P on the sphere Q, $R_{D(P)}$ and the reaction of the sphere Q on the sphere P, $R_{D(Q)}$ are equal in magnitude, opposite in direction and are collinear. So

$$R_{D(P)} = R_{D(Q)} = R_D$$

They do not appear in the combined free-body diagram of the two spheres as they form a pair of equal, opposite and collinear forces that cancel.

PROBLEM 2.16: A smooth circular cylinder of radius 1.5 metre is lying in a triangular groove, one side of which makes 15° angle and the other

40° angle with the horizontal. Find the reactions at the surfaces of contact, if there is no friction and the cylinder weighs 500 N.

Solution:

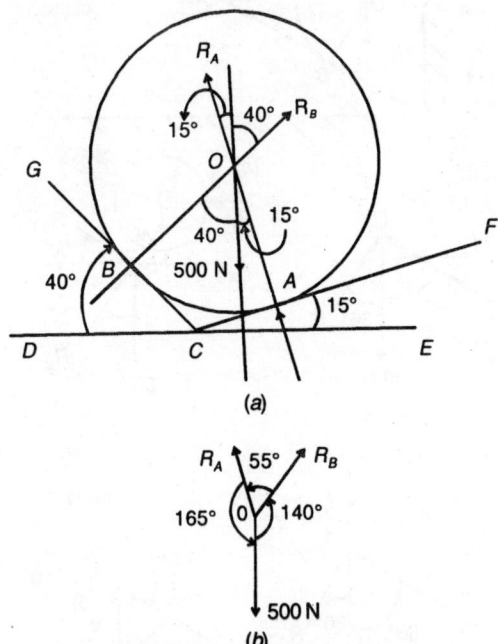

Fig. 2.22 (Illustrative Problem 2.16)

Given: Radius of the cylinder = 1.5 m

$\angle GCD = 40°$

$\angle ECF = 15°$

Let R_A = Reaction at A and

R_B = Reaction at B

The cylinder lying on a triangular groove is shown in Fig 2.22 (*a*). Since there is no friction, the reactions R_A and R_B will be normal to the surfaces as shown in Fig. 2.16 (*a*).

In order to keep the system in equilibrium, the three forces i.e., R_A, R_B and the weight = 500 N must act through θ, the c.g. of the cylinder. The angles between the forces will be as shown in Fig. 2.16 (*b*).

Applying Lami's theorem

$$\frac{R_A}{\sin 140°} = \frac{500}{\sin 55°} = \frac{R_B}{\sin 165°}$$

$$R_A = 500\frac{\sin 140°}{\sin 55°} = 392.5 \text{ N}$$

$$R_B = 500\frac{\sin 165°}{\sin 55°} = 157.98 \approx 158 \text{ N}$$

PROBLEM 2.17: Two equal heavy spheres of r units radius are in equilibrium within a smooth cup of $3r$ units radius. Show that the reaction between the cup and any one sphere is double that between the two spheres.

Solution:

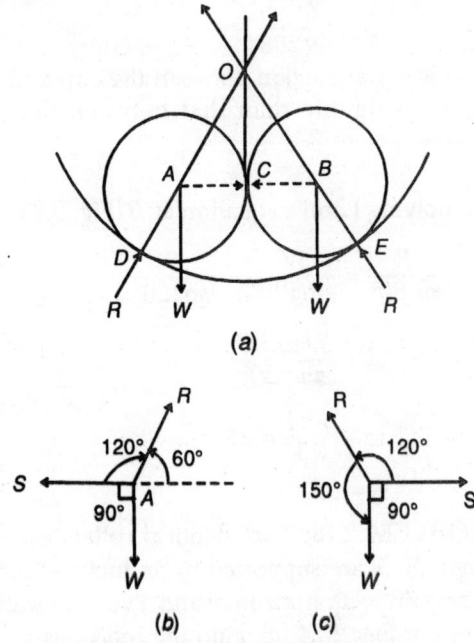

Fig. 2.23 (Illustrative Problem 2.17)

Given: Radius of spheres = 5 cm
Radius of cup = 15 cm

Let A and B be the two centres of the spheres in equilibrium in the cup with centre O, as shown in Fig. 2.23 (*a*)

Let R = Reaction between the sphere and the cup

S = Reaction between the spheres.

W = Weight of each sphere.

From the geometry of the figure, we find that

$OD = 15$ cm

$AD = 5$ cm

∴ $OA = OD - AD = 15 - 5 = 10$ cm

Similarly $OB = 10$ cm

We also know that $AB = 10$ cm
Therefore, OAB is an equilateral triangle.
Applying Lami's equation at A, (Fig. 2.23 (b)),

$$\frac{R}{\sin 90°} = \frac{W}{\sin 120°} = \frac{S}{\sin 150°}$$

$$R = \frac{S - \sin 90°}{\sin 150°}$$

$$= \frac{S \times 1}{0.5} = 2S$$

or $\qquad R = 2S$

Hence the reaction between the cup and the sphere is double than that between the two spheres.

or

Applying Lami's equation at B (Fig. 2.23 (c))

$$\frac{R}{\sin 90°} = \frac{W}{\sin 120°} = \frac{S}{\sin 150°}$$

$\therefore \qquad R = \dfrac{S \cdot \sin 90°}{\sin 150°}$

$$= \frac{S \times 1}{0.5} = 2S$$

or $\qquad R = 2S$

PROBLEM 2.18: Two identical rollers each of weight 50 N are supported by an inclined plane making 30° with horizontal and a vertical wall as shown in Fig. 2.24 (a). Find the reactions at the points of contact of the rollers with the inclined plane and the wall. All surfaces are smooth.

Solution:
Considering the free body diagram of roller 2 and its equilibrium (Fig. 2.24 (b)).

$\Sigma H = 0$

$R_D \sin 60° - R_C \sin 30° = 0$

$0.86603 R_D - 0.5 R_C = 0 \qquad \ldots(i)$

$\Sigma V = 0$

$R_D \cos 60° + R_C \cos 30° - 50 = 0$

$0.5 R_D + 0.86603 R_C = 50 \qquad \ldots(ii)$

Solving equations (i) & (ii)

$0.750 R_D - 0.433 R_C = 0$

$\underline{0.250 R_D + 0.433 R_C = 25}$

$\qquad R_D = 25$ N

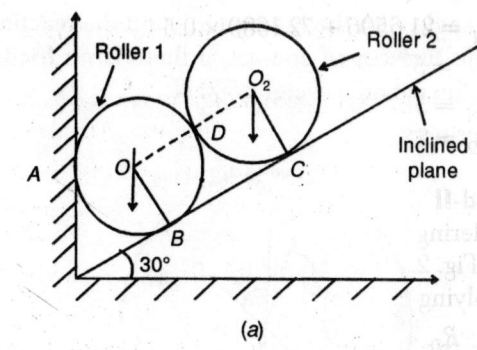

(a)

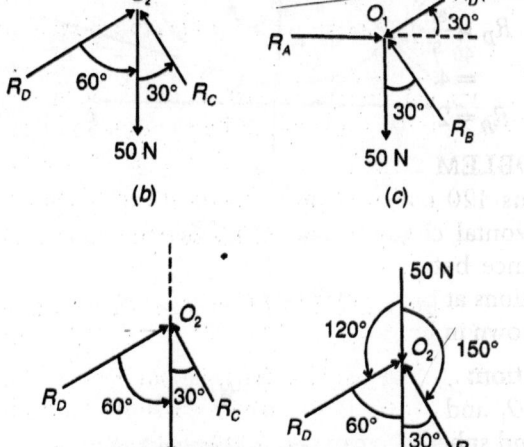

(b)   (c)

(d)   (e)

Fig. 2.24 (Illustrative Problem 2.18)

$$R_C = \frac{0.86603 \times 25}{0.5}$$

$$= 43.3 \text{ N}$$

Considering the free body diagram of roller 1 and its equilibrium (Fig. 2.24 (c)).

$\Sigma H = 0$

$R_A - R_D \cos 30° - R_B \sin 30° = 0$

$R_A - 21.6505 - 0.5 R_B = 0$

$R_A - 0.5 R_B = 21.6506 \qquad \ldots(iii)$

$\Sigma V = 0$

$-R_D \sin 30° - 50 + R_B \cos 30° = 0$

$-12.5 - 50 + R_B 0.8660254 = 0$

$$R_B = \frac{62.5}{0.660254} = 72.1689 \text{ N} \approx 72.17 \text{ N}$$

$\therefore \quad R_A = 21.6506 + 72.1689 \times 0.5$

$\qquad = 57.735$ N

$\qquad R_A = 57.735$ N; $R_B = 72.17$ N; $R_C = 43.3$ N

and $R_D = 25$ N

Method-II

Considering the free body diagram of the second roller (Fig. 2.24) (*d*) and (*e*))

Applying Lami's theorem

$$\frac{R_D}{\sin 150°} = \frac{50}{\sin 90°} = \frac{R_C}{\sin 110°}$$

$\therefore \quad R_D = \dfrac{\sin 120°}{\sin 90°} 50 = \dfrac{0866025 \times 50}{1}$

$\qquad = 43.30$ N

$R_D = 25$ N & $R_C = 43.30$ N

PROBLEM 2.19: Two smooth spheres each of radius 120 mm and weight 120 N rest in a horizontal channel having vertical walls, the distance between which is 360 mm. Find the reactions at the points of contacts *A*, *B*, *C* and *D* as shown in Fig. 2.25 (*a*)

Solution:

Let O_1 and O_2 be the centres of the first and second spheres. Drop perpendicular O_2E to the horizontal line through O_1. Since the surface of contacts are smooth, reaction of *D* is in the radial direction, i.e. in the direction of O_1O_2. Let it make angle θ with the horizontal. Then

$AO_1 = 120$ mm; $EF = 120$ mm; $AF = 360$ mm

$O_1E = AF - AO_1 - EF$

$O_1E = 360 - 120 - 120$

$\qquad = 120$ mm

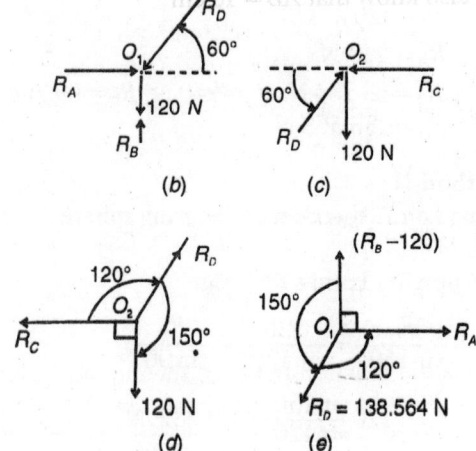

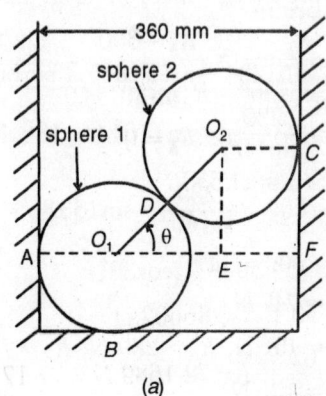

(a)

Fig. 2.25 (Illustrative Problem 2.19)

$O_1O_2 = O_1D + DO_2$

$\qquad = 120 + 120 = 240$ mm

$\therefore \quad \cos\theta = \dfrac{O_1E}{O_1O_2} = \dfrac{120}{240} = 0.5$

or $\qquad \theta = 60°$

$\sin\theta = \sin 60° = 0.866$

Method-I

Using equilibrium equation considering sphere (2) (Fig. 2.25 (*c*))

$\qquad \Sigma V = 0$

$\qquad R_D \sin 60° = 120$

$\qquad R_D \times 0.866 = 120$

or $\qquad R_D = 120/0.866 = 138.568$ N

$\qquad \Sigma H = 0$

$\qquad R_C = R_D \cos 60°$

$\qquad \quad = 138.568 \times 0.5$

$\qquad R_C = 69.284$ N

Consider sphere no. 1 (Fig. 2.25 (*b*))

$\qquad \Sigma H = 0$

$\qquad R_A = R_D \cos 60°$

$\qquad R_A = 138.568 \times 0.5$

$\qquad R_A = 69.284$ N

$\qquad \Sigma V = 0$

$\qquad R_B = 120 + R_D \sin 60°$

$\qquad \quad = 120 + 138.568 \times 0.866$

$$= 120 + 120$$

$$R_B = 240 \text{ N}$$

$$R_A = 69.284 \text{ N}; \ R_B = 240 \text{ N}; \ R_C = 69.284 \text{ N};$$
$$R_D = 138.568 \text{ N}$$

Method-II

Using Lami's theorem considering sphere (2) (Fig. 2.25 (*d*))

Applying Lami's theorem

$$\frac{R_C}{\sin 150°} = \frac{120}{\sin 120°} = \frac{R_D}{\sin 90°}$$

$$R_C = 120 \frac{\sin 150°}{\sin 120°} = 69.282 \text{ N}$$

$$R_D = 120 \frac{\sin 90°}{\sin 120°} = 138.564 \text{ N}$$

Considering sphere (1) (Fig. 2.25 (*e*))

Applying Lami's theorem

$$\frac{R_B - 120}{\sin 120°} = \frac{138.564}{\sin 90°} = \frac{R_A}{\sin 150°}$$

$$\therefore \quad R_B - 120 = 138.564 \times \frac{\sin 120°}{\sin 90°}$$

$$R_B - 120 = 120$$

$$R_B = 120 + 120 = 240 \text{ N}$$

$$R_A = 138.564 \times \frac{\sin 150°}{\sin 90°}$$

$$R_A = 69.282 \text{ N}$$

$$R_a = 69.282 \text{ N}; \ R_B = 240 \text{ N}; \ R_C = 69.282 \text{ N};$$
$$R_D = 138.564 \text{ N}$$

PROBLEM 2.20: Two cylinders A and B rest in a channel as shown in Fig. 2.26 (*a*). A has a diameter of 100 mm and weighs 240 N. B has a diameter of 180 mm and weighs 600 N. The channel is 180 mm wide at the bottom with one side vertical and other side at 120° as shown. Determine the reactions at all the four points of contact.

Solution:

$$\frac{O_1 E}{GE} = \tan 60°$$

$$\therefore \quad GE = \frac{90}{\tan 60°} = \frac{90}{1.7321} = 51.96$$

$$\approx 52 \text{ mm}$$

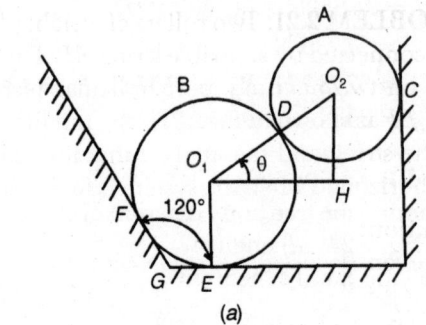

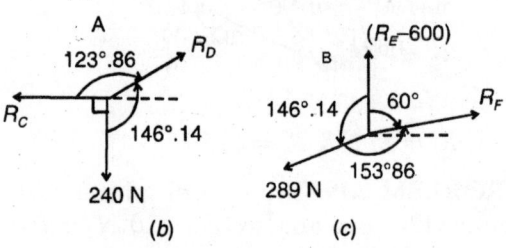

Fig. 2.26 (Illustrative Problem 2.20)

$$O_1 H = 180 - 52 - 50 = 78 \text{ mm}$$

$$\cos \theta = \frac{O_1 H}{O_1 O_2} = \frac{78}{140} = 0.557$$

$$\theta = 56.14°$$

Applying Lami's theorem (Fig. 2.26 (*b*))

$$\frac{R_C}{\sin 146.14°} = \frac{240}{\sin 123.86°} = \frac{R_D}{\sin 90°}$$

$$R_C = \frac{240 \times \sin 146.14°}{\sin 123.86°} = 161 \text{ N}$$

$$R_D = \frac{240 \sin 90°}{\sin 123.86°} = 289 \text{ N}$$

Applying Lami's theorem (Fig. 2.26 (*c*))

$$\frac{R_F}{\sin 146.14°} = \frac{R_E - 600}{\sin 153.86°} = \frac{289}{\sin 60°}$$

$$R_F = \frac{289}{\sin 60°} \times \sin 146.14° = 186 \text{ N}$$

$$R_E - 600 = \frac{289}{\sin 60°} \times \sin 153.86°$$

$$R_E - 600 = 147$$

$$\therefore \quad R_E = 747 \text{ N}$$

$$R_C = 161 \text{ N}; \ R_D = 289 \text{ N}; \ R_E = 747 \text{ N};$$
$$R_F = 186 \text{ N}$$

PROBLEM 2.21: Two rollers of weights P and Q are connected by a flexible string AB. The rollers rest on two naturally perpendicular planes DE and EF as shown in Fig. 2.21(a). Find the tension in the string and the angle θ that it makes with the horizontal when the system is in equilibrium.

Solution:

Given $P = 35$ N, $Q = 65$ N and $\alpha = 30°$

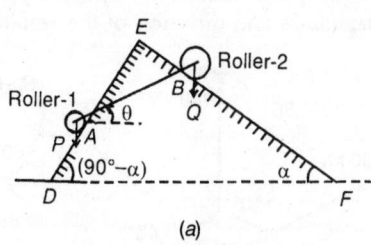

(a)

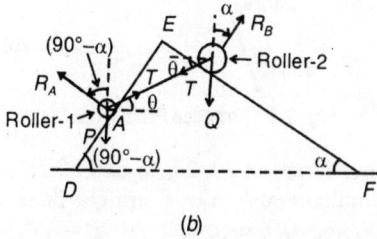

(b)

Fig. 2.27 (Illustrative Problem 2.21)

The free body diagram of each roller is shown separately in Fig. 2.21 (b).

Let T be the tension in the string R_A & R_B be the reactions of the planes on the rollers acting normal to planes DE and DF respectively.

Consider the equilibrium of the roller-1

$\Sigma H = 0$

$T\cos\theta - R_A\sin(90° - \alpha) = 0$

or $T\cos\theta = R_A\cos\alpha$ ...(i)

$\Sigma V = 0$

$T\sin\theta + R_A\cos(90° - \alpha) - P = 0$

$T\sin\theta = P - R_A\sin\alpha$ ...(ii)

Consider the equilibrium of the roller-2

$\Sigma H = 0$

$R_B\sin\alpha - T\cos\theta = 0$

$R_B\sin\alpha = T\cos\theta$ ...(iii)

$\Sigma V = 0$

$R_B\cos\alpha - T\sin\theta - Q = 0$

$-Q + R_B\cos\alpha = T\sin\theta$ ...(iv)

From (i) & (iii)

$T\sin\theta = P - T\cos\theta\cot\alpha$

or $P = T(\cos\theta\tan\alpha + \sin\theta)$ ...(v)

From (iii) & (iv)

$T\sin\theta = -Q + T\cos\theta\cot\alpha$

From (v)

or $Q = T(\cos\theta\cot\alpha - \sin\theta)$ ...(vi)

$\dfrac{(vi)}{(v)} = \dfrac{Q}{P} = \dfrac{\cos\theta\cot\alpha - \sin\theta}{\cos\theta\tan\alpha + \sin\theta}$

$Q\sin\theta + Q\cos\theta\tan\alpha = P\cos\theta\cot\alpha - P\sin\theta$

$(P + Q)\sin\theta = \cos\theta(P\cot\alpha - Q\tan\alpha)$

or $\dfrac{\sin\theta}{\cos\theta} = \tan\theta = \dfrac{(P\cot\alpha - Q\tan\alpha)}{(P+Q)}$

$$T = \frac{P}{\cos\theta\tan\alpha + \sin\theta} = \frac{Q}{\cos\theta\cot\alpha - \sin\theta}$$

Substituting $P = 35$ N, $Q = 65$ N & $\alpha = 30°$

$\tan\theta = \dfrac{35\cot30° - 65\tan30°}{35 + 65} = \dfrac{23.094}{100}$

$\tan\theta = 0.23094$

$\theta = \tan^{-1}(0.23094) = 13°$

$\therefore \quad T = \dfrac{35}{\cos13° \times \tan30° + \sin13°}$

$= \dfrac{65}{\cos13° \times \cot30° - \sin13°} = 44.44$ N

PRACTICE PROBLEM

2.1 Two forces act at an angle of 120°. The bigger force is of 500 N and the resultant is perpendicular to the smaller one. Find the smaller force.

2.2 The resultant of two forces P and $Q = R$. If Q is doubled, the new resultant is perpendicular to P. Prove that $Q = R$.

2.3 Two forces, whose magnitudes are P and $P\sqrt{2}$, act at an angle of 135° to each other. Find the magnitude and direction of the resultant.

2.4 Two forces of magnitude 10 N and 15 N are acting at a point. If the angle between the forces is 30°, determine the magnitude of the resultant force.

2.5 Two forces are acting at a point O as shown in Fig. 2.28. Determine the resultant in magnitude and direction.

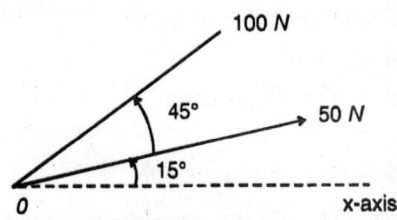

Fig. 2.28 (Practice Problem 2.5)

2.6 Two forces P and Q are acting at a point O as shown in Fig. 2.29. The resultant force is 300 N and angles β and γ are 40° and 20° respectively. Find the two forces P and Q.

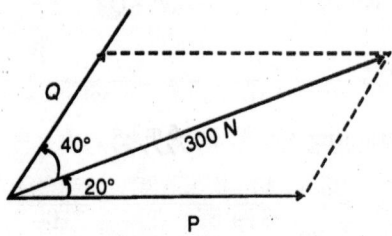

Fig. 2.29 (Practice Problem 2.6)

2.7 A small block of weight 150 N is placed on an inclined plane which makes an angle θ = 30° with the horizontal. What is the component of this weight ?

   (*i*) parallel to the inclined plane

   (*ii*) perpendicular to the inclined plane

2.8 A vector of magnitude 100 units makes an angle of 60° with the *z*-axis and its projection on the *x*-*y* plane makes an angle of 45° with the *x*-axis. Determine (*a*) the components of the vector and (*b*) the angles of the vector with the axes.

2.9 Find the magntitude and direction of the resultant of the four concurrent forces of 50 N, 100 N, 150 N and 200 N making angles of 30°, 60°, 120° and 150° respectively with a fixed line.

2.10 A particle is acted upon simultaneously by the following forces:

   (*i*) 150 N inclined 60° to the North of East

   (*ii*) 200 N towards the North.

   (*iii*) 250 N towards North West.

   (*iv*) 300 N inclined 50° to South of West

   (*v*) 350 N towards South East

Determine the magnitude and direction of the resultant.

2.11 Four forces of magnitude 10 kN, 20 kN, 30 kN and 40 kN are acting at a point O as shown in Fig. 2.30. The angles made by 10 kN, 20 kN, 30 kN and 40 kN with *x*-axis are 45°, 90°, 150° and 240° respectively. Find the magnitude and direction of the resultant.

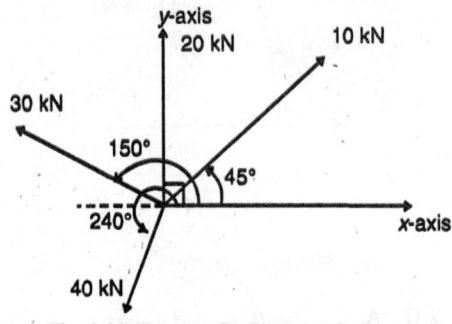

Fig. 2.30 (Practice Problem 2.11)

2.12 Four forces of 20, 25, 15 and 30 N are acting simultaneously along straight lines *OA*, *OB*, *OC* and *OD*, such that $\angle AOB = 45°$, $\angle BOC = 105°$, and $\angle COD = 120°$. Find the magnitude and direction of the resultant.

2.13 Find the resultant and equilibrant of the four collinear forces 200 N and 100 N, acting towards the right and two forces 150 N and 50 N acting to the left.

2.14 In a jib crane, jib is 5 m long, and the tie rod is horizontal having a length of 4 m. The height of the crane post is 3 m. If a load of 3 kN is suspended at crane head, find out the forces produced in the jib and tie rods.

2.15 A weight of 900 N is supported by two chains shown in Fig. 2.31. Determine the tension in each chain.

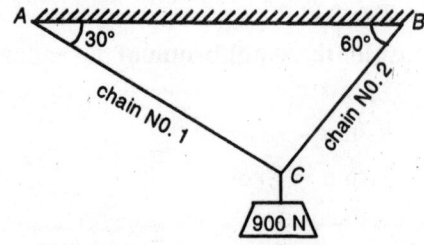

Fig. 2.31 (Practice Problem 2.15)

2.16 An electric light fixture weighing 20 N hangs from a point C, by two strings AC and BC. AC is inclined at $60°$ to the horizontal and BC at $45°$ to the vertical as shown in Fig. 2.32. Using Lami's theorem or otherwise, determine the forces in the strings AC and BC.

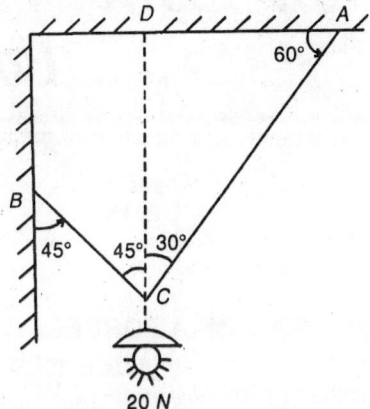

Fig. 2.32 (Practice Problem 2.16)

2.17 A uniform sphere of weight W rests between a smooth vertical plane and a smooth plane at an angle θ with the first plane. Determine the reactions at the contact surfaces.

2.18 A sphere of weight 120 N is tied to a smooth wall by a string as shown in Fig. 2.33. Find the tension in the string and reaction R of the wall.

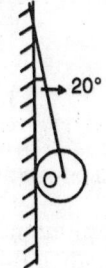

Fig. 2.33 (Practice Problem 2.18)

2.19 A roller of weight 15 kN rests on a smooth horizontal floor and is connected to the floor by the bar AC as shown in Fig. 2.34. Determine the force in the bar AC and reacton from floor if the roller is subjected to a horizontal force of 10 kN and an inclined force of 12 kN as shown in Fig. 2.34.

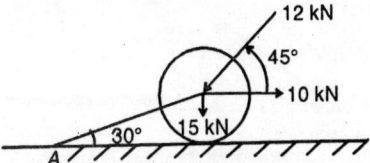

Fig. 2.34 (Practice Problem 2.19)

2.20 A cord AB 6.0 m long is connected at two points A and B at the same level 5 m apart. A load of 20 kN is suspended from a point D on the cord 1.5 m from B as shown in Fig. 2.35. A load of W is suspended from a point C on the cord 2 m from A. Determine the value of W so that CD remains horizontal.

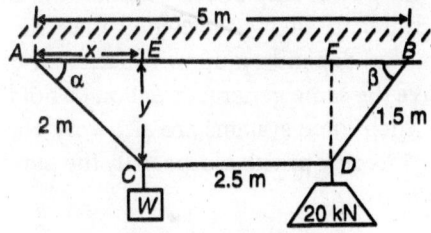

Fig. 2.35 (Practice Problem 2.20)

2.21 A machine shaft BC 1.5 m long and weighing 1000 N is supported by two strings AB and CD as shown in Fig. 2.36. Find the tensions T_1 and T_2 in the ropes AB and CD.

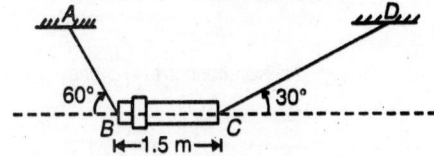

Fig. 2.36 (Practice Problem 2.21)

Chapter 3

Coplanar Non-Concurrent Forces

3.1 INTRODUCTION

If all the forces of a system lie in the same plane and the forces are not concurrent, the system is said to be a 'coplanar non-current force system'.

A special case arises when all the forces act parallel to each other; they may be 'like' or 'unlike' depending upon whether all the forces have the same general direction or not (Fig. 3.1).

Such force systems are also very common in engineering practice, especially the parallel force systems.

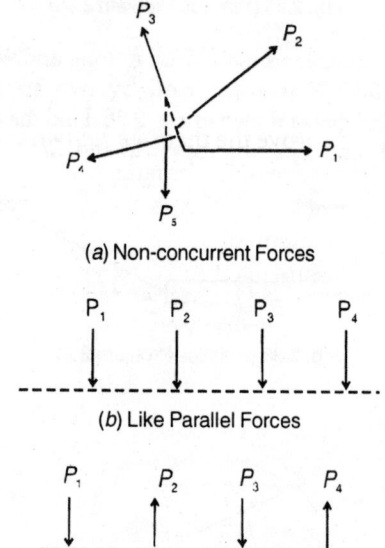

(a) Non-concurrent Forces

(b) Like Parallel Forces

(c) Unlike Parallel Forces

Fig. 3.1: Coplanar Non-Current Force Systems

3.2 MOMENT OF A FORCE

'Moment' of a force about a point is defined as the product of the magnitude of the force and the perpendicular distance of the point from the line of action of the force. It is the measure of its 'rotational' effect about the point, which is called the 'Moment Centre'. The perpendicular distance of the point from the line of action of the force is called the 'moment arm' or 'lever arm'. (Fig. 3.2).

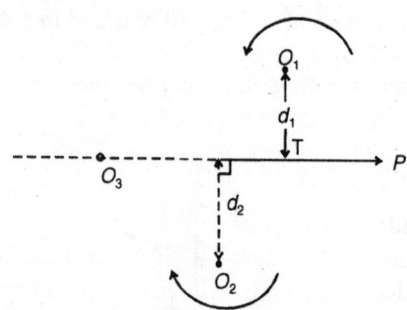

Fig. 3.2: Moment of a Force

In Fig. 3.2, the moment of the force P about O_1 is $P \cdot d_1$; the moment of the force P about O_2 is $P \cdot d_2$; and its moment about O_3 is zero (since the moment centre lies on the line of action of the force, the moment arm is zero).

3.2.1 Sign of Moment of a Force

Since force is a vector, moment of a force is also a vector, having rotational sign or sense. The sign of a moment is decided based on the tendency for rotation of the moment centre under its influence.

In Fig. 3.2, the moment of force P about O_1, is counter-clockwise and that about O_2 is clockwise. Conventionally counterclockwise moments are considered to be positive while clockwise moments are considered to be negative. The tendency for rotation will be clear if it is imagined that the moment centre is connected by a rigid rod perpendicular to the line of action of the force and pinned at the moment centre.

The cumulative effect of two opposite kinds of moments on a rigid body is obtained as the resultant moment, which is simply the algebraic sum of these moments. For example, the resultant moment in the case of two forces P_1 and P_2 about a moment centre O with moment arms d_1 and d_2 as shown in Fig. 3.3 is given by $P_1 \cdot d_1 \sim P_2 \cdot d_2$, with the sign as that of the greater one.

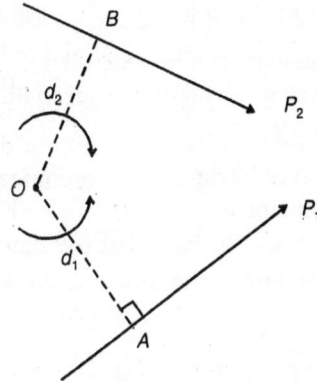

Fig. 3.3: Resultant of Several Moments

3.2.2 Unit of Moment
Since moment of a force is the product of the force and the moment arm from the moment centre, the unit of the moment is Newton-mm (N.mm), Newton-metre (N·m), or kN·m.

In engineering applications, the last two— N·m and kN·m—are very commonly used.

3.2.3 Geometrical Interpretation of Moment
Let us consider the moment of a force P about a moment centre O as shown in Fig. 3.4.

The moment of a force P about the moment centre O is given by the product $P \cdot d$, where d is the moment arm, or the perpendicular distance of O from the line of action of the force.

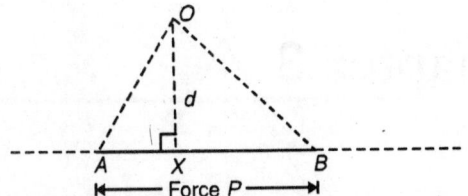

Fig. 3.4: Geometrical Interpretation of Moment of a Force

If the force P is represented vectorially by AB and the moment arm by $OX (= d)$,

$$\text{Moment} = P \cdot d = (AB)(OX) = 2(\triangle AOB)$$

Thus the area of the triangle formed by the ends of the force-vector drawn on its line of action as the base and the moment centre as the vertex represents half the moment of the force about the particular moment centre.

Or, the moment of the force is twice the area of such a triangle.

3.3 VARIGNON'S THEOREM
The following theorem is due to VARIGNON (1654-1722), a French mathematician:

The algebraic sum of the moments of a coplanar force system about any point in its plane is equal to the moment of the resultant of the system about the same point.

This is also known as the "Principle of Moments". The theorem may be easily proved for a concurrent force system (Fig. 3.5); it may be extended to a general coplanar force system in a later section.

It is easier to prove the theorem for two forces and extend it to several forces later.

In (*b*) the resultant R of two forces P_1 and P_2, alongwith their moment arms d, d_1 and d_2 with respect to the moment centre O, is shown.

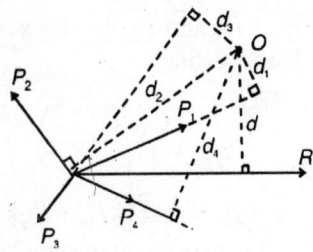

(*a*) General Coplanar Force System

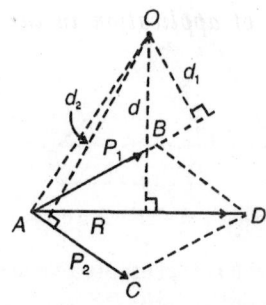

(b) Resultant of Two Forces

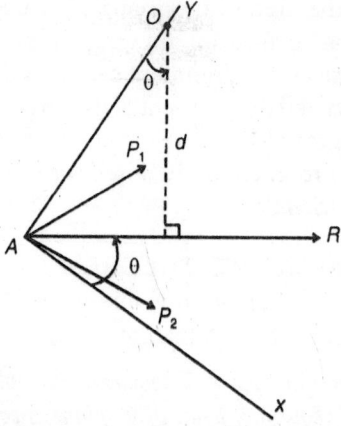

(c) Rectangular Components of the Resultant of Two Forces
Fig. 3.5: VARIGNON'S Theorem for a Concurrent Force System

In (c) the rectangular components of the resultant R with respect to the line joining the point of action of the forces and the moment centre as y-axis and an axis perpendicular to it as x-axis are shown (θ is the angle made by R with respect to the x-axis).

$$R \cdot d = R \times \overline{OA} \cdot \cos\theta$$
$$= \overline{OA} \times (R\cos\theta)$$

or $R \cdot d = \overline{OA} \times R_x$, R_x being the x-component of R
...(Eq. 3.1)

Similarly, if P_{1x} and P_{2x} are the x-components of P_1 and P_2 respectively,

$$P_1 \cdot d_1 = \overline{OA}(P_{1x})$$

and $P_2 \cdot d_2 = \overline{OA}(P_{2x})$

Hence $P_1 \cdot d_1 + P_2 \cdot d_2 = \overline{OA}(P_{1x} + P_{2x})$

But, from the theorem of resolved parts,

$$P_{1x} + P_{2x} = R_x$$

$\therefore P_1 \cdot d_1 + P_2 \cdot d_2 = \overline{OA} \cdot R_x$ ...(Eq. 3.2)

From equations 3.1 and 3.2; we have

$$R \cdot d = P_1 \cdot d_1 + P_2 \cdot d_2$$

which proves Varignon's theorem for two forces.

Now the proof can be extended to several forces as follows:

Let R_1 be the resultant of P_1 and P_2 and let its moment arm about O be d_1'.

Applying Varignon's theorem for P_1, P_2 and R_1,

$$R_1 \cdot d_1' = P_1 \cdot d_1 + P_2 \cdot d_2$$

If R_2 is the resultant of R_1 and P_3 (and so of P_1, P_2 and P_3) and its moment arm about O is d_2', from Varignon's theorem,

$$R_2 \cdot d_2' = R_1 \cdot d_1' + P_3 \cdot d_3$$
$$= P_1 d_1 + P_2 d_2 + P_3 d_3$$

Thus the logic can be extended continuously until the system of forces is covered fully.

$$R \cdot d = \Sigma P_1 d_1$$ ...(Eq. 3.3)

This proves Varignon's theorem for a system of concurrent forces.

It can be shown later that the theorem may also be extended to a general coplanar force system.

3.4 PARALLEL FORCES

A system of parallel forces is shown in Fig. 3.1 (b) and (c) – (b) represents a 'like' parallel force system in which the general direction of all the forces is same, while (c) represents an 'unlike' parallel force system. Such force systems are commonly encountered in engineering practice.

3.4.1 Resultant of Two Parallel Forces

The parallelogram law of forces cannot be applied directly to determine the resultant of two parallel forces; it can be applied indirectly.

The two cases of 'like' and 'unlike' forces will be considered separately.

Like Forces

Two like parallel forces P_1 and P_2 are shown in Fig. 3.6.

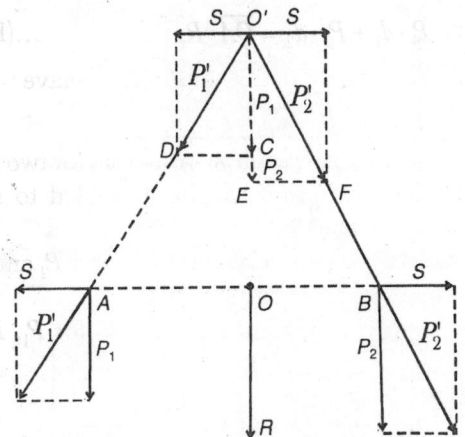

Fig. 3.6: Resultant of Two like Parallel Forces

Let two forces, S each, be introduced in opposite directions at the points of action A and B of P_1 and P_2 as shown, which do not alter the system. The resultants of S and P_1 at A (P_1') and of S and P_2 at B (P_2') are obtained by the parallelogram law. These sets of forces are transmitted to the point of intersection O' of the lines of action of P_1' and P_2' as shown.

It is obvious that the forces S directed opposite to each other cancel each other; further, the forces P_1 and P_2 which act along the same line $O'O$ and in the same direction merely add to each other.

The resultant R is therefore, given by

$$R = P_1 + P_2 \qquad \text{...(Eq. 3.4)}$$

This acts in the same direction—parallel to P_1 and P_2.

For determining the position O of the line of action of R, sets of similar triangles $O'CD$, $O'OA$ and $O'EF$ and $O'OB$ can be used:

$$\frac{O'O}{OA} = \frac{O'C}{CD} = \frac{P_1}{S} \qquad \text{...(Eq. 3.5)}$$

$$\frac{O'O}{OB} = \frac{O'E}{EF} = \frac{P_2}{S} \qquad \text{...(Eq. 3.6)}$$

Dividing (Eq. 3.5) by (Eq. 3.6),

We get $\qquad \dfrac{OB}{OA} = \dfrac{P_1}{P_2} \qquad$ ... (Eq. 3.7)

Thus it may be stated:

The resultant of two like parallel forces is equal to the sum of the forces, acts parallel to them, and its position is such that it divides the distance between their

points of application in the ratio inversely proportional to their magnitudes.

Analytically, the resultant of two or more parallel forces may be determined by summing up their rectangular components and applying the method of resolved parts; the position of the line of action of the resultant may be determined by applying Varignon's theorem.

Unlike Forces

Let two unequal parallel forces P_1 and P_2 acting at A and B be considered (Say P_1 is greater than P_2) (Fig. 3.7).

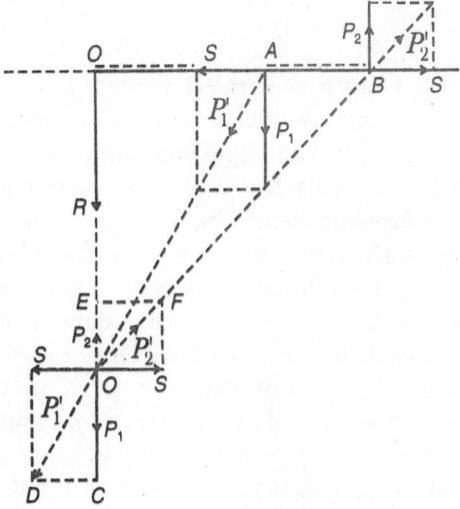

Fig. 3.7: Resultant of Two unlike Parallel Forces

A graphical procedure similar to that adopted for like forces may be adopted for this case also.

Alternatively, an analytical procedure using the method of resolved parts and Varignon's theorem may also be obtained to yield the same results.

From Fig. 3.7, the resultant R of P_1 and P_2 is given by

$$R = P_1 - P_2 \ (P_1 > P_2) \qquad \text{...(Eq. 3.8)}$$

From the sets of similar triangles $O'CD$, $O'OA$ and $O'EF$, $O'OB$,

$$\frac{O'O}{OA} = \frac{O'C}{CD} = \frac{P_1}{S}$$

$$\frac{O'O}{OB} = \frac{O'E}{EF} = \frac{P_2}{S}$$

Whence $\dfrac{OB}{OA} = \dfrac{P_1}{P_2}$ ...(Eq. 3.9)

Since $P_1 > P_2$, $\dfrac{OB}{OA} > 1$ or $OB > OA$

Thus the location of the resultant R in this case will be outside the line joining the points of action of P_1 and P_2, and on the side of the greater force; P_1, as shown.

Note: The procedure can be extended to any number of parallel forces by repeating it taking two forces at a time, while the graphical procedure is laborious, the analytical approach using Varignon's theorem.

3.4.2 Centre of Parallel Forces

If three like parallel forces act at three non-collinear points on a rigid body, the resultant may be found using the procedure stated in the previous subsection. The point of action of the resultant remains unaltered even if the body is rotated, keeping the force system constant; the same is the case if the force system is rotated keeping the body in position. This can be easily understood from the fact that only the magnitudes of the forces and their locations, and not their directions, govern the location of the point of action of the resultant of a system of forces. If the parallel forces are due to the gravitational force of the earth on the body, this centre of parallel forces is known as the 'Centre of Gravity'; this will be considered in detail in Chapter 5.

3.5 COUPLES

A system of two parallel forces, equal in magnitude and opposite in direction, constitute a 'couple' (Fig. 3.8). The perpendicular distance between these two parallel forces is called the 'lever arm' or simply the 'arm' of the couple.

The algebraic sum of the forces of a couple is obviously zero; hence, there will be no translation of a body under its influence. But there will be a rotational effect when a couple acts on a body, owing to the 'moment' caused by it.

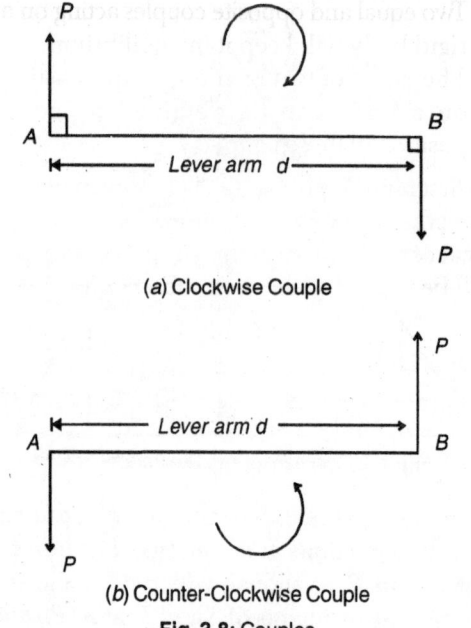

(a) Clockwise Couple

(b) Counter-Clockwise Couple

Fig. 3.8: Couples

The moment caused by the couple is the product of the force P and its lever arm $d - P \cdot d$.

The sense of the moment in Fig. 3.8 (a) is clockwise and that in (b) is counter-clockwise; since their rotational effect is in opposite directions, a clockwise couple is conventionally treated as a positive couple, a counter-clockwise one as a negative couple. These are shown by curved arrows.

3.5.1 Properties of Couples

The following are the additional properties of couples:

(i) The moment of a couple, a measure of its turning capacity or rotational effect, is a constant about any point.

(ii) The effect of a couple is unaltered when it is shifted to any other position in its own plane or into any parallel plane.

(iii) The couple may be rotated through any angle without changing its effect.

(iv) The effect of two couples is same if their moment is same in magnitude and sense that is to say, the pair of unlike parallel forces may be changed and the lever arm suitably changed to produce the same effect.

(v) Two equal and opposite couples acting on a rigid body will keep it in equilibrium.

(vi) The effect of two or more couples acting on a body can be obtained by composition of these couples.

The first may be proved easily as shown below (Fig. 3.9); then the next four automatically follow as logical consequences of this property. The last one will be treated in the next subsection.

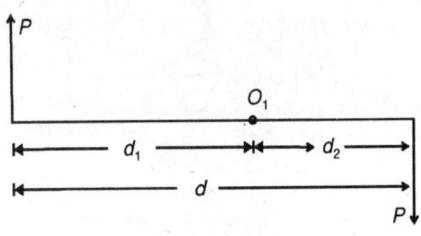

(a) Moment Centre Inside

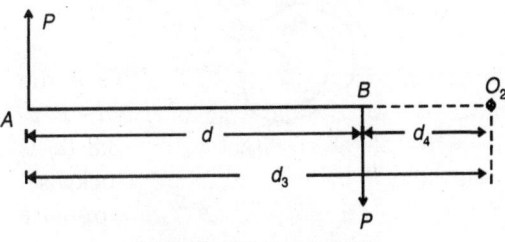

(b) Moment Centre Outside

Fig. 3.9: Moment of a Couple

The resultant moment M by the couple about the moment centre O_1 (within the lever arm) [Fig. 3.9 (a)] is given by

$M_1 = P \cdot d_1 + P \cdot d_2$ (both clockwise moments add to each other)

$= P(d_1 + d_2) = P \cdot d$ (from the geometry of the figure)

Similarly the moment M_2 about O_2 (outside the lever arm) [Fig. 3.9 (b)] is given by

$M_1 = Md_3 - Pd_4$ (the first term is clockwise moment and and second counter-clockwise one)

$= P(d_3 - d_4)$

$= P \cdot d$ (from the geometry of the figure)

$\therefore M_1 = M_2 = P \cdot d$, which is a constant value for a given couple, irrespective of the moment centre chosen. Thus, a couple can never be equivalent to a single force in view of this property, which cannot be satisfied by the latter. This proves the first property.

As already stated the next four follow as a logical consequence of the first. Hence a couple is treated as a 'free' vector.

3.5.2 Composition of Couples

When two or more couples act in the same plane their resultant can be obtained as the algebraic sum of their moments. If L_1, L_2, L_3, ... are the couples and M_1, M_2, M_3, ... are their moments, the moment M of the resultant couple L is given by

$$L = L_1 + L_2 + L_3 + \ldots = M_1 + M_2 + M_3 \ldots,$$
algebraically.

The sign convention for moments comes in handy here.

Composition of couples acting in different planes may also be achieved, but is not treated here.

3.6 RESOLUTION OF A FORCE INTO A FORCE AND A COUPLE

A force acting at a point on body can be shown to be equivalent to a system consisting of the same force acting at another point and a couple (Fig. 3.10).

In (a), force P acts at a point A as shown.

In (b), equal and opposite force P are intro-duced at another point B, parallel to the one at A; this does not alter the situation in (a).

In (c), the equivalent system is shown by taking P at B parallel to the one at A, and the equal and opposite forces P at a perpendicular distance d, acting through A and B, constituting a couple with its moment $M (= P \cdot d)$.

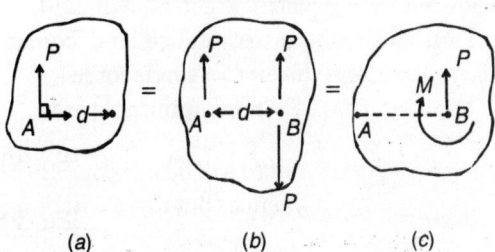

(a)   (b)   (c)

Fig. 3.10: Resolution of a Force into a Force and a Couple

Thus the given force P at A is replaced by the same force P at B and a couple with a moment $P \cdot d$, d being the perpendicular distance from B on to the line of action of P.

Note: The converse process of replacing a force-couple system by a single force at a different point is also possible.

This has many engineering applications in structural analysis.

3.7 RESULTANT OF GENERAL COPLANAR FORCE SYSTEMS

The resultant of a general coplanar force system may be found such that it has the same translational and rotational effect as the given system. *The resultant may be a single force, a pure couple, or a force and a couple.*

Let P_1, P_2, P_3, P_4, ... constitute a system of coplanar forces acting on a rigid body (Fig. 3.11).

Each of the forces can be replaced by a force of the same magnitude and acting in the same direction at a chosen point O by the principle of transmissibility and a moment about O, as seen in the previous section. Thus the system shown in Fig. 3.11 (*a*) is equivalent to that shown in (*b*)— the same system reduced to a concurrent force system and a moment, ΣM_O, the algebraic sum of the moments of all the forces about O.

The concurrent force system at O may now be combined to get the resultant force R as usual; this leads to the equivalent system shown in (*c*)— a force R at O and a moment ΣM_O.

The system shown in (*c*) is a force and a moment, or a couple, which can be now transformed into a single force R acting at a distance d from O such that the moment produced by R equals ΣM_O, as shown in (*d*).

Thus the resultant of the general coplanar system may be reduced to a single force.

The analytical equations applicable are:

$$R = \sqrt{(\Sigma P_x)^2 + (\Sigma P_y)^2} \qquad \text{... (Eq. 3.10)}$$

$$\tan \alpha = \Sigma P_y / \Sigma P_x \qquad \text{... (Eq. 3.11)}$$

$$d = \Sigma M_O / R \qquad \text{...(Eq. 3.12)}$$

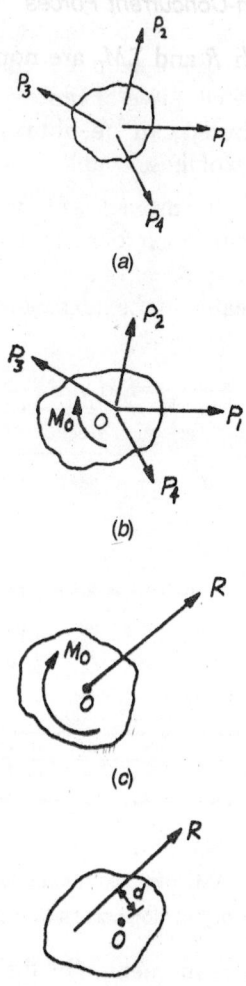

(a)

(b)

(c)

(d)

Fig. 3.11: Resolution of a General Coplanar Force System

Here ΣP_x and ΣP_y are the algebraic sums of the x-components and y-components of all the forces respectively, and is the inclination of the resultant R to the x-direction.

Note:

1. R must be located such that it produces the sense of the moment ΣM_O about O.

2. If the resultant R reduces to zero, ΣP_x and ΣP_y will be zero, but ΣM_O may exist. The system then reduces to a pure moment or a couple.

3. If the resultant R is not zero, but ΣM_O reduces to zero, the system reduces to a single force.

4. If both R and ΣM_0 are non-zero values, the system reduces to a force and a couple.

5. Sometimes, we may be interested in determining the x-and y-intercepts of the line of action of R with the origin at O.

These may be got as

$$x = \frac{d}{\sin \alpha} \quad \text{and} \quad y = \frac{d}{\cos \alpha}$$

or as $x = \Sigma M_0 / R_y = \Sigma M_0 / \Sigma P_y$

and $y = \Sigma M_0 / R_x = \Sigma M_0 / \Sigma P_x$

by applying Varignon's theorem.

3.8 EQUILIBRIUM CONDITIONS

The conditions required to be satisfied by a general coplanar force system to keep a body in equilibrium may now be easily deduced.

The resultant of the system R and the moment ΣM_0 shall reduce to zero for equilibrium. If the resultant R is imagined to be resolved into its x- and y-components R_x and R_y, both R_x and R_y shall reduce to zero.

But $R_x = \Sigma P_x$ and $R_y = \Sigma P_y$

by the theorem of resolved parts.

Thus the conditions of equilibrium for a general coplanar force system may be summarized as follows:

$$\left.\begin{array}{l} \Sigma P_x = 0 \\ \Sigma P_y = 0 \\ \text{and } \Sigma M_0 = 0 \end{array}\right\} \qquad \text{...(Eq. 3.13)}$$

Now it can be easily proved that "if a system is in equilibrium under the action of three forces, these three forces shall be concurrent" (Fig. 3.12).

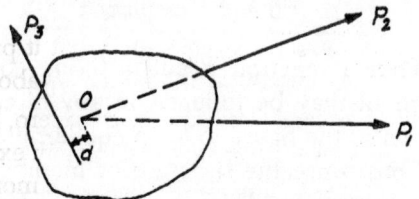

Fig. 3.12: Three forces in equilibrium

Let P_1 and P_2 intersect at the point O. Then applying the moment equilibrium condition about O, we have,

$$\Sigma M_0 = 0$$

Since P_1 and P_2 pass through O, they do not produce any moment about O. Now the moment of P_3 about O, i.e., $P_3 \cdot d$ (d being the perpendicular distance of O from the line of action of P_3) shall be zero. This is possible only if d is zero since $P_3 \neq 0$. In otherwords, P_3 also shall pass through O.

Or, this leads to the proposition that when a three-force system keeps a body in equilibrium, the three forces have to be necessarily concurrent.

3.9 REACTIONS AT SUPPORTS

Civil Engineering structures, elements or members of machines, and the like will need to be supported appropriately for providing stability and capability to serve the intended function of transmission of applied loads or forces in a safe manner.

At such supports, reactive forces or 'reactions', as they are called, are induced to keep the applied load/force system in equilibrium, so that the structure or machine member is, on the whole, kept in equilibrium.

The determination of the reactions at the supports is an essential prerequisite for the analysis of the structure, which, in turn, is an essential step in the design of the structure. As a prelude to this let us consider the types of supports, types of structures, and types of loading which occur in engineering practice.

3.9.1 Types of Supports

The types of supports are:

   (*i*) Simple Support (or knife-edge support)
   (*ii*) Roller Support
   (*iii*) Hinged Support (or Pinned Support)
   (*iv*) Fixed Support

(i) Simple Support (or Knife-edge support)

If the end of the member merely rests on the support, or on a knife-edge, it acts as a 'simple support' (Fig. 3.13).

The action of these two is one and the same in providing reaction to any applied force system on the member.

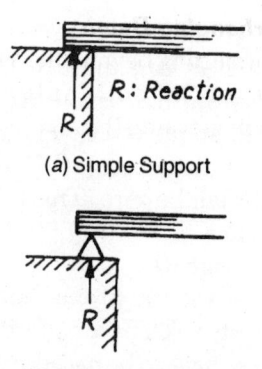

(a) Simple Support

(b) Knife-edge Support

Fig. 3.13: Simple Support

The reaction at the support will be a force at right angles to the plane of support for the end or the knife-edge. The end of the member is then restrained against any motion or displacement in a direction at right angles to the plane of the support

(It can move parallel to the plane of the support under the influence of force components in this directions and also rotate about the support).

(ii) Roller Support

The end of the member supported on rollers, which permit motion parallel to the plane of the support (Fig. 3.14). (The rollers may be treated as being smooth or frictionless).

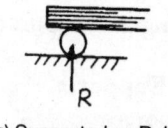

(a) Supported on Roller

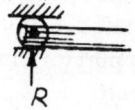

(b) Connected to Roller between Guides

Fig. 3.14: Roller Support

At the roller support, the reaction is always perpendicular to the plane in which motion is permitted, irrespective of the nature of the applied force system. (Thus the end of the member can move parallel to the plane of

support and can also rotate about the support, while it is restrained against motion normal to the plane of support).

(iii) Hinged Support (or Pinned Support)

The end of the member is connected to a frictionless 'pin' or 'hinge' firmly fixed to a rigid surface (Fig. 3.15).

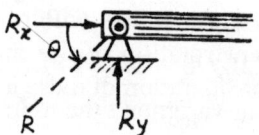

Fig. 3.15: Hinged or Pinned Support

The end of the member is thus restrained against any translatory motion in any direction, but rotation of the end is permitted through the pin or hinge. Thus the support provides a reaction R in any direction to keep equilibrium but does not provide any resisting moment.

The reaction R may always be resolved into its rectangular components R_x and R_y ($R_x = R\cos\theta$ and $R_y = R\sin\theta$).

(iv) Fixed Support

If the end of the member is built into a rigid surface so as to prevent both translation and rotation at the support, it is known as a fixed support (Fig. 3.16).

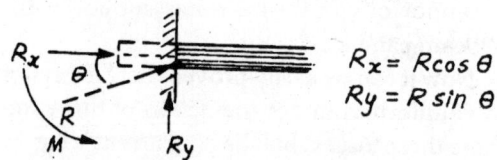

$$R_x = R\cos\theta$$
$$R_y = R\sin\theta$$

Fig. 3.16: Fixed Support

Thus a reaction R and a fixing couple or moment may be induced in any direction as shown in the figure to keep equilibrium.

Thus, since the structure or member which may be supported at one or more such supports has to be in equilibrium under an applied force system, the reactions at the supports may be found by using the principles seen in this and the previous chapter.

3.9.2 Types of Structures

Broadly speaking, the types of Civil Engineering structures may be classified as follows, from the analysis point of view:

(a) Beams,
(b) Rigid Frames, and
(c) Pin-jointed frames (or Trusses)

Rigid frames consist of two or more members joined by rigid joints which do not permit independent rotation of the members joining there; a consideration of these is out of scope of this text.

Beams are structural members with considerably more length than the cross-sectional dimensions, loaded with loads/forces applied transverse to the axis. The most common system is one which consists of only vertical loads. Since the beams are to be appropriately supported, reactions are induced at these supports to keep the beam in equilibrium under the system of applied loads. These reactions will depend upon the nature of the supports in addition to the system of loads, as seen in the last subsection.

The types of beams are as follows:

(i) Simply Supported Beam

A beam with both ends provided with simple supports or hinged/roller supports is called a 'simply supported beam' (Fig. 3.17) (both supports cannot obviously be roller supports since the beam cannot be stable).

Fig. 3.17: Simply Supported Beam

(ii) Cantilever Beam

A beam fixed at one end and free at the other is called a 'cantilever beam' (Fig. 3.18).

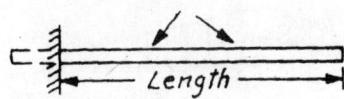

Fig. 3.18: Cantilever Beam

(iii) Overhanging Beam

A beam projecting beyond the support, either to one side or to both sides is called an 'Overhanging beam'. (Fig. 3.19).

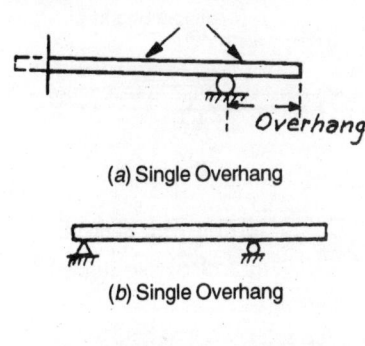

(a) Single Overhang

(b) Single Overhang

(c) Double Overhang

Fig. 3.19: Overhang Beams

The overhang portion acts like a cantilever from analysis point of view.

(iv) Propped Cantilever

A cantilever beam with an additional support (or prop) is called a propped cantilever (Fig. 3.20).

Fig. 3.20: Propped Cantilever

(v) Fixed Beam

A beam with both ends fixed or built-in is called a "fixed beam" (Fig. 3.21).

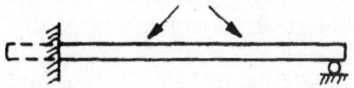

Fig. 3.21: Fixed Beam

(vi) Continuous Beam

A beam with more than two supports is called a "continuous beam"; the beam is continuous over the intermediate supports (Fig. 3.22).

The ends of continuous beams may be both fixed, both hinged/rollers, or one fixed and the other hinged; overhangs also may be there either to one side or onto both sides.

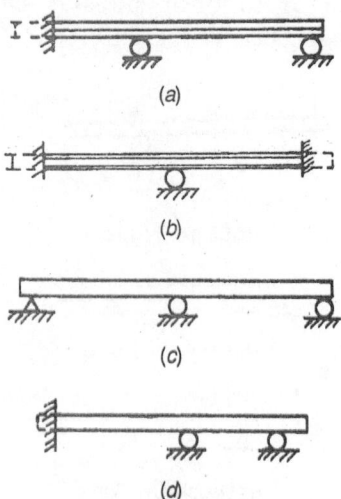

Fig. 3.22: Continuous Beams

Of the above types, (*i*), (*ii*) and (*iii*) can be solved for the reactions, etc. for any loading by using the laws of static equilibrium (Eq. 3.13) only, and hence are known as "Statically Determinate Beams". The remaining types cannot be solved using the laws of static equilibrium alone, and need additional conditions/equations for the purpose; hence such beams are known as "Statically Indeterminate Beams".

'Pin-jointed Frames' or 'Trusses' consist of several members joined by pinned ends and supported appropriately to transmit forces/loads. The members of such frames are subjected to only axial forces—tension or compression.

Beams will be treated in detail in Chapter 20 and Frames or Trusses will be treated in Chapter 4.

3.9.3 Types of Loading
The common types of loadings on beams are given below:

(i) Concentrated Loads or Point Loads
If load acts over a very small length of the beam it is considered to act at the mid-point of this small length and treated as a concentrated/point load. A point load is a hypothetical concept as it is not possible to transmit any force through a

point, and needs a finite area. It is represented by an arrow in the direction of the force. (Fig. 3.23).

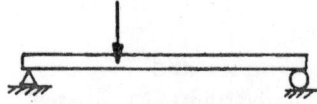

Fig. 3.23: Concentrated Load

(ii) Uniformly Distributed Load (UDL)
If the intensity of loading is the same over a significant length, it is known as "Uniformly Distributed Load". For the purpose of moments, the total load is assumed to be concentrated at the middle of its length. It is conventionally represented in one of the ways as shown in Fig. 3.24.

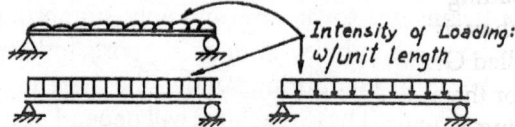

Fig. 3.24: Uniformly Distributed Load (UDL)

The loading may be on the whole or any part of the span.

(iii) Uniformly Varying Load
If the intensity of loading varies uniformly or linearly from one value to another along the length of the member, it is called a "Uniformly varying load". While the loading diagram in the case of uniformly distributed is represented by a rectangle, it is represented by a triangle or a trapezoid in this case. (A loading diagram on a member is one in which the load at any point of a member is shown as an ordinate).

Different possible cases of uniformly varying load are shown in Fig. 3.25.

In this case also the whole or any part of the span may be loaded.

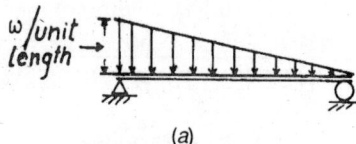

(a)

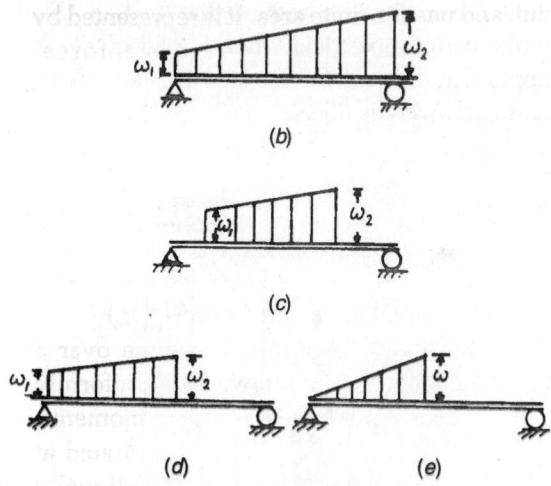

(b)

(c)

(d)   (e)

Fig. 3.25: Uniformly Varying Load

(iv) General Loading

A loading which consists of a loading diagram which varies in any specified curvilinear manner is called General Loading (Fig. 3.26).

For the sake of convenience in analysis, such loading is transformed into a set of equivalent concentrated loads.

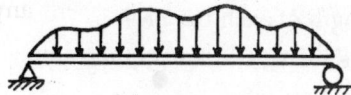

Fig. 3.26: General Loading

(v) Moment (or Couple) Loading

This consists of externally applied moments or couples at certain sections of the member (Fig. 3.27).

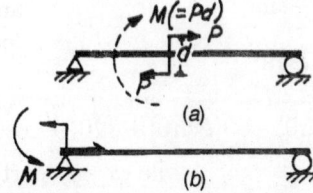

(a)

(b)

Fig. 3.27: Moment Loading (Externally Applied Couples)

In this Chapter, the determination of reactions at the supports of determinate beams under different kinds of loading will be illustrated.

ILLUSTRATIVE PROBLEMS

PROBLEM 3.1: A force of 100 N is acting at a point A as shown in Fig. 3.28 (a). Determine the moment of this force about point 'O'.

Solution:

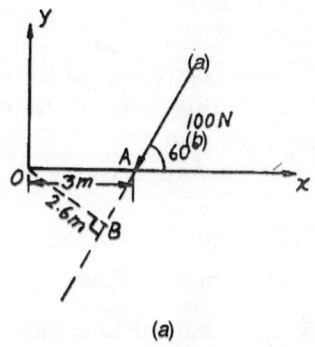

(a)

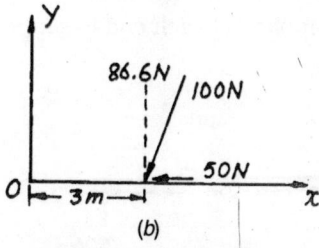

(b)

Fig. 3.28 (Illustrative Problem 3.1)

Given force at $A = 100$ N

Draw a perpendicular from O on the line of action of force 100 N. Hence OB is the perpendicular on the line of action of 100 N as shown in Fig. 3.28 (a).

1st Method

Triangle OBA is a right-angled triangle and angle $OAB = 60°$

$$\sin 60° = \frac{OB}{OA} = \frac{OB}{3}$$

$$OB = 3 \sin 60° = 3 \times 0.866 = 2.598 \text{ m}$$

Moment of the force 100 N about O.

$$= 100 \times OB$$

$$= 100 \times 2.598 = 259.8 \text{ Nm (Clockwise)}$$

2nd Method

The force of 100 N is resolved into two rectangular components i.e., the vertical component and the horizontal component.

The vertical component

$= 100 \sin 60°$

$= 100 \times 0.866 = 86.6$ N

The horizontal component

$= 100 \cos 60°$

$= 100 \times 0.5$

$= 50$ N

The horizontal component of force 100 N is passing through the point O, therefore, its moment about O is zero.

The moment of vertical component of force 100 N

$= 86.6 \times 3.0$

$= 259.8$ Nm (Clockwise)

PROBLEM 3.2: A bar 3 m long carries a weight 150 N at 2 m from left hand end. What force must be applied at each end to support the rod?

Solution:

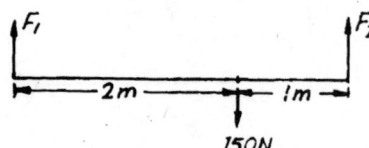

Fig. 3.29 (Illustrative Problem 3.2)

Let F_1 and F_2 be the two forces applied at the ends.

Sum of the upward forces = sum of the downward forces

∴ $F_1 + F_2 = 150$ ...(i)

By the principle of moments, algebraic sum of moments about any point in the plane is zero. Thus taking moments about A and treating anticlockwise moments as negative.

$150 \times 2 - F_2 \times 3 = 0$

$3F_2 = 300$

$F_2 = \dfrac{300}{3} = 100$ N   ...(ii)

∴ $F_1 = 150 - 100 = 50$ N

PROBLEM 3.3: Two prismatic bars AB and CD are welded together in the form of a rigid T and suspended in a vertical plane as shown in Fig.

3.30. Determine the angle θ that the bar AB will make with the vertical when a load of 120 N is applied at the end D. The two bars are identical and each weighs 60 N.

Solution:

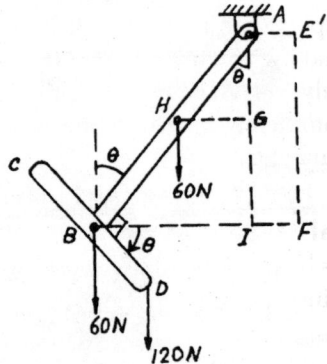

Fig. 3.30 (Illustrative Problem 3.3)

Forces acting on the bars are

  (*i*)   Reaction R_A at the hinge A

 (*ii*)   Weight of the bar $AB = 60$ N

(*iii*)   Weight of the bar $CD = 60$ N

(*iv*)   Force of 120 N acting at D

Let length of each bar be l.

Writing the equation of equilibrium

$\Sigma M_A = 0$

$-60 \times HG - 60 \times BI + 120 \times AE = 0$

(By taking moments about A, the reaction R_A has been eliminated)

$AE = IF = BF - BI = \dfrac{l}{2} \cos\theta - l\sin\theta$

$HG = \dfrac{l}{2} \sin\theta$

$BI = l\sin\theta$

∴ $-60(\dfrac{l}{2}\sin\theta) - 60(l\sin\theta) + 120(\dfrac{l}{2}\cos\theta - l\sin\theta)$

$= 0$

$-30\, l\sin\theta - 60(l\sin\theta) + 60\, l\cos\theta - 120\, l\sin\theta = 0$

$(210\sin\theta)l = 60\, l\cos\theta$

$\tan\theta = \dfrac{60}{210} = \dfrac{2}{7}$

$\theta = \tan^{-1}(2/7)$

$= 15.95°$

or $= 15°\ 57'$

PROBLEM 3.4: A horizontal line *ABCD* 15 m long, where $AB = BC = CD = 5$ m, forces 20 kN, 30 kN, 20 kN and 10 kN weight acting at *A*, *B*, *C* and *D* respectively, all downwards their lines of action, making angles of 90°, 60°, 45° an 30° respectively with *AD*. Obtain the resultant of the system completely in magnitude, direction and position analytically.

Solution:

Analytical Method

Let *R* = Resultant of the system of forces

Resolving all forces horizontally

$$\Sigma H = 20\cos90° + 30\cos60° + 20\cos45°$$
$$+ 10\cos30°$$
$$= 20 \times 0 + 30 \times 0.5 + 20 \times 0.7071 + 10$$
$$\times 0.866$$
$$= 0 + 15 + 14.142 + 8.66$$
$$= 37.802 \text{ kN}$$

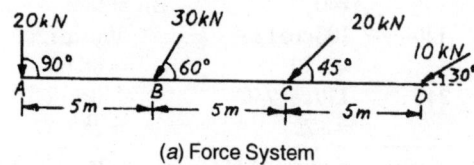

(a) Force System

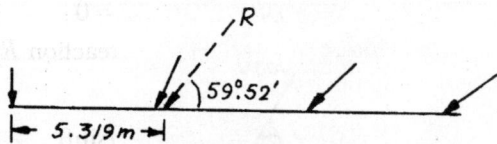

5.319 m

(b) Position of Resultant

Fig. 3.31 (Illustrative Problem 3.4)

Now resolving all the forces vertically

$$\Sigma V = 20\sin90° + 30\sin60° + 20\sin45° + 10\sin30°$$
$$= 20 \times 1 + 30 \times 0.866 + 20 \times 0.7071 + 10 \times 0.5$$
$$= 20 + 25.98 + 14.142 + 5$$
$$= 65.122 \text{ kN}$$

Using the relation,

$$R = \sqrt{(\Sigma H)^2 + (\Sigma V)^2} = \sqrt{(37.802)^2 + (65.122)^2}$$

$$= \sqrt{5669.87} = 75.3 \text{ kN}$$

Direction of Resultant

Let θ = Angle which the resultant makes with *AD*

$$\tan\theta = \frac{\Sigma V}{\Sigma H} = \frac{65.122}{37.802} = 1.7227$$

$$\theta = \tan^{-1}(1.7277) = 59.8657° \text{ or } 59°52'$$

Position of Resultant

Let *x* = distance between *A* and the line of action of the resultant

Using Varignon's theorem

Moment of the resultant about the point *A*
= Algebraic sum of the moments of all forces about *A*

Note: Only the moments of the vertical components is to be taken. The moments of horizontal components about *A* will be equal to zero.

$$65.122x = 20 \times 0 + 25.98 \times 5 + 14.142 \times 10 + 5 \times 15$$

or $\quad 65.122x = 0 + 129.9 + 141.42 + 75 = 346.32$

$$x = \frac{346.32}{65.122} = 5.319$$

Or

Method II

Applying Varignon's theorem at the point where the resultant strikes the horizontal.

Equating the sum of the anticlockwise moments to sum of the clockwise moments.

$$20x = 25.98\ (5-x) + 14.142\ (10-x) + 5(15-x)$$

$$65.122x = 129.9 + 141.42 + 75 = 346.32$$

$$x = \frac{346.32}{65.122} = 5.319 \text{ m}$$

PROBLEM 3.5: A flat plate is subjected to the coplanar system of forces shown in Fig. 3.32 (*a*). The inscribed grid with each square having a length of 1 metre locates each force and its slope. Determine the resultant and its *X* and *Y* intercepts.

Solution:

$$\tan\theta_1 = \frac{2}{3}$$

$$\theta_1 = \tan^{-1}(2/3) = 33.69°$$

$$\tan\theta_2 = \frac{2}{1} = 2$$

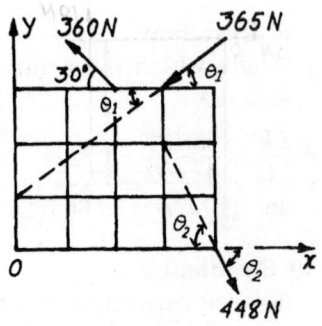

(a) Force System

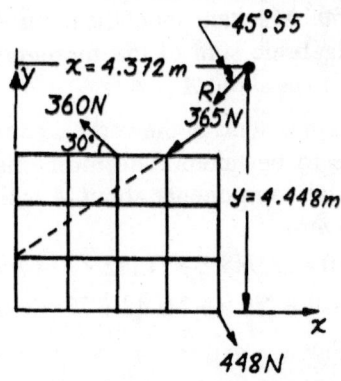

(b) Location of the Resultant

Fig. 3.32 (Illustrative Problem 3.5)

$\theta_2 = \tan^{-1}(2) = 63.435°$

$\therefore \quad \Sigma x = -360\cos 30° - 365\cos 33.69°$

$\qquad + 448\cos 63.435°$

$\qquad = -311.769 - 303.699 + 200.351$

$\qquad = -415.117 \text{ N}$

$\Sigma y = 360\sin 30° - 365\sin 33.69° - 448\sin 63.435°$

$\qquad = 180 - 202.465 - 400.704$

$\qquad = -423.169 \text{ N}$

$R = \sqrt{(\Sigma x)^2 + (\Sigma y)^2}$

$\quad = \sqrt{(-415.117)^2 + (-423.169)^2}$

$\quad = 592.785 \text{ N}$

$\tan\theta = \dfrac{\Sigma y}{\Sigma x} = \dfrac{423.169}{415.117} = 1.0194$

$\therefore \qquad \theta = \tan^{-1}(1.0194) = 45.55°$

Applying Varignon's theorem

(Choose point 'O')

$(-360\cos 30°)\,(3) - (365\cos 33.69°)\,(3)$

$\qquad = -415.117\ (Y)$

$+\,935.307 + 911.096 = 415.117\,y$

$\qquad y = \dfrac{1846.403}{415.117}$

$\qquad = 4.448 \text{ m}$

$(-360\sin 30°)2 + (365\sin 33°.69)3 + (448\sin$

$63.435°)4$

$\qquad = 423.169\ (x)$

$\qquad x = 4.372 \text{ m}$

PROBLEM 3.6: A particle is acted upon by three forces equal to 50 N, 100 N and 150 N, along the sides of an equilateral triangle, taken in order.

Find the magnitude and direction of the resultant force.

Solution:

S.No.	Force (N)	θ (in deg.)	Horizontal Component (N)	Vertical Component (N)
1	50	0	$50\cos 0°$ $= 0$	$50\sin 0°$ $= 0$
2	100	120	$100\cos 120°$ $= -50$	$100\sin 120°$ $= 86.60$
3	150	240	$150\cos 240°$ $= -75$	$150\sin 240°$ $= 129.90$
			$\Sigma H = -75 \text{ N}$	$\Sigma V = -43.3 \text{ N}$

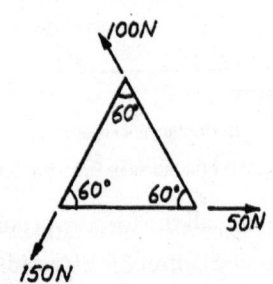

Fig. 3.33 (Illustrative Problem 3.6)

$R = \sqrt{(\Sigma H)^2 + (\Sigma V)^2}$

$\quad = \sqrt{(-75)^2 + (-43.3)^2}$

$\quad = \sqrt{7499.89}$

$\quad = 86.60 \text{ N}$

$$\theta = \tan^{-1}\left(\frac{\Sigma V}{\Sigma H}\right) = \tan^{-1}\left(\frac{43.3}{75.0}\right) = \tan^{-1}(0.5773)$$

$$= 30°$$

Actual $\theta = 180 + 30 = 210°$

PROBLEM 3.7: Four forces of magnitudes 5 N, 10 N, 15 N and 20 N are acting respectively along the four sides of a square *ABCD* as shown in Fig. 3.3 (*a*). Determine the magnitude, direction and position of the resultant force.

Solution:

Force along $AB = 5$ N

Force along $BC = 10$ N

Force along $CD = 15$ N

Force along $DA = 20$ N

The net force in the horizontal direction

$$= 5 - 15 = -10 \text{ N}$$

The net force in the vertical direction

$$= 10 - 20 = -10 \text{ N}$$

Resultant, $R = \sqrt{(\Sigma H)^2 + (\Sigma V)^2}$

$$= \sqrt{(-10)^2 + (-10)^2}$$

$$= 10\sqrt{2}$$

$$\tan\theta = \frac{\Sigma V}{\Sigma H} = \frac{10}{10} = 1$$

$$\therefore \quad \theta = 45°$$

since ΣH and ΣV are $-$ve, θ lies in the 3rd quadrant.

Actual $\theta = 180° + 45°$

$$= 225°$$

Position of the resultant force

Method I

Let $x =$ Perpendicular distance between A and line of action of the resultant force

and $a =$ the side of the square *ABCD*

Using Varignon's theorem

Algebraic sum of all the forces about A

$$= \text{Moment of resultant about } A$$

$$5 \times 0 + 10 \times a + 15 \times a + 20 \times 0 = R \times x$$

$$25a = 10\sqrt{2}x$$

$$\therefore \quad x = \frac{25a}{10\sqrt{2}} = \frac{5a}{2\sqrt{2}} = 1.7678a$$

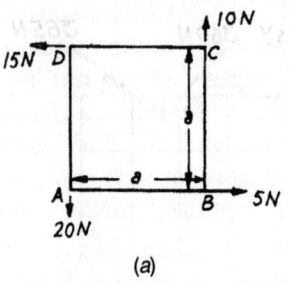

(a)

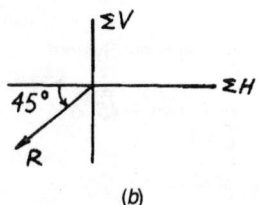

(b)

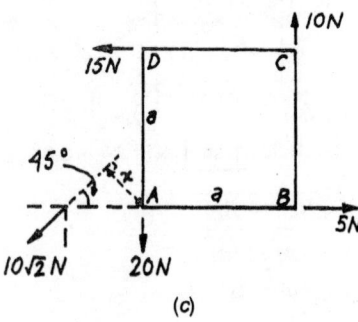

(c)

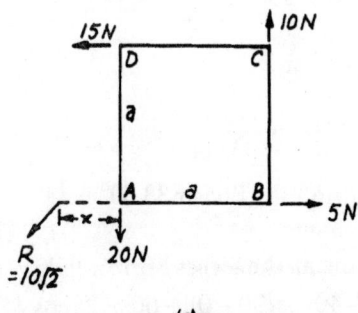

(d)

Fig. 3.34 (Illustrative Problem 3.7)

Method II

Considering the point on the line *BA* produced where the resultant cuts

$$20x + 5 \times 0 - 10 \times (x + a) - 15a = 0$$

$$10x = 25a$$

$$x = \frac{25}{10}a = \frac{5}{2}a$$

PROBLEM 3.8: *ABCD* is a square, each side being 2*a* units and *E* is the middle point of *AB* forces of 35, 40, 60, 25, 45 and 30 N act on a body in the lines of directions *AB, EC, BC, BD, CA* and *DE* respectively. Find the magnitude, direction and position with respect to *BC* of a single force, required to keep the body in equilibrium.

Solution: The system of the given forces is shown in Fig. 3.35.

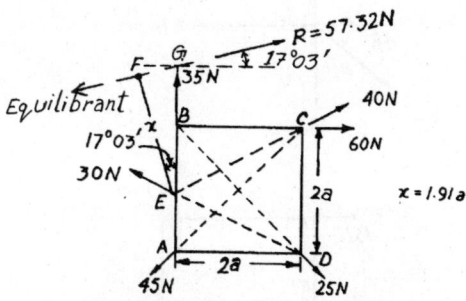

Fig. 3.35 (Illustrative Problem 3.8)

$$EC = \sqrt{BC^2 + BE^2}$$
$$= \sqrt{(2a)^2 + (a)^2}$$
$$= \sqrt{5}\,a$$

Let $\angle BEC = \theta$

$$\tan\theta = \frac{2a}{a} = 2$$

$$\sin\theta = \frac{2a}{\sqrt{5}a} = \frac{2}{\sqrt{5}}$$

$$\cos\theta = \frac{a}{\sqrt{5}a} = \frac{1}{\sqrt{5}}$$

Resolving all the forces horizontally

$$\Sigma H = 40\cos(90-\theta) + 60 + 25\cos 45°$$
$$-45\cos 45° - 30\cos(90-\theta)$$
$$= 10\sin\theta + 60 - 20\cos 45°$$

$$= 10 \times \frac{2}{\sqrt{5}} + 60 - 20 \times \frac{1}{\sqrt{2}}$$

$$= 54.8022 \text{ N}$$

Resolving all the forces vertically,

$$\Sigma V = 35 + 40\sin(90-\theta) - 25\sin 45°$$
$$-45\sin 45° + 30\sin(90-\theta)$$
$$= 35 + 70\cos\theta - 70\sin 45°$$

$$= 35 + 70 \times \frac{1}{\sqrt{5}} - 70 \times \frac{1}{\sqrt{2}}$$
$$= 35 + 31.3050 - 49.50$$
$$= 16.805 \text{ N}$$

The magnitude of the resultant force,

$$R = \sqrt{(\Sigma H)^2 + (\Sigma V)^2}$$
$$= \sqrt{(54.8022)^2 + (16.805)^2}$$
$$= \sqrt{3003.28 + 282.408}$$
$$= \sqrt{3285.688}$$
$$= 57.32 \text{ N}$$

Direction of the resultant force

Let α = Angle made by the resultant force with the horizontal

$$\tan\alpha = \frac{\Sigma V}{\Sigma H} = \frac{16.805}{54.8022} = 0.3066$$

$$= \tan^{-1}(0.30665) = 17°03'$$

Since ΣH and ΣV are +ve, therefore α lies between 0° and 90°.

$\therefore$ Actual $\alpha = 17°03'$

Taking moments about *E* and applying Varignon's theorem

$$57.32(x) = 60 \times a + (25 \times 0.7071)a$$
$$+ (45 \times 0.7071)a$$

$$57.32x = (60 + 17.6775 + 31.8195)a$$
$$= 109.497a$$

$$x = 109.497/57.32 = 1.91a$$

$$EF = 1.91a$$

$$\angle FEG = 17°03'$$

$$\cos 17°03' = \frac{EF}{EG} = \frac{1.91a}{EG}$$

$$EG = 1.91a/\cos 17°03'$$
$$= 2a$$

$\therefore$ $$BG = EG - EB$$
$$= 2a - a$$
$$= a$$

Magnitude of equilibrant = 57.32 N

Its direction $= 180 + 17°03'$

$\qquad = 197°03'$ with the +ve x-axis

Its position $= a$ from BC

PROBLEM 3.9: Three like parallel forces 100 N, 200 N, and 300 N are acting at points A, B and C respectively on a straight line ABC as shown in Fig. 3.36. The distances $AB = 40$ cm and $BC = 60$ cm. Find the resultant and also the distance of the resultant from point A on the line ABC.

Solution:

Let R be the resultant of the parallel forces 100 N, 200 N and 300 N.

Let the resultant R be acting at a distance of x cm from point A as shown in Fig. 3.36.

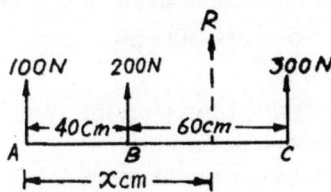

Fig. 3.36 (Illustrative Problem 3.9)

$R = 100 + 200 + 300$

$\qquad = 600$ N

Using Varignon's theorem,

The moment of the resultant R about 0

$\qquad$ = the albegraic sum of moments of all the forces about 0

$600 \times x = 100 \times 0 + 200 \times 40 + 300 \times 100$

$\qquad = 8000 + 30,000$

$x = \dfrac{38000}{600} = 63.33$ cm

PROBLEM 3.10: The three like parallel forces of magnitude 60 N, F and 120 N are shown in Fig. 3.37. If the resultant $R = 270$ N and is acting at a distance of 5 m from A, then find

(*i*) Magnitude of force F

(*ii*) Distance of F from A.

Solution:

Given forces at $A = 60$ N, at $B = F$, and at $D = 120$ N

$\qquad$ Distance $AC = 5$ m

$\qquad CD = 4$ m

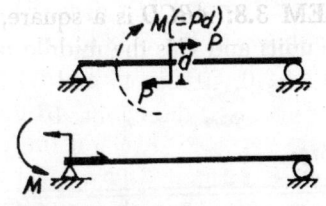

Fig. 3.37 (Illustrative Problem 3.10)

Magnitude of force F

The resultant of three like forces is given by,

$\qquad R = 60 + F + 120$

$\quad 270 = 60 + F + 120$

$\qquad F = 270 - 60 - 120 = 90$ N

Distance of F from A

Applying Varignon's theorem

$\qquad$ Taking moments about A

$\quad 60 \times 0 + F \times x + 120 \times 9 = 270 \times 5$

$\quad 90x + 1080 = 1350$

$\quad 90\,x = 1350 - 1080 = 270$ m

or $\quad x = \dfrac{270}{90} = 3$ m

PROBLEM 3.11: A system of parallel forces are acting on a rigid bar as shown in Fig. 3.38 (*a*). Reduce this system to

(*i*) a single force

(*ii*) a single force at A and a couple at A

(*iii*) a single force at B and a couple at B

Solution:

(*i*) a single force

$\qquad R = 30 - 150 + 60 - 10 = -70$ N

$\qquad R = 70$ N (downwards)

Let x be the distance of resultant from A towards right.

Moment of resultant about A

$\qquad$ = Algebraic sum of moments of all forces about A

$70x = 30 \times 0 + 150 \times 1 - 60 \times 2.2 + 10 \times 3.7$

$\qquad = 150 - 132 + 37 = 55$

$\qquad x = \dfrac{55}{70} = 0.786$ m

Hence the given system of parallel forces is equivalent to a single force 70 N acting vertically downwards at point E at a distance of 0.786 m from A as shown in Fig. 3.38 (*a*).

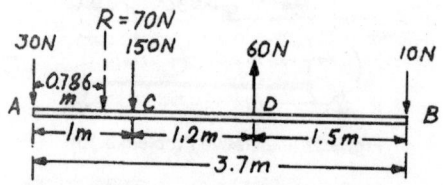

(*a*) Parallel Force System (Resultant Single Force)

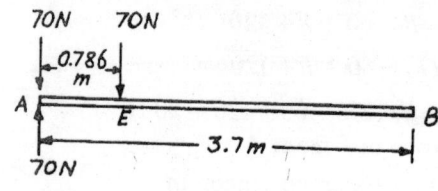

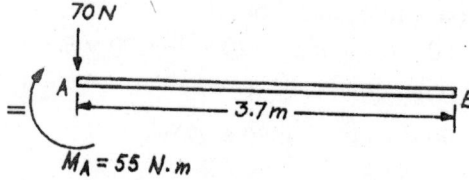

(*b*) Single Force and a Couple at A

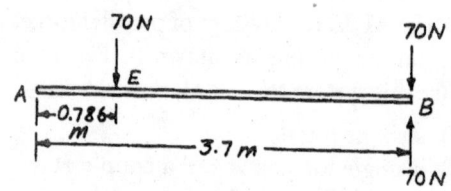

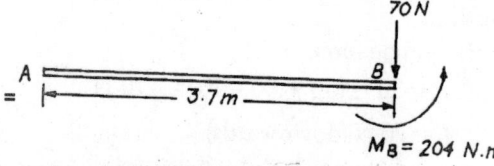

(*c*) Single Force and a Couple at B

Fig. 3.38 (Illustrative Problem 3.11)

(*ii*) A single force at A and a couple at A

$M_A = 70 \times 0.786$

$= 55$ N-m (clockwise)

(*iii*) A single force at B and a couple at B

$M_b = -70(3.70 - 0.786)$

$= -204$ N-m

$= 204$ N-m (anticlockwise)

PROBLEM 3.12: Determine the resultant of the parallel force system shown in Fig. 3.39.

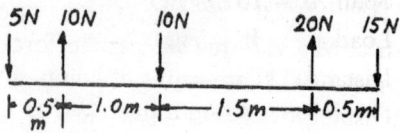

Fig. 3.39 (Illustrative Problem 3.12)

Solution:

Forces at A, B, C, D and E are 5 N, 10 N, 20 N and 15 N respectively.

Distance $AB = 0.5$ m, $BC = 1.0$ m, $CD = 1.5$ m and $DE = 0.5$ m

Since all the forces are vertical and parallel, hence their resultant is given by

$R = -5 + 10 - 10 + 20 - 15$

$= 0$

As the resultant force on the system is zero, there will be two possibilities.

The system has a resultant couple or the system is in equilibrium.

To distinguish between these two possi-bilities, take the sum of the moments of all forces about any point.

Let us take the moments about point A

$= 5 \times 0 - 10 \times 0.5 + 10 \times 1.5 - 20 \times 3 + 15 \times 3.5$

$= -5 + 15 - 60 + 52.5$

$= -65 + 67.5$

$= +2.5$ N-m (clockwise)

As the algebraic sum of moments of all forces about any point is not zero, the system will have a resultant couple of magnitude of +2.5 N-m i.e., clockwise couple.

PROBLEM 3.13: A beam of span 10 m is carrying a point load of 500 N at a distance of 4 m from A. Determine the beam reactions.

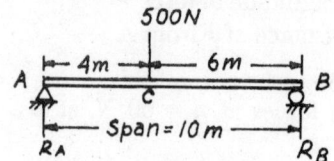

Fig. 3.40 (Illustrative Problem 3.13)

Solution:

Given:

Span $AB = 10$ m

Load at C, $W = 500$ N

Distance, $AC = 4$ m

Distance, $BC = 10 - 4 = 6$ m

Let R_A = reaction at A and R_B = reaction at B.

The beam is in equilibrium.

∴ The sum of the upward forces = the sum of the downward forces

$$R_A + R_B = 500 \text{ N} \qquad \qquad ...(i)$$

The anticlockwise moments about any point must be equal to the clockwise moments about the same point.

Taking moments about A

Anticlockwise moments = clockwise moments

$$10R_B = 500 \times 4$$

$$R_B = \frac{500 \times 4}{10} = 200 \text{ N} \qquad ...(ii)$$

Substituting (ii) in (i)

$$R_A + 200 = 500$$

or $\quad R_A = 500 - 200 = 300$ N

$R_A = 300$ N and $R_B = 200$ N

PROBLEM 3.14: The beam AB of span 6 m shown in Fig. 3.41(a) is hinged at A and is on rollers at B. Determine the reactions at A and B for the loading shown.

Solution: The reaction at A can be in any direction as the beam is hinged at A. This reaction can be represented by its components V_A and H_A as shown in Fig. 3.41 (b). At B the reaction is in vertical direction only. The beam is in equilibrium under the action of system of forces shown in Fig. 3.41 (b).

$$\Sigma H = 0$$

$$-3\cos 30° - 4\cos 60° + H_A = 0$$

$$-3 \times 0.866 - 4 \times 0.500 + H_A = 0$$

$$-2.598 - 2.00 + H_A = 0$$

$$H_A - 4.598 = 0$$

or $\quad H_A = 4.598$ kN

$$\Sigma V = 0$$

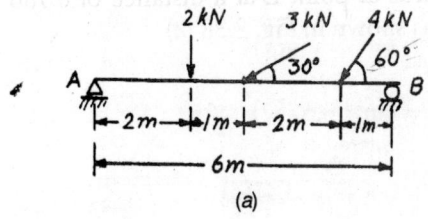

(a)

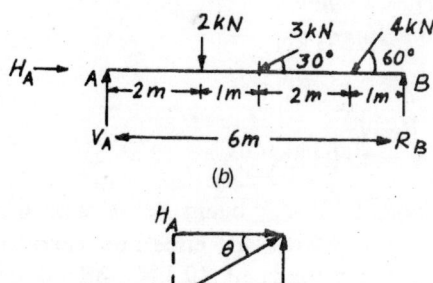

(b)

(c)

Fig. 3.41 (Illustrative Problem 3.14)

$$V_A + R_B - 2 - 3\sin 30° - 4\sin 60° = 0$$

$$V_A + R_B - 2 - 3 \times 0.5 - 4 \times 0.866 = 0$$

$$V_A + R_B = 2 + 1.5 + 3.464$$

$$V_A + R_B = 6.964 \text{ kN} \qquad ... (i)$$

$$\Sigma M_A = 0$$

$$-R_B \times 6 + 2 \times 2 + (3\sin 30°)(3) + (4\sin 60°)(5) = 0$$

$$-R_B \times 6 + 4 + 4.5 + 17.32 = 0$$

$$6R_B = 25.82$$

$$R_B = \frac{25.82}{6} = 4.303 \text{ kN} \qquad ...(ii)$$

Substituting (ii) in (i)

$$V_A + 4.303 = 6.964$$

$$V_A = 6.964 - 4.303$$

$$V_A = 2.661 \text{ kN}$$

$$R_A = \sqrt{H_A^2 + V_A^2}$$

$$= \sqrt{(4.598)^2 + (2.661)^2}$$

$$= \sqrt{21.1416 + 7.0809}$$

$$= \sqrt{28.2225}$$

$$= 5.312 \text{ kN}$$

$$\theta = \tan^{-1}\left(\frac{V_A}{H_A}\right)$$

$$= \tan^{-1}\left(\frac{2.661}{4.598}\right)$$

$$= \tan^{-1}(0.5787)$$

$$= 30°03'$$

$$\approx 30° \text{ [Fig. 3.41(c)]}$$

$H_A = 4.598$ kN

$V_A = 2.661$ kN

$R_A = 5.132$ kN

$\theta = 30°$

$R_B = 4.303$ kN

PROBLEM 3.15: A beam *AB* of span 6 m is simply supported at its ends and carries two concentrated loads of 20 kN and 40 kN at distances 1 m, and 2 m from the left hand end. It also carries a load of 20 kN/m for a length of 2 m starting at 3 m from *A*. Determine the reactions.

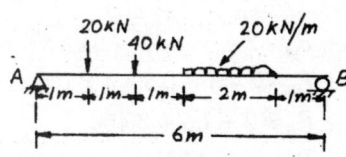

Fig. 3.42 (Illustrative Problem 3.15)

Solution:

Let R_A & R_B be the support reactions.

$$R_A + R_B = 20 + 40 + 20 \times 2$$

or $R_A + R_B = 100$ kN   ...(*i*)

$$6R_A = 20 \times 1 + 40 \times 2 + 20 \times 2 \times \left(1 \times 1 + 1 + \frac{2}{2}\right)$$

$$= 20 + 80 + 160$$

$$= 260$$

∴ $R_A = \dfrac{260}{6} = 43.33$ kN   ...(*ii*)

∴ $R_B = 100 - 43.33 = 56.67$

$R_A = 43.33$ kN & $R_B = 56.67$ kN

PROBLEM 3.16: A beam *ABC* is supported and loaded as shown in Fig. 3.43 (*a*). Find the reactions at the supports.

Solution: Free body diagram of the beam *ABC* is as shown (Fig. 3.43 (*b*)).

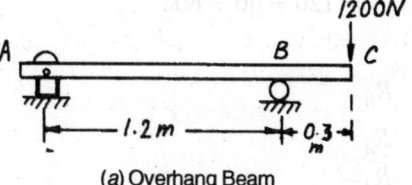

(a) Overhang Beam

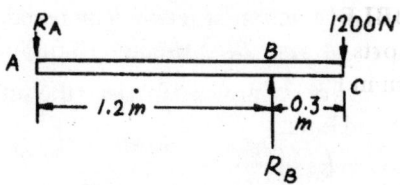

(b) Free Body of the Beam

Fig. 3.43 (Illustrative Problem 3.16)

$$\Sigma V = 0$$

$$R_B - R_A - 1200 = 0$$

$$R_B - R_A = 1200 \qquad\qquad ...(i)$$

$$\Sigma M_A = 0$$

$$-R_B \times 1.2 + 1200 \times 1.5 = 0$$

or $R_B = \dfrac{1800}{1.2} = 1500$ N   ...(*ii*)

Substituting (*ii*) in (*i*)

∴ $1500 - R_A = 1200$

or $R_A = 300$ N (down)

$R_A = 300$ N (down)

$R_B = 1500$ N (up)

PROBLEM 3.17: Fig. 3.44 shows an overhanging beam carrying loads. Determine the reactions at the supports *A* & *B*.

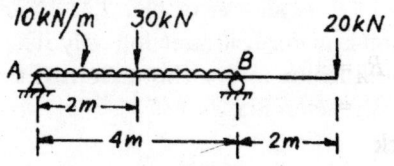

Fig. 3.44 (Illustrative Problem 3.17)

Solution:

Let R_A & R_B be reactions at the supports *A* & *B* respectively.

∴ $R_A + R_B = 30 + 20 + 10 \times 4 = 90$ kN   ...(*i*)

$$4R_B = 20 \times 6 + 30 \times 2 + 10 \times 4 \times 2$$

$$= 120 + 60 + 80$$
$$= 260$$
$$\therefore \quad R_B = \frac{260}{4} = 65 \text{ kN} \qquad ...(ii)$$
$$\therefore \quad R_A = 90 - 65 = 25 \text{ kN}$$
$$R_A = 25 \text{ kN} \ \& \ R_B = 65 \text{ kN}$$

PROBLEM 3.18: Determine the reactions at supports A and B of the overhanging beam shown in Fig. 3.45.

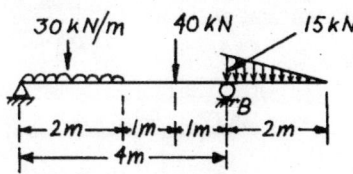

Fig. 3.45 (Illustrative Problem 3.18)

Solution:
$$\Sigma V = 0$$

$$R_A + R_B - 30 \times 2 - 40 - \frac{1}{2} \times 2 \times 15 = 0$$
$$R_A + R_B - 60 - 40 - 15 = 0$$
$$R_A + R_B - 115 = 0$$
$$R_A + R_B = 115 \qquad \qquad ...(i)$$
$$\Sigma M = 0 \qquad \qquad \text{(About point '}A\text{')}$$

$$-R_B \times 4 + 30 \times 2 \times 1 + 40 \times 3 + \frac{1}{2} \times 2 \times 15 \times$$
$$\left(\frac{2}{3} + 4\right) = 0$$
$$-4R_B + 60 + 120 + 70 = 0$$
$$4R_B = 250$$
$$R_B = \frac{250}{4}$$
$$= 62.5 \text{ kN}$$
$$\therefore \quad R_A = 115 - 62.5$$
$$= 52.5 \text{ kN}$$

Check
$$\Sigma M = 0 \qquad \qquad \text{(About point '}B\text{')}$$

$$R_A \times 4 - 30 \times 2 \times 3 - 40 \times 1 + \frac{1}{2} \times 2 \times 15 \times \frac{2}{3} = 0$$
$$4R_A - 180 - 40 + 10 = 0$$
$$4R_A = 210$$
$$R_A = \frac{210}{4} = 52.5 \text{ kN}$$

PROBLEM 3.19: A bar AB of length $2l$ and negligible weight rests on two roller supports C and D placed at a distance l apart. The bar supports two vertical loads as shown (Fig. 3.46). For the reactions at the supports to be equal find the distance x, of the end A of the bar, from the support C.

Solution:

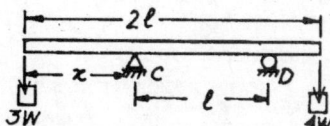

Fig. 3.46 (Illustrative Problem 3.19)

Forces acting on the bar AB are:
Reactions R_C, R_D, $3W$ and $4W$.
Writing the equations of equilibrium of the bar
$$\Sigma V = 0$$
$$R_C + R_D - 3W - 4W = 0$$
But $R_C = R_D$ (Given)

Therefore $\quad R_C = R_D = \dfrac{7W}{2}$
$$\Sigma M_C = 0$$
$$4W(2l - x) - R_D l - 3Wx = 0$$
$$8Wl - 4Wx - \frac{7W}{2}l - 3Wx = 0$$
$$7Wx = 8Wl - 3.5Wl$$
$$7Wx = 4.5Wl$$
$$x = \frac{4.5l}{7} = 0.6429 \ l$$
$$x = 0.6429 \ l$$

PROBLEM 3.20: A uniform wheel 600 mm diameter rests against rigid block 150 mm thick as shown in Fig. 3.47 (*a*). Find the least pull through the centre of the wheel to just turn the wheel over the corner of the block. All surfaces are smooth. Find the reaction of the block. The wheel weighs 30 kN.

Solution:
When the wheel is about to turn then its contact with the ground will be lost.

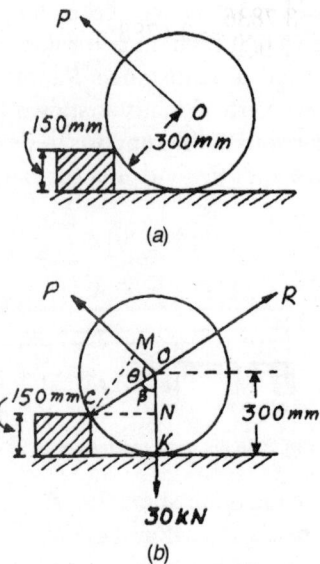

(a)

(b)

Fig. 3.47 (Illustrative Problem 3.20)

Hence under the action of two forces i.e. P and W, wheel will be in equilibrium.

From the Fig. 3.47 (b), $OK = 300$ mm radius of the wheel.

Now drop perpendiculars on the lines of actions of two forces, P and W. These are CM and CN.

$$CN = \sqrt{OC^2 - ON^2}$$

where $ON = OK - NK$

$$300 - 150 = 150 \text{ mm}$$

$$CN = \sqrt{300^2 - 150^2}$$

$$= 259.81 \text{ mm}$$

$$CM = OC\sin\theta$$

Taking moments about point C, we get

$$P \times CM = W \times CN$$

$$P \times 300\sin\theta = 30 \times 259.81$$

$$P = \frac{30 \times 259.81}{300\sin\theta} = \frac{25.981}{\sin\theta}$$

P will be minimum, when θ is maximum.

i.e. $\sin\theta = 1$, $\theta = 90°$

Hence P minimum $= 25.981$ kN & is applied perpendicular to CO.

For determining reaction R, W and P are resolved along OC.

The component of W along $OC = W\cos\beta$

$$= 30 \times \frac{150}{300} = 15 \text{ kN}$$

The component of P along $OC =$ Zero.

(Since P acts perpendicular to CO)

$\therefore \quad R = 15$ kN

PROBLEM 3.21: A plane rectangular plate of height 3.6 m and length 2.4 m is hinged at A and is subjected to water pressure from one side. Find the force P to be applied at point B of the plate to keep it in vertical position.

Solution:

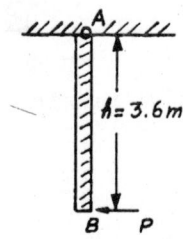

Fig. 3.48 (Illustrative Problem 3.21)

The total resultant force due to water pressure for 2.4 m length

$$\frac{9.81 \times 3.6 \times 3.6 \times 2.4}{2}$$

$$= 152.565 \text{ kN}$$

acts at a height of $\frac{3.6}{2} = 1.2$ m from B

In order to keep the plate in vertical position the moment of resultant force due to water pressure about the hinge. 'A' must balance with the moment of the force P about the same point A.

$$152.565 \times (3.6 - 1.2) = P \times 3.6$$

$\therefore \quad P = \dfrac{152.565 \times 2.4}{3.6}$

$$= 101.71 \text{ kN}$$

PROBLEM 3.22: A partition 2 m long divides a storage tank. On one side of this partition is petrol of unit weight 7.65 kN/m^3, stored to a depth of 1.2 m; and on the other side is oil of unit weight 8.65 kN/m^3 to a depth of 0.6 m. Determine the total resultant force exerted on the partition and position of its line of action.

Solution:

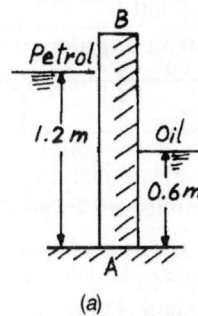

(a)

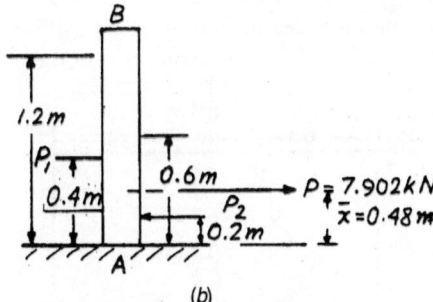

(b)

Fig. 3.49 (Illustrative Problem 3.22)

Total force due to petrol on the partition,

$$P_1 = \frac{1}{2} \times (1.2 \times 7.65) \times 1.2 \times 2$$

$$= 11.016 \text{ kN acts at } 0.4 \text{ m from } A$$

Total force due to oil on the partition,

$$P_2 = \frac{1}{2} \times (0.6 \times 8.65) \times 0.6 \times 2$$

$$= 3.114 \text{ kN acts at } 0.2 \text{ m from } A$$

∴ Net total force acting on the partition wall,

$$P = P_1 - P_2$$

$$= 11.016 - 3.114$$

$$= 7.902 \text{ kN}$$

Let net total force P be acting at a distance $\bar{x}$ from A

Applying Varignon's theorem (choosing point A)

$$P_1 \times 0.4 - P_2 \times 0.2 = P(\bar{x})$$

$$\bar{x} = \frac{11.016 \times 0.4 - 3.114 \times 0.2}{7.902}$$

$$= \frac{4.4064 - 0.6228}{7.902}$$

$$= \frac{3.7836}{7.902} = 0.4788 \text{ m}$$

or

$$P_1(0.4 - \bar{x}) + P_2(\bar{x} - 0.2) = P \times 0$$

$$(11.016)(0.4 - \bar{x}) + (3.114)(\bar{x} - 0.2) = 0$$

$$4.4064 - 11.016\bar{x} + 3.114\bar{x} - 0.6228 = 0$$

$$(11.016 - 3.114)\bar{x} = (4.4064 - 0.6228)$$

$$\bar{x} = \frac{4.4064 - 0.6228}{11.016 - 3.114}$$

$$= \frac{3.7836}{7.902}$$

$$= 0.4788 \text{ m}$$

$$P = 7.902 \text{ kN}$$

$$\bar{x} = 0.4788 \text{ from } A$$

PRACTICE PROBLEMS

3.1 A force of 20 N is applied perpendicular to the edge of a door 0.8 m as shown in Fig. 3.50. Find the moment of the force about the hinge. Also find it when the force acts at 60° to the door.

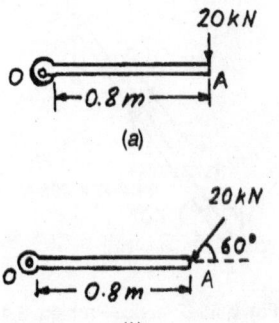

(a)

(b)

Fig. 3.50 (Practice Problem 3.1)

3.2 A three wheeler scooter of weight 2 kN and a driver weighing 0.6 kN are shown schematically in Fig. 3.51, where G and D denote the positions of C.G. of the scooter and location of the driver respectively. Find the reactions of the wheels A, B and C of the scooter for the equilibrium conditions on a level road.

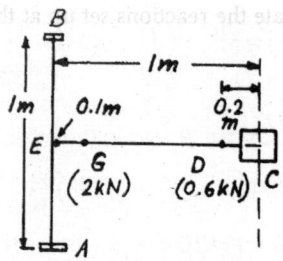

Fig. 3.51 (Practice Problem 3.2)

3.3 A uniform plank *ABC* of weight 60 N is supported at a point *B* 2.1 m from *A* as shown in Fig. 3.52. Find the value of weight *W*, that can be placed at *C*, so that the plank does not topple.

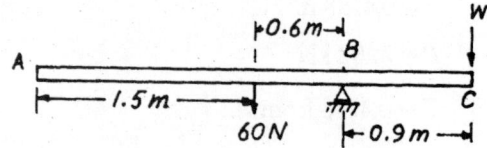

Fig. 3.52 (Practice Problem 3.3)

3.4 Two men support a weightless wooden beam *AB* with a weight of 800 N hanging from the beam as shown in Fig. 3.53. Find the load shared by the each man.

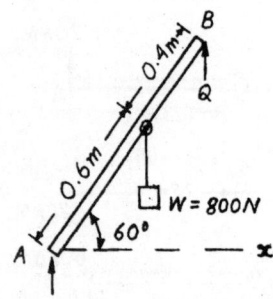

Fig. 3.53 (Practice Problem 3.4)

3.5 A horizontal line *PQRS* is 9 m long, where *PQ* = *QR* = *RS* = 3 m. Forces of 500 N, 750 N, 500 N and 250 N act at *P, Q, R* and *S* respectively with downward direction. The lines of action of these forces make angles of 90°, 60°, 45° and 30° respectively with *PS*. Find the magnitude, direction and position of the resultant force.

3.6 Find the resultant, *R* for the system shown in Fig. 3.54.

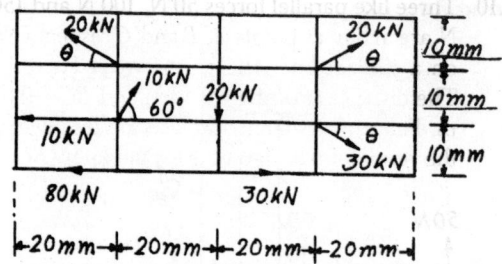

Fig. 3.54 (Practice Problem 3.6)

3.7 A rigid plate *ABCD* is subjected to force system shown in Fig. 3.55. Compute the magnitude, direction and position of the resultant from the centroid *O* of the plate.

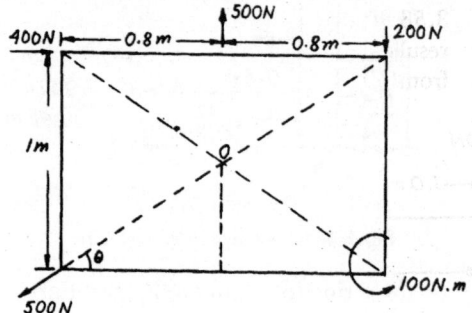

Fig. 3.55 (Practice Problem 3.7)

3.8 *ABCD* is a square. Forces of 50 N, 40 N and 20 N act at *A* in the directions *AD, AC* and *AB* respectively. Determine:
   (*i*) the resultant force in magnitude and direction;
   (*ii*) the component forces of the above resultant force along the directions *AJ* and *AH*, where *J* and *H* are the mid-points of *CD* and *BC* respectively.

3.9 A rigid bar is subjected to a system of parallel forces as shown in Fig. 3.56. Reduce this system to
   (*a*) a single force
   (*b*) a single force-moment system at *A*
   (*c*) a single force-moment system at *B*

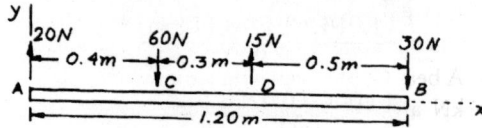

Fig. 3.56 (Practice Problem 3.9)

3.10 Three like parallel forces 50 N, 100 N and 150 N are acting at points A, B and C respectively on a straight line ABC as shown in Fig. 3.57. The distances are AB = 0.40 m and BC = 0.60 m. Find the resultant and also the distance of the resultant from point A on the line ABC.

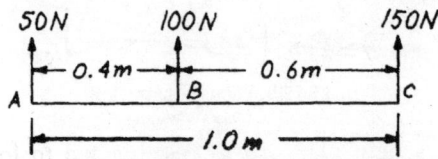

Fig. 3.57 (Practice Problem 3.10)

3.11 Four parallel forces of magnitudes 100 N, 200 N, 50 N and 300 N are shown in Fig. 3.58. Determine the magnitude of the resultant and also the distance of the resultant from point A.

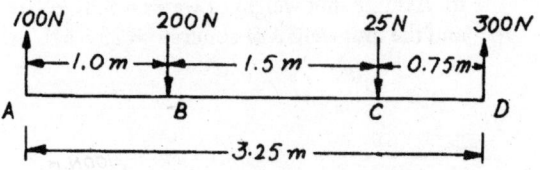

Fig. 3.58 (Practice Problem 3.11)

3.12 Determine the resultant of the parallel forces acting on a body shown in Fig. 3.59.

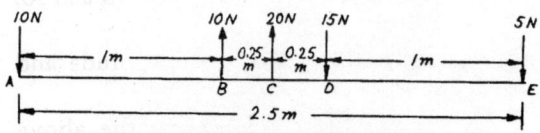

Fig. 3.59 (Practice Problem 3.12)

3.13 The cantilever beam shown in Fig. 3.60 is fixed at A and is free at B. Find the reactions at A for the loading shown in the Fig. 3.60.

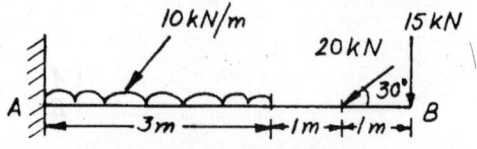

Fig. 3.60 (Practice Problem 3.13)

3.14 A beam AB of span 6 m carries two loads of 20 kN and 30 kN at 1m from either ends as shown in Fig. 3.61.

Calculate the reactions set up at the supports.

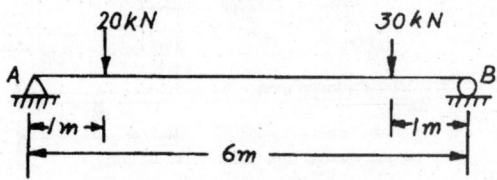

Fig. 3.61 (Practice Problem 3.14)

3.15 A beam supports a load distributed parabolically over its length. Determine the resultant of this distributed load and its line of action (Fig. 3.62). Also determine the support reactions.

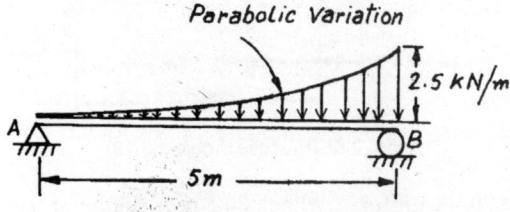

Fig. 3.62 (Practice Problem 3.15)

3.16 Three beams AB, BC and CD are hinged at their ends and supported as shown (Fig. 3.63). Determine the reactions at the points A, E, F and D.

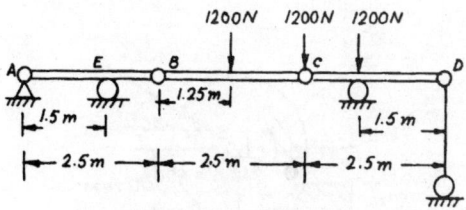

Fig. 3.63 (Practice Problem 3.16)

3.17 Determine the reactions at supports A and B of the simply supported beam shown in Fig. 3.64.

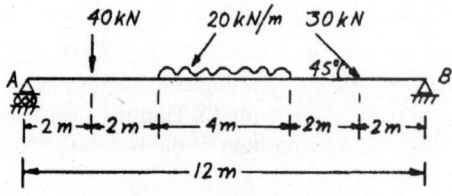

Fig. 3.64 (Practice Problem 3.17)

3.18 Determine the reactions at the supports A and B of the overhanging beam shown in Fig. 3.65.

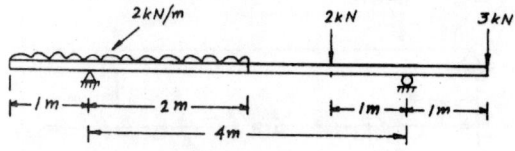

Fig. 3.65 (Practice Problem 3.18)

3.19 Determine the reactions at supports A and B of the overhanging beam shown in Fig. 3.66.

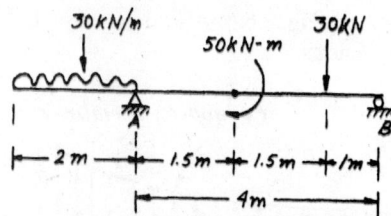

Fig. 3.66 (Practice Problem 3.19)

3.20 A uniform wheel 600 mm diameter rests against a rigid block 150 mm thick as shown in Fig. 3.67. Find the least pull through the centre of the wheel to just turn the wheel over the corner of the block. All surfaces are smooth. Find the reaction of the block. The wheel weighs 15 kN.

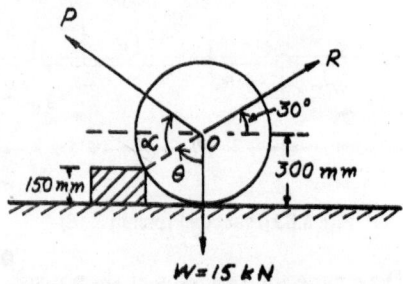

Fig. 3.67 (Practice Problem 3.20)

3.21 Determine the magnitude and direction of the smallest force P required to start the wheel (Fig. 3.68) over the block.

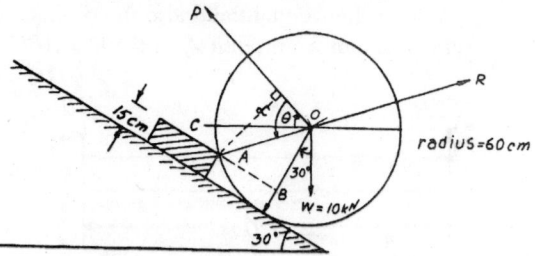

Fig. 3.68 (Practice Problem 3.21)

3.22 A concrete dam has a rectangular cross-section of height h and width b and is subjected to a water pressure on one side. Determine the minimum width b of this dam if the dam is not to overturn about the point B when $h = 5$ m. Assume unit weight of water $= 9.81$ kN/m^3 and the unit weight of concrete $= 23.5$ kN/m^3.

Chapter 4

Analysis of Pin-Jointed Frames

4.1 INTRODUCTION

A pin-jointed frame is a structure consisting of slender members, which are welded, riveted or pinned together at their ends, and supported appropriately to perform the function of transmitting static or moving loads. They are also called 'Trusses', when used to support roofs of buildings, or decks of bridges. Frames for transmission towers and other miscellaneous kinds of steel structures are other examples.

4.2 TYPES OF FRAMES

Broadly speaking, frames or trusses are classified as 'Plane frames' and 'Space frames'; the former are those in which all the members lie in a single plane, while the latter are those in which all the members do not lie in one plane. Roof trusses are examples of plane frames while transmission towers and tripods are examples of space frames.

The analysis of plane frames only will be considered here.

Frames are further subdivided into three categories:

(1) Perfect or rigid frames
(2) Deficient frames
(3) Redundant frames

4.2.1 Perfect or Rigid Frames

Perfect or rigid frames are those in which the number of members is just sufficient to transmit the loads without undergoing appreciable deformation or change of shape. The simplest of such frames is a triangular truss (Fig. 4.1). Other examples of perfect frames are shown in Fig. 4.2.

Fig. 4.1: Triangular Truss (Simplest Plane Perfect Frame)

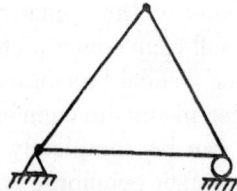

(a)

(b)

(c)

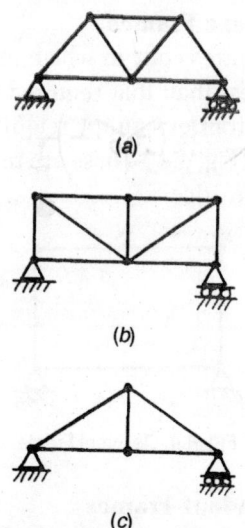

Fig. 4.2: Perfect Frames (Trusses)

With the addition of each joint to a triangular frame, two members will get added. This leads us to a relationship between the number of joints, *j*, and the number of members, *m*:

$$(m - 3) = 2(j - 3)$$
$$m = (2j - 3) \qquad \dots \text{(Eq. 4.1)}$$

This is a necessary condition for a perfect frame but not sufficient; the following frame (Fig. 4.3) satisfies this condition as does the one shown in Fig. 4.2 (*b*). But this frame is not capable of retaining its shape when loaded at the joint without a diagonal member as shown:

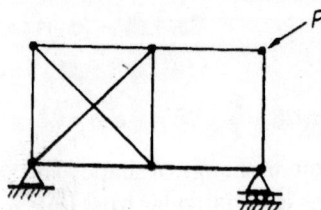

Fig. 4.3: Frame without Rigidity

This leads one to the conclusion that the necessary and sufficient condition for a frame to be perfect is that it should retain its shape when loads are applied in any direction at any joint. A perfect frame can be completely analysed by using the laws of static equilibrium alone and are said to be statically determinate.

4.2.2 Deficient Frames

A deficient frame is one in which the number of members is less than that required for a perfect frame. Such frames cannot retain their shape when loaded (Fig. 4.4). These are not adopted in engineering practice.

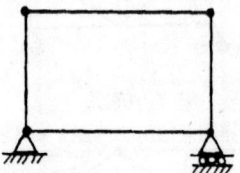

Fig. 4.4: Deficient Frame

4.2.3 Redundant Frames

A 'redundant' frame is one in which the number of members is more than that required for a perfect frame. Such frames cannot be analysed

by using the laws of static equilibrium alone, but will need additional conditions. They are therefore said to be statically indeterminate. The degree of indeterminacy increases with the number of additional members (Fig. 4.5).

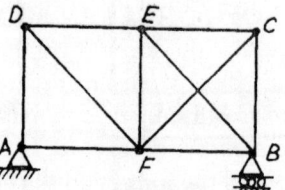

Fig. 4.5: Redundant Frame

Note: One of the diagonals in the panel *FB* is the extra member.

Only the analysis of perfect frames will be considered in this book.

4.3 ASSUMPTIONS IN THE ANALYSIS OF FRAMES

The following assumptions are relevant to the analysis of frames:

(*i*) The joints are all pin-connected or frictionless hinges, (This is an idealisation of the situation since invariably the joints are bolted, riveted, or welded in practice. However, this does not introduce any appreciable error, in view of the slenderness of the members). Consequently, the joints cannot resist moments.

(*ii*) All the loads on the truss are assumed to act or be transmitted through the nodes or joints, irrespective of their location. (Even if a load acts in between joints, its effect may be replaced by the reactions at the adjacent hinges, the local bending effect being ignored).

(*iii*) In most practical cases, the self-weight of truss is negligible when compared with the external loads, and hence the self-weight is neglected.

(*iv*) The cross-section of the members is constant throughout the length, or each member is prismatic.

(*v*) The members are all straight.

(vi) The truss is rigid so that it can deform without any appreciable change in size and shape; this is in view of the fact that the deformations and displacements are relatively small.

4.4 NATURE OF FORCES IN MEMBERS
In view of the assumptions made in the previous section, it should be obvious that the members of a perfect frame are all subjected to axial forces—either tensile or compressive.

The reactions of the hinges on the ends of the members and the internal forces in the members are depicted by arrows as shown in Fig. 4.6.

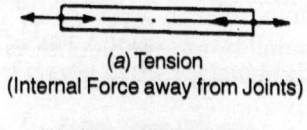

(a) Tension
(Internal Force away from Joints)

(b) Compression
(Internal Force towards the Joints)

Fig. 4.6: Forces in the Members of a Frame

Thus, a truss is completely analysed if the internal forces (tension or compression) in all the members are solved.

4.5 INDETERMINACY OF A TRUSS
Since a frame or a truss also needs to be supported like beams, they can be statically determinate or indeterminate with respect to the reactions at the supports. This indeterminacy is called 'external', and has already been seen in the last chapter.

In section 4.2, we defined determinacy with respect to the number of members and their arrangement in the frame or truss. This kind of indeterminacy is called 'internal'.

Thus a truss can have indeterminacy with respect to support reactions, or with regard to the number of members, or both.

Examples of trusses with external indeterminacy, internal indeterminacy, or both are shown in Fig. 4.7.

A truss with external indeterminacy can be solved first by solving the reactions; further, there are methods to solve the forces in members of a truss with internal indeterminacy. However, all these are out of scope of the present text.

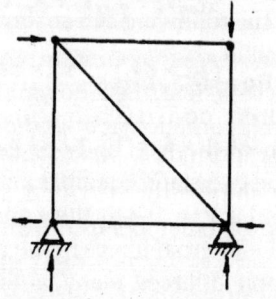

(a) Externally Indeterminate to the First Degree

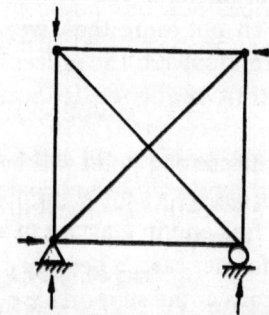

(b) Internally Indeterminate to the First Degree

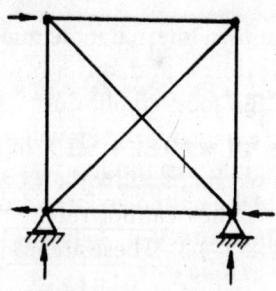

(c) Total Indeterminacy: 2
(External: 1 and Internal: 1)

Fig. 4.7: Indeterminate Trusses

4.6 ANALYSIS OF A FRAME
The following are the methods for the analysis of pin-jointed frames:

(1) Method of Joints
(2) Method of Sections
(3) Graphical Method (This is dealt with in Chapter 7)

In all these methods, the conditions of static equilibrium of a structure, studied in the last chapter, will be intelligently used along with the free body concept.

Now each of these methods will be considered in detail in the following sub-sections.

4.6.1 Method of Joints

This method consists of considering the equilibrium of the free body of each joint and applying the two static equilibrium conditions— $\Sigma P_x = 0$ and $\Sigma P_y = 0$—to the concurrent force system of the internal forces in the members and external applied forces, if any, at the joint. Since two independent equations are available, only two unknown forces can be solved. Thus, to start with, a joint at which not more than two unknowns exist is chosen and solved. The process is repeated in a convenient sequence to complete the analysis.

A convenient starting point will be one of the supports in the case of a simply supported frame, while it is the free end in the case of a cantilever frame; in the former case, it becomes imperative that the reactions at the supports are solved first by considering the equilibrium of the frame as a whole.

Any unknown internal force may be assumed as a tensile force and on completion of the analysis of the joint equilibrium, if the value is obtained as a negative one, it will be understood that the original assumption was incorrect and the nature of the unknown force is compressive.

The determinacy of a frame can also be established as follows:

Let j be the number of joints and m be the number of members.

In general, the number of unknown reaction components for a truss which is statically determinate, externally speaking, is three.

Thus the total number of unknown forces to be solved is $(m + 3)$.

Since, at each joint, two equilibrium equations can be formed, the total number of equations available becomes $2j$.

Thus, for a frame to be determinate from the analysis point of view.

$$2j = (m + 3)$$

or $m = (2j - 3)$, the same as Eq. 4.1.

If $m > (2j - 3)$, then the frame is redundant, with the degree of redundancy being equal to the number of additional unknowns.

If $m < (2j - 3)$, there can be more than one set of solutions satisfying the conditions. This indicates 'instability' of the frame, thus a deficient frame is unstable.

These principles lead us to the following corollaries:

(*i*) When only two members meet at a joint and are not collinear, the forces in both the members shall be zero, if there is no external force active at the joint. (This is because, in such a case, only one of the two equilibrium conditions— $\Sigma P_x = 0$ and $\Sigma P_y = 0$—can be satisfied, and not the other).

(*ii*) When three members only meet at a joint, two of which are collinear, there being no external load at the joint, the force in the third member shall be zero. (This is because, in such a case, the triangle of forces cannot be completed).

The method of joints will be understood from the illustrative problems 4.1 to 4.4.

4.6.2 Method of Sections

In this method, a section line is imagined so as to cut the truss into two parts, and the free body concept is applied to any of the parts, the unknown forces in the cut-members being considered as external forces along with the applied external forces and the support reactions. (The section shall be passed through members and not through joints).

Since the frame or truss as a whole is in equilibrium, any portion of the truss, considered as a free body as stated above, shall also be in equilibrium. As the number of static equilibrium conditions available is three, this is the maximum number of unknowns that can be solved at a time; thus the section shall be imagined to be passed such that the maximum number of unknown internal forces is three. The system of forces acting on any part of the truss is a non-concurrent force system, while that in the method of joints is a concurrent one.

Usually the moment equilibrium condition $\Sigma M = 0$ is applied by choosing a convenient moment centre such that number of unknowns is reduced to one, and is straight away solved. Sometimes, the condition $\Sigma V = 0$ is preferred to this in the absence of a convenient moment centre. The former is called 'moment approach' while the latter 'shear approach'. (The significance of the latter name will be appreciated after studying Chapter 24). The advantages of this method or the situation in which it is preferred are:

(*i*) The forces in only a few members of a large truss are of interest. (In such case the need to proceed from one side continuously to reach the members of interest, as is case with the method of joints, is obviated).

(*ii*) In some complex trusses, the method of joints fails to start the analysis or one gets stuck in the middle.

In certain cases, it is advantageous to apply both the methods—the method of joints to solve the forces in a few members and the method of sections for those in the others.

ILLUSTRATIVE PROBLEMS

PROBLEM 4.1: The truss *ABC* shown in Fig. 4.8 (*a*) has a span of 5 metres. It is carrying a load of 20 kN at its apex. Find the forces in the members *AB*, *BC* and *AC*, using method of joints. End *A* is hinged and *B* is supported on rollers.

Solution: A roller offers a reaction perpendicular to the plane of rolling. Let R_B is reaction at *B*. A hinge offers two reaction components one in vertical direction and other in horizontal direction. Since the load of 20 kN acts vertically downward, only vertical reaction R_A is developed and no horizontal reaction.

From the geometry of the truss, the distance of 20 kN load from *A* in horizontal direction along *AB* is *AC* cos 60°.

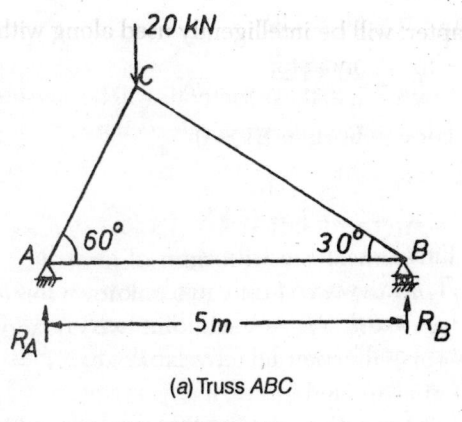

(a) Truss *ABC*

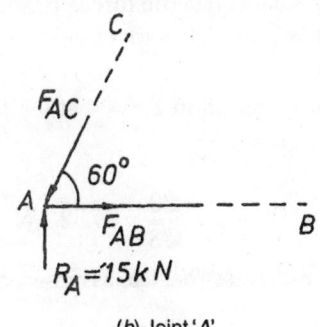

(b) Joint 'A'

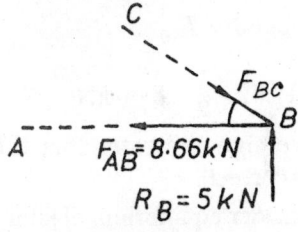

(c) Joint 'B'

Fig. 4.8 (Illustrative Problem 4.1)

In the triangle *ACB*
$$\angle ACB = 90°$$

$$AC = AB \cos 60° = 5 \times \frac{1}{2} = 2.5 \text{ m}$$

$$BC = AB \sin 60° = 5 \times \frac{\sqrt{3}}{2} \text{ m}$$

Distance of 20 kN load from $A = AC \cos 60°$

$$= 2.5 \times \frac{1}{2}$$

$$= 1.25 \text{ m}$$

Taking moments about *A*

$$R_B \times 5 = 20 \times 1.25$$

$$R_B = \frac{20 \times 1.25}{5} = 5 \text{ kN}$$

For equilibrium $\Sigma V = 0$

$$R_A + R_B = 20$$

$$R_A = 20 - 5 = 15 \text{ kN}$$

First consider equilibrium of joint A

R_A is known and only two unknown forces F_{AB} & F_{AC}, are there. At each joint two equations of static equilibrium are available i.e. $\Sigma V = 0$ & $\Sigma H = 0$.

Let F_{AC} and F_{AB} be the forces produced in the members AC & AB respectively as shown in Fig. 4.8 (*b*).

Applying condition $\Sigma V = 0$ at joint A

$$F_{AC} \sin 60° = 15$$

$$\therefore \quad F_{AC} = 15 \times \frac{2}{\sqrt{3}} = \frac{30}{\sqrt{3}} = 17.32 \text{ kN}$$

(+ve sign indicates that the assumed direction is correct)

Now applying condition $\Sigma H = 0$

$$F_{AC} \cos 60° = F_{AB}$$

$$F_{AB} = 17.32 \times \frac{1}{2} = 8.66 \text{ kN}$$

(Again +ve sign indicates that the assumed direction is correct).

Next consider equilibrium of joint B

Applying $\Sigma V = 0$ at joint B

$$F_{BC} \sin 30° = R_B = 5$$

$$F_{BC} \times \frac{1}{2} = 5$$

$$F_{BC} = 5 \times 2 = 10 \text{ kN}$$

Applying $\Sigma H = 0$ at the joint B

$$F_{BC} \cos 30° = F_{BA}$$

$$10 \times \frac{\sqrt{3}}{2} = F_{BA}$$

$$F_{BA} = 10 \times 0.866$$

$$8.66 \text{ kN} = 8.66 \text{ kN}$$

Now the forces in the various members are tabulated in the table below.

S. No.	Member	Magnitude of force (kN)	Nature of force
1.	AB	8.66	Tension
2.	BC	10.00	Compression
3.	AC	17.32	Compression

PROBLEM 4.2: A cantilever truss of 3 m span is loaded as shown in Fig. 4.9 (*a*). Find the forces in the various members of the truss, using method of joints.

Solution:

Considering Joint B

Applying $\Sigma V = 0$

$$F_{BC} \sin 60° = 20 \text{ kN}$$

$$F_{BC} \frac{\sqrt{3}}{2} = 20$$

$$F_{BC} = \frac{20 \times 2}{\sqrt{3}} = \frac{40}{\sqrt{3}} \text{ kN (Tension)}$$

Applying $\Sigma H = 0$

$$F_{BA} = F_{BC} \times \cos 60°$$

$$F_{BA} = \frac{40}{\sqrt{3}} \times \frac{1}{2} = \frac{20}{\sqrt{3}} \text{ kN (Compression)}$$

Considering Joint 'C':

Applying $\Sigma V = 0$

$$F_{CB} \cdot \sin 60° = F_{CA} \sin 60°$$

$$F_{CB} = F_{CA}$$

$$\therefore \quad F_{CA} = 40/\sqrt{3} \text{ (Compression)}$$

Applying $\Sigma H = 0$

$$F_{CD} = F_{CA} \cos 60° + F_{CB} \cos 60°$$

$$= \frac{40}{\sqrt{3}} \cdot \frac{1}{2} + \frac{40}{\sqrt{3}} \cdot \frac{1}{2}$$

$$= \frac{40}{\sqrt{3}} \text{ (Tension)}$$

S. No.	Member	Magnitude of force (kN)	Nature of force
1.	AB	20/√3	Compression
2.	BC	40/√3	Tension
3.	AC	40/√3	Tension
4.	CA	40/√3	Tension

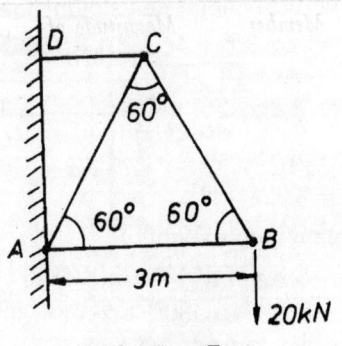

(a) Cantilever Truss

(b) Joint 'B'

(c) Joint 'C'

Fig. 4.9 (Illustrative Problem 4.2)

PROBLEM 4.3: Fig. 4.10 (a) shows a framed structure of 8 m span and 3 m height subjected to two point loads at B and D. Find the forces in all members of the structure, using method of joints.

Solution:

$$\sin\theta = \frac{3}{5} = 0.6 \; ; \; \cos\theta = 4/5 = 0.8$$

$$H_A = 10 \text{ kN } (\leftarrow)$$

Taking moments about 'C'

$$V_A \times 8 + 10 \times 3 = 15 \times 4$$

$$8V_A = 60 - 30 = 30$$

$$V_A = \frac{30}{8} = \frac{15}{4} = 3.75 \text{ kN}$$

$$\therefore \quad V_C = 15 - V_A = 15 - 3.75 = 11.25 \text{ kN}$$

Check: Taking moments about '*A*',

$$8V_C = 15 \times 4 + 10 \times 3$$

$$= 90$$

$$V_C = \frac{90}{8} = 11.25 \text{ kN}$$

Considering Joint '*C*'
Applying $\Sigma V = 0$

$$F_{CB}\sin\theta = 11.25$$

$$F_{CB} \times \frac{3}{5} = 11.25$$

$$F_{CB} = \frac{11.25 \times 5}{3} = 18.75 \text{ kN (Compression)}$$

Applying $\Sigma H = 0$

$$F_{CB}\cos\theta = F_{CD}$$

$$F_{CD} = 18.75 \times \frac{4}{5} = 15 \text{ kN (Tension)}$$

Considering Joint '*A*'
Applying $\Sigma V = 0$

$$F_{AB}\sin\theta = 3.75$$

$$F_{AB} \times \frac{3}{5} = 3.75$$

$$F_{AB} = \frac{3.75 \times 5}{3} = 6.25 \text{ kN (Compression)}$$

Applying $\Sigma H = 0$

$$F_{AB}\cos\theta + 10 = F_{AD}$$

$$6.25 \times \frac{4}{5} + 10 = F_{AD}$$

$$F_{AD} = 5 + 10 = 15 \text{ kN (Tension)}$$

Considering Joint '*D*'
Applying $\Sigma V = 0$

$$F_{CD} = 15 \text{ kN}$$

Applying $H = 0$

$$F_{DA} = F_{DC}$$

$$15 = 15 \text{ (check)}$$

S. No.	Member	Magnitude of force (kN)	Nature of force
1.	AB	6.25	Compression
2.	BC	18.75	Compression
3.	CD	15.00	Tension
4.	DA	15.00	Tension
5.	DB	15.00	Tension

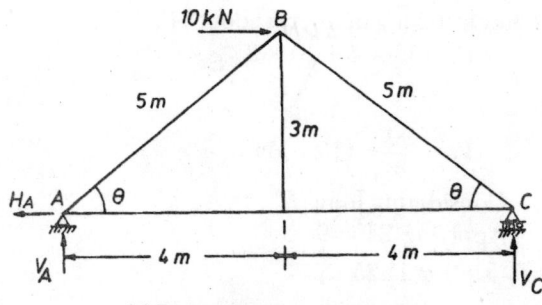

(a) Framed Structure (Pin-Jointed)

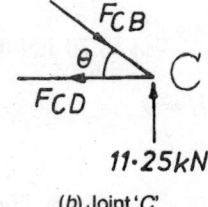

(b) Joint 'C'

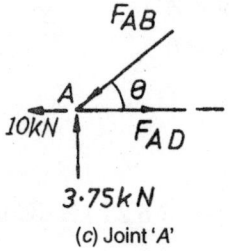

(c) Joint 'A'

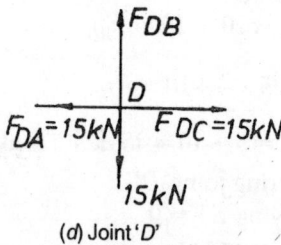

(d) Joint 'D'

Fig. 4.10 (Illustrative Problem 4.3)

PROBLEM 4.4: A truss of 12 m span is loaded as shown in Fig. 4.11 (a). Determine the forces in all the members of the truss, using method of joints.

Solution:

Total horizontal components of inclinded loads

$$= 2\sin 30° + 4\sin 30° + 2\sin 30°$$

$$= 2 \times \frac{1}{2} + 4 \times \frac{1}{2} + 2 \times \frac{1}{2}$$

$$= 1 + 2 + 1$$

$$= 4 \text{ kN}$$

Total vertical components of inclined load

$$= 2\cos 30° + 4\cos 30° + 2\cos 30°$$

$$= 8 \times \cos 30° = \frac{8\sqrt{3}}{2} = 4\sqrt{3} = 6.928 \text{ kN}$$

Total vertical load $= 6.928 + 2$

$$= 8.928 \text{ kN}$$

Taking moments about 'A'

$$V_E \times 12 = 2 \times 4 + 4(AB) + 2(AC)$$

$$= 8 + 4(4\cos 30°) + 2(6/\cos 30°)$$

$$= 8 + 13.856 + 13.856$$

$$= 35.712$$

$$V_E = \frac{35.712}{12} = 2.976 \text{ kN } (\uparrow)$$

$$V_A = 8.928 - V_E$$

$$\therefore \quad V_A = 8.928 - 2.976$$

$$= 5.952 \text{ kN } (\uparrow)$$

$$H_A = 4 \text{ kN} \quad (\leftarrow)$$

Considering Joint 'A':

Applying $\Sigma V = 0$

$$F_{AB}\sin 30° + 2 \times \cos 30° = 5.952$$

$$F_{AB}\frac{1}{2} + 2 \times 0.866 = 5.952$$

$$F_{AB} = 2(5.952 - 1.732)$$

$$F_{AB} = 8.44 \text{ kN (Compressive)}$$

Applying $\Sigma H = 0$

$$F_{AB}\cos 30° + H_A = F_{AG} + 2 \times \sin 30°$$

$$8.44 \times 0.866 + 4 = F_{AG} + 2 \times 0.5$$

$$F_{AG} = 7.309 + 4 - 1 = 10.309 \text{ kN (Tensile)}$$

Considering Joint 'B'

Resolving all forces perpendicular to line ABC

$$F_{BG} = 4 \text{ kN} \quad \text{(Compressive)}$$

Resolving all forces along the line ABC

$$F_{BC} = F_{BA} = 8.44 \text{ kN} \quad \text{(Compressive)}$$

Considering Joint 'G'

Applying $\Sigma V = 0$

$$4\sin 60° + 2 = F_{GC}\sin 60°$$

$$4 \times 0.866 + 2 = F_{GC} \times 0.866$$

$$3.464 + 2 = 0.866 \, F_{GC}$$

$$F_{GC} = \frac{5.464}{0.866} = 6.3095 = 6.310 \text{ kN} \quad \text{(Tensile)}$$

Applying $\Sigma H = 0$

$$F_{GF} + F_{GB}\cos 60° + F_{GC}\cos 60° = F_{GA}$$

$$F_{GF} + 4 \times \frac{1}{2} + 6.310 \times \frac{1}{2} = 10.309$$

$$F_{GF} = 10.309 - 2 - 3.155 = 5.154 \text{ kN}$$
$$\text{(Tensile)}$$

Considering Joint 'D'

Resolving all forces perpendicular to line CDE

$$F_{DF} = 0$$

Resolving all forces along line CDE

$$F_{DC} = F_{DE} = 5.952 \text{ kN (Compressive)}$$

Considering Joint 'E'

Applying $\Sigma V = 0$

$$F_{ED}\sin 30° = 2.976$$

$$F_{ED} = \frac{2.976 \times 2}{1} = 5.952 \text{ kN (Compressive)}$$

Applying $\Sigma H = 0$

$$F_{EF} = F_{ED}\cos 30° = 5.952 \times 0.866$$
$$= 5.154 \text{ kN (Tension)}$$

Considering Joint 'F'

Applying $\Sigma V = 0$

$$F_{FC} = F_{FD} = 0$$

Applying $\Sigma H = 0$

$$F_{FG} = F_{EF} = 5.154 \text{ kN (Tensile) (Check)}$$

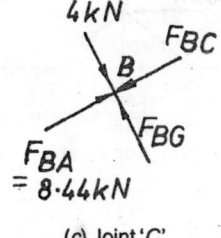

(c) Joint 'C'

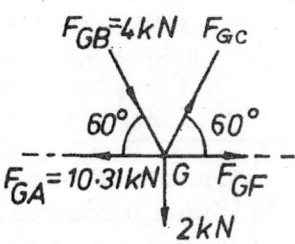

(d) Joint 'G'

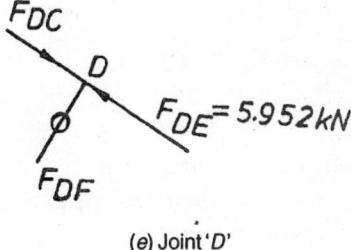

(e) Joint 'D'

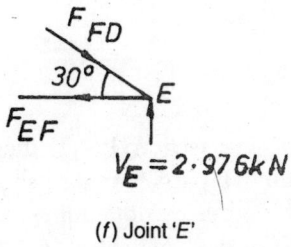

(f) Joint 'E'

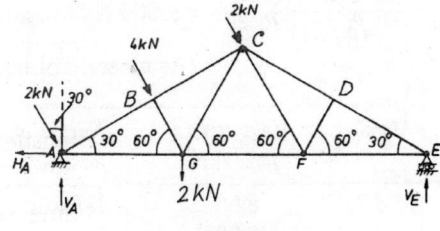

(a) Truss

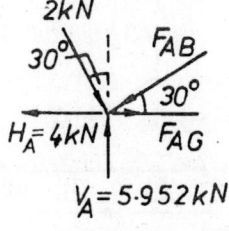

(b) Joint 'A'

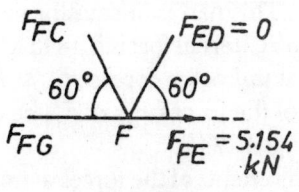

(g) Joint 'F'

Fig. 4.11 (Illustrative Problem 4.4)

S. No.	Member	Magnitude of force (kN)	Nature of force
1.	AB	8.44	Compressive
2.	BC	8.44	Compressive
3.	CD	5.952	Compressive
4.	DE	5.952	Compressive
5.	EF	5.154	Tensile
6.	FC	0	--
7.	FD	0	--
8.	FG	5.154	Tensile
9.	GA	10.309	Tensile
10.	GB	4	Compressive
11.	GC	6.310	Tensile

PROBLEM 4.5: The truss *ABC* shown in Fig. 4.12 (*a*) has a span of 5 metres. It is carrying a load of 20 kN at its apex. End *A* is hinged and *B* is supp-orted on rollers. Find the forces in the members *AB*, *BC* and *AC*, using method of sections.

Solution:

$$AC = 5 \times \cos 60° = 2.5 \text{ m}$$

$$BC = 5 \times \sin 60° = 5 \times \frac{\sqrt{3}}{2} = 4.33 \text{ m}$$

Determination of reactions

Let R_A and R_B be the reactions at *A* and *B*.
Taking moments about *A*

$$5 \times R_B = 20 \times (AC \times \cos 60°)$$
$$= 20 \times 2.5 \times 0.5$$
$$R_B = 25/5 = 5 \text{ kN}$$
$$R_A - R_B = 20$$
$$R_A = 20 - 5 = 15 \text{ kN}$$

First of all, pass section (1) – (1), thereby cutting the truss into two parts (one part shown by firm lines and the other by dotted lines) through the members *AB* and *AC* of the truss as shown in Fig. 4.12 (*b*). Now consider equilibrium of the left part of the truss. This part is in equilibrium under the action of one external force $R_A = 15$ kN and other two internal unknown forces F_{AC} & F_{AB}. Let the directions of the forces F_{AC} & F_{AB} be assumed as shown in Fig. 4.12 (*b*).

Taking moments of the forces acting in the left part of the truss about the joint *B*,

$$F_{AC} \times BC = 15 \times 5$$
$$F_{AC}(5 \times \sin 60°) = 15 \times 5$$

$$F_{AC}\left(5 \times \frac{\sqrt{3}}{2}\right) = 15 \times 5$$

$$F_{AC} = \frac{15 \times 2}{\sqrt{3}} = 17.32 \text{ kN} \quad \text{(Compressive)}$$

and now taking moments of the forces acting in the left part of the truss only about the joint *C*.

$$15 \times AC \times \cos 60° = F_{AB} \times AC \times \sin 60°$$

$$15 \times 2.5 \times \frac{1}{2} = F_{AB} \times 2.5 \times \frac{\sqrt{3}}{2}$$

$$15 = F_{AB}\sqrt{3}$$

$$F_{AB} = \frac{15}{\sqrt{3}} = 8.66 \text{ kN (Tensile)}$$

Now pass section (2) – (2) cutting the truss into two parts through the members *BC* and *AB*. Now consider the equilibrium of the right part of the truss. Let the directions of the forces F_{CB} and F_{AB} be assumed as shown in Fig. 4.12 (*c*).

Taking moments of the forces acting in the right part of the truss only about the joint *A*.

$$F_{CB} \times AC = 5 \times 5$$

$$F_{CB} = \frac{5 \times 5}{2.5} = 10 \text{ kN (Compressive)}$$

and now taking moments of the forces acting on the right part of the truss only about the joint *C*.

$$F_{AB} AC \sin 60° = 5 \times BC \cos 30°$$

$$F_{AB} 2.5 \frac{\sqrt{3}}{2} = 5 \times 4.33 \times \frac{\sqrt{3}}{2}$$

$$F_{AB} = \frac{5 \times 4.33 \times (\sqrt{3}/2)}{(\sqrt{3}/2) \times (2.5)} = 8.66 \text{ kN (Tension)}$$

(As already obtained)

S. No.	Member	Magnitude of force (kN)	Nature of force
1.	AB	8.66	Tension
2.	BC	10.00	Compression
3.	CA	17.32	Compression

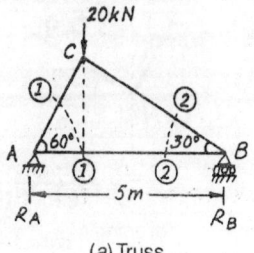

(a) Truss

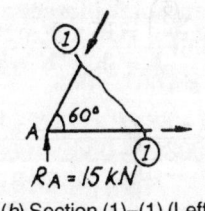

(b) Section (1)–(1) (Left)

(c) Section (2)–(2) (Right)

Fig. 4.12 (Illustrative Problem 4.5)

PROBLEM 4.6: A cantilever truss of 3 m span is loaded as shown in Fig. 4.13 (a). Find the forces in the various members of the truss using method of sections.

Solution: First of all, pass section (1) – (1) cutting the truss through the members *AB* and *BC*.

Now consider the equilibrium of the right part of the truss. Let the directions of the forces F_{BC} and F_{BA} be assumed as shown in Fig. 4.13 (b).

Taking moments of the forces acting on the right part of the truss only, about the joint *A*.

$$F_{BC} \times 3 \times \sin 60° = 20 \times 3$$

$$F_{BC} = \frac{20 \times 3}{3 \times 0.866} = 23.10 \text{ kN} \quad \text{(Tension)}$$

and now taking moments of the forces in the right part of the truss only, about the joint *C*.

$$F_{AB} \times 3 \times \sin 60° = 20 \times 3 \times \cos 60°$$

$$F_{AB} = \frac{20 \times 3 \times 0.5}{3 \times 0.866} = 1.55 \text{ kN (Compression)}$$

Now pass section (2) – (2) cutting the truss through the members *CD*, *CA* and *BA*.

Now consider the equilibrium of the right part of the truss.

Let the directions of the forces F_{CD} and F_{CA} be assumed as shown in Fig. 4.13 (c).

Taking moments of the forces acting on the right part of the truss only, about the joint *A*.

$$F_{CD} \times 3 \times \sin 60° = 20 \times 3$$

$$F_{CD} \times 3 \times 0.866 = 20 \times 3$$

$$F_{CD} = \frac{20 \times 3}{3 \times 0.866} = 23.10 \text{ kN} \quad \text{(Tension)}$$

and now taking moments of the forces in the right part of the truss only, about the joint *D*.

$$F_{AB} \times 3 \times \sin 60° + F_{AC} \times 1.5 \times \sin 60° = 20 \times 3$$

$$11.55 \times 3 \times 0.866 + F_{AC} \times 1.5 \times 0.866 = 60$$

$$30 + F_{AC} \times 1.5 \times 0.866 = 60$$

$$F_{AC} = \frac{60 - 30}{1.5 \times 0.866} = 23.10 \text{ kN (Compression)}$$

S. No.	Member	Magnitude of force (kN)	Nature of force
1.	*AB*	11.55	Compression
2.	*BC*	23.10	Tension
3.	*CA*	23.10	Compression
4.	*CD*	23.10	Tension

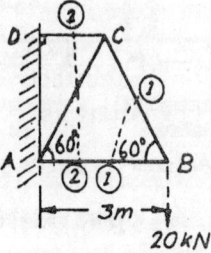

(a) Cantilever Truss

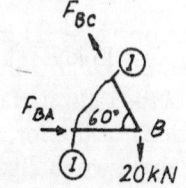

(b) Section (1)–(1) (Right)

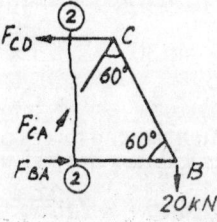

(c) Section (2)–(2) (Right)

Fig. 4.13 (Illustrative Problem 4.6)

PROBLEM 4.7: Find the forces in the members *GF*, *FC* and *CD* of the truss shown in Fig. 4.14 (*a*) by the method of sections.

Solution: Total horizontal components of inclined loads

$$= 2\sin 30° + 4\sin 30° + 2\sin 30°$$

$$= 8 \times \sin 30°$$

$$= 8 \times \frac{1}{2} = 4 \text{ kN}$$

Total vertical components of inclined loads

$$= 2\cos 30° + 4\cos 30° + 2\cos 30°$$

$$= 8 \times \cos 30° = 8\frac{\sqrt{3}}{2} = 6.928 \text{ kN}$$

Total vertical load $= 6.928 + 2 = 8.928$ kN

Taking moments about '*A*'

$$V_E \times 12 = 2 \times 4 + 4 \times AB + 2 \times AC$$

$$= 8 + 4(4\cos 30°) + 2\left(\frac{6}{\cos 30°}\right)$$

$$= 8 + 4 \times 3.464 + \frac{2 \times 6}{0.866}$$

$$= 8 + 13.856 + 13.856$$

$$= 35.712 \text{ kN}$$

$$V_E = \frac{35.712}{12} = 2.976 \text{ kN} (\uparrow)$$

$$\therefore \quad V_A = 8.928 - 2.976 = 5.952 \text{ kN} \qquad (\uparrow)$$

$$H_A = 4 \text{ kN} \qquad (\leftarrow)$$

Now pass section (1) – (1) cutting the truss through the members *GF*, *FC* and *CD*. Consider the equilibrium of the right part of the truss.

Let the directions of the forces F_{FG}, F_{FC} and F_{DC} be assumed as shown in Fig. 4.14 (*b*).

Taking moments of the forces on the right part of the truss only about *F*

$$F_{CD} \times FD = 2.976 \times 4$$

$$F_{CD} \times 4 \times \sin 30° = 2.976 \times 4$$

$$F_{CD} = 2.976 \times 2 = 5.952 \text{ kN (Compression)}$$

Taking moments of the forces on the right part of the truss only about *C*.

$$F_{FG} \times 4 \times \sin 60° = 2.976 \times 6$$

$$F_{FG} = \frac{2.976 \times 6}{4 \times 0.866} = 5.155 \text{ kN (Tension)}$$

Taking moments of the forces on the right part of truss only about *E*.

$$F_{FC} \times 4 \times \sin 60° = 0$$

$$F_{FC} = 0$$

Applying $\Sigma V = 0$

$$F_{FC} \times \sin 60° + 2.976 = F_{CD}\sin 30°$$

$$F_{FC} \sin 60° = 5.952 \times \frac{1}{2} - 2.976 = 0$$

$$\therefore \quad F_{FC} = 0$$

The results are tabulated as given below:

S.No.	Member	Magnitude of force (kN)	Nature of force
1.	DC	5.952	Compression
2.	FC	0	–
3.	FG	5.155	Tension

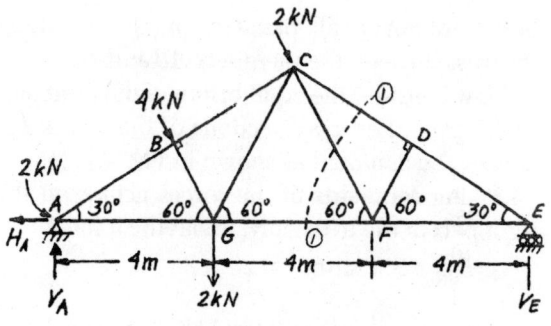

(a) Truss

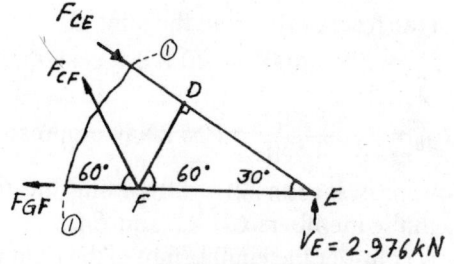

(b) Section (1) – (1) (Right)

Fig. 4.14 (Illustrative Problem 4.7)

PRACTICE PROBLEMS

4.1 Find the forces in the members of the frame shown in Fig. 4.15.

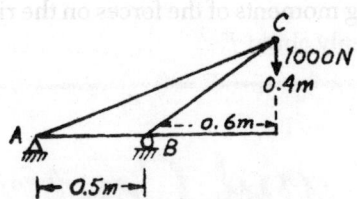

Fig. 4.15 (Practice Problem 4.1)

4.2   A warren girder consisting of seven members
each of 2 m length, is hinged or pin-jointed at
one end. The girder is supported on rollers at
other end, is shown in Fig. 4.16. The girder is
loaded at *B* and *C* as shown. Find the forces in
all the members of the girder.

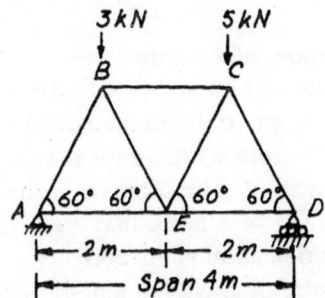

Fig. 4.16 (Warren Girder (Practice Problem 4.2)

4.3   A truss of span 5 metres is loaded as shown in
Fig. 4.17. Find the forces in all the members of
the truss.

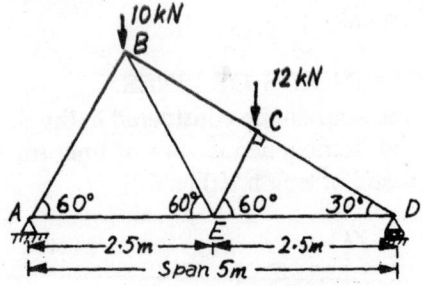

Fig. 4.17 (Practice Problem 4.3)

4.4   A king post truss of 8 m span is loaded as
shown in Fig. 4.18. Find the forces in all the
members of the truss.

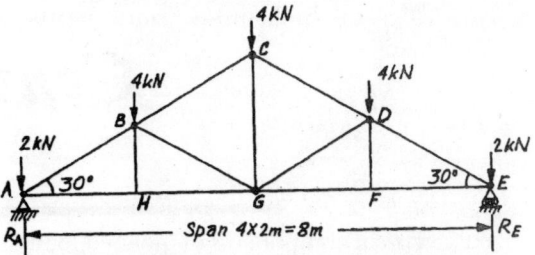

Fig. 4.18 King Post Truss (Practice Problem 4.4)

4.5   Determine the forces in the various members
of the truss loaded as shown in Fig. 4.19.

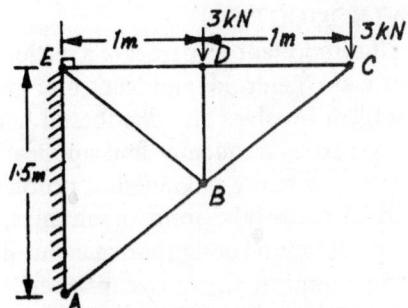

Fig. 4.19 (Practice Problem 4.5)

4.6   Determine the forces in the various members
of the truss loaded as shown in Fig. 4.20.

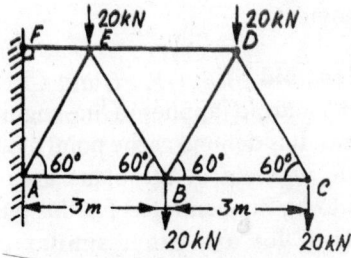

Fig. 4.20 (Practice Problem 4.6)

Chapter 5

Centroid and Centre of Gravity

5.1 INTRODUCTION

Among the important properties of a section, area, or a body are 'centroid' and 'centre of gravity'. The problem involves any distributed quantity. These properties frequently find application in the analysis of many engineering problems in strength of materials, solid mechanics, fluid mechanics, structural design and machine design.

In this chapter, these two properties are defined and their determination for lines, areas (laminas), and solid bodies is considered in detail.

5.2 DEFINITIONS

The following definitions are relevant in the present context.

5.2.1 Centroid

The term 'centroid' applies to line segments and plane areas. It is defined as the point at which the entire line segment or the plane area may be assumed to be concentrated for the purpose of moments or for any other similar purpose. Consequent to this definition, it is easily understood that the moment of the area about any axis through the centroid is zero. In the case of a line segment, one can say that it will be symmetrical about its centroid.

5.2.2 Centre of Mass

The 'Centre of Mass' of a body is the point where the whole mass of the body appears to be (or is assumed to be) concentrated.

5.2.3 Centre of Gravity

The 'Centre of Gravity' of a body is the point where its weight or the resultant force of gravity on the body appears to be (or may be taken to be) concentrated. The location of the centre of gravity (C.G.) of a body may vary depending upon the orientation of the body.

The centre of gravity of a body will coincide with its centre of mass for uniform gravitational field. However, since earth's gravitational field is not strictly uniform but shows slight variation depending upon certain factors, the centre of gravity and centre of mass are not identical, strictly speaking.

5.3 CENTROIDS OF LINES

Let a line segment be considered in the shape of a curved homogeneous wire of uniform cross-section and of length, l (Fig. 5.1).

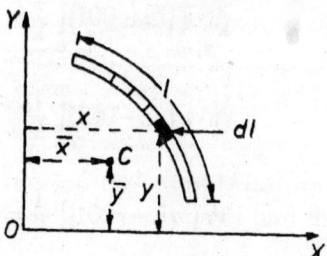

Fig. 5.1: Centroid of a Line Segment

Consider an infinitesimally small length dl of the line segment. If ρ is the mass density of the material, A the area of the cross-section of the wire, the mass of the element dm is

$$dm = \rho \cdot A \cdot dl$$

Moment of this mass about x-axis is $\rho Ay \cdot dl$
Total moment of the wire $= \int_0^l \rho Ay \cdot dl$

Since, by definition, the whole segment is assumed to be concentrated at the centroid, the total moment of the mass of the wire is $m \cdot \bar{y}$, m being the mass of the total length l and $(\bar{x}, \bar{y})$ are the coordinates of the centroid.

$$\therefore \quad m \cdot \bar{y} = \int_0^l \rho \cdot A \cdot y\,dl$$

$$\bar{y} = \frac{\int_0^l \rho \cdot A \cdot y\,dl}{m} = \frac{\int_0^l \rho \cdot A \cdot y\,dl}{\rho A l} = \frac{\int_0^l y \cdot dl}{l}$$

ρ and A constant along the length.

Similarly, $\quad \bar{x} = \dfrac{\int_0^l x\,dl}{l}$

Thus the co-ordinates of the centroid of a line segment are

$$\left. \begin{array}{l} \bar{x} = \int x\,dl / l \\ \bar{y} = \int y\,dl / l \end{array} \right\} \qquad \text{...(Eq. 5.1)}$$

When the curve does not lie in a single plane, the logic may be extended to the third dimension, and it may be shown that

$$\bar{z} = \int z\,dl / l$$

5.4 CENTROIDS OF AREAS

The concept of centroid may be extended to areas as shown in Fig. 5.2.

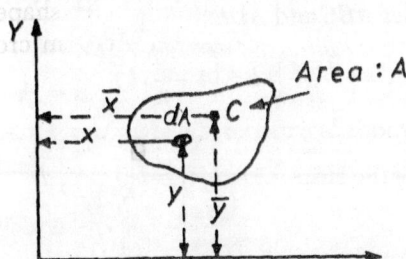

Fig. 5.2: Centroid of an Area

Let the area be A as shown. Let the centroid of the area be C with co-ordinates $(\bar{x}, \bar{y})$.

Let us consider an elemental area dA with coordinates $(x, y,)$.

If ρ is the mass density and t the uniform thickness of the area (lamina), the mass dm of the elemental area is given by

$$dm = \rho \cdot t \cdot dA$$

Since the mass density is same throughout for a homogeneous material, and the thickness is taken to be the same throughout the area, taking moments of such elemental masses about the axes, and equating these to the moments of the entire area about the respective axes, we have

$$\bar{x} = \int x \cdot dA / \int dA = \int x\,dA / A \quad \text{...(Eq. 5.2)}$$

and $\quad \bar{y} = \int y\,dA / \int dA = \int y\,dA / A$

If the area does not lie in a single plane, the logic may be extended to the third dimension, and it may be shown that

$$\bar{z} = \int z \cdot dA / A$$

Note: The numerators in equations 5.1 and 5.2 are called the first moments of the line/area about the respective axes.

Thus it may be seen that the co-ordinates of the centroid may be obtained by dividing the first moment of the area by the area itself. The first moment will have the same sign as that of the respective co-ordinate of the centroid.

Centroids of Symmetrical Areas

A line or an area may be symmetrical about an axis; in such case, every elemental area on one side of the axis of symmetry will have a mirror image on the other side. The moments of these elemental areas about this axis will be equal and opposite. Using the principle of summation/integration, it can be concluded that the first moment of the whole line/area about the axis of symmetry is zero. Hence the distance of the centroid from the axis of symmetry is zero, or the centroid lies on the axis of symmetry. Thus the work of determination of the centroid will be reduced, as one of the co-ordinates need not be obtained.

If the area has two axes of symmetry, as for example, a square or a rectangle, the centroid is obviously the point of intersection of the axes of symmetry.

5.5 CENTROIDS BY INTEGRATION

The principle of integration (for continuous functions) or summation (for discrete quantities) may be used to determine the location of the centroid of regular shaped geometrical figures. This amounts to obtaining the location of centroid from first principles.

This is illustrated by the following cases:

1. Semi-Circular Arc

This may be a uniform thin wire.

Let the semi-circular arc have the *y*-axis as the axis of symmetry as shown in Fig. 5.3.

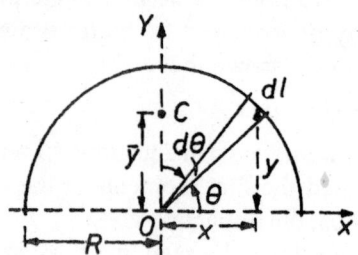

Fig. 5.3: Semi-Circular Arc-Centroid

Since there is symmetry about the *y*-axis,
$$\bar{x} = 0.$$
An element of the wire of length *dl* located at an angle θ from the *x*-axis, subtending $d\theta$ at the centre.

$$dl = R \cdot d\theta$$
$$y = R\sin\theta$$
$$\bar{y} = \frac{\int y dl}{\int dl} = \frac{\int y dl}{l}; \ l = \pi R$$

$$l\bar{y} = \int y dl = 2 \int_{0}^{\pi/2} (R\sin\theta)(Rd\theta) = 2R^2 \int_{0}^{\pi/2} \sin\theta d\theta$$

$$2R^2 \left[(-\cos\theta) \right]_{0}^{\pi/2} = 2R^2$$

$$\therefore \quad \pi R\bar{y} = 2R^2$$

or $\qquad \bar{y} = \frac{2R}{\pi}$ $\qquad\qquad$ ...(Eq. 5.3)

The centroid for a quarter-circle arc can be easily deduced from this (Fig. 5.4).

Since a quarter of a circle is one-half of a semi-circle, $\bar{y}$ will be the same as that for a semi-circular arc.

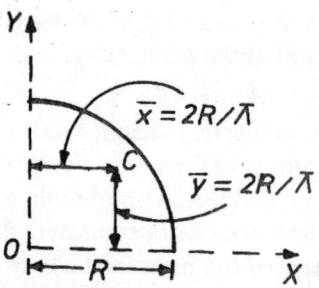

Fig. 5.4: Quarter-Circular Arc-Centroid

Once again because of symmetry, $\bar{x}$ also will be equal to

$$\bar{y}\left[= \frac{2R}{\pi} \right]$$

2. Triangle

Let the triangle be *ABC* with base *b* and altitude or height *h* as shown in Fig. 5.5. Let the distance of the centroid from the base be $\bar{y}$.

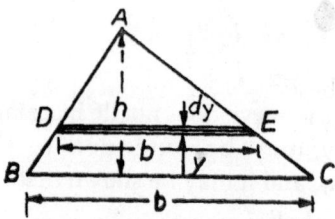

Fig. 5.5: Triangle-Centroid

Let a strip of elemental thickness *dy* at a distance *y* from the base be considered. The width of this strip *b*, may be got from the similar triangles *ABC* and *ADE*:

$$b_1 = \left(\frac{h-y}{h} \right) \cdot b = \left(1 - \frac{y}{h} \right) b$$

Area of the strip, $dA = b_1 dy = b\left(1 - \frac{y}{h} \right) dy$

Area of the triangle, $A = \frac{1}{2}bh$

$$\bar{y} = \frac{\int y dA}{A}$$

$$\int y dA = \int_{0}^{h} \frac{b}{h}(h-y)y dy = \frac{b}{h} \int_{0}^{h}(hy - y^2) dy$$

$$= \frac{b}{h}\left[\frac{hy^2}{2} - \frac{y^3}{3}\right]_0^h = \frac{b}{h}\left[\frac{h^3}{2} - \frac{h^3}{3}\right] = \frac{bh^2}{6}$$

$$\therefore \qquad \bar{y} = \frac{bh^2}{6} \times \frac{2}{bh} = \frac{h}{3} \qquad \qquad ...(\text{Eq. } 5.4)$$

Thus, the centroid of a triangle of height h will be at a distance $h/3$ from the base (or $2h/3$ from the vertex) of the triangle.

3. Sector of a Circle

Let the sector of a circle subtends an angle 2α at the centre as shown in Fig. 5.6.

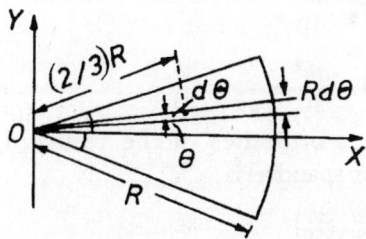

Fig. 5.6: Sector of a Circle-Centroid

Since the angle $d\theta$ is very small, the elementary sector may be treated as a triangle with centroidal distance $(2/3)R$ from the vertex O. The area dA of the elemental sector is $(1/2)\,(R)\,(R \cdot d\theta)$.

The x coordinates of the centroid of the elemental area is

$$\left(\frac{2}{3}R\right)(\cos\theta)$$

The area of the whole sector,

$$A = \frac{1 \times R}{2} \times R\alpha \times 2 = R^2\alpha$$

Now,

$$A\bar{x} = \int x\,dA$$

$$R^2\alpha\bar{x} = 2\int_0^\alpha \left(\frac{2}{3}R\cos\theta\right)\left(\frac{R^2}{2}d\theta\right)$$

$$= \frac{2}{3}R^3[\sin\theta]_0^\alpha = \frac{2}{3}R^3\sin\alpha$$

$$\therefore \qquad \bar{x} = \frac{2}{3}R^3\frac{\sin\alpha}{R^2\alpha} = \frac{2}{3}\cdot\frac{R\sin\alpha}{\alpha} \qquad ...(\text{Eq. } 5.5)$$

Special Cases:

(*i*) Semi-Circular Area

For a semi-circular area (Fig. 5.7) of radius R,

$$\alpha = \pi/2, \ \bar{y} = \frac{4R}{3\pi} \text{ (if the } y\text{-axis is the axis}$$

of symmetry) $\qquad ...(\text{Eq. } 5.6)$

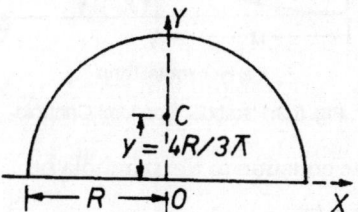

Fig. 5.7: Semi-Circular Area Centroid

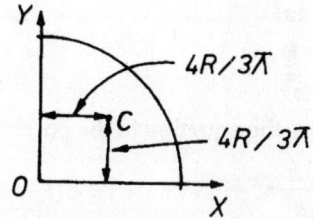

Fig. 5.8: Quarter Circular Area-Centroid

(*ii*) Quarter-Circular Area

For a quarter-circular area (Fig. 5.8) of radius R, by symmetry, it can be easily understood that $\bar{x} = \bar{y} = \frac{4R}{3\pi}$.

4. Parabolic Spandrel

The centroid of a parabolic spandrel may be found by considering a vertical strip and a horizontal strip separately, and using the principle of integration (Fig. 5.9).

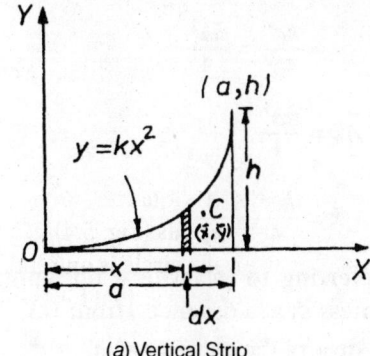

(a) Vertical Strip

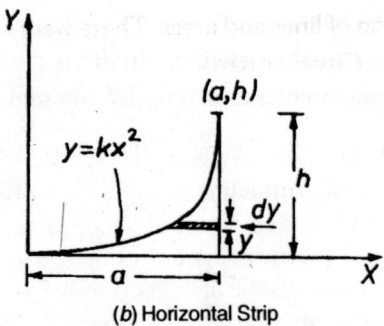

(b) Horizontal Strip

Fig. 5.9: Parabolic Spandrel-Centroid

Let the equation to the parabola be

$$y = kx^2$$

Since at $x = a$, $y = h$ (as assumed),

$$h = ka^2$$

or

$$k = \frac{h}{a^2}$$

Therefore the equation to the parabola is

$$y = \frac{h}{a^2}x^2$$

Referring to (a), with a vertical strip of thickness dx at a distance x from OY,

$$\text{area of the element} = y \cdot dx = \frac{h}{a^2}x^2 dx$$

Total area A of the spandrel

$$= \int_0^a y dx = \int_0^a \frac{h}{a^2}x^2 dx = \frac{ha^3}{3a^2} = \frac{ha}{3}$$

Moment of the area about the y-axis

$$= \int_0^a xy dx = \frac{h}{a^2}x^3 dx$$

$$= \frac{ha^4}{4a^2} = \frac{ha^2}{4}$$

$\therefore \quad A\bar{x} = \dfrac{ha^2}{4}$

or $\quad \bar{x} = \dfrac{ha^2}{4} \times \dfrac{3}{ha} = \dfrac{3a}{4}$ ...(Eq. 5.7)

Referring to (b), with a horizontal strip of thickness dy at a distance y from OX

$$\text{area of the element} = (a - x)dy$$

$$= a - \left(\frac{a^2}{h}y\right)^{\frac{1}{2}} dy = \left(a - \frac{a}{\sqrt{h}}y^{\frac{1}{2}}\right)dy$$

Moment of the area about x-axis

$$= \int_0^h (a - x)y dy$$

$$= \int_0^h \left(ay - \frac{ay^{\frac{3}{2}}}{\sqrt{h}}\right)dy$$

$$= \left(\frac{ah^2}{2} - \frac{2a}{5}h^2\right) = \frac{ah^2}{10}$$

$\therefore \quad A\bar{y} = \dfrac{ah^2}{10}$

or $\quad \bar{y} = \dfrac{ah^2}{10} \times \dfrac{3}{ah} = \dfrac{3h}{10}$ ...(Eq. 5.8)

The co-ordinates of the centroid of the parabolic spandrel is

$$\left(\frac{3a}{4}, \frac{3h}{10}\right)$$

5.6 CENTROIDS OF COMPOSITE LINES AND SECTIONS

In case the composite lines can be divided into simple lines of regular geometric shapes, the centroid may be determined by the principle of moments. The sum of the moments of each of the simple lines, the lengths of which are imagined to be concentrated at their respective centroids, about the reference axis, divided by the total length of the composite line gives the distance of the centroid of the composite line from the axis (Fig. 5.10).

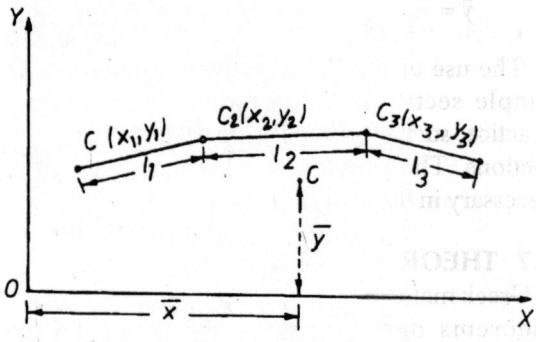

Fig. 5.10: Composite Line-Centroid

If the simpler lines have centroids with coordinates (x_1, y_1), (x_2, y_2), ..., with lengths l_1, l_2, ..., the coordinates $(\bar{x}, \bar{y})$ of the centroid C of the composite line may be obtained as follows:

$$\left.\begin{aligned}\bar{x} &= \left(\frac{l_1 x_1 + l_2 x_2 + ...}{l_1 + l_2 + ...}\right) = \frac{\Sigma l_i x_i}{\Sigma l_i} \\ \bar{y} &= \left(\frac{l_1 y_1 + l_2 y_2 + ...}{l_1 + l_2 + ...}\right) = \frac{\Sigma l_i y_i}{\Sigma l_i}\end{aligned}\right\} \quad ...\text{(Eq. 5.9)}$$

(This may not lie on the composite line).

In the case of composite sections or areas, a similar procedure is applicable for the location of the centroid. The composite area is split up into simple geometrical figures of regular shape, the centroids of each being known (Fig. 5.11). Then the coordinates of the centroid of the composite section may be got by applying the principle of moments:

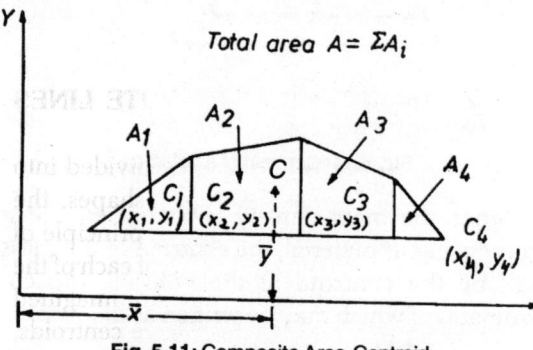

Fig. 5.11: Composite Area-Centroid

$$\left.\begin{aligned}\bar{x} &= \frac{\Sigma A_i X_i}{\Sigma A_i} \\ \bar{y} &= \frac{\Sigma A_i y_i}{\Sigma A_i}\end{aligned}\right\} \quad ...\text{(Eq. 5.10)}$$

The use of composite sections, built-up from simple sections, is common in engineering practice; such sections are also called "built-up" sections. The position of the centroid may be necessary in the analysis.

5.7 THEOREMS OF PAPPUS-GULDINUS

A Greek mathematician, Pappus, formulated two theorems during the third century AD, to determine the surface areas and volumes of revolution of lines and areas. These were restated by a Swiss mathematician, Guldinus during the seventeeth century AD, utilising the concept of centroid.

A 'surface of revolution' is the surface generated by rotating a straight line or a plane curve about an axis which does not intersect in the middle or is parallel. For example, if a straight line is rotated about an axis parallel to it, the surface of revolution is a cylinder; if a straight line is rotated about a line meeting it at one end, the surface generated is that of a cone. The surface generated when a semi-circular arc is rotated about a diametral axis is that of a sphere.

A 'body of revolution' is a solid generated by rotating a plane area about an axis not intersecting it, but lying in it in the same plane. For example, when a rectangle is rotated about one of its sides, a cylinder is the solid body generated; when a right-angled triangle is rotated about one of the adjacent sides, a cone is generated. The solid generated when a semi-circular area is rotated about its diameter is a sphere.

The theorems of Pappus-Guldinus are now stated:

First Theorem

The area of a surface of revolution is given by the product of the length of the generating line and the distance traversed by the centroid of the generating line.

The proof is simple (Fig. 5.12).

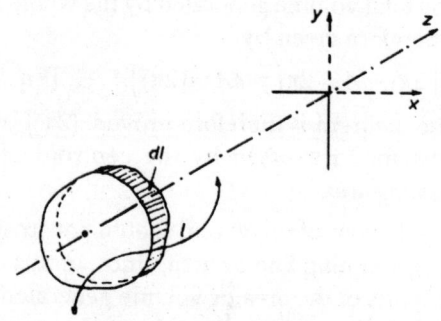

Fig. 5.12: Area of Revolution

Let us consider a small element of length dl of a generating line of length l, rotating about z-axis as shown. The area generated by it is $2\pi x \cdot dl$ The total area generated, A, is given by

$$A = \int 2\pi x dl = 2\pi \int x dl = 2\pi \bar{x} l$$

$$\therefore \quad A = (2\pi \bar{x})l \qquad \dots(\text{Eq. 5.11})$$

The theorem is therefore proved, $(2\pi \bar{x})$ being the distance traversed by the centroid.

Second Theorem

The volume of a body of revolution is given by the product of the generating area and the distance traversed by the centroid of the generating area.

The proof can be understood with reference to Fig. 5.13.

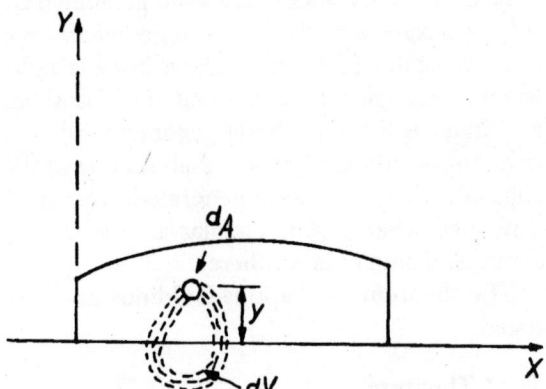

Fig. 5.13: Volume of Revolution

Let dA be a differential area of a total generating area A, revolving about the x-axis at the edge of the area as shown. The volume generated by this element is $2\pi y \cdot dA$.

The total volume generated by the whole area A is therefore given by

$$V = \int 2\pi y \cdot dA = 2\pi \int y \cdot dA = (2\pi \bar{y})A \quad \dots(\text{Eq. 5.12})$$

The theorem is therefore proved, $(2\pi \bar{y})$ being the distance traversed by the centroid of the generating area.

Note. The axis of revolution should not cross the generating line or area, since, in this case, parts of the area or volume generated will be of opposite signs. In such cases, the theorems will not apply.

Although these theorems have been propounded to obtain surface areas/volumes of

revolution, these can be applied in a converse manner, to determine the centroid of the generating line/area if the surface area/volume of revolution are known.

5.8 CENTROID OF VOLUMES

Referring to Fig. 5.14, the elemental mass dm of an elemental volume dV is given by

$$dm = \rho \cdot dV, \rho \text{ being the mass density of the material.}$$

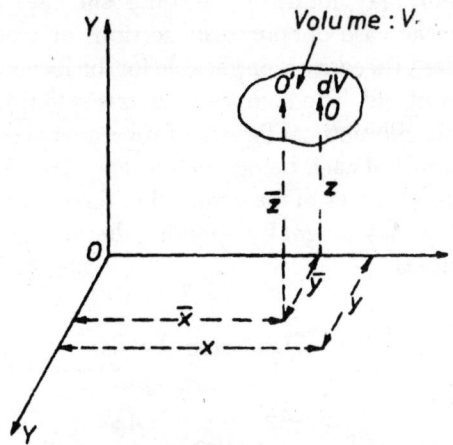

Fig. 5.14: Centroid of a Volume

Since the mass density is a constant for a homogeneous material, the centroid of the mass will be the centroid of the volume, the co-ordinates of which may be written as follows:

$$\left. \begin{array}{l} \bar{x} = \dfrac{\int x dv}{V} \\[2mm] \bar{y} = \dfrac{\int y dv}{V} \\[2mm] \bar{z} = \dfrac{\int z dv}{V} \end{array} \right\} \quad \dots(\text{Eq. 5.13})$$

For a composite body which can be divided into many simple volumes, the centroids of which are known, the co-ordinates of the whole volume may be obtained as follows:

$$\left. \begin{array}{l} \bar{x} = \dfrac{\Sigma V_i x_i}{V} \\[2mm] \bar{y} = \dfrac{\Sigma V_i y_i}{V} \\[2mm] \bar{z} = \dfrac{\Sigma V_i z_i}{V} \end{array} \right\} \quad \dots(\text{Eq. 5.14})$$

For a homogeneous body, the centroid of volume and the centre of gravity of the body will coincide; but the centre of gravity will not coincide with the centroid of the volume if the body is non-homogeneous.

If there is a plane of symmetry for the volume, the centroid of volume will lie in this plane. If there are two planes of symmetry, the centroid will lie on the intersection line of these two planes. If there are three planes of symmetry, the centroid is the point of inter-section of these planes.

5.9 CENTROIDS AND CENTRES OF GRAVITY FOR STANDARD CASES

The coordinates of the centroid and areas of some standard cases are given in Table 5.1.

TABLE 5.1: CENTROIDS AND AREAS FOR STANDARD CASES

S. N.	Shape	Figure	Coordinates of Centroid		Area
			$\bar{x}$	$\bar{y}$	
1.	Triangle		–	$h/3$	$\dfrac{bh}{2}$
2.	Semi-circle		0	$\dfrac{4R}{3\pi}$	$\dfrac{\pi R^2}{2}$
3.	Quadrant of a circle		$\dfrac{4R}{3\pi}$	$\dfrac{4R}{3\pi}$	$\dfrac{\pi R^2}{4}$
4.	Sector of a circle		$\dfrac{2R\sin\alpha}{3\alpha}$	0	αR^2
5.	Semi-parabola		$\dfrac{3a}{8}$	$\dfrac{3h}{5}$	$\dfrac{2}{3}ah$
6.	Parabolic spandrel		$\dfrac{3a}{4}$	$\dfrac{3h}{10}$	$\dfrac{1}{3}ah$

The coordinates of the centroids and the volumes of some standard cases are given in Table 5.2.

TABLE 5.2: CENTROIDS AND VOLUME FOR SOME STANDARD CASES

S. N.	Shape	Figure	Coordinates of Centroid $\bar{x}$	$\bar{y}$	Area
1.	Hemi-sphere		0	$\dfrac{3R}{8}$	$\dfrac{2}{3}\pi R^3$
2.	Right-circular Cone		0	$\dfrac{h}{4}$	$\dfrac{1}{3}\pi R^2 h$
3.	Rectangular Pyramid		0	$\dfrac{h}{4}$	$\dfrac{1}{3}lbh$
4.	Paraboloid of Revolution		0	$\dfrac{h}{3}$	$\dfrac{1}{2}\pi R^2 h$

ILLUSTRATIVE PROBLEMS

PROBLEM 5.1: Find the centroid of a line segment in the form of a circular arc shown in Fig. 5.15.

Solution:
There is symmetry about x-axis.

Therefore $\bar{y} = 0$

A differential element of arc having length

$$dL = r \cdot d\theta$$

The x-coordinate of the element $= r\cos\theta$

$$\bar{x} = \frac{\int x \cdot dL}{L}$$

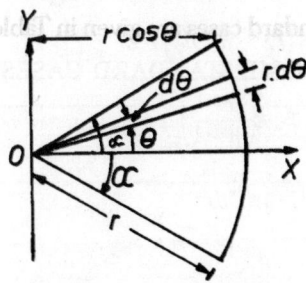

Fig. 5.15 (Illustrative Problem 5.1)

$$= \frac{\int\limits_{-\alpha}^{+\alpha} r \cdot d\theta \cdot r\cos\theta}{2\alpha r}$$

$$= \frac{r^2 \int\limits_{-\alpha}^{+\alpha} \cos\theta \cdot d\theta}{2\alpha r}$$

$$= \frac{2r^2 \int\limits_{\alpha}^{\alpha} \cos\theta \cdot d\theta}{2\alpha r}$$

$$= \frac{2r^2}{2\alpha r} [\sin\theta]_0^\alpha$$

$$= \frac{2r^2}{2\alpha r} \sin\alpha$$

$$= \frac{r\sin\alpha}{\alpha}$$

PROBLEM 5.2: Find the centroid of a right-angled triangle of base '*b*' and altitude '*h*'.

Solution:

Let *G* be the centroid
Let the centroid be at a distance of $\bar{x}$ from *AC*

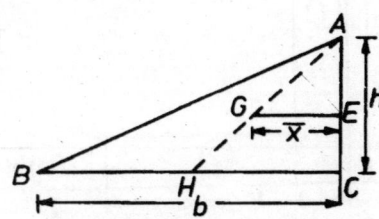

Fig. 5.16 (Illustrative Problem 5.2)

By similar triangles *AGE* and *AHC*,

$$\frac{GE}{HC} = \frac{AG}{AH} = \frac{2}{3}$$

$$\therefore \quad GE = \frac{2}{3} HC$$

$$\bar{x} = \frac{2}{3} \times \frac{b}{2} = \frac{b}{3}$$

The centroid is at $\frac{b}{3}$ from the end *C* and $\frac{2}{3}b$ from the end *B*.

Similarly the centroid is at a distance of $\frac{h}{3}$ from the end *C* and $\frac{2}{3}h$ from the end *A*.

PROBLEM 5.3: Find the centroid distance $(\bar{y})$ of the trapezium shown in Fig. 5.17 from *DC*.

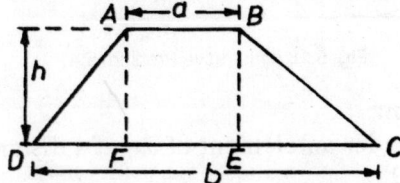

Fig. 5.17 (Illustrative Problem 5.3)

Solution:

Draw *AF* perpendicular to *DC* from *A* and *BE* perpendicular to *DC* from *B*

$$\text{Total area} = \left(\frac{a+b}{2}\right)h$$

The moment of the trapezium about *DC*
= Sum of the moments of triangle *ADF*, triangle *BEC* and rectangle *ABEF*

$$\left(\frac{a+b}{2}\right)h\bar{y} = \left[\frac{1}{2}h \times DF\right]\frac{h}{3} + \left[\frac{1}{2}h \times EC\right]\frac{h}{3} + ah\frac{h}{2}$$

$$\left(\frac{a+b}{2}\right)h\bar{y} = \frac{h^2}{6}[b - a + 3a]$$

$$\left(\frac{a+b}{2}\right)h\bar{y} = \frac{h^2}{6}[b + 2a]$$

$$\bar{y} = \frac{h^2(b+2a)2}{6(a+b)h}$$

$$\bar{y} = \frac{h}{3}\frac{(b+2a)}{(b+a)}$$

Hence centroidal distance from

$$DC = \frac{h}{3}\left(\frac{b+2a}{b+a}\right)$$

Similarly, centroidal distance from

$$AB = \frac{h}{3}\left(\frac{a+2b}{a+b}\right)$$

PROBLEM 5.4: Determine the centroid of the general parabolic spandrel having an equation of $y = kx^n$ as shown in Fig. 5.18.

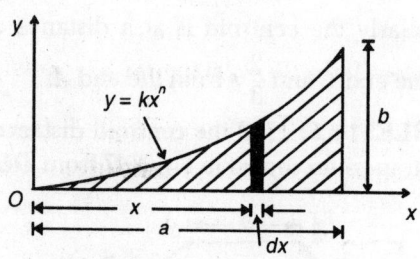

Fig. 5.18 (Illustrative Problem 5.4)

Solution:

Consider an element of dx, at a distance x from 'O'.

Height of the element at a distance x from O is
$$y = kx^n$$
$\therefore$ Area of the element $= y \cdot dx$
$$= kx^n \cdot dx$$

$$\bar{x} = \frac{\Sigma A\bar{x}}{\Sigma A}$$

$$= \frac{\int_0^a kx^n \cdot dx \cdot x}{\int_0^a kx^n \cdot dx}$$

$$= \frac{k\int_0^a x^{n+1} \cdot dx}{k\int_0^a x^n \cdot dx}$$

$$= \left[\frac{x^{n+2}}{n+2}\right]_0^a \times \left[\frac{n+1}{x^{n+1}}\right]_0^a$$

$$= \frac{a^{n+2}}{n+2} \times \frac{n+1}{a^{n+1}}$$

$$= \left(\frac{n+1}{n+2}\right)a$$

$$\bar{y} = \frac{\Sigma A\bar{y}}{\Sigma A} = \frac{\int_0^a kx^n \cdot dx \cdot y/2}{\int_0^a kx^n \cdot dx}$$

$$= \frac{\int_0^a (kx^n \cdot kx^n dx)/2}{k\left(\frac{x^{n+1}}{n-1}\right)_0^a}$$

$$y = \frac{\frac{k^2}{2}\int_0^a x^{2n}dx}{k\frac{a^{n+1}}{n+1}}$$

$$= \frac{k}{2}\frac{n+1}{a^{n+1}}\left[\frac{x^{2n+1}}{2n+1}\right]_0^a$$

$$= \frac{k}{2}\cdot\frac{n+1}{a^{n+1}}\frac{a^{2n+1}}{2n+1}$$

$$= \left(\frac{n+1}{2n+1}\right)a^n\cdot\frac{k}{2}$$

But
$$y = kx^n$$
$$b = ka^n$$
$$k = \frac{b}{a^n}$$
$$y = \left(\frac{n+1}{2n+1}\right)a^n\cdot\frac{b}{2a^n}$$
$$= \frac{(n+1)}{2(2n+1)}b$$

PROBLEM 5.5: Find the centroid of the lamina shown in Fig. 5.19.

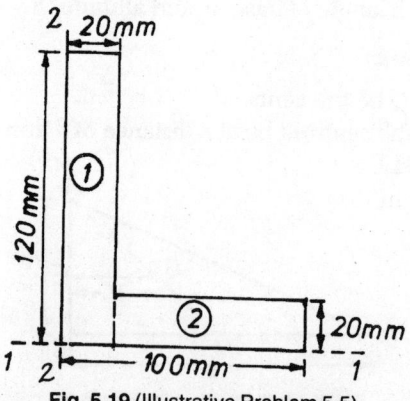

Fig. 5.19 (Illustrative Problem 5.5)

Solution:
The given lamina may be split up into two rectangles.

Rectangle (1)

$$A_1 = 120 \times 120$$

$$= 2400 \text{ mm}^2$$

$$x_1 = \frac{20}{2} = 10 \text{ mm}$$

$$y_1 = \frac{120}{2} = 60 \text{ mm}$$

Rectangle (2)

$$A_2 = 80 \times 20 = 1600 \text{ mm}^2$$

$$x_2 = 20 + \frac{80}{2} = 60 \text{ mm}$$

$$y_2 = \frac{20}{2} = 10 \text{ mm}$$

$$\bar{x} = \frac{A_1 x_1 + A_2 x_2}{A_1 + A_2}$$

$$= \frac{2400 \times 10 + 1600 \times 60}{2400 + 1600}$$

$$= \frac{24000 + 96000}{4000}$$

$$= \frac{120000}{4000} = 30 \text{ mm}$$

$$\bar{y} = \frac{A_1 y_1 + A_2 y_2}{A_1 + A_2}$$

$$= \frac{2400 \times 60 + 1600 \times 10}{2400 + 1600}$$

$$= \frac{144000 + 16000}{4000}$$

$$= \frac{160000}{4000} = 40 \text{ mm}$$

Centroid (30 mm; 40 mm)

PROBLEM 5.6: Find the centroid of the lamina shown in Fig. 5.20.

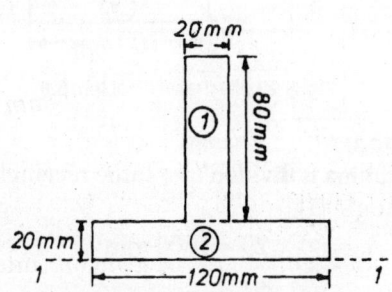

Fig. 5.20 (Illustrative Problem 5.6)

Solution:

The lamina will be split up into two rectangular areas.

Rectangle (1)

$$A_1 = 80 \times 20 = 1600 \text{ mm}^2$$

$$y_1 = 20 + \frac{80}{2} = 60 \text{ mm}$$

Rectangle (2)

$$A_2 = 120 \times 20 = 2400 \text{ mm}^2$$

$$y_2 = \frac{20}{2} = 10 \text{ mm}$$

$$\bar{y} = \frac{A_1 y_1 + A_2 y_2}{A_1 + A_2}$$

$$= \frac{1600 \times 60 + 2400 \times 10}{1600 + 2400}$$

$$= \frac{96 \times 10^3 + 24 \times 10^3}{40 \times 10^2}$$

$$= \frac{120 \times 10^3}{40 \times 10^3}$$

$$= 30 \text{ mm}$$

Above the axis (1) – (1)

PROBLEM 5.7: A rolled steel joist of I-section has the following dimensions.

Top flange   = 60 mm wide and 15 mm thick
Bottom flange = 100 mm wide and 30 mm thick
Web thickness = 15 mm
Overall depth = 120 mm

Find the C.G.

Solution:

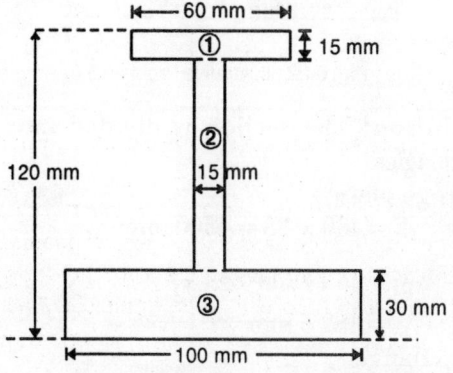

Fig. 5.21 (Illustrative Problem 5.7)

Let us divide the I-section into three rectangles

Rectangle 1 (Top flange)

$A_1 = 60 \times 15 = 900 \text{ mm}^2$

$y_1 = 120 - \dfrac{15}{2} = 112.5 \text{ mm}$

Rectangle 2 (Web)

$A_2 = (120 - 15 - 30) \times 15$

$\quad = 75 \times 15 = 1125 \text{ mm}^2$

$y_2 = 30 + \dfrac{75}{2} = 67.5 \text{ mm}$

Rectangle 3 (Bottom flange)

$A_3 = 100 \times 30 = 3000 \text{ mm}^2$

$y_3 = \dfrac{30}{2} = 15 \text{ mm}$

$\bar{y} = \dfrac{A_1 y_1 + A_2 y_2 + A_3 y_3}{A_1 + A_2 + A_3}$

$\quad = \dfrac{900 \times 112.5 + 1125 \times 67.5 + 3000 \times 15}{900 + 1125 + 3000}$

$\quad = \dfrac{101250 + 75937.5 + 45000}{5025}$

$\quad = \dfrac{222187.5}{5025} = 44.216 \text{ mm}$

PROBLEM 5.8: Determine the centroid of the section shown (Fig. 5.22).

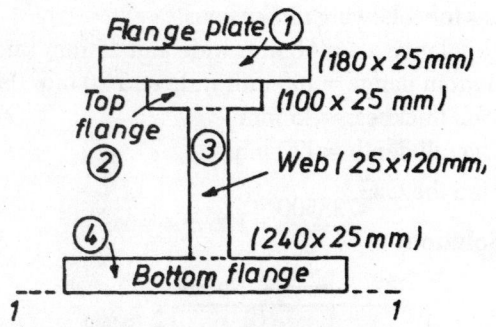

Fig. 5.22 (Illustrative Problem 5.8)

Solution: The section is divided into four rectangles.

Flange plate

$A_1 = 180 \times 25 = 4500 \text{ mm}^2$

$y_1 = 25 + 120 + 25 + \dfrac{25}{2}$

$\quad = 162.5 \text{ mm}$

Top flange

$A_2 = 100 \times 25 = 2500 \text{ mm}^2$

$y_2 = 25 + 120 + \dfrac{25}{2} = 137.5 \text{ mm}$

Web $A_3 = 120 \times 25 = 3000 \text{ mm}^2$

$y_3 = 25 + \dfrac{120}{2} = 85 \text{ mm}$

Bottom Flange

$A_4 = 240 \times 25 = 6000 \text{ mm}^2$

$y_4 = \dfrac{25}{2} = 12.5 \text{ mm}$

$\bar{y} = \dfrac{A_1 y_1 + A_2 y_2 + A_3 y_3 + A_4 y_4}{A_1 + A_2 + A_3 + A_4}$

$\quad = \dfrac{4500 \times 162.5 + 2500 \times 137.5}{4500 + 2500 + 3000 + 6000}$

$\qquad + \dfrac{3000 \times 85 + 6000 \times 12.5}{4500 + 2500 + 3000 + 6000}$

$\quad = \dfrac{2500(1.8 \times 162.5 + 137.5 + 1.2 \times 85 + 2.4 \times 12.5)}{16000}$

$\quad = \dfrac{2500 \times (292.5 + 137.5 + 102 + 30)}{16000}$

$\quad = \dfrac{2500 \times 562}{16000} = 87.8125 \text{ mm}$

PROBLEM 5.9: Determine the centroid of the lamina shown in Fig. 5.23.

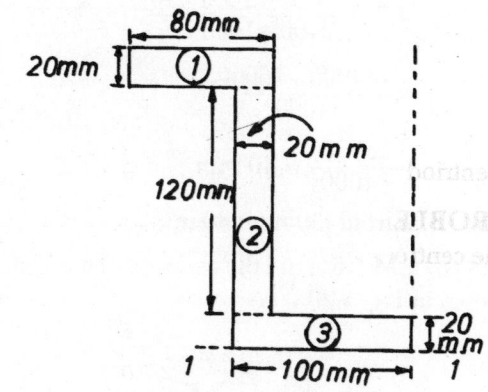

Fig. 5.23 (Illustrative Problem 5.9)

Solution:

The lamina is divided into three rectangles.

Rectangle (1)

$A_1 = 80 \times 20 = 1600 \text{ mm}^2$

$x_1 = 80 + \dfrac{80}{2} = 120 \text{ mm}$

$$y_1 = 20 + 120 + \frac{20}{2} = 150 \text{ mm}$$

Rectangle (2)

$$A_2 = 120 \times 20 = 2400 \text{ mm}^2$$

$$x_2 = 80 + \frac{20}{2} = 90 \text{ mm}$$

$$y_2 = 20 + \frac{120}{2} = 80 \text{ mm}$$

Rectangle (3)

$$A_3 = 100 \times 20 = 2000 \text{ mm}$$

$$x_3 = \frac{100}{2} = 50 \text{ mm}$$

$$y_3 = \frac{20}{2} = 10 \text{ mm}$$

$$\bar{x} = \frac{A_1 x_1 + A_2 x_2 + A_3 x_3}{A_1 + A_2 + A_3}$$

$$= \frac{1600 \times 120 + 2400 \times 90 + 2000 \times 50}{1600 + 2400 + 2000}$$

$$= \frac{(192 + 216 + 100)1000}{6000}$$

$$= \frac{508 \times 1000}{6000} = 84.67 \text{ mm}$$

$$\bar{y} = \frac{A_1 y_1 + A_2 y_2 + A_3 y_3}{A_1 + A_2 + A_3}$$

$$= \frac{1600 \times 150 + 2400 \times 80 + 2000 \times 10}{1600 + 2400 + 2000}$$

$$= \frac{240 \times 1000 + 192 \times 1000 + 20 \times 1000}{6000}$$

$$= 75.33 \text{ mm}$$

Centriod = (84.67 mm; 75.33 mm)

PROBLEM 5.10: Determine the coordinates of the centroid for the area shown in Fig. 5.24.

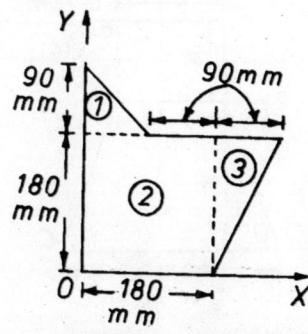

Fig. 5.24 (Illustrative Problem 5.10)

Solution:

The area is divided into two triangles and a square as shown in Fig. 5.10.

Triangle (1)

$$A_1 = \frac{1}{2} \times 90 \times 90 = 4500 \text{ mm}^2$$

$$x_1 = \frac{1}{3} \times 90 = 30 \text{ mm}$$

$$y_1 = 180 + \frac{90}{3} = 210 \text{ mm}$$

Square (2)

$$A_2 = 180 \times 180 = 32400 \text{ mm}^2$$

$$x_2 = 180/2 = 90 \text{ mm}$$

$$y_2 = 180/2 = 90 \text{ mm}$$

Triangle (3)

$$A_3 = \frac{1}{2} \times 180 \times 90 = 8100 \text{ mm}^2$$

$$x_3 = 180 + \frac{1}{3} \times 90 = 210 \text{ mm}$$

$$y_3 = \frac{2}{3} \times 180 = 120 \text{ mm}$$

$$\bar{x} = \frac{A_1 x_1 + A_2 x_2 + A_3 x_3}{A_1 + A_2 + A_3}$$

$$= \frac{4050 \times 30 + 32400 \times 90 + 8100 \times 210}{4050 + 32400 + 8100}$$

$$= \frac{121500 + 2916000 + 1701000}{44550}$$

$$= \frac{4738500}{44550} = 106.364 \text{ mm}$$

$$\bar{y} = \frac{A_1 y_1 + A_2 y_2 + A_3 y_3}{A_1 + A_2 + A_3}$$

$$= \frac{4050 \times 210 + 32400 \times 90 + 8100 \times 120}{4050 + 32400 + 8100}$$

$$= \frac{850500 + 2916000 + 972000}{44550}$$

$$= \frac{4738500}{44550}$$

$$= 106.34 \text{ mm}$$

Centroid (106.364 mm; 106.34 mm)

PROBLEM 5.11: Determine the centroid of the section of the concrete dam shown in Fig. 5.25.

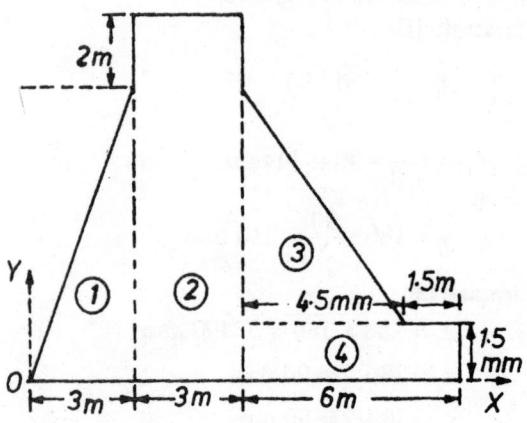

Fig. 5.25 (Illustrative Problem 5.11)

Solution:

The section of the dam is divided into triangles and rectangles as shown in Fig. 5.25.

Triangle (1)

$$A_1 = \frac{1}{2} \times 3 \times 8 = 12 \text{ m}^2$$

$$x_1 = \frac{2}{3} \times 3 = 2 \text{ m}$$

$$y_1 = \frac{1}{3} \times 8 = \frac{8}{3} \text{ m}$$

Triangle (2)

$$A_2 = 3 \times 10 = 30 \text{ m}^2$$

$$x_2 = 3 + \frac{3}{2} = 4.5 \text{ m}$$

$$y_2 = \frac{10}{2} = 5 \text{ m}$$

Triangle (3)

$$A_3 = \frac{1}{2} \times 4.5 \times 6.5 = 14.625 \text{ m}^2$$

$$x_3 = 3 + 3 + \frac{1}{3} \times 4.5 = 7.5 \text{ m}$$

$$y_3 = 1.5 + \frac{1}{3} \times 6.5 = 11/3 \text{ m}$$

Rectangle (4)

$$A_4 = 6 \times 1.5 = 9.0 \text{ m}^2$$

$$x_4 = 3 + 3 + \frac{6}{2} = 9 \text{ m}$$

$$y_4 = \frac{1.5}{2} = 0.75 \text{ m}$$

$$\bar{x} = \frac{A_1 x_1 + A_2 x_2 + A_3 x_3 + A_4 x_4}{A_1 + A_2 + A_3 + A_4}$$

$$\bar{x} = \frac{12 \times 2 + 30 \times 4.5 + 14.625 \times 7.5 + 9 \times 9}{12 + 30 + 14.625 + 9}$$

$$= \frac{24 + 135 + 109.6875 + 81}{65.625}$$

$$= 349.6875/65.625 = 5.329 \text{ m}$$

$$\bar{y} = \frac{A_1 y_1 + A_2 y_2 + A_3 y_3 + A_4 y_4}{A_1 + A_2 + A_3 + A_4}$$

$$= \frac{12 \times 8/3 + 30 \times 5 + 14.625 \times 11/3 + 9 \times 0.75}{12 + 30 + 14.625 + 9}$$

$$= (32 + 150 + 53.625 + 6.75) / (65.625)$$

$$= \frac{242.375}{65.625}$$

$$= 3.693 \text{ m}$$

Centroid = (5.329 m; 3.693 m)

PROBLEM 5.12: Find the centre of gravity of a uniform plate in the form of a symmetrical trapezium whose parallel sides are 1.5 metres and 0.75 m in length and 1 metre apart.

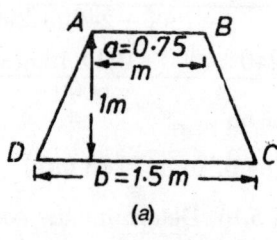

(a)

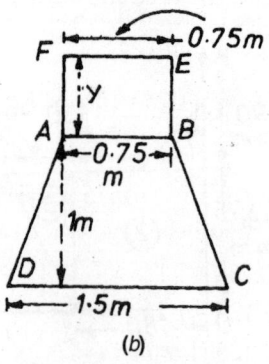

(b)

Fig. 5.26 (Illustrative Problem 5.12)

It has a rectangular extension of the same weight per square metre attached to the 0.75 m edge and y metres long so as just to fit that edge, find what should be the height of the rectangular piece if the centre of gravity of the whole is on the 0.75 m edge of the trapezium.

Solution:
Case 1
Trapezoidal plate

Distance of the centroid from the side AB

$$= \left(\frac{a+2b}{a+b}\right)\left(\frac{h}{3}\right)$$

$$= \frac{(0.75+2\times1.5)}{(0.5+1.5)}\times\frac{1}{5}$$

$$= 0.5556 \text{ m}$$

Case 2
Let the size of the rectangular piece be 0.75 m $\times\ y$ m.

Since AB is the centroidal axis,

Moment of the area $FEBA$ about AB

$= $ Moment of the area $ABCD$ about AB

$$\therefore \quad 0.75\times y\times\frac{Y}{2} = \left(\frac{1.5+0.75}{2}\right)(1)(0.5556)$$

$$Y^2 = \left(\frac{2.25}{2}\right)(1)(0.5556)\left(\frac{2}{0.75}\right)$$

$$Y = \sqrt{1.6668}$$

$$= 1.291 \text{ m}$$

or

Moment of the rectangle $ABEF$ about CD + Moment of the trapezium about $CD=$ Moment of the whole area about CD

$$\left(0.75\times y\right)\left(\frac{Y}{2}+1\right)+\left(\frac{0.75+1.5}{2}\right)(1)\left[\left(\frac{1.5+2\times0.75}{1.5+0.75}\right)\frac{1}{3}\right]$$

$$= \left[0.75\times y+\left(\frac{0.75+1.50}{2}\right)1\right]\times1$$

$0.375y^2 + 0.75y + 1.125\times0.4444 = 0.75y + 1.125$

$0.375y^2 = 1.125 - 1.25\times0.4444$

$$= 1.125\,(1 - 0.4444)$$

$$= 1.125\times0.5556$$

$$y^2 = \frac{1.125\times0.5556}{0.375}$$

$$y^2 = 1.6668$$

$$y = \sqrt{1.6668}$$

$$y = 1.291 \text{ m}$$

PROBLEM 5.13: Find the centroid for the area shown in Fig. 5.27.

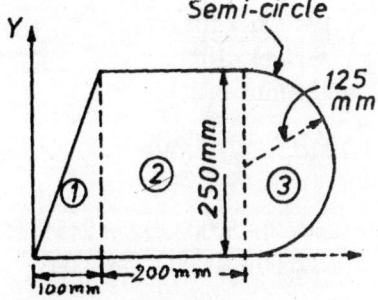

Fig. 5.27 (Illustrative Problem 5.13)

Solution:
The given area is divided into a triangle, a rectangle and a semi-circle.

Triangle (1)

$$A_1 = \frac{1}{2}\times100\times250 = 12500 \text{ m}^2$$

$$x_1 = \frac{2}{3}\times100 = \frac{200}{3} \text{ mm}$$

$$y_1 = \frac{1}{3}\times250 = \frac{250}{3} \text{ mm}$$

Triangle (2)

$$A_2 = 200\times250 = 50000 \text{ m}^2$$

$$x_2 = 100 + \frac{200}{2} = 200 \text{ mm}$$

$$y_2 = \frac{250}{2} = 125 \text{ mm}$$

Semi-circle

$$\text{Area, } A_3 = \frac{\pi\times125^2}{2} = 24543.693 \text{ mm}^2$$

$$x_3 = 100 + 200 + \frac{4\times125}{3\pi}$$

$$= 300 + 53.052$$

$$= 353.052 \text{ mm}$$

$$y_3 = \frac{250}{2} = 125 \text{ mm}$$

$$\bar{x} = \frac{A_1 x_1 + A_2 x_2 + A_3 x_3}{A_1 + A_2 + A_3}$$

$$= \frac{\dfrac{12500 \times 200}{3} + 50000 \times 200 + 24543.693 \times 353.052}{12500 + 50000 + 24543.693}$$

$$= \frac{83.33 \times 10^4 + 1000 \times 10^4 + 866.52 \times 10^4}{8.7044 \times 10^4}$$

$$= \frac{1949.85 \times 10^4}{8.7044 \times 10^4}$$

$$= 224 \text{ mm}$$

$$\bar{y} = \frac{A_1 y_1 + A_2 y_2 + A_3 y_3}{A_1 + A_2 + A_3}$$

$$= \frac{12500 \times (250/3) + 50000 \times 125 + 24543.693 \times 125}{12500 + 50000 + 24543.693}$$

$$= \frac{104.17 \times 10^4 + 625 \times 10^4 + 306.80 \times 10^4}{8.7044 \times 10^4}$$

$$= \frac{1035.97 \times 10^4}{8.7044 \times 10^4}$$

$$= 119 \text{ mm}$$

PROBLEM 5.14: Determine the coordinates of the shaded area as shown in Fig. 5.28.

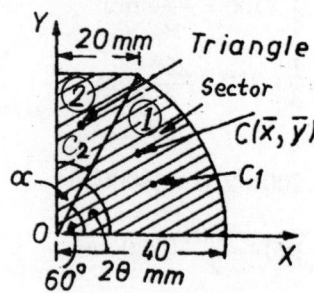

Fig. 5.28 (Illustrative Problem 5.14)

Solution:

$$\sin \alpha = \frac{20}{40} = \frac{1}{2}$$

or $\quad \alpha = \sin^{-1}(1/2) = 30°$

∴ $\quad 2\theta = 90° - \alpha = 90° - 30° = 60°$

∴ $\quad \theta = 30°$

Sector

$$A_1 = \frac{\pi(40)^2 \times 60°}{360°} = 837.758 \text{ mm}^2$$

$$OC_1 = \frac{2r \sin(\theta)}{3(\theta)}$$

$$= \frac{2 \times 40 \times \sin 30°}{3 \times \left(\dfrac{30}{180} \times \pi\right)}$$

$$= 25.465 \text{ mm}$$

$$x_1 = 25.465 \times \cos 30°$$

$$= 22.053 \text{ mm}$$

$$y_1 = 25.465 \times \sin 30°$$

$$= 12.7325 \text{ mm}$$

Triangle

$$A_2 = \frac{1}{2} \times 20 \times (40 \times \cos 30°)$$

$$= 346.41 \text{ mm}^2$$

$$x_1 = \frac{20}{3} = 6.6667 \text{ mm}$$

$$y_1 = \frac{2}{3} \times 40 \times \cos 30° = 23.094 \text{ mm}$$

$$\bar{x} = \frac{A_1 x_1 + A_2 x_2}{A_1 + A_2}$$

$$= \frac{837.758 \times 22.053 + 346.41 \times 6.6667}{837.758 + 346.41}$$

$$= \frac{20784.49}{1184.168}$$

$$\bar{x} = 17.552 \text{ mm}$$

$$\bar{y} = \frac{A_1 y_1 + A_2 y_2}{A_1 + A_2}$$

$$= \frac{837.758 \times 12.7325 + 346.41 \times 23.094}{837.758 + 346.41}$$

$$= \frac{18666.75}{1184.168}$$

$$\bar{y} = 15.764$$

c.g. (17.552 mm; 15.764 mm)

PROBLEM 5.15: Find the height of the centroid above the axis (1) – (1) for the lamina shown in Fig. 5.29.

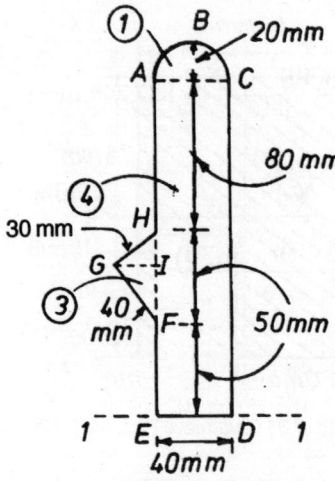

Fig. 5.29 (Illustrative Problem 5.15)

Solution:

$$GF^2 + GH^2 = 40^2 + 30^2$$

$$= 1600 + 900 = 2500 \text{ mm}^2$$

$$HF^2 = 50^2 = 2500 \text{ mm}^2$$

i.e., $GF^2 + GH^2 = HF^2$

$\therefore$ The triangle FGH is a right angled triangle.

$\therefore \quad \angle FGH = 90°$

$$FI = FG\cos\angle GFH$$

$$= 40 \times \frac{40}{50} = 32 \text{ mm}$$

$$IH = 50 - 32 = 18 \text{ mm}$$

The lamina can be split up into a number of components (Semi-circle, rectangle & triangles).

Semi-circle

$$A_1 = \frac{1}{2}\pi(20)^2 = 628.32 \text{ mm}^2$$

$$y_1 = 50 + 50 + 80 + \frac{4 \times 20}{3\pi}$$

$$= 180 + 8.488 = 188.488 \text{ mm}$$

Rectangle

$$A_2 = 40 \times 180 = 7200 \text{ mm}^2$$

$$y_2 = \frac{180}{2} = 90 \text{ mm}$$

Triangle (3)

$$GI = \sqrt{(30^2 - 18)^2}$$

$$= 24 \text{ mm}$$

or

$$GI = \sqrt{(40)^2 - (32)^2}$$

$$= 24 \text{ mm}$$

$$A_3 = \frac{1}{2} \times 32 \times 24 = 384 \text{ mm}^2$$

$$y_3 = 50 + \frac{2}{3} \times 32$$

$$= \frac{214}{3} = 71.3333$$

$$A_4 = \frac{1}{2} \times 18 \times 24 = 216 \text{ mm}^2$$

$$y_4 = 50 + 32 + \frac{18}{3}$$

$$= 88 \text{ mm}$$

$$\bar{y} = \frac{628.32 \times 188.488 + 7200 \times 90}{628.32 + 7200 + 384 + 216}$$

$$+ \frac{384 \times 71.3333 + 216 \times 88}{628.32 + 7200 + 384 + 216}$$

$$= \frac{(11.843 + 64.8 + 2.74 + 1.9) \times 10^4}{8428.32}$$

$$= \frac{81.283 \times 10^4}{8428.32}$$

$$= 96.44 \text{ mm}$$

PROBLEM 5.16: Determine the centre of gravity of the plane uniform lamina shown in Fig. 5.30.

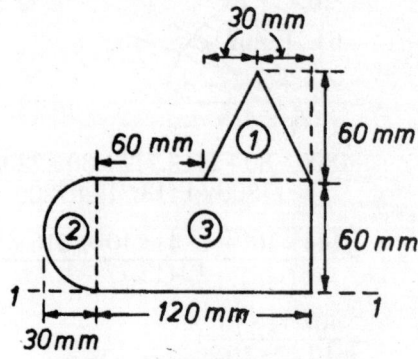

Fig. 5.30 (Illustrative Problem 5.16)

Solution:

Divide the lamina into a triangle, semi-circle and a rectangle.

Triangle (1)

$$A_1 = \frac{1}{2} \times 60 \times 60 = 1800 \text{ mm}^2$$

$$x_1 = 30 \text{ mm}$$

$$y_1 = 60 + \frac{60}{3} = 80 \text{ mm}$$

Semi-circle (2)

$$A_2 = \frac{1}{2}\pi(30)^2 = 1413.717 \text{ mm}^2$$

$$x_2 = 120 + \frac{4 \times 30}{3 \times \pi} = 132.732 \text{ mm}$$

$$y_2 = \frac{60}{2} = 30 \text{ mm}$$

Rectangle

$$A_3 = 120 \times 60 = 7200 \text{ mm}^2$$

$$x_3 = \frac{120}{2} = 60 \text{ mm}$$

$$y_3 = \frac{60}{2} = 30 \text{ mm}$$

$$\bar{x} = \frac{A_1 x_1 + A_2 x_2 + A_3 x_3}{A_1 + A_2 + A_3}$$

$$= \frac{1800 \times 30 + 1413.717 \times 132.732 + 7200 \times 60}{1800 + 1413.717 + 7200}$$

$$= \frac{54 \times 10^3 + 187.65 \times 10^3 + 432 \times 10^3}{10413.717}$$

$$= \frac{673.65 \times 10^3}{10.41 \times 10^3}$$

$$= 64.71 \text{ mm}$$

$$\bar{y} = \frac{A_1 y_1 + A_2 y_2 + A_3 y_3}{A_1 + A_2 + A_3}$$

$$= \frac{1800 \times 80 + 1413.717 \times 30 + 7200 \times 30}{1800 + 1413.717 + 7200}$$

$$= \frac{144 \times 10^3 + 42.41 \times 10^3 + 216 \times 10^3}{10413.77}$$

$$= \frac{402.41 \times 10^3}{10.41 \times 10^3}$$

$$= 38.656 \text{ mm}$$

Centroid (64.71 mm; 38.656 mm)

PROBLEM 5.17: Determine the centroid of the section shown in Fig. 5.31.

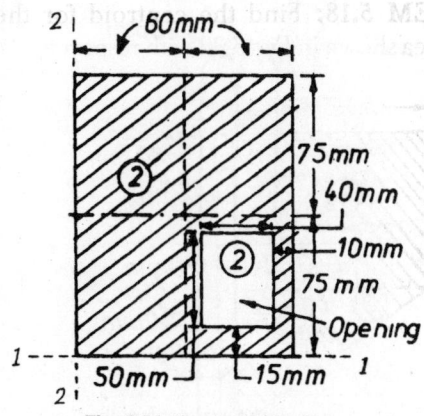

Fig. 5.31 (Illustrative Problem 5.17)

Solution:
Rectangle (1)

$$A_1 = 120 \times 150 = 18000 \text{ mm}^2$$

$$x_1 = 120/2 = 60 \text{ mm}$$

$$y_1 = 150/2 = 75 \text{ mm}$$

Rectangle (2)

$$A_2 = 40 \times 50 = 2000 \text{ mm}^2$$

$$x_2 = 120 - \left(10 + \frac{40}{2}\right) = 90 \text{ mm}$$

$$y_2 = 15 + \frac{50}{2} = 40 \text{ mm}$$

$$\bar{x} = \frac{A_1 x_1 - A_2 x_2}{A_1 - A_2}$$

$$= \frac{18000 \times 60 - 2000 \times 90}{18000 - 2000}$$

$$= \frac{1080000 - 180000}{16000}$$

$$= \frac{900000}{16000}$$

$$= 56.25 \text{ mm}$$

$$\bar{y} = \frac{A_1 y_1 - A_2 y_2}{A_1 - A_2}$$

$$= \frac{18000 \times 75 - 2000 \times 40}{18000 - 2000}$$

$$= \frac{1350000 - 80000}{16000}$$

$$= \frac{1270000}{16000}$$

$$= 79.375 \text{ mm}$$

PROBLEM 5.18: Find the centroid for the shaded area shown in Fig. 5.32 with respect to 'O'

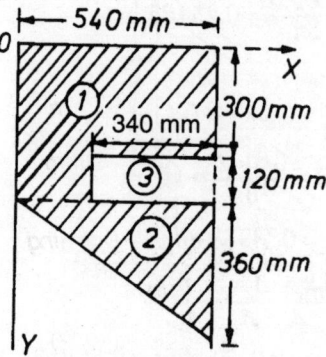

Fig. 5.32 (Illustrative Problem 5.18)

Solution:
Rectangle (1)

$$A_1 = 540 \times 420 = 226800 \text{ mm}$$

$$x_1 = \frac{540}{2} = 270 \text{ mm}$$

$$y_1 = \frac{-420}{2} = -210 \text{ mm}$$

Triangle (2)

$$A_1 = \frac{1}{2} \times 540 \times 360 = 97200 \text{ mm}^2$$

$$x_2 = \frac{2}{3} \times 540 = 360 \text{ mm}$$

$$y_2 = -420 - \frac{1}{3} \times 360$$
$$= -540 \text{ mm}$$

Triangle (3)

$$A_3 = 340 \times 120 = 40800 \text{ mm}^2$$

$$x_3 = 200 + \frac{340}{2} = 370 \text{ mm}$$

$$y_3 = -300 - \frac{120}{2} = -360 \text{ mm}$$

$$\bar{x} = \frac{A_1 x_1 + A_2 x_2 - A_3 x_3}{A_1 + A_2 - A_3}$$

$$= \frac{(226800 \times 270) + (97200 \times 360) - (40800 \times 370)}{226800 + 97200 - 40800}$$

$$= \frac{(6123.6 + 3499.2 - 1509.6) \times 10^4}{28.32 \times 10^4}$$

$$= \frac{8113.2 \times 10^4}{28.32 \times 10^4} = 286.48 \text{ mm}$$

$$\bar{y} = \frac{A_1 y_1 + A_2 y_2 - A_3 y_3}{A_1 + A_2 - A_3}$$

$$= \frac{(226800) \times (-210) + (97200) \times (-540)}{226800 + 97200 - 40800}$$

$$+ \frac{(40800) \times (-360)}{226800 + 97200 - 40800}$$

$$= \frac{(-4762.8 - 5248.8 + 1468.8) \times 10^4}{28.32 \times 10^4}$$

$$= \frac{-8542.8 \times 10^4}{28.32 \times 10^4} = -301.65 \text{ mm}$$

c.g. (286.48 mm; –301.65 mm)

PROBLEM 5.19: A square hole is punched of a circular lamina as shown (Fig. 5.33). The diagonal of the square which is punched out is equal to the radius of the circle. Find the centroid of the remaining lamina.

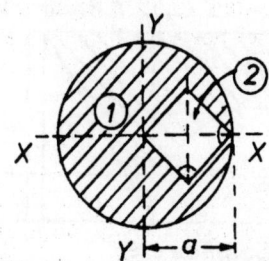

Fig. 5.33 (Illustrative Problem 5.19)

Solution:
Let 'a' be the radius of the circle.
    The diagonal of the square $= a$

$\therefore$ Side of the square $= \sqrt{\dfrac{a^2}{2}}$

$$= \frac{a}{\sqrt{2}}$$

The figure is symmetrical about x-axis
Circle

$$A_1 = \pi a^2$$
$$x_1 = 0$$

Square

$$A_2 = \frac{a}{\sqrt{2}} \times \frac{a}{\sqrt{2}} = \frac{a^2}{2}$$

$$x_2 = \frac{a}{2}$$

$$\therefore \quad \bar{x} = \frac{A_1 x_1 - A_2 x_2}{A_1 - A_2}$$

$$= \frac{\pi a^2 \times 0 - (a^2/2) \times (a/2)}{\pi a^2 - (a^2/2)}$$

$$= -\frac{a^3}{4} \times \frac{2}{(2\pi - 1)a^2}$$

$$= -\frac{a}{2} \frac{1}{(6.283 - 1)}$$

$$= -0.09464a$$

say $-0.095a$

The negative sign indicates that the centroid lies on the left side of the origin 'O'.

PROBLEM 5.20: Find the centroid of the shaded area as shown in Fig. 5.34 with respect to 'O'

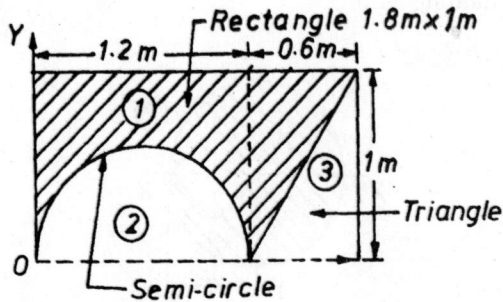

Fig. 5.34 (Illustrative Problem 5.20)

Solution:
The shaded area = Area of rectangle (1)
$\qquad$ = –Area of semi-circle (2)
$\qquad$ = –Area of triangle (3)
Rectangle (1)

$$A_1 = (1.2 + 0.6) \times (1) = 1.8 \text{ m}^2$$

$$x_1 = \frac{1.8}{2} = 0.9 \text{ m}$$

$$y_1 = \frac{1}{2} = 0.5 \text{ m}$$

Semi-circle (2)

$$A_2 = \frac{1}{2}\left(\frac{\pi}{4} \times 1.2^2\right) = 0.5655 \text{ m}^2$$

$$x_2 = \frac{1.2}{2} = 0.6 \text{ m}$$

$$y_2 = \frac{4 \times 0.6}{3\pi} = 0.25465 \text{ m}$$

Triangle (3)

$$A_3 = \frac{1}{2} \times 0.6 \times 1 = 0.3 \text{ m}^2$$

$$x_3 = 1.2 + \frac{0.6 \times 2}{3} = 1.6 \text{ m}$$

$$y_3 = \frac{1}{3} = 0.33333 \text{ m}$$

$$\bar{x} = \frac{A_1 x_1 - A_2 x_2 - A_3 x_3}{A_1 - A_2 - A_3}$$

$$= \frac{1.8 \times 0.9 - 0.5655 \times 0.6 - 0.3 \times 1.6}{1.8 - 0.5655 - 0.3}$$

$$= \frac{1.62 - 0.3393 - 0.48}{0.9345}$$

$$= \frac{0.8007}{0.9345} = 0.8568 \text{ m or } 856.8 \text{ mm}$$

$$\bar{y} = \frac{A_1 y_1 - A_2 y_2 - A_3 y_3}{A_1 - A_2 - A_3}$$

$$= \frac{1.8 \times 0.5 - 0.5655 \times 0.25465 - 0.3 \times (1/3)}{1.8 - 0.5655 - 0.3}$$

$$= \frac{0.9 - 0.144 - 0.1}{0.9345}$$

$$= \frac{0.656}{0.9345}$$

$$= 0.702 \text{ m or } 702 \text{ mm}$$

c.g, (856.8 mm; 702 mm)
or $\quad$ (0.8568 m; 0.702 m)

PROBLEM 5.21: Locate the centroid of the shaded area obtained by removing a semi-circle of diameter 'a' from a quadrant of a circle of radius 'a'.

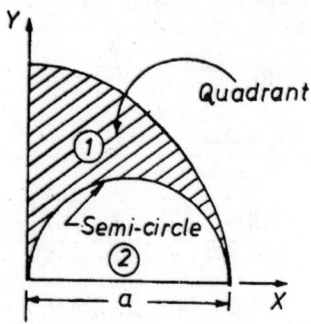

Fig. 5.35 (Illustrative Problem 5.21)

Solution:

Quadrant of a circle

$$A_1 = \frac{\pi}{4}a^2$$

$$x_1 = \frac{4a}{3\pi}$$

$$y_1 = \frac{4a}{3\pi}$$

Semi-circle

$$A_2 = \frac{1}{2} \times \frac{\pi}{4}a^2 = \frac{\pi}{8}a^2$$

$$x_2 = \frac{a}{2}$$

$$y_2 = \frac{4(a/2)}{3\pi} = \frac{2a}{3\pi}$$

$$\bar{x} = \frac{A_1x_1 - A_2x_2}{A_1 - A_2}$$

$$= \frac{\dfrac{\pi}{4}a^2 \times \dfrac{4a}{3\pi} - \dfrac{\pi}{8}a^2 \times \dfrac{a}{2}}{\dfrac{\pi}{4}a^2 - \dfrac{\pi}{8}a^2}$$

$$= \frac{\dfrac{a^3}{3} - \dfrac{\pi a^3}{16}}{\dfrac{\pi a^2}{8}}$$

$$= \frac{16a^3 - 3\pi a^3}{3 \times 16 \times \dfrac{\pi a^2}{8}}$$

$$= \frac{a^3(16 - 3\pi)}{6\pi a^2}$$

$$= 0.3488\,a$$

$$\bar{y} = \frac{\dfrac{\pi}{4}a^2 \times \dfrac{4a}{3\pi} - \dfrac{\pi}{8}a^2 \times \dfrac{2a}{3\pi}}{\dfrac{\pi}{4}a^2 - \dfrac{\pi}{8}a^2}$$

$$= \frac{\dfrac{a^3}{3} - \dfrac{a^3}{12}}{\dfrac{\pi a^2}{8}}$$

$$= \frac{3a^3 \times 8}{12 \times \pi a^2}$$

$$= 0.6366a$$

Centroid $(0.3488a;\ 0.6366a)$

PROBLEM 5.22: From a circular plate of diameter 80 mm is cut out a circle whose diameter is a radius of the plate. Find the c.g. of the remainder.

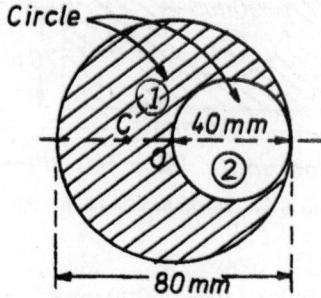

Fig. 5.36 (Illustrative Problem 5.22)

Solution:

Circle (1)

$$A_1 = \frac{\pi}{4} \times 80^2 = 1600\pi \text{ mm}^2$$

$$x_1 = 0$$

Circle (2)

$$A_2 = \frac{\pi}{4} \times 40^2 = 400\pi \text{ mm}^2$$

$$x_2 = \frac{40}{2} = 20 \text{ mm}$$

$$\bar{x} = \frac{A_1x_1 - A_2x_2}{A_1 - A_2}$$

$$= \frac{(1600 \times \pi) \times (0) - (400\pi) \times (20)}{1600\pi - 400\pi}$$

$$= \frac{-400 \times 20 \times \pi}{1200 \times \pi}$$

$$= -6.67 \text{ mm}$$

C.g. of the remainder is at a distance of 6.67 mm from 'O' (left side)

PROBLEM 5.23: Determine the co-ordinates x and y of the centre of a 90 mm diameter circular hole cut in thin plate so that this point will be the centroid of the remaining shaded area shown in Fig. 5.37.

Solution:

The shaded area = Area of rectangle (1)
      − Area of circle (2)
      − Area of triangle (3)

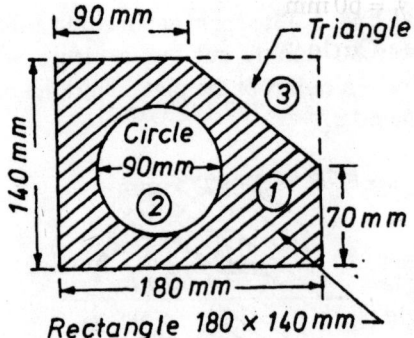

Fig. 5.37: (Illustrative Problem 5.23)

Rectangle (1)

$$A_1 = 180 \times 140 = 25200 \text{ mm}^2$$

$$x_1 = \frac{180}{2} = 90 \text{ mm}$$

$$y_1 = \frac{140}{2} = 70 \text{ m}$$

Circle (2)

$$A_2 = \frac{\pi}{4} \times (90)^2 = 6361.725 \text{ mm}^2$$

$$x_2 = \bar{x}$$

$$y_2 = \bar{y}$$

Triangle (3)

$$A_3 = \frac{1}{2} \times 90 \times 70 = 3150 \text{ mm}^2$$

$$x_3 = 90 + \frac{2}{3} \times 90 = 150 \text{ mm}$$

$$y_3 = 70 + \frac{2}{3} \times 70 = 116.67 \text{ mm}$$

$$\bar{x} = \frac{A_1 x_1 - A_2 x_2 - A_3 x_3}{A_1 - A_2 - A_3}$$

$$\bar{x} = \frac{25200 \times 90 - 6361.725 \times \bar{x} - 3150 \times 150}{25200 - 6361.725 - 3150}$$

$$= \frac{2268000 - 6361.625 \bar{x} - 472500}{15688.275}$$

$$15688.275\bar{x} = 1795500 - 6361.725\bar{x}$$

$$22050\bar{x} = 1795500$$

$$\bar{x} = \frac{1795500}{22050} = 81.429 \text{ mm}$$

$$\bar{y} = \frac{A_1 y_1 - A_2 y_2 - A_3 y_3}{A_1 - A_2 - A_3}$$

$$\bar{y} = \frac{25200 \times 70 - 6361.725 \times \bar{y} - 3150 \times 116.67}{25200 - 6361.725 - 3150}$$

$$= \frac{1764000 - 6361.725\bar{y} - 367510.5}{15688.275}$$

$$15688.275\bar{y} = 1396489.5 - 6361.725\bar{y}$$

$$22050\bar{y} = 1396489.5$$

$$\bar{y} = \frac{1396489.5}{22050} = 63.333 \text{ mm}$$

c.g [81.429 mm; 63.333 mm]

PROBLEM 5.24: For the lamina shown in Fig. 5.38, find the centroidal axis parallel to the base.

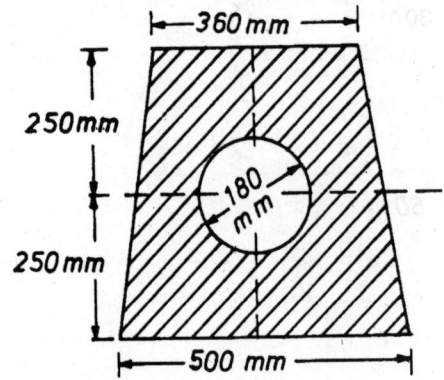

Fig. 5.38 (Illustrative Problem 5.24)

Solution:
Trapezium

$$A_1 = \frac{1}{2}(500 + 360) \times 500$$

$$= 215000 \text{ mm}^2$$

$$y_1 = \frac{500}{3}\left[\frac{500 + 2 \times 360}{500 + 360}\right]$$

$$= \frac{500}{3} \times \frac{1220}{860}$$

$$= 236.4341 \text{ mm}$$

Circle

$$A_2 = \frac{\pi}{2}(180)^2 = 25446.9 \text{ mm}^2$$

$$y_2 = 250 \text{ mm}$$

$$\bar{y} = \frac{A_1 y_1 - A_2 y_2}{A_1 - A_2}$$

$$= \frac{(215000 \times 236.4341) - (25446.9 \times 250)}{(215000 - 25446.9)}$$

$$= \frac{50.83 \times 10^6 - 6.36 \times 10^6}{0.18955 \times 10^6}$$

$$= \frac{44.47}{0.18955}$$

$$= 234.608 \text{ mm or } 23.46 \text{ cm}$$

PROBLEM 5.25: The surface of a plate of uniform thickness is given by the shaded area in Fig. 5.39. If this plate is suspended from a hinge at A, what angle will line AB make with the vertical?

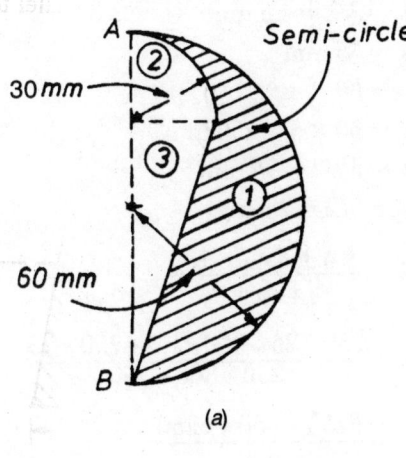

(a)

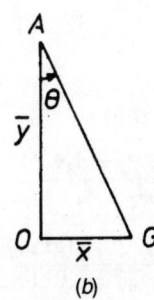

(b)

Fig. 5.39 (Illustrative Problem 5.25)

Solution:

Take 'A' as the origin

Semi-Circle (1)

$$A_1 = \frac{\pi(60)^2}{2} = 5657.1427 \text{ mm}^2$$

$$x_1 = \frac{4 \times 60}{3\pi} = 25.455 \text{ mm}$$

$$y_1 = 60 \text{ mm}$$

Quarter-Circle (2)

$$A_2 = \frac{\pi(30)^2}{4} = 707.1429 \text{ mm}^2$$

$$x_2 = \frac{4 \times 30}{3\pi} = 12.7273 \text{ mm}$$

$$y_2 = 30 - \frac{4 \times 30}{3\pi} = 17.2727 \text{ mm}$$

Triangle (3)

$$A_3 = \frac{1}{2} \times 90 \times 90 = 1350 \text{ mm}^2$$

$$x_3 = \frac{1}{3} \times 30 = 10 \text{ mm}$$

$$y_3 = 30 + \frac{1}{3} \times 90 = 60 \text{ mm}$$

$$\bar{x} = \frac{A_1 x_1 - A_2 x_2 - A_3 x_3}{A_1 - A_2 - A_3}$$

$$= \frac{(5657.1427 \times 25.4545) - (707.1429 \times 12.7273)}{5657.1427 - 707.1429 - 1350}$$

$$- \frac{(1350 \times 10)}{5657.1427 - 707.1429 - 1350}$$

$$= \frac{144 \times 10^3 - 9 \times 10^3 - 13.5 \times 10^3}{3600}$$

$$= \frac{121.5 \times 10^3}{3.6 \times 10^3}$$

$$= 33.75 \text{ mm}$$

$$\bar{y} = \frac{A_1 y_1 - A_2 y_2 - A_3 y_3}{A_1 - A_2 - A_3}$$

$$= \frac{(5657.1427 \times 60) - (707.1429 \times 17.2727) - (1350 \times 60)}{5657.1427 - 707.1429 - 1350}$$

$$= \frac{339.429 \times 10^3 - 12.214 \times 10^3 - 81 \times 10^3}{3600}$$

$$= \frac{246.215 \times 10^3}{3.6 \times 10^3}$$

$$= 68.39 \text{ mm}$$

$$\tan\theta = \frac{GO}{AO} = \frac{\bar{x}}{\bar{y}} = \frac{33.75}{68.39} = 0.4935$$

$$\theta = \tan^{-1}(0.4935) = 26.27° \text{ or } 26°16'$$

The angle made by the line *AB* with the vertical $= 26°16'$

PROBLEM 5.26: Find the c.g. of the channel section whose flanges are 50 mm × 5 mm, and web is 100 mm × 3 mm. The total height of the channel being 110 mm.

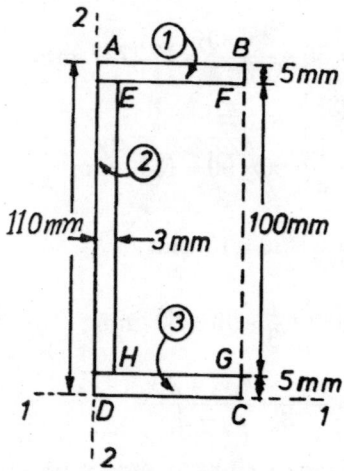

Fig. 5.40 (Illustrative Problem 5.26)

Solution:
I Method
Rectangle *ABCD*

$$A_1 = 50 \times 110 = 5500 \text{ mm}^2$$

$$x_1 = 50/2 = 25 \text{ mm}$$

$$y_1 = 110/2 = 55 \text{ mm}$$

Rectangle *EFGH*

$$A_2 = 47 \times 100 = 4700 \text{ mm}^2$$

$$x_2 = 3 + \frac{47}{2} = 26.5 \text{ mm}$$

$$y_2 = 5 + \frac{100}{2} = 55 \text{ mm}$$

$$\bar{x} = \frac{A_1 x_1 - A_2 x_2}{A_1 - A_2}$$

$$= \frac{5500 \times 25 - 4700 \times 26.5}{5500 - 4700}$$

$$= \frac{12950}{800} = 16.1875 \text{ mm}$$

$$\bar{y} = \frac{A_1 y_1 - A_2 y_2}{A_1 - A_2}$$

$$= \frac{5500 \times 55 - 4700 \times 55}{5500 - 4700}$$

$$= \frac{55 \times 800}{800} = 55 \text{ mm}$$

II Method
Rectangle (1)

$$A_1 = 50 \times 5 = 250 \text{ mm}^2$$

$$x_1 = 50/2 = 25 \text{ mm}$$

$$y_1 = 100 + 5 + \frac{5}{2} = 107.5 \text{ mm}$$

Rectangle (2)

$$A_2 = 100 \times 3 = 300 \text{ mm}^2$$

$$x_2 = 3/2 = 1.5 \text{ mm}$$

$$y_2 = 55 \text{ mm}$$

Rectangle (3)

$$A_3 = 50 \times 5 = 250 \text{ mm}^2$$

$$x_3 = 50/2 = 25 \text{ mm}$$

$$y_3 = 5/2 = 2.5 \text{ mm}$$

$$\bar{x} = \frac{A_1 x_1 + A_2 x_2 + A_3 x_3}{A_1 + A_2 + A_3}$$

$$= \frac{250 \times 25 + 300 \times 1.5 + 250 \times 25}{250 + 300 + 250}$$

$$= \frac{6250 + 450 + 6250}{800}$$

$$= \frac{12950}{800}$$

$$= 16.1875 \text{ mm}$$

$$\bar{y} = \frac{A_1 y_1 + A_2 y_2 + A_3 y_3}{A_1 + A_2 + A_3}$$

$$= \frac{250 \times 107.5 + 300 \times 55 + 250 \times 2.5}{250 + 300 + 250}$$

$$= \frac{44000}{800}$$

$$= 55 \text{ mm}$$

c.g. [16.1875 mm; 55 mm]

PROBLEM 5.27: Determine the centre of gravity of a right circular cone.

Fig. 5.41 shows a right circular cone of base radius *r* and height *h*. Obviously the centre of gravity lies on the vertical line through *O*.

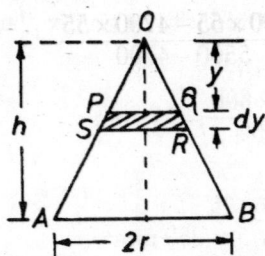

Fig. 5.41 (Illustrative Problem 5.27)

Solution: Consider the cone as split into an infinite number of horizontal discs. Consider one such disc of thickness dy at a depth y. Let x be the radius of this disc. We have

$$\frac{x}{r} = \frac{y}{h}$$

$\therefore \qquad x = \frac{y}{h}r$

Volume of the disc $= \pi x^2 dy$

$$= \frac{\pi y^2}{h^2}r^2 dy$$

Let ρ be the density of the material

$\therefore \quad dm = \frac{\rho \pi r^2}{h^2}y^2 dy$

$\therefore \quad \bar{y} = \frac{\int y \, dm}{\int dm}$

$$= \frac{\displaystyle\int_0^h y \cdot \frac{\rho \pi r^2}{h^2}y^2 dy}{\displaystyle\int_0^h \frac{\rho \pi r^2}{h^2}y^2 dy}$$

$$= \frac{\displaystyle\int_0^h Y^3 dy}{\displaystyle\int_0^h Y^2 dy} = \frac{\left(\dfrac{y^4}{4}\right)_0^h}{\left(\dfrac{y^3}{3}\right)_0^h}$$

$$= \frac{h^4}{4} \times \frac{3}{h^3}$$

$$\bar{y} = \frac{3}{4}h$$

$\therefore$ Distance of centre of gravity of a solid cone

from the vertex $= \frac{3}{4}h$.

$\therefore$ Distance of the centre of gravity of the cone from the base $= \frac{1}{4}h$.

PROBLEM 5.28: Determine the centre of gravity of a hollow cone

Solution: Let the base radius be r and let the height be h (Fig. 5.42).

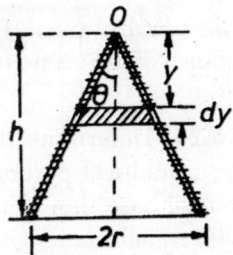

Fig. 5.42 (Illustrative Problem 5.28)

Let $\theta =$ Semi-vertical angle
ρ be the surface density.

The cone can be split up into an infinite number of small rings. Consider one such elemental ring between depths y and $(y + dy)$ from the vertex. Let the radius at this level be x.

Area of the surface ring $= 2\pi x \, dy \sec\theta$

Mass of the elemental part, $dm = \rho 2\pi x \, dy \sec\theta$

But $\quad \dfrac{x}{r} = \dfrac{y}{h}$

$\therefore \qquad x = \frac{y}{h}r$

$\therefore \quad dm = \rho 2\pi \frac{y}{h} r \, dy \sec\theta$

or $\quad dm = \rho 2\pi \frac{r}{h} y \, dy \sec\theta$

Distance of centre of gravity from the vertex,

$$\bar{y} = \frac{\int y \, dm}{\int dm}$$

$\therefore \quad \bar{y} = \dfrac{\displaystyle\int_0^h y \cdot \rho 2\pi \frac{r}{h} y \, dy \sec\theta}{\displaystyle\int_0^h \rho 2\pi \frac{r}{h} y \, dy \sec\theta}$

$$\bar{y} = \frac{\displaystyle\int_0^h y^2 dy}{\displaystyle\int_0^h y \, dy}$$

$$\bar{y} = \left[\frac{y^3}{3}\right]_0^h \Bigg/ \left[\frac{y^2}{2}\right]_0^h$$

$$\bar{y} = \frac{h^3}{3} \times \frac{2}{h^2}$$

$$\bar{y} = \frac{2}{3} h$$

∴ Distance of the centre of gravity from the vertex is $(2/3h)$ or distance of the centre of gravity from the base = $(h/3)$.

PROBLEM 5.29: Determine the maximum height h of the cylindrical portion of the body with hemispherical base shown in Fig. 5.43, so that it is in stable equilibrium on its base.

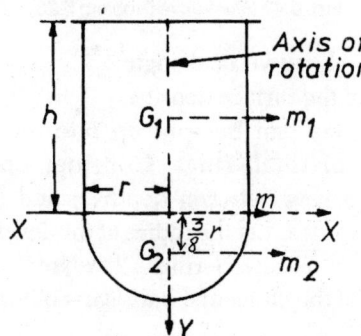

Fig. 5.43 (Illustrative Problem 5.29)

Solution:
The body will be stable on its base as long as its centre of gravity is in hemi-spherical base. The limiting case is when it is on the plane x-x shown in the figure. Centroid lies on the axis of rotation.

Let ρ be the density of the material.

Mass of the cylindrical portion,

$$m_1 = \pi r^2 h \rho$$

Its centre of gravity g_1 is at a height,

$$y_1 = \frac{h}{2} \text{ from x-axis.}$$

Mass of hemispherical portion,

$$m_2 = \frac{2}{3} \pi r^3 \text{ and its centre of gravity is at a distance}$$

$$y_2 = \frac{3r}{8} \text{ from x-x plane}$$

Since centroid is to be on x-x plane $\bar{y} = 0$

i.e. $\Sigma m_i y_i = 0$

∴ $$m_1 \frac{h}{2} - m_2 \frac{3}{8} r = 0$$

$$\pi r^2 h \rho \frac{h}{2} - \frac{2}{3} \pi r^3 \rho \frac{3}{8} r = 0$$

or $$h^2 - \frac{r^2}{2} = 0$$

$$h^2 = \frac{r^2}{2}$$

$$h = \frac{r}{\sqrt{2}} = 0.7071 \, r$$

PROBLEM 5.30: A steel ball of diameter 120 mm rests centrally over a concrete cube of size 120 mm. Determine the centre of gravity of the system, taking weight of concrete = 25 kN/m³ and that of steel 80 kN/m³.

Solution:
Weight of concrete cube, $W_1 = (0.12)^3 \times 25$

$$= 0.04320 \text{ kN}$$

Distance of centre of gravity of concrete cube from base, $y_1 = \frac{0.12}{2} = 0.06$ m

Weight of steel ball, $W_2 = \frac{4\pi}{3}(0.06)^3 \times 80$

$$= 0.07238 \text{ kN}$$

Distance of centre of gravity of steel ball from base, $y_2 = 0.12 + \frac{0.12}{2}$

$$= 0.18 \text{ m}$$

∴ $$\Sigma W_i Y_i = W_1 Y_1 + W_2 Y_2$$

$$= 0.0432 \times 0.06 + 0.07238 \times 0.18$$

$$= 2.592 \times 10^{-3} + 0.0130284$$

$$= 0.0156204 \text{ kN-m}$$

$$\Sigma W_i = W_1 + W_2$$

$$= 0.04320 + 0.07238$$

$$= 0.11558 \text{ m}$$

∴ $$\bar{y} = \frac{\Sigma W_i y_i}{\int W_i} = \frac{0.0156204}{0.11558}$$

$$= 0.13515 \text{ m}$$

or $$= 135.15 \text{ mm}$$

PROBLEM 5.31: A frustum of a solid circular cone has an axial hole of 250 mm diameter as shown in Fig. 5.44 (*a*).

Determine the centre of gravity of the body.

Solution:

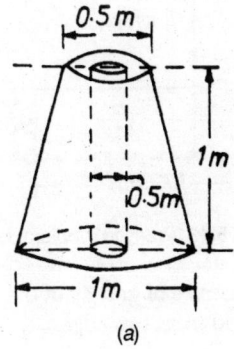

(a)

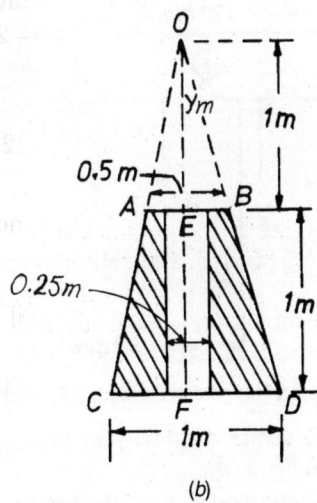

(b)

Fig. 5.44 (Illustrative Problem 5.31)

Referring to Fig. 5.44 (*b*)

$$AE = EB = \frac{0.5}{2} = 0.25 \text{ m}$$

$$CF = FD = \frac{1}{2} = 0.5 \text{ m}$$

Let $OE = y$ metres

The triangles OBE and ODF are similar triangles.

$$\therefore \quad \frac{OE}{OF} = \frac{EB}{FD}$$

$$\frac{OE}{OE + EF} = \frac{0.25}{0.50}$$

$$\frac{y}{y+1} = \frac{1}{2}$$
$$2y = y + 1$$
or $\quad y = 1$ m

i.e., $OE = 1$ m

Right circular cone OCD:

$$V_1 = \frac{1}{3}\pi r^2 h = \frac{1}{3} \times \pi \times (0.5)^2 \times 2 = 0.5236 \text{ m}^3$$

$$y_1 = \frac{h}{4} = \frac{2}{4} = 0.5 \text{ m}$$

Right Circular cone OAB:

$$V_1 = \frac{1}{3}\pi r^2 h = \frac{1}{3} \times \pi \times (0.5)^2 \times 1$$
$$= 0.06545 \text{ m}^3$$

$$y_1 = 1 + \frac{1}{4} = 1.25 \text{ m}$$

Circular hole

$$V_3 = \frac{\pi}{4} d^2 h = \frac{\pi}{4} \times 0.25^2 \times 2$$
$$= 0.04909 \text{ m}$$

$$y_3 = \frac{h}{2} = \frac{1}{2} = 0.5 \text{ m}$$

Using the relation,

$$\bar{y} = \frac{V_1 y_1 - V_2 y_2 - V_3 y_3}{V_1 - V_2 - V_3}$$

$$= \frac{0.5236 \times 0.5 - 0.06545 \times 1.25 - 0.04909 \times 0.5}{0.5236 - 0.06545 - 0.04909}$$

$$= \frac{0.2618 - 0.0818 - 0.0245}{0.40906}$$

$$= \frac{0.1555}{0.40906}$$

$$= 0.38 \text{ m}$$

PRACTICE PROBLEMS

5.1 Calculate the centroid of a line segment shown in Fig. 5.45.

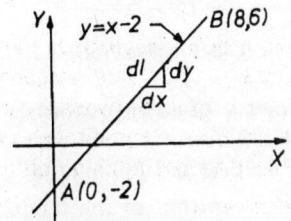

Fig. 5.45 (Practice Problem 5.1)

5.2   Find the position of the centroid of **any** triangle from the left end or the right end. (Fig. 5.46)

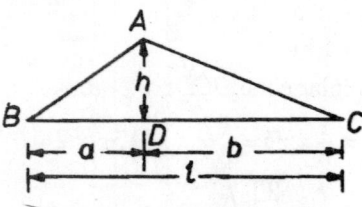

Fig. 5.46 (Practice Problem 5.2)

5.3   Fig. 5.47 shows the cross-section of the masonry dam. Determine the distance of the centroid from the vertical face.

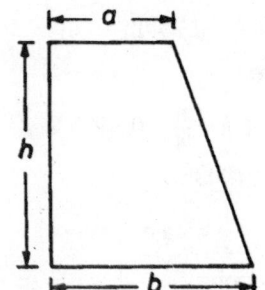

Fig. 5.47 (Practice Problem 5.3)

5.4   Determine the centroid of parabolic spandrel having an equation of $y = kx^2$ as shown in Fig. 5.48.

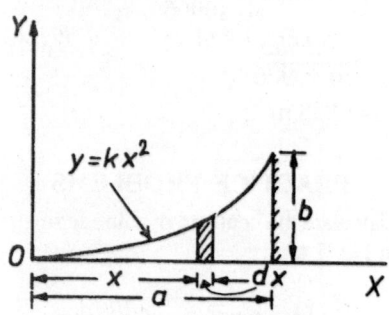

Fig. 5.48 (Practice Problem 5.4)

5.5   Find the c.g. of an unequal angle section 120 mm × 100 mm × 20 mm with its longer leg placed vertical and 100 mm side at the top.

5.6   Find the centroid of the lamina shown in Fig. 5.49.

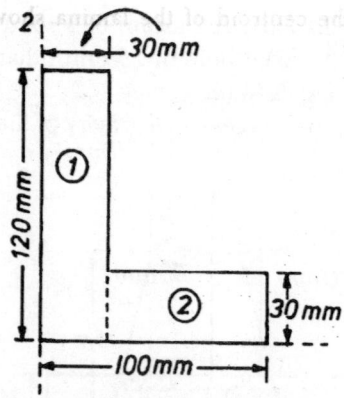

Fig. 5.49 (Practice Problem 5.6)

5.7   Find the centre of gravity of the lamina shown in Fig. 5.50 from top edge.

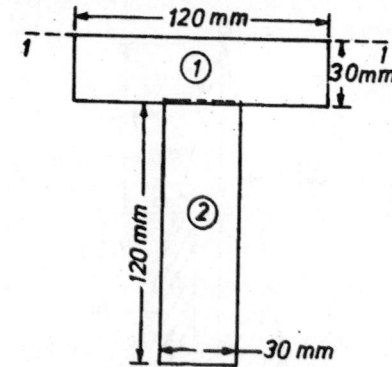

Fig. 5.50: (Practice Problem 5.7)

5.8   Find the centre of gravity of the *T*-section shown in Fig. 5.51.

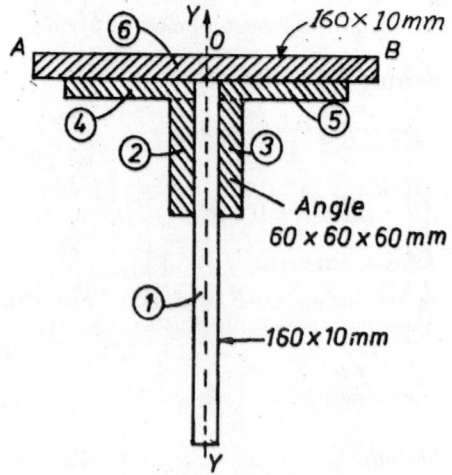

Fig. 5.51 (Practice Problem 5.8)

5.9 Find the centroid of the lamina shown in Fig. 5.52.

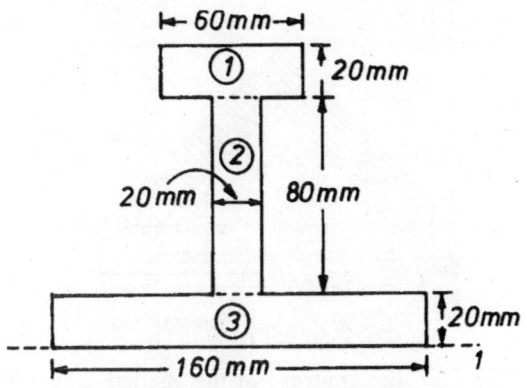

Fig. 5.52 (Practice Problem 5.9)

5.10 Find the centroid for the shaded area shown in Fig. 5.53 with respect to 'O' as the origin.

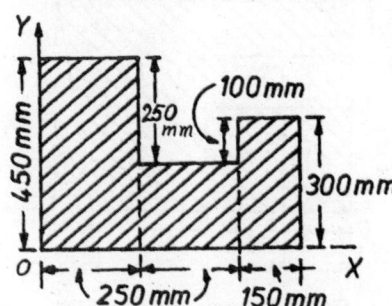

Fig. 5.53 (Practice Problem 5.10)

5.11 Determine the centroid for the area shown in Fig. 5.54 with respect to 'O'.

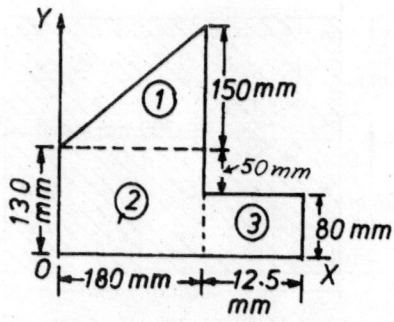

Fig. 5.54 (Practice Problem 5.11)

5.12 Determine the centroid of the dam section shown in Fig. 5.55

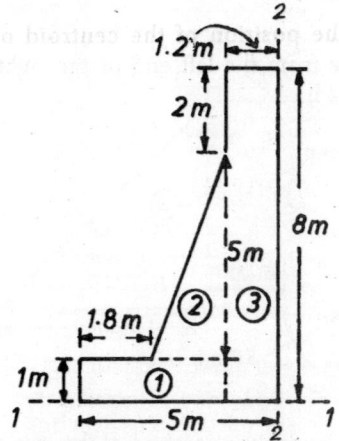

Fig. 5.55 (Practice Problem 5.12)

5.13 Determine the position of the centroid of the area as shown in Fig. 5.56.

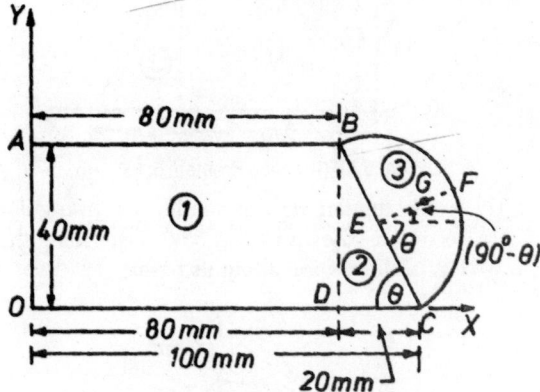

Fig. 5.56 (Practice Problem 5.13)

5.14 Find the centroid of the area shown in Fig. 5.57.

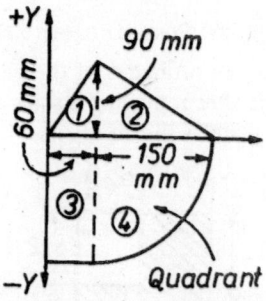

Fig. 5.57 (Practice Problem 5.14)

5.15 Find the centroid for the area shown in Fig. 5.58.

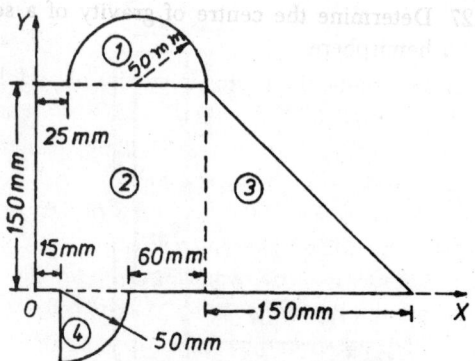

Fig. 5.58 (Practice Problem 5.15)

5.16 Determine the centroid of the section shown in Fig. 5.59.

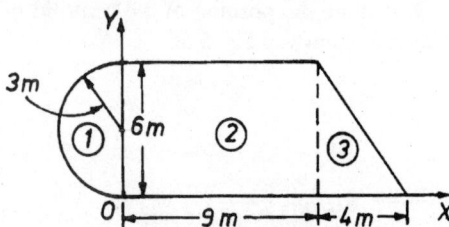

Fig. 5.59 (Practice Problem 5.16)

5.17 A rectangular hole is made in a triangular section as shown in Fig. 5.60. Determine the c.g. of the section about its base.

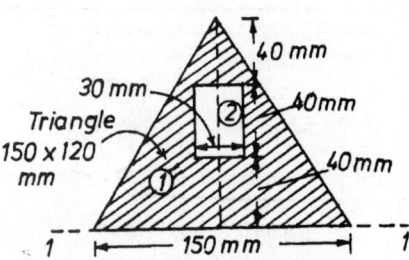

Fig. 5.60 (Practice Problem 5.17)

5.18 Find the co-ordinates of the centroid of the area left after removing a square area from a square plate as shown in Fig. 5.61.

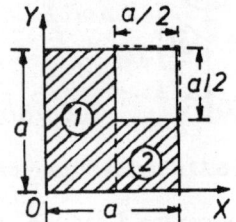

Fig. 5.61 (Practice Problem 5.18)

5.19 Find the centroidal axis of the lamina shown in Fig. 5.62 parallel to the base.

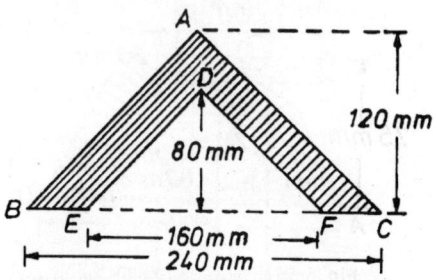

Fig. 5.62 (Practice Problem 5.19)

5.20 Find the centroid of the shaded area with respect to '*O*' (Fig. 5.63).

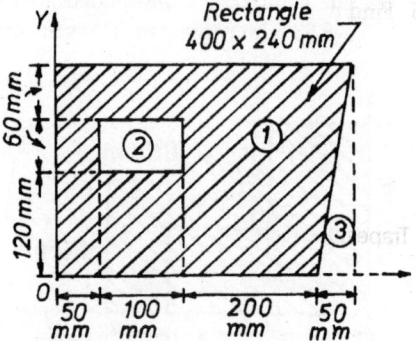

Fig. 5.63 (Practice Problem 5.20)

5.21 A rectangular plate of uniform thickness of size 400 mm × 150 mm has a circular hole as shown in Fig. 5.64. Find the centroid.

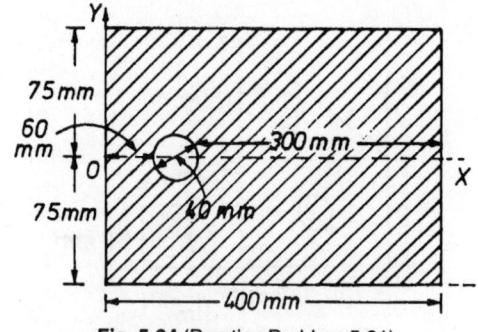

Fig. 5.64 (Practice Problem 5.21)

5.22 Find the centre of gravity of a semi-circular section having the outer and inner radii of 150 mm and 120 mm respectively.

5.23 Find the position of centre of gravity of the plane figure, shown in Fig. 5.65.

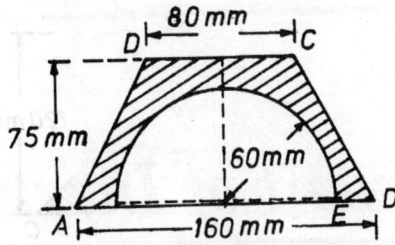

Fig. 5.65 (Practice Problem 5.23)

5.24 Find the coordinates of the centroid of the area obtained after removing a semi-circle of radius 80 mm from a quadrant of a circle of radius 160 mm.

5.25 Find the c.g. of the shaded area in Fig. 5.66.

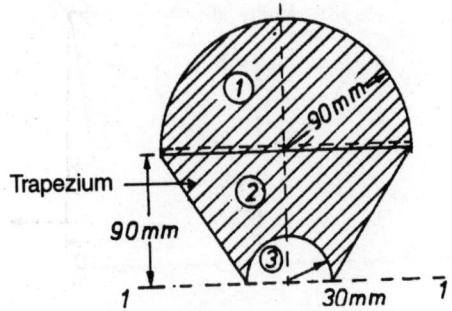

Fig. 5.66 (Practice Problem 5.25)

5.26 Find the centroid of a channel section as shown in Fig. 5.67.

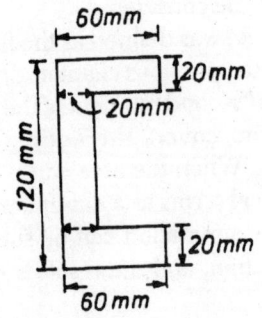

Fig. 5.67 (Practice Problem 5.26)

5.27 Determine the centre of gravity of a solid hemisphere.

5.28 Determine the centre of gravity of a Hollow Hemisphere.

5.29 A right circular cylinder of 120 mm diameter is joined with a hemisphere of the same diameter face to face. Find the greatest height of the cylinder, so that centre of gravity of the composite section coincides with the plane of joining the two sections. The density of the material of hemisphere is twice that of the material of cylinder.

5.30 A uniform solid of revolution is formed by joining the base of a right circular cone of height H to the equal base of a right circular cylinder of height h. Calculate the distance of the centre of mass of the solid, when $H = 60$ mm and $h = 15$ mm.

5.31 A concrete block of size 0.9 m × 0.72 m × 0.6 m is cast with a hole of diameter 0.24 m and depth 0.36 m as shown in Fig. 5.68. Locate the centre of gravity of the body. Take the weight of concrete = 25 kN/m³.

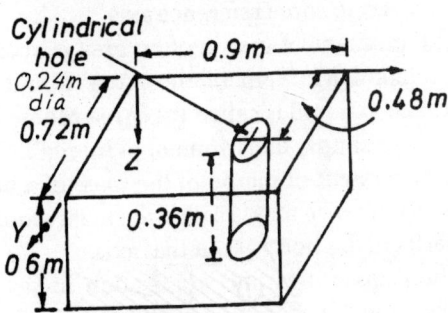

Fig. 5.68 (Practice Problem 5.31)

Chapter 6

Moment of Inertia

6.1 INTRODUCTION

The distribution and orientation of an area (like the surface area of a body) and of the mass of a body are often of as much significance as of their absolute values.

The concept of 'moment of inertia' of an area or of a mass provides a quantitative idea of the relative distribution of these quantities of a body with respect to some reference axis.

The moment of inertia of an area is actually the 'Second Moment of the Area' about the reference axis and is called the 'Area Moment of Inertia', or simply the 'Moment of Inertia'.

The moment of inertia of the mass of a body about a reference axis is called the mass moment of inertia of the body about that axis.

These quantities find application in several problems in Mechanics of Solids, Mechanics of Fluids, Structural Analysis, and so on. Thus the importance of the concept of moment of inertia in engineering practice cannot be overemphasized.

6.2 MOMENT OF INERTIA

Let us understand the moment of inertia of an area A about any specified axis as shown in Fig. 6.1.

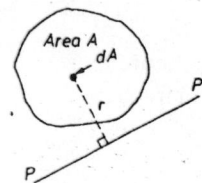

(a) About an axis PP'

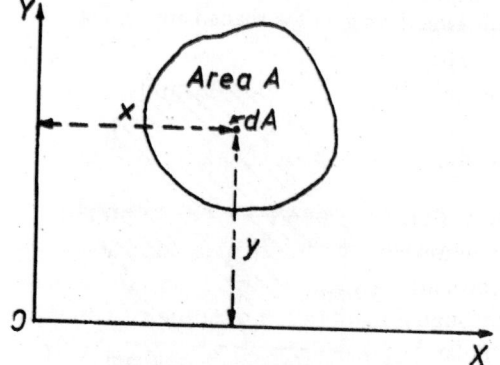

(b) About Rectangular Axes OX & OY

Fig. 6.1: Moment of Inertia of an Area about Specified Axis

With reference to (a), let an infinitesimally small area dA, at a perpendicular distance r from an axis $P-P'$, be considered.

Just as $\Sigma r \cdot dA$ was defined as the first moment of the area in the previous chapter, $\Sigma r^2 \cdot dA$ is now defined as the "second moment of the area" when the summation covers the entire area under consideration. When the area can be expressed in mathematical terms as a continuous function, the process of summation can be substituted by that of integration, and the result is denoted by the symbol I_p.

$$I_P = \Sigma r^2 \cdot dA = \int r^2 \cdot dA \qquad \text{...(Eq. 6.1)}$$

I_p is the second moment of area of the area A about the axis $P-P'$.

Similarly, with respect to Fig. 6.1 (b), the following are easily understood.

$$I_x = \int y^2 \cdot dA \\ I_y = \int x^2 \cdot dA \Bigg\} \quad ...(Eq.\ 6.2)$$

The second moment of area is also called the 'Area Moment of Inertia'; in fact the term 'inertia' is appropriate only when 'mass' is involved. That is the reason for specifically calling it 'Area Moment of Inertia' to differentiate it from 'Mass Moment of Inertia', dealt with in Section 6.9.

However, the term 'moment of inertia' is freely used even when referring to that for an area. Since the square of the distance of the elemental area is involved, the summation is always positive and hence the moment of inertia is always positive for an area (unless the area itself is treated as being negative, as in the case of the inner portions of hollow sections).

The units are obviously fourth power of linear units–m⁴ in the S.I. units. In the MKS units, either m⁴ or mm⁴ may be used according as the linear units used.

6.3 POLAR MOMENT OF INERTIA

The moment of inertia of an area about an axis perpendicular to the plane of the area and passing through a point called the 'Pole', is known as the "Polar Moment of Inertia" of area at that pole (Fig. 6.2).

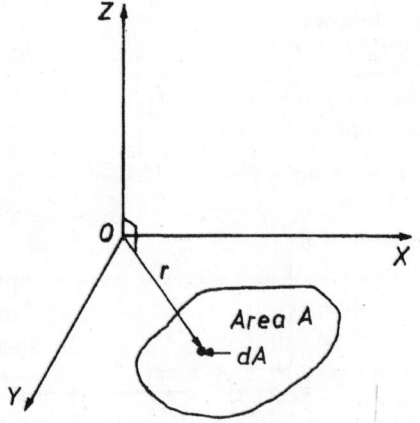

Fig. 6.2: Polar Moment of Inertia of an Area

With reference to the figure,

$$I_Z = \Sigma r^2 dA = \int r^2 \cdot dA \qquad ...(Eq.\ 6.3)$$

The concept of polar moment of inertia finds application in the analysis of numbers subjected to 'torsion' or twisting.

6.4 RADIUS OF GYRATION

'Radius of gyration' of an area or a section is defined as the square root of the ratio of its moment of inertia to the area.

$$\text{Radius of gyration, } k = \sqrt{\frac{I}{A}} \qquad ...(Eq.\ 6.4)$$

Since I is to be taken about a reference axis, the corresponding suffix is used for k as for I.

$$k_x = \sqrt{\frac{I_x}{A}} \\[2mm] k_y = \sqrt{\frac{I_y}{A}} \\[2mm] k_z = \sqrt{\frac{I_z}{A}} \Bigg\} \quad ...(Eq.\ 6.5)$$

Eq. 6.4 may also be written in the form:

$$I = A \cdot k^2 \qquad ...(Eq.\ 6.6)$$

From this, a geometric interpretation may be given to radius of gyration as the distance at which the whole area is taken to be concentrated as thin sheet of negligible width such that there is no change in moment of inertia (Fig. 6.3).

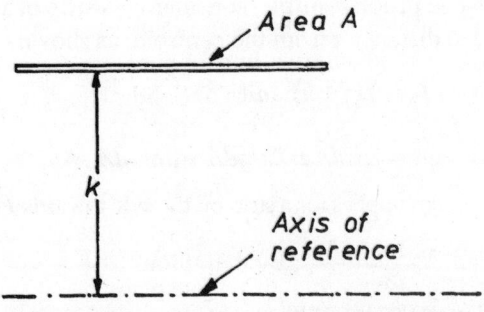

Fig. 6.3: Radius of Gyration—Physical Interpretation

6.5 THEOREMS OF MOMENT OF INERTIA

There are two theorems involving moment of inertia which find frequent application. These are:

(*i*) Parallel Axis Theorem
(*ii*) Perpendicular Axis Theorem

These are dealt with in the following sub-sections.

6.5.1 Parallel Axis Theorem

This theorem enables one to obtain the moment of inertia about any axis parallel to a centroidal axis in terms of that about the centroidal axis and the perpendicular distance between these two axes.

With reference to Fig. 6.4, the theorem may be stated mathematically as follows:

$$I_p = I_g + A \cdot h^2 \qquad \qquad \text{...(Eq. 6.7)}$$

where I_p = moment of inertia of the area A about an axis P-P, parallel to a centroidal axis G-G,

I_g = moment of inertia of the area A about an axis G-G passing through the centroid,

h = perpendicular distance between the axes G-G & P-P

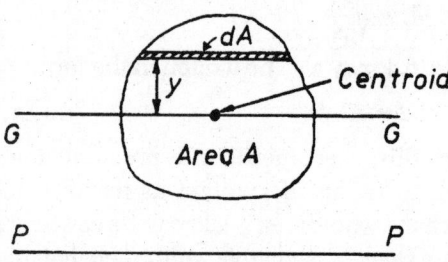

Fig. 6.4: Parallel Axis Theorem for Moment of Inertia

The proof is simple, considering a strip of area dA at distance y from the centroid, as shown.

$$I_p = \Sigma(y \pm h)^2 \cdot dA$$

$$= \Sigma y^2 dA \pm \Sigma 2hy dA + \Sigma h^2 \cdot dA$$

(–ve sign applies for a strip on the side towards P-P)

$$\Sigma y^2 \cdot dA = I_g, \text{ by definition.}$$

Also, $\Sigma h^2 \cdot dA = h^2 \Sigma dA = A \cdot h^2$

$$\Sigma 2yh dA = 2h \Sigma y \cdot dA$$

$\Sigma y \cdot dA$ is the first moment of the area about the centroidal axis, which has been shown to be zero in the previous chapter.

Hence $I_p = I_g + A \cdot h^2$, which proves the theorem.

[It may be noted that the theorem is applicable only if one of the axes is a centroidal axis; it cannot be applied to any two parallel axes].

For the radius of gyration, we can rewrite Eq. 6.7 as follows:

$A \cdot k_p^2 = A \cdot k_g^2 + A \cdot h^2$ by definition.

or, $\quad k_p^2 = k_g^2 + h^2 \qquad \qquad \text{...(Eq. 6.8)}$

(Note that k_p is always greater than h)

Thus it is observed that both the moment of inertia and the radius of gyration about any axis away from the centroid are always greater than those with respect to a centroidal axis.

Similarly, it can be easily shown that the parallel axis theorem is applicable for the polar moment of inertia too:

$$I_{z1} = I_z + A \cdot h^2 \qquad \qquad \text{...(Eq. 6.9)}$$

and $\quad k_{z1}^2 = k_z^2 + h^2 \qquad \qquad \text{...(Eq. 6.10)}$

If z_1-axis is an axis normal to the plane of the area located at a perpendicular distance h from z-axis, normal to the plane and passing through the centroid.

6.5.2 Perpendicular Axis Theorem

This theorem relates the Polar Moment of Inertia with the centroid as the pole and the moments of inertia about the other two perpendicular axes in the plane with the centroid as the origin.

With reference to Fig. 6.5, the theorem may be stated as follows:

$$I_z = I_x + I_y \qquad \qquad \text{...(Eq. 6.11)}$$

with the usual notation.

The proof is obvious:

$$I_z = \int r^2 dA = \int (x^2 + y^2) dA$$

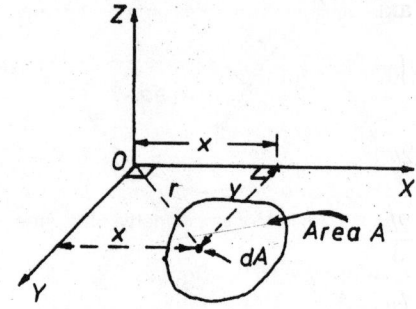

Fig. 6.5: Perpendicular Axis Theorem

$$= \int x^2 dA + \int y^2 dA$$

$\therefore \quad I_z = I_x + I_y$, which proves the theorem.

6.6 MOMENT OF INERTIA BY INTEGRATION

The moment of inertia for simple geometrical figures can be obtained by considering an elemental area or strip, as the case be, and integrating the general expression for it between the appropriate limits for the variables so as to cover the entire area. Depending upon the shape of the figure and the element, double integration may also be used, if necessary.

Further, the moment of inertia about a centroidal axis may be determined first and that about any other axis parallel to it may be obtained by using the parallel axis theorem. Some times, however, it would be easier to determine the value about an axis situated away from the centroid, and that about a centroidal axis from the parallel axis theorem

6.6.1 Moment of Inertia of Standard Sections

Now the principle of integration will be applied to obtain the moment of inertia of standard sections like rectangle, triangle, circle, semi-circle and hollow sections.

There is also a graphical method (for irregular figures, especially) which is given in the next chapter.

(1) Rectangle

With reference to Fig. 6.6, let us consider a thin strip of thickness dy at a distance y from the centroidal axis XX.

$$I_x = \int y^2 dA = \int_{-d/2}^{d/2} y^2 b \cdot dy = b \int_{-d/2}^{d/2} y^2 \cdot dy$$

$$= 2b \int_0^{d/2} y^2 dy \qquad \text{(owing to symmentry)}$$

$$= \frac{2b}{3} \Big[y^3 \Big]_0^{d/2} = \frac{bd^3}{12}$$

$$\therefore \quad I_x = \frac{bd^3}{12} \qquad \qquad \dots(\text{Eq. } 6.12)$$

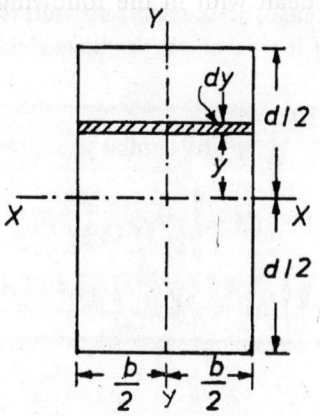

Fig. 6.6: Moment of Inertia of a Rectangle

Similarly,

$$I_y = \frac{db^3}{12} \qquad \qquad \dots(\text{Eq. } 6.13)$$

If the moment of inertia about the base BB is required, the parallel axis theorem may be applied:

$$I_b = I_x + A(d/2)^2 = \frac{bd^3}{12} + \frac{bd \times d^2}{4} = \frac{bd^3}{3}$$

$$\therefore \quad I_b = \frac{bd^3}{3} \qquad \qquad \dots(\text{Eq. } 6.14)$$

Note. The moment of inertia of a parallelogram of base b and height d will also be the same as that for the rectangle shown in Fig. 6.6, since the strip will be only laterally shifted with the same distance from the axis and the same area.

(2) Triangle

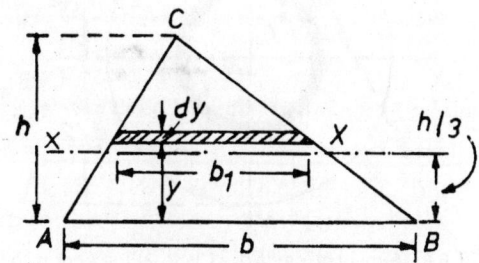

Fig. 6.7: Moments of Inertia of a Triangle

Referring to Fig. 6.7,

Let the triangle ABC be of base b and altitude (height) h as shown. Let us consider a strip of width b_1 at a distance y and of thickness dy.

In this case, it is easier to determine the moment of inertia about the base AB from first principles.

$$b_1 = \left(\frac{h-y}{h}\right)b, \text{ by similar triangles.}$$

$$I_{AB} = \int y^2 \cdot dA = \int_0^h y^2 b_1 dy = \frac{b}{h}\int_0^h y^2(h-y)dy$$

$$= \frac{b}{h}\int_0^h (hy^2 - y^3)dy = \frac{b}{h}\left[\frac{hy^3}{3} - \frac{y^4}{4}\right] = \frac{bh^3}{12}$$

$$\therefore \quad I_{AB} = \frac{bh^3}{12} \qquad\qquad \text{...(Eq. 6.15)}$$

If a centroidal axis parallel to AB is XX,

$$I_x = I_{AB} - A \cdot \left(h/3\right)^2 \text{ (by the parallel axis theorem)}$$

$$= \frac{bh^3}{12} - \frac{bh}{2} \times \frac{h^2}{9} = \frac{bh^3}{36}$$

$$\therefore \quad I_x = \frac{bh^3}{36} \qquad\qquad \text{...(Eq. 6.16)}$$

(3) Circle

Let the circle be of diameter d and let it be required to determine the moment of inertia about a diametral axis XX. For this, we can apply three different approaches.

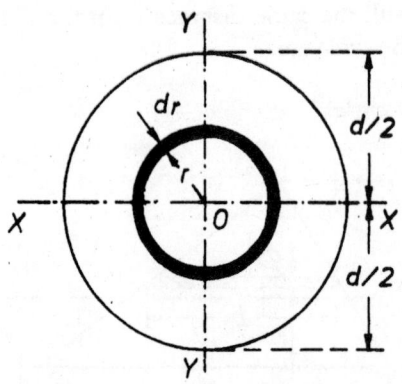

Fig. 6.8: Moment of Inertia of a Circle—First Approach

First Approach

Let us consider an elemental ring of thickness dr at a radial distance r as shown in Fig. 6.8.

The area dA of the ring is given by

$$dA = 2\pi r \cdot dr$$

The polar moment of inertia of this ring about a centroidal axis zz normal to the plane of the figure is given by

$$dI_z = r^2 dA = 2\pi r^3 \cdot dr$$

The polar moment of inertia for the entire area, I_z is given by

$$I_z = \int_0^{d/2} dI_z = \int_0^{d/2} 2\pi r^3 dr = \frac{2\pi}{4}\left[r^4\right]_0^{d/2}$$

$$= \frac{\pi}{2} \times \frac{d^4}{16} = \frac{\pi d^4}{32}$$

Applying the perpendicular axis theorem,

$$I_z = I_x + I_y = 2I_x \text{ by symmetry.}$$

$$\therefore \quad I_x = \frac{I_z}{2} = \frac{\pi d^4}{64} \qquad\qquad \text{...(Eq. 6.17)}$$

Second Approach

Let us consider a thin strip of thickness dy at a distance y from the centroidal axis xx as shown in Fig. 6.9.

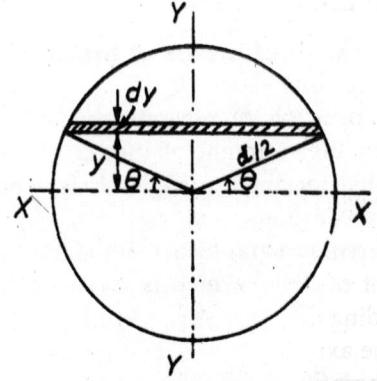

Fig. 6.9: Moment of Inertia of a Circle—Second Approach

Area of the elemental strip dA may be found as follows:

Let the ends of the strip be at an angular distance θ with reference to axis xx.

Then $\quad y = \frac{d}{2}\sin\theta \quad dy = \frac{d}{2}\cos\theta \cdot d\theta$

Width of the strip $= 2 \times \frac{d}{2}\cos\theta = d\cos\theta$

Area of the strip $dA = d\cos\theta \cdot \frac{d}{2}\cos\theta d\theta$

$$= \frac{d^2}{2}\cos^2\theta d\theta$$

$$\therefore \quad I_x = 2\int_0^{\pi/2} y^2 dA = 2\int_0^{\pi/2}\left(\frac{d^2}{4}\sin^2\theta\right)\left(\frac{d^2}{2}\cos^2 d\theta\right)$$

$$= \frac{d^4}{4}\int_0^{\pi/2}(\sin^2\theta\cdot\cos^2\theta)d\theta = \frac{d^4}{4}\int_0^{\pi/2}\sin^2\theta(1-\sin^2\theta)d\theta$$

$$= \frac{d^4}{4}\int_0^{\pi/2}(\sin^2\theta-\sin^4\theta)d\theta = \frac{d^4}{4}\left[\frac{1}{2}\frac{\pi}{2}-\frac{3}{4}\times\frac{1}{2}\times\frac{\pi}{2}\right]$$

$$= \frac{d^4}{4}\times\frac{1}{4}\times\frac{1}{2}\times\frac{\pi}{2}$$

or $\quad I_x = \dfrac{\pi d^4}{64}$, the same as Eq. 6.17.

Third Approach

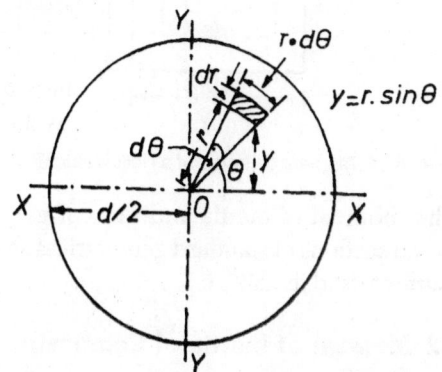

Fig. 6.10: Moment of Inertia of a Circle—Third Approach

Referring to Fig. 6.10, let us consider an element of thickness dr at a radial distance r, subtending an angle $d\theta$ at an angular distance θ from the axis xx.

Its moment of inertia about xx is

$$dI_x = y^2 dA$$
$$= r^2\sin^2\theta \cdot r d\theta\, dr$$
$$= r^3\sin^2\theta\, d\theta\, dr$$

So, I_x may be obtained by double integration of dI_x between the appropriate limits for the variables r and θ:

$$I_x = 4\int_0^{d/2}\int_0^{\pi/2} r^3\sin^2\theta\, d\theta\, dr$$

$$= 4\int_0^{d/2}\int_0^{\pi/2} r^3\left(\frac{1-\cos 2\theta}{2}\right)d\theta\, dr$$

$$= 4\int_0^{d/2}\frac{r^3}{2}\left[\theta-\frac{\sin 2\theta}{2}\right]_0^{\pi/2} dr$$

$$= 2\int_0^{d/2}\frac{\pi}{2}r^3 dr = \pi\left[\frac{r^4}{4}\right]_0^{d/2} = \frac{\pi d^4}{64}$$

or $\quad I_x = \dfrac{\pi d^4}{64}$, the same as Eq. 6.17.

Thus, the moment of inertia of a circle about any centroidal axis is the same value—$(\pi d^4/64)$—owing to perfect symmetry about the centre.

(4) Semi-Circle

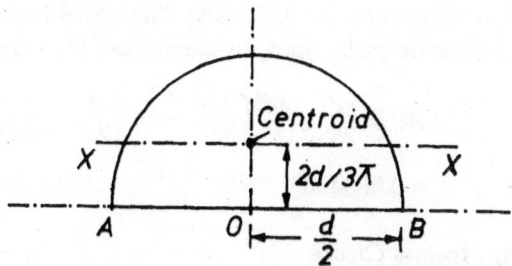

Fig. 6.11: Moment of Inertia of a Semi-Circle

Referring to Fig. 6.11, the moment of inertia of the semi-circle about its base is given by half that of the circle about the same axis, owing to symmetry.

or $\quad I_{ab} = \dfrac{1}{2}\times\dfrac{\pi d^4}{64} = \dfrac{\pi d^4}{128}$ ...(Eq. 6.18)

The centroid will be at a height of $(2d/3\pi)$ above the base as shown. Therefore, the moment of inertia about an axis XX through the centroid and parallel to the base may be obtained by using the parallel axis theorem:

$$I_x = I_{ab} - Ak^2 = \frac{\pi d^2}{128} - \frac{\pi d^2}{8}\left(\frac{2d}{3\pi}\right)^2 = \pi d^4\left[\frac{1}{128}-\frac{1}{18\pi}\right]$$

or $\quad I_x \approx 0.00686\, d^4$ ...(Eq. 6.19)

(5) Quadrant of a Circle

Referring to Fig. 6.12, the moment of inertia of a quadrant of a circle about its base OA is given by one-fourth of that of a semi-circle about the base, because of symmetry.

or $\quad I_{0a} = \dfrac{1}{4}\left(\dfrac{\pi d^4}{64}\right) = \dfrac{\pi d^4}{256}$ ...(Eq. 6.20)

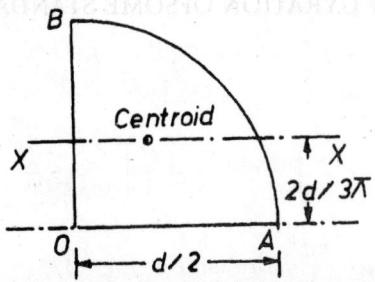

Fig. 6.12: Moment of Inertia of a Quadrant of a Circle

The centroid of the quadrant is also at the same height—$\frac{2d}{3\pi}$—as that of the semi-circle.

Therefore, the moment of inertia of the quadrant about a centroidal axis XX parallel to the base may be got by using the parallel axis theorem:

$$I_x = I_{0a} - Ak^2 = \frac{\pi d^4}{256} - \frac{\pi d^2}{16}\left(\frac{2d}{3\pi}\right)^2 = \pi d^4\left(\frac{1}{256} - \frac{1}{36\pi}\right)$$

or $\quad I_x = 0.00343d^4 \qquad$ ...(Eq. 6.21)

(6) Hollow Circle

With reference to Fig. 6.13, for a hollow circular or annular area with outer diameter D and inner diameter d, the moment of inertia about a centroidal axis may be obtained by subtracting the value for the area bounded by the inner circle from that for the area bounded by the outer circle:

$$I_x = \frac{\pi D^4}{64} - \frac{\pi d^4}{64} = \frac{\pi}{64}\left(D^4 - d^4\right) \quad \text{...(Eq. 6.22)}$$

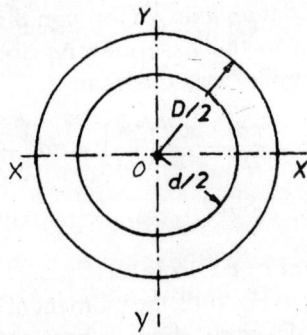

Fig. 6.13: Moment of Inertia of a Hollow Circle

(7) Hollow Rectangle

With reference to Fig. 6.14, similarly, the moments of inertia of a hollow rectangle about the centroidal axes parallel to the sides may be written as follows:

$$I_x = \frac{1}{12}\left(BD^3 - bd^3\right) \qquad \text{...(Eq. 6.23)}$$

$$I_y = \frac{1}{12}\left(DB^3 - db^3\right) \qquad \text{...(Eq. 6.24)}$$

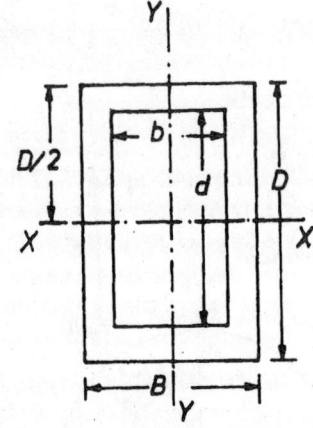

Fig. 6.14: Moment of Inertia of a Hollow Rectangle

The moment of inertia and radii of gyration for some sections of standard geometrical shapes are summarised in Table 6.1.

6.6.2 Moment of Inertia of Composite Sections

If an area/section is composed of several simple areas such as rectangle, triangle, circle, and so on, the moment of inertia of such a section can be found by merely summing up the values for these simple areas about the particular axis under consideration. If any area is to be deducted, the moment of inertia for it is also to be deducted. Some standard rolled steel sections used in engineering practice are I-section, channel section ([), angle section (L), T-section, and so on; for such built-up sections the above procedure is used for the determination of the moment of inertia and radius of gyration.

6.7 PRODUCT OF INERTIA

With respect to the set of rectangular axes OX and OY, the 'product of inertia' of an area, I_{xy}, is defined as the summation or the integral

$$I_{xy} = \int xy \cdot dA \qquad \text{...(Eq. 6.25)}$$

TABLE 6.1: MOMENTS OF INERTIA AND RADII OF GYRATION OFSOME STANDARD SECTIONS

S.No.	Shape of Section	Moment of Inertia	Radius of Gyration
1.	Rectangle	$I_x = \dfrac{bd^3}{12}$ $I_y = \dfrac{db^3}{12}$ $I_{ab} = \dfrac{bd^3}{3}$	$k_x = \dfrac{d}{\sqrt{12}}$ $k_y = \dfrac{b}{\sqrt{12}}$ $k_{ab} = \dfrac{d}{\sqrt{3}}$
2.	Triangle	$I_x = \dfrac{bh^3}{36}$ $I_{ab} = \dfrac{bh^3}{12}$	$k_x = \dfrac{h}{3\sqrt{2}}$ $k_{ab} = \dfrac{h}{\sqrt{b}}$
3.	Circle	$I_x = I_y = \dfrac{\pi d^4}{64}$	$k_x = k_y = \dfrac{d}{4}$
4.	Semi-circle	$I_{ab} = \dfrac{\pi d^4}{128}$ $I_x = 0.00686\, d^4$	$k_{ab} = \dfrac{d}{4}$ $k_x = 0.132\, d$
5.	Quadrant	$I_{oa} = \dfrac{\pi d^4}{256} = I_{ob}$ $I_x = 0.00343\, d^4 = Iy$	$k_{oa} = \dfrac{d}{4} = k_{ob}$ $k_x = 0.132\, d$
6.	Hollow circle	$I_x = I_y = \dfrac{\pi}{64}\left(D^4 - d^4\right)$	$k_x = k_y = \dfrac{\sqrt{D^2 + d^2}}{4}$

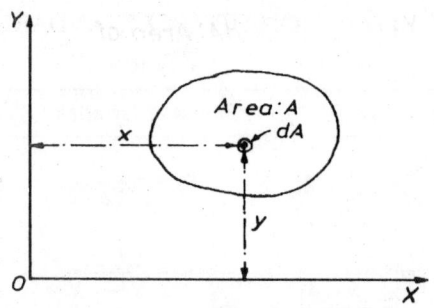

Fig. 6.15: Product of Inertia of an Area

The integral shall be taken over the entire area under consideration. While the elemental area dA is positive, the product of x and y can be positive or negative according as both x and y are of the same sign or of opposite sign. Thus the contribution of the portions of the area lying in the first and third quadrants will be positive and of those lying in the second and fourth quadrants will be negative. Hence the product of inertia of the whole area may be *positive or negative*. This is in contrast to the fact that the moments of inertia about any axis is always positive for an area, which is always positive, except for holes or voids.

(The units of product of inertia are the same as those for moment of inertia—m^4 or mm^4).

Another interesting point about product of inertia is that if the area/section has an axis of symmetry, the product of inertia about this axis and an axis perpendicular to it is *zero*; this should be obvious by virtue of the fact that for every elemental area on one side of the axis of symmetry there lies an identical area as a mirror image on the other side of this axis, the products of inertia being of opposite sign for these two elemental areas. Thus the product of inertia for the whole area reduces to zero. (Note that it is adequate if one of the axes with respect to which the product of inertia is being considered is an axis of symmetry for the area).

6.7.1 Product of Inertia of Certain Areas about Specified Axes

Now the product of inertia of certain regular geometrical areas about specified axes will be obtained.

(1) Rectangle about axes coinciding with two adjacent edges

Referring to Fig. 6.16, let us consider an elemental area dA as shown $\{(dx\,dy)$ at $(x, y)\}$

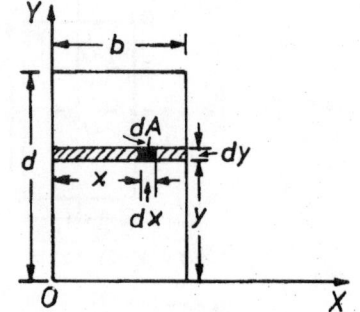

Fig. 6.16: Product of Inertia of a Rectangle

$$I_{xy} = \int xy\,dA$$

$$= \int_{y=0}^{y=d}\int_{x=0}^{x=b} xy\,dx\,dy = \int_{y=0}^{y=d} y\,dy \int_{x=0}^{x=b} x\,dx$$

$$= \frac{d^2}{2}\frac{b^2}{2} = \frac{b^2 d^2}{4}$$

$$I_{xy} = \frac{b^2 d^2}{4} \qquad\qquad \text{...(Eq. 6.26)}$$

(2) Right-angled triangle, base b and height h, with reference to axes along these sides

Referring to Fig. 6.17, let us consider an elemental area dA as shown $[(dx \times dy)$ at $(x, y)]$.

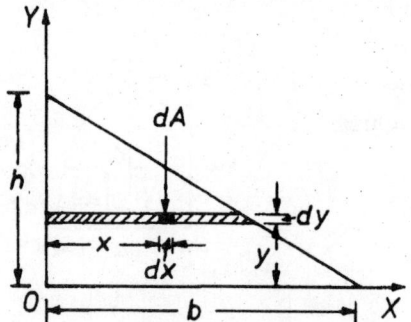

Fig. 6.17: Product of Inertia of a Triangle

$$I_{xy} = \int xy\,dA$$

$$= \int_{y=0}^{y=h}\int_{x=0}^{x=((h-y)/h)b} (xy\,dx \cdot dy)$$

Width at height y

$$= \frac{(h-y)}{h} \times b$$

$$\therefore \quad I_{xy} = \int\limits_{y=0}^{y=h} y \cdot dy \int\limits_{x=0}^{x=\left(\frac{h-y}{h}\right)b} x\,dx = \int\limits_{y=0}^{y=h} \frac{y}{2} dy \left[x^2\right]_0^{(h-y)b/h}$$

$$= \frac{b^2}{2h^2} \int\limits_{y=0}^{y=h} y(h-y)^2 dy$$

$$= \frac{b^2}{2h^2} \int\limits_{y=0}^{y=h} y(h^2 - 2hy + y^2)dy$$

$$= \frac{b^2}{2h^2} \int\limits_0^h (h^2 y - 2hy^2 + y^3)dy = \frac{b^2}{2h^2} \left[\frac{h^2 y^2}{2} - \frac{2hy^3}{3} + \frac{y^4}{4}\right]_0^h$$

$$= \frac{b^2 h^4}{2h^2} \left[\frac{1}{2} - \frac{2}{3} + \frac{1}{4}\right] = \frac{b^2 h^2}{24}$$

or $\quad I_{xy} = \dfrac{b^2 h^2}{24} \qquad \qquad$ ...(Eq. 6.27)

(3) Quadrant of a circle with the reference axes along the bounding radii

Let the radius of the quadrant be R.

Let the reference axes be OX and OY, the bounding radii of the quadrant.

Let us consider an element bounded by radii r and $(r + dr)$ between angles θ and $(\theta + d\theta)$ as shown in Fig. 6.18.

$$x = r\cos\theta$$
$$y = r\sin\theta$$
$$dA = (rd\theta)\,dr$$

$$I_{xy} = \int\limits_{r=0}^{R} \int\limits_{\theta=0}^{\pi/2} (rd\theta)(dr)(r\cos\theta)(r\sin\theta)$$

$$= \frac{1}{2}\int\limits_0^R r^3 dr \int\limits_0^{\pi/2} \sin 2\theta = \frac{1}{4}\int\limits_0^R r^3 dr[\cos 2\theta]_{\pi/2}^0$$

$$= \frac{1}{2}\int\limits_0^R r^3 dr = \frac{1}{8}[r^4]_0^R = \frac{R^4}{8}$$

$$\therefore \quad I_{xy} = \frac{R^4}{8} \qquad \qquad \text{...(Eq. 6.28)}$$

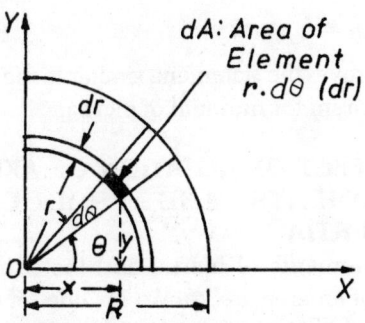

Fig. 6.18: Product of Inertia of Quadrant of a Circle

6.7.2 Parallel Axis Theorem for Product of Inertia

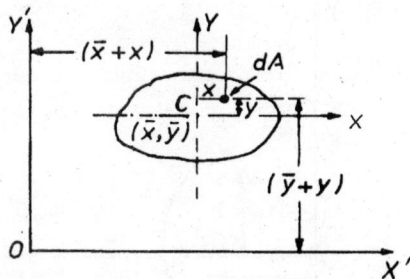

Fig. 6.19: Parallal Axis Theorem for Product of Inertia

We have seen the Parallel Axis Theorem for moment of inertia; a similar one can be enunciated for product of inertia too:

If the product of inertia about the centroidal axes is known, that about axes with respect to which the coordinates of the centroid of the area is given by the former plus the product of the area and coordinates of the centroid.

The proof of this is simple and may be understood with respect to Fig. 6.19.

$$I_{x'y'} = \int (\bar{x} + x)(\bar{y} + y)dA$$

$$= \int \bar{x}\bar{y} \cdot dA + \int \bar{x}y\,dA + \int x\bar{y}\,dA + \int xy\,dA$$

$$= \int \bar{x}\bar{y}\,dA + \bar{x}\int y\,dA + \bar{y}\int x\,dA + \int xy\,dA$$

$\int y\,dA$ and $\int x\,dA$ are first moments of the area about the centroidal axes and hence are zero.

$$\int \bar{x}\bar{y}\,dA = \bar{x}\bar{y}\int dA = A\bar{x}\,\bar{y}$$

$$\int xy\,dA = I_{xy}$$

$$I_{x'y'} = I_{xy} + A\bar{x}\,\bar{y} \qquad \dots\text{(Eq. 6.29)}$$

which proves the statement, similar to the parallel axis theorem for moment of inertia.

6.8 EFFECT OF ROTATION OF AXES ON MOMENTS AND PRODUCT OF INERTIA

If the moments of inertia and the product of inertia of an area are known for one set of axes, those for another set of axes through the same origin may be determined from first principles (Fig. 6.20).

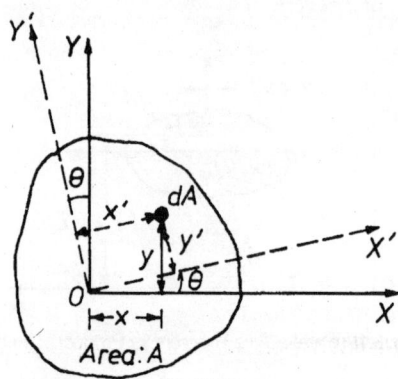

Fig. 6.20: Effect of Rotation of Axes

Let the coordinates of an elemental area dA with respect to the original set of axes OX and OY (with respect to which the moments of inertia I_x, I_y and I_{xy}, are known for an area A) and another set of axes OX' and OY' (at an angle θ with respect to OX and OY) be (x, y) and $(x'\ y')$, respectively. This is simply a rotation of reference axes.

From the geometry of the figure,

$$\left. \begin{array}{l} x' = x\cos\theta + y\sin\theta \\ y' = -x\sin\theta + y\cos\theta \end{array} \right\} \qquad \dots\text{(Eq. 6.30)}$$

$$I_{x'} = \int_A (y')^2 dA = \int_A (-x\sin\theta + y\cos\theta)^2 \cdot dA$$

$$= \sin^2\theta \int_A x^2 dA - 2\sin\theta\cos\theta \int_A xy\,dA + \cos^2\theta \int_A y^2 dA$$

$$\therefore \quad I_x = I_y\sin^2\theta + I_x\cos^2\theta - 2I_{xy}\sin\theta\cdot\cos\theta$$

This can be rewritten as following using trigonometry:

$$I_{x'} = \left(\frac{I_x + I_y}{2}\right) + \left(\frac{I_x - I_y}{2}\right)\cos 2\theta - I_{xy}\sin 2\theta$$

$$\dots\text{(Eq. 6.31)}$$

$I_{y'}$ may be got by replacing θ by $(\theta + \pi/2)$ in Eq. 6.31.

$$I_{y'} = \left(\frac{I_x + I_y}{2}\right) + \left(\frac{I_y - I_x}{2}\right)\cos 2\theta + I_{xy}\sin 2\theta$$

$$\dots\text{(Eq. 6.32)}$$

The product of inertia $I_{x'y'}$ may also be obtained in a similar manner.

$$I_{x'y'} = \int_A x'\,y'\,dA = \int_A (x\cos\theta + y\sin\theta)(-x\sin\theta + y\cos\theta)dA$$

or $\quad I_{x'y'} = \sin\theta\cos\theta(I_x - I_y) + (\cos^2\theta - \sin^2\theta)I_{xy}$

Once again, using trigonometry,

$$I_{x'y'} = \left(\frac{I_x - I_y}{2}\right)\sin 2\theta + I_{xy}\cos 2\theta \quad \dots\text{(Eq. 6.33)}$$

Another corollary is that, since the polar moment of inertia of the area at the pole is a constant, the sum of the moments of inertia about any set of rectangular axes through the pole is also a constant.

$$I_x + I_y = I_{x'} + I_{y'} = \text{a constant} = I_z \quad \dots\text{(Eq. 6.34)}$$

6.8.1 Principal Axes and Principal Moments of Inertia

We have seen that once the moments and product of inertia for an orthogonal reference axes are known, those for any other orthogonal reference axes through the same origin can be determined.

It is observed from equations 6.31 to 6.33, that moments and products of inertia vary with the angle θ, specifying the orientation of the reference axes; it may now be investigated for what value of θ the moments of inertia are the maximum or minimum. Since the sum of the moments of inertia for any set of rectangular reference axes is constant for any pole or origin, the minimum moment of inertia must correspond to an axis perpendicular to that of maximum moment of inertia.

The orientation of these axes may be readily determined by equating the partial derivative of I_x with respect to θ to zero:

$$\frac{\partial I_{x'}}{\partial \theta} = (I_x - I_y)(-\sin 2\theta) - 2I_{xy}\cos 2\theta = 0$$

If the value of θ which satisfies this equation is denoted as $\overline{\theta}$,

$$(I_x - I_y)(-\sin 2\overline{\theta}) = 2I_{xy}\cos 2\overline{\theta}$$

$$\text{or} \quad \tan 2\overline{\theta} = \frac{2I_{xy}}{(I_x - I_y)} \qquad \text{...(Eq. 6.35)}$$

$\overline{\theta}$ corresponds to extreme values of $I_{x'}$, i.e., to a maximum or a minimum value. There are two possible values of $\overline{\theta}$ which satisfy this equation:

$$\theta = \overline{\theta} \text{ and } \theta = \overline{\theta} + \pi/2$$

These two axes are orthogonal to each other; the moment of inertia with respect to one of these axes is the maximum and that with respect to the other is the minimum.

Such a set of orthogonal axes are known as the 'Principal Axes' through the pole or origin chosen.

Substituting the value of θ as $\overline{\theta}$ or $\left\{\frac{1}{2}\tan^{-1}\left(\frac{2I_{xy}}{I_y - I_x}\right)\right\}$ and $(\overline{\theta} + \pi/2)$ in equation 6.31 and 6.32, and simplifying,

We obtain the maximum and minimum values of the moments of inertia:

$$I_{\max} = \left(\frac{I_x + I_y}{2}\right) + \sqrt{\left(\frac{I_x - I_y}{2}\right)^2 + \left(I_{xy}\right)^2}$$

$$\text{...(Eq. 6.36)}$$

$$I_{\min} = \left(\frac{I_x + I_y}{2}\right) - \sqrt{\left(\frac{I_x - I_y}{2}\right)^2 + \left(I_{xy}\right)^2}$$

$$\text{...(Eq. 6.37)}$$

Further, substituting the value of $\overline{\theta}$ for θ in Eq. 6.33, and simplifying, we observe that

$$I_{x'y} = 0$$

Thus it is interesting to note that the product of inertia with respect to the principal axes is zero.

If the principal axes, the maximum and minimum values of the moments of inertia are known as the principal centroidal moments of inertia, and such principal axes as principal centroidal axes. These have very great significance in mechanics of solids and structural analysis.

6.8.2 Mohr's Circle of Inertia

If the moments of inertia and product of inertia about any set of rectangular axes are known, these may be represented along two perpendicular axes as co-ordinates, thus obtaining two points for each such set of axes—I_x and I_{xy} for one point and I_y and I_{xy} for the other point. These two points will always lie on opposite sides of the axis representing moments of inertia (say x-axis); the product of inertia is represented on the other orthogonal axis. If these two points are taken as the ends of a diameter and a circle is described as shown in Fig. 6.21, it is called the "Mohr's Circle of Inertia".

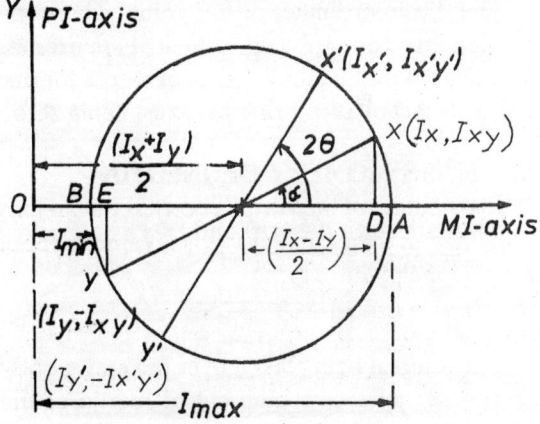

Fig. 6.21: Mohr's Circle of Inertia

If the values of $I_{x'}$, I_y, and $I_{x'y}$ corresponding to another set of axis inclined at θ with respect to the first set of axes are required, one has to draw their radius vector Cx' from the centre of the circle C at an angle 2θ with respect to the vector Cx and produce it on to the other side to meet the circle in y'. Then the MI -coordinates of x' and y'

represent $I_{x'}$ and $I_{y'}$ respectively, and the *PI* - coordinate of x' gives the value of the product of inertia, $I_{x'y'}$.

This can be easily deduced from the geometry of the circle by deriving equations for the co-ordinates of x' and y':

$$I_{x'} = \left(\frac{I_x + I_y}{2}\right) + \left(\frac{I_x - I_y}{2}\right)\cos 2\theta - I_{xy}\sin 2\theta$$

$$I_{y'} = \left(\frac{I_x + I_y}{2}\right) + \left(\frac{I_y - I_x}{2}\right)\cos 2\theta + I_{xy}\sin 2\theta \quad \Bigg\} \ *$$

$$I_{x'y'} = \left(\frac{I_x - I_y}{2}\right)\sin 2\theta + I_{xy}\cos 2\theta$$

*These are the same as equations 6.31 to 6.33.

Since the M.I's are always positive, the figure will always in the first and fourth quadrants. The centre of the circle is $\left[\left(\frac{I_x + I_y}{2}\right), 0\right]$ and the radius is $\left[\left(\frac{I_x - I_y}{2}\right)^2 + I_{xy}^2\right]^{1/2}$.

The *MI* co-ordinates of the points where the circle cuts the *MI*-axis give the principal moments of inertia, the product of inertia for this particular set of axes (principal axes) being zero.

6.9 MASS MOMENT OF INERTIA

'Mass moment of inertia' of a body about an axis is defined in a way similar to the area moment of inertia for a plane area:

$$I_{pp} = \Sigma r^2 \cdot dm = \int r^2 \cdot dm \qquad ...(\text{Eq. } 6.38)$$

The integral being taken over the entire body (Fig. 6.22). I_{pp}: Mass moment of inertia of the body of mass m about axis pp.

When rotation of rigid bodies is involved, the resistance to rotation may be shown to be proportional to the mass and the square of its perpendicular distance from the axis of rotation. Thus the mass moment of inertia is a measure of the resistance of the body to rotation about an axis. It is also merely called moment of inertia and can be easily understand from the context.

Dimensionally speaking, the units of mass moment of inertia in the S.I. units are those of (mass × square of distance) or $\frac{N}{m/\sec^2} \times m^2$ or N-m-$\sec^2$. No special name has been assigned to it. [In the MKS-units, it is (kgm) · (m²)].

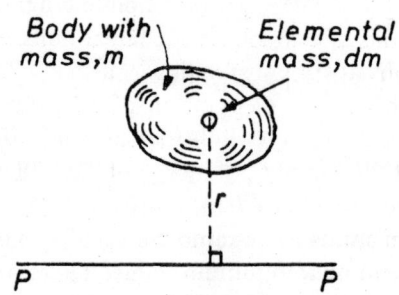

Fig. 6.22: Mass Moment of Inertia

With respect to a set of three mutually perpendicular axes—x-, y- and z-axes (Fig. 6.23)

$$I_{xx} = \int (y^2 + z^2)dm$$
$$I_{yy} = \int (x^2 + z^2)dm \quad \Bigg\} \quad ...(\text{Eq. } 6.39)$$
$$I_{zz} = \int (x^2 + y^2)dm$$

[Double subscripts are used to differentiate it from moment of inertia of areas].

$$I_0 = \int R^2 \cdot dm$$

I_0 is the mass moment of inertia about the point O.

$$I_0 = \int (x^2 + y^2 + z^2)dm$$

$$= \frac{1}{2}(I_{xx} + I_{yy} + I_{zz}) \qquad ...(\text{Eq. } 6.40)$$

by virtue of Equation 6.39.

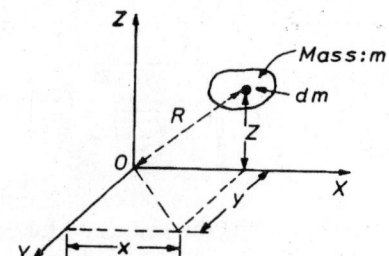

Fig. 6.23: Mass Moment of Inertia of a Body about Three Mutually Perpendicular Directions

6.9.1 Mass Radius of Gyration

The 'Mass Radius of Gyration' of a body with respect to a specified axis is defined in a much similar way to the radius of gyration of an area:

$$k = \sqrt{\frac{I_m}{m}} \qquad \text{...(Eq. 6.41)}$$

$(I_m:$ Mass moment of inertia)

The units of k are obviously linear units.

The physical interpretation is also similar to that of the radius of gyration of an area—the distance at which the whole mass is concentrated such that the moment of inertia of the body and that of the concentrated mass are the same.

The mass radius of gyration is also simply called the radius of gyration of the body and may be easily understood from the context.

6.9.2 Mass Product of Inertia

The 'Mass Product of Inertia' of a body with respect to a pair of axes may be defined similar to the product of inertia of an area:

$$\left.\begin{array}{l} I_{xy} = \int xy \cdot dm \\[2mm] I_{yz} = \int yz \cdot dm \\[2mm] I_{zx} = \int zx \cdot dm \end{array}\right\} \qquad \text{...(Eq. 6.42)}$$

6.9.3 Parallel Axis Theorem for Mass Moment and Mass Product of Inertia

The parallel axis theorem is applicable for mass moment of inertia too (Fig. 6.24).

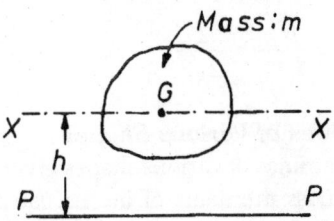

Fig. 6.24: Parallel Axis Theorem for Mass Moment of Inertia

If I_{xx} is the mass moment of inertia of the body of mass m about an axis through the centre of gravity G of the body, that about an axis pp parallel to it and located at a distance h from it, is given by

$$I_{pp} = I_{xx} + mh^2 \qquad \text{... (Eq. 6.43)}$$

This principle is applicable even for mass product of inertia:

$$I_{x'y'} = I_{xy} + mh_1h_2 \qquad \text{...(Eq. 6.44)}$$

(h_1 and h_2 are the shifts or translations needed to get the new axes from the original set of axes).

Note: The proofs are much similar to those for moments of inertia for areas and hence are not given here.

6.9.4 Perpendicular Axes Theorem for Mass Moment of Inertia of a Thin Lamina of Uniform Thickness

The statement of the "Perpendicular Axes Theorem" for mass moment of inertia of a uniformly thin lamina is as follows:

The mass moment of inertia of uniformly thin plane lamina with respect to an axis perpendicular to its plane is equal to the sum of its values with respect to two rectangular axes taken in the central plane of the lamina through the point where the former intersects that plane.

$$I_{zz} = I_{xx} + I_{yy} \qquad \text{...(Eq. 6.45)}$$

The proof of this is also similar to the perpendicular axes theorem for areas, and hence is not given here.

It can also be shown that the mass radius of gyration of a uniformly thick homogeneous lamina will be the same as the radius of gyration of its face area.

6.9.5 Principal Moments of Inertia of a Body

For plane laminas of uniform thickness, with any origin, there is always a pair of in-plane orthogonal axes with respect to which the mass *P.I.*, I_{xy}, is zero. These are called the principal axes for that origin and *M.I.* values for these axes are called the principal mass moments of inertia (one being the maximum and the other the minimum).

For three-dimensional bodies, it can be shown that there are three mutually perpendicular axes x, y and z with respect to which I_{xy}, I_{yz} and I_{zx} are all zero. They are called the principal axes through the point, and the mass moments of inertia I_{xx}, I_{yy} and I_{zz} are called the principal moments of inertia.

In dynamic analysis of bodies, the principal moments of inertia and principal radii of gyration find extensive application.

6.9.6 Mass Moment of Inertia by Integration

Mass moment of inertia of simple bodies of regular geometric shape can be determined from first principles by integration. Appropriate expressions for a general element of mass dm and its distance r, from the axis, are written down, and the term $r^2 \cdot dm$ is integrated between suitable limits such that the entire mass of the body is covered.

6.9.7 Mass Moment of Inertia of Thin Plates

Let us consider a homogeneous thin plate of uniform thickness lying in the *XY*-plane as shown in Fig. 6.25.

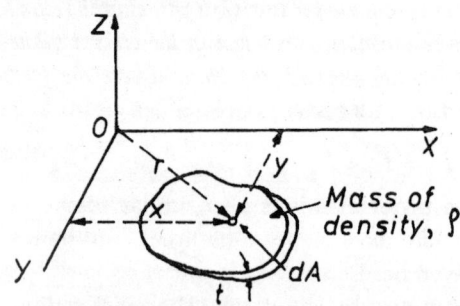

Fig. 6.25: Mass Moment of Inertia of a Thin Plate

Let the uniform thickness be t and the mass density of the material, ρ. Let a small area dA be considered at a distance r from the z-axis. The mass dm of the elemental area is $dm = \rho \cdot t \cdot dA$. Therefore, the mass moment of inertia of the plate about x-axis is given by

$$I_{xx} = \int y^2 dm = \rho t \int y^2 dA = \rho t \cdot I_x \quad ...(\text{Eq. } 6.46)$$

Similarly,

$$I_{yy} = \rho t \cdot I_y$$

$$I_{zz} = \rho t \cdot I_z$$

Since $r^2 = x^2 + y^2$,

$I_{zz} = I_{xx} + I_{yy}$; this is the same as equation 6.45.

Thus, the mass moment of inertia of a thin plate about any axis contained in the plane of the plate is obtained by its product of the area

moment of inertia about the same axis, the uniform thickness, and the mass density of the material.

6.9.8 Mass Moment of Inertia of Standard Bodies

The mass moment of inertia of a few bodies of standard geometrical shape may now be derived.

(1) Thin Uniform Rod

Let the rod be of a total mass m, and length L (Fig. 6.26).

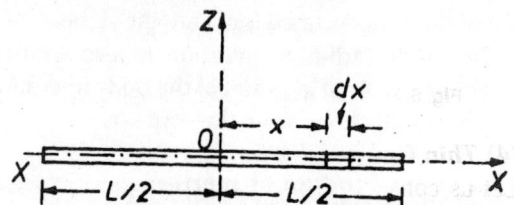

Fig. 6.26: Mass Moment of Inertia fo a Thin Uniform Road

Let us consider a small length dx at a distance x from the mid point of the rod, O. With O as origin, let the axes be as shown.

$$I_{zz} = 2 \int_0^{L/2} x^2 \cdot \frac{m}{L} \cdot dx = \frac{2m}{3L} \left(x^3 \right)_0^{L/2}$$

$$= \frac{2m}{3L} \times \frac{L^3}{8} = \frac{mL^2}{12}$$

$$\therefore \quad I_{zz} = \frac{mL^2}{12} \quad ...(\text{Eq. } 6.47)$$

Also, $k_{zz} = \sqrt{\dfrac{I_{zz}}{m}} = \dfrac{L}{2\sqrt{3}} \quad ...(\text{Eq. } 6.48)$

(2) Laminas of Various Shapes

For thin laminas of various shapes given in Table 6.1, the mass moments of inertia can be easily obtained from the value of the radius of gyration, as stated in sub-section 6.9.7; hence these are not given here.

(3) Thin Ring

Let the thin ring of mass m and diameter d (Fig. 6.27).

Let us consider an element of mass dm of the thin ring.

$$I_{zz} = \int \left(\frac{d}{2}\right)^2 \cdot dm$$

or $\quad I_{zz} = \dfrac{md^2}{4} \qquad \qquad \text{...(Eq. 6.49)}$

$\qquad k_{zz} = \dfrac{d}{2} \qquad \qquad \text{...(Eq. 6.50)}$

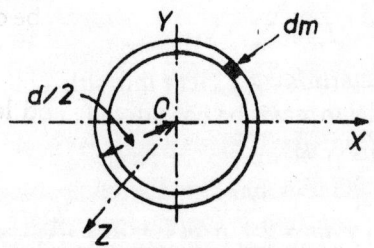

Fig. 6.27: Mass Moment of Inertia of Thin Ring

(4) Thin Circular Plate

Let us consider a thin circular plate of mass m and diameter d (Fig. 6.28).

From equation 6.46,

$\quad I_{zz} = \rho t \cdot I_z$, with the usual notation for ρ.

It is known that

$$I_z = \frac{\pi d^4}{32}$$

$\therefore \quad I_{zz} = \rho t \cdot \dfrac{\pi d^4}{32} = \left(\rho t \cdot \dfrac{\pi d^2}{4}\right)\left(\dfrac{d^2}{8}\right)$

or $\quad I_{zz} = \dfrac{md^2}{8} \qquad \qquad \text{...(Eq. 6.51)}$

$\qquad k_{zz} = \dfrac{d}{2\sqrt{2}} \qquad \qquad \text{...(Eq. 6.52)}$

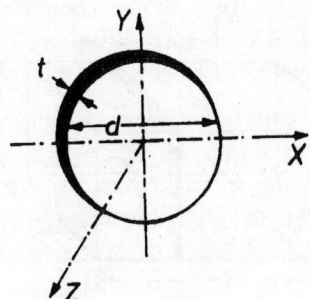

Fig. 6.28: Mass Moment of Inertia of Thin Circular Plate

(5) Thin Hollow Circular Plate

Let the outer diameter be D and inner diameter be d.

$$I_{zz} = \rho t I_z$$

$$= \rho t \frac{\pi}{32}\left(D^4 - d^4\right)$$

$$= \left\{\rho t \frac{\pi}{4}\left(D^2 - d^2\right)\right\}\frac{\left(D^2 + d^2\right)}{8}$$

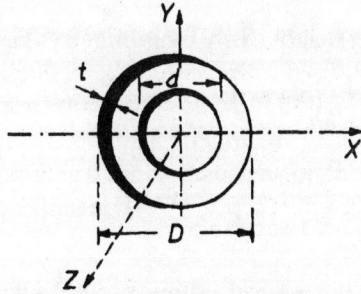

Fig. 6.29: Mass Moment of Inertia of Thin Hollow Circular Plate

or $\quad I_{zz} = \dfrac{m\left(D^2 + d^2\right)}{8} \qquad \text{...(Eq. 6.53)}$

$\qquad k_{zz} = \dfrac{\sqrt{D^2 + d^2}}{2\sqrt{2}} \qquad \text{...(Eq. 6.54)}$

(6) Right Circular Cylinder

Let a circular slice of small thickness dz at a distance z from the base.

Let the diameter be d and height h, with the total mass m.

I'_{zz} for the thin slice $= \dfrac{m' \cdot d^2}{8}$, where m' is the mass of ring.

I_{zz} for the cylinder

$$= \int_0^h I'_{zz} = \int_0^h \frac{m' d^2}{8}$$

or $\quad I_{zz} = \dfrac{m \cdot d^2}{8} \qquad \qquad \text{...(Eq. 6.55)}$

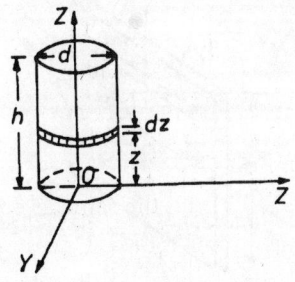

Fig. 6.30: Mass Moment of Inertia of Right Circular Cylinder

(7) Sphere

We can use (Eq. 6.40) for this case conveniently.

$$I_{xx} = I_{yy} = I_{zz}$$

But $I_{xx} + I_{yy} + I_{zz} = 2I_0$

or $I_{zz} = \dfrac{2}{3} I_0$

I_0 may be got easily by symmetry (Fig. 6.31)

$$I_0 = \int_0^R \{\rho(4\pi r^2)dr\}r^2 \text{ considering a spherical}$$

shell of thickness dr at a radius r.

$$= 4\pi\rho \int_0^R r^4 \cdot dr = \frac{4\pi\rho R^5}{5}$$

But the mass of sphere $m = \rho \dfrac{4}{3}\pi R^3$

$\therefore \quad I_0 = \dfrac{m \cdot 3R^2}{5}$

or $\quad I_{zz} = \dfrac{2}{3} I_0 = \dfrac{2}{5} mR^2 = \dfrac{mD^2}{10}$ ...(Eq. 6.56)

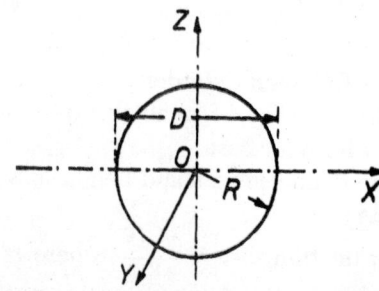

Fig. 6.31: Mass Moment of Inertia of a Sphere

(8) Thin Rectangular Plate

Let the rectangular plate be $b \times d$, and of a small thickness t (Fig. 6.32).

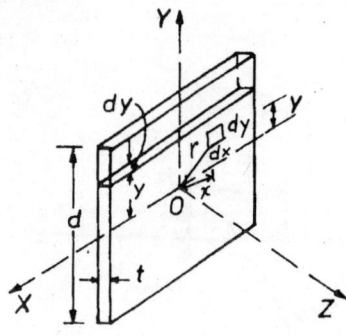

Fig. 6.32: Mass Moment of Inertia of a thin Rectangular Plate

Let an element of size $dx \cdot dy$ and thickness 't' be considered at (x, y).

$$r^2 = x^2 + y^2$$

$$I_{zz} = \int r^2 dm = \int (x^2 + y^2)dm$$

$$= \int x^2 dm + \int y^2 dm$$

$$= I_{xx} + I_{yy}$$

For determining I_{xx}, let a thin strip of thickness dy parallel to x-axis be considered at a distance y.

$$dm = \rho \cdot bt \cdot dy$$

$$I_{xx} = 2\int_0^{d/2} y^2 dm = 2\int_0^{d/2} \rho \cdot bt \cdot y^2 \cdot dy = \frac{2\rho bt}{3}(y^3)_0^{d/2}$$

$$= \frac{\rho \cdot btd^3}{12}$$

But the mass of the plate, $m = \rho btd$

$$I_{xx} = \frac{md^2}{12} \qquad\qquad \text{...(Eq. 6.57)}$$

Similarly, $I_{yy} = \dfrac{mb^2}{12}$ (by considering a strip parallel to y-axis) ...(Eq. 6.59)

Hence $I_{zz} = I_{xx} + I_{yy} = \dfrac{m}{12}(b^2 + d^2)$...(Eq. 6.59)

(9) Rectangular Prism

Let the rectangular prism be $l \times b \times d$ in size.

Let a thin rectangular slice of thickness dz at a distance z from the xy-plane be considered (Fig. 6.33).

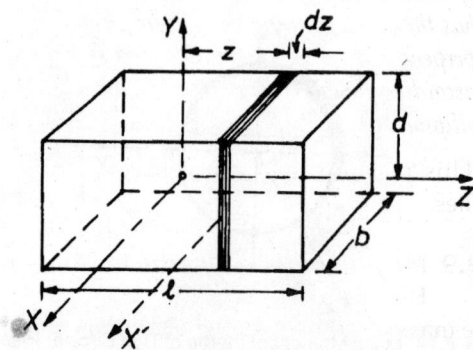

Fig. 6.33: Mass Moment of Inertia of a Rectangular Prism

The mass moment of inertia of this element about OX' parallel to OX.

$$dI_{x'x'} = \frac{1}{12} \cdot dm \cdot d^2$$

Using the parallel axis theorem,

$$dI_{xx} = \frac{1}{12} \cdot dm \cdot d^2 + dm \cdot z^2 = \left(\frac{1}{12}d^2 + z^2\right)dm$$

$$= \left(\frac{1}{12}d^2 + z^2\right)\rho bd \cdot dz$$

$$\therefore \quad I_{xx} = 2\int_0^{l/2} \left(\frac{1}{12}d^2 + z^2\right)\rho bd \cdot dz$$

$$= 2\rho bd\left[\frac{1}{12}d^2 z + \frac{z^3}{3}\right]_0^{d/z} = \frac{\rho bdl}{12}\left(d^2 + l^2\right)$$

$$= \frac{m}{12}\left(d^2 + l^2\right)$$

or $\quad I_{xx} = \dfrac{m}{12}\left(d^2 + l^2\right) \qquad \text{...(Eq. 6.60)}$

Similarly, $\quad I_{yy} = \dfrac{m}{12}\left(b^2 + l^2\right) \qquad \text{...(Eq. 6.61)}$

$$I_{zz} = \frac{m}{12}\left(b^2 + d^2\right) \qquad \text{...(Eq. 6.62)}$$

Note: In this case $I_{zz} \neq I_{xx} + I_{yy}$, because thickness of the block is not small.

If l is zero, it becomes a plate in the *xy*-plane; substituting $l = 0$ in the above equations, we get the results already obtained for a thin plate.

There is a rule enunciated by Routh as follows:

If a body has three axes of symmetry, then the square of the radius of gyration about an axis of symmetry has the value given by $k^2 =$ (Sum of squares of the perpendicular semi-axes) $\div D$, where $D = 3, 4,$ or 5, according as the body is rectangular, elliptical, or ellipsoidal.

This may be used conveniently for some bodies.

6.9.9 Mass moment of Inertia of Composite Bodies

The mass moment of inertia of a composite body may be determined by dividing it into simple parts of regular geometrical shape with known M.I. about their centroidal axes. The parallel axis theorem may be used judiciously to obtain the required result by the process of summation for the simple parts.

ILLUSTRATIVE PROBLEMS

PROBLEM 6.1: Calculate the moment of inertia of the area enclosed by a line $y = 6x$ and a curve $y = 3x^2$ about both *x*-axis and *y*-axis as shown in Fig. 6.34.

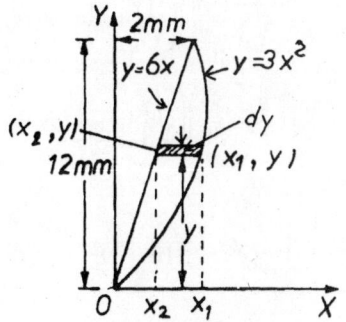

Fig. 6.34 (Illustrative Problem 6.1)

Solution: Consider an elemental area parallel to *x*-axis. The elemental area intersects the line and the curve at the arbitrary points (x_2, y) and (x_1, y) respectively.

The elemental area,

$$dA = (x_1 - x_2)\, dy$$

$$I_{xx} = \int_A y^2 \cdot dA$$

$$= \int_0^{12} y^2(x_1 - x_2)\, dy$$

$$= \int_0^{12} y^2\left(\sqrt{\frac{y}{3}} - \frac{y}{6}\right) dy$$

$$= \int_0^{12}\left(\frac{y^{5/2}}{\sqrt{3}} - \frac{y^3}{6}\right) dy$$

$$= \left[\frac{y^{7/2}}{\sqrt{3}(7/2)} - \frac{y^4}{6 \times 4}\right]_0^{12}$$

$$= \frac{12^{7/2}}{(\sqrt{3})(7/2)} - \frac{12^4}{6 \times 4}$$

$$= 987.429 - 864$$

$$= 123.429 \text{ mm}^4$$

$$I_{yy} = \int_A dA \left(\frac{x_1 + x_2}{2}\right)^2$$

$$= \int_0^{12} (x_1 - x_2) dy \left(\frac{x_1 + x_2}{2}\right)^2$$

$$= \frac{1}{4} \int_0^{12} \left(\sqrt{\frac{y}{3}} - \frac{y}{6}\right)\left(\sqrt{\frac{y}{3}} + \frac{y}{6}\right)^2 dy$$

$$= \frac{1}{4} \int_0^{12} \left(\frac{y}{3} - \frac{y^2}{36}\right)\left(\sqrt{\frac{y}{3}} + + \frac{y}{6}\right) dy$$

$$= \frac{1}{4} \int_0^{12} \left(\frac{y^{3/2}}{3\sqrt{3}} - \frac{y^3}{216} + \frac{y^2}{18} - \frac{y^{5/2}}{36\sqrt{3}}\right) dy$$

$$= \frac{1}{4}\left[\frac{y^{5/2}}{3\sqrt{3}} \times \frac{2}{5} - \frac{y^4}{216 \times 4} + \frac{y^3}{18 \times 3} - \frac{y^{7/2}}{36\sqrt{3}} \frac{2}{7}\right]_0^{12}$$

$$= \frac{1}{4}\left[\frac{12^{5/2}}{3\sqrt{3}} \frac{2}{5} - \frac{12^4}{216 \times 4} + \frac{12^3}{18 \times 3} - \frac{12^{7/2}}{36\sqrt{3}} \frac{2}{7}\right]$$

$$= \frac{1}{4}[38.40 - 24.0 + 32.0 - 27.4286]$$

$$= \frac{1}{4}(18.9714)$$

$$= 4.743 \text{ mm}^4$$

PROBLEM 6.2: Find the moment of inertia of a rectangular section shown in Fig. 6.35 about the faces AB and BC.

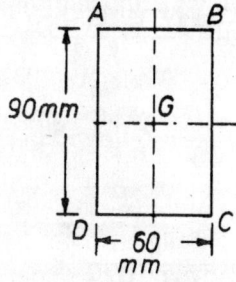

Fig. 6.35 (Illustrative Problem 6.2)

Solution: Moment of inertia about AB

$$I_{AB} = I_G + Ah^2$$

$$I_G = \frac{bd^3}{12}$$

$$= \frac{60 \times 90^3}{12} = 364.5 \times 10^4 \text{ mm}^4$$

$$A = bd = 60 \times 90 = 5400 \text{ mm}^2$$

$$h = \frac{d}{2} = \frac{90}{2} = 45 \text{ mm}$$

$$\therefore \quad I_{AB} = 364.5 \times 10^4 + 5400 \times (45)^2$$

$$= 364.5 \times 10^4 + 1093.5 \times 10^4$$

$$= 1458 \times 10^4 \text{ mm}^4$$

or $\qquad 1458 \text{ cm}^4$

Moment of inertia about BC

$$I_{BC} = I_G + Ah^2$$

$$I_G = \frac{db^3}{12} = \frac{90 \times 60^3}{12}$$

$$= 162 \times 10^4 \text{ mm}^4$$

$$A = bd = 60 \times 90 = 5400 \text{ mm}^2$$

$$h = \frac{b}{2} = \frac{60}{2} = 30 \text{ mm}$$

$$\therefore \quad I_{BC} = 162 \times 10^4 + 5400 \times (30)^2$$

$$= 162 \times 10^4 + 486 \times 10^4$$

$$= 648 \times 10^4 \text{ mm}^4$$

$$= 648 \text{ cm}^4$$

PROBLEM 6.3: Compute the moment of inertia of the 50 mm × 75 mm rectangle shown in Fig. 6.36 about x-x axis to which it is inclined at angle $\theta = \sin^{-1}(4/5)$.

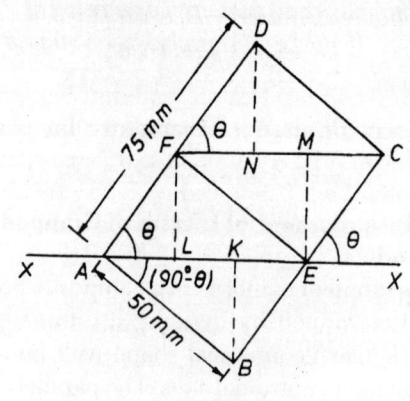

(a) Rectangle

$$I_{AEF} = \frac{62.5 \times 30^3}{12} = 140625 \text{ mm}^4$$

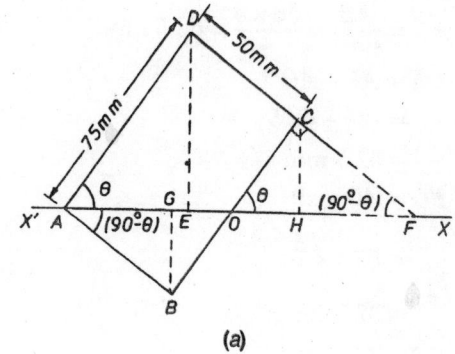

$$\theta = \sin^{-1}\left(\frac{4}{5}\right)$$

$$\therefore \sin\theta = \frac{4}{5}; \quad \cos\theta = \frac{3}{5}$$

(b) Inclination of the Rectangle with x-axis

Fig. 6.36 (Illustrative Problem 6.3)

Solution:
Method-I
Moment of inertia of the section about X-X axis
= sum of the moments of inertia of individuals
triangular areas about x-x axis.

$$= I_{CDF} + I_{CEF} + I_{AEF} + I_{ABE}$$

FC is parallel to X-X axis

$$BK = AB\sin(90 - \theta) = AB\cos\theta$$

$$= 50 \times \frac{3}{5}$$

$$= 30 \text{ mm}$$

$$ND = BK = 30 \text{ mm}$$

$$FD = \frac{DN}{\sin\theta} = \frac{30}{(4/5)} = \frac{30 \times 5}{4} = 37.50 \text{ mm}$$

$$AF = 75 - 37.50 = 37.50 \text{ mm}$$

Hence $FL = ME = 37.5\sin\theta$

$$= 37.5 \times \frac{4}{5} = 30 \text{ mm}$$

$$AE = FC = \frac{AB}{\cos(90 - \theta)} - \frac{50}{\sin\theta} = \frac{50}{4} \times 5$$

$$= 62.5 \text{ mm}$$

$$I_{CDF} = \frac{62.5 \times 30^3}{36} + \frac{1}{2} \times 62.5 \times 30\left(30 + \frac{30}{3}\right)^2$$

$$= 46875 + 1500000$$

$$= 1546875 \text{ mm}^4$$

$$I_{CEF} = \frac{62.5 \times 30^3}{4} = 421875 \text{ mm}^4$$

$$I_{AEF} = \frac{62.5 \times 30^3}{12} = 140625 \text{ mm}^4$$

$$I_{ABE} = \frac{62.5 \times 30^3}{12} = 140625 \text{ mm}^4$$

$$I_{xx} = 1546875 + 421875 + 140625 + 140625$$

$$= 2250000 \text{ mm}^4$$

$$= 225 \times 10^4 \text{ mm}^4$$

$$= 225 \text{ cm}^4$$

Method-II

$$\theta = \sin^{-1}\left(\frac{4}{5}\right)$$

or $\sin\theta = \frac{4}{5}$

$$\cos\theta = \frac{3}{5}$$

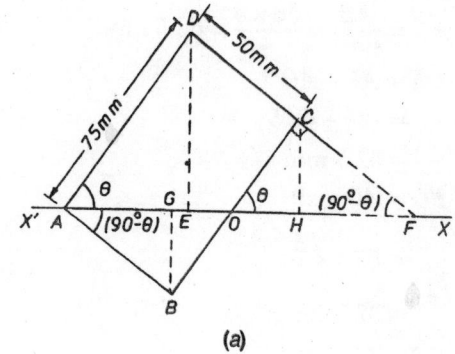

(a)

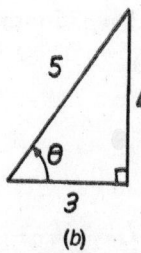

(b)

Fig. 6.37 (Illustrative Problem 6.3)

Moment of inertia of the section about X-X axis
= M.I. of triangle ADF about X-X axis
+ M.I. triangle ABO about X-X axis
− M.I. triangle COF about X-X axis
$= I_{ADF} + I_{ABO} - I_{COF}$

$$= \sin(90 - \theta) = \frac{AD}{AF}$$

or $AF = \dfrac{AD}{\sin(90-\theta)} = \dfrac{AD}{\cos\theta}$

$= \dfrac{75 \times 5}{3} = 125 \text{ mm}$

$\sin\theta = \dfrac{DE}{AD}$

or $\quad = DE = AD\sin\theta$

$= 75 \times \dfrac{4}{5}$

$= 60 \text{ mm}$

$\sin(90-\theta) = BG/AB$

or $\quad BG = AB\sin(90-\theta)$

$= 50 \times \cos\theta = 50 \times \dfrac{3}{5} = 30 \text{ mm}$

$\sin\theta = \dfrac{AB}{AO}$

$AO = \dfrac{AB}{\sin\theta} = \dfrac{50 \times 5}{4} = 62.5 \text{ mm}$

$OF = AF - AO$

$= 125 - 62.5$

$= 62.5 \text{ mm}$

$BO = AO\cos\theta$

$= 62.5 \times \dfrac{3}{5}$

$= 37.5 \text{ mm}$

$\therefore \quad CO = BC - BO$

$= 75 - 37.5 = 37.5 \text{ mm}$

$CH = OC\sin\theta$

$= \dfrac{37.5 \times 4}{5}$

$= 30 \text{ mm}$

$I_{XX} = I_{ADF} + I_{ABO} - I_{COF}$

$= \dfrac{AF \times DE^3}{12} + \dfrac{AO \times BG^3}{12} - \dfrac{OF \times CH^3}{12}$

$= \dfrac{125 \times 60^3}{12} + \dfrac{62.5 \times 30^3}{12} - \dfrac{62.5 \times 30^3}{12}$

$= 2250000 + 0$

$= 225 \times 10^4 \text{ mm}^4$

$= 225 \text{ cm}^4$

PROBLEM 6.4: Find the second moment of the area shown in Fig. 6.38 about x-x & y-y.

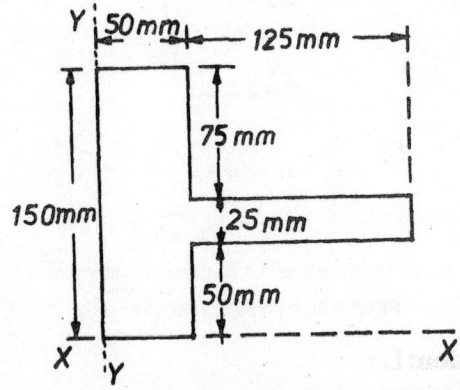

Fig. 6.38 (Illustrative Problem 6.4)

Solution: Let us divide section into two rectangles as shown in Fig. 6.38.

$I_{xx} = (I_{xx})_1 + (I_{xx})_2$

$(I_{xx})_1 = \dfrac{1}{3} \times (50) \times (150)^3$

$= 5625 \times 10^4 \text{ mm}^4$

$(I_{xx})_2 = \dfrac{1}{13} \times (125) \times (25)^3 + (125)(25)\left(50 + \dfrac{25}{2}\right)^2$

$= 162.76 \times 10^4 + 1220.70 \times 10^4$

$= 1236.976 \times 10^4$

$= 1237 \times 10^4 \text{ mm}^4$

$\therefore \quad I_{xx} = 5625 \times 10^4 + 1237 \times 10^4$

$= 6862 \times 10^4 \text{ mm}^4$

or $\quad 6862 \text{ cm}^4$

$I_{yy} = (I_{yy})_1 + (I_{yy})_2$

$(I_{yy})_1 = \dfrac{1}{3} \times (150) \times (50)^3$

$= 625 \times 10^4 \text{ mm}^4$

$(I_{yy})_2 = \dfrac{1}{3} \times (25) \times (125)^3 + (25)(125)\left(50 + \dfrac{125}{2}\right)^2$

$= 406.90 \times 10^4 + 3955.08 \times 10^4$

$= 4361.98 \times 10^4 \text{ mm}^4$

$I_{yy} = (625 + 4361.98) \times 10^4 \text{ mm}^4$

$= 4986.98 \times 10^4 \text{ mm}^4$

$= 4987 \times 10^4 \text{ mm}^4$

or $\quad 4987 \text{ cm}^4$

PROBLEM 6.5: For the area shown in Fig. 6.39, find the moment of inertia about the line *AB*.

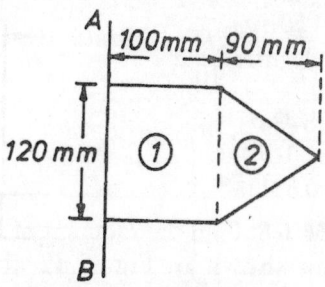

Fig. 6.39 (Illustrative Problem 6.5)

Solution: Let us divide the figure into a rectangle and a triangle as shown in Fig. 6.39.

$$I_{AB} = (I_{AB})_1 + (I_{AB})_2$$

Moment of inertia of the rectangle about *AB*

$$(I_{AB})_1 = \frac{1}{3} \times (120) \times (100)^3$$

$$= 4000 \times 10^4 \text{ mm}^4$$

Moment of inertia of the triangle about *AB*

$$(I_{AB})_2 = \frac{1}{36}(120)(90)^3 + \frac{1}{2}(120)(90)\left(100 + \frac{90}{3}\right)^2$$

$$= 243 \times 10^4 + 9126 \times 10^4$$

$$= 9369 \times 10^4 \text{ mm}^4$$

$$\therefore \quad I_{AB} = (4000 + 9369) \times 10^4$$

$$= 13369 \times 10^4 \text{ mm}^4$$

or $\qquad$ 13369 cm^4

PROBLEM 6.6: Determine the radius of gyration for the area shown in Fig. 6.40 with respect to *YY*.

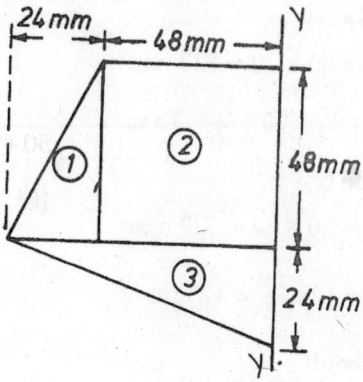

Fig. 6.40 (Illustrative Problem 6.6)

Solution: Let us divide the given area into triangles and square.

$$I_{yy} = (I_{yy})_1 + (I_{yy})_2 + (I_{yy})_3$$

$$(I_{yy})_1 = \frac{48 \times 24^3}{36} + \left(\frac{48 \times 24}{2}\right)\left(48 + \frac{24}{3}\right)^2$$

$$= 1.8432 \times 10^4 + 180.6336 \times 10^4$$

$$= 182.4768 \times 10^4 \text{ mm}^4$$

$$= 182.4768 \text{ cm}^4$$

$$(I_{yy})_2 = \frac{48 \times 48^3}{3} = 176.9472 \times 10^4 \text{ mm}^4$$

$$= 176.9472 \text{ cm}^4$$

$$(I_{yy})_3 = \frac{24 \times 72^3}{3} = 74.6496 \times 10^4 \text{ mm}^4$$

or $\qquad$ 74.6496 cm^4

$$I_{yy} = 182.4768 \times 10^4 + 176.9472 \times 10^4 + 74.6496 \times 10^4$$

$$= 434.0736 \times 10^4 \text{ mm}^4$$

or $\qquad$ 434.0736 cm^4

$$K_{yy} = \sqrt{\frac{I_{yy}}{A}}$$

$$A = A_1 + A_2 + A_3$$

$$= \frac{1}{2} \times 24 \times 48 + 48 \times 48 + \frac{1}{2} \times 24 \times 72$$

$$= 5.76 \times 10^2 + 23.04 \times 10^2 + 8.64 \times 10^2$$

$$= 37.44 \times 10^2 \text{ mm}^2$$

$$\therefore \quad K_{yy} = \sqrt{\frac{434.0736 \times 10^4}{37.44 \times 10^2}}$$

$$= \sqrt{11.59385} \times 10$$

$$= 3.405 \times 10$$

$$= 34.05 \text{ mm}$$

or $\qquad$ 3.405 cm

PROBLEM 6.7: Compute the moment of inertia with respect to an axis passing through two opposite apexes of regular hexagon of side '*a*'.

Solution: Fig. 6.41 shows the given regular hexagon of side '*a*'. All interior angles in case of a regular hexagon are of 120°.

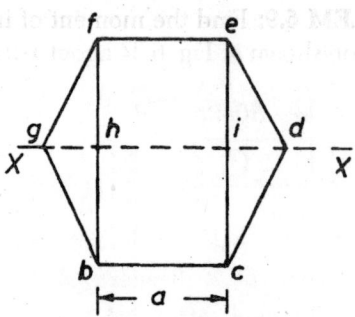

Fig. 6.41 (Illustrative Problem 6.7)

Therefore $\angle fgh = 60°$

$$gh = fg \cdot \cos 60° = \frac{a}{2}$$

$$fh = fg \cdot \sin 60° = \frac{a\sqrt{3}}{2}$$

M.I. about xx for the given hexagon is desired to be determined. The hexagon is divided into a rectangle $bcef$ and four triangles as shown in Fig. 6.41.

Rectangle $bcef$
Centroid of this rectangle is same as that of the hexagon

$$(I_{xx})_1 = \frac{(bc) \times (bf)^3}{12}$$

$$= \frac{(a) \times (a\sqrt{3})^3}{12} \qquad (\because bf = 2 \times fh)$$

$$= \frac{\sqrt{3}}{4} a^4$$

Triangle ghf

M.I. of one triangle about $xx = \frac{gh \times (fh)^3}{12}$

$$= \left(\frac{a}{2}\right)\left(\frac{a\sqrt{3}}{2}\right)^3 / 12$$

$$= \frac{\sqrt{3}\, a^4}{64}$$

Moment of inertia of four triangles about xx,

$$(I_{xx})_2 = \frac{4 \times \sqrt{3}a^4}{64}$$

$$= \frac{\sqrt{3}}{16} a^4$$

Moment of inertia of the hexagon about xx

$$= (I_{xx})_1 + (I_{xx})_2$$

$$= \frac{\sqrt{3}}{4} a^4 + \frac{\sqrt{3}}{16} a^4$$

$$= \frac{5\sqrt{3}a^4}{16}$$

$$= 0.5413 a^4$$

PROBLEM 6.8: Find the moment of inertia of the section shown in Fig. 6.42 about the centroidal axis x-x perpendicular to the web.

Solution: First of all, let us find out the c.g. of the section. Since the section is symmetrical about y-y axis, therefore c.g. will be on this axis.

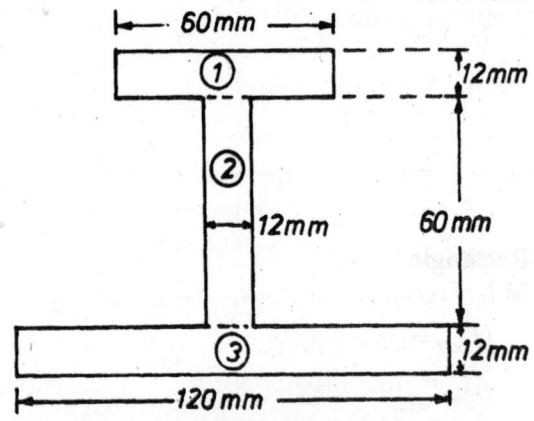

Fig. 6.42 (Illustrative Problem 6.8)

Let $\bar{y}$ be the distance between the c.g. of the section and the bottom face.

Split up the whole section into three rectangles.

Rectangle (1)
$$A_1 = 60 \times 12 = 720 \text{ mm}^2$$

$$y_1 = 12 + 60 + \frac{12}{2} = 78 \text{ mm}$$

Rectangle (2)
$$A_2 = 60 \times 12 = 720 \text{ mm}^2$$

$$y_2 = 12 + \frac{60}{2} = 42 \text{ mm}$$

Rectangle (3)
$$A_3 = 120 \times 12 = 1440 \text{ mm}^2$$

$$y_3 = \frac{12}{2} = 6 \text{ mm}$$

$$\therefore \quad \bar{y} = \frac{A_1 y_1 + A_2 y_2 + A_3 y_3}{A_1 + A_2 + A_3}$$

$$= \frac{720 \times 78 + 720 \times 42 + 1440 \times \iota}{720 + 720 + 1440}$$

$$= \frac{56.16 \times 10^3 + 30.24 \times 10^3 + 8.640 \times 10^3}{2.88 \times 10^3}$$

$$= \frac{95.04 \times 10^3}{2.88 \times 10^3} = 33 \text{ mm}$$

Determination of M.I.
Rectangle (1)
M.I. of rectangle (1) about x-x axis,

$$(I_{xx})_1 = (I_G)_1 + A_1 h_1^2$$

$$= \frac{60 \times (12)^3}{12} + (60 \times 12)(78 - 33)^2$$

$$= 0.864 \times 10^4 + 145.8 \times 10^4$$

$$= 146.664 \times 10^4 \text{ mm}^4$$

Rectangle (2)
M.I. of rectangle (2) about x-x axis,

$$(I_{xx})_2 = (I_G)_2 + A_2 h_2^2$$

$$= \frac{12 \times 60^3}{12} + (12 \times 60)(42 - 33)^2$$

$$= (21.6 \times 10^4 + 5.832 \times 10^4) \text{ mm}^4$$

$$= 27.432 \times 10^4 \text{ mm}^4$$

Rectangle (3)
M.I. of rectangle (3) about x-x axis,

$$(I_{xx})_3 = (I_G)_3 + A_3 h_3^2$$

$$= \frac{120 \times (12)^3}{12} + (120 \times 12)(33 - 6)^2$$

$$= 1.728 \times 10^4 + 104.976 \times 10^4$$

$$= 106.704 \times 10^4 \text{ mm}^4$$

Now M.I. of whole section about x-x axis,

$$I_{xx} = (I_{xx})_1 + (I_{xx})_2 + (I_{xx})_3$$

$$= (146.664 + 27.432 + 106.704) \times 10^4$$

$$= 280.8 \times 10^4 \text{ mm}^4$$

or $\qquad 280.8 \text{ cm}^4$

PROBLEM 6.9: Find the moment of inertia of the section shown in Fig. 6.34 about x-x axis.

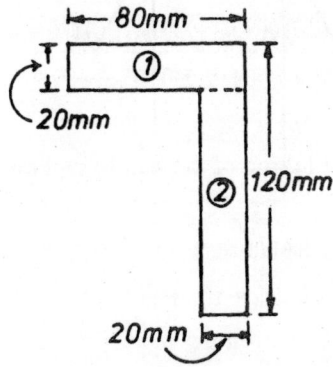

Fig. 6.34 (Illustrative Problem 6.9)

Solution: The section is divided into two rectangles.

Let $\bar{y}$ be the distance between c.g. of the section and the top face.

Rectangle (1)
$$A_1 = 80 \times 20 = 1600 \text{ mm}^2$$

$$y_1 = \frac{20}{2} = 10 \text{ mm}$$

Rectangle (2)
$$A_2 = (120 - 20)(20) = 2000 \text{ mm}^2$$

$$y_2 = \left(20 + \frac{120 - 20}{2}\right)$$

$$= 70 \text{ mm}$$

$$\therefore \quad \bar{y} = \frac{A_1 y_1 + A_1 y_2}{A_1 + A_2} = \frac{1600 \times 10 + 2000 \times 70}{1600 + 2000}$$

$$= \frac{16000 + 140000}{3600}$$

$$= \frac{156000}{3600}$$

$$= 43.33 \text{ mm}$$

Moment of inertia of rectangle (1) about x-x axis,

$$(I_{xx})_1 = (I_G)_1 + A_1 h_1^2$$

$$= \frac{1}{12}(80)(20)^3 + (80)(20)\left(43.33 - \frac{20}{2}\right)^2$$

$$= (5.3333 + 177.7422) \times 10^4 \text{ mm}^4$$

$$= 183.0755 \times 10^4 \text{ mm}^4$$

Moment of inertia of rectangle (2) about x-x axis,

$$(I_{xx})_2 = (I_G)_2 + A_2 h_2^2$$

$$= \frac{1}{12}(20)(100)^3 + (20)(100)(70 - 43.33)^2$$

$$= (166.67 + 142.26) 10^4 \text{ mm}^4$$

$$= 308.93 \times 10^4 \text{ mm}^4$$

∴ Moment of inertia of the whole section about x-x axis,

$$I_{xx} = (I_{xx})_1 + (I_{xx})_2$$

$$I_{xx} = 183.0755 \times 10^4 + 308.93 \times 10^4$$

$$= 492 \times 10^4 \text{ mm}^4$$

or 492 cm^4

PROBLEM 6.10: A strut consists of a T–bar of 60 mm × 40 mm × 6 mm section as shown in Fig. 6.44. Find the moment of inertia and radius of gyration of the cross-section of the bar about a neutral axis (i.e., through centroid parallel to the top face/flange).

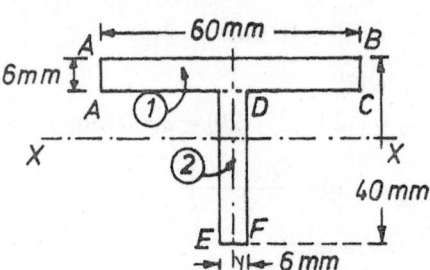

Fig. 6.44 (Illustrative Problem 6.10)

Solution: The section is symmetrical about the y-y axis.

Let $\bar{y}$ be the distance of the centre of gravity of the section from the top AB.

Divide the section into two rectangles.

Rectangle (1)

$$A_1 = 60 \times 6 = 360 \text{ mm}^2$$

$$y_1 = \frac{6}{2} = 3 \text{ mm}$$

Rectangle (2) .

$$A_2 = (40 - 6) \times 6$$

$$= 34 \times 6$$

$$= 204 \text{ mm}^2$$

$$y_2 = 6 + \left(\frac{40 - 6}{2}\right) = 6 + 17 = 23 \text{ mm}$$

$$\bar{y} = \frac{A_1 y_1 + A_2 y_2}{A_1 + A_2}$$

$$= \frac{360 \times 3 + 204 \times 23}{360 + 204}$$

$$= \frac{1080 + 4692}{564}$$

$$= \frac{5772}{564} = 10.234 \text{ mm}$$

M.I. rectangle (1) about x-x axis,

$$(I_{xx})_1 = (I_G)_1 + A_1 h_1^2$$

$$= \frac{1}{12}(60)(6)^3 + (60 \times 6)(10.234 - 3)^2$$

$$= 1080 + 18839.07$$

$$= 19919.07 \text{ mm}^4$$

Moment of inertia of the rectangle (2) about x-x axis,

$$(I_{xx})_2 = (I_G)_2 + A_2 h_2^2$$

$$= \frac{1}{12}(6)(40 - 6)^3 + (40 - 6) \times 6(23 - 10.234)^2$$

$$= 19652 + 33246.03$$

$$= 52898.03 \text{ mm}^4$$

Moment of inertia of the whole section about x-x axis,

$$I_{xx} = (I_{xx})_1 + (I_{xx})_2$$

$$I_{xx} = 72817.10 \text{ mm}^4$$

$$A = A_1 + A_2$$

$$= 360 + 204$$

$$= 564 \text{ mm}^2$$

∴ $$K_{xx} = \sqrt{\frac{I_{xx}}{A}}$$

$$= \sqrt{\frac{72817.10}{564}}$$

$$= \sqrt{129.1083}$$

$$= 11.3626 \text{ mm}$$

PROBLEM 6.11: Find the moment of inertia of a channel section shown in Fig. 6.45 (*a*) about the centroidal x-x and y-y axes.

Solution: The channel section is divided into three rectangles.

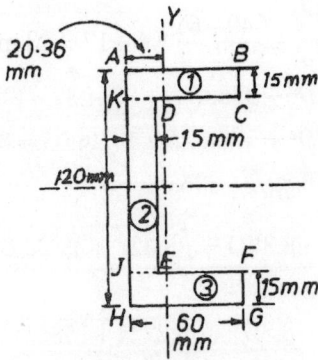

(a) Channel Section (for I_{yy})

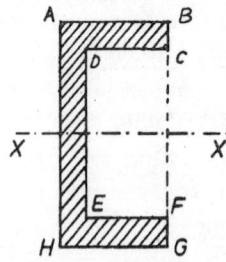

(b) Channel Section (for I_{xx})

Fig. 6.45 (Illustrative Problem 6.11)

Rectangle (1)

$$A_1 = 60 \times 15 = 900 \text{ m}^2$$

$$x_1 = \frac{60}{2} = 30 \text{ mm (from face } AH)$$

Rectangle (2)

$$A_2 = (120 - 15 - 15)(15)$$

$$= 90 \times 15$$

$$= 1350 \text{ mm}^2$$

$$x_2 = \frac{15}{2} = 7.5 \text{ mm (from face } AH)$$

Rectangle (3)

$$A_3 = 60 \times 15 = 900 \text{ mm}^2$$

$$x_3 = \frac{60}{2} = 30 \text{ mm (from face } AH)$$

$$\therefore \quad \bar{x} = \frac{A_1 x_1 + A_2 x_2 + A_3 x_3}{A_1 + A_2 + A_3}$$

$$= \frac{900 \times 30 + 1350 \times 7.5 + 900 \times 30}{900 + 1350 + 900}$$

$$= \frac{27000 + 10125 + 27000}{3150}$$

$$= \frac{64125}{3150}$$

$$= 20.357 \text{ mm}$$

Moment of inertia of rectangle (1) about y-y axis, $(I_{yy})_1$

$$= (I_G)_1 + A_1 h_1^2$$

$$= \frac{15 \times (60)^3}{12} + (15 \times 60) \times (30 - 20.357)^2$$

$$= 27 \times 10^4 + 8.37 \times 10^4$$

$$= 35.37 \times 10^4 \text{ mm}^4$$

M.I. of rectangle (2) about y-y axis,

$$(I_{yy})_2 = (I_G)_2 + A_2 h_2^2$$

$$= \frac{90 \times 15^3}{12} + 90 \times 15 \times (20.357 - 7.5)^2$$

$$= 2.53 \times 10^4 + 22.32 \times 10^4$$

$$= 24.85 \times 10^4 \text{ mm}^4$$

M.I. of rectangle (3) about y-y axis,

$$(I_{yy})_3 = (I_G)_3 + A_3 h_3^2$$

$$= \frac{15 \times (60)^3}{12} + (15 \times 60) \times (30 - 20.357)^2$$

$$= 27 \times 10^4 + 8.37 \times 10^4$$

$$= 35.37 \times 10^4 \text{ mm}^4$$

M.I. of the whole section about y-y axis,

$$I_{yy} = (I_{yy})_1 + (I_{yy})_2 + (I_{yy})_3$$

$$= (35.37 + 24.85 + 35.37)10^4$$

$$= 95.59 \times 10^4 \text{ mm}^4$$

or $\qquad 95.59 \text{ cm}^4$

Moment of inertia about x-x axis

I_{xx} = M.I. of solid rectangle $ABGH$ – M.I. of rectangle $CDFE$

$$= \frac{60 \times (120)^3}{12} - \frac{(60-15)(120-15-15)^3}{12}$$

$$= \frac{60 \times (120)^3}{12} - \frac{45 \times (90^3)}{12}$$

$$= 864 \times 10^4 - 273.375 \times 10^4$$

$$= 590.625 \times 10^4 \text{ mm}^4$$

or $\qquad 590.625 \text{ cm}^4$

PROBLEM 6.12: A plate girder section is built up of two flange plates 360 mm × 10 mm, two web plates 240 mm × 10 mm and four angles 80 mm × 80 mm × 10 mm as shown in Fig. 6.46. Calculate the distance 'D' between the web plates so that the moment of inertia of the section about the axes x-x and y-y which pass through the centre of gravity will be equal. Properties of 80 mm × 80 mm × 10 mm angle are, area = 1505 mm^2; distance of centre of gravity from back of angle = 23.4 mm and $I = 87.7 \times 10^4$ mm^4.

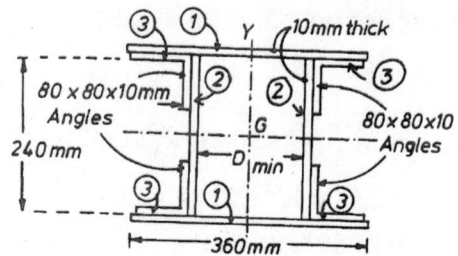

Fig. 6.46 (Illustrative Problem 6.12)

Solution:

$$I_{xx} = 2\left[\frac{1}{12}(36)(1)^3 + (36 \times 1)(13 - 0.5)^2\right]$$

$$+ 2\left[\frac{1 \times 24^3}{12}\right] + 4\left[87.7 + 15.05(12 - 2.34)^2\right]$$

$$= 2[3 + 5625] + 2(1152) + 4(87.7 + 1404.4)$$

$$= 2 \times 5628 + 2 \times 1152 + 4 \times 1492.1$$

$$= 11256 + 2304 + 5968.4$$

$$= 19528.4 \text{ cm}^4$$

$$I_{yy} = 2\left[\frac{1}{12}(1)(36)^3\right] + 2\left[\frac{24 \times 1^3}{12} + (24 \times 1)\left(\frac{D}{2} + 0.5\right)^2\right]$$

$$+ 4\left[87.7 + 15.05\left(\frac{D}{2} + 1 + 2.34\right)^2\right]$$

$$= 7776 + 2[2 + 6(d+1)^2][350.8 + 15.05(D+6.88)^2]$$

$$= 7776 + 16 + 24D + 12D^2 + 1022.37$$
$$+ 201.07D + 15.05D^2$$

$$= 27.05D^2 + 225.07D + 8814.37$$

Since $I_{xx} = I_{yy}$

or $\quad I_{yy} - I_{xx} = 0$

$$27.05D^2 + 225.07D + 8814.37 - 19528.4 = 0$$

$$27.05D^2 + 225.07D - 10714.03 = 0$$

$$D^2 + 8.32D - 396.08 = 0$$

$$D = \frac{-8.32D \pm \sqrt{(8.32)^2 + 4(396.08)}}{2}$$

$$= \frac{-8.32 \pm \sqrt{69.2224 + 1584.32}}{2}$$

$$= \frac{-8.32 \pm \sqrt{1653.5424}}{2}$$

$$= \frac{-8.32 \pm 40.66}{2}$$

Considering +ve sign,

$$D = \frac{-8.32 + 40.66}{2}$$

$$= \frac{32.24}{2}$$

$$= 16.17 \text{ cm}$$

or $\qquad 161.7 \text{ mm}$

PROBLEM 6.13: Find the moment of inertia of hollow rectangular section about its centre of gravity, if the external dimensions are breadth 80 mm and depth 100 mm with 20 mm thickness throughout.

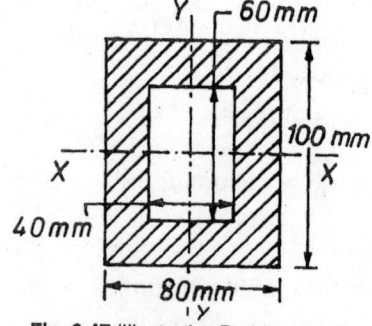

Fig. 6.47 (Illustrative Problem 6.13)

Solution:

$$B = 80 \text{ mm}; D = 100 \text{ mm}$$
$$b = 40 \text{ mm}; d = 60 \text{ mm}$$

$$I_{xx} = \frac{BD^2}{12} - \frac{bd^3}{12}$$

$$= \frac{80 \times 100^3}{12} - \frac{40 \times 60^3}{12}$$

$$= 666.67 \times 10^4 - 72 \times 10^4$$

$$= 594.67 \times 10^4 \text{ cm}^4$$

or $\qquad 594.67 \text{ cm}^4$

$$I_{yy} = \frac{BD^3}{12} - \frac{db^3}{12}$$

$$= \frac{100 \times 80^3}{12} - \frac{60 \times 40^3}{12}$$

$$= 426.67 \times 10^4 - 32 \times 10^4$$

$$= 394.67 \times 10^4 \text{ mm}^4$$

or $\qquad 394.67 \text{ cm}^4$

PROBLEM 6.14: Find the moment of inertia of the shaded (hatched) area about the line *AB*.

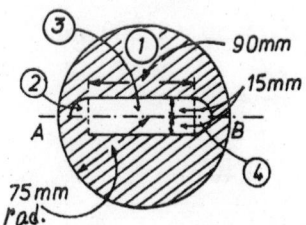

Fig. 6.48 (Illustrative Problem 6.14)

Solution:

M.I. of the hatched portion about *AB*

$\qquad$ = M.I. of the circle about *AB*

$\qquad$ – M.I. of the quarter circle about *AB*

$\qquad$ – M.I. of the rectangle about *AB*

$\qquad$ – M.I. of the semi-circle about *AB*

$$= \left[\frac{\pi}{64}(150)^4 \right] - \left[\frac{\pi}{256}(30)^4 \right]$$

$$- \left[\frac{1}{12} \times 90 \times 30^3 \right] - \left[\frac{\pi}{128}(30)^4 \right]$$

$$= 2485 \times 10^4 - 0.994 \times 10^4 - 20.25$$
$$\times 10^4 - 1.988 \times 10^4$$

$$= 2461.768 \times 10^4 \text{ mm}^4$$

or $\qquad 2461.768 \text{ cm}^4$

PROBLEM 6.15: Find the moment of inertia of the area shown shaded in Fig. 6.49 about the edge *AB*.

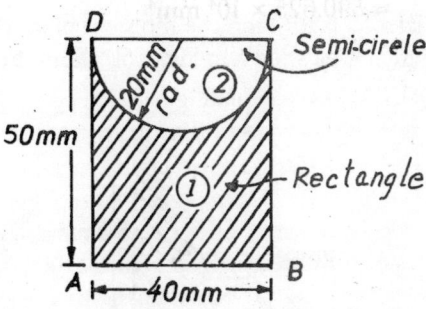

Fig. 6.49 (Illustrative Problem 6.15)

Solution: Moment of inertia of the shaded area about *AB*

= Moment of inertia of the rectangle about *AB*

$\qquad$ –Moment of inertia of the semi-circle about *AB*

Moment of inertia of the rectangle about the base *AB*,

$$= \frac{bd^3}{3} = \frac{40 \times 50^3}{3}$$

$$= 166.67 \times 20^4$$

Moment of the semi-circle about its centre of gravity,

$$I_G = 0.11 r^4$$

$$= 0.11 \times 20^4$$

$$= 1.76 \times 10^4 \text{ mm}^4$$

Distance of centre of gravity of a semi-circle about the base, *AB*

$$h = 50 - \frac{4r}{3\pi}$$

$$= 50 - \frac{4 \times 20}{3\pi}$$

$$= 41.5117 \text{ mm}$$

Moment of inertia of the semi-circle about the base *AB*

$$= I_G + Ah^2$$

$$= 1.76 \times 10^4 + \frac{\pi}{2}(20)^2 (41.5117)^2$$

$$= 1.76 \times 10^4 + 108.27 \times 10^4$$

$$= 110.03 \times 10^4 \text{ mm}^4$$

∴ Moment of inertia of the shaded area about *AB*

$$= 166.67 \times 10^4 - 110.03 \times 10^4$$

$$= 56.64 \times 10^4 \text{ mm}^4$$

PROBLEM 6.16: Find the moment of inertia of the shaded area shown in Fig. 6.50 about the axis *AB*.

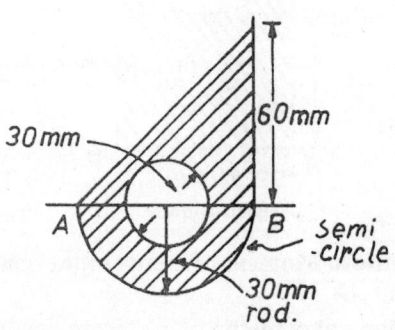

Fig. 6.50 (Illustrative Problem 6.16)

Solution: The section is divided into a triangle, a semi-circle having base on axis *AB* and a circle having its centre on axis *AB*.

Moment of inertia of the shaded area about axis *AB*

= Moment of inertia of the triangle about *AB*

+ Moment of inertia of semi-circle about *AB*

– Moment of inertia of circle about *AB*

$$= \frac{60 \times 60^3}{12} + \frac{\pi \times 60^4}{128} - \frac{\pi}{64}(30)^4$$

$$= 108 \times 10^4 + 31.809 \times 10^4 - 3.976 \times 10^4$$

$$= 135.833 \times 10^4 \text{ mm}^4$$

$$= 135.833 \text{ cm}^4$$

PROBLEM 6.18: Find the second moment of the shaded portion shown in Fig. 6.51 about its centroidal axes.

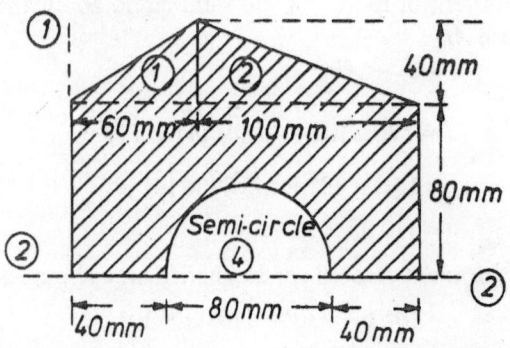

Fig. 6.51 (Illustrative Problem 6.17)

Solution:
Triangle 1

$$A_1 = \frac{1}{2} \times 60 \times 40 = 1200 \text{ mm}^2$$

$$x_1 = \frac{2}{3} \times 60 = 40 \text{ mm}$$

$$y_1 = 80 + \frac{40}{3} = 93.33 \text{ mm}$$

Triangle 2

$$A_2 = \frac{1}{2} \times 100 \times 40 = 2000 \text{ mm}^2$$

$$x_2 = 60 + \frac{100}{3} = 93.33 \text{ mm}$$

$$y_2 = 80 + \frac{40}{3} = 93.33 \text{ mm}$$

Triangle 3

$$A_3 = 160 \times 80 = 12800 \text{ mm}^2$$

$$x_3 = \frac{160}{2} = 80 \text{ mm}$$

$$y_3 = \frac{80}{2} = 40 \text{ mm}$$

Semi-circle (4)

$$A_4 = \frac{1}{2} \times \frac{\pi(80)^2}{4}$$

$$= 2513.27 \text{ mm}^2$$

$$x_4 = 40 + \frac{80}{2} = 80 \text{ mm}$$

$$y_4 = \frac{4 \times 40}{3\pi} = 16.98 \text{ mm}$$

$$\bar{x} = \frac{A_1 x_1 + A_2 x_2 + A_3 x_3 - A_4 x_4}{A_1 + A_2 + A_3 - A_4}$$

$$= \frac{1200 \times 40 + 2000 \times 93.33 + 12800 \cdots}{1200 + 2000 + 12800}$$

$$\frac{\times 80 - 2513.27 \times 80}{-2513.27}$$

$$= \frac{(48 + 186.66 + 1024 - 201.06)10^3}{13.49 \times 10^3}$$

$$= \frac{1057.6 \times 10^3}{13.49 \times 10^3}$$

$$= 78.40 \text{ mm}$$

$$\bar{y} = \frac{A_1 y_1 + A_2 y_2 + A_3 y_3 - A_4 y_4}{A_1 + A_2 + A_3 - A_4}$$

$$= \frac{1200 \times 93.33 + 2000 \times 93.33 + 12800 \times 40}{1200 + 2000 + 12800} \cdots$$
$$\frac{-2513.27 \times 16.98}{-2513.27}$$

$$= \frac{(111.996 + 186.66 + 512 - 42.675)10^3}{13.49 \times 10^3}$$

$$= \frac{767.981}{13.49} = 56.93 \text{ mm}$$

$$I_{xx} = \left[\frac{160 \times 40^3}{36} + \frac{1}{2} \times 160 \right.$$
$$\left. \times 40\left(80 + \frac{40}{3} - 56.93\right)^2 \right]$$
$$+ \left[\frac{160 \times 80^3}{12} + 160 \times 80(56.93 - 40)^2 \right]$$
$$- \left[0.00686 \times 80^4 + \frac{\pi}{2}40^2\left(56.93 - \frac{4 \times 40}{3\pi}\right)^2 \right]$$

$$= (28.44 \times 10^4 + 424.06 \times 10^4) + (682.67 \times 10^4 + 366.88 \times 10^4)$$
$$- [28.10 \times 10^4 + 401.19 \times 10^4]$$
$$= (452.5 + 1049.55 - 429.29) \ 10^4$$
$$= 1072.76 \times 10^4 \text{ mm}^4$$
$$= 1072.76 \text{ cm}^4$$

$$I_{yy} = \left[\frac{40 \times 60^3}{36} + \frac{1}{2} \times 40 \times 60\left(78.40 - \frac{2}{3} \times 60\right)^2 \right]$$
$$+ \left[\frac{40 \times 100^3}{36} + \frac{1}{2} \times 40 \times 100\left(60 + \frac{100}{3} - 78.40\right)^2 \right]$$
$$+ \left[\frac{80 \times 160^3}{12} + 80 \times 160(80 - 78.40)^2 \right]$$
$$- \left[\frac{\pi \times 80^4}{128} + \frac{\pi 40^2}{2}(80 - 78.40)^2 \right]$$
$$= (24 \times 10^4 + 176.95 \times 10^4) + (111.11 \times 10^4 + 44.58 \times 10^4)$$

$$+ (2730.67 \times 10^4 + 3.280 \times 10^4) - (100.53 \times 10^4 + 0.64 \times 10^4)$$
$$= 200.95 \times 10^4 + 155.69 \times 10^4 + 2733.95 \times 10^4 - 101.17 \times 10^4$$
$$= 2989.42 \times 10^4 \text{ mm}^4$$

or $\qquad 2989.42 \text{ cm}^4$

PROBLEM 6.18: Determine the product of inertia of a rectangular area about the x- and y-axes.

Solution: Consider an element of area dA situated at a distance y from the x-axis (Fig. 6.52(a)).

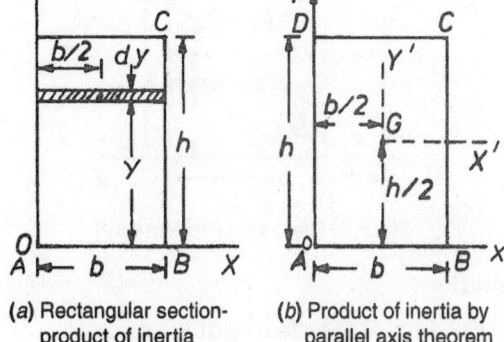

(a) Rectangular section-
product of inertia

(b) Product of inertia by
parallel axis theorem

Fig. 6.52 (Illustrative Problem 6.18)

Area of the element $= b \cdot dy$

$$dI_{xy} = (b \cdot dy)\left(\frac{b}{2}\right)y$$

$$= \frac{b^2}{2} \cdot y \cdot dy$$

$$I_{xy} = \int_0^h \frac{b^2}{2} y \, dy$$

$$= \frac{b^2}{2}\left[\frac{y^2}{2}\right]_0^h$$

$$= \frac{b^2 h^2}{4}$$

Alternatively, we can use parallel axes theorem. Referring to Fig. 6.52 (b)

$$I_{xy} = I_{x'y'} + A\left(\frac{h}{2}\right)\left(\frac{b}{2}\right)$$

$$= 0 + (b \times h)\frac{hb}{4}$$

$$= \frac{b^2 h^2}{4}$$

$[\because I_{x'y'} = 0, x' \& y'$ being the axes of symmetry$]$

PROBLEM 6.19: Determine the product of inertia of the right angled triangle shown in Fig. 6.53 with respect to (i) x, y axis and (ii) a pair of centroidal axes parallel to x, y axes.

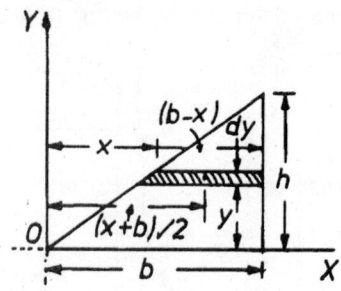

Fig. 6.53 (Illustrative Problem 6.19)

Solution:

(i) **Product of inertia about x, y axes**

Consider an elemental strip of thickness dy at a distance of y- from x-axis.

Length of the strip is $(b - x)$, where by similar

triangles, $x = \dfrac{b}{h} y$

$dA = (b - x)\, dy$

$$= \left(b - \frac{b}{h} y\right) dy$$

$$= b\left(1 - \frac{y}{h}\right) dy$$

The centroid of elemental area is situated at

$\left(x + \dfrac{b-x}{2},\ y\right)$, i.e. $\left(\dfrac{x+b}{2},\ y\right)$

dI_{xy} = Product of inertia of elemental area about

its own axes $+ dA\left(\dfrac{x+b}{2}\right)(y)$

The first term is zero because the elemental area is symmetrical about its centroidal axes.

$$dI_{xy} = 0 + dA\left(\frac{x+b}{2}\right)(y)$$

Substituting for dA & x

$$dI_{xy} = b\left(1 - \frac{y}{h}\right) dy\left(\frac{b}{h}y + b\right)(y)/2$$

$$= \frac{1}{2}b^2\left(y - \frac{y^3}{h^2}\right) dy$$

$$\therefore \quad I_{xy} = \frac{b^2}{2}\int_0^h \left(y - \frac{y^3}{h^2}\right) dy$$

$$= \frac{b^2}{2}\left[\frac{y^2}{2} - \frac{y^4}{4h^2}\right]_0^h$$

$$= \frac{b^2}{2}\left[\frac{h^2}{2} - \frac{h^4}{4h^2}\right]$$

$$= \frac{b^2 h^2}{8}$$

(ii) **Product of inertia about centroidal axes**

$$I_{xy} = \bar{I}_{xy} + Aab$$

$$\frac{b^2 h^2}{8} = \bar{I}_{xy} + \left(\frac{1}{2}bh\right)\left(\frac{2b}{3}\right)\left(\frac{h}{3}\right)$$

$$\frac{b^2 h^2}{8} = \bar{I}_{xy} + \frac{b^2 h^2}{9}$$

$$\bar{I}_{xy} = \frac{b^2 h^2}{8} - \frac{b^2 h^2}{9}$$

$$= \frac{9b^2 h^2 - 8b^2 h^2}{72}$$

$$= \frac{b^2 h^2}{72}$$

PROBLEM 6.20: Determine the product of inertia of a semi-circle shown in Fig. 6.54 about X- and Y-axes.

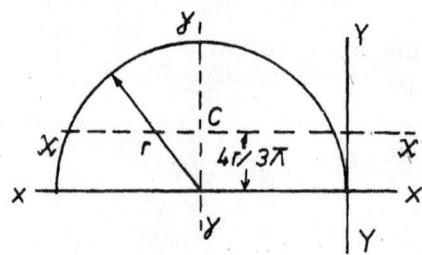

Fig. 6.54 (Illustrative Problem 6.20)

Solution: By the parallel axis theorem for products of inertia.

$$I_{XY} = I_{xy} + Aab$$

$I_{xy} = 0$ as the section is symmetrical about y-y axis.

$$A = \frac{\pi r^2}{2}$$

$$a = -r$$

$$b = \frac{4r}{3\pi}$$

$$\therefore \quad I_{XY} = 0 + \left(\frac{\pi r^2}{2}\right)(-r)\left(\frac{4r}{3\pi}\right)$$

$$= -\frac{2r^4}{3}$$

PROBLEM 6.21: Find the product of inertia of a quarter of a circle with respect to the x- and y-axes.

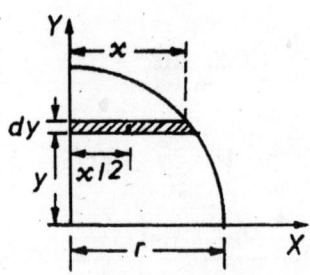

Fig. 6.55 (Illustrative Problem 6.21)

Solution:

Method-I

Consider an element of area dA situated at a distance y from the x-axis (Fig. 6.55).

$$dA = x \cdot dy$$

$$dI_{xy} = (x \cdot dy)\left(\frac{x}{2}\right) y$$

$$= \frac{x^2}{2} y \cdot dy$$

Using the equation of circle

$$x^2 + y^2 = r^2$$

$$x^2 = r^2 - y^2$$

$$\therefore \quad I_{xy} = \int_0^r \left(\frac{r^2 - y^2}{2}\right) y \cdot dy$$

$$= \left[\frac{r^2 y^2}{4} - \frac{y^4}{8}\right]_0^r$$

$$= \frac{r^4}{4} - \frac{r^4}{8}$$

$$= \frac{2r^4 - r^4}{8} = \frac{r^4}{8}$$

$$I_{xy} = \frac{r^4}{8}$$

Method-II

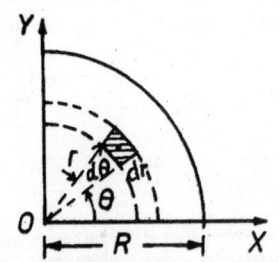

Fig. 6.56 (Illustrative Problem 6.21)

$$I_{xy} = \int_0^{\pi/2} \int_0^R (r d\theta)\, dr\, (r \sin\theta)(r \cos\theta)$$

$$= \int_0^{\pi/2} \int_0^R (r^3 dr)\left(\frac{\sin\theta}{2} d\theta\right)$$

$$= \frac{R^4}{8} \int_0^{\pi/2} \sin 2\theta \cdot d\theta$$

$$= \frac{R^4}{8}\left[-\frac{\cos 2\theta}{2}\right]_0^{\pi/2}$$

$$= \frac{R^4}{8}\left[\frac{1+1}{2}\right]$$

$$= \frac{R^4}{4}$$

$$I_{xy} = \frac{R^4}{8}$$

PROBLEM 6.22: Calculate the product of inertia I_{xy} of the area of a three-quarter circular sector as shown in Fig. 6.57.

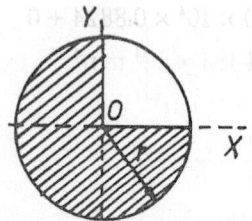

Fig. 6.57 (Illustrative Problem 6.22)

Solution: Area of the given three-quarter circular sector

= Area of the circle – area of the quarter circle

I_{xy} of the three-quarter circular sector

= I_{xy} of the circle – I_{xy} of the quarter circle.

$$= 0 - \frac{r^4}{8}$$

$$= -\frac{r^4}{8}$$

PROBLEM 6.23: Find the I_{xy}, I_{xy} & I_{xy} for the area under the parabolic curve shown in Fig. 6.58.

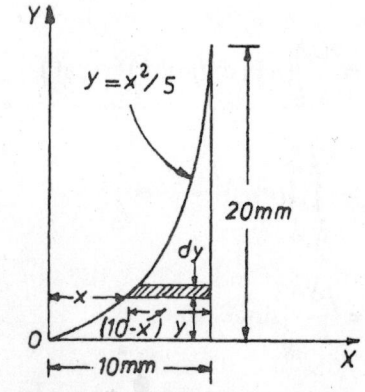

Fig. 6.58 (Illustrative Problem 6.23)

Solution: Consider an elemental area to x-axis as shown in Fig. 6.58. The elemental area,

$$dA = (10 - x)\,dy$$

$$I_{xx} = \int_A y^2\,dA$$

$$= \int_0^{20} y^2(10 - x)\,dy$$

$$= \int_0^{20} y^2\left(10 - \sqrt{5y}\right)dy \qquad \left(\because x = \sqrt{5y}\right)$$

$$= \int_0^{20}\left(10y^2 - \sqrt{5}\,y^{5/2}\right)dy$$

$$= \left[10\frac{y^3}{3} - \sqrt{5}\,\frac{y^{7/2}}{7/2}\right]_0^{20}$$

$$= \left[10 \times \frac{20^3}{3} - 5\frac{20^{7/2}}{7/2}\right]$$

$$= 26667 - 22857$$

$$= 3810 \text{ mm}^4$$

$$I_{yy} = \int_0^{20}(10 - x)\,dy\left(\frac{x+10}{2}\right)^2$$

$$= \frac{1}{4}\int_0^{20}(100 - x^2)\,(10 + x)\,dy$$

$$= \frac{1}{4}\int_0^{20}\left[1000 - 10x^2 + 100x - x^3\right]dy$$

$$= \frac{1}{4}\int_0^{20}\left[1000 - 10 \times 5y + 100\sqrt{5y} - 5y\sqrt{5y}\right]dy$$

$$= \frac{1}{4}\left[1000y - \frac{50y^2}{2} + \frac{100\sqrt{5}\,y^{3/2}}{3/2} - 5\sqrt{5}\,\frac{y^{5/2}}{5/2}\right]_0^{20}$$

$$= \frac{1}{4}[20000 - 10000 + 13333 - 8000]$$

$$= 3833 \text{ mm}^4$$

$$I_{xy} = \int_0^{20}(10 - x)\,dy \cdot y\left(\frac{x+10}{2}\right)$$

$$= \frac{1}{2}\int_0^{20}(100 - x^2)\,y \cdot dy$$

$$= \frac{1}{2}\int_0^{20}(100 - 5y)\,y \cdot dy$$

$$= \frac{1}{2}\int_0^{20}(100y - 5y^2)\,dy$$

$$= \frac{1}{2}\left[100\frac{y^2}{2} - \frac{5y^3}{3}\right]_0^{20}$$

$$= \frac{1}{2}\left[\frac{50 \times 20^2}{2} - \frac{5 \times 20^3}{3}\right]$$

$$= \frac{400}{2}\left[25 - \frac{100}{3}\right]$$

$$= \frac{200 \times (-25)}{3}$$

$$= -\frac{5000}{3}$$

$$= -1667 \, \text{mm}^4$$

PROBLEM 6.24: Find the product of inertia I_{xy} of the rectangle shown in Fig. 6.59.

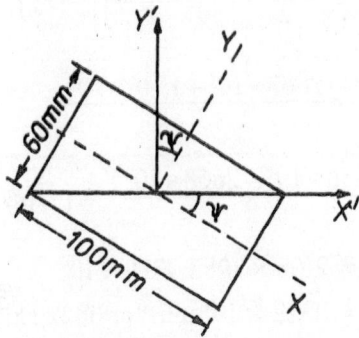

Fig. 6.59 (Illustrative Problem 6.24)

Solution: $I_x = \frac{1}{12} \times 100 \times 60^3$

$$= 180 \times 10^4 \, \text{mm}^4$$

$$I_y = \frac{1}{12} \times 60 \times 100^3$$

$$= 500 \times 10^4 \, \text{mm}^4$$

$$I_{xy} = 0$$

(Since, the rectangle is symmetrical about both x- & y-axes)

$$\tan \psi = \frac{30}{50}, \, \psi = 30°58'$$

$$I_{x'y'} = \frac{I_x - I_y}{2} \sin 2y + I_{xy} \cos 2\psi$$

$$= \frac{180 \times 10^4 - 500 \times 10^4}{2} \sin (2 \times 30°58')$$

$$+ 0 \times \cos (2 \times 30°58')$$

$$= 160 \times 10^4 \times 0.8824 + 0$$

$$= 141.184 \times 10^4 \, \text{mm}^4$$

or $141.184 \, \text{cm}^4$

PROBLEM 6.25: Calculate the angle θ_m defining the direction of principal axes x' and y' through point O for the angle section shown in Fig. 6.60. Each leg of the angle is 12 mm wide.

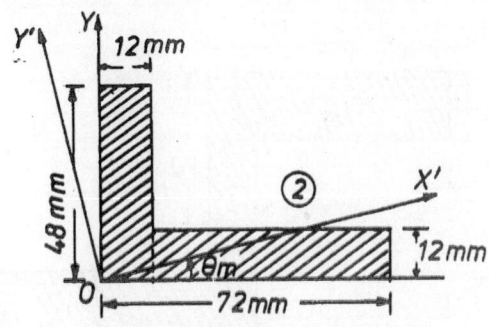

Fig. 6.60 (Illustrative Problem 6.25)

Solution:

$$I_{xx} = \frac{60 \times 12^3}{3} + \frac{12 \times 48^3}{3}$$

$$= 3.456 \times 10^4 + 44.237 \times 10^4$$

$$= 47.797 \times 10^4 \, \text{mm}^4$$

$$I_{yy} = \left[\frac{48 \times 12^3}{3}\right] + \left[\frac{12 \times 60^3}{12} + 60 \times 12\left(12 + \frac{60}{2}\right)^2\right]$$

$$= (2.765 \times 10^4) + (21.6 \times 10^4 + 127.008 \times 10^4)$$

$$= 2.765 \times 10^4 + 148.608 \times 10^4$$

$$= 151.373 \times 10^4 \, \text{mm}^4$$

$$I_{xy} = 48 \times 12 \times 24 \times 6 + 60 \times 12 \times \left(12 + \frac{60}{2}\right) \times 6$$

$$= 8.294 \times 10^4 + 18.144 \times 10^4$$

$$= 26.438 \times 10^4 \, \text{mm}^4$$

$$\tan 2\theta_m = -\frac{2 \times I_{xy}}{I_{xx} - I_{yy}}$$

$$= -\frac{2 \times 26.438 \times 10^4}{47.797 \times 10^4 - 151.373 \times 10^4}$$

$$= +\frac{52.876 \times 10^4}{103.576 \times 10^4}$$

$$= 0.51$$

or $\quad 2\theta_m = \tan^{-1}(0.51) = 27°01' \; \& \; 207°01'$

or $\quad\; \theta_m = 13°30'30'' \; \& \; 103°30'30''$

PROBLEM 6.26: For a z-section shown (Fig. 6.61) the moments of inertia with respect to the x- and y-axes are given to be $I_x = 1548 \times 10^4 \; \text{mm}^4$ and $I_y = 2668 \times 10^4 \; \text{mm}^4$. Determine the principal axes of the section about the O and the values of the principal of the moment of inertia.

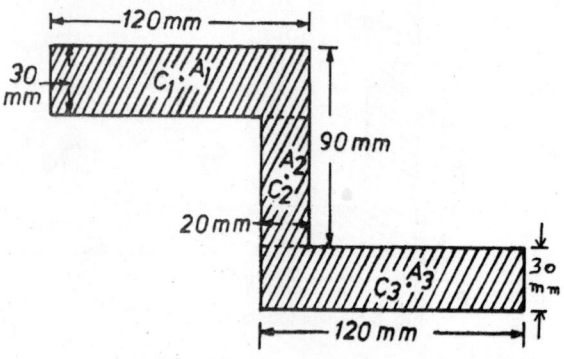

Fig. 6.61 (Illustrative Problem 6.26)

Solution: Area of z-section can be considered to be made up of three rectangles A_1, A_2 and A_3 with three centroids C_1, C_2 and C_3 respectively. Note that C_2 and 0 are coincident points.

Area (mm²)	Distance of the Centroid from x- and y-axes (mm)
$A_1 = 120 \times 30$ = 3600	$\bar{x}_1 = -50$ $\bar{y}_1 = +45$
$A_2 = 20 \times 60$ = 1200	$\bar{x}_2 = 0$ $\bar{y}_2 = 0$
$A_3 = 120 \times 30$ = 3600	$\bar{x}_3 = +50$ $\bar{y}_3 = -45$

Product of inertia of the total area,

$I_{xy} = [0 + (3600)(-50)(45)] + [0+0] + [0 + (3600)(50)(-45)]$

$\qquad = -8100000 - 8100000$

$\qquad = -16200000 \; \text{mm}^4$

(Using parallel axis theorem and the concept that product of inertia vanishes if any one of the axes is the axis of symmetry).

$$\left.\begin{array}{l} I_x = 1548 \times 10^{+4} \; \text{mm}^4 \\ I_y = 2668 \times 10^{+4} \; \text{mm}^4 \end{array}\right\} \quad \text{(given)}$$

For finding the directions of the principal axes,

$$\tan 2\theta_m = \frac{2I_{xy}}{I_y - I_x}$$

$$= \frac{-2(1620 \times 10^4)}{2668 \times 10^4 - 1548 \times 10^4}$$

$$= -2.893$$

$$2\theta_m = -70.93°$$

$$\theta_m = -35.46° \; \& \; \theta_m = 54.54°$$

$$I_{\text{max, min}} = \frac{I_x + I_y}{2} \pm \sqrt{\left(\frac{I_x - I_y}{2}\right)^2 + (I_{xy})^2}$$

$$= \frac{2668 \times 10^4 + 1548 \times 10^4}{2}$$

$$\pm \sqrt{\left(\frac{1548 \times 10^4 - 2668 \times 10^4}{2}\right)^2 + (-1620 \times 10^4)^2}$$

$$= 2108 \times 10^4 \pm 1714 \times 10^4$$

$$I_{\text{max}} = 3822 \times 10^4 \; \text{mm}^4 \text{ or } 3822 \; \text{cm}^4$$

$$I_{\text{min}} = 394 \times 10^4 \; \text{mm}^4 \text{ or } 394 \; \text{cm}^4$$

$$2\theta_m = -70.93° + 360°$$

$$= 289.07° \; \& \; 109.07°$$

$$\therefore \quad \theta_m = 143.535° \; \& \; 54.535°$$

$$I_x = \left[\frac{1}{12}(120)(30)^3 + (120 \times 30 \times 45^2)\right] \times 2 + \frac{1}{12} 20 \times 60^3$$

$$= 1548 \times 10^4 \; \text{mm}^4$$

$$I_y = \left[\frac{1}{12} 30 \times 120^3 + 120 \times 30 \times 50^2\right] \times 2 + \frac{1}{12} 60 \times 20^3$$

$$= 2668 \times 10^4 \; \text{mm}^4$$

PROBLEM 6.27: Determine the radius of gyration of the body shown in Fig. 6.62 about the centroidal x-axis. The grooves are semi-circular with radius 60 mm.

Solution: The composite body may be divided into (1) A solid block of $120 \times 180 \times 150$ mm and (2) Two semi-circular grooves each of radius 60 mm and length 120 mm.

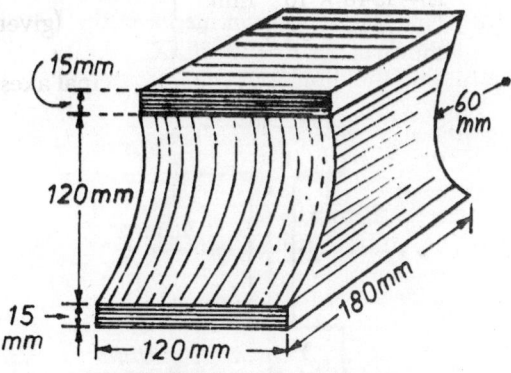

Fig. 6.62 (Illustrative Problem 6.27)

Mass of solid block,

$$M_1 = 120 \times 180 \times 150\rho$$
$$= 3.24 \times 10^6 \rho$$

where, ρ is mass of 1 mm^3 of material.
Its moment of inertia about x-axis.

$$(I_{xx})_1 = \frac{M_1(l^2 + d^2)}{12}$$
$$= \frac{3.24 \times 10^6 \rho \left(180^2 + 150^2\right)}{12}$$
$$= 14.8323 \times 10^9 \rho$$

Semi-circular groove

$$\text{Mass, } M_2 = \frac{\pi r^2 b}{2}\rho$$
$$= \frac{\pi}{2} \times 60^2 \times 120 \times \rho = 0.6786 \times 10^6 \rho$$

Moment of inertia about the axis parallel to x-axis through the centre of semi-circular $= 1/2$ of that of cylinder

$$= \frac{1}{2}\left[2M_2 \times \frac{r^2}{2}\right]$$
$$= \frac{M_2 r^2}{2}$$

Centre of gravity of semi-circular groove from this axis is at a distance,

$$d' = \left(\frac{3}{8}r\right)$$
$$= \frac{3}{8} \times 60 = 22.5 \text{ mm}$$

$\therefore$ Moment of inertia about the axis through centre of gravity I_g is given by

$$\frac{M_2 r^2}{2} = I_g + M_2 \cdot d'^2$$

$$\therefore \quad I_g = M_2\left(\frac{r^2}{2} - d'^2\right)$$

$$= 0.6786 \times 10^6 \rho\left(\frac{60^2}{2} - 22.5^2\right)$$

$$= 8.78 \times 10^8 \rho$$

The distance of this centroid from x-axis is

$$d'' = 90 - 22.5 = 67.5 \text{ mm}$$

$$\therefore \quad (I_{xx})_2 = I_g + M_2 \cdot d''^2$$
$$= 8.78 \times 10^8 \rho + 0.6786 \times 10^6 \rho \times (67.5)^2$$
$$= 8.78 \times 10^8 \rho + 30.92 \times 10^8 \rho$$
$$= 39.70 \times 10^8 \rho = 3.97 \times 10^9 \rho$$

(I_{xx}) of composite body

$$= (I_{xx})_1 - 2(I_{xx})_2$$
$$= 14.823 \times 10^9 \rho - 2(3.970 \times 10^9)\rho$$
$$= 14.823 \times 10^9 \rho - 7.940 \times 10^9 \rho$$
$$= 6.883 \times 10^9 \rho$$

$$M = M_1 - 2M_2$$
$$= 3.24 \times 10^6 \rho - 2 \times 0.6786 \times 10^6 \rho$$
$$= 3.24 \times 10^6 \rho - 1.36 \times 10^6 \rho$$
$$= 1.88 \times 10^6 \rho$$

$$K_{xx} = \sqrt{\frac{I_{xx}}{M}} = \sqrt{\frac{6.883 \times 10^9 \rho}{1.88 \times 10^6 \rho}}$$
$$= 60.47 \text{ mm}$$

PRACTICE PROBLEMS

6.1 Compute the moment of inertia of the shaded area as shown in Fig. 6.63 about x-axis.

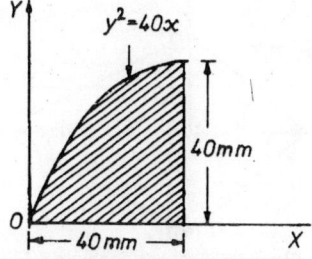

Fig. 6.63 (Practice Problem 6.1)

6.2 Determine the moment of inertia of the shaded area about the x-axis as shown in Fig. 6.64.

6.3 Find the moment of inertia of a rectangular section, 40 mm wide and 60 mm deep, about *x-x* and *y-y* axes passing through the centroid of the section.

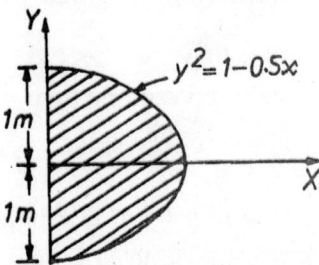

Fig. 6.64 (Practice Problem 6.2)

6.4 Find the moment of inertia of an isosceles triangular section *ABC* with base 60 mm and height 45 mm about *x-x* axis passing through the centre of gravity of the section and parallel to the base *AB*.

6.5 For the rectangle shown in Fig. 6.65, find the moment of inertia about *AA* and *BB*.

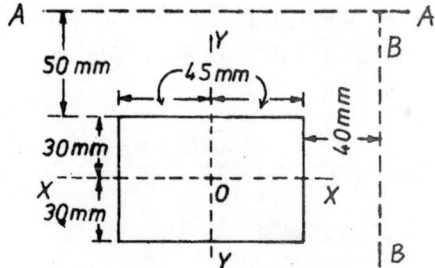

Fig. 6.65 (Practice Problem 6.5)

6.6 For the area shown in Fig. 6.66, determine the second moment area (moment of inertia) about *yy*.

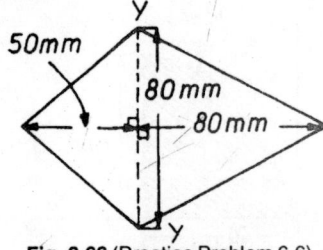

Fig. 6.66 (Practice Problem 6.6)

6.7 Find the polar moment of inertia of the cross-section of the cylinder of 0.60 m radius, about the longitudinal centroidal axis.

6.8 Find the moment of inertia of the I-section shown in Fig. 6.67 about *XX*.

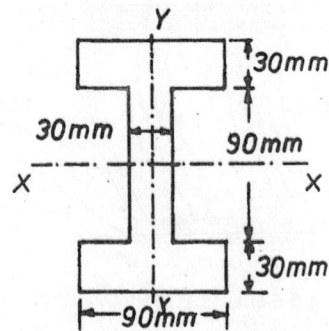

Fig. 6.67 (Practice Problem 6.8)

6.9 Find the second moment of area given in Fig. 6.68 about *XX*

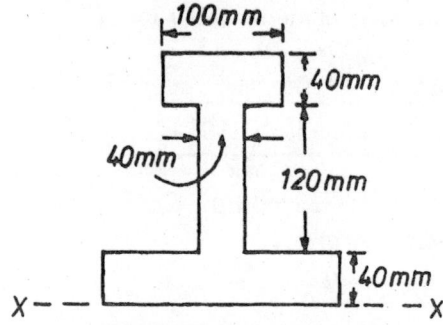

Fig. 6.68 (Practice Problem 6.9)

6.10 Determine the moment of inertia of the L-section shown in Fig. 6.69 about the vertical centroidal axis.

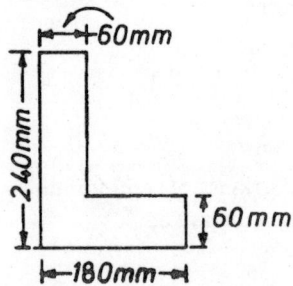

Fig. 6.69 (Practice Problem 6.9)

6.11 An inverted T-section is $120 \times 120 \times 16$ mm as shown in Fig. 6.70. Calculate the moment of inertia of the section about *X-X* axis parallel to the base of the Tee passing through its centre of gravity.

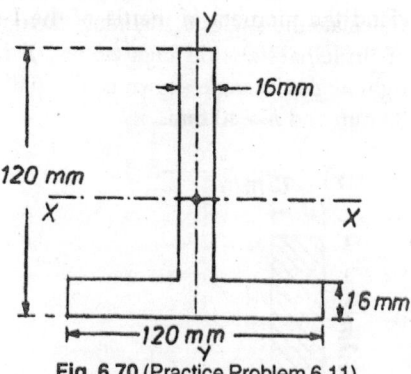

Fig. 6.70 (Practice Problem 6.11)

6.12 Find the moment of inertia of the section given in Fig. 6.71 below, about the axis passing through its c.g. and parallel to the base.

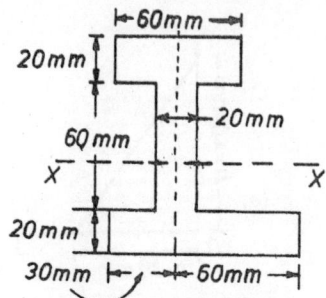

Fig. 6.71 (Practice Problem 6.12)

6.13 A hollow circular section has an external diameter of 60 mm and internal diameter of 40 mm. Find its moment of inertia about the horizontal axis passing through its centre.

6.14 Determine the moment of inertia of the shaded area shown in Fig. 6.72 about the line (1) – (1).

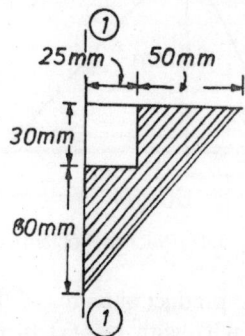

Fig. 6.72 (Practice Problem 6.14)

6.15 Find the moment of inertia of the shaded area about the base *AB* (Fig. 6.73).

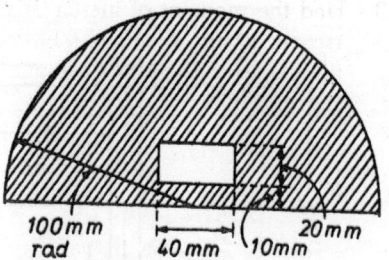

Fig. 6.73 (Practice Problem 6.15)

6.16 Find the moment of inertia of a hollow section shown in Fig. 6.74 about an axis passing through its centre of gravity and parallel to the base.

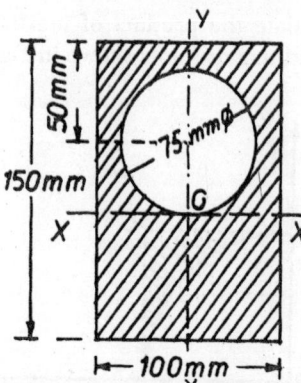

Fig. 6.74 (Practice Problem 6.16)

6.17 Determine the product of inertia of right angle triangle (Fig. 6.75) (*i*) with respect to *x*- and *y*-axis (*ii*) with respect to centroidal axes parallel to *x*- and *y*-axes.

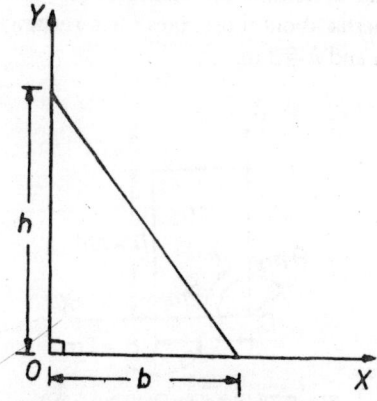

Fig. 6.75 (Practice Problem 6.17)

6.18 Determine the product of inertia I_{xy} of the *z* section shown in Fig. 6.76.

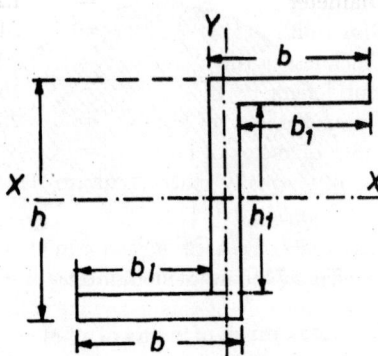

Fig. 6.76 (Practice Problem 6.18)

6.19 Calculate the product of inertia I_{xy} of the shaded spandrel area shown in Fig. 6.77.

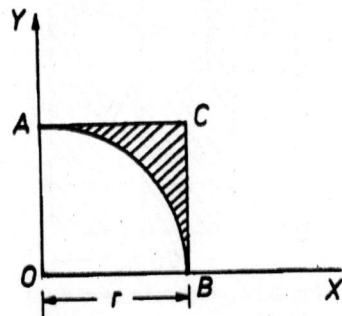

Fig. 6.77 (Practice Problem 6.19)

6.20 Locate the axes of minimum and maximum moments of inertia at the corner of a rectangle and determine the values of the moments of inertia about those axes for a rectangle $b = 2$ m and $h = 3$ m.

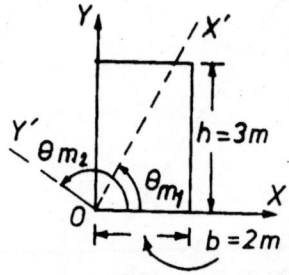

Fig. 6.78 (Practice Problem 6.20)

6.21 Determine the product of inertia of the L-section with respect to the x- and y-axes.

6.22 Calculate the angles θ_m defining the direction of principal axes through point 'O' for the right angled triangle shown in Fig. 6.80 if $b = 30$ mm and $h = 40$ mm.

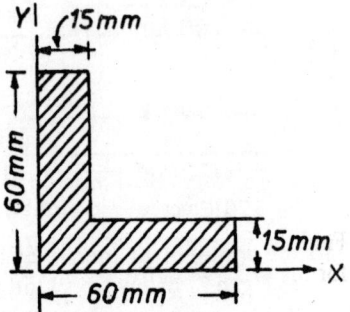

Fig. 6.79 (Practice Problem 6.21)

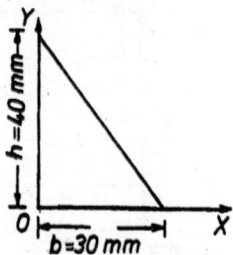

Fig. 6.80 (Practice Problem 6.22)

6.23 Find the product of inertia of the triangle OAB shown in Fig. 6.81 about (i) the axes, OX, OY, (ii) the parallel axes through its centroid.

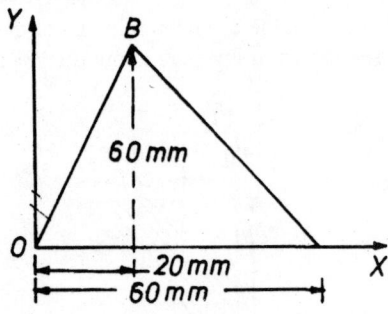

Fig. 6.81 (Practice Problem 6.23)

6.24 Find the product of inertia of the area shown in Fig. 6.82 with respect to the centroidal x- and y-axes. Also find the angle θ defining the directions of principal axes through the centroid and the principal moment of inertia.

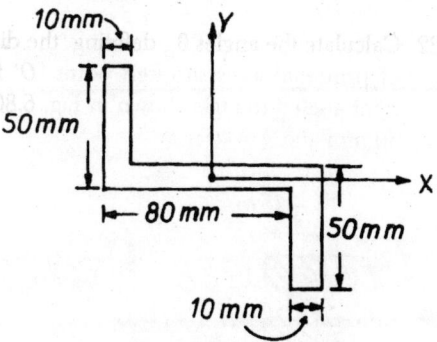

Fig. 6.82 (Practice Problem 6.24)

6.25 Find the moment of inertia about the axis *LL* for the section shown in Fig. 6.83.

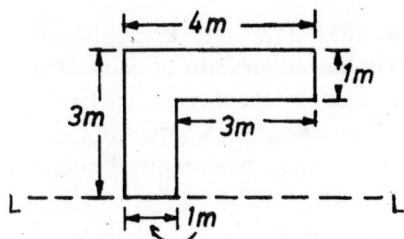

Fig. 6.83 (Practice Problem 6.25)

6.26 A cast iron fly wheel has the following dimensions:

Diameter	=	1.2 m
Rim width	=	240 mm
Thickness of rim	=	40 mm
Hub length	=	160 mm
Outer diameter of hub	=	200 mm
Inner diameter of hub	=	80 mm

Arms: 6 equally spaced uniform slender rods of length 0.46 m

Cross-sectional area of each arm = 5120 mm²
Determine the moment of inertia of the wheel about the axis of rotation. Take mass of cast iron as 7200 kg/m³

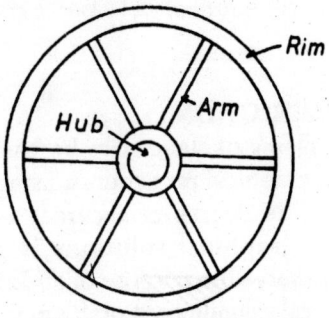

Fig. 6.84 (Practice Problem 6.26)

Chapter 7

Graphic Statics

7.1 INTRODUCTION

Many problems of statics can be easily solved by elegant graphical procedures, using practical geometry. The degree of accuracy obtainable from such graphical solutions is generally adequate to serve the purpose of an Engineer. In case analytical solution for a problem is tedious or is not available, graphical solution will provide a quick and mechanical answer.

Problems involving coplanar force systems, reactions at the supports of structures, internal forces in the members of pin-jointed frames or trusses, location of the centroid/centre of gravity, and determination of moment of inertia of plane areas are amenable to graphical procedures/methods.

All these are collectively treated under the head—"Graphic Statics"—in this chapter.

7.2 BOW'S NOTATION

For designating loads/forces (external or internal), a two letter notation for each force is adopted in graphic statics; this is based on the fact that the line of action of each force divides the plane in which it acts into two 'spaces' one on each of its sides. Each of the spaces is marked by a letter symbol like A, B, ..., etc. In the vector representation of the force-diagram, also called the vector-diagram, each force is represented by a line parallel to its direction with a length to a convenient vector scale (like, say 1 cm = p newtons), the ends of the vector being named a, b, ..., etc.,

corresponding to the letter symbols used to mark the spaces on either side of each of the forces (Fig. 7.1). This system of graphical representation of the forces is known as 'Bow's notation', which is very convenient in drawing force diagrams/vector diagrams.

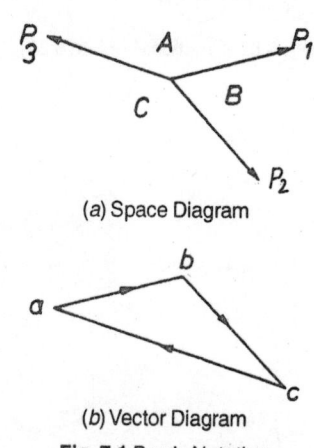

(a) Space Diagram

(b) Vector Diagram

Fig. 7.1 Bow's Notation

Fig. 7.1 (a) in which the forces are represented in their correct orientation/positions is called the 'space diagram', and (b) shows the 'vector diagram', with appropriate Bow's notation in each of these, drawn to convenient scales, called 'space scale' and 'vector scale'.

(Space scale will be some thing like 1 cm = s metres and vector scale will be some thing like 1 cm = p newtons).

[Fig. 7.1 shows the case of three concurrent forces in equilibrium].

7.3 FUNICULAR POLYGON

It was seen in Chapter 2 that the resultant of a system of coplanar concurrent forces may be determined completely by the polygon of forces i.e., the magnitude, direction as well as the line of action of the resultant can be obtained.

However, the line of action of resultant of a coplanar, nonconcurrent force system cannot be determined, but only its magnitude and direction, by using the principle of force polygon /vector diagram with the aid of Bow's notation. For this purpose, another concept is necessary— that of the funicular polygon.

Let the force system be as shown in Fig. 7.29 (*a*), drawn to a suitable space scale (1 cm = s m), acting on rigid body. Bow's notation is given to represent the forces.

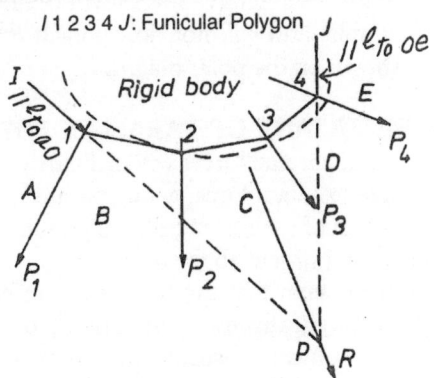

(*a*) Force System (Space Diagram) and Funicular Polygon

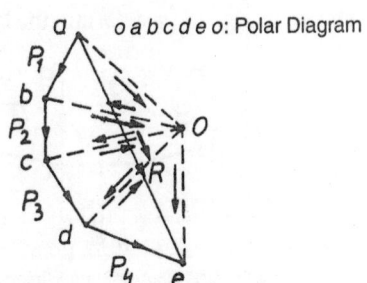

(*b*) Force Diagram and Polar Diagram

Fig. 7.2: Resultant of a Non-Concurrent Force System Funicular Polygon

The force diagram is drawn to a convenient vector or force scale (1 cm = p newtons, say) as usual as shown in Fig. 7.2 (*b*). The resultant of the force system is the vector $\overline{ae}$, which gives its magnitude and direction. A direct and straightforward approach to determine the line of action (or point of application) of the resultant R is as follows:

Let R_1 be the resultant of P_1 and P_2, which is given by $\overline{ac}$, its line of action shall pass through the point of intersection of P_1 and P_2. This line of action of R_1 and that of P_3 may be extended to meet to give the point through which the resultant R_2 of R_1 and P_3 shall pass; R_2 is given by vector $\overline{ad}$. Similarly, the resultant R of the system is given by that of R_2 and P_4. The corresponding vector is $\overline{ae}$; the line of action of R shall pass through the inter-section of those of R_2 and P_4.

Obviously this approach is laborious and the graphical errors tend to accumulate.

To obviate this, an ingenious method has been devised to obtain the line of action of the resultant R as given below:

The method consists in replacing the given force system by an equivalent system of just two forces which are known completely; the resultant R must necessarily pass through the point of intersection of the lines of action of these two force components. This is achieved by means of what is called the 'Polar diagram'.

Any point O, outside the force diagram, is chosen (Fig. 7.2) and joined to the vertices of the force diagram–a, b, c, d and e. The point O is called the 'Pole' and Oa, Ob, Oc, Od and Oe the 'rays'; this part of the diagram is called the 'Polar diagram'. Inspection of this reveals that each of the forces is resolved into two components– $\overline{ab}$ into $\overline{ao}$ and $\overline{ob}$, $\overline{bc}$ into $\overline{bo}$ and $\overline{oc}$ and so on.

Now lines I-1, 1-2, 2-3, 3-4 and 4-J are drawn respectively parallel to rays Oa, Ob, Oc, Od and Oe; that is, these lines are drawn in the corresponding spaces A, B, C, D, and E. Lines I-1, 1-2, ... are called 'strings' and the figure I-1-2-3-4-J is called, the 'string polygon', 'link polygon', or 'funicular polygon'. The end strings I-1 and J-4 are produced to meet at a point p, which is a point on the line of action of the resultant R. This can be understood from the physical significance of the polar diagram and the funicular polygon as follows:

The polar diagram is just a device to split each force into two oblique components. For example, force P_1, represented by vector $\vec{ab} = \vec{ao} + \vec{ob}$.

Force P_2, represented by $\vec{bc}$ is

$$\vec{bc} = \vec{bO} + \vec{Oc}$$

Similarly, P_3, represented by $\vec{cd}$ is

$$\vec{cd} = \vec{cO} + \vec{Od}$$

and P_4, represented by $\vec{de}$ is

$$\vec{de} = \vec{dO} + \vec{Oe}$$

Thus it may be observed that, in the polar diagram:

(*i*) The force system is split up into concurrent vectors, double the number of forces; and

(*ii*) except the first and the last vector, all others cancel each other; the resultant therefore is that of the first and the last vectors, i.e., $\vec{ae}$ in this case, which represents the resultant R in magnitude, direction and sense.

In the funicular polygon, each of the forces is split into its oblique components at the respective vertices. The resultant R, being the sum of vectors $\vec{ao}$ and $\vec{oe}$, is seen to be reduced to a statically equivalent one of only two forces, parallel to $a\vec{O}$ and $O\vec{e}$, in the funicular polygon, as the first and last rays–*I*-1 and 4-*J*. (The force system is reduced to a system of eight components at 1, 2, 3 and 4, of which the force pairs acting along the intermediate strings cancel out each other). The intersection of the first and last strings is therefore a point on the line of action of the resultant.

The characteristics of the polar diagram are:

(1) Each force of the system is split up into two oblique components.

(2) Any two components when shown on the space have to intersect.

(3) For every pair of adjacent forces of the system has a pair of components, represented on the polar diagram by a single ray, only the sense of the vector differing for the two components; these two act on the same string of the funicular polygon which lies between those two forces, and cancel each other.

The characteristics of the funicular polygon are:

(1) Each vertex of it lies on the line of action of a force of the given system.

(2) Every string is parallel to the corresponding ray of the polar diagram.

(3) The given force system reduces to two forces, one acting along each end string.

(4) The locations of the lines of action of the various forces and their components are given by the corresponding vertices and strings of the funicular polygon.

Note: Unless the strings are continued in one sequence, the resultant of the components of two consecutive forces will reduce to a couple, and will not cancel each other as shown on the polar diagram.

7.4 RESULTANT OF PARALLEL FORCES

A system of parallel forces will ordinarily reduce to a single resultant force, acting parallel to each of the forces of the system, the magnitude being the algebraic sum of all the forces.

We have seen that the line of action can be located analytically using Varignon's theorem; now the graphical approach using force polygon and funicular polygon will be studied.

The system may be 'like' or 'unlike'; the former is shown in Fig. 7.3 (*a*) & (*b*) and the latter in (*c*) and (*d*).

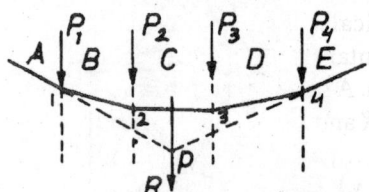

(a) Space Diagram and Funicular Polygon

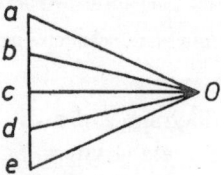

(b) Force Diagram and Polar Diagram

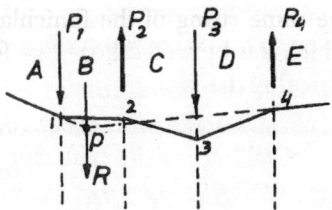

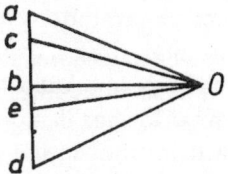

(*c*) Space Diagram and Funicular Polygon

(*d*) Force Diagram and Polar Diagram

Fig. 7.3: Resultant of Parallel Force Systems; (*a*) & (*b*) Like parallel forces (*c*) & (*d*) Unlike parallel forces

In both cases, the force diagram is a straight line parallel to the forces. The vector $\overline{ae}$ gives the resultant in magnitude, direction and sense. (Obviously, $\overline{ae}$ is the algebraic sum of all the forces).

The line of action of the resultant is obtained by the polar diagram and funicular polygon, as usual; the intersection of the first and last strings of the funicular polygon gives a point on the line of action of the resultant, as shown in the figure.

7.5 LOCATION OF CENTROID AND CENTRE OF GRAVITY—GRAPHICAL METHOD

The centroid of a plane figure may be located graphically by dividing it into a number of elementary areas, the centroids of which are known. A suitable reference set of axes is chosen— say OX and OY. Let the elementary areas be A_1, A_2, A_3, ..., and the coordinates of their centroids be (x_1, y_1), (x_2, y_2), (x_3, y_3) and so on, with respect to those axes.

Now, if $(\bar{x}, \bar{y})$ are the co-ordinates of the centroid of the whole area A, it has been shown analytically that

$$\bar{x} = \frac{\Sigma A_1 x_1}{A} \quad \text{and} \quad \bar{y} = \frac{\Sigma A_1 y_1}{A}$$

In the graphical method, the concepts of force diagram and funicular polygon are used, treating

areas A_1, A_2, A_3, ..., as coplanar forces acting through their respective centroids. For locating the x-axis through the centroid of the area, the elementary areas are treated as forces acting in the positive direction of the x-axis. The force diagram and the funicular polygon are drawn to convenient scales, and the intersection of the first and last string of the latter gives a point on the x-axis, through the centroid, which is now drawn parallel to the force vectors.

The y-axis through the centroid of the area is similarly obtained by repeating the above procedure, treating the elementary areas as forces acting in the positive direction of the y-axis.

The intersection of the x- and y-axes locates the centroid of the whole area (Fig. 7.4).

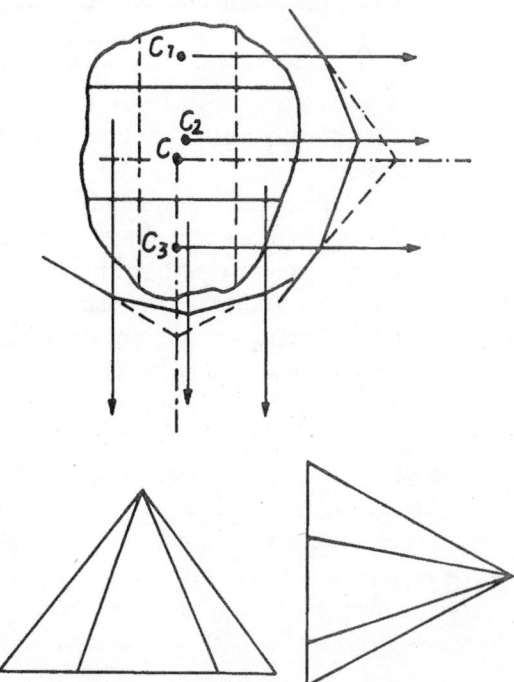

Fig. 7.4: Graphical Determination of Centroid

If there is an axis of symmetry for the area, the centroid lies on it, and the work is reduced to the other direction alone.

The centre of gravity of a body may be located graphically by locating the line of action of the resultant of a number of parallel forces, which are the weights of the elements of which the body is

146 Statics

composed; obviously, the force diagram and the funicular polygon will be the aids for this purpose too.

For plane laminar bodies, this procedure has to be carried out twice, once taking the forces of individual parts acting vertically through their centres of gravity, and again acting horizontally. The intersection of these two lines of action will be the centre of gravity.

For a linear body with uniform cross-section, its weight may be taken being distributed uniformly along its length, or its weight per unit length is constant. Similarly for a laminar body made of homogeneous material and of constant thickness, its weight per unit area may be taken as constant. If these constants are taken as unity for convenience, the length of elements of linear bodies and area of laminar bodies automatically represent their weights. In such a case, the centre of gravity of any such body will be the location of the centroid of the figure defined by its geometrical shape.

The body may be split up judiciously into as few parts as possible, so that the weight and location of C.G. for each part is known from standard values. For bodies of irregular shape, it is necessary to split into a number of small parts of approximately regular shape; a certain degree of error is inevitable, but it can be minimised by adopting greater number of parts.

If the body has any cut away portions, these can be taken as parts with negative weights (acting opposite to the other parts), and the procedure can be applied in the normal way.

The procedure becomes clear from the illustrative problem Nos. 7.8 to 7.9.

7.6 GRAPHICAL CONDITIONS OF EQUILIBRIUM

If a system of coplanar forces keep a body in equilibrium, the conditions to be fulfilled is that the resultant of the system shall reduce to zero.

In general, a coplanar force system will reduce either to

(i) a single resultant R, or
(ii) a single couple L, or

(iii) a zero resultant, leading to equilibrium of the body on which the system acts.

Graphically, the condition of equilibrium means the closure of the force diagram/polygon, the last point falling on the first, thus ensuring that the possibility (i) does not occur. But this does not ensure that the possibility (ii) does not occur, since a couple consists of equal and opposite forces separated by a certain perpendicular distance, thus satisfying (i), but it cannot keep the body in equilibrium.

Let us see what further is required to ensure equilibrium, with respect to the coplanar force system shown in Fig. 7.5.

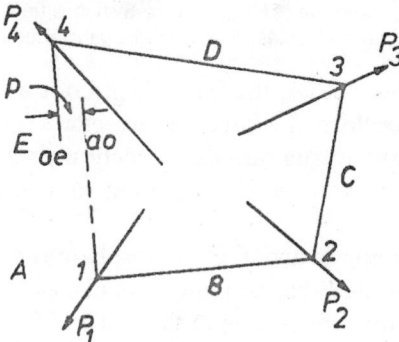

(a) Space Diagram and Funicular Polygon

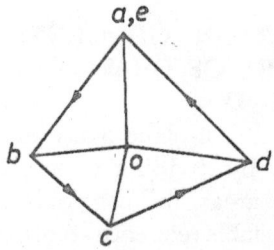

(b) Force Polygon and Plan Diagram

Fig. 7.5: Graphical Conditions of Equilibrium—Closure of Funicular Polygon

For the system shown, the force polygon closes as shown in (b), e coinciding with a. Next the polar diagram is drawn with an arbitrary pole O. The funicular polygon is started from an arbi-trarily chosen print 1 on the force P_1 and drawn as 1-2-3-4. But the string parallel to $\overline{ao}$ through 1 and that parallel to $\overline{oe}$ through 4 do not coincide, but appear separated by a perpendicular distance p shown in (a).

Recalling the discussion of the polar diagram and funicular polygon in Sec. 7.3 (Fig. 7.2), the resultant is the vector sum of the first and last strings of the funicular polygon. In Fig. 7.5 (a), therefore, the resultant appears as a couple (equal to opposite forces, separated by a lever arm), which cannot keep the system in equilibrium. Thus the closure of the force polygon/diagram does not ensure equilibrium as can be seen from this situation.

It becomes necessary that the two components $\overline{ao}$ and $\overline{oe}$ into which the whole system reduces (by the polar diagram); besides being equal and opposite (as seen from the polar diagram), must act along the same line in the funicular polygon to balance each other and keep equilibrium.

Hence the second condition necessary for equilibrium is that the funicular polygon must *close*, the first and last strings falling on each other, providing a common line of action for the equal and opposite force components $\overline{ao}$ and $\overline{oe}$.

To summarise—the graphical conditions of equilibrium are:

(1) the force diagram/polygon must close, and

(2) the funicular diagram must *also* close.

7.7 SUPPORT REACTIONS

An important problem involving equilibrium of a coplanar force system is the determination of the reactions at the supports of beams and frames. The external load/force system along with the reactions induced at the supports shall keep the beam or frame in equilibrium. The reactions induced will depend upon the nature of the external loads (vertical or inclined) and the kind of supports (roller, hinge, or fixed support). As seen in the previous section, the force diagram as well as the funicular polygon must close for the force system including the reactions.

The reaction at a roller support is always normal to the plane in which motion is permitted and so the string of the funicular polygon can start or end any where on its line of action.

The reaction at a hinge can be in any direction through the hinge, depending upon the external load system; however, the reaction shall always pass through the hinge. Hence the string of the funicular polygon there should be made to pass through it. This will obviously obviate the need to know the direction of the reaction before hand.

Now a few specific types of beam/frame problems involving the graphical determination of reactions will be seen:

Case 1: A simple beam with only vertical loads (Fig. 7.6)

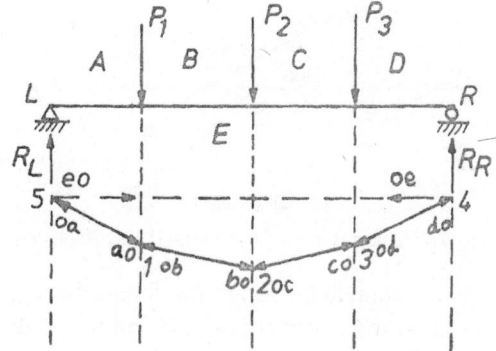

(a) Space Diagram and Funicular Polygon

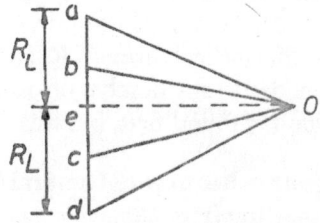

(b) Force Diagram and Polar Diagram

Fig. 7.6: Simple Beam with Vertical Loads-Reactions

The closing line of the funicular polygon is obtained and Oe is drawn parallel to it; e then divides the total load into the reactions R_L $(\overline{ea})$ and R_R $(\overline{de})$. The first string of the funicular polygon may be drawn from any point on the vertical through the left support L (since the reactions are known to be vertical) and allowed to end on a vertical through the right support R.

Case 2: A simple beam with an inclined load (Fig. 7.7)

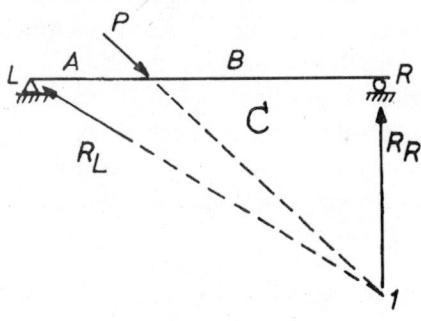

(a) Space Diagram

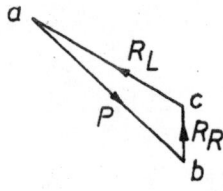

(b) Force Diagram

Fig. 7.7: Simple Beam with an Inclined Load -Reactions

The reaction at the hinge L will pass through L and that at the roller support R will be vertical. Since only three forces are involved, the force diagram will be a triangle as shown, the direction of the reaction at the hinge being got by joining L to the intersection 1 of the line of action of force P and the vertical through R (The principle may be extended to any number of loads provided their resultant is found first; but this is tedious).

Case 3: Simple Beam with General Coplanar Loading (Fig. 7.8)

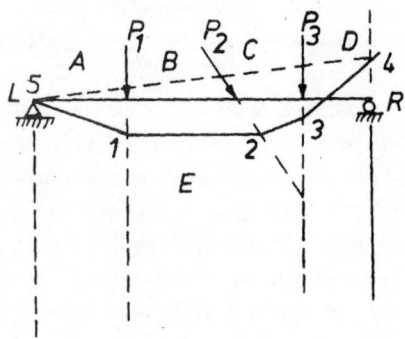

(a) Space Diagram and Funicular Polygon

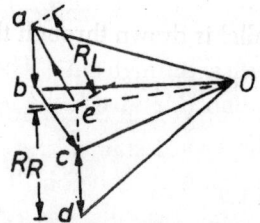

(b) Force Diagram and Polar Diagram

Fig. 7.8: Simple Beam with General Coplanar Loading-Reactions

Since the reaction R_L is known to pass through the hinge L, the funicular polygon is started from L itself, and allowed to end on the vertical through R at 4. The closing line of the funicular polygon 4-L.

A parallel to this closing line is drawn through O to meet the vertical through d in e; ea is joined, then the vector $\vec{de}$ is R_R and $\vec{ea}$ is R_L.

Case 4: A pin-jointed plane frame with general loading (Fig. 7.9)

Since L is a hinge, the first string is made to pass through it and the last string to intersect a vertical through the horizontal roller R, and the closing line of the funicular polygon is 3-L, to

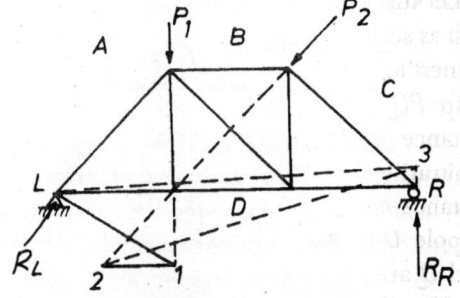

(a) Space Diagram and Funicular Polygon

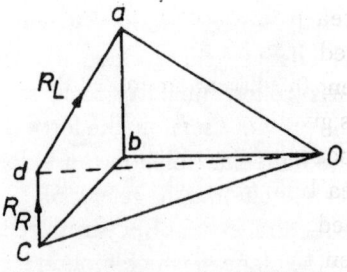

(b) Force Diagram and Polar Diagram

Fig. 7.9: A Pin-Jointed Plane Frame with General Loading-Reactions

which a parallel is drawn through the pole O to intersect a vertical through c in d. Then the vector $\vec{da}$ is the reaction R_L and $\vec{cd}$ is the reaction R_R.

7.8 GRAPHICAL METHOD FOR MOMENT OF INERTIA

A graphical method is available for the determination of the first moment as well as the second moment of area (area moment of inertia) of a plane lamina (or areas about an axis in its plane). This is explained below (Fig. 7.10).

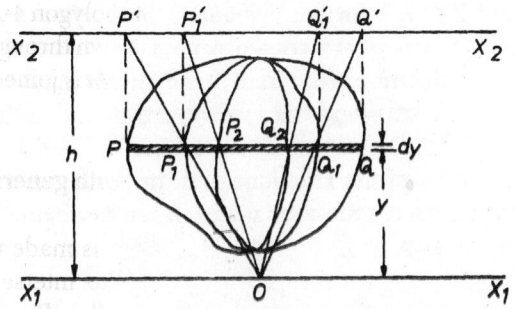

Fig. 7.10: Graphical Method for Moment of Inertia

Let the area be A and let its first moment as well as second moment of area (or area moment of inertia) be required about the axis X_1X_1. Let a strip PQ of thickness dy be considered at a distance y. Let P and Q be projected on to an arbitrarily chosen axis X_2X_2, parallel to X_1X_1, at a distance h from it. Let these points be P' and Q'. A pole O is chosen on X_1X_1 and points like P' and Q' are joined to O, to intersect PQ in P_1 and Q_1. Thus for every strip PQ there will be a pair of points P_1 and Q_1. By considering several strips, the area bounded by points like P_1, Q_1, is obtained; let it be A_1.

Then, the first moment of the area A about X_1X_1 is given by $A_1 \cdot h$.

Repeating the same procedure on the area A_1, an area bounded by points like P_2 and Q_2 is obtained; let it be A_2.

Then, the second moment of the area A about the axis X_1X_1 is given by $A_2 \cdot h^2$.

The proof is simple.

Referring to Fig. 7.10,

$\dfrac{P_1Q_1}{PQ} = \dfrac{P_1Q_1}{P'Q'} = \dfrac{y}{h}$ (from similar triangles OP_1Q_1 and $OP'Q'$).

or $PQ \cdot y = P_1Q_1 \cdot h$

Multiplying both sides by dy,

$PQ \cdot y\,dy = P_1Q_1 \cdot h \cdot dy$

or $y - da = h\,da_1$ (da and da_1, are the areas of elemental strips PQ and P_1Q_1 of thickness dy).

The first moment of the area A about X_1X_1 is the algebraic sum of the moments of the elemental area da about X_1X_1.

$\therefore \quad A\bar{y} = \sum y \cdot da = \sum h \cdot da_1 = h\sum da_1 = A_1 h$

($\bar{y}$ is the distance of centroid of the area A from AB)

Similarly,

$\dfrac{P_2Q_2}{P_1Q_1} = \dfrac{P_2Q_2}{P_1'Q_1'} = \dfrac{y}{h}$

similar triangles OP_2Q_2 and $OP_1'Q_1'$)

$\therefore \quad P_2Q_2 = \dfrac{y}{h}P_1Q_1 = \dfrac{y}{h} \cdot \dfrac{y}{h} \cdot PQ = \dfrac{y^2}{h^2} \cdot PQ$

$Y^2 \cdot PQ = h^2 \cdot P_2Q_2$

Multiplying both sides by dy,

or $y^2 \cdot da = h^2 \cdot da_2$, da_2 being the area of the elemental strip P_2Q_2 of thickness dy.

But

$I_{x_1x_1} = \sum y^2 da = \sum h^2 da_2 = h^2 \sum da_2 = A_2 \cdot h^2$

Thus the proposition is proved.

(What is achieved by the graphical construction is to reduce the width of the strip in the ratio y/h).

A_1 is called the 'Modulus Figure' of area A, and the area A_2, that of area A_1.

It is rare that the graphical method is necessary for regular geometrical figure; so it is considered useful in the case of irregular areas.

Even for regular figures like rectangle and triangle one can apply the graphical approach and derive the moments of inertia about any axis partly analytically, in place of totally analytical approach.

7.9 GRAPHICAL METHOD FOR FORCES IN MEMBERS OF TRUSSES— MAXWELL DIAGRAMS

In Chapter 4, we have seen how to determine the forces in the members of pin-jointed frames/ trusses analytically, either by the method of joints or the method of sections.

In the method of joints, the equilibrium of each joint, treated as a free body, under the action of external loads and internal forces at the joint, is considered; the unknown internal forces (not more than two at a time) at the joints are solved by applying the laws of static equilibrium.

$$\Sigma P_x = 0 \text{ and } \Sigma P_y = 0$$

Alternatively, each joint can be solved graphically by drawing the force polygon for the forces acting at the joint.

This graphical solution for each joint may be sketched in a combined manner for the entire frame/truss, drawn in an appropriate sequence. The resulting force diagram for the whole truss is known its "Maxwell Diagram".

A prerequisite step is to evaluate the reactions at the supports of the truss, so that the number of unknown forces at any joint does not exceed two. This may be done either analytically, or graphically. Three types of trusses will be illustrated:

(1) Simply Supported Truss
(2) Cantilever Truss
(3) A truss with more than two unknown forces at a joint. (Compound Truss).

(1) Simply Supported Truss

A Warren Truss and the Maxwell Diagram for it are shown in Fig. 7.11 and Fig. 7.12.

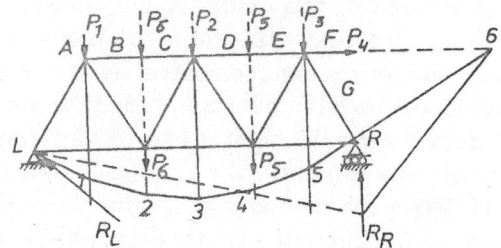

(a) Space Diagram and Funicular Polygon

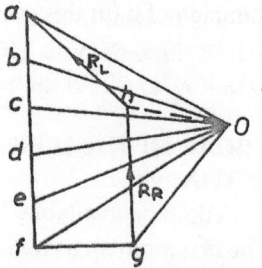

(b) Force Diagram and Polar Diagram

Fig. 7.11: Warren Truss—Graphical Determination of Reactions

Fig. 7.11 for the determination of the reactions and Fig. 7.12 for the evaluation of the forces in the members are shown separately, since the Bow's notation is different in these two cases.

If the truss and the loading are such that the same Bow's notation is convenient for the determination of the reactions and the forces in the members, the Maxwell diagram can be obtained in one step.

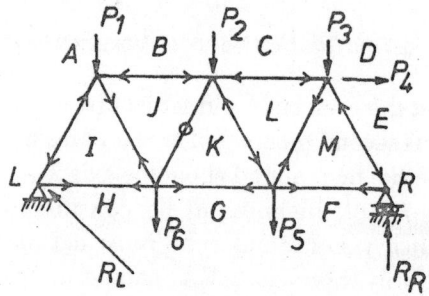

(a) Truss with Nature of Forces in the Members

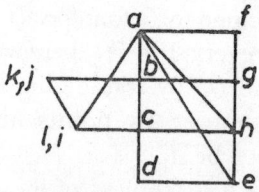

(b) Maxwell Diagram for Forces in Members

Fig. 7.12: Warren Truss—Maxwell Diagram

Note: Bow's notation is given clockwise.

The important point to be noted in the Maxwell diagram is that, to ascertain the nature of the force in any member, the member shall be read in the Maxwell diagram in the direction in which Bow's notation is given around one of the joints on either side of it (in this case, clock-wise), and the arrow head marked in that direction at the particular joint. When the equilibrium of the other joint is considered, the same internal force appears in the reverse direction-leading, in both cases, to an internal force towards the joint (compression), or away from the joint (tension). It is for this reason that no arrow heads are shown on the Maxwell diagram.

The magnitudes of the internal forces are scaled off from the Maxwell diagram.

(2) Cantilever Truss

A cantilever truss—a bracket with truss–support—with vertical loads—is shown along with the Maxwell diagram in Fig. 7.13. Determination of reactions is not necessary here.

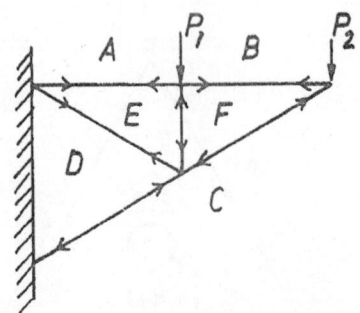

(*a*) Space Diagram with Bow's Notation and Nature of Forces in Members

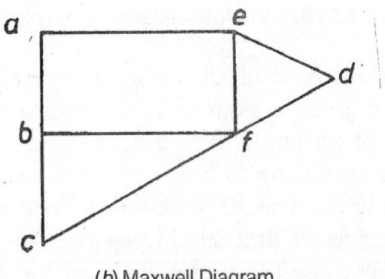

(*b*) Maxwell Diagram

Fig. 7.13: Cantilever Truss—Maxwell Diagram

(3) Compound Truss (Method of Substitution)

In some trusses, the members are so arranged that there are more than two unknown forces at a joint. Such problems are solved by substituting or replacing some members by an imaginary member so as to reduce the unknown forces at a

joint to two only. After solving this joint and reaching a certain stage of the Maxwell diagram, the imaginary member is replaced by the actual members and worked back to obtain relevant points and the forces in these. The procedure is then continued in the usual way to complete the Maxwell diagram for the whole truss.

An example is shown in Fig. 7.14.

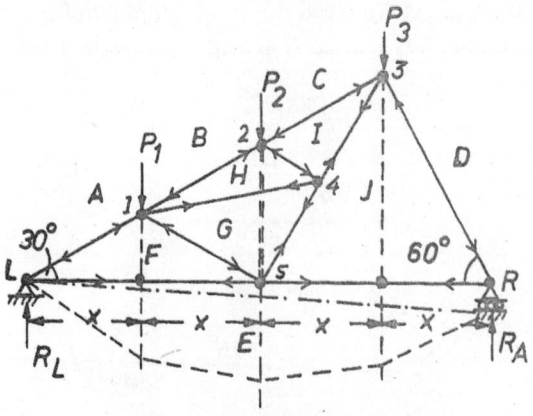

(*a*) Space Diagram and Funicular Polygon

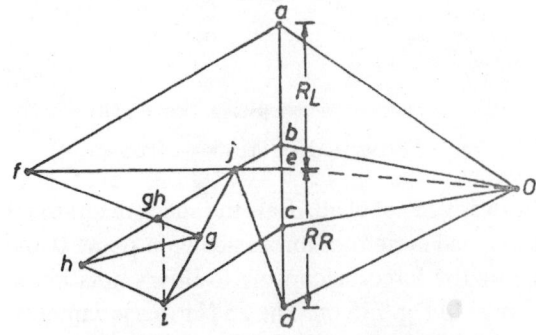

(*b*) Polar Diagram and Maxwell Diagram

Fig. 7.14: Compound Truss (Method of Substitution)

After joint L, we encounter the difficulty of more than two unknowns both at joint 1 and joint 5. Now the two members 1-4 and 2-4 are imagined to be substituted by a single member 2-5 temporarily, and the spaces G and H by a single space $g\,h$. Now the Maxwell diagram can be continued and the points g, h and e obtained from joints 1 and 2. Now the original members are imagined to be put back in place instead of the imaginary member and by working back, the points *g* and *h* are obtained. The work is continued and the Maxwell diagram is completed as usual.

Note: In the case of this truss, the problem will not require substitution if one proceeds in the reverse direction from joint 2 in a continuous manner!

ILLUSTRATIVE PROBLEMS

PROBLEM 7.1: Find the resultant of two forces equal to 50 N and 25 N acting at an angle of 60°.

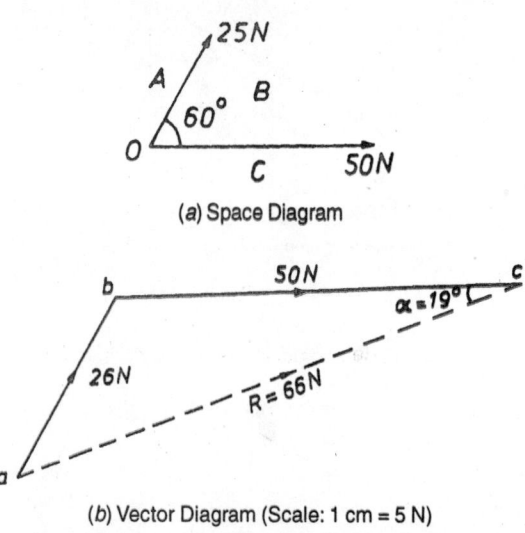

(*a*) Space Diagram

(*b*) Vector Diagram (Scale: 1 cm = 5 N)

Fig. 7.15 Bow's Notation (Illustrative Problem 7.1)

Solution: First of all, draw the space diagram for the given system of forces acting at point *O* and name the forces according to Bow's notation as shown in Fig. 7.15 (*a*). The 25 N force is named as *AB* and 50 N force as *BC*. Now draw the vector diagram for the given system of forces as shown in Fig. 7.15 (*b*). Select a suitable scale (1 cm = 5 N) and draw vectors $\overline{ab}$ and $\overline{bc}$ to represent the forces *AB* and *BC*, respectively, in magnitude and direction. It can be observed that the force polygon does not close, which means that there is a resultant unbalanced force. Line joining the initial point, *a*, to the last point, *c*, gives the resultant (*ac*) in magnitude (66 N), direction (at an angle of 19° with 50 N force) and sense. In space diagram it must pass through *O*, the point of concurrence of forces.

PROBLEM 7.2: A particle is acted upon by three forces equal to 10 N, 20 N and 30 N, along the three sides of an equilateral triangle taken in order. Find the magnitude and direction of the resultant force.

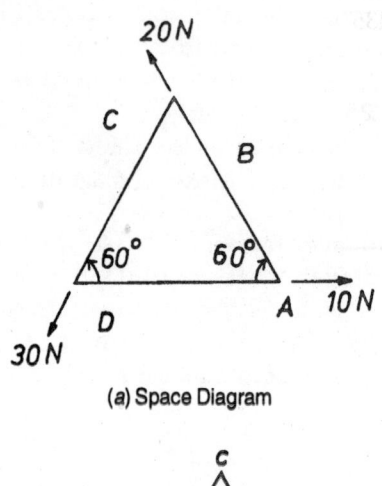

(*a*) Space Diagram

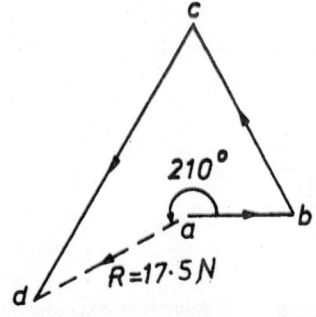

(*b*) Vector Diagram (Scale: 1 cm = 5 N)

Fig. 7.16 Bow's Notation (Illustrative Problem 7.2)

Solution: First of all, draw the space diagram for the given system of forces acting along the sides of an equilateral triangle and name the forces according to Bow's notation as shown in Fig. 7.16 (*a*). The 10 N force is named as *AB*, 20 N force as *BC* and 30 N force as *CD*.

Now draw the vector diagram for the given system of forces as shown in Fig. 7.16 (*b*). Select a suitable scale (1 cm = 5 N) and draw vectors *ab*, *bc* and *cd* to represent the forces *AB*, *BC* and *CD*, respectively in magnitude and direction. It can be observed that the force polygon does not close, which means that there is a resultant unbalanced force. Line joining the initial point, *a*, to the last point, *d*, gives the resultant (*ad*) in magnitude

(27.5 N), direction (acting at an angle of 207° with 10 N force) and sense.

PROBLEM 7.3: Find the magnitude and direction of the resultant of the concurrent forces of 15 N, 20 N, 25 N and 30 N making angles of 30°, 90°, 135° and 240° respectively with a fixed line.

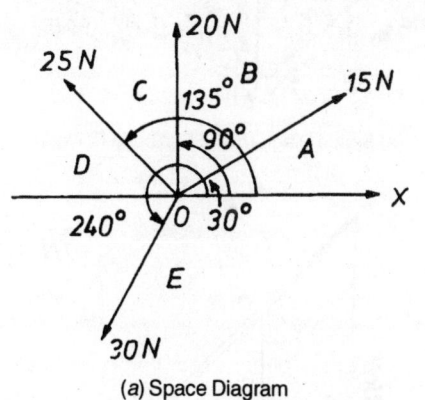

(*a*) Space Diagram

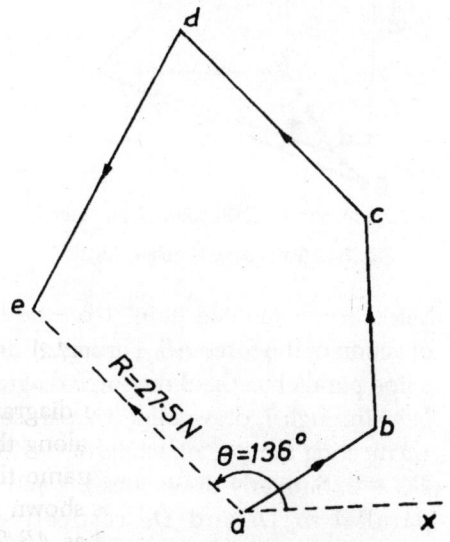

(*b*) Vector Diagram (Scale: 1 cm = 5 N)

Fig. 7.17 Bow's Notation (Illustrative Problem 7.3)

Solution: First of all draw the space diagram for the given system of forces acting at point *O* and the name the forces according to Bow's notation as shown in Fig. 7.17 (*a*). The 15 N force is named as *AB* the 20 N force as *BC*, the 25 N force as *CD* and 30 N force as *DE*.

Now draw the vector diagram for the given system of forces as shown in Fig. 7.17 (*b*). Select a

suitable scale (1 cm = 5 N) and draw vectors *ab*, *bc*, *cd* and *de* to represent the forces *AB*, *BC*, *CD* and *DE*, respectively, in magnitude and direction. It can be observed that there is a resultant unbalanced force. Line joining the initial point, *a*, to the last point, *e*, gives the resultant (*ae*) in magnitude (27.5 N), direction (at an angle of 136° with the horizontal) and sense. In space diagram it must pass through *O*, the point of concurrence of forces.

PROBLEM 7.4: Four like parallel forces 50 N, 75 N, 25 N and 75 N are acting at point *P*, *Q*, *R* and *S* respectively on a straight line *PQRS*. The distances are *PQ* = 3 m, *QR* = 2 m and *RS* = 4 m. Find the resultant and also the distance of the resultant from point *P*.

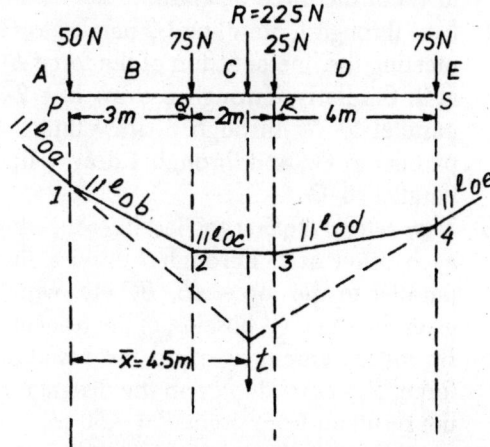

(*a*) Space Diagram and Funicular Polygon

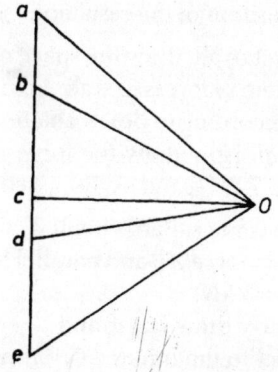

(*b*) Force Diagram and Polar Diagram
(Scale: 1 cm = 25 N)

Fig. 7.18 (Illustrative Problem 7.4)

Solution: First of all, draw the space diagram for the given system of forces to scale 1 cm = 1 m and name them according to Bow's notation as shown in Fig. 7.18 (*a*). Now draw the force diagram, as shown in Fig. 7.18 (*b*) and as discussed below:

(1) Take some suitable point *a*, and draw *ab* equal and parallel to the force *AB* (i.e. 50 N) to scale 1 cm = 25 N. Similarly draw *bc* equal to the force *BC* (i.e. 75 N), *cd* equal to the force *CD* (i.e. 25 N), and *de* equal to the force *DE* (i.e. 75 N), respectively.

(2) Now select some suitable point *O*, and join *Oa, Ob, Oc, Od* and *Oe*.

(3) Now take some suitable point 1 on the line of action of the force *AB* of the space diagram. Through 1 draw a line parallel to *Oa* of the force diagram.

(4) Now through 1, draw line 1-2 parallel to *Ob* meeting the line of action of the force *BC* at 2. Similarly, through 2 draw line 2-3 parallel to *Oc*, through 3 draw line 3-4 parallel to *Od*, and through 4 draw a line parallel to *Oe*.

(5) Now extend the first and last lines meeting each other at *t*. Through *t* draw a line parallel to the force *AB, BC* etc. which gives the required position of the resultant.

(6) By measurement, we find the resultant force, $R = ae = 225$ N and the distance of the resultant from point $P = 4.50$ m.

PROBLEM 7.5: Four forces act on a rigid bar as shown in Fig. 7.19 (*a*). Find the magnitude, direction and position of the resultant.

Solution: First of all, draw the space diagram for the given system of forces to scale 1 cm = 1 m and name them according to Bow's notation as shown in Fig. 7.19 (*a*). Now draw the force diagram as shown in Fig. 7.19 (*b*) and as discussed below:

(1) Select some suitable point *a*, and draw *ab* equal to force *AB* and parallel to it to scale 1 cm = 5 kN.

(2) Similarly draw *bc, cd,* and *de* equal to and parallel to the forces *BC, CD* and *DE* respectively.

(3) Now select some suitable point *O*, and join *Oa, Ob, Oc* and *Od*.

(4) Extend the lines of action of the forces *AB, BC, CD* and *DE*.

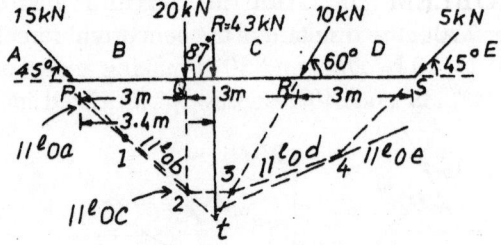

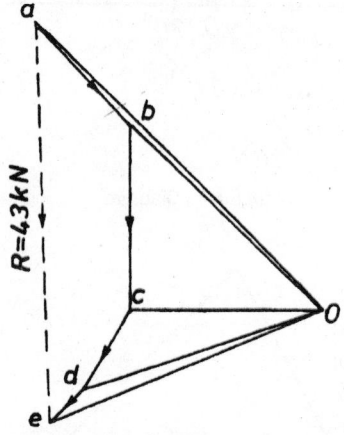

(*a*) Space Diagram and Funicular Polygon

(*b*) Force Diagram and Polar Diagram (Scale: 1 cm = 5 N)

Fig. 7.19 (Illustrative Problem 7.5)

(5) Select some suitable point 1 on the line of action of the force *AB*. Through 1 draw a line parallel to *Oa* of the force diagram.

(6) Now through 1, draw a line 1-2 parallel to *Ob* meeting the line of action of the force *BC* at 2. Similarly draw lines 2-3 and 3-4 parallel to *Oc* and *Od* respectively. Through point 4 draw a line parallel to *Oe*.

(7) Now extend the first and last lines to meet at '*t*'.

(8) Through '*t*' draw a line parallel to *ea* which gives the inclination and position of the resultant force.

(9) By measurement, we find that the resultant force, $R = ae = 43$ kN, direction of the resultant = 87° with respect to *QP*

distance between P and line of action of resultant force = 3.40 m.

PROBLEM 7.6: Given two forces P and Q (Fig. 7.20 (*a*)) acting at O, find their equilibrant.

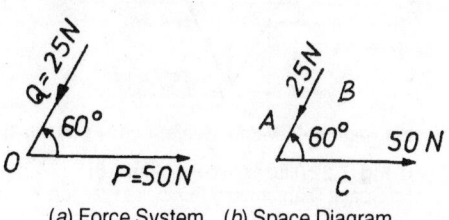

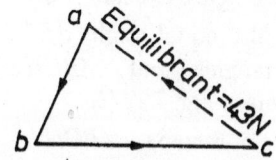

(*a*) Force System   (*b*) Space Diagram

(*c*) Vector Diagram (Scale: 1 cm = 10 N)

Fig. 7.20 (Illustrative Problem 7.6)

Solution: First of all, let us draw the space diagram, for the given system of forces and name them according to Bow's notation as shown in Fig. 7.20 (*b*).

Now draw the vector diagram for the given forces as shown in Fig. 7.20 (*c*) and as discussed below:

1. Select some suitable point a and draw a line equal and parallel to the force AB (i.e., 25 N).
2. Through b draw a line bc equal and parallel to the force BC (i.e., 50 N).
3. Join the last point 'c' to the initial point 'a'.
4. Then ca will represent the equilibrant (E).
5. Now a draw a line parallel to ca through point 'O' on the space diagram as shown in Fig. 7.22 (*b*).
6. By measurement, we find that the magnitude of equilibrant, $E = 43$ N and its direction with the line of action of force $P = 30°$.

PROBLEM 7.7: Five strings are tied at a point and are pulled in all the directions, equally spaced, from one another. If the magnitude of the pulls on three consecutive strings is 50 N, 100 N and 75 N respectively, find the magnitude of the pulls on two other strings.

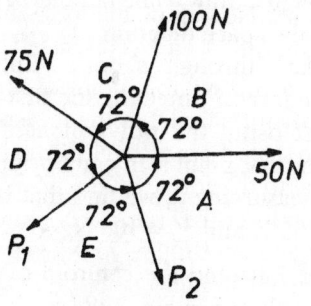

(*a*) Space Diagram

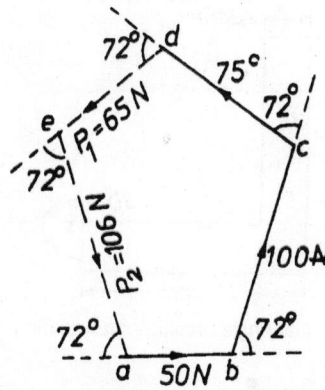

(*b*) Vector Diagram (Scale: 1 cm = 25 N)

Fig. 7.21 (Illustrative Problem 7.7)

Solution:

Given: Pulls = 50 N, 100 N and 75 N;

Angle between all the forces $= \dfrac{360}{5} = 72°$

Let P_1 and P_2 = Pulls in the two strings.

First of all, let us draw the space diagram for the given system of forces and name them according to Bow's notation as shown in Fig. 7.21 (*a*).

Now draw the vector diagram for the given forces as shown in Fig. 7.21 (*b*) and as discussed below:

1. Select some suitable point a and draw a horizontal line ab equal to 50 N to some suitable scale representing the force AB.
2. Through B draw a line bc equal to 100 N to the scale and parallel to BC.

3. Similarly through C, draw cd equal to 75 N to the scale and parallel to CD.

4. Through d draw a line parallel to the force P_1 of the space diagram.

5. Similarly through 'a' draw a line parallel to the force P_2 meeting the first line at e, thus closing the polygon $abcde$, which means that point is in equilibrium.

6. By measurement, we find that the forces $P_1 = 65$ N and $P_2 = 106$ N.

PROBLEM 7.8: Find the centroid of an equal angle section shown in Fig. 7.22 (a).

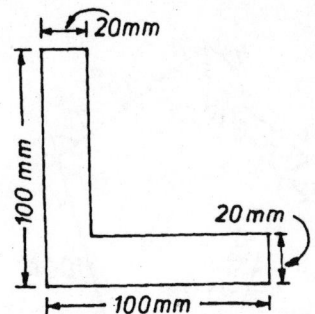

(a) Equal Angle (Scale: 1 cm = 20 mm)

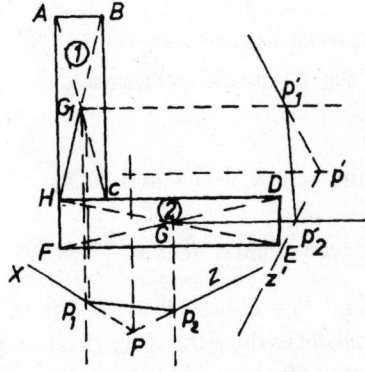

(b) Centroid from Funicular Polugon in Perpendicular Directions

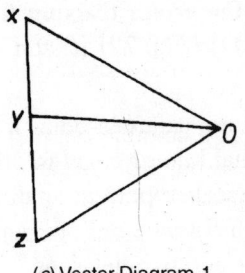

(c) Vector Diagram-1

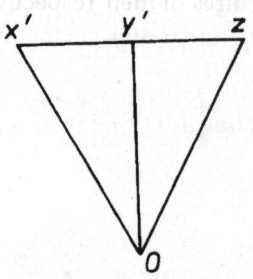

(d) Vector Diagram-2 (Vector Scale: 1 cm = 800 mm²)

Fig. 7.22 (Illustrative Problem 7.8)

Soultion:

1. Draw the figure to scale (1 cm = 20 mm).

2. Divide the equal angle $ABCDEFH$ into two rectangles 1 and 2 ($ABCH$ and $DEFH$) as shown in Fig. 7.22 (b).

   a_1 = area of rectangle $ABCH$
   $= 80 \times 20 = 1600$ mm²

   a_2 = area of rectangle $DEFH$
   $= 100 \times 20 = 2000$ mm²

3. First of all treat graphically a_1 and a_2 are two parallel vertical forces acting through the centres of their respective rectangles. Now set off vertical vector xy and yz to represent a_1 and a_2 respectively to a convenient/suitable scale (1 cm = 800 mm²) (Fig. 7.22 (c)).

4. Select a suitable point O, and join ox, oy and oz.

5. Draw vertical lines through the centres of gravity of the two rectangles. The centres of gravity of these two rectangles G_1 and G_2 are the intersection points of diagonals of the relevant rectangles.

6. Select a suitable point p_1 on the vertical line through the centre of gravity of rectangle 1 (G_1) and draw a line p_1x parallel to ox and p_1p_2 parallel to oy meeting the vertical line through the centre of gravity of rectangle 2 (G_2) at p_2.

7. Through p_2 draw a line p_2z parallel to oz produce this line (zp_2) to meet the line xp_1 at p. Now draw a vertical line through p.

8. Now treat graphically a_1 and a_2 as two parallel horizontal forces acting through

the centres of their respective rectangles. Set off horizontal vector $x'y'$ and $y'z'$ to represent a_1 and a_2, respectively to a suitable or convenient scale (1 cm = 800 mm²). (Fig. 7.25 (d)).

9. Select a suitable point O' and joint $O'x'$, $O'y'$ and $O'z'$.

10. Draw horizontal lines through the centres of gravity of the two rectangles.

11. Select a suitable point p_1' on the horizontal line through the centre of gravity of the rectangle 1 and draw a line $p_1'x'$ parallel to $o'x'$ and $p_1'p_2'$ parallel to $o'y'$ meeting the horizontal line through the centre of gravity of rectangle 2 at p_2'.

12. Through p_2' draw a line $p_2'z'$ parallel to oz'. Produce this line $(z'p_2')$ to meet the line $x'p_1'$ at p'.

13. Now draw a horizontal line through p' meeting the vertical line through p at G, which is the required centre of gravity of the section.

The coordinates of centre of gravity of the figure G are (32 mm; 32 mm) from the vertical line AF and from the horizontal line FE respectively by measurement.

PROBLEM 7.9: Determine the position of the centroid of the I-section shown in Fig. 7.23 (a).

Solution:

1. Draw the I-section to scale (1 cm = 20 mm).
2. Divide the I-section *ABCDEFGHIJKL* into three rectangles 1, 2, and 3 (*ABCJKL, CDIJ* and *DEFGHI*) as shown in Fig. 7.23 (b).

a_1 = area of rectangle 1
   $= 100 \times 20 = 2000$ mm²
a_2 = area of rectangle 2
   $= 60 \times 20 = 1200$ mm²
a_3 = area of rectangle 3
   $= 50 \times 20 = 1000$ mm²

3. Treat a_1, a_2 and a_3 are three vertical forces acting through the centres of their respective rectangles. Now set off vertical vector ux, xy and yz to represent a_1, a_2 and

a_3 respectively to a convenient scale (1 cm = 1000 mm²) (Fig. 7.23 (c)).

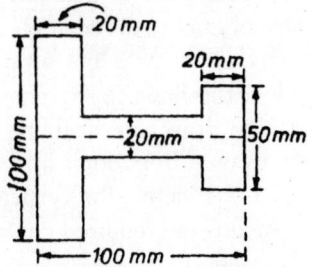

(a) I-Section (Scale: 1 cm = 20 mm)

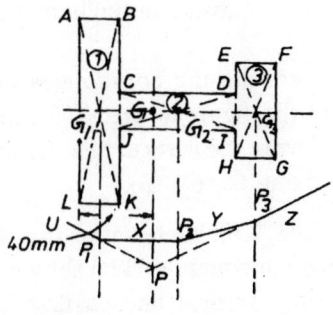

(b) Centroid from Funicular Polygon

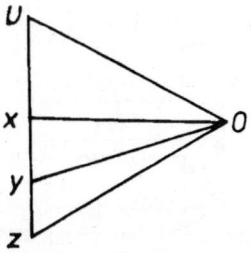

(c) Vector Diagram (Scale: 1 cm = 1000 mm²)

Fig. 7.23 (Illustrative Problem 7.9)

4. Select a suitable point O and join ou, ox, oy and oz.

5. Draw vertical lines through the centres of gravity of the three rectangles. The centres of gravity of these two rectangles G_1, G_2 and G_3 are the intersection points of diagonals of the relevant rectangles.

6. Select a suitable point p_1 on the vertical line through the centre of gravity of rectangle 1 (G_1) and draw a line p_1U parallel to ou, p_1p_2 parallel to ox meeting the

vertical line through the centre of gravity of rectangle 2 (G_2) at p_2 and $p_2 p_3$ parallel to *oy* meeting the vertical line through the centre of gravity of rectangle 3 (G_3) at p_3.

7. Through p_3 draw a line $p_3 z$ parallel to *oz*. Produce this line ($Z p_3$) to meet the line $U p_1$ at p. Now draw a vertical line through p.

8. Now draw a horizontal line through G_1, G_2 and G_3 meeting the vertical line at p at G, which is the required centre of gravity of the section.

(Since the I-section is symmetrical about the horizontal axis passing through the points G_1, G_2 and G_2).

Therefore, the the coordinates of centre of gravity of the figure G are 40.5 mm & 0 mm measured from the vertical line AL and from the horizontal line G_1, G_2, G_3.

PROBLEM 7.10: Determine the moment of inertia of an unequal I-section shown in Fig. 7.24 (*a*) about the vertical line passing through the centroid.

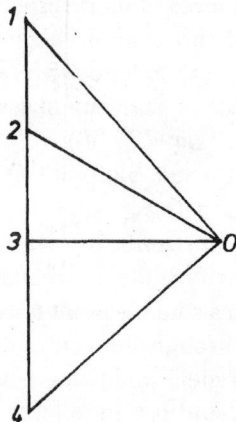

(*c*) Vector Diagram (Scale: 1 cm = 1000 mm²)

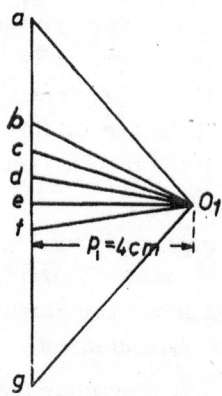

(*d*) Vector & Polar Diagrams (Scale: = 1 cm = 1000 mm²)

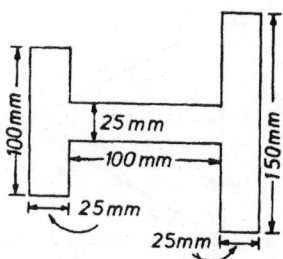

(*a*) Unequal I-Section
(Scale: 1 cm = 25 mm)

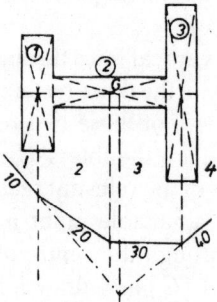

(*b*) Funicular Polygon (For the Location of Centroid, *G*)

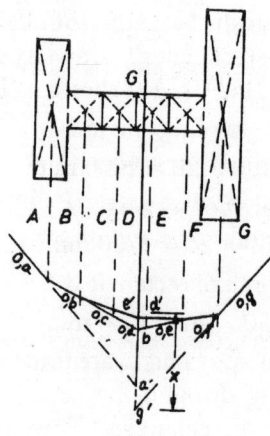

(*e*) Funicular Polygon from (*d*)

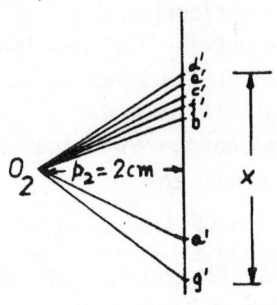

(*f*) Vector & Polar Diagrams Obtained from (*e*)

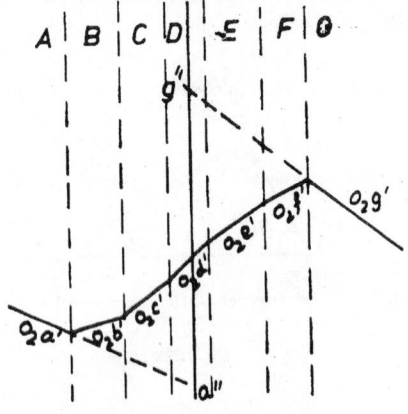

(*g*) Funicular Polygon from (*f*)

Fig. 7.24 (Illustrative Problem 7.10)

Solution:

1. Draw the space diagram and determine the vertical line passing through the C.G., with the usual procedure [Fig. 7.24 (*b*) and (*c*)].

2. Divide the section into smaller areas and convert it into an equivalent parallel force system, treating area of smaller sections as equivalent forces acting through their respective centroids, in a direction parallel to the vertical line passing through the centroid of the section.

3. Name the spaces according to Bow's notation and draw the vector diagram, Fig. 7.24 (*d*), to an area scale 1 cm = 1000 mm².

4. Select pole O_1 at a known perpendicular pole $P_1 = 40$ mm distance from the resultant area vector *ag*. Join *O* to *a, b, c, d, e, f* and *g* to complete the polar diagram.

5. Draw the funicular polygon, Fig. 7.24 (*e*) for the polar diagram of Fig. 7.24 (*d*).

6. Get the points of intersection of strings O_1a, O_1b, O_1c, O_1d, O_1e, O_1f and O_1g of the funicular polygon with the vertical line, drawn through C.G. of the section, at points *a', b', c', d', e', f'* and *g'* respectively marked (Fig. 7.24 (*e*)).

7. Transfer the intercept (X) to a new (parallel) as in Fig. 7.24 (*f*), and select a new pole O_2 at a known distance $P_2 = 20$ mm. Join points *a', b', c', d', e', f'* and *g'* to O_2 to complete the polar diagram, Fig. 7.24 (*f*).

8. Draw the corresponding funicular polygon for this polar diagram, Fig. 7.24 (*g*) and obtain the intercept *a'' g''*, on the vertical line through C.G., where *a''* is the point of intersection of the first funicular string O_2a' and *g''* is the point of intersection of the last string O_2g' with the vertical line draw through C.g. about which the second moment of area is to be found.

9. The required moment of area of the section

$$= a''g'' \text{ (mm)} \times \text{(Linear scale)}^2$$
$$= \times \text{Area scale} \times \text{(mm}^2/\text{mm)}$$
$$\times p_1 \text{(mm)} \times p_2 \text{ (mm)}$$
$$= 2612.53 \times 10^4 \text{ mm}^4$$

PROBLEM 7.11: A horizontal beam *PQ*, hinged at *P* and supported on rollers at *Q* is loaded as shown in Fig. 7.25 (*a*). Determine the magnitude of the vertical reaction at *Q* and the magnitude and line of action of reaction at the hinge *P*.

1. First of all, draw the space diagram of the beam to a scale 1 cm = 1 m, and name all the loads and reactions according to Bow's notations.

2. Select some suitable point *a* and draw *ab*, *bc, cd,* and *de* parallel and equal to the loads 20 kN, 30 kN, 15 kN and 20 kN to a scale 1 cm = 10 kN.

3. Select any point *O* and joint *Oa, Ob, Oc, Od,* and *Oe.*

4. Now extent the lines of the loads *AB, BC, CD, DE* and the reaction *RQ*. Through *P* or (6) draw (6) (1) parallel to *Oa* intersecting the line of action of 20 kN load at (1).

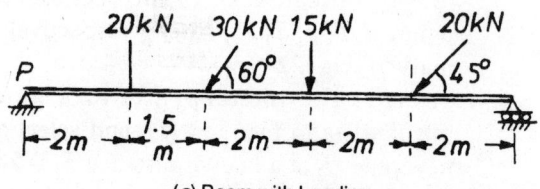

(a) Beam with Loading

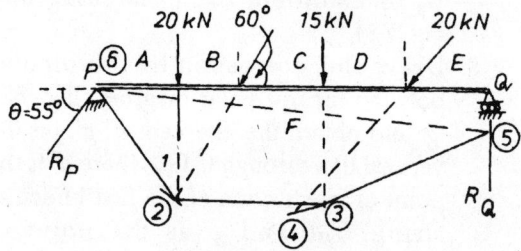

(b) Space Diagram and Funicular Polygon
(Scale: 1 cm = 1 m)

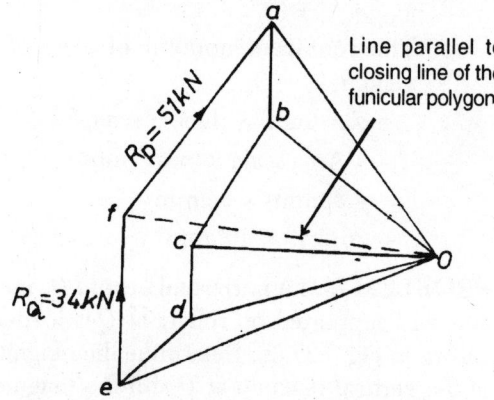

(c) Force Diagram and Polar Diagram
(Scale: 1 cm = 10 kN)

Fig. 7.25 (Illustrative Problem 7.11)

5. Similarly, draw (1)(2), (2)(3), (3)(4) and (4)(5) parallel to *Ob, Oc, Od* and *Oe* respectively. Join (6) and (5). Through *O*, draw a line parallel to this line. Now through *e*, draw a vertical line (as the reaction R_Q is vertical) meeting the line through *O* at *f*. Join *fa*.

6. Now the lengths *fa* and *ef*, in the vector diagram, give the magnitudes and direction of the reaction R_P and R_q respectively to the scale.

By measurement, we find that

$R_P = fa = 51$ kN;

$R_Q = ef = 34$ kN and $\theta = 55°$

PROBLEM 7.12: The truss *ABC* shown in Fig. 7.26 (*a*) a span of 5 m. It is carrying a load of 20 kN at its apex. End *A* is hinged and *B* is supported on rollers. Find the forces in the members *AB, BC* and *AC*.

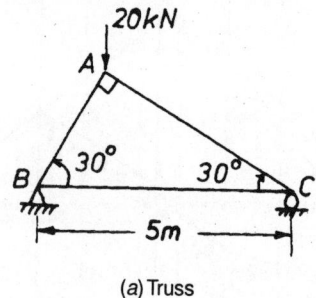

(a) Truss

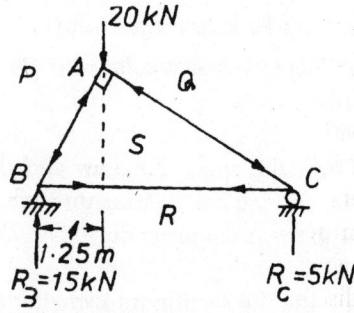

(b) Space Diagram (Scale: 1 cm = 1 m)

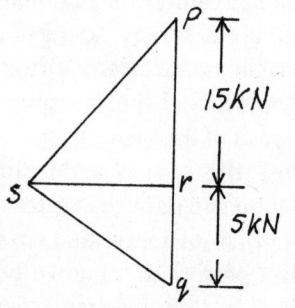

(c) Vector Diagram (Scale: 1 cm = 5 kN)

Fig. 7.26 (Illustrative Problem 7.12)

From the geometry of the truss

$$\frac{h}{x} = \tan 60° \quad \therefore \quad h = x \tan 60°$$

$$\frac{h}{5-x} = \tan 30° \quad \therefore \quad h = (5-x) \tan 30°$$

$$\therefore \quad x \tan 60° = (5-x) \tan 30°$$

$$x(\tan 30° + \tan 60°) = 5 \tan 30°$$

$$x = \frac{5 \tan 30°}{\tan 30° + \tan 60°}$$

$$= \frac{2.88675}{0.57735 + 1.73205}$$

$$= \frac{2.88675}{2.3094}$$

$$= 1.25 \text{ m}$$

The load of 10 kN is acting at a distance of 1.25 m from the left hand support i.e. B and 3.75 m from C.

Taking moments of all forces including reactions about C.

$$5R_B = 20 \times 3.75$$

$$R_B = 15 \text{ kN}$$

But $\quad R_B + R_C = 20$

$$R_C = 20 - R_B = 20 - 15 = 5 \text{ kN}$$

First of all, draw the space diagram for the truss along with the load at its apex and the reactions R_B and R_C as shown in Fig. 7.26 (*b*). Name the members AB, BC, and AC according to Bow's notation as PS, RS and SQ respectively. Now draw the vector diagram as shown in Fig. 7.26 (*c*) and as discussed below:

1. Select some suitable point p and draw a vertical line pq equal to 20 kN to some suitable scale (1 cm = 5 kN) to represent the load PQ at joint A.
2. Now cut off qr equal to 5 kN to the scale to represent the reacton R_C at C. Thus rp will represent the reaction R_B at the scale.
3. Now draw the vector diagram for the joint B. For doing so, through p draw ps parallel to PS and through r draw rs parallel to RS meeting the first line at S. Now psr is the vector diagram for the joint B, whose directions follow p-s, s-r and r-p.

4. Similarly, draw vector diagram for the joint C, whose directions follow q-r, r-s and s-q as shown in Fig. 7.26 (*b*) and (*c*). Now check the vector diagram for the joint A, whose directions follow p-q, q-s and s-p.

Now measuring the various sides of the vector diagram and keeping due note of the directions of the arrow heads, the results are tabulated below:

Sl.No.	Member	Magnitude of force in kN	Nature of force
1.	AB (PS)	17.4	Compression
2.	BC (SR)	8.6	Tension
3.	AC (SQ)	10.0	Compression

PROBLEM 7.13: Fig. 7.27 (*a*) shows a framed structure of 8 m span and 3.0 height subjected to two points load at B and D. Find the forces in all members of the structure.

Solution: Since the structure is supported on rollers at the right hand support (C), therefore the reaction at this support will be vertical (because of horizontal support). The reaction at the left hand support (A) will be the resultant of vertical and horizontal forces and thus will be inclined with the vertical.

Taking moments of all forces about A including reactions.

$$V_c \times 8 = (10 \times 3) + (15 \times 4)$$

$$8V_c = 30 + 60 = 90$$

$$V_c = \frac{90}{8} = 11.25 \text{ kN } (\uparrow)$$

and $\quad V_A = 15 - V_c = 15 - 11.25$

$$V_A = 3.75 \text{ kN } (\uparrow)$$

Horizontal reaction at the left hand support A,

$$H_A = 10 \text{ kN } (\leftarrow)$$

First of all, draw the space diagram and name the members and forces according to Bow's notation as shown in Fig. 7.27 (*b*).

Now draw the vector diagram as shown in Fig. 7.27 (*c*). The diagram is self instructive. Measuring the various sides of the vector diagram the results are tabulated in the table below.

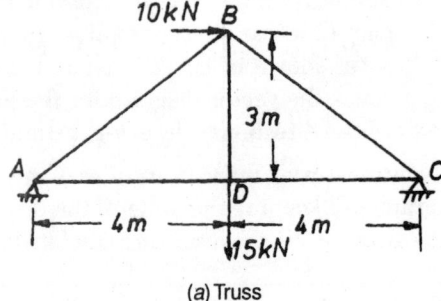

(*a*) Truss

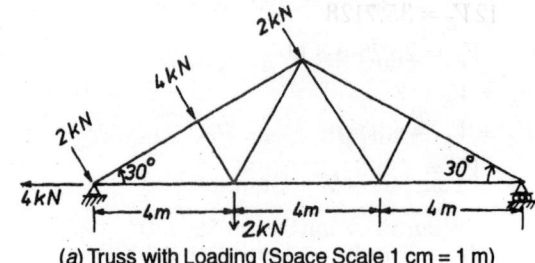

(*a*) Truss with Loading (Space Scale 1 cm = 1 m)

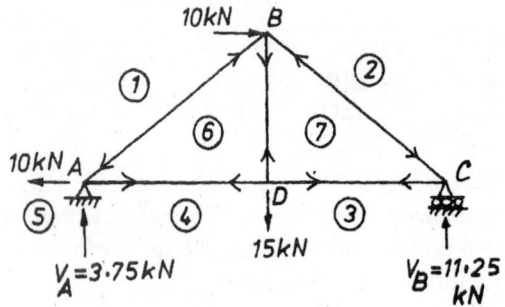

(*b*) Space Diagram

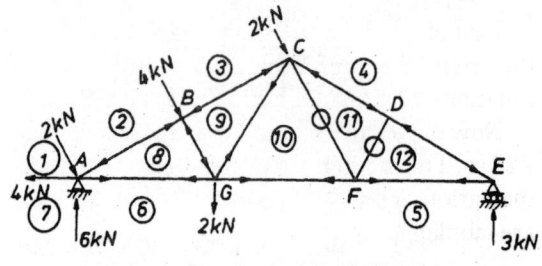

(*b*) Space Diagram (with Bow's Notation)

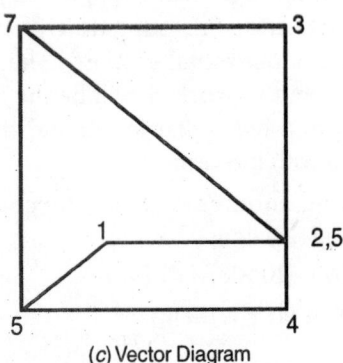

(*c*) Vector Diagram

Fig. 7.27 (Illustrative Problem 7.13)

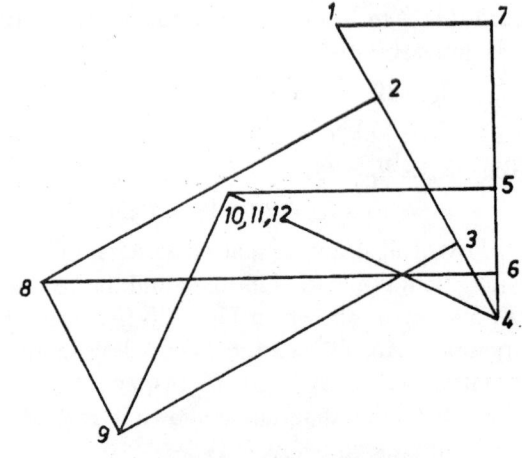

(*c*) Vector Diagram (Maxwell Diagram)
(Scale 1 cm = 1 kN)

Fig. 7.28 (Illustrative Problem 7.14)

Sl.No.	Member	Magnitude of force in kN	Nature of force
1.	AB (1-6)	6.25	Compression
2.	BC (2-7)	18.75	Compression
3.	CD (3-7)	15.00	Tension
4.	DA (4-6)	15.00	Tension
5.	DB (6-7)	15.00	Tension

PROBLEM 7.14: A truss of 12 m span is loaded as shown in Fig. 7.28 (*a*). Determine the forces in all the members of the truss.

Solution: Since the structure is supported on rollers at the right hand support (*E*), therefore the reactions at this support will be vertical (because of horizontal support).

The reaction at the left hand support (*A*) will be the resultant of vertical and horizontal forces and thus will be inclined with the vertical.

Taking moments of all forces about *A* including reactions

$$12 V_E = 2 \times 4 + 4 \times 4 \cos 30^{\text{go}} + 2 \times 8 \cos 30°$$

$$12V_E = 35.7128$$
$$V_E = 2.976 \approx 3 \text{ kN}$$
$$V_A + V_E = 2\cos 30° + 4\cos 30° + 2\cos 30° + 2$$
$$V_A + V_E = 6.928 + 2 = 8.928$$
$$V_A = 8.928 - 2.976$$
$$= 5.952 \text{ kN} \approx 6 \text{ kN}$$
$$H_A = 2\sin 30° + 4\sin 30° + 2\sin 30°$$
$$H_A = 4 \text{ kN}$$

First of all, draw the space diagram and name the members and forces according to Bow's notations at shown in Fig. 7.28 (*b*).

Now draw the vector diagram as shown in Fig. 7.28 (*c*). The diagram is self instructive. Measuring the various sides of the vector diagram the results are tabulated here.

S.No.	Member	Magnitude of force in kN	Nature of force
1.	AB (2-8)	8.4	Compression
2.	BC (3-9)	8.4	Compression
3.	CD (4-11)	6.0	Compression
4.	DE (4-12)	6.0	Compression
5.	EF (5-12)	5.2	Tension
6.	FC (10-11)	0	–
7.	FD (11-12)	0	–
8.	FG (5-10)	5.2	Tension
9.	GA (6-8)	10.3	Tension
10.	GB (8-9)	4.0	Compression
11.	GC (9-10)	6.3	Tension

PRACTICE PROBLEMS

7.1 Two forces are acting at a point *O* as shown in the Fig. 7.29. Determine the magnitude and direction of the resultant.

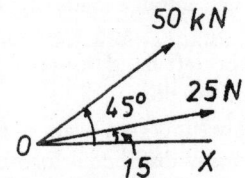

Fig. 7.29 (Practice Problem 7.1)

7.2 Find the resultant of coplanar concurrent forces acting at the point *O* as shown in the Fig. 7.30.

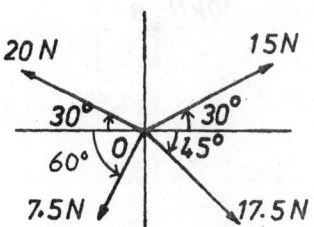

Fig. 7.30 (Practice Problem 7.2)

7.3 A rigid bar *PQRS* is subjected to a system of parallel forces as shown in Fig. 7.31. Find the magnitude, direction and position of the resultant force.

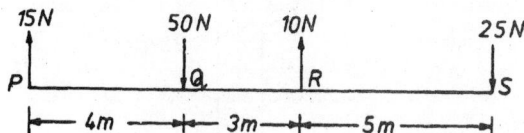

Fig. 7.31 (Practice Problem 7.3)

7.4 A rigid bar is subjected to a system of forces as shown in Fig. 7.32. Find the magnitude, direction and position of the resultant force.

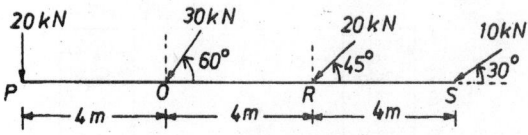

Fig. 7.32 (Practice Problem 7.4)

7.5 Three forces F_1, F_2 and F_3 are acting at a point *O* as shown in Fig. 7.33. If $F_1 = 5$ N, find the other two forces (F_2 and F_3), if the system of forces keeps the point *O* in equilibrium.

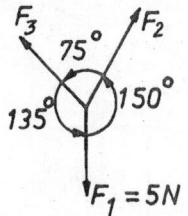

Fig. 7.33 (Practice Problem 7.5)

7.6 A body is acted upon by five forces F_1, F_2, F_3, F_4 and F_5 as shown in Fig. 7.34 and the body is in equilibrium. If $F_1 = 20$ N, $F_2 = 25$ N, $F_3 = 15$ N and $F_4 = 30$ N, find the force F_5, both in magnitude and direction.

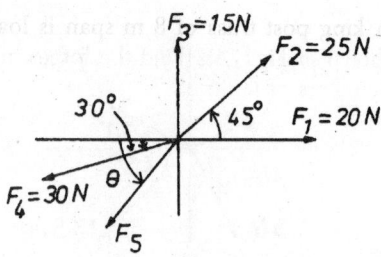

Fig. 7.34 (Practice Problem 7.6)

7.7   Find the centroid of 100 mm × 150 mm × 25 mm T-section shown in Fig. 7.35.

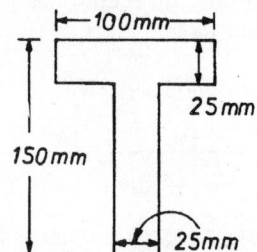

Fig. 7.35 (Practice Problem 7.7)

7.8   Find the centroid of the channel section shown in Fig. 7.36.

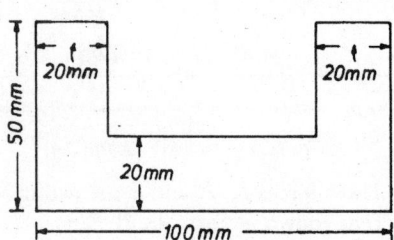

Fig. 7.36 (Practice Problem 7.8)

7.9   Find the centroid of the Z-section shown in the Fig. 7.37.

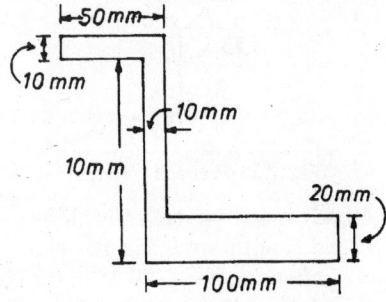

Fig. 7.37 (Practice Problem 7.9)

7.10   Find the moment of inertia of a T-section, flange 100 mm × 25 mm and web 100 mm × 25 mm as shown in Fig. 7.38 about the vertical axis yy passing through its centroid G.

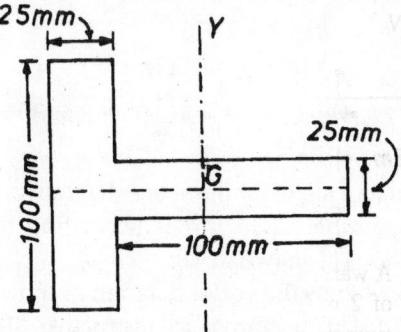

Fig. 7.38 (Practice Problem 7.10)

7.11   Find the moment of inertia of an equal angle section 125 × 125 × 25 mm as shown in Fig. 7.39 about the vertical line YY passing through the centroid.

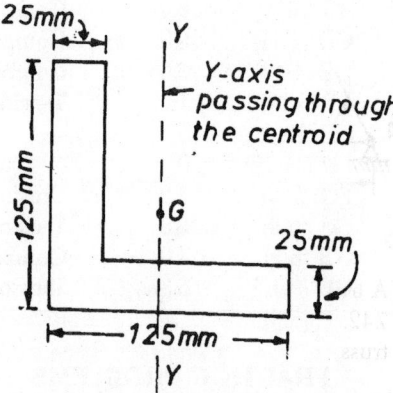

Fig. 7.39 (Practice Problem 7.11)

7.12   A simply supported beam of 6 m span carries three concentrated loads of 30 kN, 15 kN and 25 kN at 1, 2.5 and 4.5 metres respectively from the left hand support. Determine the support reactions.

7.13   A simply supported beam of 6 m span carries a uniformly distributed load of 15 kN/m run over 2 m length, starting at the left hand support. In addition there are two point loads each of 30 kN at 1.5 m and 3.0 m from the right hand support. Determine the support reactions.

7.14 A simply supported beam, overhung at each end, has four concentrated loads applied as shown in Fig. 7.40. Determine the support reactions.

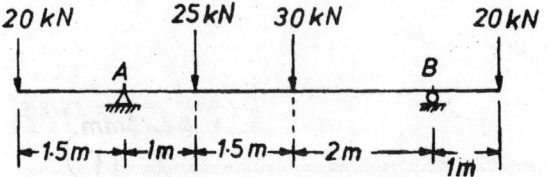

Fig. 7.40 (Practice Problem 7.14)

7.15 A warren girder consisting of 7 members each of 2 m length hinged at one end and freely supported on rollers at the other end is shown in Fig. 7.41. The girder is loaded at *B* and *C* as shown. Find the forces in all the members of the girder.

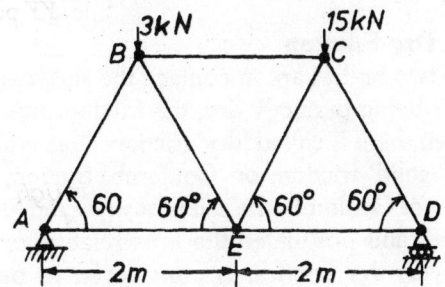

Fig. 7.41 (Practice Problem 7.15)

7.16 A truss of span 5 m is loaded as shown in Fig. 7.42. Find the forces in all the members of the truss.

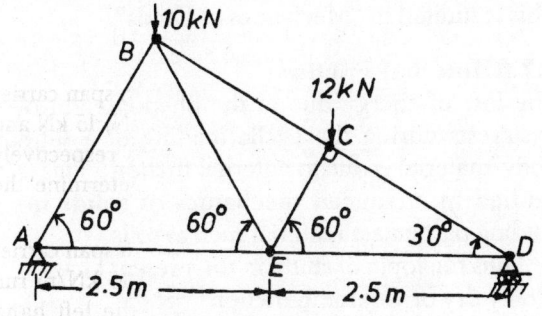

Fig. 7.42 (Practice Problem 7.16)

7.17 A king post truss of 8 m span is loaded as shown in Fig. 7.43. Find the forces in all the members of the truss.

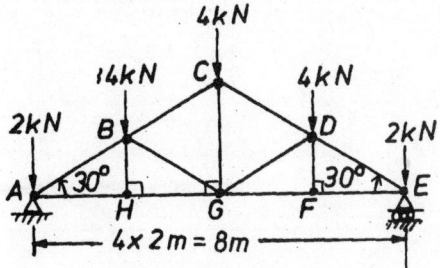

Fig. 7.43 (Practice Problem 7.17)

7.18 A cantilever truss of 3 m span is loaded as shown in Fig. 7.44. Find the forces in the various members of the truss.

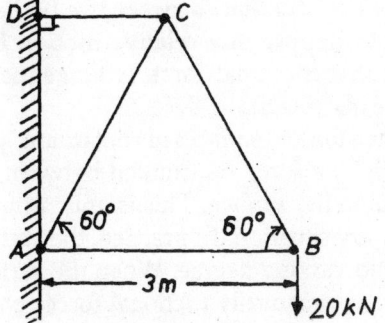

Fig. 7.44 (Practice Problem 7.18)

7.19 Determine the forces in the various members of the truss loaded as shown in Fig. 7.45.

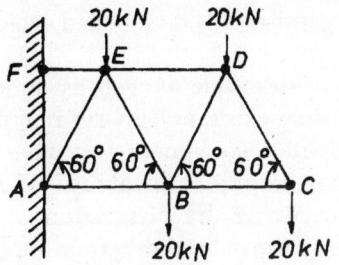

Fig. 7.45 (Practice Problem 7.19)

Chapter 8

Friction

8.1 INTRODUCTION

'Friction' may be defined as the force distribution at a surface of contact between two bodies which tends to oppose any relative motion between them; this frictional force is tangential to the surface of contact.

When the two surfaces in contact are "perfectly smooth" the force transmitted between them is normal to the surface. This is only an ideal and imaginary situation. In practice, the surfaces are rough to varying degree. When the surfaces are rough, automatically frictional forces come into play as stated above.

There are always microscopic protruberances at the surfaces of bodies, the interlocking of which causes friction, tending to oppose relative motion. When ultimately motion occurs, the protruberances may get sheared off or melted, which result in "wear".

There are desirable as well as undesirable effects if friction exists, as it always is in practice. Human activities would not be possible without friction—for example, "braking" of moving vehicles involving the dissipation of kinetic energy in the form of heat. In certain applications of statics, frictional forces may be needed to keep equilibrium of a body. However, undesirable effects of friction are also common as the "wear and tear" of machine parts such as bearings and gears.

In this chapter, the concepts relating to friction, laws of friction, and different applications of friction are presented.

8.2 TYPES OF FRICTION

Broadly speaking, three types of friction are recognised:

8.2.1 Dry Friction

When two bodies are in contact, the surfaces of contact being perfectly dry, the friction present between them is called 'dry' friction. This is also called 'solid' friction, or 'Coulomb' friction, in honour of Coulomb, the scientist who studied it in detail and postulated the laws relating to it. However, dry friction is considered to be a complex phenomenon, actually speaking.

8.2.2 Fluid Friction

The friction which exists between adjacent layers of a fluid when relative motion occurs between them is called 'fluid friction' or 'viscous friction'. This is studied in "Mechanics of Fluids".

8.2.3 Internal Friction

The loss of energy due to the phenomenon of hysteresis during the cyclic loading of a solid body/material is due to 'internal friction'. This is studied in advanced mechanics of solids including particulate material such as soils.

Thus the topic of study in the present chapter is only dry or Coulomb friction.

8.3 LAWS OF COULOMB FRICTION

Coulomb enunciated the laws of dry friction after a lot of experimental work, which is typified by the following (Fig. 8.1).

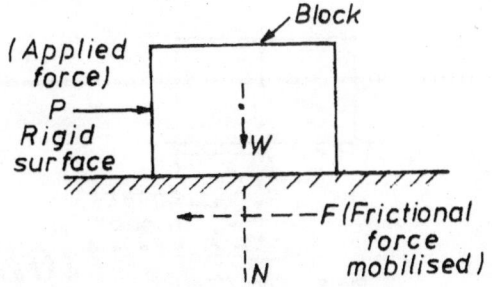

(a) Block on a Rough Rigid Surface under the Action of an Applied Force

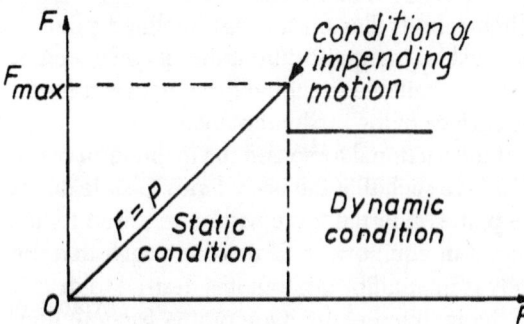

(b) Relation between Applied Friction and Friction Mobilised

Fig. 8.1: Block Experiment for Dry Friction

Let a block be at rest on rough rigid horizontal surface as in (a). Let a horizontal force P be applied to the block. As P is gradually increased, a frictional force F, which is just equal to it in magnitude and opposite in direction will get mobilised at the surface of contact to keep the block in equilibrium.

However, a limiting condition will be reached when the maximum possible frictional force (or resistance) has been mobilized. Motion of the block is then said to be impending, or about to occur and the maximum frictional force developed is known as 'Limiting Friction'.

As the force P is increased beyond the value for which the limiting friction has been mobilised, there will be a sudden and slight drop in the frictional force, and the block is set in motion. For uniform motion of the block, this frictional force remains constant at low velocities.

The frictional force mobilised when the block is in static equilibrium, including that at the condition of impending motion, is known as 'static friction', while that mobilised under motion is known as 'dynamic friction' or 'kinetic friction'.

The following is the summary of what are known as the 'Laws of Friction', enunciated first by Coulomb (1781) and later extended and confirmed by Morin (1831):

(1) Frictional force always acts in a direction opposite to that in which there is tendency for motion, or there is motion.

(2) The frictional force is exactly equal in magnitude to the force tending to cause motion, until the limiting frictional force is reached; then motion is impending.

(3) Under the condition of impending motion, the magnitude of the limiting friction is proportional to the normal force transmitted across the surface of contact.

(4) The available limiting friction depends upon the roughness of the surfaces of contact.

(5) The frictional force is independent of the area of contact between the two surfaces/bodies.

(6) Once motion is established, the frictional force developed is practically independent of velocity, for low values of velocity; however, it is less than the limiting friction at impending motion (or the static friction). Condition (3), expressed mathematically, is:

$F \propto W$ (= N, the normal reaction force equal and opposite to the weight W of the block)

or $F = \mu \cdot N$ ...(Eq. 8.1)

μ is called the 'Coefficient of Friction'. For static friction, we may write

$F_s = \mu_s \cdot N$ ...(Eq. 8.2)

Similarly for dynamic or kinetic friction,

$F_k = \mu_k \cdot N$ ...(Eq. 8.3)

μ_k here is called the 'coefficient of dynamic friction' and is always less than the 'coefficient of static friction', or simply the 'coefficient of friction' (about 25% less).

Eqs. 8.1 to 8.3 are, it should be remembered, applicable only to the condition of impending motion, or when the body is actually in motion.

The value of μ_s is about 0.40 for steel on cast iron, 0.57 for mild steel on mild steel, and 0.70 for rope on wood.

8.4 ANGLE OF FRICTION AND CONE OF FRICTION

Let us consider the case of impending motion of the block in Fig. 8.1 (*a*), the frictional force mobilised being the maximum or the limiting value (Fig. 8.2).

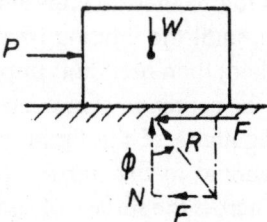

Fig. 8.2: Angle of Friction

The limiting frictional force $F (= P)$ and the normal reaction $V (= N)$ will yield a resultant reaction, R, as shown, which will be equal and opposite to the resultant force acting on the block–the resultant of the self-weight, W and the applied horizontal force, P.

In the corresponding triangle of forces, the angle between the normal reaction N and the resultant reaction R is called the "angle of friction" (ϕ). This value corresponds to the condition of impending motion. Obviously, the angle which corresponds to any condition prior to that of impending motion will be less than ϕ.

It is obvious that $\tan\phi = F/N = \mu$ ...(Eq. 8.4)

If the direction of motion is imagined to be changed, the resultant force R also will correspondingly get altered. Thus if this direction is changed through 360°, the direction of the resultant R generates a right circular cone with the semi-vertex angle equal to the angle of friction, ϕ (Fig. 8.3).

This right circular cone is called the 'Cone of friction'. If the angle corresponds to the static condition, the cone is 'static cone of friction' and that corresponding to motion is 'kinetic' or 'dynamic cone of friction'.

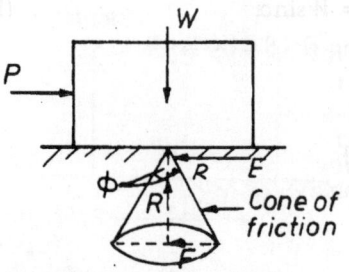

Fig. 8.3: Cone of Friction

8.5 EQUILIBRIUM OF A BODY ON A ROUGH INCLINED PLANE

When a body lies on a rough inclined plane it may just be in equilibrium under its own weight and the available frictional force between it and the surface of the inclined plane.

If the frictional force and the inclination of the plane are such that the body tends to slide down the plane, external force will be required to just keep it in equilibrium. If these are such that the body is in equilibrium, but it is desired to drag it up the inclined plane overcoming friction, then also a force will be required, but in a direction opposite to that in the previous case.

All these possibilities will be considered in the following sub-sections.

8.5.1 Equilibrium of a Body under its Own Weight

Let a block of weight W rest on an inclined plane which makes an angle α with the horizontal (Fig. 8.4);

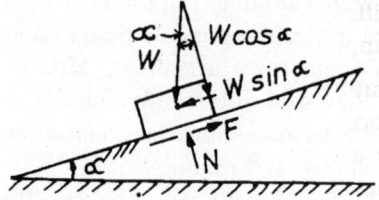

Fig. 8.4: Body on an Inclined Plane—Equilibrium under its Own Weight

Considering the equilibrium of the block under the normal reaction N and the frictional force F directed up the plane, and the components of W parallel to and perpendicular to the plane, we have

$$N = W\cos\alpha \qquad ...(Eq. 8.5)$$

and $\quad F = W\sin\alpha \qquad \ldots$(Eq. 8.6)

Dividing Eq. 8.6 by Eq. 8.5,

$$\tan\alpha = \frac{F}{N} \qquad \ldots\text{(Eq. 8.7)}$$

If θ is the value of α when motion is impending, frictional force will be the limiting value:

$$\tan\alpha = \frac{F}{N} = \mu$$

μ being the coefficient of friction (by Eq. 8.4).

This leads to

$$\alpha = \mu_s \qquad \ldots\text{(Eq. 8.8)}$$

Thus the value of the maximum slope of the inclined plane with the horizontal to just keep the body in equilibrium (or when motion down the plane is impending) is equal to the static coefficient of friction. For any slope less than this limiting value, the block will be in equilibrium.

This is actually known as the 'Angle of Repose' (angle of 'sleep' or 'rest') under the conditions.

It is a well known fact that when particulate materials such as food grains, sand, and soil are poured in the form of a heap, there will be a limiting slope beyond which the grains start rolling down; this limiting slope is also called the 'angle of repose' of the material.

If the frictional force and the inclination of the plane are such that the body tends to slide down the plane, an external force is required to just keep it in equilibrium. This situation occurs when the inclination of the plane is more than the angle of repose.

The different ways in which this external force may be applied are:

(*i*) parallel to the plane,
(*ii*) horizontally, and
(*iii*) in an appropriate direction other than these two.

8.5.2 Equilibrium under a Force Parallel to the Inclined Plane

Let us consider a block of a weight W resting on an inclined plane with an angle of inclination α to the horizontal. Let a force P be applied parallel to this plane directed upwards as shown in Fig. 8.5. Let N be the normal reaction of the plane, F

be the limiting frictional force directed upwards when there is impending downward motion of the block. Under these conditions, the minimum value of P required may be determined as follows:

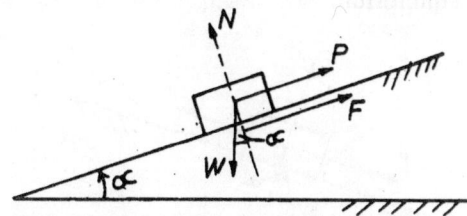

Fig. 8.5: Equilibrium under a Force Parallel to the Inclined Plane—Impending Downward Motion

Resolving parallel to and perpendicular to the inclined plane,

$$N = W\cos\alpha \qquad \ldots\text{(Eq. 8.9)}$$
$$P + F = W\sin\alpha \qquad \ldots\text{(Eq. 8.10)}$$
or $\quad P + \mu N = W\sin\alpha \qquad \ldots$(Eq. 8.11)

Eliminating N between Eqns. 8.9 and 8.11,

$$P + \mu W\cos\alpha = W\sin\alpha$$
$$P = W(\sin\alpha - \mu\cos\alpha) \qquad \ldots\text{(Eq. 8.12)}$$

Substituting $\tan\phi$ for μ, and simplifying, we get

$$P = \frac{W\sin(\alpha - \phi)}{\cos\phi} \qquad \ldots\text{(Eq. 8.13)}$$

Here ϕ is the angle of repose or angle of friction. P given by Eq. 8.12 is the minimum external force necessary up the inclined plane. If a force more than this is applied, the block will still be in equilibrium; but the frictional force mobilised will be less than the limiting value.

From Eq. 8.10, we have

$$F = (W\sin\alpha - P) \qquad \ldots\text{(Eq. 8.14)}$$

From Eq. 8.14, it is obvious that if P is continuously increased, the necessary friction to be mobilised will go on decreasing, although the tendency for the block is still to slide down. If we apply a force $P_o = W\sin\alpha$ up the plane, no frictional resistance is called for and the block will be in equilibrium with no tendency to slide downwards or move upwards.

If P is still increased, the tendency of the block will now be to move *upwards*, and the frictional force F will then act *downwards* as shown in Fig. 8.6. Let us consider the case of impending upward motion, when the limiting force is mobilised, P being the minimum force necessary to keep equilibrium.

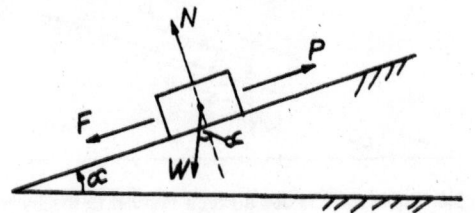

Fig. 8.6: Equilibrium under a Force Parallel to the Inclined Plane—Impending upward Motion

Resolving parallel and perpendicular to the inclined plane,

$$N = W\cos\alpha$$

$$P = W\sin\alpha + F \qquad ...(Eq.\ 8.15)$$

or $\quad P = W\sin\alpha + \mu N$

$$= W\sin\alpha + \mu W\cos\alpha$$

$$= W(\sin\alpha + \mu\cos\alpha)$$

$$= W(\sin\alpha + \tan\phi \cdot \cos\alpha)$$

$$= \frac{W(\sin\alpha\cos\phi + \cos\alpha\sin\phi)}{\cos\phi}$$

$$P = \frac{W\sin(\alpha + \phi)}{\cos\phi} \qquad ...(Eq.\ 8.16)$$

If a force larger than this acts, the block starts moving up. For any value of P between those given by Equations 8.13 and 8.16, the block will be in equilibrium with the necessary frictional force acting *up* the plane if $P < P_0$ and acting *down* the plane if $P > P_0$.

In fact even a graphical approach can be used to determine the limiting values of P (Eq. 8.13 & 8.16), by combining F and N vectorially to get the direction of the resultant. This resultant force along with P and W keep the block in equilibrium. Of these three concurrent forces, W is completely known and the directions of the other two are also known. The force triangle may be completed to determine the vector P.

8.5.3 Equilibrium under a Horizontal Force

This case is shown in Fig. 8.7.

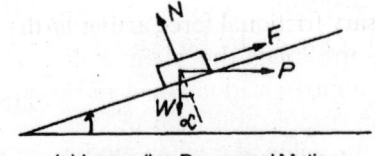

(a) Impending Downward Motion

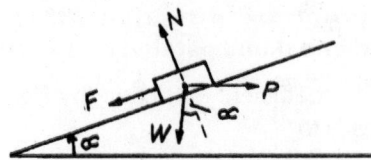

(b) Impending Upward Motion

Fig. 8.7: Equilibrium under a Horizontal Force

Referring to Fig. 8.7 (*a*), for impending downward motion, and resolving parallel to and perpendicular to the inclined plane,

$$N = W\cos\alpha + P\sin\alpha \qquad ...(Eq.\ 8.17)$$

$$F = W\sin\alpha - P\cos\alpha \qquad ...(Eq.\ 8.18)$$

(F is the limiting friction)

or $\quad \mu N = W\sin\alpha - P\cos\alpha \qquad ...Eq.\ 8.19$

Eliminating N from Eqns. 8.17 and 8.19, and putting $\mu = \tan\phi$, we get, after simplification,

$$P = W\tan(\alpha - \phi) \qquad ...(Eq.\ 8.20)$$

This is the minimum value of P required to keep the block in equilibrium. If P is increased, the block will still be in equilibrium with the value of frictional force mobilised being less than the limiting value.

In fact no friction need be mobilised to keep equilibrium when

$$P = P_0 = W\tan\alpha \qquad ...(Eq.\ 8.21)$$

if P is increased still further, the tendency for motion will be upwards, with the direction of F reversed as in Fig. 8.7 (*b*).

The equations are

$$N = W\cos\alpha + P\sin\alpha$$

$$F = P\cos\alpha - W\sin\alpha \qquad ...(Eq.\ 8.22)$$

The minimum force P for impending upward motion can be shown to be

$$P = W\tan(\alpha + \phi) \qquad ...(Eq.\ 8.23)$$

For a value of P between those given by Eqns. 8.20 & 8.23, the block will be in equilibrium with the necessary frictional force acting *up* the plane if $P < P_o$ and *down* the plane if $P > P_o$. The graphical approach mentioned in the previous sub-section is applicable here also.

8.5.4 Equilibrium under a force acting in a appropriate direction other than horizontal and parallel to the inclined plane

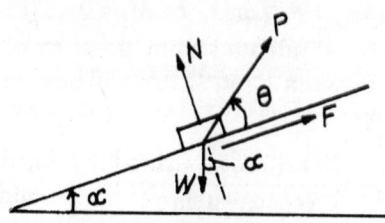

(a) Impending Downward Motion

(b) Impending upward motion

Fig. 8.8: Equilibrium under a force acting in a appro-priate direction other than horizontal and parallel to the inclined plane

(a) is for impending downward motion and
(b) for impending upward motion.

Let θ be the angle between the direction P to that of the plane, P being directed upwards.

The equations for case (a) are:

$$N = W\cos\alpha - P\sin\theta \qquad ...(Eq. 8.24)$$

$$\mu N = F = W\sin\alpha - P\cos\theta \qquad ...(Eq. 8.25)$$

P can be shown to be

$$P = W\frac{\sin(\alpha - \phi)}{\cos(\theta + \phi)} \qquad ...(Eq. 8.26)$$

The equations for case (b) are:

$$N = W\cos\alpha - P\sin\theta$$

$$\mu N = F = P\cos\theta - W\sin\alpha \qquad ...(Eq. 8.27)$$

P can be shown to be

$$P = W\frac{\sin(\alpha + \phi)}{\cos(\theta - \phi)} \qquad ...(Eq. 8.28)$$

Similarly,

$$P_0 = W\frac{\sin\alpha}{\cos\theta} \cdot \qquad ...(Eq. 8.29)$$

By substituting $\theta = -\alpha$, we get the equations for those in sub-section 8.5.2 and $\theta = 0°$, we get those in sub-section 8.5.3, as there are special cases of the general case of this sub-section.

The graphical approach is applicable for this case too.

8.6 WEDGE

The 'Wedge' is a simple contrivance used for raising or lowering heavy loads through small heights; this is achieved by the application of a small force in a convenient direction after interposing a wedge-shaped body or a wedge between the load and its supporting surface. The load is constrained from displacing laterally in the process.

The weight of the wedge is very small compared to the load sought to be lifted, and is consequently neglected in all the problems involving wedges.

Invariably, problems on wedges will involve frictional force on inclined plane, and can better be appreciated by considering a few illustrative examples.

8.7 SCREW

Screws, nuts and bolts are basic fastening accessories used in components of machines and structures. The rotational motion of an axially loaded nut over a bolt is equivalent to pushing up or down of a load along an inclined plane which is wrapped around a cylindrical surface. The axial force between them is the load and the mean cylindrical surface of the screw-threads of the bolt is the surface on which the inclined plane is wrapped.

To appreciate this, it is necessary to know the geometry of the screw-threads which are cut externally on the bolt and internally for the nut.

Let us consider the action of a nut on a screw that has square threads (Fig. 8.9).

The pitch, p, is the distance between similar points in adjacent threads, and lead, L, is the distance that a nut advances in the axial direction of the screw in one revolution. For a single-threaded screw, $L = p$, and for multi-threaded screw with n-threads, $L = np$; here we consider only the former.

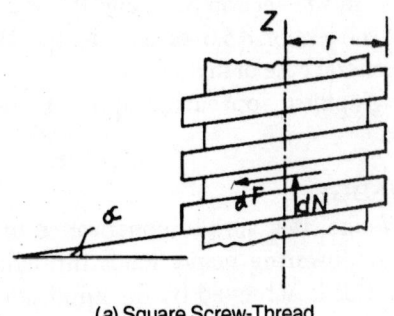

(a) Square Screw-Thread

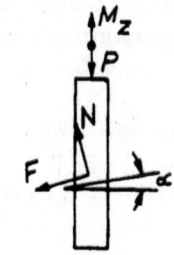

(b) Forces on the Screw-Thread

Fig. 8.9: Action of a Nut on a Square Screw-Thread

Forces are transmitted from the screw to the nut and we have a certain distribution of normal and frictional forces.

But, in view of the narrow nature of the thread, the distribution may be confined at a radial distance r from the axis, thus forming a strip-winding around the centre line of the screw.

In Fig. 8.9 (a), the infinitesimal normal and frictional forces, dN and dF, on the strip are shown. Angle α is given by

$$\tan \alpha = \frac{p}{2\pi r} \qquad \text{...(Eq. 8.30)}$$

The force distribution may be replaced by a single normal force N and a single frictional force F at the inclination shown in Fig. 8.9 (b), at a position anywhere along the thread. The inclination as well as the moment arm about the axis are constant for the elemental forces; so the moments of the concentrated forces N and F may be taken about the axis. The force distribution on the screw and the forces N and F have 'limited'

equivalence. The other forces on the screw will be taken as an axial load P and a torque M_z collinear with P as shown.

For equilibrium when motion is impending, we have:

$$\Sigma F_z = 0$$

or $\quad N\cos\alpha - P - \mu N\sin\alpha = 0 \qquad \text{...(Eq. 8.31)}$

$$\Sigma M_z = 0$$

or $\mu N\cos\alpha \cdot r + N\sin\alpha \cdot r - M_z = 0$ ...(Eq. 8.31)

N may be eliminated from these to obtain a relation between P and M_z that will be of practical significance; we get

$$M_z = \frac{P \cdot r \cdot (\mu\cos\alpha + \sin\alpha)}{(\cos\alpha - \mu\sin\alpha)} \qquad \text{...(Eq. 8.32)}$$

An important application of this is the "Screw-jack", one of the lifting machines considered in the next chapter.

8.8 DYNAMIC FRICTION

In Section 8.3, differentiation between static friction and dynamic friction has been brought out. It has also been brought out that dynamic friction is a little less than the limiting friction in the static condition.

Dynamic friction is classified into two groups:

   (*i*) Sliding friction
   (*ii*) Rolling friction

8.8.1 Sliding Friction

It is the friction experienced by a body when it tends to slide over a surface/a body. The friction that has been considered till now in the previous sections is of this type.

8.8.2 Rolling Friction

It is the friction experienced by a body which rolls on a surface. Ideally speaking, the contact of a wheel/roller with a plane surface is a point; however, in practice the contact will be through a very small area owing to the deformation of the surface upon which rolling occurs.

Let us consider the situation where a hard wheel of weight W rolls without slipping along a horizontal surface (Fig. 8.10).

A horizontal force P is obviously required to maintain uniform motion; so, a resistance—frictional resistance F—as shown, has to be mobilised to keep dynamic equilibrium, or static equilibrium, if the position at any instant is considered as in the figure.

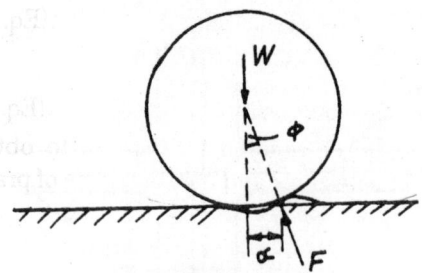

Fig. 8.10: Rolling Friction

All the three must be concurrent at the centre of the wheel or roller. Since the area of contact is small, ϕ must be a small angle.

Considering the equilibrium of the wheel,

$$W = F\cos\phi$$
$$P = F\sin\phi$$

or $\quad P/W = \tan\phi \qquad \qquad ...(\text{Eq. 8.33})$

Since ϕ is small, $\tan\phi \approx \sin\phi$

But $\sin\phi = a/r$

$$\therefore \quad \frac{P}{W} = \frac{a}{r}$$

or $\quad P = \dfrac{Wa}{r} \qquad \qquad ...(\text{Eq. 8.34})$

The distance 'a' is called the 'coefficient of rolling resistance'.

8.9 ROPE AND BELT FRICTION

Transmission of power by means of rope or belt drives involves friction which exists between the wheels and the rope or belt. Tension in the rope or belt is more on the side on which it is pulled and less on the other side—these two are the tight side and slack side, respectively.

Let the impending motion of the belt be clockwise relative to the drum, the tension, T_1, being greater than the tension T_2 (Fig. 8.11).

Let an infinitesimal length of the belt ds be considered as a free body; this subtends an angle $d\beta$ at the centre of the drum. The forces acting on this portion of the belt are tension T and $T + dT$ at the ends, normal reaction dN and frictional force $\mu \cdot dN$ as shown.

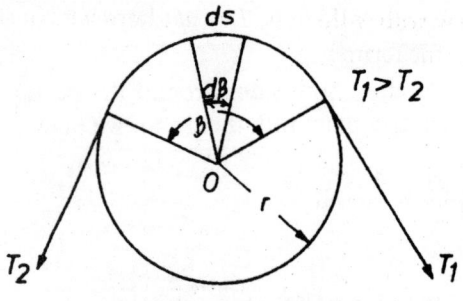

(a) Rope or Belt around Drum

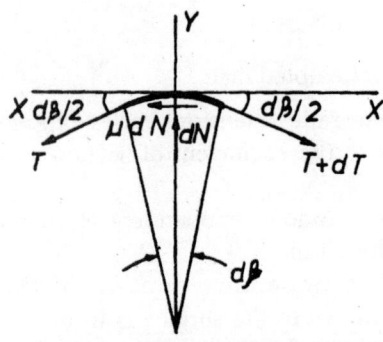

(b) Free Body of a Small Length of Belt

Fig. 8.11: Rope or Belt Friction

Resolving in the radial and tangential directions, we have:

$$\Sigma F_x = 0$$

$$-T\cos\frac{d\beta}{2} + (T + dT)\cos\frac{d\beta}{2} - \mu dN = 0$$

$$\therefore \quad dt\cos\frac{d\beta}{2} = \mu dN$$

$$\Sigma F_y = 0$$

$$-T\sin\frac{d\beta}{2} - (T + dT)\sin\frac{d\beta}{2} + dN = 0$$

$$\therefore \quad -2T\sin\frac{d\beta}{2} - dT\sin\frac{d\beta}{2} + dN = 0$$

Since $d\beta$ is very small,

$$\left.\begin{array}{l} dT = \mu dN \\[6pt] -TdB - dT\dfrac{d\beta}{2} + dN = 0 \end{array}\right\} \quad ...(\text{Eq. 8.35})$$

Neglecting the second term, a product of two infinitesimals,

$$Td\beta = dN \qquad \qquad ...(\text{Eq. 8.36})$$

Eliminating dN from these equations, we have:

$$dT = \mu T \cdot d\beta$$

$$\therefore \quad \frac{dT}{T} = \mu d\beta$$

Integrating both sides around the portion of the belt in contact with the drum, we obtain

$$\int_{T_2}^{T_1} \frac{dT}{T} = \int_0^B \mu \cdot d\beta$$

or $\quad \log \dfrac{T_1}{T_2} = \mu \cdot \beta$

$$\therefore \quad \frac{T_1}{T_2} = e^{\mu\beta} \qquad \qquad \ldots\text{(Eq. 8.37)}$$

It may be noted that:

(*i*) The ratio of tensions varies exponentially with the coefficient of friction and angle of lap;

(*ii*) This ratio is independent of the radius of the drum;

(*iii*) The impending motion of the rope relative to the surface is from the "slack side" to the "tight side";

(*iv*) The impending motion of the drum relative to the rope is naturally opposite to that of the rope;

(*v*) The angle β must be measured in radians; and

(*vi*) The equation 8.37 is valid only at impending motion of the belt.

Generally the lap of the rope or belt may be varied at will round a cylindrical pulley or drum. The ratio of tensions may be increased very fast with the lap angle of the belt over the drum as seen from the following Table.

TABLE 8.1: LAP ANGLE AND TENSION-RATIO

No. of Rounds	β in Radians	Tension ratio $T_1/T_2 = e^{\mu\beta}$ ($\mu = 0.2$)
1/2 round	3.142	1.874
1	6.283	3.514
2	12.568	12.345
3	18.852	43.376
4	25.136	152.406

8.10 DISC AND BEARING FRICTION

Power can be transmitted between two shafts by virtue of the friction present between the two discs at the ends of the shafts (Fig. 8.12) (The friction between two circular surfaces is called 'Disc friction').

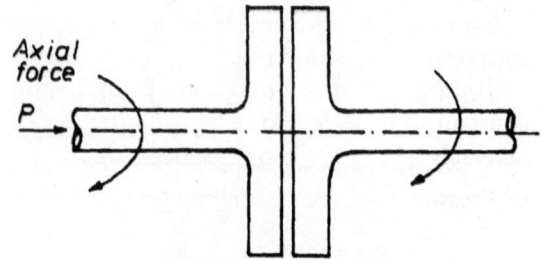

(a) Discs at the Ends of Shafts

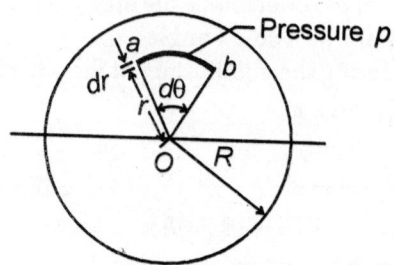

(b) Free Body of a Small Element

Fig. 8.12: Disc Friction

Let the radius of the discs be R, area be A, and let it be subjected to an axial force P. The free body of an element ab of the disc at a radial distance r from the centre and a central angle $d\theta$ as shown in Fig. 8.12 (*b*) is considered.

$$dA = rdr\, d\theta$$

Let p be the pressure at this radial distance.

Then, force on the element $= p \cdot dA$

Frictional force on the element

$$dF = \mu(p\, dA)$$

$$= \mu(pr \cdot dr \cdot d\theta)$$

Moment of this force about the axis of the disc

$$d\mu = \mu(pr \cdot d \cdot r \cdot d\theta)r$$

$$= \mu pr^2 \cdot dr \cdot d\theta$$

Total moment of the frictional force at the contact surface

$$M = \int_0^{2\pi} \int_0^R \mu \cdot pr^2 \cdot d\theta \cdot dr = \int_0^R \int_0^{2\pi} \mu pr^2 \cdot dr \cdot d\theta$$

$$= 2\pi\mu \int_0^R pr^2 dr \qquad ...(Eq.\ 8.38)$$

For evaluating this integral, the variation of p with respect to r should be known.

Two possible assumptions are that pressure is constant with respect to radius and that it varies inversely with respect to the radius.

(a) Pressure is constant with respect to the radius:

$$p = \frac{P}{\pi R^2} \qquad ...(Eq.\ 8.39)$$

Therefore Eq. 8.38 may be written as

$$M = 2\pi p\mu \int_0^R r^2 dr = \frac{2\pi\mu pR^3}{3} \qquad ...(Eq.\ 8.40)$$

In terms of P,

$$M = \frac{2}{3}\mu PR \qquad ...(Eq.\ 8.41)$$

(b) Pressure varies inversely with respect to the radius:

$$p \propto \frac{1}{r}$$

or $\quad p = \dfrac{k}{r}$, k being a constant $\quad ...(Eq.\ 8.42)$

$$P = \int_0^R 2\pi rp\, dr \int_0^R 2\pi r \frac{k}{r} \cdot dr = 2\pi kR \quad ...(Eq.\ 8.43)$$

where $k = \dfrac{P}{2\pi R}$

From Eq. 8.42, $p = \left(\dfrac{P}{2\pi R}\right)\dfrac{1}{r}$

Substituting in Eq. 8.38,

$$M = 2\pi\mu \int_0^R \left(\frac{P}{2\pi Rr}\right) r^2 dr$$

$$= \frac{\mu P}{R} \int_0^R r\, dr = \frac{\mu P}{R} \cdot \frac{R^2}{2}$$

$$M = \frac{\mu PR}{2} \qquad ...(Eq.\ 8.44)$$

Bearing Friction

Journal bearing gives lateral support to the rotating shaft, whereas the thrust bearing gives an axial support to the shaft. A thrust bearing can be an end bearing or a collar bearing. Further, an end bearing may be of the flat pivot type or of the conical pivot type.

The moment M of the frictional force may be can be obtained by the same approach as for the disc clutch even in the case of the end bearings (Fig. 8.13).

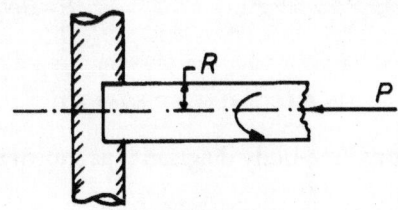

(a) Flat Pivot Type of End Bearing

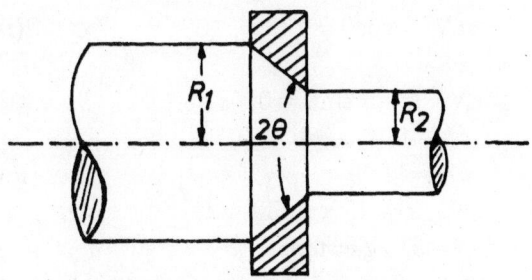

(b) Conical Pivot Type of End Bearing

Fig. 8.13: Bearing Friction (Thrust Bearing)

Flat pivot type of end bearing

$$M = \frac{2}{3}\mu PR \text{ (for uniform pressure)} \quad ...(Eq.\ 8.45)$$

$$M = \frac{\mu PR}{2} \text{ (for pressure varying inversely with}$$
$$\text{the radius)} \qquad ...(Eq.\ 8.46)$$

Conical Pivot Bearing

$$M = \frac{2}{3}\frac{\mu P}{\sin\theta}\left(\frac{R_1^3 - R_2^3}{R_1^2 - R_2^2}\right) \text{ (for uniform pressure)}$$
$$...(Eq.\ 8.47)$$

ILLUSTRATIVE PROBLEMS

PROBLEM 8.1: A body of weight 50 N is placed on a rough horizontal plane. To just move the body on the horizontal plane, a push of 15 N inclined at 30° to the horizontal plane is required. Find the coefficient of friction.

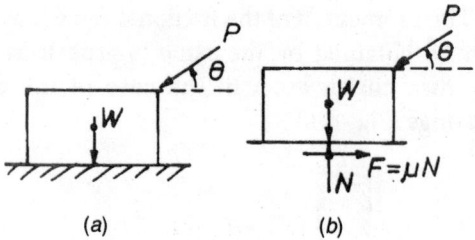

(a)   (b)

Fig. 8.14 (Illustrative Problem 8.1)

Solution: Free-body diagram is as shown in Fig. 8.14 (*b*)

Applying equilibrium equations

$\Sigma H = 0$

$\mu N - P\cos\theta = 0$ ...(i)

$\Sigma V = 0$

$N - W - P\sin\theta = 0$ ...(ii)

$\mu N - 15 \times \cos 30° = 0$

$\mu N = 12.99$ ...(iii)

$N - 50 - 15 \times \sin 30° = 0$

$N = 57.5$ Newtons   ...(iv)

(iii)/(iv) gives

$\mu = 12.99/57.5 = 0.2259$

PROBLEM 8.2: Find the least friction required to pull a body of weight W placed on a rough horizontal plane, when the force is applied at an angle θ with the horizontal.

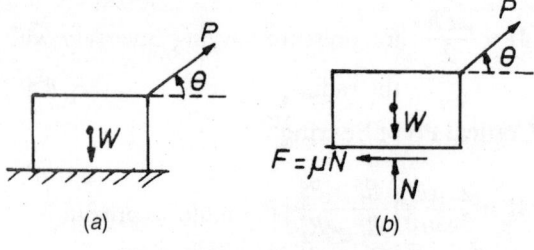

(a)   (b)

Fig. 8.15 (Illustrative Problem 8.2)

Solution: Free body diagram is shown in Fig. 8.15 (*b*)

Writing the equations of equilibrium

$\Sigma H = 0$; $P\cos\theta - \mu N = 0$

where $\mu = \tan\phi$

     ϕ = angle of friction

     $P\cos\theta - (\tan\phi)N = 0$ ...(i)

     $\Sigma V = 0$

     $P\sin\theta + N - W = 0$

or $P\sin\theta + N = W$ ...(ii)

Let us eliminate 'N' from equations (i) & (ii)

$$P = \frac{W\tan\phi}{\cos\theta + (\sin\theta)(\tan\phi)}$$

$$= \frac{W}{\sin\theta + \cos\theta \cdot \cot\phi}$$

$$= \frac{W \cdot \sin\phi}{\sin\phi \cdot \sin\theta + \cos\phi \cdot \cos\theta}$$

$$P = \frac{W \cdot \sin\phi}{\cos(\theta - \phi)}$$

Now in this expression ϕ & W are constants. Hence, numerator $W \cdot \sin\phi$ is a constant quantity. When the denominator is maximum, the quantity p is minimum. Denominator $\cos(\theta - \phi)$ is maxi-mum, when it attains a value of 1.

$\cos(\theta - \phi) = 1$

i.e. $\cos 0° = 1$

$\therefore$ $\theta - \phi = 0$

or $\theta = \phi$

i.e. the angle of inclination of force P with the horizontal plane is equal to angle of friction.

Hence least value of $P = W\sin\theta$

PROBLEM 8.3: A wooden block rests on a rough horizontal plane. Determine the force required to (*a*) pull it (*b*) push it. Assume the weight of the block to be 100 N and the coefficient of friction μ = 0.3. Also assume that the force (pull or push) is inclined at 20° to (*a*) pull the horizontal plane.

Solution: Let P_1 be the force required to just pull it. Free-body diagram is as shown in Fig. 8.16 (*b*).

Writing the equation of equilibrium

$\Sigma H = 0$

$P_1\cos 20° - \mu N = 0$

$0.9397 P_1 - 0.3 N = 0$ ...(i)

$\Sigma V = 0$

$P_1 \sin\theta - W + N = 0$

$P_1 \sin 20° - 100 + N + 0$

$0.3420 P_1 + N = 100$ ...(ii)

$(i) + (0.3) \times (ii)$

$(0.9397 P_1 - 0.3N) + 0.3(0.3420 P_1 + N)$

$= 0.3 \times 100$

$(0.9397 + 0.1026)P_1 = 30$

$P_1 = \dfrac{30}{(0.9397 + 0.1026)}$

$P_1 = \dfrac{30}{1.0423} = 28.7825 \text{ N}$

Check

$N = \dfrac{0.9397 \times 28.7825}{0.3} = 90.1564 \text{ N}$

$0.3420 \times 28.7825 + 90.1564 = 100$

$9.843615 + 90.1564 = 100$

$100 = 100 \text{ OK}$

(b) **Push:** Let P_2 be the force required

Free-body diagram is as shown in Fig. 8.16 (c)

Writing the equations of equilibrium

$\Sigma H = 0$

$\mu N - P_2 \cos 20° = 0$

$0.3 N - 0.9397 P_2 = 0$ ...(i)

$\Sigma V = 0$

$N - W - P_2 \sin 20° = 0$

$N - 100 - 0.3420 P_2 = 0$

$N - 0.3420 P_2 = 100$ ...(ii)

$0.3 \times (ii) - (i)$ gives

$0.3 N - 0.3 \times 0.3420 P_2 - 0.3N + 0.9397 P_2$

$= 0.3 \times 100$

$-0.1026 P_2 + 0.9397 P_2 = 30$

$P_2 = \dfrac{30}{0.9397 - 0.1026}$

$\therefore \quad P_2 = \dfrac{30}{0.8371} = 35.838 \text{ N}$

Check

$N = \dfrac{0.9397 \times 35.838}{0.3}$

$= 112.2566 \text{ N}$

$112.2566 - 0.3420 \times 35.838 = 100$

$112.2566 - 12.2566 = 100$

$100 = 100 \text{ (OK)}$

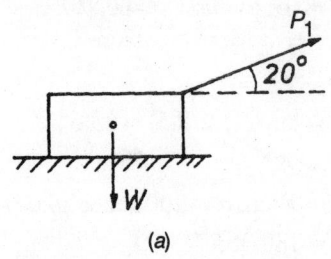

(a)

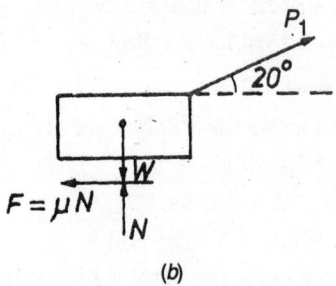

(b)

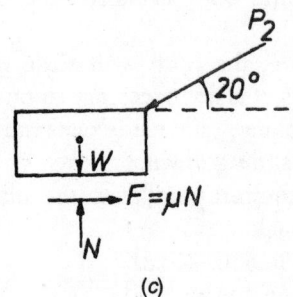

(c)

Fig. 8.16 (Illustrative Problem 8.3)

PROBLEM 8.4: An inclined plane as shown in Fig. 8.17 is used to unload slowly a heavy safe body weighing 200 N from a truck 1.2 m high onto the ground. The angle of friction is 15°. State whether it is necessary to push the body down the plane or hold it back from sliding down.

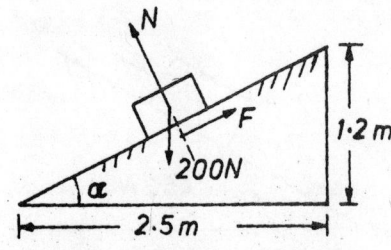

Fig. 8.17 (Illustrative Problem 8.4)

Solution:

$$W = 200 \text{ N}$$

$$\mu = \tan\phi = \tan 15° = 0.26795$$

$$\tan\alpha = \frac{1.2}{2.5}$$

$$\alpha = \tan^{-1}(1.2/2.5) = 25.641°$$

or $= 25°38'$

$$N = W\cos\alpha = 200 \times \cos 25.641°$$

$$= 180.305 \text{ N}$$

∴   Force of friction, $F = \mu N$

$$= 0.26795 \times 180.305$$

$$= 48.313 \text{ N}$$

Now resolving the 200 N force along the plane

$$= 200 \sin\alpha$$

$$= 200 \times \sin 25.641°$$

$$= 86.546 \text{ N} < 48.313 \text{ N}$$

The force along the plane which is responsible for sliding the body is more than the force of friction.

Therefore, the body will slide down or in other words, it is not necessary to push the body down the plane, rather it is necessary to hold it back from sliding down.

Force required parallel to the plane to hold the body back

$$= 86.546 - 48.313$$

$$= 38.233 \text{ N}$$

PROBLEM 8.5: A body of weight 450 Newtons is lying on a rough plane inclined at an angle of 30° with the horizontal. It is supported by an effort (P) parallel to the plane as shown in Fig. 8.18. Determine the maximum and minimum values of P, for which the equilibrium can exist, if the angle of friction is 20°.

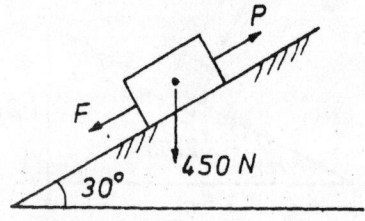

Fig. 8.18 (Illustrative Problem 8.5)

$$W = 450 \text{ N}$$

$$\alpha = 30°$$

$$\phi = 20°$$

Maximum value of P:

For the maximum value of P, the body is at the point of sliding upwards.

$$P = W \times \frac{\sin(\alpha + \phi)}{\cos\phi}$$

$$= 450 \times \frac{\sin(30° + 20°)}{\cos 20°}$$

$$= \frac{450 \times \sin 50°}{\cos 20°}$$

$$= \frac{450 \times 0.7660}{0.9397}$$

$$= 366.82 \text{ N}$$

Minimum Value of P:

For the minimum value of P, the body is at the point of sliding downwards.

$$P = W\frac{\sin(\alpha - \phi)}{\cos\phi}$$

$$= 450 \times \frac{\sin(30° - 20°)}{\cos 20°}$$

$$= \frac{450 \times \sin 10°}{\cos 20°}$$

$$= \frac{450 \times 0.17365}{0.9397}$$

$$= 83.16 \text{ N}$$

PROBLEM 8.6: A chord connects two bodies of weight 250 N and 500 N. The two bodies are placed on an inclined plane and chord is parallel to the inclined plane. The coefficient of friction for the weight of 250 N is 0.15 and that for 500 N is 0.3. Determine the inclination of the plane to the horizontal and tension in the chord when the motion is about to take place, down the inclined plane. The body weighing 250 N is below the body weighing 500 N.

Solution: Let A and B are the two bodies placed on inclined plane having inclination α with horizontal as shown in Fig. 8.19 (a). Chord AB is in tension, carrying pull T. When the motion is about to take place down the plane friction forces act upward the plane.

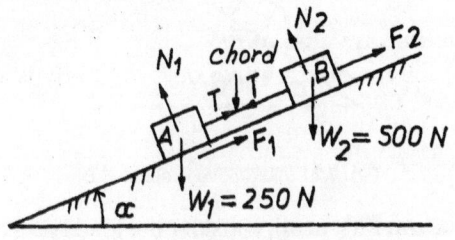

(a) Two Connected Bodies on an Inclined Plane

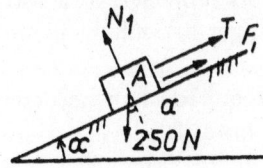

(b) Free Body Diagram for 'A'

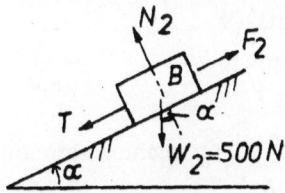

(c) Free Body Diagram for 'B'

Fig. 8.19 (Illustrative Problem 8.6)

Consider the free body diagram for 'A'
Resolving the forces perpendicular to the plane.

$$N_1 = 250\cos\alpha$$

Resolving the forces along the plane

$$T + F_1 = 250\sin\alpha$$
$$T + \mu_1 N_1 = 250\sin\alpha$$
$$T + \mu_1 250\cos\alpha = 250\sin\alpha$$
$$T + 0.15 \times 250\cos\alpha = 250\sin\alpha$$
$$T + 37.5\cos\alpha = 250\sin\alpha \qquad \ldots(i)$$

Consider the free body diagram for 'B'
Resolving the forces perpendicular to the plane

$$N_2 = 500\cos\alpha$$

Resolving the forces along the plane

$$T + 500\sin\alpha = F_2$$

$$T + 500\sin\alpha = \mu_2 N_2$$
$$T + 500\sin\alpha = 0.3 \times 500 \times \cos\alpha$$
$$T + 500\sin\alpha = 150\cos\alpha \qquad \ldots(ii)$$

Solving (i) & (ii)

$$T + 500\sin\alpha = 150\cos\alpha$$
$$(-)\,T + 37.5\cos\alpha = 250\sin\alpha$$
$$500\sin\alpha - 37.5\cos\alpha = 150\cos\alpha - 250\sin\alpha$$
$$750\sin\alpha = 187.5\cos\alpha$$

$$\tan\alpha = \frac{187.5}{750} = 0.25$$

$$\alpha = \tan^{-1}(0.25)$$
$$= 14.04° \approx 14°$$
$$T = 150\cos(14.04°) - 500\sin(14.04°)$$
$$T = 145.52 - 121.30$$
$$T = 24.22 \text{ N}$$

Check

$$T = 250\sin\alpha - 37.5\cos\alpha$$
$$= 250\sin(14.04°) - 37.5\cos(14.04°)$$
$$= 60.65 - 36.38$$
$$= 24.27 \text{ N}$$

PROBLEM 8.7: A load of 300 N, resting on an inclined rough plane, can be moved up the plane by a force of 400 N applied horizontally or by a force 250 N applied parallel to the plane. Find the inclination of the plane and the coefficient of friction.

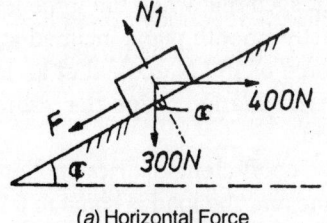

(a) Horizontal Force

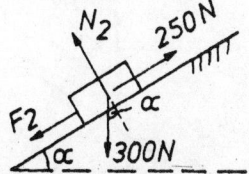

(b) Force Parallel to the Plane
Fig. 8.20 (Illustrative Problem 8.7)

Solution: Let α = Inclination of the plane, and

ϕ = Angle of friction

First of all, consider the load of 300 N subjected to a horizontal force of 400 N as shown in Fig. 8.20 (*a*).

The magnitude of the force, which can move the load up the plane,

$$P = W \tan(\alpha + \phi)$$

$$400 = 300 \tan(\alpha + \phi)$$

$$(\alpha + \phi) = 53°08'$$

Now consider the load of 300 N subjected to a force of 250 N along the plane as shown in Fig. 8.20 (*b*).

The magnitude of the force, which can move the load up the plane.

$$P = W \times \frac{\sin(\alpha + \phi)}{\cos\phi}$$

$$250 = 300 \times \frac{\sin 53°08'}{\cos\phi}$$

$$\cos\phi = \frac{300}{250}(0.8)$$

$$\phi = \cos^{-1}(0.96) = 16°16'$$

$$\alpha = 53°08' - 16°16' = 36°52'$$

$\therefore$ $\mu = \tan\phi = \tan 16°16' = 0.292$

PROBLEM 8.8: Find the force required to move a load of 400 N up a rough plane, the force being applied parallel to the plane. The inclination of the plane is such that when the same load is kept on a perfectly smooth plane inclined at the same angle, a force of 100 N applied at an inclination of 30° to the plane, keeps the same load in equilibrium.

Assume coefficient of friction between the rough plane and the load is equal to 0.25.

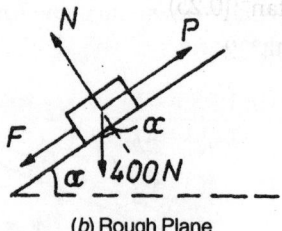

(*a*) Smooth Surface

(*b*) Rough Plane

Fig. 8.21 (Illustrative Problem 8.8)

Solution: First of all, consider the load lying on a smooth plane inclined at an angle α with the horizontal and subjected to a force of 100 N acting at an angle 30° with the plane as shown in Fig. 8.21 (*a*).

The force required when the load is at the point of sliding downwards, P

$$P = W\frac{\sin(\alpha + \phi)}{\cos(\theta - \phi)}$$

$$P = 100 \text{ N}$$

$$W = 400 \text{ N}$$

$$\alpha = ?$$

$\phi = 0°$ ($\therefore$ the plane is smooth)

$$\theta = 30°$$

$$100 = 400\frac{\sin(\alpha + \theta°)}{\cos(30° - 0°)}$$

$$\sin\alpha = \frac{100 \times \cos 30°}{400} = \frac{100 \times 0.866}{400}$$

$$\sin\alpha = 0.2165$$

$$\alpha = \sin^{-1}(0.2165) = 12°30'$$

Now consider the load lying on the rough plane inclined at an angle of 12°30' with the horizontal as shown in Fig. 8.21(*b*).

The force required to move the load up the plane, P

$$P = \frac{W\sin(\alpha + \phi)}{\cos\phi}$$

$$P = ?$$

$$W = 400 \text{ N}$$

$$\alpha = 12°30'$$

$$\mu = 0.25 = \tan\phi$$

$\phi = \tan^{-1}(0.25)$

$\phi = 14°02'$

$\therefore \quad P = \dfrac{400\sin(12°30' + 14°02')}{\cos 14°02'} = \dfrac{400\sin(26°32')}{\cos 14°02'}$

$P = \dfrac{400 \times 0.4467}{0.9702}$

$P = 184.17$ N

The force required to move the load up the plane, P is 184.17 N.

PROBLEM 8.9: A block weighing 2000 N overlying a 10° wedge on a horizontal floor and leaning against a vertical wall, is to be raised by applying a horizontal force to the wedge. Assuming coefficient of friction between all the surfaces in contact to be 0.3, determine the minimum horizontal force to be applied to raise the block.

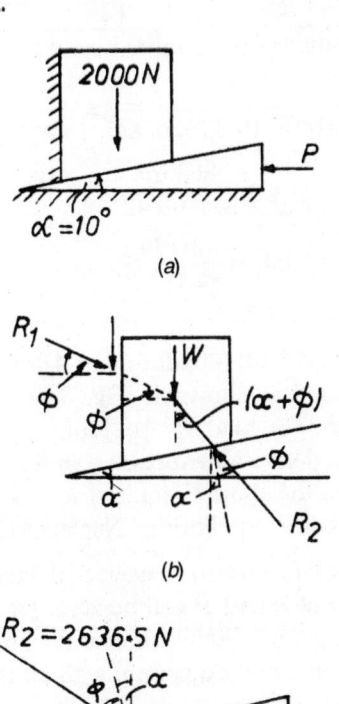

(a)

(b)

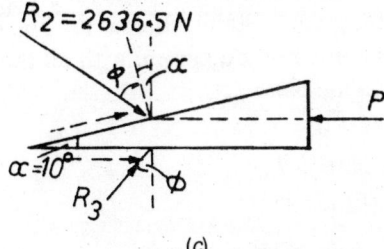

(c)

Fig. 8.22 (Illustrative Problem 8.9)

Solution: Fig. 8.22 (a) shows a block which is to be raised up by applying minimum force P on the wedge.

$\mu = \tan\phi = 0.3$

$\phi = \tan^{-1}(0.3) = 16°42'$

$W = 2000$ N

$\alpha = 10°$

Let $\quad P =$ Minimum horizontal force to be applied to raise the block.

Equilibrium of block should be considered first because the load is given. As the block moves up, frictional force developed between wall and block acts downward. Also the angle of friction is known, therefore a reactive force R_1 making an angle ϕ with the normal to the wall is shown in Fig. 8.22(b). R_1 is the resultant of frictional force and normal reaction. The block moves up the inclined plane, therefore frictional force acts down the plane. Normal reaction perpendicular to inclined plane is also there. These two will give a resultant force R_2 making an angle ϕ with the normal. Block is in equilibrium under the action of forces, W, R_1 and R_2.

Similarly, wedge is in equilibrium under the action of forces P, R_2 and R_3 as shown in Fig. 22 (c).

Method-I

Consider the equilibrium of block

$\Sigma H = 0$

i.e. $\quad R_1\cos\phi = R_2\sin(\alpha + \phi)$

$R_1\cos 16°42' = R_2\sin(10 + 16°42')$

$R_1\cos 16°42' = R_2\sin 26°42'$

$0.9578 R_1 = 0.4493 R_2$

$R_2 = \dfrac{0.9578}{0.4493} R_1$

$R_2 = 2.132 R_1$

$\Sigma V = 0$

$R_1\sin\phi + W = R_2\cos(\alpha + \phi)$

$R_2\sin 16°42' + 2000 = R_2\cos 26°42'$

$0.2874 R_1 + 2000 = R_2 0.8934$

$0.2874 R_1 + 2000 = 2.132 \times R_1 \times 0.8934$

$0.2874 R_1 + 2000 = 1.9047 R_1$

$R_1(1.9047 - 0.2874) = 2000$

$1.6173 R_1 = 2000$

$R_1 = \dfrac{2000}{1.6173} = 1236.63$ N

$R_2 = 2.132 \times 1236.63 = 2636.49$ N

Consider the equilibrium of the wedge

$\Sigma H = 0$

$P = R_2\sin(\alpha + \phi) + R_3\sin\phi$

$P = 2636.49\sin 26°42' + R_3\sin 16°42'$

$P = 1184.63 + 0.2874 R_3$

$\Sigma V = 0$

$R_2\cos(\alpha + \phi) = R_3\cos\phi$

$R_3 = \dfrac{\cos 26°42'}{\cos(16°42')} R_2$

$R_3 = 2636.49 \times \dfrac{0.8934}{0.9578} = 2459.22$ N

$\therefore \quad P = 1184.63 + 0.2874(2459.22)$

$= 1184.63 + 706.78 = 1891.41$ N

Method-II

By applying Lami's theorem to the block

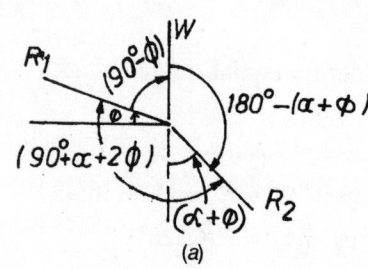

(a)

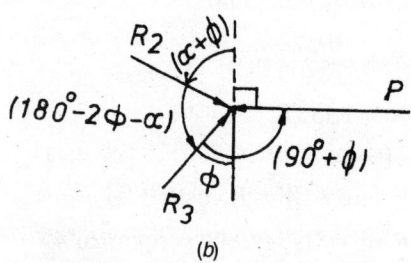

(b)

Fig. 8.23 (Illustrative Problem 8.9)

$\dfrac{R_2}{\sin(90 - \phi)} = \dfrac{W}{\sin(90 + \alpha + 2\phi)}$

$= \dfrac{R_1}{\sin 180 - (\alpha + \phi)}$

$\dfrac{R_2}{\sin(90 - 16°42')} = \dfrac{2000}{\sin(90° + 10° + 2 \times 16°42')}$

$= \dfrac{R_1}{\sin(180° - (10° + 16°42'))}$

$\dfrac{R_2}{\sin 73.3°} = \dfrac{2000}{\sin 133.4°} = \dfrac{R_1}{\sin 153.3°}$

$R_2 = 2000\dfrac{\sin 73.3°}{\sin 133.4°} = 2636.54$ N

$R_1 = 2000\dfrac{\sin 153.3°}{\sin 133.4°} = 1236.81$ N

Now applying Lami's theorem to the wedge

$\therefore \quad \dfrac{P}{\sin[180° - \phi - (\alpha + \phi)]}$

$= \dfrac{R_2}{\sin(90 + \phi)} = \dfrac{R_3}{\sin(90 + \alpha + \phi)}$

$\dfrac{P}{\sin(180 - 16°42' - 26°42')} = \dfrac{2636.54}{\sin(90 + 16°42')}$

$\dfrac{P}{\sin 136.6°} = \dfrac{2636.54}{\sin 106°42'}$

$P = 2636.54\dfrac{\sin 136.3°}{\sin 106.70°}$

$= 1891.30$ N

PROBLEM 8.10: A uniform bar AB of weight W is supported as shown in Fig. 8.24 (a) and is subjected to a load P. Applying equilibrium equations derive an expression in terms of P, θ and W, for the effort F required at B as shown to keep the bar in equilibrium. Neglect friction.

Solution: Friction is to be neglected. Hence force of friction at B and D will be zero. Fig. 8.24 (a) shows the given system, whereas the Fig. 8.24 (b) shows its free body diagram, with all the forces acting on the bar.

$\Sigma H = 0$

$R_D\sin\theta = F$ ...(1)

$\Sigma V = 0$

$R_B + R_D\cos\theta = W + P$

$R_D\cos\theta = W + P - R_B$ ...(2)

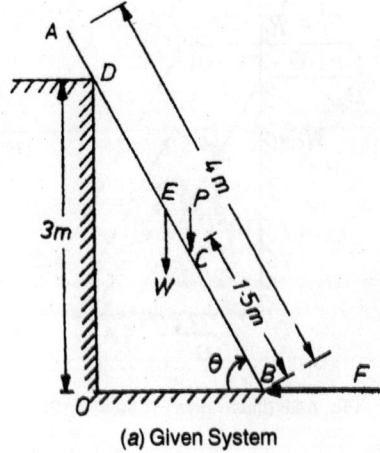

(a) Given System

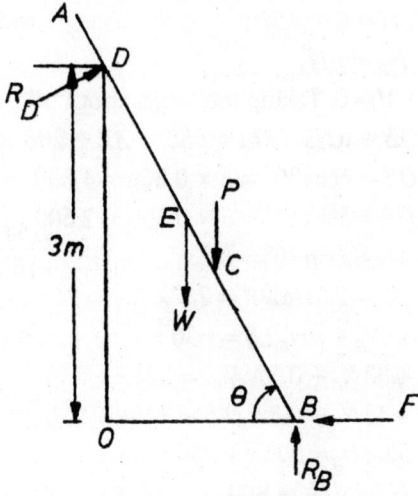

(b) Free Body Diagram

Fig. 8.24 (Illustrative Problem 8.10)

$$\Sigma M = 0 \ 'D'$$

$$R_B 3\cot\theta = (3\cot\theta - 1.5\cos\theta)P$$
$$+ (3\cot\theta - 2\cos\theta)W + 3F \qquad ...(3)$$

From (1) & (2)

$$\tan\theta = F/(W + P - R_B)$$

or $\quad F = (W + P)\tan\theta - R_B\tan\theta \qquad ...(4)$

$$3F = 3R_B\cot\theta - 3\cot\theta(P + W)$$
$$+ \cos\theta(2W + 1.5P) \qquad ...(3)$$

$(4) \times 3\cot\theta$ gives $(3\cot\theta)F = 3(W+P) - 3R_B \quad ...(5)$

$(3) \times \tan\theta$ gives $(3\tan\theta)F = 3R_B - 3(P+W)$
$$+ \sin\theta(2W+1.5P) \ ...(6)$$

$(5) + (6)$ gives $3F(\cot\theta + \tan\theta) = \sin\theta(2W + 1.5P)$

$$3F\left(\frac{\cos\theta}{\sin\theta} + \frac{\sin\theta}{\cos\theta}\right) = \sin\theta(2W + 1.5P)$$

$$3F\left(\frac{\cos^2\theta + \sin^2\theta}{\sin\theta \cdot \cos\theta}\right) = \sin\theta(2W + 1.5P)$$

$$\frac{3F}{\sin\theta \times \cos\theta} = \sin\theta(2W + 1.5P)$$

$$(\because \ \cos^2\theta + \sin^2\theta = 1)$$

$$F = \frac{(2W + 1.5P)\cos\theta\sin^2\theta}{3}$$

PROBLEM 8.11: A ladder of length 5 m weighing 300 N is placed against a vertical wall making an angle of $\alpha°$ with the floor as shown in Fig. 8.25. The coefficient of friction between the wall and ladder is 0.20 and that between the floor and the ladder is 0.30. The ladder in addition to its own weight has to support a man weighing 700 N at a distance of 3.5 m from *A*. Calculate the minimum value of α so that there is no slipping of the ladder.

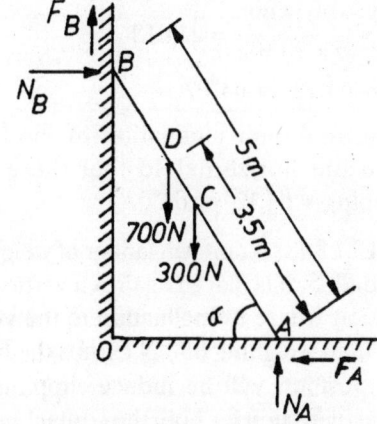

Fig. 8.25 (Illustrative Problem 8.11)

Solution: Applying equilibrium equation

$$\Sigma H = 0 \qquad \therefore \ F_A = N_B$$

But $\quad F_A = \mu_f N_A = 0.3 N_A$

$$\therefore \ 0.3 N_A = N_B \qquad ...(1)$$

$$\Sigma V = 0$$

$N_A + F_B = 700 + 300 = 1000$

But $F_B = \mu_W N_B = 0.2 N_B$

$\therefore$ $N_A + 0.2 N_B = 1000$ ...(2)

$\Sigma M = 0$ Taking moments about 'A'

$N_B \times 5 \times \sin\alpha + F_B \times 5 \times \cos\alpha$

$= 300 \times 2.5\cos\alpha + 700 \times 3.5\cos\alpha$

$= (750 + 2450)\cos\alpha$

$5 N_B \sin\alpha + 5 \times 0.2 N_B \cos\alpha = 3200\cos\alpha$

$5 N_B \sin\alpha + N_B \cos\alpha = 3200\cos\alpha$ ...(3)

From (1) & (2)

$N_A + 0.2 \times 0.3 \times N_A = 1000$

$N_A (1 + 0.06) = 1000$

$\therefore$ $N_A = \dfrac{1000}{1.06} = 943.4$ N   is

$N_B = 0.3 \times 943.4$

$N_B = 283.02$ N

Substituting the value of N_B in Eqn. (3)

$5 \times 283.02 \times \sin\alpha + 283.02 \times \cos\alpha = 3200 \times \cos\alpha$

$(5 \times 283.02)\sin\alpha = (3200 - 283.02)\cos\alpha$

$\tan\alpha = \dfrac{2916.98}{5 \times 283.02} = 2.0613$

$= 64.12°$ or $64°07'$

$\therefore$ The minimum inclination of the ladder with the horizontal so that there is no slipping $= 64.12°$ or $64°07'$

PROBLEM 8.12: A uniform ladder of weight 200 N and length 5 m is placed against a vertical wall in a position where its inclination to the vertical is 30°. A man weighing 650 N climbs the ladder. At what position will he induce slipping. The coefficient of friction for both the contact surfaces of the ladder is 0.3.

Solution: Fig. 8.26 shows the ladder and the forces acting on it. Let D be the position of the man, where he will induce slipping.

Let AD be x metres

For equilibrium of the ladder

$\Sigma V = 0$ ——

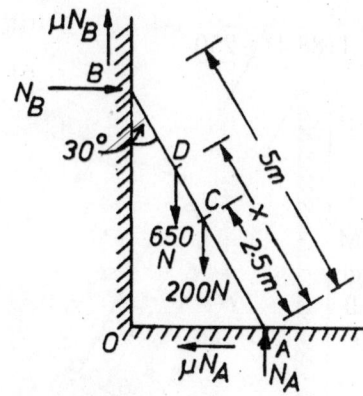

Fig. 8.26 (Illustrative Problem 8.12)

$\therefore$ $N_A + \mu N_B = 200 + 650 = 850$ ...(i)

$\Sigma H = 0$

$\therefore$ $N_B = \mu N_A$ ...(ii)

$\Sigma M = 0$, Taking moments about 'A'

$N_B \cdot OB + \mu N_B \times OA = 650 \times AE + 200 \times AF$

$OB = 5\cos30° = 5 \times 0.866 = 4.330$ m

$OA = 5\sin30° = 5 \times 0.500 = 2.500$ m

$AE = x\sin30° = 0.5x$

$AF = 2.5\sin30° = 2.5 \times 0.5 = 1.250$ m

$\therefore$ $4.33 N_B + \mu N_B 2.5 = 650 \times 0.5x + 200 \times 1.25$

$4.33 N_B + 0.3 N_B 2.5 = 325x + 250$

$4.33 N_B + 0.3 N_B 2.5 = 325x + 250$

$5.08 N_B = 325x + 250$ ...(iii)

$N_A + \mu N_B = 850$ ...(i)

$N_B = \mu N_A$ ...(ii)

$5.08 N_B = 325x + 250$ ...(iii)

From (i) & (ii)

$N_A + \mu^2 N_a = 850$

$N_A (1 + \mu^2) = 850$

$N_A = \dfrac{850}{1 + \mu^2} = \dfrac{850}{1 + (0.3)^2} = \dfrac{850}{1.09}$

$= 779.82$ N

$N_B = 0.3 \times 779.82$

$= 233.945$ N

Substituting in equation (iii)

$x = \dfrac{(5.08 \times 233.945) - 250}{325}$

$$= \frac{1188.44 - 250}{325}$$

$$= \frac{938.44}{325}$$

$$= 2.8875 \text{ m}$$

PROBLEM 8.13: A uniform ladder of length 5 m and weighing 300 N is placed against a smooth vertical wall with its lower end 4 m from the wall. If the ladder is just to slip, find:

(a) the coefficient of friction between ladder and floor,

(b) frictional force acting on the ladder at the point of contact between ladder and floor.

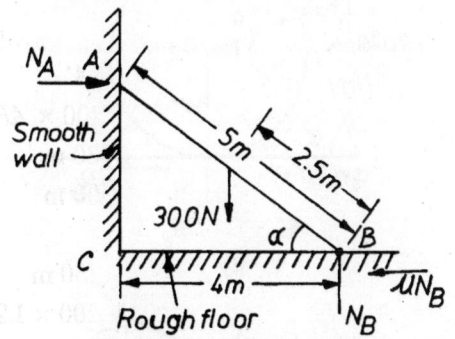

Fig. 8.27 (Illustrative Problem 8.13)

Solution: In the Fig. 8.27, AB is a ladder of 5 m length. At the end A, N_A is normal reaction. Since wall is smooth, coefficient of friction is zero. Therefore no frictional force will be at point A. At point B, frictional force μN_B will be developed, because the ladder slips to the right. Hence, frictional force will be towards left as shown. N_B is normal reaction. The whole ladder is in equilibrium under the action of the forces, N_A, μN_B, weight of ladder and N_B.

From Fig. 8.27

$$AB^2 = AC^2 + CB^2$$

$$5^2 = AC^2 + 4^2$$

$$AC = \sqrt{25 - 19} = \sqrt{9} = 3 \text{ m}$$

$$\sin\alpha = \frac{3}{5}; \qquad \cos\alpha = \frac{4}{5}$$

$$\sin\alpha = 0.6; \ \cos\alpha = 0.8$$

Resolving all forces horizontally,

$$N_A = \mu N_B \qquad \qquad \ldots(1)$$

Resolving all forces vertically,

$$N_B = 300 \text{ N} \qquad \qquad \ldots(2)$$

Taking moments of all forces about 'B'

$$N_A \times 3 = 300 \times \frac{5}{2}\cos\alpha$$

$$3N_A = 300 \times 2.5 \times 0.8$$

$$3N_A = 600$$

$$N_A = 200 \text{ N} \qquad \qquad \ldots(3)$$

From equations (1), (2) & (3)

$$200 = \mu 300$$

$$\mu = \frac{200}{300} = 0.6667$$

Frictional force $= \mu N_B$

$$= \frac{2}{3} \times 300$$

$$= 200 \text{ N}$$

PROBLEM 8.14: A short semicircular cylinder of radius r and weight W rests on a horizontal surface. It is pulled at right angles to its geometrical axis by a horizontal force P applied at the middle of the front edge. Find the angle θ that the flat surfaces will make with the horizontal plane just before sliding begins. Assume the coefficient of friction at the point of contact as μ and all the forces to be coplanar.

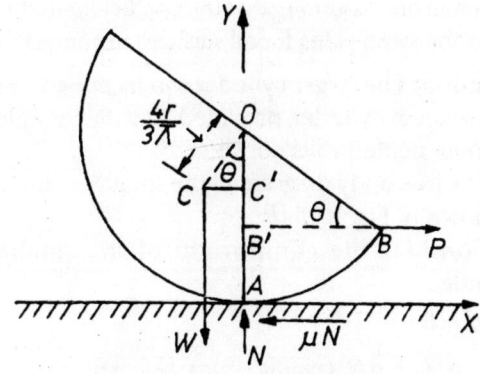

Fig. 8.28 (Illustrative Problem 8.14)

Free body diagram of the semicircular cylinder is as shown in Fig. 8.28.

The normal reaction N at A would pass through the centre O and weight W through the centroid C.

Writing the equations of equilibrium

$\Sigma H = 0$

i.e. $\quad P = \mu N$ ...(1)

$\Sigma V = 0$

i.e. $\quad N = W$ ...(2)

From (1) & (2)

$P = \mu W$

$\Sigma M = 0 \quad$ Taking moments about A

i.e. $\quad W \times CC' = P \times AB'$

$W(OC\sin\theta) = P(AO - B'O)$

$W(OC\sin\theta) = P(r - r\sin\theta)$

$W\left(\dfrac{4r}{3\pi}\sin\theta\right) = Pr(1 - \sin\theta)$ ...(4)

Substituting the value P from eqn. (3) in equation (4)

$$W = \frac{4r}{3\pi}\sin\theta = \mu Wr(1 - \sin\theta)$$

$$4\sin\theta = 3\pi\mu(1 - \sin\theta)$$

$$(4 + 3\pi\mu)\sin\theta = 3\pi\mu$$

$$\sin\theta = \frac{3\pi\mu}{4 + 3\pi\mu}$$

PROBLEM 8.15: The heavy circular cylinders of radii R and r rest on a rough horizontal plane as shown in Fig. 8.29 (a). The larger cylinder has a string wound round it to which a horizontal force P can be applied. Determine the necessary condition so that the larger cylinder can be pulled over the small one assuming that the coefficient friction μ has the same value for all surfaces in contact.

Solution: The larger cylinder can be pulled over the smaller cylinder provided that the smaller cylinder neither rolls nor slides.

The free body diagram of the smaller cylinder is shown in Fig. 8.29 (b).

Consider the equilibrium of the smaller cylinder.

$\Sigma H = 0$

i.e. $\quad \mu N_B + \mu N_D\cos(90 - \theta) = N_D\cos\theta$

$\mu N_B + \mu N_D\sin\theta = N_D\cos\theta$ ...(1)

$\Sigma V = 0$

$N_D\sin\theta + w + \mu N_D\cos\theta = N_B$ ...(2)

$\Sigma M = 0$ Taking moments about C_2

$\mu N_B r = \mu N_D r$

i.e. $\quad N_B = N_D$ ...(3)

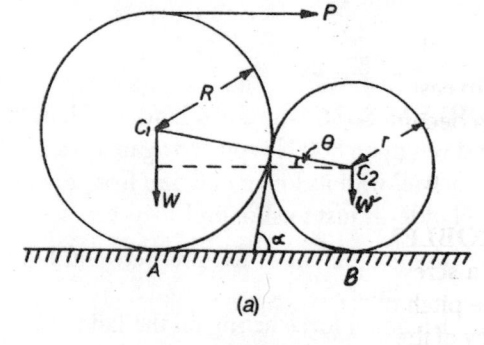

(a)

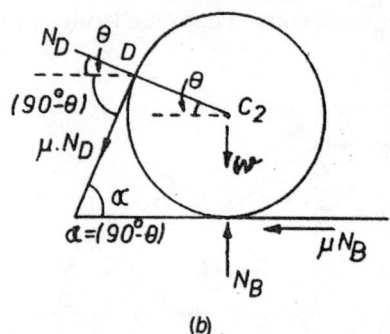

(b)

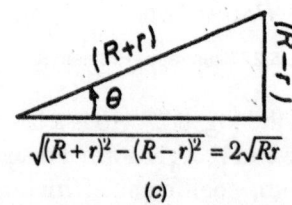

$$\sqrt{(R+r)^2 - (R-r)^2} = 2\sqrt{Rr}$$

(c)

Fig. 8.29 (Illustrative Problem 8.15)

Substituting (3) in (1)

$$\mu N_D + \mu N_D\sin\theta = N_D\cos\theta$$

$$\mu(1 + \sin\theta) = \cos\theta$$

$$\mu = \frac{\cos\theta}{1 + \sin\theta}$$

But $\sin\theta = \dfrac{R - r}{R + r}$

$\therefore \quad \cos\theta = \dfrac{2\sqrt{Rr}}{R + r}$

$$1 + \sin\theta = 1 + \frac{R - r}{R + r}$$

$$= \frac{2R}{R + r}$$

$$\therefore \quad \mu = \frac{2\sqrt{Rr}}{(R+r)} \times \frac{(R+r)}{2R}$$

$$\mu = \sqrt{\frac{r}{R}}$$

In case of limiting friction the necessary condition becomes

$$\mu > \sqrt{\frac{r}{R}}$$

PROBLEM 8.16: A load of 2 kN is to be raised by a screw jack with mean diameter of 50 mm. The pitch of the thread is 6 mm. Find the efficiency of the screw jack, if the coefficient of friction between the screw and nut is 0.10. Is it self-locking?

$P = 6$ mm; $d = 50$ mm; $W = 2$ kN $= 2000$ N,

$\tan\phi = 0.10$

$$\therefore \quad \tan\alpha = \frac{P}{\pi d} = \frac{6}{\pi \times 50} = 0.0382$$

$$\alpha = 2.1875°$$

$$\eta = \frac{\tan\alpha}{\tan(\alpha+\phi)}$$

$$= \frac{\tan\alpha}{\dfrac{\tan\alpha+\tan\phi}{1-\tan\alpha\cdot\tan\phi}}$$

$$= \frac{0.0382}{\dfrac{0.0382+0.10}{1-0.0382\times0.10}}$$

$$= \frac{0.0382}{\dfrac{0.1382}{0.99618}}$$

$$= \frac{0.0382}{0.13873}$$

$$= 0.2754 \quad \text{or} \quad 27.54\%$$

$\tan\phi = 0.10$

$\phi = \tan^{-1}(0.10) = 5.71° > 2.1875°$

i.e. $\phi > \alpha$

It is a self-locking machine

Check

The efficiency of the jack is 27.54% which is less than 50%.

$\therefore$ It is a self-locking machine.

PROBLEM 8.17: The screw of a jack is square threaded with two threads to a centimeter. The outer diameter of the screw is 5 cm. If the co-efficient of friction is 0.12, find the force required to be applied at the end of the level, which is 60 cm length (a) to lift a load of 5 kN, and (b) to lower it.

Solution: There are two threads in a cm.

i.e. $n = 2$

Therefore pitch of the screw, $p = \frac{1}{2} = 0.5$ cm

Internal diameter of the screw

$= 5 - 2 \times 0.5 = 4$ cm

Mean diameter of the screw, $d = \dfrac{5+4}{2} = 4.5$ cm

Let α be helix angle

$$\tan\alpha = \frac{p}{\pi d}$$

$$= \frac{0.5}{\pi \times 4.5} = 0.0354$$

(a) Force required to be applied at the end of 60 cm length (long) lever to lift the load.

Let P_1 be the force required to be applied at the end of the lever to lift the load.

Force required to be applied at the end of the radius to lift the load,

$$P = W\tan(\alpha+\phi)$$

$$= \frac{W(\tan\alpha+\tan\phi)}{1-\tan\alpha\tan\phi}$$

$$= \frac{5000(0.0354+0.12)}{1-0.0354\times0.12}$$

$$= \frac{5000\times0.1554}{0.9958}$$

$$= 780.31 \text{ N}$$

$$\therefore \quad P_1 = \frac{780.31\times60}{(4.5/2)} = 20,808.3 \text{ N or } 20.81 \text{ kN}$$

(b) Force required to be applied at the end of 60 cm long lever to lower the load.

Force required to be applied at the mean radius to lower the load,

$$P = W\tan(\phi-\alpha)$$

$$= W \left[\frac{\tan\phi - \tan\alpha}{1 + \tan\phi \cdot \tan\alpha} \right]$$

$$= 5000 \left[\frac{0.12 - 0.0354}{1 + 0.12 \times 0.0354} \right]$$

$$= \frac{5000 \times 0.0846}{1.004248}$$

$$= 421.21 \text{ N}$$

The force required to be applied at the end of the lever to lower the load,

$$P_2 = \frac{421.21 \times 60}{(4.5/2)} = 11232.27 \text{ N or } 11.23 \text{ kN}$$

PROBLEM 8.18: A screw press is used to compress books. The thread is a double thread (square head) with a pitch of 6 mm and a mean radius of 25 mm. The coefficient of friction for the contact surface of the thread is 0.25. Find the torque for a pressure of 600 N.

Solution:

$$\tan\alpha = \frac{np}{2\pi r}$$

$$= \frac{2 \times 6}{2 \times \pi \times 25} = 0.0764 \quad \text{or } 4.3689°$$

Effort required at the mean radius of the screw to press the books.

$$P = W \tan(\alpha + \phi)$$

$$= W \left[\frac{\tan(\alpha) + \tan(\phi)}{1 - \tan(\alpha) \cdot \tan(\phi)} \right]$$

$$= 600 \left[\frac{0.0764 + 0.25}{1 - 0.0764 \times 0.25} \right]$$

$$= \frac{600 \times 0.3264}{(1 - 0.0191)}$$

$$= 199.65 \text{ Newtons}$$

Torque required to press the books,

$$T = P \times r$$

$$= 199.65 \times 25$$

$$= 4991.25 \text{ N-mm}$$

PROBLEM 8.19: A belt supports two weights W_1 and W_2 over a pulley as shown in Fig. 8.30. If W_1 = 1500 N, find the minimum weight W_2 to keep W_1 in equilibrium. Assume that the pulley is locked and coefficient of friction between the pulley and the belt is 0.30.

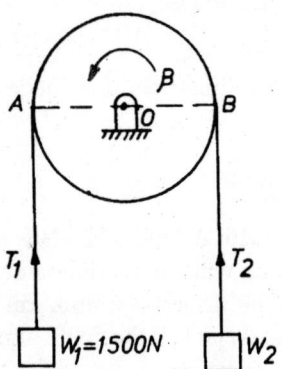

Fig. 8.30 (Illustrative Problem 8.19)

Solution: Let the tension in the belt be T_1 and T_2 as shown in Fig. 8.30. Since the weight W_2 just checks the tendency of weight W_1 to move down, tension on the side of W_1 is larger.

That is $T_1 > T_2$

$$\mu = 0.30; \beta = \pi \text{ radians}$$

$$W_1 = 1500 \text{ N}$$

$$\frac{T_1}{T_2} = e^{\mu\beta}$$

$$\frac{T_1}{T_2} = e^{0.30 \times \pi} = 2.5663$$

But $\quad T_1 = W_1 = 1500 \text{ N}$

$\therefore \quad T_2 = 1500 \times \dfrac{1}{2.5663}$

$$T_2 = 584.5 \text{ Newtons}$$

But $\quad W_2 = T_2 = 584.5 \text{ N}$

The weight, $W_2 = 584.5 \text{ N}$

PROBLEM 8.20: Find out the number of turns a hauling rope must be wound round a rotating capstan in order to haul a load of 2.1 MN up to a gradient of 1 in 30. Resistance due to rolling is 0.005 N per Newton and pull on the free end of the rope is 161 N. Coefficient of friction between the rope and the drum is 0.33.

Solution:

$T_1 =$ Weight × Resistance per Newton + Weight component

$$= 2.1 \times 10^6 \times 0.005 + 2.1 \times 10^6 \times \frac{1}{30}$$

$$= 10500 + 70000$$

$$= 80500 \text{ N}$$

$$T_2 = 161 \text{ N}, \ \mu = 0.33$$

$$\frac{T_1}{T_2} = e^{\mu\beta}$$

$$\frac{80500}{161} = e^{0.33 \times \beta}$$

$$e^{0.33 \times \beta} = 500$$

$$0.33 \times \beta = 6.2146$$

$$\beta = \frac{6.2146}{0.33} = 18.83212$$

$$\therefore \quad \text{Number of turns} = \frac{\beta}{2\pi}$$

$$= \frac{18.83212}{2\pi}$$

$$= 2.9972 \text{ say } 3$$

Number of turns required is 3

PROBLEM 8.21: The end of a vertical shaft of 300 mm diameter and weighing 25 kN rests on a flat surface. If the coefficient of friction between the surfaces is 0.10, find the frictional torque assuming constant pressure.

Solution:

Frictional Torque, $T = M = \frac{2}{3} PR\mu$

$$P = 25 \text{ kN} = 25 \times 1000 = 25000 \text{ N}$$

$$R = \frac{300}{2} = 150 \text{ mm or } 0.15 \text{ m}$$

$$\mu = 0.10$$

$$\therefore \quad T = \frac{2}{3} \times 25000 \times 0.15 \times 0.10$$

$$= 250 \text{ N-m}$$

PROBLEM 8.22: A shaft supported on a collar whose external diameter is 750 mm and the shift diameter is 500 mm. The coefficient of friction between the surfaces is 0.05. Find the power absorbed in overcoming friction when the shaft runs at 150 r.p.m. and carries an axial load of 100 kN. Assume uniform intensity of pressure.

Solution:

$$T = M = \frac{2}{3} \mu P \left(\frac{R_1^3 - R_2^3}{R_1^2 - R_2^2} \right)$$

$$\mu = 0.05$$

$$P = 100 \text{ kN} = 100 \times 1000 \text{ N}$$

$$R_1 = \frac{750}{2} = 375 \text{ mm} = 0.375 \text{ m}$$

$$R_2 = \frac{500}{2} = 250 \text{ mm} = 0.250 \text{ m}$$

$$\therefore T = \frac{2}{3} \times 0.05 \times 100 \times 1000 \times \left(\frac{0.375^3 - 0.250^3}{0.375^2 - 0.250^2} \right)$$

$$= 3333.33 \times \left(\frac{0.05273 - 0.015625}{0.625 \times 0.125} \right)$$

$$= \frac{3333.33 \times 0.037105}{0.625 \times 0.125}$$

$$= 1583.15 \text{ N-m}$$

Power absorbed in overcoming friction

$$= \frac{2\pi NT}{60} \text{ Watts}$$

$$= \frac{2 \times \pi \times 150 \times 1583.15}{60 \times 1000} \text{ kW} = 24.87 \text{ kW}$$

PROBLEM 8.23: A shaft of 75 mm diameter rests in a conical bearing of cone angle 60°. Calculate the frictional torque and the power required to rotate the shaft at 1020 r.p.m., if the axial load on the shaft is 6 kN and the coefficient of friction is 0.25. Assume normal pressure to be uniform.

Solution:

$$T = M = \frac{2}{3} \frac{\mu PR}{\sin \theta}$$

$$\mu = 0.25$$

$$P = 6 \text{ kN} = 6000 \text{ N}$$

$$R = \frac{75}{2} = 37.5 \text{ mm or } 0.0375 \text{ m}$$

$$2\theta = 60°$$

$$\theta = 30°$$

$$\therefore \quad T = \frac{2}{3} \times \frac{0.25 \times 6000 \times 0.0375}{\sin 30°}$$

$$= 75 \text{ N-m}$$

Power required $= \dfrac{2\pi NT}{60}$ Watts

$= 8011$ Watts

$= 8.011$ kW

≈ 8 kW

PRACTICE PROBLEMS

8.1. Determine the minimum force P to cause the 600 N block of Fig. 8.31 to move. The coefficient of friction is 0.25.

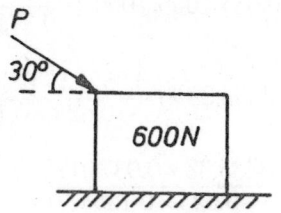

Fig. 8.31 (Practice Problem 8.1)

8.2. A body resting on a rough horizontal plane required a pulley of 200 N inclined at 25° to the plane just to move it. It was found that a push of 250 N inclined at 25° to the plane just moved the body. Determine the weight of the body and th coefficient of friction.

8.3. The apex angle of a cone of friction is 45°. Find the coefficient of friction.

8.4. A body of weight of 500 N is pulled up an inclined plane, by a force of 300 N. The inclination of the plane is 25° to the horizontal and the force is applied parallel to the plane. Find the coefficient of friction.

8.5. A pull of 250 N is required just to make a certain body up an inclined plane of angle 20°, the force acting parallel to the plane. If the angle of inclination of the plane is made 30°, the pull required again parallel to the plane, is found to 300 N. Find the weight of the body and the angle of friction.

8.6. An object of weight 200 N is kept in position on a plane inclined to 25° to the horizontal by a force(P) applied horizontally. The coefficient of friction of the surface of the inclined plane is 0.30. Determine the minimum magnitude of the force(P).

8.7 Block weighing 20 N is a rectangular prism resting on a rough inclined plane as shown in

Fig. 8.32. The block is tied up with a horizontal string which has a tension of 5 N, find

(*i*) Normal reaction of the inclined plane.
(*ii*) The coefficient of friction between the surfaces of contact.
(*iii*) The frictional force on the block.

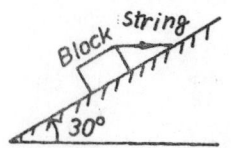

Fig. 8.32 (Practice Problem 8.7)

8.8. A 15° wedge of negligible weight is to be driven to tighten a body B which is supporting a vertical load of 1500 N. If the angle of friction for all the surfaces is 14°, find the minimum force P required to drive the wedge (Fig. 8.33).

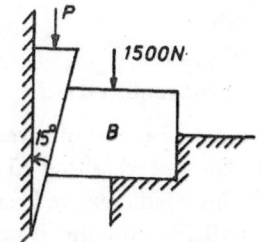

Fig. 8.33 (Practice Problem 8.8)

8.9. A uniform ladder of weight W and length L is held in equilibrium by a horizontal force P as shown in Fig. 8.34 below. Using the equilibrium equations, express P in terms of W and θ where θ is the angle made by ladder with the vertical, the contact surfaces are smooth.

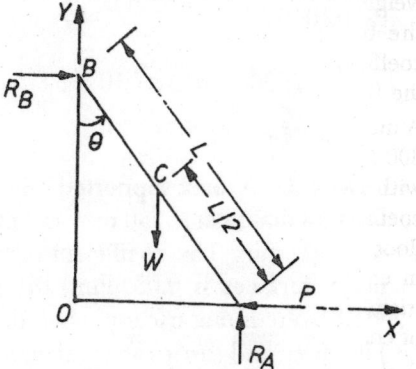

Fig. 8.34 (Practice Problem 8.9)

8.10. A ladder of length 5 m weighing 300 N is placed against a vertical wall making an angle of 60° with the floor as shown in Fig. 8.35. The coefficient of friction between the wall and the ladder is 0.20 and the that between the floor and the ladder is 0.30. The ladder in addition to its own weight has to support a man weighing 700 N at a distance of 3.5, from A. Calculate the minimum horizontal force to be applied at A to prevent slipping.

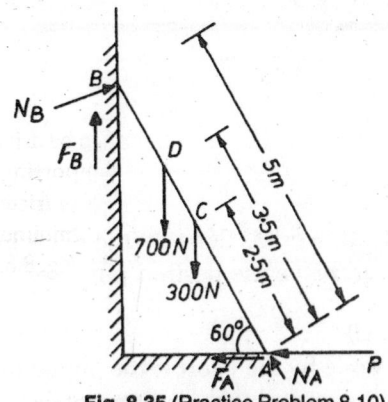

Fig. 8.35 (Practice Problem 8.10)

8.11. A uniform ladder of 5 m rests against a vertical wall with which it makes an angle of 45°, the coefficient of friction between the ladder and the wall is 0.25 and that between ladder and the floor is 0.5. If a man, whose weight is one half of that of the ladder, ascends it, how high will he be when the ladder slips.

8.12. A ladder 5 m long rests on a horizontal ground and leans against a smooth vertical wall at an angle 60° with the horizontal. The weight of the ladder is 600 N and acts at its middle. The ladder is at the point of sliding when a man weighing 650 N stands on a rung 2 m from the bottom of the ladder. Calculate the coefficient of friction between the ladder and the floor.

8.13. A uniform ladder of length 13 m and weighing 300 N is placed against a smooth vertical wall with its lower end 5 m from the wall. The coefficient of friction between the ladder and floor is 0.25. Show that the ladder will remain in equilibrium in this position. What is the frictional force acting on the ladder at the point of contact between the ladder and the floor.

8.14. A short semi-circular cylinder of radius r and weight W rests on a horizontal surface. It is pulled at right angles to its geometrical axis by a horizontal force P applied at the middle of the front edge. Find the angle θ that the flat surface will make with the horizontal plane just before sliding begins. Assume the coefficient of friction at the point of contact as 0.25 and all the forces to be coplanar.

8.15. A homogenous cylinder of weight W rests on a horizontal floor in contact with a wall if the coefficient of friction for all contact surfaces be μ, determine the couple M acting on the cylinder which start counter clockwise rotation.

8.16. The mean radius of a screw of the square threaded screw jack is 25 mm. The pitch of the thread is 6 mm. If the coefficient of friction is 0.10, find the effort to be applied at the end of a lever 50 cm length, need to lower a weight of 1.5 kN.

8.17. A screw has a square thread of 75 mm mean diameter and 15 mm pitch. The coefficient of friction at the screw thread is 0.05. A load of 6 kN is to be lifted by the screw jack. Find the efficiency of the screw jack. State whether the jack is self-locking.

8.18. The mean radius of a screw of a square threaded screw jack is 25 mm. The pitch of thread is 6 mm. If the coefficient of friction is 0.10. Find the effort to be applied at the end of a lever 50 cm length, needed to raise a weight of 1.5 kN.

8.19. A weight of 625 N is held by passing a rope around a horizontal post and exerting a pull of 50 N. How many turns of the rope around the post are necessary if the coefficient of friction between the post and the rope is 0.2.

8.20. Find the force required to hold a weight of 15 kN suspended on a rope wrapped twice around a post. Assume the coefficient of friction between the rope and the post is 0.25.

8.21. A flat foot step bearing supporting an axial load of 20 kN is rotating at 100 r.p.m. The diameter of the bearing is 22.5 cm (225 mm) and the coefficient of friction may be assumed to be 0.05. Find the frictional torque and power loss due to bearing friction assuming uniform wear.

8.22. A cone clutch has an effective diameter of 60 mm, and semi-cone angle of 15°. If the coefficient of friction is 0.25, find the frictional torque when axial force applied is 15 kN.

Chapter 9

Lifting Machines

9.1 INTRODUCTION

'Lifting Machines' are devices used to lift heavy objects by applying relatively small effort in a convenient direction. Most of these are of simple nature and are, therefore, called 'Simple machines'.

Common examples in daily life are a crowbar for moving heavy objects, a simple pulley to draw water from a well, and a screw-jack to lift a car for replacing a flat tyre.

In this chapter, some common terms and concepts used with lifting machines are defined, and some important machines like pulleys, wheel and axle, inclined plane, screw-jack, worm-wheel, and winch crab are studied.

9.2 DEFINITIONS AND CERTAIN RELATIONSHIPS

Some terms commonly used with lifting machines are defined and explained briefly.

(i) Simple Machine

A device by means of which a load can be lifted by a relatively small effort, or the direction can be changed, or the speed of motion can be increased is called a 'simple machine'.

(ii) Load (W)

The load lifted or the resistance overcome by the machine—it is also called 'resistance'.

(iii) Effort (P)

The force required to overcome the resistance or

to lift the load; work is done by the effort in the machine to achieve the desired purpose.

(iv) Input (I)

The work done by the effort is the 'input' to the machine; this is equal to the product of the effort and the distance moved in its direction.

(v) Output (O)

This is the useful work done by the machine; this is therefore the work done by the load or the product of the load and the distance through which it moves.

(vi) Mechanical Advantage (A_m)

This is the ratio of the load lifted to the effort applied.

(vii) Velocity Ratio (V_r)

This is the ratio of the distance moved by the effort (D) to that moved by the load (d) in a given time interval.

(viii) Efficiency (η)

The efficiency of the machine is defined as the ratio of the output to the corresponding input; this is usually expressed as per cent.

(ix) Ideal Machine

A machine for which the efficiency is the maximum of one hundred per cent is an 'ideal machine'; the output equals the input for this.

(x) Ideal Effort (P_i)

This is the effort required to lift a given load by the machine, assuming the latter to be ideal.

(xi) Ideal Load (W_i)

This is the load that can be lifted using a given effort by the machine, assuming the latter to be ideal.

(xii) Compound Machine

A machine made up of a number of simple machines.

The mechanical advantage, the velocity ratio, and the efficiency of a machine are interrelated as shown below:

Mechanical advantage,

$$A_m = \frac{W}{P} \qquad \text{...(Eq. 9.1)}$$

Velocity ratio,

$$V_r = \frac{D}{d} \qquad \text{...(Eq. 9.2)}$$

Efficiency,

$$\eta = \frac{\text{Output}}{\text{Input}} = \frac{W \cdot d}{P \cdot D}$$

$$= \left(\frac{W}{P}\right) \times \left(\frac{d}{D}\right)$$

$$\eta = \frac{A_m}{V_r} \qquad \text{...(Eq. 9.3)}$$

Thus efficiency is the ratio of the mechanical advantage to the velocity ratio.

Hence for an ideal machine ($\eta = 1$ or 100%), the mechanical advantage equals the velocity ratio.

$$A_m = V_r$$

If P_i is the ideal effort, then

$$V_r = \frac{W}{P_i}$$

or $\qquad P_i = \frac{W}{V_r} \qquad \text{...(Eq. 9.4)}$

If W_i is the ideal load, then

$$V_r = \frac{W_i}{P}$$

or $\qquad W_i = V_r \cdot P \qquad \text{...(Eq. 9.5)}$

9.3 PRACTICAL MACHINE

It is impossible to obtain an ideal machine in practice. There always exists friction at all contact surfaces of moving parts, and it is impossible to get a frictionless machine. Some of work done by the effort is utilised in overcoming frictional resistance, which results in reduction of the output and consequent decrease in efficiency. Such a machine is termed the 'practical machine'.

As already pointed out, in a simple machine, a large load is lifted by a relatively small effort; thus, the mechanical advantage is invariably greater than unity. Since the work done by the effort is utilised in lifting the load, except for the friction losses, "a small force applied through a large distance overcomes a large load through a small distance".

In view of the friction losses, the mechanical advantage of a practical machine is always less than that of an ideal one. But the velocity ratios are practically the same as the geometrical features determine it, and it is assumed that the slip between the motion of the load and effort is negligible.

The mechanical advantage, being a ratio of forces, can be determined by the equations of equilibrium. The velocity ratio is obtained by giving a unit displacement to the effort and checking the corresponding displacement of the load.

The fact that the difference in velocity ratios of a practical machine and its ideal model is negligible is utilised in the estimation of the friction losses in a machine. This is illustrated below.

Friction losses may be estimated in two different ways; first, as the difference between the actual effort required and ideal effort for lifting a given load; this represents the effort wasted in overcoming friction. Second, as the difference between the ideal load and the actual load lifted by a given effort; this represents the loss of load that could be lifted owing to friction.

In the first approach, the ideal effort P_i from Eq. 9.4 is

$$P_i = \frac{W}{V_r}, \text{ and}$$

The effort loss in friction, $P_f = (P - P_i)$, P being the actual effort needed to lift a given load W.

Hence the efficiency, η, of the machine is given by

$$\eta = \frac{P_i}{P} \qquad ...(Eq.\ 9.6)$$

In the second approach, the ideal load W_i from Eq. 9.5 is

$$W_i = V_r \cdot P, \text{ and}$$

The loss of load that could be lifted by a given effort, owing to friction, $W_f = (W_i - W)$, W being the actual load lifted by a given effort P.

Hence the efficiency may be put down as

$$\eta = \frac{W}{W_i} \qquad ...(Eq.\ 9.7)$$

Thus

$$\eta = \frac{P_i}{P} = \frac{W}{W_i} \qquad ...(Eq.\ 9.8)$$

In terms of input and output,

Friction losses = Input – Output

$$F = (I - O) \qquad ...(Eq.\ 9.9)$$

9.4 LAW OF MACHINES

The relationship between the load or resistance overcome (W) and the corresponding effort required (P) is called the "Law of Machines". This can be established experimentally. In practice, this relationship approximates a straight line with an intercept on the effort-axis as shown in Fig. 9.1.

This relationship may be expressed mathematically as

$$P = mW + C \qquad ...(Eq.\ 9.10)$$

Here C is the intercept on the P-axis and m is the slope of the line or tangent of the angle made with the W-axis.

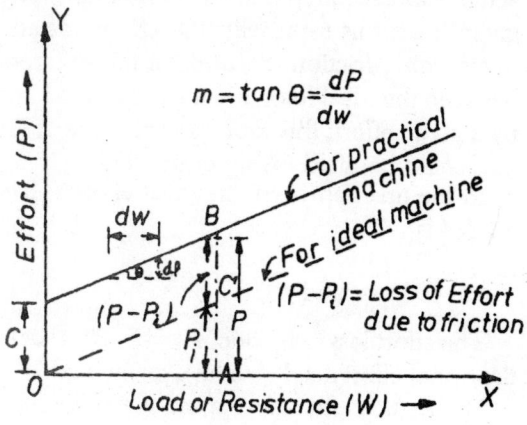

Fig. 9.1: Law of Machines

For an ideal machine, however, this straight line will pass through the origin, since the friction losses are not there. It is obvious that the intercept between the straight lines for the practical machine and its ideal counterpart represents the loss of effort due to friction in the former.

Efficiency, $\eta = \dfrac{P_i}{p} = \dfrac{AC}{AB}$, with reference to the plot.

9.4.1 Variation of Mechanical Advantage
Mechanical advantage, A_m, is given by

$$A_m = \frac{W}{P}$$

From the law of machines (Eq. 9.10),

$$P = mW + C$$

$$A_m = \frac{W}{mW + C}$$

or

$$A_m = \frac{1}{\left(m + \dfrac{C}{W}\right)} \qquad ...(Eq.\ 9.11)$$

As W increases, C/W decreases, and A_m increases.

The maximum possible mechanical advantage for very large load is $1/m$. Variation of the mechanical advantage with load is shown in Fig. 9.2:

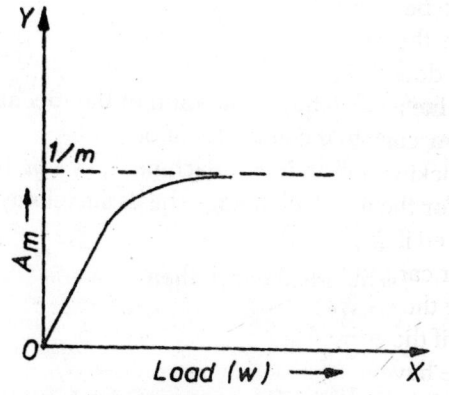

Fig. 9.2: Variation of Mechanical Advantage with Load

9.4.2 Variation of Efficiency
Efficiency of a machine is given by

$$\eta = \frac{A_m}{V_r} \qquad ...(\text{From Eq .9.3})$$

Substituting for A_m from Eq. 9.11,

$$\eta = \frac{1}{V_r\left(m + \dfrac{C}{W}\right)} \qquad ...(Eq.\ 9.12)$$

Since the velocity ratio, V_r, for a given machine is a constant for all practical purposes, the variation of efficiency, η, with load is similar to the variation of mechanical advantage with load (Fig. 9.3).

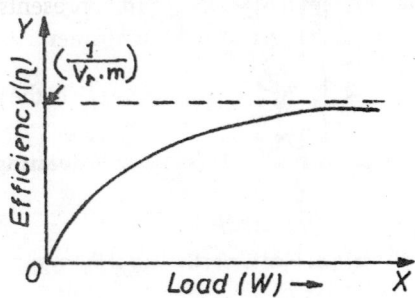

Fig. 9.3: Variation of Efficiency with Load

The maximum efficiency for a very large value of load approaches a value $\left(\dfrac{1}{V_r \cdot m}\right)$, as shown in the figure.

9.5 REVERSIBILITY AND SELF-LOCKING OF A MACHINE

If, while lifting a load, the withdrawal of the effort results in lowering of the load, the machine is said to be 'reversible'.

On the other hand, if the withdrawal of the effort does not result in lowering of the load, the machine is said to be 'self-locking'.

Two common examples of reversibility and self-locking of machines are lifting water from a well for the former (since the pot or bucket gets lowered if the effort is withdrawn), and lifting of a motor car by means of a screw-jack for the latter (since the car will be held in a particular position even if the effort is withdrawn).

We have seen that input (I) for the machine is given by

$$I = P \times D,$$

and the output (O) by

$$O = W \times d$$

Work lost in friction $= I - O$

$$= P \cdot D - W \cdot d$$

When the effort is withdrawn, the load can start moving down if the latter can overcome the frictional resistance.

That is to say, the condition for reversibility is:

$$Wd > (PD - Wd)$$

or $2Wd > PD$

or $\left(\dfrac{W}{P}\right)\left(\dfrac{d}{D}\right) > \dfrac{1}{2}$

or $\left[(A_m)/(V_r)\right] > \dfrac{1}{2}$

or $\eta > \dfrac{1}{2}$ or 50%

Thus, a machine will be reversible if its efficiency is greater than 50%. Obviously, the condition for self-locking of a machine is that its efficiency should be less than 50%, since the two ideas of reversibility and self-locking are of opposite nature.

9.6 PULLEYS

A 'pulley' is one of the simplest of the lifting machines. A system of pulleys arranged in a certain manner provides a lifting machine with a higher mechanical advantage than that for a single pulley.

The following assumptions are relevant to pulleys:

 (*i*) The weight of the pulley is considered very small compared to the load lifted, and hence neglected.

 (*ii*) The friction at the pivot of the pulley as well as that between the pulley and the rope passing over it is considered negligibly small. Consequently the tensions on either side are considered to be the same.

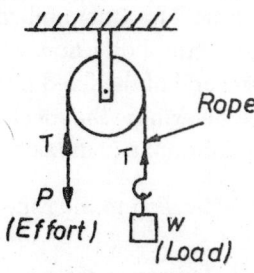

Fig. 9.4: Single Pulley

For a single pulley (Fig. 9.4), the velocity ratio is unity since the distance moved by the load is just equal to that moved by the effort.

If T is the tension in the string, considering the equilibrium of the load as a free body,

$$T = W$$

But T is also equal to P.

$$P = W$$

Mechanical Advantage $A_m = \dfrac{W}{P} = 1$

For an ideal case where friction losses are negligible,

Velocity ratio, $V_r = A_m = 1$, as stated earlier.

Thus the only advantage of a single pulley is that the direction of the effort is conveniently altered; a classic example is a pulley and rope used to draw a bucket of water from a well.

When a system of pulleys is used, it may fall under one of the following categories, depending upon the arrangement:

(1) First order system of pulleys
(2) Second order system of pulleys
(3) Third order system of pulleys

Direct determination of the velocity ratio in these or other cases may be some times difficult; then, the mechanical advantage may be determined first by using the equations of equilibrium. The velocity ratio is then taken to be the same as the mechanical advantage under ideal conditions of the absence of friction; thus the velocity ratio is got indirectly. This approach is used, in general, in the following treatment.

9.6.1 First-Order System of Pulleys

In this system there are as many number of ropes as the number of pulleys. The first pulley is fixed in position and merely serves to change the direction of the effort. The other pulleys are movable and move in a vertical direction, when an effort is applied at one end of the fixed pulley (Fig. 9.5).

Let us first determine the mechanical advantage using equilibrium equations:

Pulley 4: Tension in the rope $= \dfrac{W}{2}$

Pulley 3: Tension in the rope $= \dfrac{W}{4}$

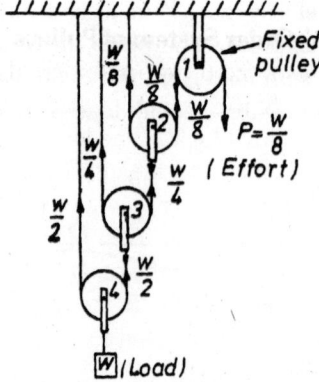

Fig. 9.5: First-Order System of Pulleys

Pulley 4: Tension in the rope $= \dfrac{W}{8}$

Effort, $P = \dfrac{W}{8}$

Hence, mechanical advantage,

$$A_m = \dfrac{W}{P} = \dfrac{W}{W/8} = 8$$

But $A_m = 8 = 2^3 = 2^n$ ...(Eq. 9.13)

where n is the number of movable pulleys. (It is easy to understand that $A_m = 2^2$, if the number of movable pulleys is 2 and so on)

Since friction is neglected,

The velocity ratio, $V_r = A_m = 2^n$ ...(Eq. 9.14)

Now the alternative approach of direct determination of the velocity ratio may be seen:

Let the effort move through a distance y. Since there are two segments of the rope supporting the pulley 2, each segment shortens by a vertical distance $y/2$; this moves the pulley 3 by a vertical distance $1/2\,(y/2)$ or $y/2^2$. Similarly the pulley 4 moves up by a distance $y/2^3$.

Therefore, the velocity ratio,

$$V_r = \frac{\text{Distance moved by effort } (D)}{\text{corresponding distance moved by load } (d)}$$

$$= \frac{y}{\left(y/2^3\right)} = 2^3$$

Thus, the velocity ratio, $V_r = 2^n$, n being the number of movable pulleys. (Same as Eq. 9.14).

9.6.2 Second-Order System of Pulleys

This consists of a fixed (upper) block of pulleys and a movable (lower) block of pulleys (This is also called a 'pulley block'). The load is attached to the movable block and the effort is applied to the end of the rope passing over all the pulleys at the top most pulley of the fixed block (Fig. 9.6) (The number of pulleys in the fixed and movable blocks need not necessarily be the same. Fig. 9.7 shows an unsymmetrical block).

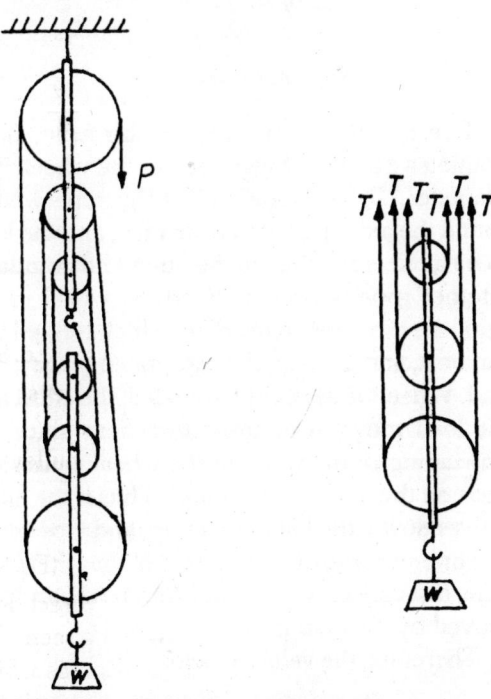

(a) Pulley Block with 3   (b) Equilibrium of Movable
   Pulleys Each   Block

Fig. 9.6: Symmetrical Pulley Block

Let n be the number of the parts of the string supporting the lower block.

The tension, T, is the same throughout and is equal to P. Considering the equilibrium of the lower block [Fig. (b)].

$$nT = W = nP$$

or $\quad A_m = \dfrac{W}{P} = n \qquad \qquad$ ...(Eq. 9.15)

Since for an ideal counterpart, efficiency is 1,

$$\eta = \frac{A_m}{V_r} = 1$$

$\therefore$ The velocity ratio, $V_r = A_m = n \quad$ ...(Eq. 9.16)

If the weight of the lower block (w) is also considered

$$w + W = nT = nP \qquad \qquad \text{...(Eq. 9.17)}$$

However, the velocity ratio of the system remains the same and is equal to n.

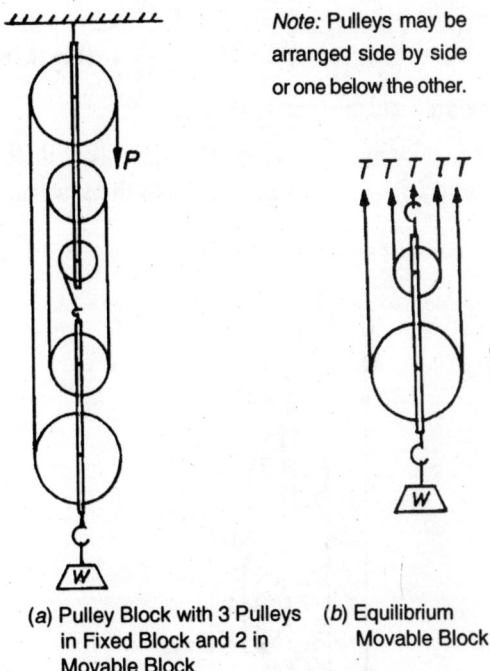

Note: Pulleys may be arranged side by side or one below the other.

(a) Pulley Block with 3 Pulleys   (b) Equilibrium
   in Fixed Block and 2 in   Movable Block
   Movable Block

Fig. 9.7: Unsymmetrical Pulley Block

9.6.3 Third-Order System of Pulleys

In this system, there are as many ropes as there are number of pulleys; one end of each rope is attached to a common rigid block which supports the load to be lifted, the other end is fixed to the next pulley. The effort is applied to the free end of the lowermost pulley, as shown in Fig. 9.8.

Considering the equilibrium of pulley 1,

Tension, $T_1 = P$

Similarly, from the equilibrium of pulleys 2, 3 and 4,

$$T_2 = P + P = 2P$$
$$T_3 = 2P + 2P = 4P = 2^2 \cdot P$$
$$T_4 = 4P + 4P = 8P = 2^3 \cdot P \text{ and so on}$$

If there are n pulleys,

$$T_n = 2^{(n-1)} \cdot P$$

Considering the equilibrium of the rigid block,

$$W = T_1 + T_2 + T_3 + \dots T_n$$
$$P + 2P + 2^2 \cdot P + \dots + 2^{(n-1)} \cdot P$$

Mechanical Advantage, A_m is given by

$$A_m = \frac{W}{P} = 1 + 2 + 2^2 + \dots + 2^{n-1}$$

or $\quad A_m = \frac{(2^n - 1)}{(2 - 1)} = (2^n - 1) \qquad \dots \text{(Eq. 9.18)}$

For an ideal system $V_r = A_m$,

and so, velocity ratio, $V_r = (2^n - 1)$ ...(Eq. 9.19)
where n is the number of pulleys in the system.

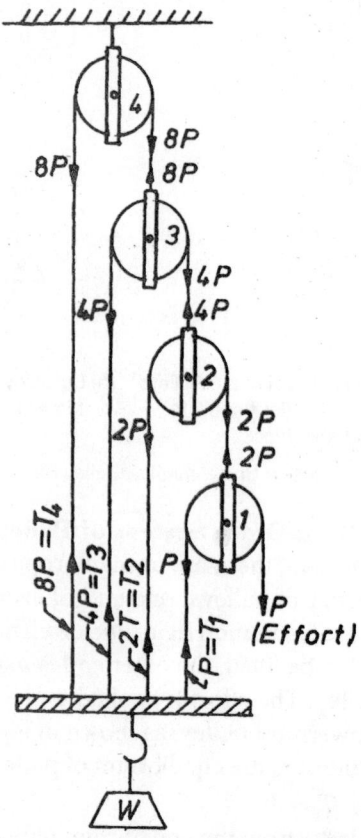

Fig. 9.8: Third-order System of Pulleys

9.7 WHEEL AND AXLE

This consists of an axle (a shaft) and a wheel of a bigger diameter fitted coaxially, and the assembly is supported on bearings; the wheel and axle can be rotated about their common axis (Fig. 9.9).

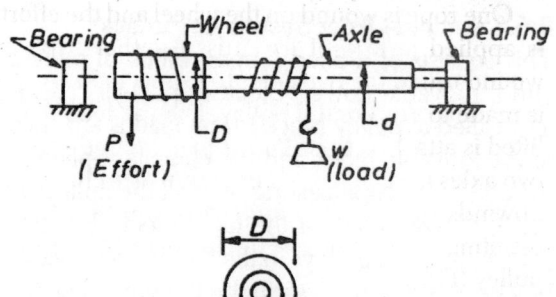

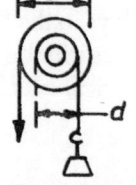

Fig. 9.9: Wheel and Axle

The smaller cylinder, or the axle, is of diameter d and the bigger one or the wheel, is of diameter, D, say. One end of the rope wound round the axle is fixed to a pin on it, and the load to be lifted is attached to the other end. Similarly another rope is wound round the wheel in the opposite direction, one end of which is fixed to a pin on it, and the effort being applied to the other end. When the assembly is rotated, the effort and the load move in opposite directions since the unwinding of the rope on the wheel causes the rope on the axle to get wound. Thus if the effort moves down, the load moves up, and vice versa.

For one complete revolution of the wheel and axle, distance moved by the effort is πD, and that moved by the load is πd.

Therefore, the velocity ratio,

$$V_r = \frac{D}{d} \qquad \dots \text{(Eq. 9.20)}$$

Mechanical advantage,

$$A_m = \frac{W}{P}$$

Efficiency, $\eta = \dfrac{A_m}{V_r} = \dfrac{W/P}{D/d}$

9.8 WHEEL AND DIFFERENTIAL AXLE

An improvement over the wheel and axle can be achieved by adding another wheel of a bigger diameter; both the original wheel and the axle may now be treated as axles of different diameters. All the three are supported on bearings and fitted coaxially so as to rotate together.

One rope is wound on the wheel and the effort is applied at one of its ends. Another rope is wound round the two axles to form a loop which is made to go round a pulley, and the load to be lifted is attached to it. This rope is wound on the two axles in such a way that as the axles rotate, it unwinds on one axle and winds on the other, resulting in the lifting of the load attached to the pulley (Fig. 9.10).

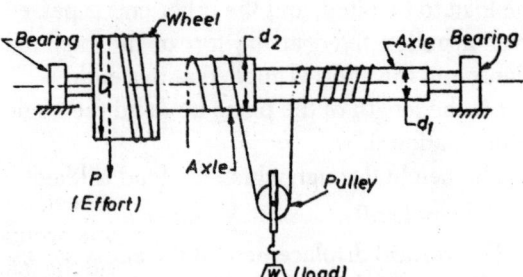

Fig. 9.10: Wheel and Differential Axle

Let the assembly make one complete revolution.

Distance moved by effort $= \pi D$

Length of rope winding on the bigger axle
$$= \pi d_2$$

Length of rope unwinding on the smaller axle
$$= \pi d_1$$

Net winding of rope $= \pi(d_2 - d_1)$

Since this goes round the pulley,
the height through which the pulley (and hence the load) is lifted

$$= \frac{\pi}{2}(d_2 - d_1)$$

The velocity ratio,

$$V_r = \frac{\pi D}{\frac{\pi}{2}(d_2 - d_1)}$$

or $\quad V_r = \frac{2D}{(d_2 - d_1)} \qquad \ldots\text{(Eq. 9.21)}$

For increasing the velocity ratio an obvious way is to make the difference in diameter of the axles small.

9.9 WESTON DIFFERENTIAL PULLEY BLOCK

This is a special, but simple, pulley system, which consists of two pulley blocks. One block of two coaxial pulleys of different diameters is fixed to a rigid horizontal support, and the other at the bottom is hung in the loop of an endless rope. The load to be lifted is attached to the movable bottom pulley. A single rope is made to pass over the pulleys as shown in Fig. 9.11 (All the pulleys may be provided with teeth to engage with the links if a chain is used in place of a rope. This prevents slipping).

The effort pulls one part of the loop unwinding the rope from the bigger pulley on the top block and winding it at the same time on the smaller pulley of this block. The load attached to the lower movable pulley gets lifted.

For one revolution, the length of the rope pulled by the effort is πD, D being the diameter of the bigger pulley of the top block. This is the displacement of the effort. Also, the length of the rope released from the smaller pulley is πd, d being its diameter.

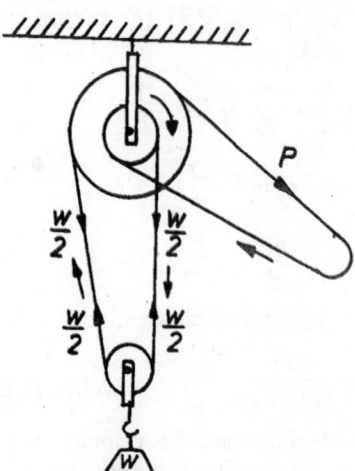

(a) Differential Pulley Block

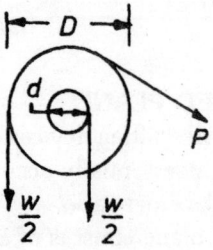

(b) Free Body of Top Block

Fig. 9.11: Weston Differential Pulley Block

The rope on the load-side shortens by $(\pi D - \pi d)$. This is divided equally between the two segments of the rope supporting the lower pulley to which the load is attached.

Therefore, the displacement of the load

$$= \frac{\pi}{2}(D - d)$$

Hence, the Velocity Ratio, V_r, of the system is:

$$V_r = \frac{\pi D}{(\pi/2)(D-d)} = \frac{2D}{(D-d)} \quad ...(\text{Eq. 9.22})$$

For an ideal system of this type, the mechanical advantage,

$$A_m = V_r = \frac{2D}{(D-d)} \quad ...(\text{Eq. 9.23})$$

Now, we can obtain the mechanical advantage by considering the equilibrium of the forces on the top block.

The tension in the chain loop of the lower pulley to which the load is attached is $W/2$.

Referring to Fig. 9.11 (b), taking moments about the axis,

$$\frac{W}{2} \cdot \frac{D}{2} = \frac{W}{2} \cdot \frac{d}{2} + P \cdot \frac{D}{2}$$

or $\qquad \dfrac{W}{4}(D - d) = P \cdot \dfrac{D}{2}$

or $\qquad \dfrac{W}{P} = \dfrac{2D}{D - d}$

The mechanical advantage,

$$A_m = \frac{2D}{(D - d)} \quad ...(\text{same as Eq. 9.23})$$

For an ideal system, the velocity ratio will be the same as the mechanical advantage.

Hence the Velocity Ratio, V_r is given by

$$V_r = \frac{2D}{(D - d)} \quad ...(\text{same as Eq. 9.22})$$

9.10 INCLINED PLANE

One of the simplest lifting devices is the "inclined plane"; the lift is essentially accompanied by a horizontal displacement also.

An inclined plane consists of a plane surface at a certain angle over which a load is to be lifted, made to slide, or rolled (Fig. 9.12).

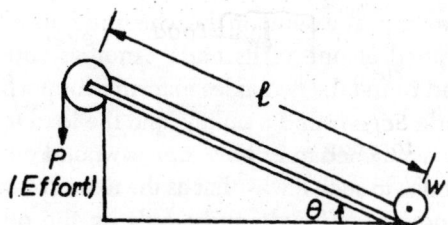

Fig. 9.12: Inclined Plane

One end of the rope is tied to the weight W, the load to be lifted, and the other end is passed over a pulley fixed at the top of the inclined plane, and the effort is applied at this end.

Let the length of the plane be l and the angle of inclination θ.

The height through which the load is lifted

$$= l \sin \theta$$

The vertical displacement of the effort corresponding to this lift $= l$

Velocity Ratio,

$$V_r = \frac{l}{l \sin \theta} = \frac{1}{\sin \theta} \quad ...(\text{Eq. 9.24})$$

If the friction is negligible, the mechanical advantage, A_m, is

$$A_m = V_r = \frac{1}{\sin \theta} \quad ...(\text{Eq. 9.25})$$

9.11 SCREW-JACK

This is a lifting device used to lift heavy loads like the body of a car, bus, truck, and so on. This consists of a square-threaded screw turning in a fixed nut, usually forming the body of the device. The load is carried by the head at the top of the screw, and the effort is applied at the end of a handle fitted to the head of the screw, by rotating it, thus accomplishing the task of causing relative motion between the fixed nut and the movable screw. The axial distance of this relative motion for one complete revolution is known as the 'lead of the screw head'. The distance between consecutive threads of the screw is the 'pitch', as defined earlier. For a single-threaded screw, the lead is equal to the pitch, and for a double-threaded screw, the lead equals double the pitch. The screw-jack is schematically shown in Fig. 9.13.

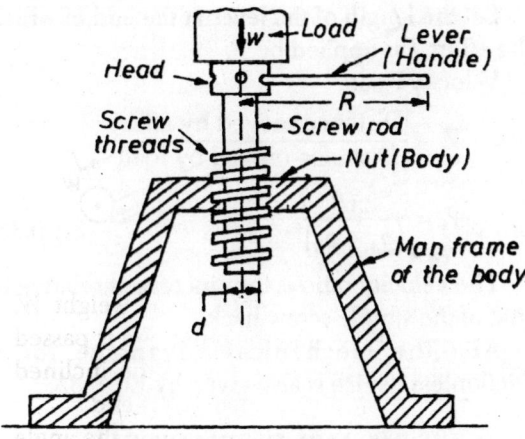

(a) Schematic of Screw-Jack

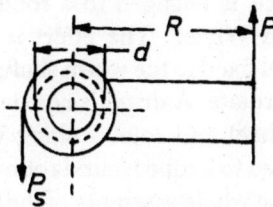

(b) Effort at the Screw

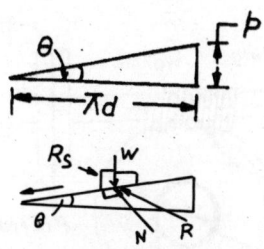

(c) Equivalent Inclined Plane

Fig. 9.13: Screw-Jack

Let the effort applied at the end of the lever or handle be P and its equivalent at the screw be P_s. Let the length of the lever from the axis of the jack be R and the mean diameter of the screw be d.

From (b), we have

$$P \times R = P_s \times \frac{d}{2}$$

$$P_s = \frac{2PR}{d} \qquad \ldots (\text{Eq. 9.26})$$

Let us consider one complete revolution of the lever. The load is lifted by the pitch p of the screw or the lead for a single-threaded screw.

This is analogous to an inclined plane with an angle of inclination θ equal to $\tan^{-1}(p/\pi d)$, with an effort P_s applied on this inclined plane.

Resolving horizontally, or parallel to P_s,

$$P_s = R \sin(\theta + \phi) \qquad \ldots (\text{Eq. 9.27})$$

R here is the resultant the normal reaction N and the frictional resistance F (directed downwards) for ascending load), and ϕ is the limiting friction angle.

Resolving vertically,

$$W = R \cos(\theta + \phi) \qquad \ldots (\text{Eq. 9.28})$$

Dividing equation (9.27) by (9.28)

$$\frac{P_s}{W} = \tan(\theta + \phi)$$

or $\qquad P_s = W \tan(\theta + \phi)$

But by Equation (9.26),

$$\frac{2PR}{d} = W \tan(\theta + \phi)$$

$$\therefore \quad P = \frac{d}{2R} W \tan(\theta + \phi) \qquad \ldots (\text{Eq. 9.29})$$

Using $\tan \phi = \mu$, the coefficient of friction,

$$P = \frac{d}{2R} \cdot W \frac{(\mu + \tan \theta)}{(1 - \mu \tan \theta)} \qquad \ldots (\text{Eq. 9.30})$$

where $\tan \theta = \dfrac{p}{\pi d}$

For the case of descent of the load, the friction F will be act in the expand direction along the equivalent inclined plane, and the resultant reaction R shifts to the other direction.

Then Eq. 9.29 changes to

$$P = \frac{d}{2R} W \tan(\theta - \phi) \qquad \ldots (\text{Eq. 9.31})$$

Torque required, $T = P \cdot R$.

$\therefore$ Torque required during ascent of the load,

$$T = \frac{d}{2} W \tan(\theta + \phi) \qquad \ldots (\text{Eq. 9.32})$$

The value required during descent of the load,

$$T = \frac{d}{2} W \tan(\theta - \phi) \qquad \ldots (\text{Eq. 9.33})$$

Now, the velocity ratio

$$V_r = \frac{\text{Distance moved by the effort}}{\text{Distance moved by the load}} = \frac{2\pi R}{p}$$

$$(\ldots \text{Eq. 9.34})$$

For an ideal device with negligible friction, mechanical advantage is the same as the velocity ratio.

9.12 DIFFERENTIAL SCREW JACK
This is an improvement over the simple screw-jack considered in the previous section. A typical differential screw-jack is shown schematically in Fig. 9.14.

It consists of two screw-threads of different diameters, both threaded in the same direction (right-handed). The bigger screw-shaft is threaded both inside and outside; the threads on the outer surface fit into the big nut below which also serves as the base of the device. The threads on the smaller screw fit those on the inner surface of the bigger one. Thus the bigger screw moves in the nut (base) below, and also acts as a nut for the smaller screw. A lever (handle) is fixed to the block fitted to this smaller screw, and the block can be rotated, thereby rotating the bigger screw.

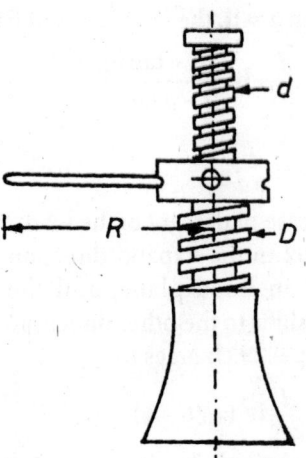

Fig. 9.14: Differential Screw-Jack

Let D and d be the means diameters of the bigger and smaller screws respectively. Let their pitches be p_A and p_B, p_A being greater than p_B.

For one complete revolution, the height through which the bottom bigger screw moves up is say p_A; but the top smaller screw moves down through p_B.

So, the net height through which the load gets lifted = $(p_A - p_B)$.

Let the length of the lever at the end of which the effort P is applied be R.
Velocity Ratio,

$$V_r = \frac{\text{Distance moved by effort}}{\text{Distance moved by load}}$$

or $\quad V_r = \dfrac{2\pi R}{(p_A - p_B)}$ ...(Eq. 9.35)

The velocity ratio is thus increased relative to that of the simple screw hack.

Also, the mechanical advantage for a frictionless device is also given by Eq. 9.35.

9.13 WORM AND WORM-WHEEL
It consists of a square threaded-screw, called 'Worm', which is engaged to a toothed wheel, called 'Worm Wheel'. The effort is applied to another wheel fixed to the worm shaft; this causes the worm to rotate. A drum is coaxially fixed to the worm wheel and rotates along with it; the load is attached to a rope wound round this drum (Fig. 9.15). The whole assembly of the worm shaft and effort wheel is supported in bearings.

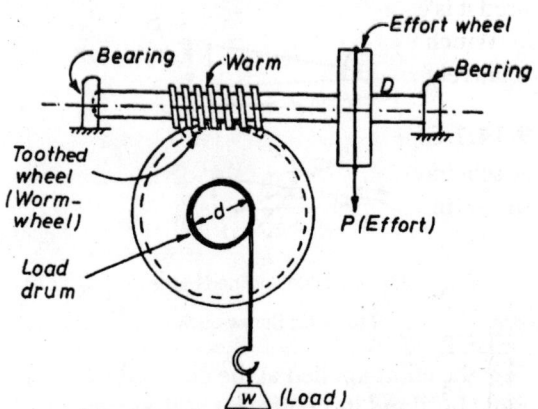

Fig. 9.15: Worm and Worm-Wheel

Let D be the diameter of the effort wheel,
d be the diameter of the load drum, and
T be the number of teeth on the worm wheel.
Let the effort wheel, be given one complete revolution.

By this the worm wheel is moved by one tooth if the worm is of single-threaded type.

Distance moved by the effort $= \pi D$

Distance moved by the load in one revolution of the worm, which advances the load drum by one tooth,

$$= \frac{\pi d}{T}$$

Velocity Ratio,

$$V_r = \frac{\pi D}{(\pi d / T)} = \frac{T \cdot D}{d} \qquad ...\text{(Eq. 9.36)}$$

For double-threaded drum, one revolution of the worm wheel moves the load drum by two teeth.

Hence, the Velocity Ratio, V_r for this is given by,

$$V_r = \frac{\pi D}{(2 \pi d / T)} = \frac{T \cdot D}{2d} \qquad ...\text{(Eq. 9.37)}$$

9.14 WINCH CRAB

A 'Winch Crab' is a lifting machine in which the velocity ratio is increased by a system of gears.

If just one set of gears is used, it is called a 'Single-Purchase Winch Crab', and if two sets are used it is called a 'Double-Purchase Winch Crab'.

Winch crabs are used to lift heavy loads by a relatively small effort.

9.14.1 Single-Purchase Winch Crab

A schematic of a single purchase winch crab is shown in Fig. 9.16.

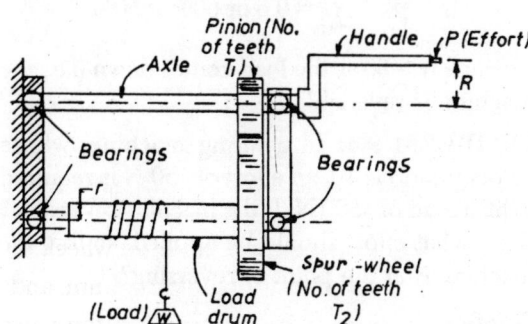

Fig. 9.16: Single Purchase Winch Crab

Let R be the radius the crank (handle) and r that of the load drum over which a rope is wound and the load attached to it. Let the number of teeth on the pinion be T_1 and that on the spur

wheel (main gear) be T_2. The pinion is always smaller than the spur wheel and the same is true of the number of teeth as well (i.e., $T_2 > T_1$).

When the handle at the end of which the effort is applied makes a complete revolution, the pinion makes one complete revolution. The load drum moves by T_1/T_2 of a revolution.

Distance moved by the effort $= 2\pi R$

Then the load moves by a distance $2\pi r \cdot \dfrac{T_1}{T_2}$

Hence the velocity ratio,

$$V_r = \frac{2\pi R}{2\pi r \dfrac{T_1}{T_2}} = \left(\frac{R}{r}\right)\left(\frac{T_2}{T_1}\right) \qquad ...\text{(Eq. 9.38)}$$

For a frictionless machine, the mechanical advantage equals this velocity ratio.

9.14.2 Double-Purchase Winch Crab

A schematic of a double-purchase winch crab is shown in Fig. 9.17.

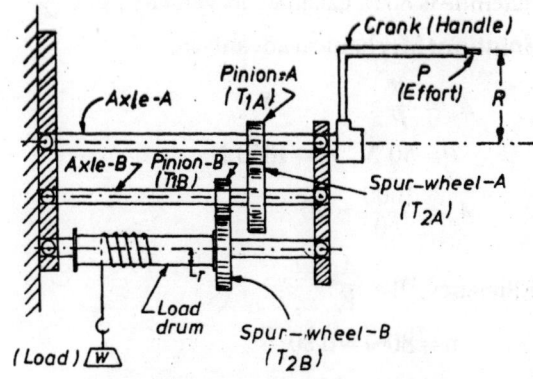

Fig. 9.17: Double-Purchase Winch Crab

Since two sets of pinions and spur wheels with two axles are used, it is known as a 'double-purchase winch crab'.

Let the number of teeth on the pinion and spur wheel of set A be T_{1A} and T_{2A}, and that on those of set B be T_{1B} and T_{2B}, respectively.

For one complete revolution of the handle,

The distance moved by the effort $= 2\pi R$

(R is the radius of the crank)

When axle-*A* makes one revolution, axle-*B* makes (T_{1A}/T_{2A}) of a revolution; the load drum, similarly, makes

$$\left(\frac{T_{1A}}{T_{2A}}\right)\left(\frac{T_{1B}}{T_{2B}}\right) \text{ of a revolution.}$$

Hence the distance moved by the load is

$$(2\pi r)\left(\frac{T_{1A}}{T_{2A}}\right)\left(\frac{T_{1B}}{T_{2B}}\right)$$

∴ The velocity ratio,

$$V_r = \frac{2\pi R}{(2\pi r)(T_{1A}/T_{2A})(T_{1B}/T_{2B})}$$

or $\quad V_r = \left(\frac{R}{r}\right)\left(\frac{T_{2A}}{T_{1A}}\right)\left(\frac{T_{2B}}{T_{1B}}\right)$ ...(Eq. 9.39)

For a friction less machine, the mechanical advantage equals this velocity ratio.

ILLUSTRATIVE PROBLEMS

PROBLEM 9.1: In a machine, 50 N effort is applied to raise a load of 1000 N. Determine the mechanical advantage. If the efficiency of the machine is 80%, calculate its velocity ratio.

Solution: Mechanical advantage,

$$A_m = \frac{W}{P}$$
$$P = 50 \text{ N}; \ W = 1000 \text{ N}$$
∴ $\quad A_m = \frac{1000}{50} = 20$

Efficiency, $\eta = \frac{A_m}{V_r}$

$$\eta = 80\% = 0.80$$

∴ $\quad V_r = \frac{A_m}{\eta} = \frac{20}{0.8} = 25$

PROBLEM 9.2: In a certain weight lifting machine, a weight of 800 N is lifted by an effort of 20 N. While the weight moves up by 10 cm, the point of application of effort moves by 6m. Determine:

(a) Mechanical advantage of the machine.
(b) Velocity ratio of the machine.
(c) Efficiency of the machine and
(d) Ideal effort required.

Solution: Actual effort, $P = 20$ N

Load lifted, $W = 800$ N

Distance moved by effort, $D = 6$ m

Distance moved by load, $d = 10$ cm $= 0.1$m

(a) Mechanical advantage, $A_m = \frac{W}{P}$

$$= \frac{800}{20} = 40$$

(b) Velocity ratio, $V_r = \frac{D}{d} = \frac{6}{0.1} = 60$
(c) Efficiency of the machine, $\eta = \frac{A_m}{V_r}$

$$= \frac{40}{60} \times 100 = 66.67\%$$

(d) Ideal effort required, $P_i = \frac{W}{V_r}$

$$= \frac{800}{60} = 13.33 \text{ N}$$

PROBLEM 9.3: A lifting machine having a velocity ratio 25 starts raising a load of 6000 N on applying a an effort of 400 N to it. If suddenly the effort is removed find whether load comes down or not?

Solution:

$$W = 6000 \text{ N}$$
$$P = 400 \text{ N}$$
$$A_m = \frac{W}{P} = \frac{6000}{400} = 15$$
$$V_r = 25$$
$$\eta = \frac{A_m}{V_r} = \frac{15}{25} = 0.6 \text{ or } 60\% > 50\%$$

Since $\eta > 50\%$, the load comes down (i.e., the machine is reversible).

PROBLEM 9.4: In a lifting machine, whose velocity ratio is 40, an effort of 150 N is required to lift a load of 4500 N. Is the machine reversible? If so, what effort should be applied, so that the machine is at the point of reversing?

Solution:

$$W = 4500 \text{ N}, P = 150 \text{ N}, V_r = 40$$

Mechanical advantage, $A_m = \frac{W}{P}$

$$= \frac{4500}{150} = 30$$

Efficiency of the machine, $\eta = \dfrac{A_m}{V_r}$

$$= \dfrac{30}{40} = 75\%$$

Since the efficiency of the machine is more than 50% the machine is reversible.

The machine will be at the point of reversing, when its efficiency $= 50\%$

$$= 0.5$$

Let P_1 be the effort required to lift a load of 4500 N when the machine is at the point of reversing.

$$A_m = \dfrac{W}{P_1} = \dfrac{4500}{P_1}$$

$$\eta = 0.5 = \dfrac{A_m}{V_r} = \dfrac{(4500 / P_1)}{40}$$

$$\dfrac{4500}{P_1} = 0.5 \times 40 = 20$$

$$\therefore \quad P_1 = \dfrac{4500}{20} = 225 \text{ N}$$

PROBLEM 9.5: In a lifting machine an effort of 400 N is to be moved through a distance of 15 m to raise a load of 8 kN through a distance of 60 cm. Determine:

(a) Mechanical advantage
(b) Velocity ratio
(c) Efficiency of the machine
(d) Ideal effort
(e) Effort lost in friction
(f) Ideal load and
(g) Frictional resistance

Solution:

(a) **Mechanical Advantage**

$$A_m = \dfrac{W}{P} = \dfrac{8 \times 1000}{400} = 20$$

(b) **Velocity Ratio**

$$V_r = \dfrac{D}{d} = \dfrac{15}{(60/100)} = 25$$

(c) **Efficiency**

$$\eta = \dfrac{A_m}{V_r} = \dfrac{20}{25} = 0.8 \text{ or } 80\%$$

(d) **Ideal Effort**

$$P_c' = \dfrac{W}{V_r} = \dfrac{8000}{400} = 320 \text{ N}$$

(e) **Effort Lost in Friction**

$$= P - P_i = 400 - 320 = 80 \text{ N}$$

(f) **Ideal Load**

$$W_i = P \times V_r$$
$$= 400 \times 25 = 10{,}000 \text{ N}$$

(g) **Frictional Resistance**

$$= W_i - W$$
$$= 10{,}000 - 8000$$
$$= 2000 \text{ N or 2 kN}$$

PROBLEM 9.6: In a lifting machine an effort of 15 N raised a load of 750 N. What is the mechanical advantage ? Find the velocity ratio if the efficiency at this load is 50%.

If on the machine an effort of 25 N raised a load of 1500 N, what is the efficiency? What will be the effort to raise a load of 500 N? Calculate the maximum mechanical advantage and maximum efficiency.

Solution:

(a) $W = 750 \text{ N}$

$P = 15 \text{ N}$

$$A_m = \dfrac{W}{P} = \dfrac{750}{15} = 50$$

$$\eta = 50\% = 0.50$$

$$V_r = \dfrac{A_m}{\eta} = \dfrac{50}{0.50} = 100$$

(b) $P = 25 \text{ N}$

$W = 1500 \text{ N}$

$$A_m = \dfrac{W}{P} = \dfrac{1500}{25} = 60$$

$$\eta = \dfrac{A_m}{V_r} = \dfrac{60}{100} = 0.60 \text{ or } 60\%$$

(c) $P = ?$

$W = 500 \text{ N}$

$$P = mW + C$$
$$15 = 750m + C \qquad \qquad \dots(1)$$
$$25 = 1500m + C \qquad \qquad \dots(2)$$

(2) – (1) gives

$$m(1500 - 750) = 25 - 15 = 10$$

$$m = \frac{10}{750} = \frac{1}{75} = 0.0133333$$

$$C = -750 \times \frac{1}{75} + 15 = 5$$

$$P = 500 \times \frac{1}{75} + 5 = 6.67 + 5 = 11.67 \text{ N}$$

Maximum Mechanical Advantage,

$$(A_m)_{max} = \frac{1}{m} = \frac{1}{(1/75)} = 75$$

Maximum efficiency,

$$\eta_{max} = \frac{1}{V_r \cdot m} = \frac{1}{100 \times (1/75)}$$

$$= \frac{75}{100}$$

$$= 0.75 \text{ or } 75\%$$

PROBLEM 9.7: The velocity ratio of a lifting machine is 25, from which loads of 125 N and 25 N are raised by applying efforts of 10 N and 5 N respectively. Find the effort lost in friction in raising a load of 75 N. Find also the maximum efficiency of the machine.

Solution:

$$P = mW + C$$
$$P_1 = 10 \text{ N}; W_1 = 125 \text{ N}$$
$$\therefore \quad 10 = 125m + C \qquad \qquad ...(1)$$
$$P_2 = 5 \text{ N}; W_2 = 15 \text{ N}$$
$$\therefore \quad 5 = 25m + C \qquad \qquad ...(2)$$

(1) – (2) gives

$$m(125 - 25) = 10 - 5 = 5$$
$$100m = 5$$
$$\therefore \quad m = \frac{5}{100} = \frac{1}{20} = 0.05$$

Substituting the value of m in equation (2)

$$5 = 25 \times 0.05 + C$$
$$C = 5 - 1.25 = 3.75$$
$$W = 75 \text{ N}$$
$$P = ?$$
$$P = 0.05 \times 75 + 3.75$$

$$= 3.75 + 3.75 = 7.5 \text{ N}$$

Ideal effort, $P_i = \dfrac{W}{V_r} = \dfrac{75}{25} = 3.0 \text{ N}$

Effort lost in friction $= P - P_i$

$$= 7.5 - 3.0$$
$$= 4.5 \text{ N}$$

$$\eta_{max} = \frac{1}{25 \times 0.05} = 0.80 \text{ or } 80\%$$

PROBLEM 9.8: In a system of pulleys of the first order type, there are four movable pulleys. If a load of 1280 N is lifted by an effort of 100 N, find:

(a) efficiency of the machine,
(b) effort lost in friction, and
(c) frictional load

(a) **Efficiency of the machine**

$$n = 4$$
$$V_r = 2^n = 2^4 = 16$$
$$W = 1280 \text{ N}; P = 100 \text{ N}$$

$$A_m = \frac{W}{P} = \frac{1280}{100} = 12.8$$

$$\eta = \frac{A_m}{V_r} = \frac{12.8}{16} = 0.80 \text{ or } 80\%$$

(b) **Effort lost in friction**

Ideal effort, $P_i = \dfrac{W}{V_r} = \dfrac{1280}{16}$

$$P_i = 80 \text{ N}$$

Effort lost in friction $= P - P_i$

$$= 100 - 80 = 20 \text{ N}$$

(c) **Frictional load**

Ideal load, $W_i = P \times V_r$

$$W_i = 100 \times 16$$
$$W_i = 1600 \text{ N}$$

Frictional load $= W_i - W$

$$= 1600 - 1280 = 320 \text{ N}$$

PROBLEM 9.9: In a first order system of pulleys, there are five movable pulleys. If the efficiency of the machine is 80% what effort can lift a load of 5 kN.

Solution:

$$W = 5 \text{ kN} = 5000 \text{ N}; \eta = 80\% = 0.80;$$
$$n = 5$$

$$V_r = 2^n = 2^5 = 32$$

$$\eta = \frac{A_m}{V_r}$$

$$0.80 = \frac{A_m}{32}$$

$$A_m = 0.80 \times 32 = 25.60$$

But $A_m = \dfrac{W}{P}$

$$25.60 = \frac{5000}{P}$$

$$\therefore \quad P = \frac{5000}{25.60}$$

$$P = 195.31 \text{ N}$$

PROBLEM 9.10: A weight of 1 kN is lifted by an effort of 250 N, by second-order system of pulleys, having three pulleys in upper block and two pulleys in the lower block. Find:

(a) Efficiency of the machine,
(b) Effort lost in friction, and
(c) Frictional load

Solution:

(a) $W = 1 \text{ kN} = 1000 \text{ N}$
$P = 250 \text{ N}$

$$\therefore \quad A_m = \frac{W}{P} = \frac{1000}{250} = 4$$

Total number of pulleys, $n = 3 + 2 = 5$

$$V_r = n = 5$$

$$\therefore \quad \eta = \frac{A_m}{V_r} = \frac{4}{5} = 0.80 \text{ or } 80\%$$

(b) Ideal effort, $P_i = \dfrac{W}{V_r} = \dfrac{1000}{5}$

$$= 200 \text{ N}$$

Effort lost in friction $= P - P_i$
$$= 250 - 200 = 50 \text{ N}$$

(c) Ideal load, $W_i = P \cdot V_r$
$$= 250 \times 5 = 1250 \text{ N}$$

$$\therefore \quad \text{Frictional load} = W_i - W$$
$$= 1250 - 1000 = 250 \text{ N}$$

PROBLEM 9.11: There are four pulleys in a third-order system of pulleys. An effort of 500 N is required to lift an unknown weight. If the efficiency of the machine is 80%, find the weight lifted. Also calculate the frictional load.

Solution:

$$n = 4$$
$$P = 500 \text{ N}$$
$$\eta = 80\% = 0.80$$
$$V_r = 2^n - 1 = 2^4 - 1 = 15$$
$$\eta = \frac{A_m}{V_r}$$

$$0.80 = \frac{A_m}{15}$$

$$A_m = 0.80 \times 15 = 12$$

But $A_m = \dfrac{W}{P}$

$$12 = \frac{W}{500}$$

$$W = 12 \times 500 = 6000 \text{ N} = 6 \text{ kN}$$

Ideal load, $W_i = P \times V_r$
$$= 500 \times 15 = 7500 \text{ N} = 7.5 \text{ kN}$$

Frictional load $= W_i - W$
$$= 7.5 - 6.0 = 1.5 \text{ kN}$$

PROBLEM 9.12: A weight of 360 N is to be raised by means of a simple wheel and axle. The axle is 100 mm diameter and wheel is 400 mm diameter. If a force of 120 N has to be applied to the wheel, find (a) mechanical advantage, (b) velocity ratio and (c) efficiency of the machine.

Solution:

$$W = 360 \text{ N}$$
$$P = 120 \text{ N}$$
$$D = 400 \text{ mm}$$
$$d = 100 \text{ mm}$$

(a) **Mechanical advantage**

$$A_m = \frac{W}{P} = \frac{360}{120} = 3$$

(b) **Velocity ratio**

$$V_r = \frac{D}{d} = \frac{400}{100} = 4$$

(c) **Efficiency of the machine**

$$\eta = \frac{A_m}{V_r} = \frac{3}{4} = 0.75 \text{ or } 75\%$$

PROBLEM 9.13: In a differential wheel and axle, the diameter of the effort wheel is 500 mm. The diameters of the axles are 300 mm and 250 mm respectively. The diameter of the rope is 10 mm. The efficiency of the machine is 80%. Find the load which can be lifted by an effort of 150 N.

Solution: Let W = Load that can be lifted
Effective diameter of effort wheel,

$$D = 500 + 10$$
$$= 510 \text{ mm}$$

Effective diameters of axles,

$$d_1 = 300 + 10 = 310 \text{ mm}$$
$$d_2 = 250 + 10 = 260 \text{ mm}$$

$$\therefore \quad V_r = \frac{2D}{d_1 - d_2} = \frac{2 \times 510}{310 - 260} = \frac{2 \times 510}{50}$$
$$V_r = 20.4$$

$$\eta = \frac{A_m}{V_r}$$

$$0.80 = \frac{A_m}{20.4}$$

$$\therefore \quad A_m = 0.80 \times 20.4$$
$$A_m = 16.32$$

But $\quad A_m = \dfrac{W}{P}$

$$16.32 = \frac{W}{150}$$

or $\quad W = 16.32 \times 150$

$$W = 2448 \text{ N} = 2.448 \text{ kN} \approx 2.45 \text{ kN}$$

PROBLEM 9.14: In a differential wheel and axle, the diameter of the effort wheel is 450 mm. A load of 2.1 kN is raised by an applying an effort of 100 N. The efficiency of the machine at this load is 70%. If the sum of diameters of the axles is 270 mm, determine the diameter of each axle.

Let d_1 & d_2 are the diameters of the large and smaller axles respectively.

$$D = 450 \text{ mm}$$
$$W = 2.1 \text{ kN} = 2100 \text{ N}$$
$$P = 100 \text{ N}$$
$$\eta = 70\% = 0.70$$

$$d_1 + d_2 = 270 \text{ mm} \qquad \qquad \dots(1)$$

$$A_m = \frac{W}{P} = \frac{2100}{100} = 21$$

$$\eta = \frac{A_m}{V_r}$$

$$0.70 = \frac{21}{V_r}$$

or $\quad V_r = \dfrac{21}{0.70}$

$$V_r = 30$$

But $\quad V_r = \dfrac{2D}{d_1 - d_2}$

$$30 = \frac{2 \times 450}{d_1 - d_2}$$

or $\quad d_1 - d_2 = \dfrac{2 \times 450}{30}$

$$d_1 - d_2 = 30 \qquad \qquad \dots(2)$$

(1) + (2) gives

$$2d_1 = 270 + 30 = 300$$

$$d_1 = \frac{300}{2} = 150 \text{ mm}$$

Substituting in equation (1)

$$150 + d_2 = 270$$

$$d_2 = 270 - 150 = 120 \text{ mm}$$

The diameter of the larger axle = 150 mm
The diameter of the smaller axle = 120 mm

PROBLEM 9.15: The efficiency of a Weston's Differential Pulley block is 40%. The upper pulley block has two pulleys of diameters 300 mm and 250 mm respectively. Find the load lifted by this machine if effort applied is 25 N.

Solution:

$$\eta = 40\% = 0.40$$
$$D = 300 \text{ mm}$$
$$d = 250 \text{ mm}$$
$$P = 25 \text{ N}$$
$$W = ?$$

$$V_r = \frac{2D}{D - d} = \frac{2 \times 300}{(300 - 250)} = \frac{2 \times 300}{50} = 12$$

$$\eta = \frac{A_m}{V_r}$$

$$0.40 = \frac{A_m}{12}$$

or $A_m = 0.40 \times 12 = 4.80$

But $A_m = \dfrac{W}{P}$

$$4.8 = \frac{W}{25}$$

or $W = 4.8 \times 25 = 120$ N

PROBLEM 9.16: In a Weston's Differential Pulley block, the difference in the number of teeth of the two pulleys is 1. The efficiency of the machine is 75%. If an effort of 12 N just lifts a load of 180 N, find the number of teeth of the two pulleys.

Solution: Let T_1 and T_2 are the number of teeth of the larger and smaller pulleys respectively.

Difference in the number of teeth of the two pulleys = 1

i.e. $T_1 - T_2 = 1$

or $T_1 = 1 + T_2$

$$V_r = \frac{2T_1}{T_1 - T_2} = \frac{2 \times (1 + T_2)}{1}$$

$V_r = 2(1 + T_2)$ ...(1)

$W = 180$ N

$P = 12$ N

$$A_m = \frac{W}{P} = \frac{180}{12} = 15$$

$\eta = 75\% = 0.75$

But $\eta = \dfrac{A_m}{V_r}$

$$0.75 = \frac{15}{V_r}$$

or $V_r = \dfrac{15}{0.75} = 20$ ...(2)

Equating (1) and (2)

$2(1 + T_2) = 20$

$1 + T_2 = \dfrac{20}{2} = 10$

or $T_2 = 10 - 1 = 9$

$T_1 = 1 + 9 = 10$

$T_1 = 10; \; T_2 = 9$

PROBLEM 9.17: A screw jack has a thread of 10 mm pitch. Determine the effort applied at the end of a handle of 450 mm length to lift a load of 2.5 kN, if the efficiency of the jack at this load is 40%.

Solution:

$p = 10$ mm

$R = 450$ mm

$W = 2.5$ kN $= 2500$ N

$\eta = 40\% = 0.40$

$$V_r = \frac{2\pi R}{p} = \frac{2 \times \pi \times 450}{10} = 282.74$$

$$\eta = \frac{A_m}{V_r}$$

$$0.4 = \frac{A_m}{482.74}$$

$\therefore \quad A_m = (282.74)0.4$

$\quad\quad = 113.096$

But $A_m = \dfrac{W}{P}$

$$113.096 = \frac{2500}{P}$$

or $P = \dfrac{2500}{113.096} = 22.105$ N

PROBLEM 9.18: A screw jack has square threads 50 mm mean diameter and 12 mm pitch. The coefficient of friction between the screw and nut is 0.10. Calculate the torque required to a raise a load of 5 kN. Also calculate the efficiency of the screw. The length of the handle is 500 mm.

Solution:

$p = 12$ mm; $W = 5$ kN $= 5000$ N

$d = 50$ mm; $R = 500$ mm; $\mu = 0.10$

$\therefore \quad \tan\theta = \dfrac{p}{\pi d} = \dfrac{12}{\pi \times 50} = 0.07639$

$$P = \frac{d}{2R} W \left(\frac{\mu + \tan\theta}{1 - \mu \cdot \tan\theta} \right)$$

$$= \frac{50}{2 \times 500} \times 5000 \left(\frac{0.10 + 0.07639}{1 - 0.10 \times 0.07639} \right)$$

$$= 44.44 \text{ N}$$

Torque, $T = P \times R$

$$= 44.44 \times 500$$

$$= 22220 \text{ N}$$

$$= 22.22 \text{ kN-mm}$$

$$A_m = \frac{W}{P} = \frac{5000}{44.44} = 112.51$$

$$V_r = \frac{2\pi R}{p} = \frac{2 \times \pi \times 500}{12}$$

$$V_r = 261.8$$

$\therefore \qquad \eta = \dfrac{A_m}{V_r} = \dfrac{112.51}{261.8} = 0.4298$

or   42.98% say 43 %

PROBLEM 9.19: In a differential screw jack, the screw threads have pitches of 10 mm and 6 mm. If the efficiency of the machine is 25%, find the effort required at the end of an arm of 400 mm long to lift a load of 4.5 kN.

Solution:

$$p_A = 10 \text{ mm}$$

$$p_B = 6 \text{ mm}$$

$$\eta = 25\% = 0.25$$

$$R = 400 \text{ mm}$$

$$W = 4.5 \text{ kN} = 4500 \text{ N}$$

$$V_r = \frac{2\pi R}{p_A - p_B}$$

$$= \frac{2 \times \pi \times 400}{10 - 6}$$

$$= 628.32$$

$$\eta = \frac{A_m}{V_r}$$

$$0.25 = \frac{A_m}{628.32}$$

$$A_m = 0.25 \times 628.32$$

$$= 157.08$$

But $A_m = \dfrac{W}{P}$

$$157.08 = \frac{4500}{P}$$

$$P = \frac{4500}{157.08}$$

$$P = 28.65 \text{ N}$$

PROBLEM 9.20: In a worm and worm-wheel, the diameter of the load drum is 150 mm and that of the effort wheel is 450 mm. The number of teeth on the worm wheel are 50 and the worm has double start threads. A load of 3000 N could just be lifted by an effort of 200 N. Find the efficiency of the machine.

Solution: Number of teeth on the worm wheel,

$$T = 50$$

Diameter of the effort wheel, $D = 450$ mm

Diameter of the load drum, $d = 150$ mm

The worm has double start

$\therefore \qquad V_r = \dfrac{TD}{2d} = \dfrac{50 \times 450}{2 \times 150} = 75$

Load to be lifted, $W = 3000$ N

Effort, $P = 200$ N

$\therefore$ Mechanical Advantage,

$$A_m = \frac{W}{P}$$

$$A_m = \frac{3000}{2000} = 15$$

Efficiency, $\eta = \dfrac{A_m}{V_r} = \dfrac{15}{75}$

$$= 0.20 \text{ or } 20\%$$

PROBLEM 9.21: A single purchase crab winch has the following specifications:

   Diameter of the load drum = 200 mm

   Length of lever arm = 1.2 m

   Number of teeth on pinion = 15

   Number of teeth on spur wheel = 90

   Find the velocity ratio of the machine.

   On this machine efforts of 50 N and 80 N are required to lift the loads of 1500 N and 4500 N respectively. Find the law of the machine and the efficiencies in the two cases.

Solution: Velocity of ratio of the single purchase crab winch is given by

$$V_r = \frac{R}{r} \times \frac{T_2}{T_1}$$

Length of the lever arm, $R = 1.2$ m $= 1200$ mm

Radius of the drum,

$$r = \frac{200}{2} = 100 \text{ mm}$$

Number of teeth on spur wheel, $T_2 = 90$

Number of teeth on pinion, $T_1 = 15$

$$V_r = \frac{1200}{100} \times \frac{90}{15}$$

$$V_r = 72$$

Let the law of machine be $P = mW + C$

$$P_1 = 50 \text{ N}; \ W_1 = 1500 \text{ N}$$

$$P_2 = 80 \text{ N}; \ W_2 = 4500 \text{ N}$$

$$50 = 1500m + C \qquad \ldots(1)$$

$$80 = 4500m + C \qquad \ldots(2)$$

$(2) - (1)$ gives

$$(4500 - 1500)m = (80 - 50)$$

$$m = \frac{30}{3000} = 0.01$$

Substituting the value of m in equation (1)

$$50 = 1500 \times 0.01 + C$$

$$C = 50 - 15 = 35$$

$$P = 0.01 W + 35$$

$$W_1 = 1500 \text{ N}; \ P_1 = 50 \text{ N}$$

$$\left(A_m\right)_1 = \frac{W_1}{P_1} = \frac{1500}{50} = 30$$

$$\eta_1 = \frac{\left(A_m\right)_1}{V_r} = \frac{30}{72} = 0.4167$$

$$= 41.67\%$$

$$W_2 = 4500 \text{ N}; \ P_2 = 80 \text{ N}$$

$$\left(A_m\right)_2 = \frac{W_2}{P_2} = \frac{4500}{80} = 56.25$$

$$\eta_2 = \frac{\left(A_m\right)_2}{V_r} = \frac{56.25}{72}$$

$$= 0.78125 \text{ or } 78.125 \%$$

PROBLEM 9.22: In a double purchase crab winch, teeth of spur wheels are 75 and 90. Length of the handle is 800 mm and radius of the load drum is 200 mm. If the efficiency of the machine is 50%, find the effort required by a load of 900 N.

Solution:

Let $P =$ Effort required to lift

$$W = 900 \text{ N}$$

$$\eta = 50\% = 0.5$$

$$T_{1A} = 30$$

$$T_{2A} = 90$$

$$T_{1B} = 25$$

$$T_{2B} = 75$$

$$R = 800 \text{ mm}$$

$$r = 200 \text{ mm}$$

$$\therefore \quad V_r = \left(\frac{R}{r}\right)\left(\frac{T_{2A}}{T_{1A}}\right)\left(\frac{T_{2B}}{T_{1B}}\right)$$

$$= \left(\frac{800}{200}\right)\left(\frac{75}{25}\right)\left(\frac{90}{30}\right)$$

$$= 4 \times 3 \times 3 = 36$$

$$\eta = \frac{A_m}{V_r}$$

$$0.5 = \frac{A_m}{36}$$

$$A_m = 0.5 \times 36 = 18.0$$

But $\quad A_m = \dfrac{W}{P}$

$$18 = \frac{900}{P}$$

or $\quad P = \dfrac{900}{18} = 50 \text{ N}$

PRACTICE PROBLEMS

9.1 The velocity ratio of a machine is 20 and its efficiency is 50%. Find the load which can be raised on application of effort of 40 N.

9.2 A load of 150 N is raised by means of a certain weight lifting machine through a distance of 150 mm. If the effort applied is 25 N and has moved through a distance of 1.2 m, find the efficiency of the machine.

9.3 In a weight lifting machine, whose velocity ratio is 20, a weight 1.5 kN can be raised by an effort of 100 N. If the effort is removed, show that the machine can work in the reverse direction.

9.4 In a lifting machine in which the velocity ratio is 30, a load of 5000 N is lifted with an effort of 400 N. Determine whether it is self-locking or reversible machine. Find the frictional resistance.

9.5 In a lifting machine whose velocity ratio is 5, an effort of 25 N was able to lift a load of 100 N. Find:

(a) mechanical advantage
(b) efficiency of the machine
(c) ideal effort
(d) effort lost in friction
(e) ideal load and
(f) frictional resistance

9.6 In a lifting machine, whose velocity ratio is 25, a load of 2000 N is lifted by an effort of 130 N and a load of 3400 N is lifted by an effort of 200 N. Find the law of the machine and calculate the load that could be lifted by a force of 180 N. Calculate also:

(a) the amount of effort lost in friction,
(b) mechanical advantage, and
(c) the efficiency

9.7 In a lifting machine efforts of 10 N and 15 N are required to lift loads of 250 N and 750 N respectively. Find out the effort required to raise a load of 600 N.

9.8 In a system of pulleys of the first order type, there are three movable pulleys. If a load of 720 N is lifted by an effort of 100 N, find the

(a) efficiency of the machine
(b) effort lost in friction, and
(c) frictional load

9.9 In a first order system of pulleys, there are three movable pulleys. If the efficiency of the machine is 75%, what effort can lift a load of 3 kN.

9.10 A weight of 1.5 kN is lifted by an effort of 200 N, by second order system of pulleys, having five pulleys in each block. Calculate:

(a) efficiency of the machine
(b) effort lost in friction
(c) frictional load

9.11 There are four pulleys in a third-order system of pulleys. Find out the effort required to lift a load of 3 kN, if the efficiency of the machine is

80%. Also calculate the amount of effort wasted in friction.

9.12 In a simple wheel and axle machine, the diameter of the axle is 50 mm. If a load of 480 N is raised by applying an effort of 120 N, at the efficiency of 80%, find the diameter of the wheel.

9.13 The larger and smaller diameters of a differential axle are 150 mm and 120 mm respectively. The diameter of the effort wheel is 450 mm. Find the efficiency of the machine, if a load of 2.1 kN is lifted by applying an effort of 100 N.

9.14 A load of 2 kN is to be lifted by a differential wheel and axle. It consists of differential axle of 200 mm and 250 mm and the wheel diameter is 600 mm. Find the effort required if the efficiency of the machine is 70%.

9.15 In a Weston's differential pulley block, diameter of the concentric pulleys are 250 mm and 200 mm respectively. It is found that an effort of 20 N just lifts a load of 120 N calculate

(a) efficiency of the machine;
(b) effort lost in friction and
(c) frictional load.

9.16 In a Weston's differential pulley block a load of 180 N is raised by an effort of 12 N. The number of teeth on the larger and smaller blocks are 24 and 22 respectively. Find the (a) velocity ratio; (b) mechanical advantage; and (c) efficiency of the machine.

9.17 A screw jack raises a load of 54 kN. The screw is square threaded having three threads per 30 mm length and 50 mm in mean diameter. Find the effort required at the end of a lever of 500 mm long measured from the axis of the screw if the coefficient of friction between screw and nut is 0.10.

9.18 A screw jack has square threads 40 mm mean diameter and 10 mm pitch. The coefficient of friction between screw and nut is 0.10. Determine the effort applied at the end of a lever of 400 mm long to lift a load of 4 kN. Also determine the efficiency of the screw.

9.19 A differential screw jack has threads of 10 mm and 6 mm. An effort of 15 kN applied at the end of a handle of 500 mm length can lift a load of 3.5 kN. Find the efficiency of the machine (screw).

9.20 In a worm and worm-wheel, the number of teeth in the worm-wheel is 30. The effort handle is 200 mm long (radius of the effort wheel) and the load drum is 100 mm diameter. Find the efficiency of the machine, if an effort of 50 N can lift a load of 750 N and the worm is double threaded.

9.21 A single purchase crab winch has the following specifications:

Diameter of the load drum = 200 m
Length of lever arm = 0.75 m
Number of teeth on pinion = 12
Number of teeth on spur wheel = 96.

Find the velocity ratio of the machine. On this machine efforts of 60 N and 100 N are required to lift the loads of 1800 N and 4800 N, respectively. Find the law of the machine and efficiencies in the two cases. Also determine the loss of load and the loss of effort in friction when lifting the load of 4800 N.

9.22 In a double purchase crab winch, the pinions have 20 and 25 teeth. While the spur wheels have 60 and 50 teeth. The effort of handle is 600 mm while the effective diameter is 160 mm. If the efficiency of the winch is 60%, find the load that will be lifted by an effort of 100 N applied at the end of the handle.

Chapter 10

Virtual Work

10.1 INTRODUCTION

Till now, in all problems involving the static equilibrium of a particle or a rigid body, the three equations of static equilibrium have been used to arrive at the solution.

Alternate methods of expressing the conditions of equilibrium are available—the method of "virtual work" and that of "minimum potential energy"; the former will be presented in this chapter and the latter in Chapter 17.

These methods provide simpler means of solving certain problems than by the conventional approach seen earlier. These concepts lead us into the realm of the "Variational principles of mechanics" which include equations of Lagrange, Hamilton, Jacobi, etc., that form the base of advanced classical mechanics.

Even for our immediate needs in mechanics, the method of virtual work is of great significance.

Before we can discuss this method we need to understand what is meant by the "work done" by a force.

If a particle is subjected to a force F and the particle gets displaced by an infinitesimal displacement ds in the direction of the force, the products of these two is called the 'work done (dW)' by the force:

$$dW = F \cdot ds$$

The work done, W, under a given finite movement or displacement in the direction of the force F, is:

$$W = \int F \cdot ds \qquad \dots \text{(Eq. 10.1)}$$

If the directions of the force and the displacement are different, with an angle θ between them, this equation gets modified as:

$$W = \int (F \cos\theta) \cdot ds \qquad \dots \text{(Eq. 10.2)}$$

The work done by a force is a scalar quantity with magnitude and sign, but with no direction.

The unit of work is N.m (newton-metre) (or a Joule). Work done by a force is zero if either the displacement is zero or the force is in a perpendicular direction to that of the displacement. Work done is positive if the direction of the force and direction of the displacement are the same. There can be exceptions as in the case of motion against gravity.

10.2 CONCEPT OF VIRTUAL WORK

Having understood the concept of work done by a force, it is easy to understand that no work is done by a system forces in equilibrium; however, if it is assumed that an infinitesimal displacement is *imagined* to be given to such a system, some work is imagined to be done. The imaginary displacement is called "Virtual displacement", and the imaginary work done is known as "Virtual work".

The adjective 'Virtual' is used to emphasize the hypothetical nature of the displacement and the work. We would only imagine what would happen, if the virtual displacement is given to the system.

The concept of virtual work is very useful in determining the unknown forces in certain problems of statics.

10.3 PRINCIPLE OF VIRTUAL WORK

The principle of virtual work may be stated as follows:

When an initially stationary particle, rigid body, or an ideally connected system of rigid bodies, is in equilibrium under a system of forces, the virtual work done by the system under any virtual displacement consistent with the constraints is zero. (This is a necessary and sufficient condition for equilibrium).

Note: The "initially stationary" condition is imposed because this will not be a sufficient condition for equilibrium for a particle moving at a constant speed in a circular path although the condition of zero virtual work will be satisfied in the absence of active forces; obviously the particle will have an acceleration towards the centre of curvature, and so cannot be considered to be in equilibrium.

The external forces which do work are known as "active" forces.

From the definition of work and/or the nature of the forces, it should be obvious that the following types of forces do not do any work under the stated conditions:

(*i*) Self-weight of a body with horizontal motion.

(*ii*) Reaction at a frictionless roller permitting motion parallel to the rigid surface of support.

(*iii*) Reaction at a frictionless hinge permitting rotation at the hinge.

(*iv*) Normal reaction on a rolling body.

(*v*) Internal forces of action and reaction–such as the tension in an inextensible spring and the internal forces in the members of trusses with negligible deformations.

(*vi*) In general, all forces acting normal to the direction of motion or displacement.

A simple way of verification of the principle of virtual work is as follows:

Let a particle or a body be held in equilibrium by a system of forces P_1, P_2, ..., P_n as shown in Fig. 10.1.

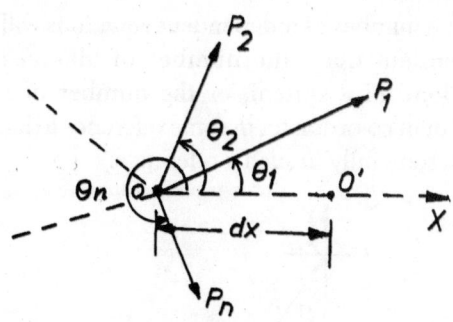

Fig. 10.1: Verification of the Principle of Virtual Work

Let each of the forces make an angle θ_1, θ_2, ..., θ_n with respect to the positive direction of the *x*-axis passing through the particle or the centre of gravity of the body, *O*. The components of the forces parallel to the *x*-axis are $P_1\cos\theta_1$, $P_2\cos\theta_2$, ..., $P_n\cos\theta_n$.

Now let us imagine a virtual displacement *dx* for the particle/body in the direction *OX*.

The work done by the force system during this virtual displacement is:

$$P_1\cos\theta_1\,dx + P_2\cos\theta_2\,dx + ... + P_n\cos\theta_n \cdot dx$$

or $(P_1\cos\theta_1 + P_2\cos\theta_2 + ... + P_n\cos\theta_n) \cdot dx.$

But the term in the brackets is zero since the force system keeps the particle/body in equilibrium. Thus the virtual work done by the force system during any arbitrary virtual displacement is zero when the system is in equilibrium (The sufficiency condition may also be proved easily).

The sign conventions that are used herein are considered to be widely acceptable in the application of the principle of virtual work:

(1) Vertical forces: Upward forces are positive.

(2) Horizontal forces: Forces towards the right are positive.

(3) Internal forces: Tensile forces are positive.

(4) Moments of forces: Clockwise moments are positive.

10.4 DEGREES OF FREEDOM

The equations of equilibrium by the method of virtual work do not involve reactions; thus when these are not of interest, the method is very useful. The number of unknown 'active' forces that can be solved is equal to the number of independent equations.

The number of independent equations will be dependent upon the number of 'degrees of freedom' of a system, or the number of independent co-ordinates in some reference to locate the system fully in such a reference.

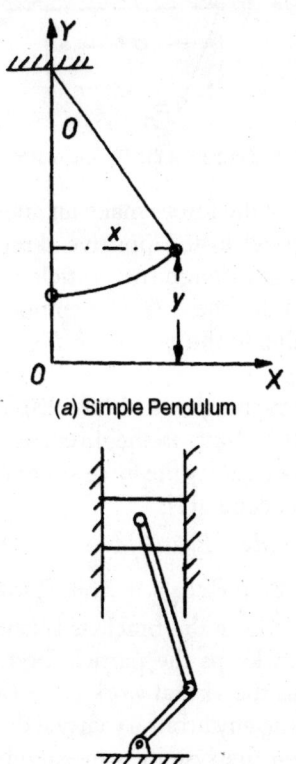

(a) Simple Pendulum

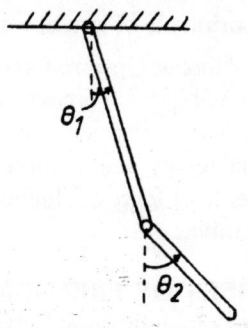

(b) Piston & Crank

Fig. 10.2: Systems with Single-Degree Freedom

(a) Plane Double Pendulum (two-degree freedom)

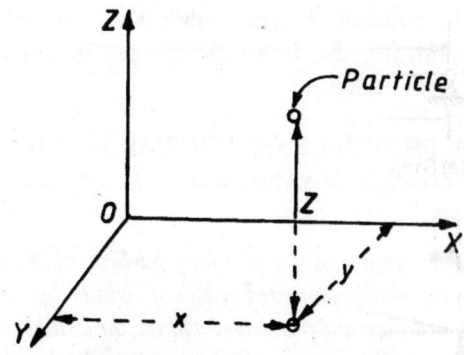

(b) Particle in Space (three-degree freedom)

Fig. 10.3: Systems with Multi-Degree Freedom

A few examples of systems with a single-degree freedom and two-degree freedom are shown in Figs. 10.2 and 10.3.

The number of unique virtual displacements that may be given to a system equals its degree of freedom, and this is the number of unknown active forces that may be solved.

10.5 APPLICATIONS OF THE PRINCIPLE VIRTUAL WORK

The principle of virtual work has very wide and versatile applications.

While writing the equation of virtual work either a virtual angular displacement, or a linear displacement, one at a time is to be given so that the unknown force to be determined can do virtual work and it appears in the equation of virtual work.

However, in the case of a fully constrained body, one of the constraints is removed and replaced by a suitable force acting at the point.

The following few important applications will now be considered, and the approach will be given in brief:

    I   Beams
    II Friction problems—ladders
    III Lifting machines
    IV Framed structures

10.5.1 Application to Beams

Let us consider a simple beam carrying a non-central concentrated load as shown in Fig. 10.4.

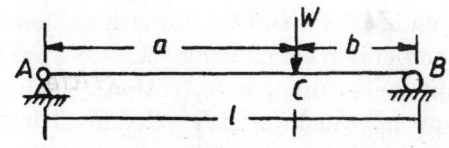

(a) Simple Beam with Non-Central Point Load

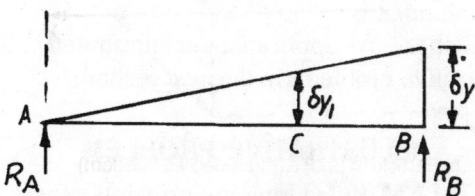

(b) Virtual Displacement of the System

Fig. 10.4: Virtual Work for a Beam with a Point Load

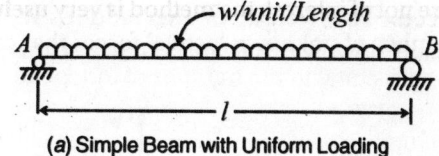

(a) Simple Beam with Uniform Loading

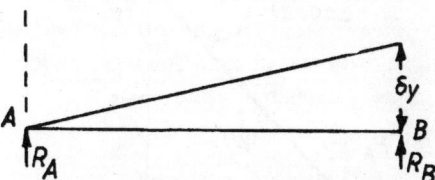

(b) Virtual Displacement of the System

Fig. 10.5: Virtual Work for a Beam with Uniform Loading

Let R_A and R_B be the vertical reactions at A and B. Let the system be given a virtual displacement δy be given upward at B, imagining the constraint at B to have been removed.

The corresponding upward displacement at the load W be δy_1 upward, and it is given by similar triangles as

$$\delta y_1 = \frac{a}{l} \cdot \delta y$$

Total virtual work done by the reactions,

$$= R_A \times O + R_B \times \delta y = R_b \cdot \delta y$$

Virtual work done by the point load W

$$= -W \cdot \delta y_1$$

(negative sign because the displacement is in opposite direction to that of the load).

By the principle of virtual work,

$$R_b \cdot \delta y - W \cdot \delta y_1 = 0$$

or $\quad R_b = W \cdot \dfrac{\delta y_1}{\delta y} = W \dfrac{a}{l} \dfrac{\delta y}{\delta y} = \dfrac{Wa}{l}$

Similarly, R_A can be shown to be Wb/l by removing the constraint at A and giving a virtual displacement at A.

[R_A may also be got by applying $\Sigma V = 0$ after obtaining R_B].

A beam subjected to uniformly distributed load may also be treated in a similar way. As a simple example, let us consider a simple beam subjected to uniform load throughout (Fig. 10.5).

With a virtual displacement δy upward at B, imagining the constraint at B to have been removed, the virtual work by the reactions

$$= (R_A \times 0 + R_B \times \delta y) = R_B \cdot \delta y$$

Virtual work done by the uniformly distributed load,

$$= -w \left[\left(\frac{0 + \delta y}{2} \right) \times l \right] = -\frac{wl}{2} \cdot \delta y$$

(Negative sign because the virtual displacement of the beam under the uniform downward load is upward, or opposite in direction to the load).

By the principle of virtual work,

$$R_B \cdot \delta y - \frac{wl}{2} \cdot \delta y = 0$$

or $\quad R_B = \dfrac{wl}{2}$

Similarly R_A is also equal to $\dfrac{wl}{2}$.

10.5.2 Friction Problems—Ladders

When a 'ladder' is used for climbing by placing it against a rough wall and or a rough floor, the frictional forces mobilised on the wall and the floor oppose impending slippage of the ladder (Fig. 10.6).

When the ladder tends to slip and slide on the floor and slide down the wall limiting frictional forces F_f and F_W will be mobilised on the floor and the wall respectively at the point of impending motion. Normal reactions against the floor and wall, R_f and R_W respectively, are induced. The weight of the ladder, W, acts vertically downwards from its centre of gravity.

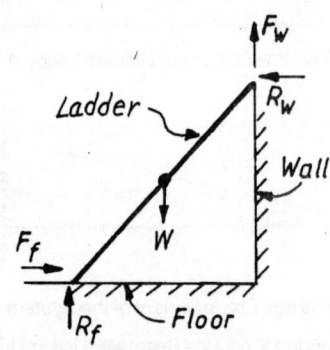

Fig. 10.6: Ladder against a Wall—Friction Problem

It is obvious that the reactions R_f and R_W do not do any work; the frictional forces, F_f and F_W and the weight of the ladder, W, will do virtual work when a virtual displacement of a small value is imagined for the foot of the ladder, since this is automatically accompanied by some virtual displacement for the top of the laddder and also for the centre of gravity of the ladder.

Applcation of the principle of virtual work to the system yields the unknown force.

Note: If the wall is considered to be smooth and frictionless, no friction force will be mobilised on the wall, and F_W vanishes; obviously, the frictional force mobilised on the floor at the foot of ladder and the self-weight of the ladder will tend to keep the ladder in equilibrium; the principle of virtual work can be applied to this problem as usual.

10.5.3 Lifting Machines
In the case of lifting machines, while the effort moves downwards, the load moves upwards. The virtual work done by the effort and by the load are determined and the principle of virtual work is applied to the force system as usual.

10.5.4 Framed Structures
In the case of framed structures, the member in which the internal force is to be detemined is assumed to be removed, and replaced by the internal force in the member acting at the joints

at the end of the member—towards the joint (if the member is in compression) or away from the joint (if the member is in tension). Then the principle of virtual work is applied involving all the external forces and this unknown internal force under an assumed virtual displacement at this member.

All the above applications are illustrated in the Illustrative Problems in the next section.

ILLUSTRATIVE PROBLEM

PROBLEM 10.1: Using the principle of virtual work, determine the reactions of a beam AB of span 10 m. The beam carries a point load of 50 kN at a distance of 4 m from A.

Solution:

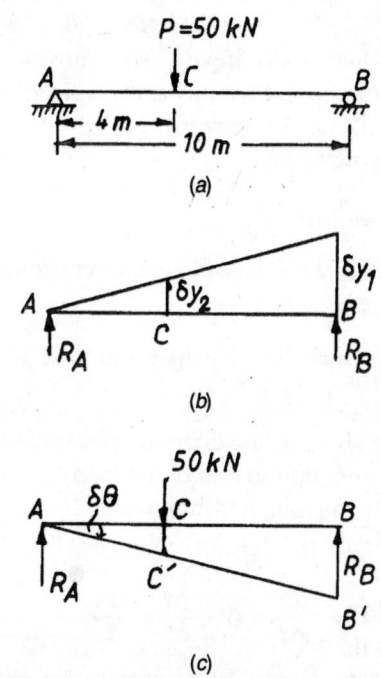

Fig. 10.7 (Illustrative Problem 10.1)

Method-I
Span of AB, $L = 10$ m

Point load, $P = 50$ N

Distance of load from A, $a = 4$ m

Distance of load from B, $b = L - a$

$$= 10 - 4 = 6 \text{ m}$$

Let the beam AB be given virtual displacement at B in the upward direction, keeping the end A intact.

Let

δY_1 = Virtual displacement at B in the upward direction.

δY_2 = Virtual displacement at C in the upward direction.

R_B = reaction at B in the upward direction.

R_A = reaction at A in the upward direction.

From the similar triangles ABE and ACD, we have

$$\frac{CD}{BE} = \frac{AC}{AB}$$

$$\frac{\delta Y_2}{\delta Y_1} = \frac{4}{10}$$

or $\delta Y_2 = 0.4\,\delta Y_1$

Algebraic sum of the virtual work done by all force

$= $ Virtual work done by reaction R_B
  $+$ Virtual work done by reaction R_A
  $+$ Virtual work done by point load

$= R_B\delta Y_1 + R_A \times 0 - 50\,\delta Y_2$

$= R_B\delta Y_1 - 50 \times 0.4\delta Y_1$

$= R_B\delta Y_1 - 20\delta Y_1$

$= \delta Y_1(R_B - 20)$

But from the principle of virtual work, the algebraic sum of the virtual work should be zero.

$\therefore \quad \delta Y_1(R_B - 20) = 0$

or $\quad R_B - 20 = 0$

$\quad R_B = 20$ kN

and $\quad R_A = 50 - 20 = 30$ kN

Method-II

Imagine that the beam rotates through an angle $\delta\theta$ about A to assume the position AB'.

The displacement of B is BB' and that of C is CC'. Since the rotation is very small, the displacement BB' and CC' are supposed to be perpendicular to AB so that we take

$BB' = 10\delta\theta$ & $CC' = 4\delta\theta$

Work done by R_B is $-R_B \times 10 \times \delta\theta$ and the work done by P is $50 \times 4 \times \delta\theta$, so that the virtual work is $50 \times 4\delta\theta - R_B \times 10$.

This must be equal to zero

$10R_B\delta\theta - 200\delta\theta = 0$

$\therefore \quad R_B = \frac{200}{10} = 20$ kN

$R_A = 50 - 20 = 30$ kN

PROBLEM 10.2: A beam AB of span 10 m carries two point loads of 20 kN and 30 kN at 4 m and 6 m from the end A respectively. Determine the reactions of the beam by the principle of virtual work.

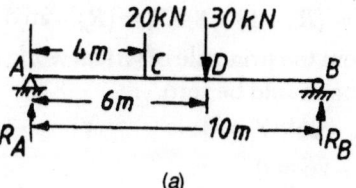

(a)

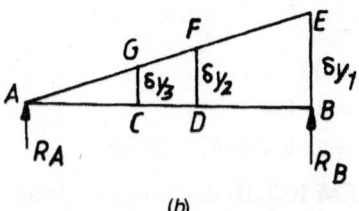

(b)

Fig. 10.8 (Illustrative Problem 10.2)

Solution:

Let the beam AB be given displacement at B in the upward keeping the end A intact.

Let

δY_1 = Virtual displacement at B in the upward direction.

δY_2 = Virtual displacement at D in the upward direction.

δY_3 = Virtual displacement at C in the upward direction

R_B = reaction at B in the upward direction.

R_A = reaction at A in the upward direction.

From the similar triangles ADF and ABE, we have

$$\frac{DF}{BE} = \frac{AD}{AB}$$

$$\frac{\delta Y_2}{\delta Y_1} = \frac{6}{10}$$

$$\delta Y_2 = 0.6\,\delta Y_1$$

From the similar triangles AGC and ABE, we have

$$\frac{CG}{BE} = \frac{AC}{AB}$$

$$\frac{\delta Y_3}{\delta Y_1} = \frac{4}{10}$$

$$\delta Y_3 = 0.4\delta Y_1$$

The virtual work done

$$= R_B\delta Y_1 - 30\delta Y_2 - 20\delta Y_3 + R_A \times 0$$
$$= R_B\delta Y_1 - 30 \times 0.6\delta Y_1 - 29 \times 0.4\delta Y_1$$
$$= (R_B - 18 - 8)\delta Y_1 = (R_B - 26)\delta Y_1$$

But from the principle of virtual work, the total work done should be zero.

$$(R_B - 26)\delta Y_1 = 0$$
$$R_B - 26 = 0$$
$$R_B = 26 \text{ kN}$$
$$R_A = 30 + 20 - 26$$
$$= 50 - 26$$
$$= 24 \text{ kN}$$
$$R_A = 24 \text{ kN } \& R_B = 26 \text{ kN}$$

PROBLEM 10.3: By the principle of virtual work, determine the reactions for the beam shown in Fig. 10.9 (*a*).

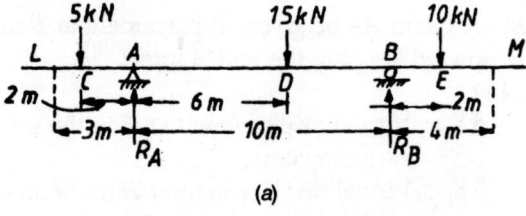

(a)

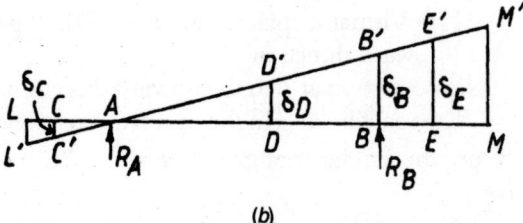

(b)

Fig. 10.9 (Illustrative Problem 10.3)

Solution:

Given:

Point load at $C = 5$ kN

Point load at $D = 15$ kN
Point load at $E = 10$ kN
Distance $AC = 2$ m
$AD = 6$ m
$BD = 4$ m
$BE = 2$ m

Let R_A = Reaction at A

R_B = Reaction at B

To determine the reaction R_B, let the beam be given a virtual displacement at B in the upward direction, keeping the end A intact. All the points lying towards right from A will have upward virtual displacements whereas the points lying towards left from A will have downward virtual displacements.

Total virtual work done

> = Virtual work done by point load at E
> + Virtual work done by point load at D
> + Virtual work done by point load at C
> + Virtual work done by R_A
> + Virtual work done by R_B

$$= -10\delta_E - 15\delta_D + 5\delta_C + R_B\delta_B + R_A \times 0$$

From similar triangles

$ADD^1, ABB^1, AEE^1, ACC^1$, we get

$$\frac{\delta_D}{6} = \frac{\delta_B}{10} = \frac{\delta_E}{12} = \frac{\delta_C}{2}$$

$$\delta_E = 1.2\delta_B$$
$$\delta_D = 0.6\delta_B$$
$$\delta_C = 0.2\delta_B$$

Total virtual work done

$$= -10\delta_E + R_B\delta_B - 15\delta_D + R_A \cdot 0 + 5 \times \delta_C$$
$$= -10 \times 1.2\delta_B + R_B \cdot \delta_B - 15 \times 0.6\delta_B + R_A \times 0 + 5 \times 0.2\delta_B$$
$$= \delta_B(R_B - 12 - 9 + 1)$$
$$= \delta_B(R_B - 20)$$

But from the principle of virtual work, the total work done should be zero

i.e., $\delta_B(R_B - 20) = 0$

$$R_B = 20 \text{ kN}$$

But $R_A + R_B = 5 + 15 + 10$

$$R_A + R_B = 30$$
$$R_A = 30 - R_B = 30 - 20 = 10 \text{ kN}$$
$$R_A = 10 \text{ kN and } R_B = 20 \text{ kN}$$

PROBLEM 10.4: Two beams AC and CD of lengths 9 m and 10 m respectively are hinged at C. These are supported on rollers at left and right ends (A and D). A hinged support is provided at B, 7 m from A. Using the principle of virtual work, determine the reaction at the support B when a load of 7000 N acts at a point 6 m from D.

Also determine the reactions at the supports A and D.

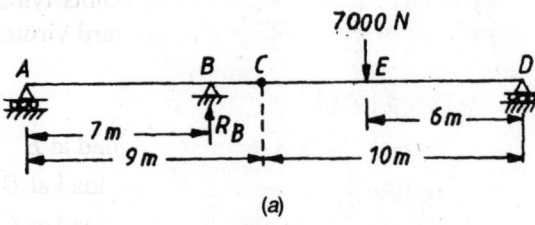

(a)

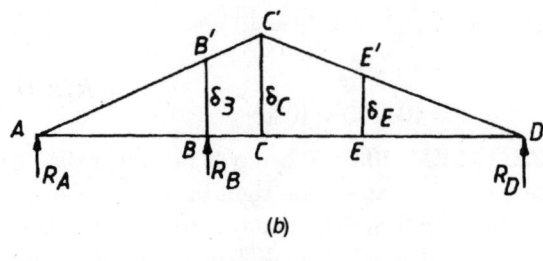

(b)

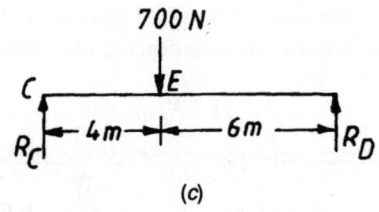

(c)

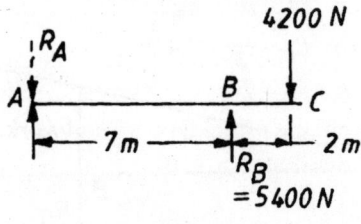

(d)

Fig. 10.10 (Illustrative Problem 10.4)

Solution:
Let a virtual displacement at hinge C be given in the upward direction, keeping ends A and D intact as shown in Fig. 10.10 (b).

Total virtual work done
$$= R_A \times 0 + R_B \times \delta_B - 7000 \times \delta_E + R_D \times 0$$
$$= R_B \delta_B - 7000 \delta_E$$

But from the principle of virtual work, the total work done should be zero.
$$R_B \delta_B - 7000 \delta_E = 0 \qquad \ldots(i)$$

From similar triangles ABB' and ACC', we get
$$\frac{\delta_B}{\delta_C} = \frac{7}{9} \quad \text{or} \quad \delta_B = \frac{7}{9} \delta_C \qquad \ldots(ii)$$

From similar triangles DEF' and DCC', we get
$$\frac{\delta B}{\delta_C} = \frac{6}{10} \quad \text{or} \quad \delta_C = \frac{10}{6} \delta_E \qquad \ldots(iii)$$

From (ii) and (iii), we get
$$\delta_B = \frac{7}{9} \delta_C = \frac{7}{9} \times \frac{10}{6} \delta_E$$
$$\delta_B = \frac{70}{54} \delta_E \qquad \ldots(iv)$$

Substituting the value of δ_B from (iv) in (i)
$$R_B = \frac{70}{54} \delta_E - 7000 \, \delta_E = 0$$
$$= \left(\frac{70}{54} R_B - 7000\right) \delta_E = 0$$

or
$$= \frac{70}{54} R_B - 7000 = 0, \quad \frac{70 R_B}{54} = 7000$$
$$R_B = \frac{7000 \times 54}{70} = 5400 \text{ kN}$$
$$R_B = 5400 \text{ N}$$
$$R_C + R_D = 7000 \text{ N}$$

Taking moments of all forces about D, we get
$$10 R_C = 7000 \times 6$$

or
$$R_C = \frac{7000 \times 6}{10} = 4200 \text{ N}$$

∴
$$R_D = 7000 - 4200 \text{ N} = 2800 \text{ N}$$
$$R_A + R_B = 4200 \text{ N}$$
$$R_A + 5400 = 4200$$

$R_A = 4200 - 5400$

$\quad = -1200$ N

The -ve sign indicates that R_A acts downwards.

PROBLEM 10.5: A simply supported beam of span 8 m is carrying a uniformly distributed load of 10 kN/m as shown in Fig. 10.11 (*a*) using the principle of virtual work, determine the reactions at *A* and *B*.

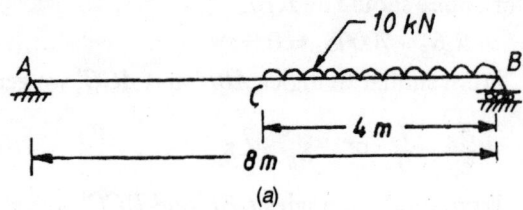

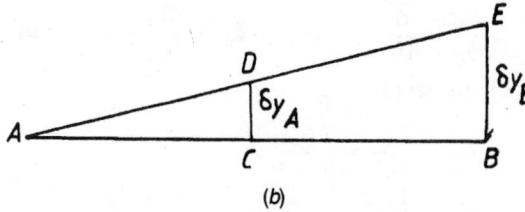

Fig. 10.11 (Illustrative Problem 10.5)

Solution: Let the beam is given a virtual displacement at *B*, keeping the point *A* fixed.

Let

δY_B = Virtual displacement at *B* in the upward direction.

δY_A = Virtual displacement at *C* in the upward direction.

R_B = Reaction at *B* in the upward direction.

R_A = Reaction at B in the upward direction.

Algebraic sum of the virtual work done by all the forces

$\quad$ = Virtual work done by reaction R_B
$\quad$ + Virtual work done by reaction R_A
$\quad$ + Virtual work done by uniformly distributed load between *C* & *B*

$\quad = R_B \delta Y_B + R_A \times 0 - 10 \times 4 \left(\dfrac{\delta Y_A + \delta Y_B}{2} \right)$

$\quad = R_B \delta Y_B - 20 \left(\delta Y_A + \delta Y_B \right) Y_A$

$\quad = \left(R_B - 20 - 20 - 20 \dfrac{\delta Y_A}{\delta Y_B} \right) \delta Y_B$

But from the principle of virtual work, the algebraic sum of the virtual work should be zero.

$$\left(R_B - 20 - 20 \dfrac{\delta Y_A}{\delta Y_B} \right) \delta Y_B = 0$$

$$R_B - 20 - 20 \, \dfrac{\delta Y_A}{\delta Y_B} = 0$$

$$R_B = 20 \left(1 + \dfrac{\delta Y_A}{\delta Y_B} \right)$$

From the similar triangles *ACD* and *ABF*

$$\dfrac{CD}{BE} = \dfrac{AC}{AB}$$

$$\dfrac{\delta Y_A}{\delta T_B} = \dfrac{4}{8} = \dfrac{1}{2}$$

$\therefore \quad R_B = 20 \left(1 + \dfrac{1}{2} \right)$

$\quad = 20 \times \dfrac{3}{2}$

$\quad = 30$ kN

$R_A + R_B = 4 \times 10 = 40$ kN

$R_A = 40 - R_B$

$\quad = 40 - 30 = 10$ kN

PROBLEM 10.6: A beam 8 m long rests on supports 6 m apart, the right hand end is overhanging by 2 m. The beam carries a uniformly distributed load of 15 kN/m over the entire length of the beam. Using the principle of virtual work, determine the reactions of the beam.

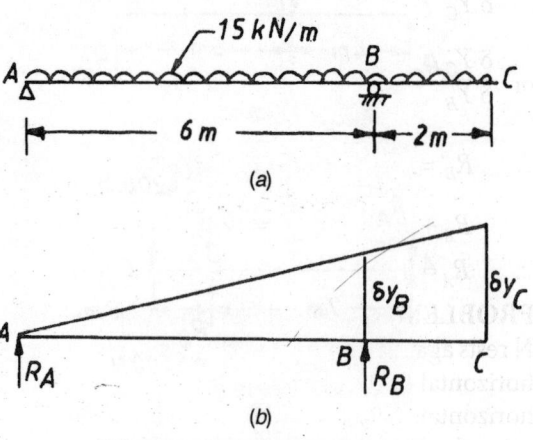

Fig. 10.12 (Illustrative Problem 10.6)

Solution: Let the beam be given a virtual displacement at B, keeping the point A fixed.

Let

δY_B = Virtual displacement at B in the upward direction.

δY_C = Virtual displacement at C in the upward direction.

R_A = reaction at A in the upward direction.

R_B = reaction at B in the upward direction.

Algebraic sum of the virtual work done by all the forces

= Virtual work done by reaction R_A
+ Virtual work done by reaction R_B
+ Virtual work done by uniformly distributed load between A & C

$$= R_A \times 0 + R_B \times \delta Y_B - 15 \times 8 \left(\frac{0 + \delta Y_C}{2} \right)$$

$$= R_B \delta Y_B - 60 \delta Y_C$$

But from the principle of virtual work, the algebraic sum of the virtual work should be zero.

$$(R_B \times \delta Y_B - 60 \delta Y_C) = 0$$

$$R_B = 60 \frac{\delta Y_C}{\delta Y_B}$$

From the similar triangles ABD and ACE

$$\frac{BD}{CE} = \frac{AB}{AC}$$

$$\frac{\delta Y_B}{\delta Y_C} = \frac{6}{8}$$

or $\frac{\delta Y_C}{\delta Y_B} = \frac{8}{6}$

$$\therefore \quad R_B = \frac{60 \times 8}{6} = 10 \times 8 = 80 \text{ kN}$$

$$R_A + R_B = 15 \times 8 = 120$$

$$\therefore \quad R_A = 120 - 80 = 40 \text{ kN}$$

PROBLEM 10.7: A uniform ladder of weight 500 N rests against a smooth vertical wall and a rough horizontal floor making an angle of 60° with the horizontal. Find the force of friction at the floor using the method of virtual work.

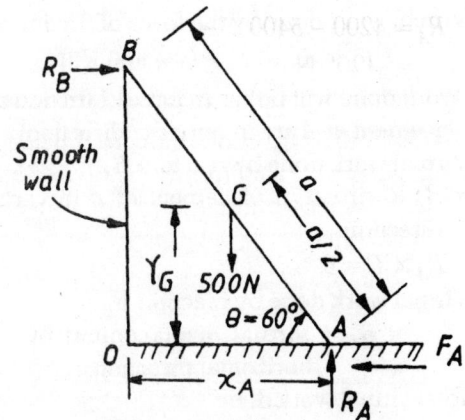

Fig. 10.13 (Illustrative Problem 10.7)

Solution: Weight of ladder, $W = 500$ N

Angle made by ladder with horizontal, $\theta = 60°$

Let a = Length of ladder AB

R_B = Reaction at point A

F_A = Force of friction at A

R_B = Reaction at B

The force of friction at B will be zero as vertical wall is smooth.

The weight of ladder is acting at G, the middle point of AB.

$$\therefore \quad AG = \frac{AB}{2} = \frac{a}{2}$$

Let Y_G = Vertical distance of G above OA,

x_A = Horizontal distance of A from OB a

$$Y_G = AG \sin\theta = \frac{a}{2} \sin\theta$$

$$x_A = Ao = AB\cos\theta = a\cos\theta$$

$$\therefore \quad dY_G = \frac{a}{2} \cos\theta \, d\theta$$

$$dx_A = -a\sin\theta \, d\theta$$

Let the ladder be given a small virtual displacement in such a way that the point A moves to the right by dx_A, keeping the length of ladder fixed. Then the point B will move downwards. Also the point G will move downwards.

Let dY_G = downward displacement of G

Virtual work done by the weight of the ladder

$$= W \times dY_G$$

$$= 500 \times \frac{a}{2} \cos\theta \times d\theta$$

Virtual work done by the force of friction at A

$$= F_A \times dx_A = F_A \times (-a\sin\theta \, d\theta)$$

[Work done will be -ve as force of friction and displacement at A are in opposite direction]

Virtual work done by reaction R_A

= $R_A \times$ virtual displacement of A in vertical direction

$= R_A \times 0 = 0$

Virtual work done by reaction R_B

= $R_B \times$ virtual displacement of B in horizontal direction

Total virtual work done

$$= 500 \times \frac{a}{2} \times \cos\theta \times d\theta - F_A \times a \times \sin\theta \times d\theta$$

But according to the principle of virtual work, total virtual work done should be zero.

$$500 \times \frac{a}{2} \times \cos\theta \times d\theta - F_A \times a \times \sin\theta \times d\theta = 0$$

$$250 \times \cos\theta - F_A \sin\theta = 0$$

$$F_A = 250\frac{\cos\theta}{\sin\theta} = 250 \times \cot\theta$$

$$\sin\theta = 250\cot 60° \qquad\qquad (\because \theta = 60°)$$

$$= 144.34 \text{ kN}$$

PROBLEM 10.8: A uniform bar AB of weight W is supported as shown in Fig. 10.14 (a) and is subjected to a load P. Applying the principle of virtual work derive an expression in terms of P, θ and W, for the effort F required at B as shown to keep the bar in equilibrium. Neglect friction.

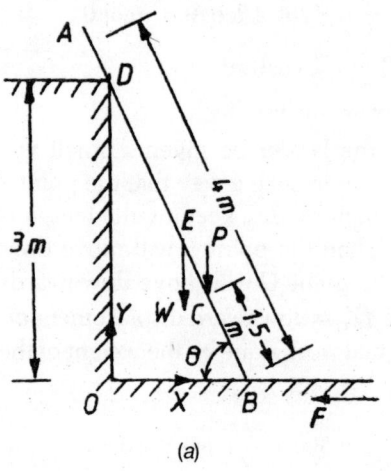

(a)

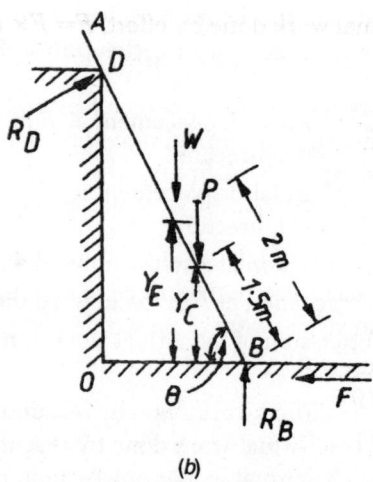

(b)

Fig. 10.14 (Illustrative Problem 10.8)

Solution: Friction is to be neglected. Hence force of friction at B and D will be zero. Fig. 10.14 (a) shows the given system, whereas the Fig. 10.14 (b) shows its free body diagram, with all the forces acting on the bar. The virtual work done by normal reactions R_B and R_D will be zero. The virtual work will be done by W, P and effort F applied at B. The weight W and force P are acting vertically downwards, whereas the effort F is acting horizontally at B.

Hence for virtual work, virtual vertical displacement at E and C and virtual horizontal displacement at B are required.

Let $dx_B =$ Virtual horizontal displacement at B

$dY_E =$ Virtual vertical displacement at E

$dY_C =$ Virtual vertical displacement at C

$$\tan\theta = \frac{OD}{OB} = \frac{3}{x_B}$$

$$x_B = 3\cot\theta$$

$$\therefore \quad dx_B = -3 \cdot \text{cosec}^2\theta \cdot d\theta$$

$$Y_E = 2\sin\theta$$

$$dY_E = 2\cos\theta \cdot d\theta$$

$$Y_C = 1.5\sin\theta$$

$$dY_C = 1.5\cos\theta \cdot d\theta$$

Virtual work done by weight, $W = W \times dy_E$

$$= W \times 2\cos\theta \cdot d\theta$$

Virtual work done by load, $P = P \times dy_C$

$$= P \times 1.5\cos\theta \cdot d\theta$$

Virtual work done by effort, $F = F \times dx_B$

$$= F(-3\operatorname{cosec}^2\theta \cdot d\theta)$$

[Work will be −ve as force of friction and displacement at B are in opposite direction]

Total virtual work done

$$= F(-3\operatorname{cosec}^2\theta \cdot d\theta) + W(2\cos\theta)\,d\theta$$
$$+ P(1.5\cos\theta)\,d\theta$$

But according to the principle of virtual work, the total virtual work done should be zero.

$$F(-3\operatorname{cosec}^2\theta \cdot d\theta) + W(2\cos\theta)d\theta + P(1.5\cos\theta)$$
$$d\theta = 0$$

or $\quad 2W\cos\theta + 1.5P\cos\theta = 3F\operatorname{cosec}^2\theta$

$$F = \frac{2W\cos\theta + 1.5P\cos\theta}{3\operatorname{cosec}^2\theta}$$

$$= \frac{(2W + 1.5P)\cos\theta}{3(1/\sin^2\theta)}$$

$$= \frac{(2W + 1.5P) \times \cos\theta \times \sin^2\theta}{3}$$

PROBLEM 10.9: Using the principle of virtual work, determine the effort P required to hold the weight 800 N in equilibrium in a system of two frictionless pulleys of the same diameter as shown in Fig. 10.15.

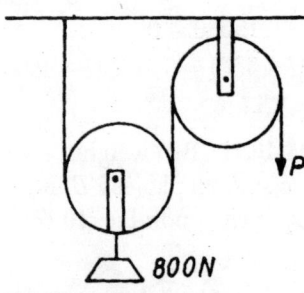

Fig. 10.15 (Illustrative Problem 10.9)

Solution:

Weight, $W = 800$ N

Effort applied $= P$

Let $\quad \delta y =$ Virtual displacement of effort P in the downward direction.

Then the weight W will be lifted by a distance of $(\delta y/2)$ in the upward direction.

$\therefore$ Virtual displacement of $W = \dfrac{\delta y}{2}$

Now the virtual work done by effort

$$= P \times \text{virtual displacement of } P$$
$$= P\delta y$$

(virtual work done will be positive as effort and virtual displacement are acting in the same direction)

Virtual work done by load $= W \times$ virtual displacement of load

$$= -800 \times \frac{\delta y}{2}$$

(Virtual work done will be −ve, as weight is acting downward, but virtual displacement is upward)

Total virtual work done $= P\delta y - 800\dfrac{\delta y}{2}$

But according to the principal of virtual work, the total virtual work done should be zero.

$$P\delta y - 800\,\frac{\delta y}{2} = 0$$

$$\delta y\left(P - \frac{800}{2}\right) = 0$$

or $\quad \delta y\,(P - 400) = 0$

$$P - 400 = 0$$

or $\quad P = 400$ N

PROBLEM 10.10: The diameters of the two steps of the pulley of a Weston's differential pulley block are 40 cm and 30 cm respectively. Determine the value of the effort required to lift a load of 60 kN using the principle of virtual work. Neglect the frictional forces.

Solution:

Diameter of the large pulley, $D = 40$ cm

Radius of the larger pulley, $R = 20$ cm

Diameter of the smaller pulley, $d = 30$ cm

Radius of smaller pulley, $r = 15$ cm

Load lifted $= 60$ kN

Let P be effort applied to lift the load.

Suppose the pulley block undergoes a small virtual angular dislacement $\delta\theta$ as shown in Fig. 10.16 (b)

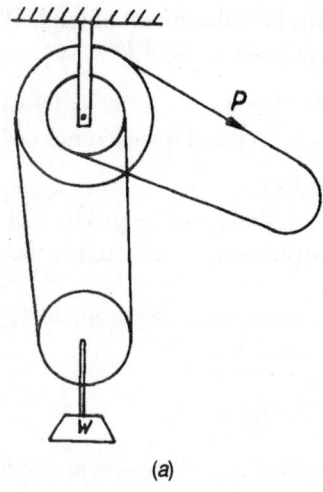

(a)

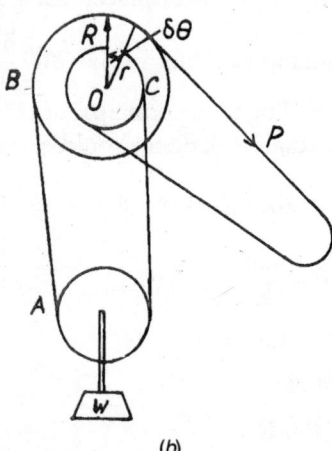

(b)

Fig. 10.16 (Illustrative Problem 10.10)

Then virtual displacement of the effort

$$= R\delta\theta$$
$$= 20\delta\theta$$

Virtual work done by effort

$$= P \times 20 \times \delta\theta$$
$$= 20P\delta\theta$$

(The virtual work done will be positive as effort and the virtual displacement of effort are in the same direction.)

From Fig. 10.16 (*b*) it is clear that the string *AB* will be moving upward whereas string *CA* will be moving downward.

Now distance moved by *AB* in the upward direction

$$= \text{Distance moved by effort}$$

$$= 20\delta\theta$$

Distance moved by *CA* in the downward direction = Radius of smaller pulley × δθ

$$= r\delta\theta = 15\delta\theta$$

As the distance moved by *AB* in the upward direction is more than the distance moved by *CA* in the downward direction. Hence load will be moving in the upward direction.

Virtual distance moved by load in the upward direction

$$= \frac{1}{2}\{(\text{Distance moved by AB}) - (\text{Distance moved by CA})\}$$

$$= \frac{1}{2}(20\delta\theta - 15\delta\theta) = 2.5\delta\theta$$

Virtual work done by load

$$= \text{Load} \times \text{Virtual displacement of load}$$
$$= -60 \times 2.5 \times \delta\theta$$
$$= -150\delta\theta$$

(The virtual work done by load will be negative as the load is acting downward but the virtual displacement is upward).

Now according to principle of virtual work, the total virtual work done should be zero.

∴ Virtual work done by effort + Virtual work done by load = 0

$$20P\delta\theta - 150\delta\theta = 0$$
$$\delta\theta(20P - 150) = 0 \text{ or } 20P - 150 = 0$$
$$\text{or } P = 7.5 \text{ kN}$$

PROBLEM 10.11: Two weights W_1 and W_2 are acting on two smooth planes *AB* and *BC* at angles of O_1 and O_2 as shown in Fig. 10.17.

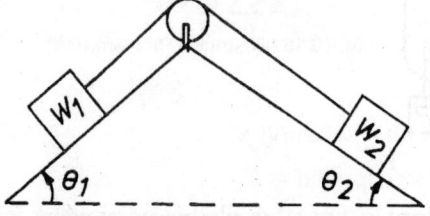

Fig. 10.17 (Illustrative Problem 10.11)

With the help of principle of virtual work, find the ratio of W_1 and W_2.

Solution: Provide a small displacement δs for the block W_1 up the plane. The block W_2 will move down the plane by δs

The vertical displacement of load, W_1
$$= \delta s \sin\theta_1$$
The vertical displacement of load, W_2
$$= \delta s \sin\theta_2$$
Virtual work done by the load, W_1
$$= W_1 \times \delta s \sin\theta_1$$
(Virtual work done will be −ve, as weight is acting downward, but virtual displacement is upward).

Virtual work done by the load, W_2
$$= W_2 \times \delta s \sin\theta_2$$
(Virtual work done will be positive, as weight and virtual displacement are acting in the same direction)

∴ Total virtual work done
$$= W_2 \times \delta s \sin\theta_2 - W_1 \times \delta s \sin\theta_1$$
$$= (W_2 \sin\theta_2 - W_1 \sin\theta_1)\delta s$$

But according to the principle of virtual work, the total work done should be zero.

∴ $W_2\sin\theta_2 - W_1\sin\theta_1 = 0$

or $W_2\sin\theta_2 = W_1\sin\theta_1$

$$\frac{W_1}{W_2} = \frac{\sin\theta_2}{\sin\theta_1}$$

PROBLEM 10.12: A block of weight W rests on the smooth surface inclined at 15° with the horizontal. The block is supported by an effort (P) hung from a pulley as shown in Fig. 10.18.

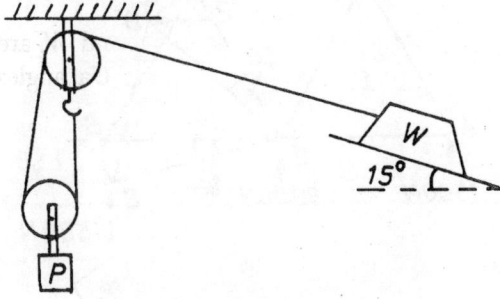

Fig. 10.18 (Illustrative Problem 10.12)

Solution: Provide a small displacement δs for the weight W up the splane. The effort P will move vertically down by $\frac{\delta s}{2}$

The vertical displacement of load, W
$$= \delta s \sin 15°$$
Virtual work done by the load, W
$$= -W(\delta s)\sin 15°$$
(Virtual work done will be −ve, as weight is acting downward but virtual displacement is upward).

Virtual work done by the effort, P
$$= P \cdot \frac{\delta s}{2}$$
(Virtual work done will be +ve, as the effort and virtual displacement are acting in the same direction)

∴ Total virtual work done
$$= \left(P \cdot \frac{\delta s}{2} - W\delta s \sin 15°\right)$$

But according to the principle of virtual work, the total work done should be zero.

∴ $\left(P \cdot \frac{\delta s}{2} - W \cdot \delta s \cdot \sin 15°\right) = 0$

$$\left(\frac{P}{2} - W\sin 15°\right)\delta s = 0$$

or $\dfrac{P}{2} - W\sin 15° = 0$

$$P = 2W\sin 15°$$
$$P = 2W\sin 15°$$
$$= 2(W)(0.258819)$$
$$= 0.517638\,W$$
$$P = 0.518\,W$$

PROBLEM 10.13: A structure with pin-connected members is shown in Fig. 10.19 (a). Determine the value of θ for equilibrium using the principle of virtual work. Neglect friction and weight of the members.

Solution: Let E and F be the projections of C and A on the horizontal line BD
$$DE = 1000\cos\theta$$
$$AF = 500\sin\theta$$
$$d(DE) = -1000\sin\theta \cdot d\theta$$
$$d(AF) = 500\cos\theta \cdot d\theta$$

Virtual work done by W

$$= W(500\cos\theta \cdot d\theta)$$

(Virtual work done will be +ve, as weight and virtual displacement are acting in the same direction).

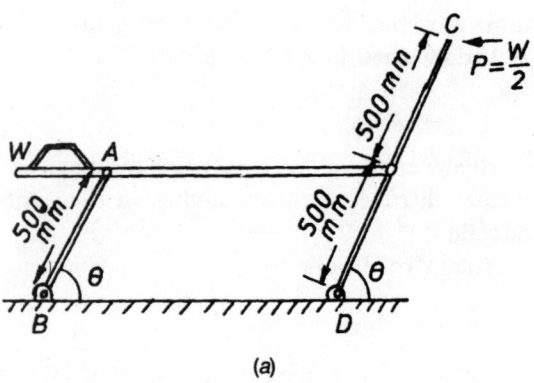

(a)

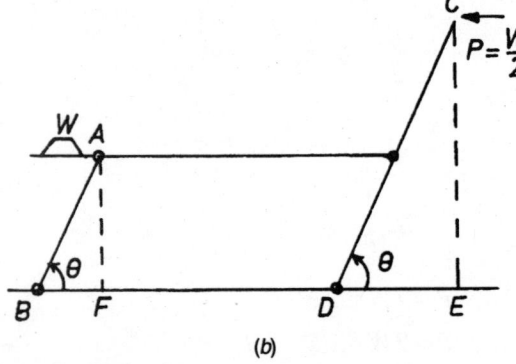

(b)

Fig. 10.19 (Illustrative Problem 10.13)

Virtual work done by $P(W/2)$

$$= \frac{W}{2}(-1000\sin\theta\,d\theta)$$

(Virtual work done will be –ve, as $P(W/2)$ and virtual displacement are acting in the opposite direction).

Total virtual work done

$$= \frac{W}{2}(-1000\sin\theta\,d\theta) + W(500\cos\theta\,d\theta)$$

But according to the principle of virtual work, the total work done should be zero.

$$= \frac{W}{2}(-1000\sin\theta\,d\theta) + W(500\cos\theta\,d\theta)$$

$$(-500\,W\sin\theta + W\,500\cos\theta)\,d\theta = 0$$

or $500\,W\sin\theta = 500\,W\cos\theta$

$$\frac{\sin\theta}{\cos\theta} = \tan\theta = \frac{500W}{500W} = 1$$

$$\tan\theta = 1$$

$$\therefore \qquad \theta = \tan^{-1}(1)$$

$$\theta = 45°$$

PROBLEM 10.14: Determine the force in the member CD of the truss in Fig. 10.20 (a) by virtual work principle.

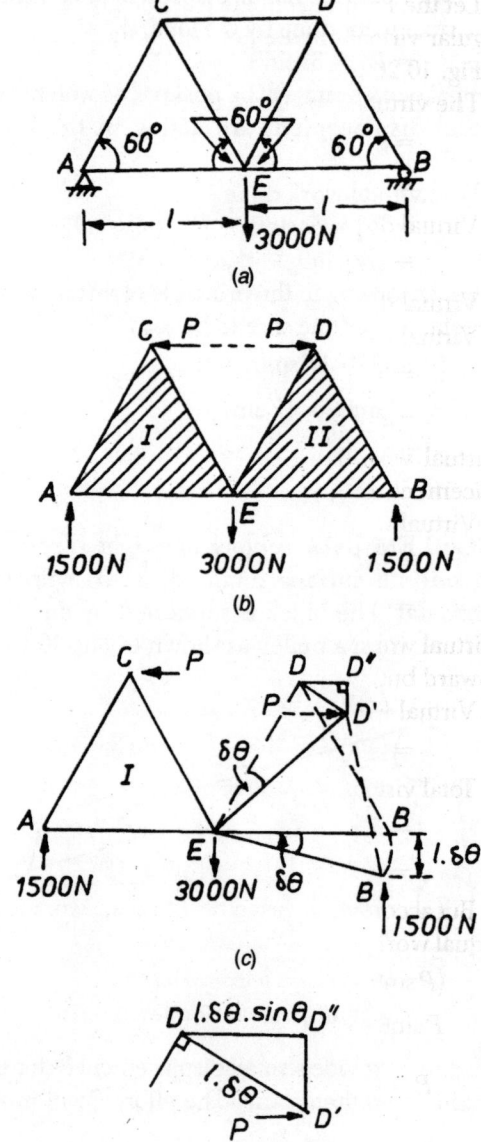

(a)

(b)

(c)

Fig. 10.20 (Illustrative Problem 10.14)

Solution:

Point load at $E = 3000\ N$

Length of each member $= l$

The reactions R_A and R_B at A and B are each equal to 1500 N.

Fig. 10.20 (*a*) shows the given truss. Remove the member CD of the truss in which force is to be determined. Replace the member CD of truss by two forces each equal to P and acting at C and D as shown in Fig. 10.20 (*b*).

Let the Portion *II* of the truss be given a small angular virtual movement O at point E as shown in Fig. 10.20 (*c*).

The virtual displacement of $P = DD''$

$$= DD' \sin\theta$$

$$= l\delta\theta \sin\theta \qquad (\because DD' = l\theta)$$

Virtual displacement of reaction at $B = BB'$

$$= l\delta\theta$$

Virtual displacement of point load at $E = 0$

Virtual work done by the force P

$$= P \times \text{virtual displacement of } P$$

$$= P \times l \times \delta\theta \times \sin\theta$$

(Virtual work done is +ve as force, P and displacement, DD'' are in the same direction).

Virtual work done by reaction R_B.

$$= -R_B \times \text{Virtual displacement at } B$$

$$= -1500 \times l \times \delta\theta$$

(Virtual work done will be –ve, as reaction, R_B' is upward but displacement, BB' is downward).

Virtual work done by the load 3000 N at E

$$= (3000) \times \text{Virtual displacement at } E$$

Total virtual work done

$$= P \times l \times \delta\theta \times \sin\theta - 1500 \times l \times \theta$$

$$= (P\sin\theta - 1500)\ (l\delta\theta)$$

But according to principle of virtual work, total virtual work done should be zero.

$$(P\sin\theta - 1500)\ l\delta\theta = 0$$

or $\quad P\sin\theta - 1500 = 0$

$$P = \frac{1500}{\sin\theta}$$

$$P = \frac{1500}{\sin 60°} \qquad (\because \theta = 60°)$$

$$P = \frac{1500 \times 2}{\sqrt{3}} = \frac{3000}{\sqrt{3}}$$

$$P = \frac{3 \times 1000}{\sqrt{3}} = \sqrt{3} \times 1000$$

$$P = 1.732 \times 1000$$

$$P = 1732\ \text{N}$$

PROBLEM 10.15: Five rods, AB, BC, CD, DA and BD each of equal length and cross-section are pin-jointed together so as to form a plane frame $ABCD$. The frame $ABCD$ has a rhombus shape with horizontal diagonal BD. The frame is suspended from the top most joint A. A weight $3W$ is attached at the lower most joint C. Neglecting the self weight of the frame, and using the method of virtual work find the magnitude of the thrust in the member BD.

Solution:

$$AB = BC = CD = DA = a\ (\text{say})$$

ABD is an equilateral triangle

$\therefore \qquad \theta = 60°$

Remove the member BD of the frame in which the force is to be determined. Replace the member BD by two forces equal to acting at B & D as shown in Fig. 10.21(*b*).

Let

y_B = Vertical distance of point B from fixed point A (AO)

x_B = Horizontal distance of point B from axis AOC as shown in Fig. 10.21(*a*) (BO)

y_C = Vertical distance of point C from fixed point A (AC)

From the triangle ABO

$$y_B = AO = a\sin\theta$$

$$x_B = BO = a\cos\theta$$

$$y_C = CO + AO = a\sin\theta + a\sin\theta$$

$$= 2a\sin\theta$$

Let the frame be given a small virtual displacement $\delta\theta$ as shown in Fig. 10.21 (*c*). Keeping the end A fixed and the point C exactly below A. The points B and D will move inward and point C will move downward.

Now virtual horizontal displacement of point B is given by

$$\delta x_B = dx_B = d\ (a\cos\theta)$$

$$= -a\sin\theta\, d\theta$$

$$= -a\sin\theta\, \delta\theta$$

(For virtual displacement change $d\theta$ to $\delta\theta$)

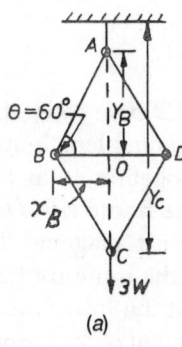

(a)

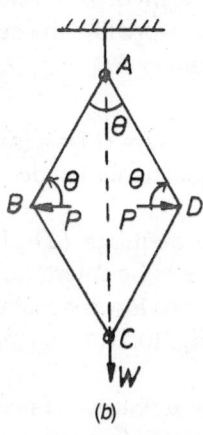

(b)

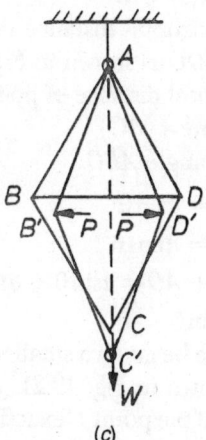

(c)

Fig. 10.21 (Illustrative Problem 10.15)

Also the virtual horizontal displacement of point D is equal to virtual horizontal displacement of point B.

$\therefore\quad \delta x_D = -a\sin\theta \cdot \delta\theta$

Virtual vertical displacement of point C is given by

$$\delta y_C = dy_C = d\,(2a\sin\theta)$$
$$= 2a\cos\theta \cdot d\theta$$
$$= 2a\cos\theta \cdot \delta\theta$$

Virtual work done by force P at B
$$= P\,(-a\sin\theta \cdot \delta\theta)$$

Virtual work done by force P at D
$$= P\,(-a\sin\theta \cdot \delta\theta)$$

Virtual work done by weight $3W$ at C
$$= 3W\,(2a\cos\theta \cdot \delta\theta)$$

Total virtual work done
$$= P\,(-a\sin\theta \cdot \delta\theta) + P\,(-a\sin\theta \cdot \delta\theta)$$
$$+ 3W\,(2a\cos\theta \cdot \delta\theta)$$

But according to the principle of virtual work, total virtual work done should be zero.

$$(P)\,(-a\sin\theta \cdot \delta\theta) + (P)\,(-a\sin\theta \cdot \delta\theta)$$
$$+ 3W\,(2a\cos\theta \cdot \delta\theta) = 0$$
$$(W2a\cos\theta - P2a\sin\theta)\,\delta\theta = 0$$
$$(W\cos\theta - P\sin\theta)\,2a\delta\theta = 0$$
or $3W\cos\theta = P\sin\theta$

$$P = \frac{\cos\theta}{\sin\theta} = 3W\cot\theta$$

$$P = 3W\cot 60°$$

$$P = \frac{3W}{\sqrt{3}}$$

$$P = \sqrt{3}\,W$$

PROBLEM 10.16: Six equal heavy bars are freely joined at their extremities; one is fixed on a horizontal plane, and the system lies in a vertical plane; the middle points of the two upper non-horizontal bars are connected by a rope. Show that the tension of this rope is $6W\cot\theta$, W being the weight of each bar and θ the inclination of the non-horizontal bars to the horizon.

Solution: Let $ABCDEF$ be the hexagon formed by the bars with AB fixed to a horizontal plane. L, the middle point of EF, and M, the middle point of DC, are connected by a rope.

Let each side of the hexagon be $2a$.

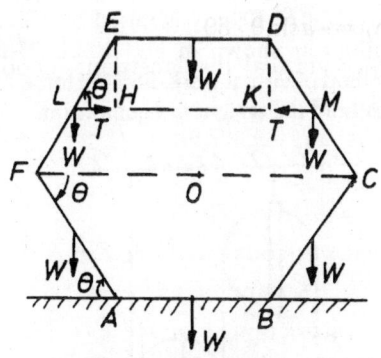

Fig. 10.22 (Illustrative Problem 10.16)

Draw *EH* and *DK* perpendiculars to *LM.* Then

$LM = LH + HK + KM$

$\quad = a\cos\theta + 2a + a\cos\theta$

$\quad = 2a\cos\theta + 2a$

$d(LM) = -2a\sin\theta \cdot d\theta$

$\quad y$ = Height of the middle point of *AF* or *BC* above *AB.*

$\quad = \left(\dfrac{1}{2} \times 2a\right)\sin\theta = a\sin\theta$

$\therefore \quad dy = a\cos\theta\, d\theta$

$\quad z$ = Height of *L* or *M* above *AB*

$\quad = 2a\sin\theta + a\sin\theta = 3a\sin\theta$

$\therefore \quad dz = 3a\cos\theta\, d\theta$

$\quad x$ = height of the middle point of *AB* above $AB = 0$

$\therefore \quad dx = 0$

$\quad h$ = Height of the middle point of *Ed* above *AB*

$\quad = 2a\sin\theta + 2a\sin\theta = 4a\sin\theta$

$dh = 4a\cos\theta\, d\theta$

The weight *W* of *AB* acts at height *x*, the weights of *AF* and *BC* act at height *y*, the weights of *FE* and *CD* act at height *z* and the weight of *ED* acts at height *h*. Let a displacement be given in which θ becomes θ + *d*θ. Then the equation of virtual work is

$-W \cdot dx - 2W \cdot dy - 2W \cdot dz - W \cdot dh - T \cdot d(LM) = 0$

or $\quad 0 - 2W(a\cos\theta\, d\theta) - 2W(3a\cos\theta\, d\theta)$
$\quad\quad - W(4a\cos\theta\, d\theta) - T \cdot (-2a\sin\theta\, d\theta) = 0$

or $\quad (12W a\cos\theta) - T\,2a\sin\theta)\, d\theta = 0$

or $\quad 12Wa\cos\theta - T\,2a\sin\theta = 0$

or $\quad 6W\cos\theta = T\sin\theta$

or $\quad T = 6W = \dfrac{\cos\theta}{\sin\theta} = 6W\cot\theta$

$\quad T = 6W\cot\theta$

PRACTICE PROBLEMS

10.1 Using the principle of virtual work, determine the reactions of a beam *AB* of span 10 m. The beam carries a point load of 60 kN at a distance of 6 m from *A*.

10.2 A beam *AB* of span 8 m carries two point loads of 30 kN and 20 kN at 3 m and 6 m from the end *A* respectively. Determine the reactions of the beam by the principle of virtual work.

10.3 By the principle of virtual work, determine the reactions for the beam shown in Fig. 10.23.

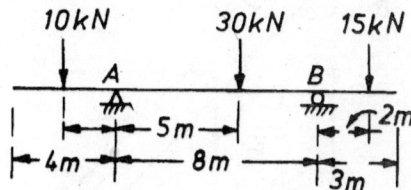

Fig. 10.23 (Practice Problem 10.3)

10.4 Two beams *AC* and *CD* of lengths 4 m and 6 m respectively are hinged at *C*. These are supported on rollers at the left and right ends (*A* and *D*). A hinged support is provided at *B*, 4 m from *D*. Two loads of 60 kN and 35 kN are acting at distances of 3 m and 2 m from *A* and *D* respectively on the two beams. Using the principle of virtual work, find the reaction at *B*.

10.5 A simply supported beam *AB* of span 10m is loaded as shown in Fig.10.24. Using the principle of virtual work, find the reactions at *A* and *B*.

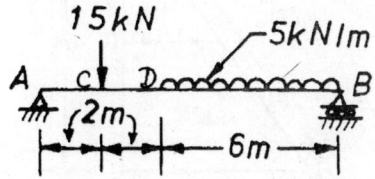

Fig. 10.24 (Practice Problem 10.5)

10.6   A beam 10 m long carries load as shown in Fig. 10.25. Using the principle of virtual work, find the reactions at *A* and *B*.

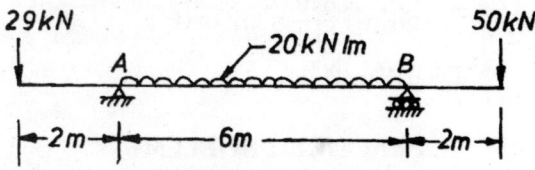

Fig. 10.25 (Practice Problem 10.6)

10.7   A uniform ladder of weight *W* and length *L* is held in equilibrium by a horizontal force *P* as shown in Fig. 10.26 below. Using the virtual work method, express *P* in terms of *W* and θ, where θ is the angle made by ladder with the vertical. The contact surfaces are smooth.

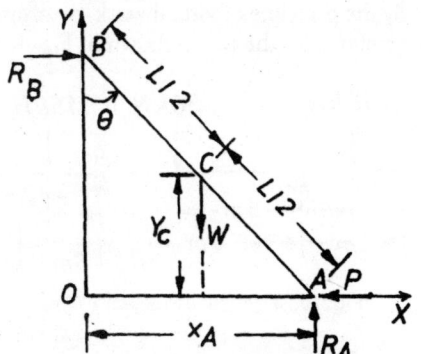

Fig. 10.26 (Practice Problem 10.7)

10.8   A uniform ladder, 5 metres long long and weighing 300 N, rests on a smooth floor at *A* and against a smooth wall at *B* as shown in Fig. 10.27. A horizontal rope *PQ* prevents the ladder from slipping. Using the method of virtual work, determine the tension in the rope if *BP* = 2.5 m.

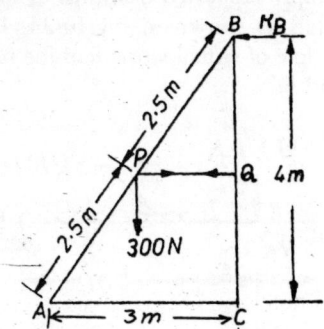

Fig. 10.27 (Practice Problem 10.8)

10.9   A weight of 4000 N is raised by a system of pulleys as shown in Fig. 10.28. Using the method of virtual work, find the force *P*, which can hold the weight in equilibrium.

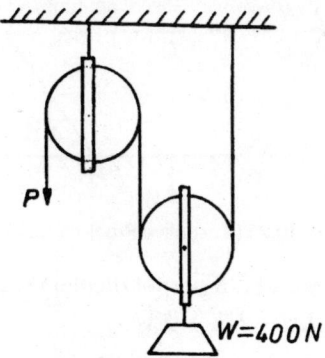

Fig. 10.28 (Practice Problem 10.9)

10.10   The diameter of the pulleys in a Weston's differential pulley block are 30 cm and 25 cm respectively. Using the principle of virtual work and neglecting friction find the value of the effort required to lift a load of 42 kN.

10.11   A weight of 1400 N resting over a smooth surface inclined at 30° with the horizontal, is supported by an effort (*P*) resting on a smooth surface inclined at 45° with the horizontal as shown in Fig. 10.29. By using the principle of virtual work, calculate the value of the effort *P*.

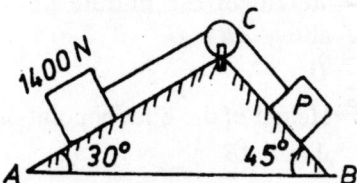

Fig. 10.29 (Practice Problem 10.11)

10.12   Two balls *D* and *E* of weights *P* and *Q* can slide freely along the bars *AC* and *BC*. The balls are

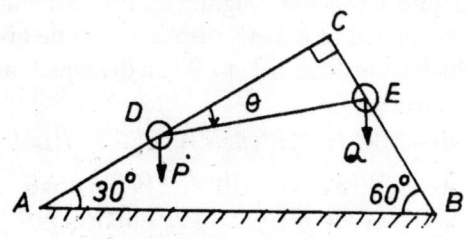

Fig. 10.30 (Practice Problem 10.12)

connected by an inextensible string *DE* (Fig. 10.30). Find the value of the angle θ defining the position of equilibrium.

10.13 A beam *AB* of 2 m length is held in equilibrium by the application of an effort (*P*) as shown in Fig. 10.31. Using the principle of virtual work, find the value of the effort (*P*), when a weight of 1000 N is hung at its middle point (*C*).

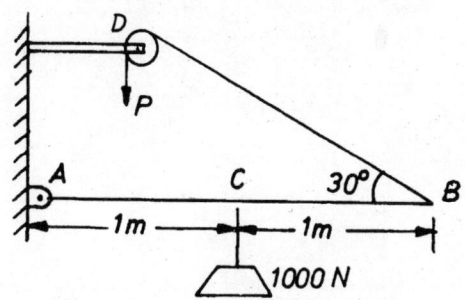

Fig. 10.31 (Practice Problem 10.13)

10.14 Fig. 10.32 shows a simple truss. Determine the force in member *CD*, using principle of virtual work.

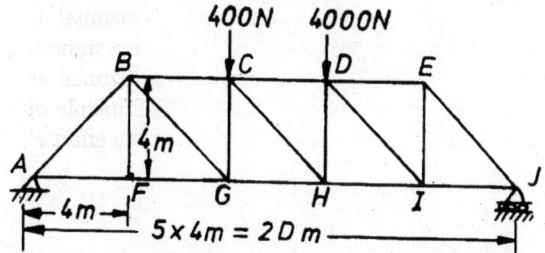

Fig. 10.32 (Practice Problem10.14)

10.15 A hexogonal frame is made up of six bars of identical lengths and cross-sections. Each bar has a weight of *W* Newton. The rod *AB* is fixed in a horizontal position. A string joins the midpoints of the bars *AB* and *DE*. Using the principle of virtual work find the tension in the string. (Fig. 10.33).

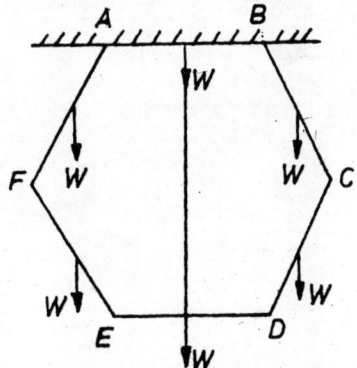

Fig. 10.33 (Practice Problem 10.15)

10.16 Determine the horizontal and vertical components of the reactions at *A* and *B* of the frame which is made up of three squares with hinged joints as shown in Fig. 10.34.

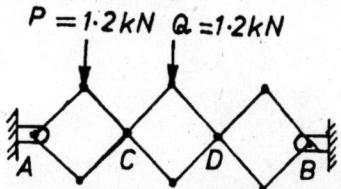

Fig. 10.34 (Practice Problem 10.16)

DYNAMICS

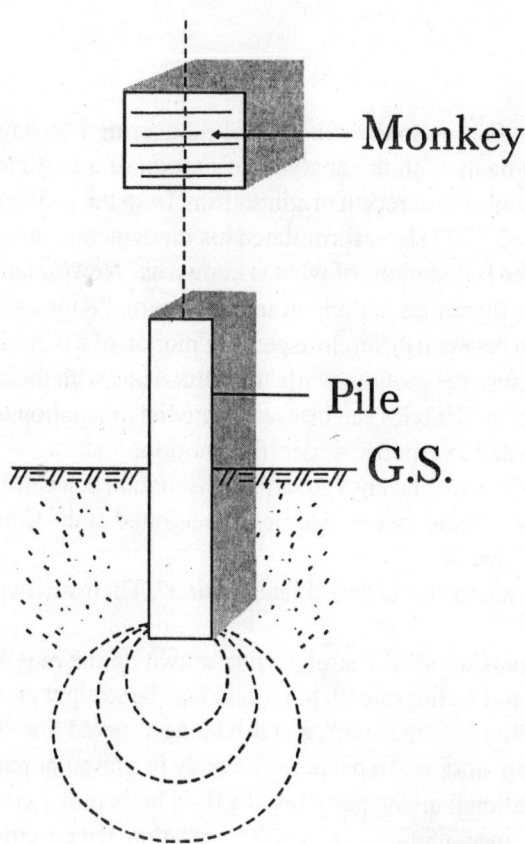

Monkey

Pile

G.S.

PART -II

Introduction to Dynamics

As already stated in the introduction to the subject of Engineering Mechanics, "Dynamics" is that branch of Mechanics which deals with the analysis of motion of a particle or a rigid body. While Statics is relatively old, Dynamics is of recent origin starting from the contributions of Galileo (1564-1642) onwards; Newton (1642-1727) later formulated his fundamental laws of motion, and his laws of gravitation and thus laid the foundations of what is known as "Newtonian Mechanics".

For convenience of study, dynamics is divided into two parts-"Kinematics" and "Kinetics", the former deals with the relation between different aspects of motion of a particle or a rigid body without reference to the forces that cause the motion, while the latter deals with these with specific reference to the forces causing the motion. Kinetics can be used to predict the motion caused by a known force, or to determine the force needed to produce a specified motion.

Dynamics is based on the fundamental laws governing the motion of a particle, which is a convenient idealization of a body, its mass being assumed to be concentrated at the Centroid, the rotation about the Centroid, if any, being neglected.

If the motion is in a single plane, it is called "Plane motion". There are two basic types of motion—"translation" and "rotation".

A motion is said to be 'translation' if a straight line drawn on the moving body always remains parallel to its original position. The line or path of motion may be straight or curved. Translation along a straight line is called "Rectilinear translation", and that along a curved line "Curvilinear Translation", an example of the latter being motion of a particle or a body in a circular path.

A motion is said to be rotation if all the particles of a rigid body move in *concentric circular paths* about a fixed axis. Thus rectilinear motion can be only translation, but curvilinear motion can be either translation or rotation.

A combination of both translation and rotation is possible in general plane motion with an instantaneous centre of rotation. The different types of motion will be considered in detail at the appropriate places in the chapters to follow.

Chapter 11

Rectilinear Motion

11.1 INTRODUCTION

A particle is said to be in rectilinear motion if the path traced by it is a straight line; a rigid body is said to be in rectilinear motion if the paths traced by different points on it are parallel straight lines.

In this chapter, only kinematics of rectilinear motion will be considered. Kinetics of rectilinear motion will be treated in a later chapter.

11.2 BASIC DEFINITIONS

Definitions of certain basic terms used in dynamics are given below.

11.2.1 Motion

A particle or a body is said to be in "motion" if it goes on changing its position with respect to a reference point. Thus, motion is purely relative—a person sitting and travelling in a vehicle is said to be in motion with respect to a reference point on the road, but is considered to be at rest with respect to a point inside the vehicle.

The reference for engineering problems is any fixed point on the earth; however, the centre of the earth is the reference point for motion of satellites, the centre of the sun for motion of planets in the solar system, and so on.

11.2.2 Displacement

The linear distance between the two positions of a particle/body in motion at the beginning and end of a specified time interval is known as its 'displacement'; it has both magnitude and direction, and so is a vector quantity.

The total length of the actual path of the motion is called the "distance travelled" by the body; this may not be the same as the displacement even in rectilinear motion, unless the motion has been in the same direction throughout the time interval considered. This is only a scalar quantity since it has only magnitude.

11.2.3 Velocity

Rate of change of displacement with respect to time is called "velocity"; it has both magnitude and direction, and so is a vector quantity. If the magnitude alone is of interest, it is called the 'speed', which is a scalar.

If the displacement in a given time interval is denoted by s, the average velocity v_{av} in the given time interval is given by

$$v_{av} = \frac{s}{t} \qquad \text{...(Eq. 11.1)}$$

The velocity v at any instant of time is called the 'instantaneous velocity', and is given by the limiting value of s/t when s and t are infinitesimally small.

$$v = \underset{\delta t \to 0}{\text{Lt}} \frac{\delta s}{\delta t} = \frac{ds}{dt} \qquad \text{...(Eq. 11.2)}$$

In S.I. units, the units of velocity are metres/second (m/s) or kilometres/hour (kmph).

$$\left[1 \text{ kmph} = \frac{5}{18} \text{ m/s} \right]$$

11.2.4 Acceleration

Rate of change of velocity with respect to time is called "acceleration"; it has both magnitude and direction, and hence is a vector quantity.

Average acceleration a_{av} over a time interval t in which the velocity changes from v_1 to v_2 is given by

$$a_{av} = \frac{(v_2 - v_1)}{t} \qquad \text{...(Eq. 11.3)}$$

The acceleration 'a' at any instant of time is called the 'instantaneous acceleration', and is given by

$$a = \frac{dv}{dt} \qquad \text{...(Eq. 11.4)}$$

If the acceleration is negative (i.e., the velocity decreases with time), it is called "retardation" or "deceleration".

From Eqs. 11.2 and 11.4, we get

$$a = \frac{d^2s}{dt^2} \qquad \text{...(Eq. 11.5)}$$

The units of acceleration are metres/second/second or metres/(second)² (m/sec²); it may also be kilometres/hour/second (kmph/s), wherever it is convenient.

In the case of rectilinear motion, both velocity and acceleration may be expressed as positive or negative quantities depending upon whether the motion is in the positive direction or negative direction along the path of motion.

11.3 GRAPHICAL REPRESENTATION OF MOTION

The motion of a body may be described with the aid of a table, a functional relationship with respect to time, or a diagram or graph.

The most accurate form is expression of the functional relationship of a particular parameter of motion (such as displacement, velocity, or acceleration) with time, although this may not always be easy.

The next best is a diagram or a graph with time interval as the abscissa and the parameter of motion as the ordinate.

The least preferable form is a table containing the values of the parameter of motion at specified instants of time with discrete intervals; the information for other instants of time has to be obtained by interpolation, which may not yield accurate results.

11.3.1 Time-Displacement Graph

This is nothing but the graph between time and the corresponding displacement of the particle/body (Fig. 11.1).

The velocity at any instant of time t is given by the slope of the curve or the tangent of the angle made by the tangent to the curve at that point.

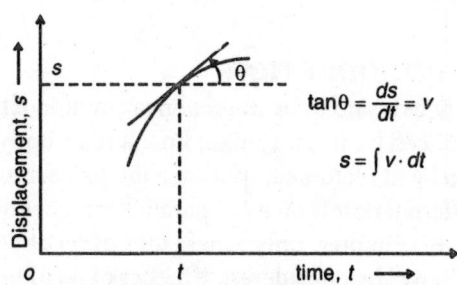

$$\tan\theta = \frac{ds}{dt} = v$$

$$s = \int v \cdot dt$$

Fig. 11.1: Time-Displacement Graph

11.3.2 Time-Velocity Graph

This graph is plotted with time as abscissa and the corresponding velocity as the ordinate (Fig. 11.2).

The slope of the graph at any point gives the acceleration at that instant of time.

$$v = \frac{ds}{dt}$$

$$\therefore \quad ds = v \cdot dt$$

$$(s_2 - s_1) = \int_{t_1}^{t_2} v \, dt \qquad \text{...(Eq. 11.6)}$$

$\therefore$ Area under the curve represents displacement.

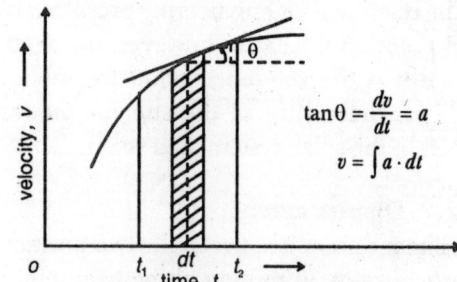

$$\tan\theta = \frac{dv}{dt} = a$$

$$v = \int a \cdot dt$$

Fig. 11.2: Time-Velocity Graph

11.3.3 Time-Acceleration Graph

This graph is plotted with time as abscissa and the corresponding acceleration as the ordinate (Fig. 11.3).

$$a = \frac{dv}{dt}$$

$$dv = a\,dt$$

$$\int_{v_1}^{v_2} dv = \int_{t_1}^{t_2} a\,dt$$

$$\therefore \ (v_2 - v_1) = \int_{t_1}^{t_2} a\,dt \qquad \ldots(\text{Eq. } 11.7)$$

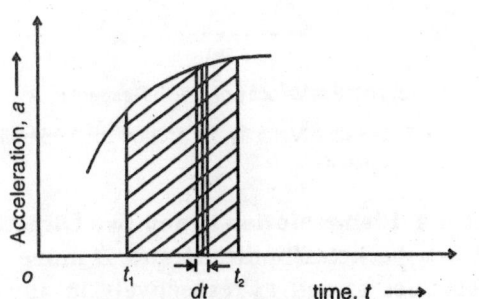

Fig. 11.3: Time–Acceleration Graph

$\therefore$ Area under the curve represents the change of velocity.

11.4 RECTILINEAR MOTION WITH CONSTANT VELOCITY

This is the simplest motion that one can think of, and is called "uniform motion".

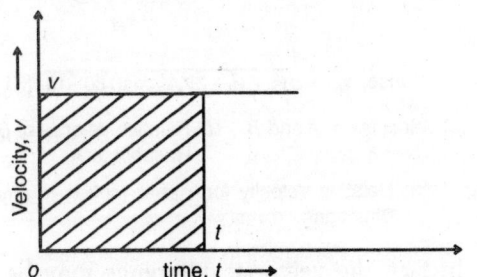

Fig. 11.4: Time-Velocity Graph for Uniform Motion

Let the constant velocity be, v

We know, $v = \dfrac{ds}{dt}$

$$ds = v\,dt$$

$$s = \int v \cdot dt$$

$$= v \int dt, \text{ since } v \text{ is a constant.}$$

This leads to $s = v \cdot t$ \qquad ...(Eq. 11.8)

The time-velocity graph is a straight line parallel to the time-axis as shown in Fig. 11.4.

The displacement (or distance travelled, in this case) is represented by the shaded area for a particular value of t.

11.5 RESULTANT VELOCITY

Since velocity is a vector quantity like force, if a particle/rigid body is simultaneously subjected to the influence of more than one velocity, the velocity to which it will be subjected will be obtained based on the concept of the 'resultant'.

The resultant velocity in a given case may be obtained by procedures similar to those for forces. In view of this, the concepts of components in two specified directions, the triangle law, the polygon law, and the theorem of resolved parts, apply equally well to velocity.

11.6 RELATIVE VELOCITY

The motion of a particle with respect to a fixed point or a fixed frame of reference is called the absolute motion of the particle.

The motion relative to a moving set of axes or a moving frame of reference is known as "relative motion".

If two trains move in parallel paths in the same direction with the same speed, an observer in either train does not feel the motion of the other. But if the train in which the observer is located moves faster than the other train, he will experience some motion; in fact, if the two trains move in opposite directions, the observer feels that the trains are approaching each at a much greater speed than that of either train. This is due to the concept of what is known as "Relative velocity". The relative velocity of B with respect to A is the vector difference of the velocities of B and A.

11.6.1 Motion in Parallel Paths in the Same Direction

If two particles/bodies A and B move with

velocities v_A and v_B respectively in parallel paths in the same direction, the relative velocity of B with respect to A, v_{BA} is given by the vector difference $(v_B - v_A)$, as shown in Fig. 11.5.

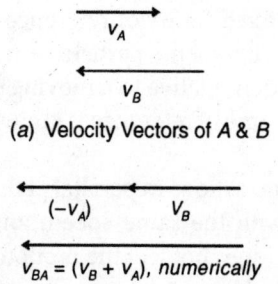

(a) Velocity Vectors of A & B

(b) Relative Velocity of B with Respect to A

Fig. 11.5: Relative Velocity for Motion in Parallel Paths in the Same Direction

11.6.2 Motion in Parallel Paths in Opposite Directions

If two particles/bodies A and B move with velocities v_A and v_B in parallel paths in opposite directions also, the relative velocity of B with respect to A, v_{BA} is once again given by the vector difference $(v_B - v_A)$, as shown in Fig. 11.6.

(a) Velocity Vectors of A & B

(b) Relative Velocity of B with Respect to A

Fig. 11.6: Relative Velocity for Motion in Parallel Paths in the Opposite Direction

$(v_B - v_A)$ is $v_B - (-v_A) = (v_B + v_A)$, numerically, in this case.

11.6.3 Motion in Perpendicular Directions

If two particles/bodies A and B move with velocities v_A and v_B in two perpendicular directions, the relative velocity of B with respect to A is once again given by the vector difference $(v_B - v_A)$, as shown in Fig. 11.7.

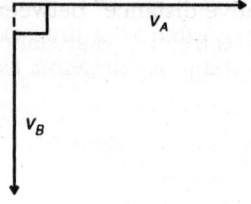

(a) Velocity of A and B

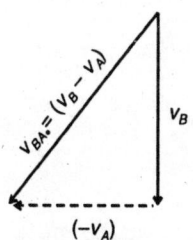

(b) Relative Velocity of B with Respect to A

Fig. 11.7: Relative Velocity for Motion in Perpendicular Directions

11.6.4 Planar Motion in any Two Directions

If two particles/bodies A and B move with velocities v_A and v_B respectively in any two directions in a plane, the relative velocity of B with respect to A is once again given by $(v_B - v_A)$, vectorially, as shown in Fig. 11.8.

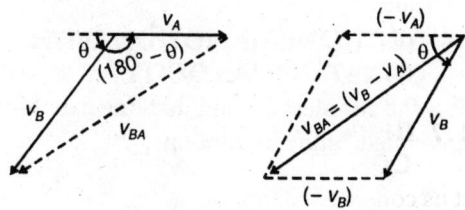

Also, $v_{BA} = \sqrt{v_A^2 + v_B^2 - 2V_A V_B \cos(180 - \theta)}$

(a) Velocities of A and B   (b) Relative Veloctiy of B with Respect to A

Fig. 11.8: Relative Velocity for Planar Motion in any two Directions

In fact, the vectorial difference may be obtained by resolving v_A and v_B in two perpendicular directions x and y, and combining the x- and y-components of v_A and v_B, and then obtaining the vector difference $(v_B - v_A)$.

The case given in 11.6.3 may be viewed as a special case of this general case.

The "relative distance" between A and B may be obtained as treating A as stationary and B as moving at its relative velocity v_{BA} with respect to A in the given time interval. (However, it may be remembered that A would have moved in the direction of its velocity v_A by $v_A \cdot t$ in the specified time interval t.

11.6.5 Relative Velocity versus Resultant Velocity

The combined effect of any number of velocities induced by several forces on a body may be obtained by the vectorial addition of such velocity vectors; the resulting velocity vector is the "Resultant Velocity" of the body.

The following differences between relative velocity and resultant velocity are obvious:

(*i*) Resultant velocity is the combined effect of more than one force causing motion of a single body, while relative velociy is the velocity of a moving body with respect to another body in motion.

(*ii*) Resultant velocity is obtained as the *vector addition* of the velocities caused by all the forces acting on the body, while the relative velocity of one body with respect to another is obtained as the vector difference of the velocity of the first body and that of the second body, taken in that order.

11.7 RECTILINEAR MOTION UNDER CONSTANT ACCELERATION

Let us consider motion of a body under constant or uniform acceleration a.

Let u be the initial velocity;

v the final velocity;

t the time interval;

and s the distance travelled.

Acceleration is the rate of change of velocity.

$\therefore \quad a = \dfrac{(v - u)}{t}$

or $\quad v = u + at$ ...(Eq. 11.9)

Alternatively,

$a = \dfrac{dv}{dt}$, a being constant

$dv = a \cdot dt$

Integrating both sides

$$\int_u^v dv = \int_0^t a.dt$$

$v - u = at$

or $\quad v = u + at$, as before

Displacement s is given by

$$s = v_{av} \cdot t = \left(\frac{u + v}{2} \right) \cdot t \qquad \text{...(Eq. 11.10)}$$

Substituting for v from Eq. 11.9,

$$s = \left(\frac{u + u + at}{2} \right) \cdot t$$

or $\quad s = ut + \dfrac{1}{2} at^2$ ...(Eq. 11.11)

Alternatively,

$\dfrac{ds}{dt} = v$

$ds = v \cdot dt$

Using Eq. 11.9 for v,

$ds = (u + at)\, dt$

Integrating both sides,

$$\int_0^s ds = \int_0^t (u + at)\, dt$$

or $\quad s = ut + \dfrac{1}{2} at^2$, as before

From Eq. 11.9,

$$t = \frac{(v - u)}{a}$$

Substituting this value in Eq. 11.10,

$$s = \left(\frac{u + v}{2} \right)\left(\frac{v - u}{a} \right)$$

$$= \frac{v^2 - u^2}{2a}$$

or $\quad v^2 - u^2 = 2as$ ...(Eq. 11.12)

Alternatively,

$$a = \frac{dv}{dt}$$

$$= \frac{dv}{ds} \cdot \frac{ds}{dt}$$

$$= v \cdot \frac{dv}{ds}$$

$a \cdot ds = v \cdot dv$

Integrating both sides,

$$\int_0^s a\, ds = \int_u^v v \cdot dv$$

$$a \cdot s = \left(\frac{v^2 - u^2}{2} \right)$$

or $v^2 - u^2 = 2as$, as before.

The following three equations are relevant for solving motion under constant aceleration:

$v = u + at$ ...(Eq. 11.9)

$s = ut + \frac{1}{2}at^2$ ...(Eq. 11.11)

$v^2 - u^2 = 2as$ ...(Eq. 11.12)

11.8 RECTILINEAR MOTION UNDER GRAVITY

All bodies are attracted towards the centre of the earth with an acceleration, which is *practically constant* everywhere. It is called "Acceleration due to Gravity", and its value is found to be 9.81 metres/second^2 very nearly. While for vertical downward motion, it is acceleration, for vertical upward motion, it is negative acceleration, or deceleration. (More about gravity will be seen in the Chapter on "Laws of Motion"). Thus, rectilinear motion under the influence of gravity is the most common case of motion under uniform acceleration, dealt with in the previous section.

11.8.1 Downward Motion under Gravity

Let the acceleration due to gravity be designated 'g'. Eqs. 11.9, 11.11, and 11.12 are applicable for this case, except that 'a' shall be replaced by 'g'

$v = u + gt$ ...(Eq. 11.13)

$s = ut + \frac{1}{2}gt^2$ ...(Eq. 11.14)

$(v^2 - u^2) = 2gs$ ...(Eq. 11.15)

In a general case, we have

$$\ddot{s} = \frac{d^2s}{dt^2} = g$$

Integrating with respect to time,

$v = \dot{s} = gt + C_1$ (C_1 is the constant of integration)
    ...(Eq. 11.16)

Integrating again with respect to t,

$$s = \frac{1}{2}gt^2 + C_1 t + C_2$$

(C_2 is another constant of integration) ...(Eq. 11.17)

Constants C_1 and C_2 can be determined from known boundary/initial conditions.

Eqs. 11.16 and 11.17 give the velocity and the position of the body at any instant of time.

11.8.2 Motion of Body Projected Vertically Upward

When a body is projected vertically upwards from a ground point with a certain initial velocity it is subjected to a deceleration equal to 'g'; thus it reaches a certain height where the velocity gets reduced to zero, and again starts falling under the influence of gravity with an acceleration g. It can be shown that the time of descent is the same as that for ascent. Numerical examples will illustrate these ideas.

11.9 MOTION WITH VARYING ACCELERATION

In practice, the acceleration of a particle/body may vary with time–for example, a vehicle may start with zero acceleration, but the acceleration may be increased until the desired speed is nearing, it may be decreased and brought to zero as soon as the desired speed is picked up. Thus it becomes a problem with varying acceleration. Such problems, as well as problems in which the velocity or displacement varies with time, may be solved provided the specific manner of variation is expressible mathematically. In this context, the following equations, already given, come in handy.

$$v = \frac{ds}{dt}$$

$$a = \frac{dv}{dt} = \frac{d^2s}{dt^2} = v \cdot \frac{dv}{ds}$$

The solution may involve differentiation or integration; in the latter case, the constants of integration may be obtained from known boundary/initial conditions of motion.

Motion of a particle in a viscous medium like oil, is motion in a resisting medium, since it offers resistance to the motion of the particle/body; it is known from the principles of fluid mechanics that the negative acceleration (retardation/deceleration) produced by the resisting force is proportional to its velocity at any instant.

$$a = \frac{dv}{dt} = -k.v. \qquad ...(\text{Eq. } 11.18)$$

k being a constant, and the negative sign indicates that it is a retardation.

Thus, this is a special case of motion with varying acceleration.

These and other related concepts will be clear from the illustrative problems given below.

ILLUSTRATIVE PROBLEMS

PROBLEM 11.1: A particle moves along a straight line in such a way that its distance in meters from a fixed point on the line after t seconds is given by $s = 4t^3 + 3t^2 + 2t + 1$. Find the distance, velocity and acceleration at the end of 2 seconds.

Solution:

$$s = 4t^3 + 3t^2 + 2t + 1$$

$$t = 2 \text{ seconds.}$$

$$\therefore \quad s = 4 \times 2^3 + 3 \times 2^2 + 2 \times 2 + 1$$

$$4 = 32 + 12 + 4 + 1$$

$$= 49 \text{ m}$$

$$v = \frac{ds}{dt} = 12t^2 + 6t + 2$$

Put $t = 2$ sec

$$\therefore \quad v = 12 \times 2^2 + 6 \times 2$$

$$= 48 + 12$$

$$= 60 \text{ m/sec}$$

$$a = \frac{dv}{dt} = 24\,t + 6$$

Put $t = 2$

$$a = 24 \times 2 + 6$$

$$= 54 \text{ m/sec}^2$$

PROBLEM 11.2: A particle moves along a straight line according to the law $S^2 = at^2 + 2bt + c$. Prove that the acceleration varies inversely as the cube of the distance.

Solution:

$$S^2 = at^2 + 2bt + c$$

$$S = (at^2 + 2bt + c)^{1/2} \qquad ...(i)$$

Differentiating (i) with respect to 't',

$$\frac{dS}{dt} = \frac{1}{2}\left(at^2 + 2bt + c\right)^{-1/2}(2at + 2$$

$$\frac{dS}{dt} = \frac{at + b}{(at^2 + 2bt + c)^{1/2}}$$

Again differentiating with respect to 't'.

$$\frac{d^2S}{dt^2} = \frac{(at^2 + 2bt + c)^{1/2}a}{(at^2 + 2bt + c)}$$

$$- \frac{(at + b)\frac{1}{2}(at^2 + 2bt + c)^{-1/2}(2at + 2b)}{(at^2 + 2bt + c)}$$

$$= \frac{\left(at^2 + 2bt + c\right)^{1/2}(a) - \dfrac{\left(at + b\right)^2}{\left(at^2 + 2bt + c\right)^{1/2}}}{\left(at^2 + 2bt + c\right)}$$

$$= \frac{\left(at^2 + 2bt + c\right)a - \left(at + b\right)^2}{\left(at^2 + 2bt + c\right)\left(at^2 + 2bt + c\right)^{1/2}}$$

$$= \frac{a^2t^2 + 2abt + ac - a^2t^2 - 2at \cdot b - b^2}{S^2 \cdot S}$$

$$= \frac{ac - b^2}{S^3} \propto \frac{1}{S^3} \quad (a,\, b,\, c \text{ are constants})$$

$$\therefore \quad \text{acceleration} \propto \frac{1}{S^3}$$

i.e., acceleration varies inversely as the cube of the distance S.

PROBLEM 11.3: A body starting with some initial velocity and moving with uniform acceleration acquires a velocity of 20 m/sec after moving through 10 m and a velocity of 30 m/sec after moving through a further 15 m. When and where will its velocity be 40 m/sec?

Solution: Let u be the initial velocity and a, the uniform acceleration.

$$OA = 10 \text{ m}$$

$$AB = 15 \text{ m}$$

Velocity at $A = 20$ m/sec

Velocity at $B = 30$ m/sec

Velocity at $C = 40$ m/sec

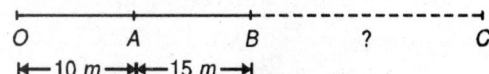

Fig. 11.9 (Illustrative Problem 11.3)

Consider the motion from A to B

$$v^2 - u^2 = 2aS$$

$$30^2 - 20^2 = 2 \times a \times 15$$

$$a = \frac{(30^2 - 20^2)}{2 \times 15} = \frac{50 \times 10}{2 \times 15}$$

$$a = \frac{50}{3} \text{ m/sec}$$

Consider the motion from B to C

$$40^2 - 30^2 = 2 \times \frac{50}{3} \times S$$

$$S = \frac{(40^2 - 30^2) \times 3}{2 \times 50} = \frac{70 \times 10 \times 3}{2 \times 50}$$

$$S = 21 \text{ m}$$

∴ Velocity becomes 40 m/sec at 21 m from the last position (i.e., from B).

$$v = u + at$$

$$t = \frac{v - u}{a} = \frac{(40 - 30)}{(50/3)} = \frac{10 \times 3}{50}$$

$$t = 0.6 \text{ sec}$$

PROBLEM 11.4: Three marks A, B and C at a distance of 120 m each are made along a straight road. A car starting from rest and with a uniform acceleration passes the mark A and takes 12 seconds to reach B and further 8 seconds to reach the mark C. Calculate (*i*) the magnitude of acceleration of the car, (*ii*) the velocity of car at A, (*iii*) the velocity of car at B, and (*iv*) the distance of the mark C from the starting point.

Solution: Car starts from 'O'

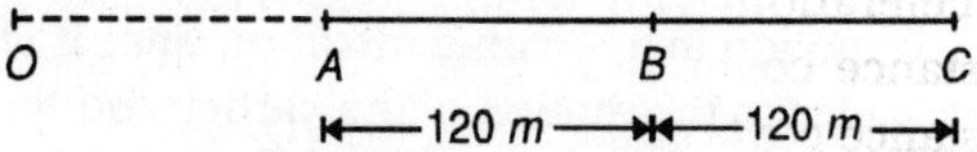

Fig. 11.10 (Illustrative Problem 11.4)

Time taken from A to $B = 12$ sec

Time taken from B to $C = 8$ sec

Considering motion from A to B

$$s = ut + \frac{1}{2}at^2$$

$$120 = v_A \times 12 + \frac{1}{2} \times a \times 12^2$$

$$v_A + 6a = 10 \qquad \ldots(i)$$

Considering motion from A to C

$$240 = v_A \times 20 + \frac{1}{2} \times a \times 20^2$$

$$v_A + 10a = 12 \qquad \ldots(ii)$$

$(ii) - (i)$ gives

$$4a = 2$$

∴ $$a = 0.5 \text{ m/sec}^2$$

Substituting the value of 'a' in Eqn. (*i*)

$$v_A + 6 \times 0.5 = 10$$

or $$v_A = 7 \text{ m/sec}$$

Considering motion from B to C

$$120 = v_B \times 8 + \frac{1}{2} \times 0.5 \times 8^2$$

$$v_B + 2 = 15$$

$$v_B = 13 \text{ m/sec}$$

Considering the motion from O to A

$$v^2 - u^2 = 2aS$$

$$7^2 - u^2 = 2 \times 0.5 \times S$$

$$S = \frac{49}{2 \times 0.5} = 49 \text{ m}$$

∴ The distance of the mark C from the starting point 'O'.

$$= 49 + 120 + 120$$

$$= 289 \text{ m}$$

PROBLEM 11.5: A particle is moving with uniform acceleration. In the fifth and ninth seconds from the beginning, it travels 16 m and 20 m respectively. Find the initial velocity and the acceleration with which it moves. Find also distance covered by it in 12 seconds and the distance covered in 12th second.

Solution: Let u be the initial velocity and a, the uniform acceleration.

$$S_n = u + \frac{1}{2}a(2n - 1)$$

$$S_5 = 16 \text{ m} \quad n = 5$$

$$16 = u + \frac{1}{2}a(2 \times 5 - 1)$$

$$16 = u + \frac{9}{2}a \qquad \qquad \ldots(i)$$

$$S_8 = 20 \text{ m}, \; n = 9$$

$$20 = u + \frac{1}{2}a(2 \times 9 - 1)$$

$$20 = u + \frac{17}{2}a \qquad \qquad \ldots(ii)$$

$(ii) - (i)$ gives

$$4 = \left(\frac{17}{2} - \frac{9}{2}\right)a = 4a$$

$$a = 1 \text{ m/sec}^2$$

Substituting in equation (i)

$$16 = u + \frac{9}{2}(1)$$

$$u = 16 - 4.5 = 11.5 \text{ m/sec}$$

Distance covered in 12 seconds

$$= ut + \frac{1}{2}at^2$$

$$= 11.5 \times 12 + \frac{1}{2} \times 1 \times 12^2$$

$$= 138 + 72$$

$$= 210 \text{ m}$$

$$S_{12} = u + \frac{1}{2}a(2 \times 12 - 1)$$

$$= 11.54 + \frac{1}{2} \times 1 \times 23$$

$$= 23 \text{ m}$$

Check

Distance covered in 11 seconds

$$= 11.5 \times 11 + \frac{1}{2} \times 1 \times 11^2$$

$$= 126.5 + 60.5$$

$$= 187 \text{ m}$$

$\therefore$ Distance covered in 12th second

$$= \text{Distance covered in 12 seconds}$$

$$- \text{Distance covered in 11 seconds}$$

$$= 210 - 187 = 23 \text{ m}$$

PROBLEM 11.6: Two motor cars are moving uniformly along two straight roads making an angle of 60° with each other, with velocities of 60 kmph and 45 kmph. If the first car be moving towards and the second away from the junction of the roads, find the relative velocity of the first with respect to the second.

Solution: Let the second car move along OX with velocity 45 kmph.

Take OX as x-axis

Let O be the junction of the road.

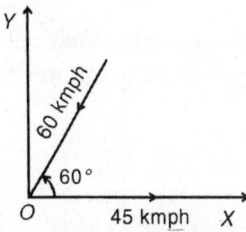

Fig. 11.11 (Illustrative Problem 11.6)

Actual velocity of second car along OX
$$= 45 \text{ kmph}$$

Actual velocity of second car along $OY = 0$

Reversed velocity of second car along OX
$$= -45 \text{ kmph}$$

Reversed velocity of second car along OY
$$= 0$$

Actual velocity of first car along OX
$$= -60\cos 60° = -30 \text{ kmph}$$

Actual velocity of first car along OY
$$= -60\sin 60°$$

$$= -30\sqrt{3} \text{ kmph}$$

Let v be the relative velocity of the first car with respect to the second making an angle θ with OX

$$v\cos\theta = -30 - 45 = -75 \text{ kmph} \qquad ...(1)$$
$$v\sin\theta = -30\sqrt{3} + 0 = -30\sqrt{3} \text{ kmph} \qquad ...(2)$$

Squaring (1) and (2) and adding

$$v^2 = (-75)^2 + \left[(-30)\sqrt{3}\right]^2$$
$$v^2 = 5625 + 2700 = 8325$$
$$\therefore \quad v = \sqrt{8325} = 91.24 \text{ kmph}$$

Dividing (2) by (1),

$$\tan\theta = \frac{-30\sqrt{3}}{-75} = 0.6928$$
$$\theta = \pi + 0.6928$$

($\cos\theta$ and $\sin\theta$ are both $-$ve $\therefore \theta$ lies in the third quadrant).

PROBLEM 11.7: A person travelling eastwards with a velocity of 6 kmph finds that the wind seems to blow directly from the north; on doubling his speed it seems to come from the north-east. Find the true direction of wind and its velocity.

Solution: Let v be the actual velocity of wind and let it blow in a direction making an angle θ south of east.

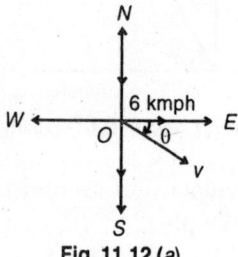

Fig. 11.12 (a)

Case 1

v is the resultant velocity of the person along OE and the apparent velocity of wind along OS.

$\therefore$ Resolving along OE

$$v\cos\theta = 6 \qquad ...(1)$$

Case 2

The velocity of the person is 12 kmph along OE.

Let u be the apparent velocity of wind along OA.

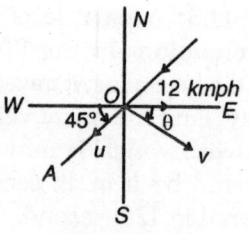

Fig. 11.12 (b)

$\therefore$ Resolving along OE

$$v\cos\theta = 12 - u\cos 45° = 12 - \frac{u}{\sqrt{2}} \qquad ...(2)$$

Resolving along OS

$$-v\sin\theta = -u\sin 45°$$

or $v\sin\theta = \dfrac{u}{\sqrt{2}} \qquad ...(3)$

Eliminate u by adding (2) & (3)

$$v\cos\theta + v\sin\theta = 12 \qquad ...(4)$$

By substituting (1) in (4)

$$v\sin\theta = 6 \qquad ...(5)$$

Squaring (1) and (5) and adding

$$v^2 = 6^2 + 6^2 = 72$$
$$v = 6\sqrt{2} \text{ kmph}$$

Dividing (5) by (1)

$$\tan\theta = \frac{6}{6} = 1$$
$$\therefore \qquad \theta = 45°$$

PROBLEM 11.8: Two ships move from a port at the same time. Ship A has a velocity of 25 kmph and is moving in north-west direction while ship B moves in S 30° W with a velocity of 60 kmph. Determine the relative velocity of A with respect to B.

Solution: Actual velocity of ship A along x-axis

$$= -v_A\sin 45°$$
$$= -25\sin 45°$$
$$= -17.68 \text{ kmph}$$

Actual velocity of ship A along y-axis

$$= +v_A\cos 45°$$
$$= +25\cos 45°$$
$$= +17.68 \text{ kmph}$$

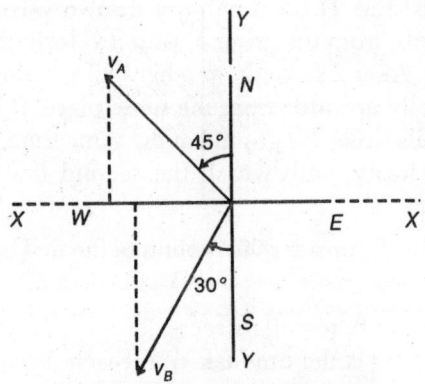

Fig. 11.13 (Illustrative Problem 11.8)

Actual velocity of ship B along x-axis

$$= -v_B \sin 30°$$

$$= -60 \times \frac{1}{2} = -30 \text{ kmph}$$

Actual velocity of ship B along y-axis

$$= -v_B \cos 30°$$

$$= -60 \times 0.866$$

$$= -51.96 \text{ kmph}$$

Reversed velocity of ship B along x-axis

$$= +30 \text{ kmph}$$

Reversed velocity of ship B along y-axis

$$= +51.96 \text{ kmph}$$

Let v be the relative velocity of ship A with respect to the second making an angle O with OX

$$v \cos\theta = -17.68 + 30$$

$$= +12.32 \text{ kmph}$$

$$v \sin\theta = +17.68 + 51.96$$

$$= 69.64 \text{ kmph}$$

$$\therefore \quad v = \sqrt{(12.32)^2 + (69.64)^2}$$

$$v = \sqrt{5001.512}$$

$$v = 70.72 \text{ kmph}$$

$$\theta = \tan^{-1}\left(\frac{69.64}{12.32}\right) = \tan^{-1}(5.6526)$$

$$= 79.97°$$

$$\approx 80°$$

The relative velocity of A with respect to B is 70.72 kmph at N 10° E (10° East of North).

PROBLEM 11.9: A stone is dropped from the top of a tower, 40 m high. At the same time another stone is thrown upwards from the foot of the tower with a velocity of 20 m/sec. When and where will the two stones cross each other and with what relative velocity?

Solution: Let the two stones cross each other at a height h from the ground after t seconds.

Height of the tower = 40 m

$\therefore$ Distance traversed by the first stone

$$= (40 - h) \text{ m}$$

$$\therefore 40 - h = u_1 t + \frac{1}{2} g t^2$$

$$40 - h = 0 \times t + \frac{1}{2} \times 9.81 \times t^2 \qquad ...(1)$$

Consider the upward motion of second stone

$$h = u_2 t - \frac{1}{2} g t^2$$

$$h = 20 \times t - \frac{1}{2} \times 9.81 \times t^2 \qquad ...(2)$$

Adding (1) and (2)

$$40 - h + h = \frac{1}{2} \times 9.81 \times t^2 + 20t - \frac{1}{2} \times 9.81 \times t^2$$

$$40 = 20t$$

or $\quad t = 2$ sec.

Substituting the value of 't' in Eqn. (1)

$$40 - h = \frac{1}{2} \times 9.81 \times 2^2$$

$$\therefore \quad h = 40 - 19.62$$

$$h = 20.38 \text{ m}$$

At $t = 2$ sec

Downward velocity of first stone, $v_1 = u_1 + gt$

$$v_1 = 0 + 9.81 \times 2$$

$$= 19.62 \text{ m/sec}$$

Upward velocity of second stone, $v_2 = u_2 - gt$

$$= 20 - 9.81 \times 2$$

$$= 0.38 \text{ m/sec}$$

$\therefore$ Relative velocity $= v_1 + v_2$

$$= 19.62 + 0.38$$

$$= 20 \text{ m/sec}$$

PROBLEM 11.10: A stone is thrown vertically upwards with a velocity of 50 m/sec. Find its position after 4 seconds.

Solution:

$$S = ut - \frac{1}{2}gt^2$$

$$u = 50 \text{ m/sec}$$

$$t = 4 \text{ sec}$$

$$g = 9.81 \text{ m/sec}^2$$

$$S = 50 \times 4 - \frac{1}{2} \times 9.81 \times 4^2$$

$$S = 200 - 78.48$$

$$S = 121.52 \text{ m}$$

PROBLEM 11.11: A stone is thrown vertically upwards, from the top of a tower with a certain velocity. It reaches the ground in 6 seconds. A second stone, thrown down from the same tower with the same initial velocity reaches ground in 4 seconds. Determine (*i*) the height of the tower, and (*ii*) the initial velocity of the stones.

Solution: Let $u =$ initial velocities of the stones

$h =$ height of the tower

Consider the motion of the first stone.

The time taken by the first stone to reach maximum height and then to reach the top of the tower from where it was thrown,

$$= 6 - 4 = 2 \text{ sec}$$

∴ The time taken by the first stone to reach maximum height

$$= \frac{2}{2} = 1 \sec$$

$$v = u - gt$$

$$0 = u - 9.81 \times 1 \qquad (\because v = 0)$$

or $\quad u = 9.81 \text{ m/sec}$

$$h = ut + \frac{1}{2}gt^2$$

$$h = 9.81 \times 4 + \frac{1}{2} \times 9.81 \times (4)^2$$

$$h = 39.24 + 78.48$$

$$h = 117.72 \text{ m}$$

PROBLEM 11.12: A ball was thrown vertically upwards, from the ground, with a velocity of 50 m/sec. After 2 seconds, another ball was thrown vertically upwards from the same place. If both the balls strike the ground at the same time, find the velocity, with which the second ball was thrown.

Solution: Consider the motion of the first ball

$$v = u - \text{gt}$$

$$v = 0; \ u = 50 \text{ m/sec};$$

$t =$ is the time taken to reach the maximum height

$$\theta = 50 - 9.81 \times t$$

$$t = \frac{50}{9.81} = 5.097 \text{ sec}$$

It means the ball will take 5.097 sec to reach the maximum height and 5.097 sec to come back to the ground.

Total time of flight $= 5.097 + 5.097$

$$= 10.194 \text{ sec}$$

Consider the motion of the second ball.

The time taken by the second ball for going upwards and coming back to the ground

$$= 10.194 - 2.000$$

$$= 8.194 \text{ sec}$$

∴ Time taken by the second ball to reach maximum height

$$= \frac{8.194}{2} = 4.097 \text{ sec}$$

Consider the upward motion of the second ball

$$v = u - gt$$

$$v = 0; \ t = 4.097 \text{ sec}$$

$$0 = u - 9.81 \times 4.097$$

$$u = 9.81 \times 4.097$$

$$u = 40.19 \text{ m/sec}$$

PROBLEM 11.13: A stone is dropped into a well and the sound of the splash is heard in 6 seconds; if the velocity of the sound be 340 m/sec, find the depth of the surface of water from the top.

Solution: Let h be the depth of the well

t_1 be the time taken by stone to strike water

t_2 be the time taken by sound to travel the height h

$\therefore \quad t_1 + t_2 = 6 \text{ sec}$ ...(1)

For downward motion of the stone,

$$h = 0 \times t_1 + \frac{1}{2} g t_1^2$$

$$h = \frac{1}{2} g t_1^2 \qquad \qquad ...(2)$$

For the upward motion of sound

$$h = 340 \, t_2 \ (\because \text{Sound moves with uniform velocity})$$

$$h = 340 \, t_2 \qquad \qquad ...(3)$$

Equating equations (2) and (3)

$$\frac{1}{2} \times 9.81 \times t_1^2 = 340 \, t_2$$

$$t_1^2 = 69.317 \, t_2 \qquad \qquad ...(4)$$

From Eqn. (1) $t_2 = 6 - t_1$ substituting in (4)

$$t_1^2 = 69.317 (6 - t_1)$$

$$t_1^2 + 69.317 \, t_1 - 415.902 = 0$$

$$t_1 = \frac{69.317 \pm \sqrt{(69.317)^2 + 4(415.902)(1)}}{2}$$

$$= \frac{-69.317 \pm \sqrt{4804.8464 + 1663.608}}{2}$$

$$= \frac{-69.317 \pm \sqrt{6468.4544}}{2}$$

$$= \frac{-69.317 + 80.427}{2}$$

$$= \frac{11.11}{2} = 5.555 \text{ sec}$$

$\therefore \quad h = \frac{1}{2} \times 9.81 \times (5.555)^2 = 151.36 \text{ m}$

PROBLEM 11.14: The motion of a particle along a straight line is given by the equation

$$a = t^2 - 3t + 5$$

where a is the acceleration in m/sec^2 and t is the time in seconds. The velocity of the particle and the displacement are 6 m/sec and 12 m after one second respectively. Find the (*i*) displacement, (*ii*) velocity and (*iii*) acceleration of the particle after 3 seconds.

Solution:

$$a = t^2 - 3t + 5$$

$$a = \frac{dv}{dt} = t^2 - 3t + 5$$

$\therefore \quad dv = (t^2 - 3t + 5) \, dt$

Integrating both sides

$$v = \frac{t^3}{3} - \frac{3}{2} t^2 + 5t + C_1$$

When $t = 1$ sec; $v = 6$ m/sec

$\therefore \quad 6 = \frac{1}{3} - \frac{3}{2} + 5 + C_1$

or $\quad C_1 = \frac{13}{6}$

$\therefore \quad v = \frac{t^3}{3} - \frac{3}{2} t^2 + 5t + \frac{13}{6}$

$$v = \frac{ds}{dt} = \frac{t^3}{3} - \frac{3}{2} t^2 + 5t + \frac{13}{6}$$

$$ds = \left(\frac{t^3}{3} - \frac{3}{2} t^2 + 5t + \frac{13}{6} \right) dt$$

Integrating both sides

$$S = \frac{t^4}{12} - \frac{t^3}{2} + \frac{5}{2} t^2 + \frac{13}{6} t + C_2$$

When $t = 1$ sec; $s = 12$ m

$$12 = \frac{1}{12} - \frac{1}{2} + \frac{5}{2} + \frac{13}{6} + C_2$$

$$C_2 = 12 - \frac{1}{12} + \frac{1}{2} - \frac{5}{2} - \frac{13}{6}$$

$$= \frac{93}{12} = \frac{31}{4}$$

$\therefore \quad s = \frac{t^4}{12} - \frac{t^3}{2} + \frac{5}{2} t^2 + \frac{13}{6} t + \frac{31}{4}$

$$v = \frac{t^3}{3} - \frac{3}{2} t^2 + 5t + \frac{13}{6}$$

$$a = t^2 - 3t + 5$$

When $t = 3$ seconds

$$s = \frac{3^4}{12} - \frac{3^3}{2} + \frac{5}{2} \times 3^2 + \frac{13}{6} \times 3 + \frac{31}{4}$$

$$s = \frac{81}{12} - \frac{27}{2} + \frac{45}{2} + \frac{39}{6} + \frac{31}{4}$$

$$s = \frac{81 - 162 + 270 + 78 + 93}{12}$$

$$s = \frac{360}{12} = 30\,\text{m}$$

$$v = \frac{3^3}{3} - \frac{3}{2} \times 3^2 + 5 \times 3 + \frac{13}{6}$$

$$v = 9 - \frac{27}{2} + 15 + \frac{13}{6}$$

$$v = \frac{54 - 81 + 90 + 13}{6}$$

$$v = 12.67 \text{ m/sec}$$

$$a = 3^2 - 3 \times 3 + 5$$

$$a = 9 - 9 + 5$$

$$a = 5 \text{ m/sec}^2$$

PROBLEM 11.15: A particle, starting from rest, moves in a straight line, whose acceleration is given by the equation

$$a = 12 - 0.01 s^2$$

where a is in m/sec^2 and s is in meters. Determine (i) the velocity of the particle, when it has travelled 25 m, (ii) the distance travelled by the particle, when it comes to rest.

Solution:

$$a = 12 - 0.01 s^2$$

$$v \cdot \frac{dv}{ds} = 12 - 0.01\, s^2$$

$$v \cdot dv = (12 - 0.01 s^2)\, ds$$

$$\frac{v^2}{2} = 12 s - \frac{0.01}{3} s^3 + C_1$$

$$s = 0; \quad v = 0$$

$$\therefore \quad C_1 = 0$$

$$\therefore \quad \frac{v^2}{2} = 12 s - \frac{0.01}{3} s^3$$

or $$v^2 = 24 s - \frac{0.02}{3} \times s^3$$

(i) When $s = 25$ m

$$v^2 = 24 \times 25 - \frac{0.02}{3} \times 25^3$$

$$v^2 = 600 - 104.17$$

$$v^2 = 495.83$$

$$v = 22.27 \text{ m/sec}$$

(ii) When $v = 0$

$$0 = 24\, s - \frac{0.02\, s^3}{3}$$

$$s \times \left(24 - \frac{0.02}{3} s^2\right) = 0$$

$$s = 0$$

or $$s = \sqrt{\frac{24 \times 3}{0.02}}$$

$$= 60 \text{ m}$$

PROBLEM 11.16: A cage goes down a mine shaft 700 m deep in 48 seconds. For the first quarter of the distance, the speed is being uniformly accelerated and during the last quarter uniformly retarded, the acceleration and retardation being equal. Find the uniform speed of the cage, while traversing the central portion of the shaft.

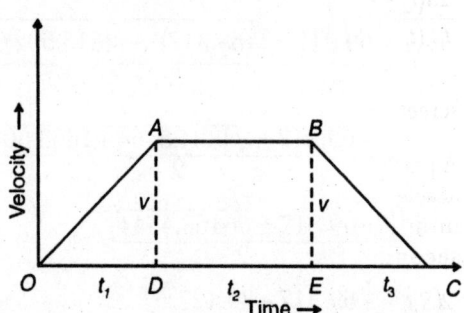

Fig. 11.14 (Illustrative Problem 11.16)

Solution: Total distance covered = 700 m

$$= \text{area of the trapezium } OABC$$

$$= \frac{1}{2}(OC + AB) \times AD$$

$$= \frac{1}{2}(48 + t_2) \times v$$

or $$(48 + t_2)\, v = 700 \times 2 = 1400 \qquad \text{...(1)}$$

Distance covered with uniform speed

$$= 700 - \frac{700}{4} - \frac{700}{4}$$

$$= 350 \text{ m}$$

$$350 \text{ m} = \text{Area of rectangle } ABCD$$

$$350 = t_2 \times v \qquad \qquad ...(2)$$

From (1) & (2)

$$48 \times v + 350 = 1400$$

$$48v = 1400 - 350 = 1050$$

$$v = \frac{1050}{48} = 21.875 \text{ m/sec}$$

PRACTICE PROBLEMS

11.1 If a particle be moving in a straight line and its distance in m from a fixed point O in the line after t seconds is given by $S = 1 + 2t - 3t^2 + 4t^3$, find the distance, velocity and acceleration at the end of 3 seconds. Also find its average velocity during the third second.

11.2 A bullet travelling horizontally pierces in succession three screens placed at equal distances 'S' apart. If the time taken from the first screen to the second is t_1 and from the second to the third be t_2, prove that the retardation (assume to be constant) is $\frac{2S(t_2 - t_1)}{t_1 t_2 (t_1 + t_2)}$ and the velocity at the middle screen is $\frac{S(t_1^2 + t_2^2)}{t_1 t_2 (t_1 + t_2)}$.

11.3 A particle moves in a straight line with uniform acceleration and describes distances S_1 and S_2 m in successive intervals of time t_1 and t_2 seconds. Prove that the acceleration is $\frac{2(S_2 t_1 - S_1 t_2)}{t_1 t_2 (t_1 + t_2)}$.

11.4 The speed of a train is reduced from 60 kmph to 20 kmph whilst it travels a distance of 125 m. If the retardation be uniform, find how much further will it travel before coming to rest.

11.5 Two trains A and B leave the same station in parallel lines. Train A starts with a uniform acceleration of 0.20 m/sec^2 and attains a speed of 30 kmph, when the steam is reduced to keep speed constant. Train B leaves 50 seconds later with uniform acceleration of 0.40 m/sec^2 to attain a maximum speed of 60 kmph. When and where will B overtake A?

11.6 Velocity of a particle is given by $v = 5\cos 2x$ where x is distance from a fixed point. Find its acceleration.

11.7 To a man walking at the rate of 6 kmph, rain appears to fall vertically. If its real velocity is 12 kmph, find its real direction.

11.8 To a cyclist travelling at 16 kmph due east, the wind seems to blow from a direction 60° south of west at 8 kmph. Find the true direction and velocity of wind.

11.9 A ship is sailing due east at 12 kmph, and another ship is sailing due south at 16 kmph. Find the velocity of the second ship relative to the first.

11.10 A ball is dropped from the top of a tower, 60 m high. At the same time another ball is thrown upwards from the foot of the tower with a velocity of 30 m/sec. When and where the two balls cross each other and with what relative velocity.

11.11 A body was thrown vertically downward from a tower and traverses a distance of 50 meters during its 5th second of its flight. Find the initial velocity of the body.

11.12 A body is projected vertically upwards with a velocity of 24 m/sec.

(i) How long will it go?

(ii) How long will it take to return to the point of projection?

(iii) When will its velocity be 4.8 m/sec?

(iv) At what time will it be 28 m above the point of projection?

(v) At what time will it be 28 m below the point of projection?

11.13 A stone is dropped into a well and reaches the bottom with a velocity of 30 m/sec and the sound of the splash on water reaches the top of the well in the $3\frac{5}{26}$ seconds from the time stone starts; find the velocity of the sound.

11.14 The equation of motion of a particle moving in a straight line is given by the expression

$$S = t^3 - 2t^2 + t + 4$$

where S is the displacement in metres and t is the time in second. Determine

(i) the velocity and acceleration at start

(ii) also the velocity and acceleration after 5 seconds

(iii) maximum or minimum velocity of the particle

(iv) time at which velocity is zero.

11.15 An automobile is moving with a velocity of 60 kmph. After seeing an animal on the road brakes are applied and the automobile is stopped in a distance of 12 m. If the retardation produced is proportional to the distance from the point where brakes are applied, find the expression of retardation.

11.16 Two stations A and B are 5 km apart. An electric train starts from rest from the station A and accelerates uniformly for the first 30 seconds during which period it covers 200 m. It then runs with constant speed, until it is finally retarded in the last 100 m. Calculate the maximum speed and the time taken over the journey.

Chapter 12

Projectile Motion

12.1 INTRODUCTION

Any object projected or thrown in a direction other than vertical with an initial velocity is called a "Projectile", and its resulting motion is called "projectile motion". The motion of a projectile has both vertical and horizontal components, and hence traces a *curvilinear* path. The vertical component of the motion is influenced by gravity, the object being accelerated or retarded according as the direction of motion at the instant is downwards or upwards; the horizontal motion, however, remains constant, uninfluenced by acceleration due to gravity, or any other acceleration/retardation. This is so under the assumption that the resistance due to air in the atmosphere is negligible. (Otherwise, the treatment becomes extremely complex). A further assumption is that the range of the projectile is very small compared to the size of the earth, the curvature of which is therefore neglected. The ground is thus assumed to be a horizontal plane.

Common examples of projectile motion are a stone or a ball thrown in the air, and a bullet fired from a gun.

12.2 CERTAIN DEFINITIONS

Definitions of certain terms used with reference to projectile motion are given below.

(i) Point of Projection

The point from which the particle is projected is called the "Point of projection".

(ii) Angle of Projection

The angle between the direction of projection and the horizontal is called the "Angle of Projection".

(iii) Velocity of Projection

The velocity with which the object is projected is called the "velocity of projection".

(iv) Trajectory

The path traced by the projectile is called its "trajectory".

(v) Time of Flight

The time interval during which the object projected travels is known as the 'Time of flight".

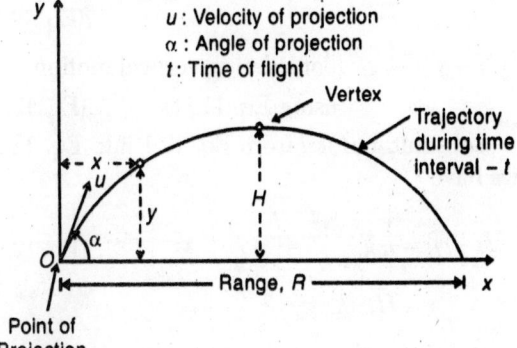

Fig.12.1: Projectile Motion—Certain Basic Definitions

(vi) Range

The horizontal distance through which the projectile travels during its flight or the distance from the point of projection and the point where the projectile strikes the ground is called the "horizontal range" or simply the "range" of the projectile. These are shown marked in Fig. 12.1.

12.3 HORIZONTAL PROJECTION

Let us consider an object thrown horizontally from A with a velocity u as shown in Fig. 12.2, and let it strike the ground at the point B.

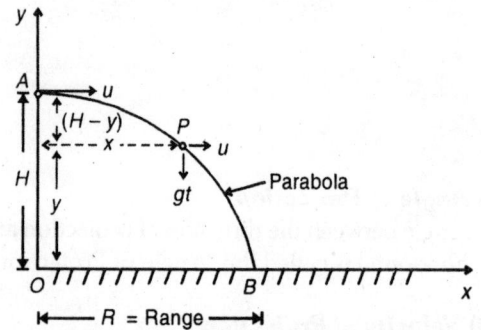

Fig. 12.2: Horizontal Projectile Motion

The motion of the particle at any instant may be resolved into the two components—horizontal and vertical. The horizontal motion is uniform with a constant velocity u, and the vertical motion is with zero initial velocity and an acceleration equal to that due to gravity, g.

Let the object be at a point $P(x, y)$ after a time interval t from the start.

$x = ut$ (considering horizontal motion)

$$\text{...(Eq. 12.1)}$$

$(H - y) = \frac{1}{2} gt^2$ (considering vertical motion and using Eq. 11.11)   ...(Eq. 12.2)

Substituting for t from Eq. 12.1 into Eq. 12.2, we have

$$(H - y) = \frac{1}{2} g \left(\frac{x}{u} \right)^2 = \left(\frac{1}{2} \frac{g}{u^2} \right) \cdot x^2 \quad \text{...(Eq. 12.3)}$$

or $y = H - k \cdot x^2$

where $k = \frac{1}{2} \cdot \frac{g}{u^2}$, a constant for the given problem.

Thus the equation of the path or trajectory of the object is a *parabola* as seen from Eq. 12.3.

The range, R, may be determined as mentioned below.

Let T be the time required for the object to travel from A, the point of projection, to the point B on the ground. From Eqs. 12.1 and 12.2 for point B,

$$R = uT \qquad \text{...(Eq. 12.4)}$$

$$H = \frac{1}{2} gT^2 \qquad \text{...(Eq. 12.5)}$$

Time of flight, $T = \dfrac{R}{u}$ ...(Eq. 12.6)

Substituting for T from Eq. 12.6 into Eq. 12.5,

$$H = \frac{1}{2} g \cdot \left(\frac{R}{u} \right)^2 = \frac{1}{2} \frac{g}{u^2} \cdot R^2$$

or   Range, $R = \sqrt{\dfrac{2h}{g}} \cdot u$ ...(Eq. 12.7)

Comparing Eqs. 12.4 and 12.7

Time of flight,

$$T = \sqrt{\frac{2H}{g}} \qquad \text{...(Eq. 12.8)}$$

12.4 INCLINED PROJECTION ON LEVEL GROUND

Let us consider the motion of an object projected from a point A with a velocity u at an angle of projection α, falling onto level ground as shown in Fig. 12.3.

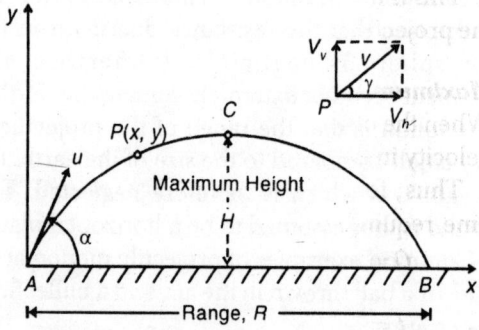

Fig.12.3: Inclined Projection on Level Ground

The horizontal and vertical components of the motion may be considered separately, for convenience in analysis.

Horizontal Component of the Motion

Velocity component = $u\cos \alpha$

This is considered constant during the flight, neglecting air resistance.

Vertical Component of the Motion

Initial velocity = $u\sin\alpha$ (upward)

Acceleration due to gravity = g = 9.81 m/sec^2

(Acts as a retardation for the upward part of the flight from A to C, the highest point reached)

Velocity at C becomes zero and the object starts moving downward with acceleration g.

Trajectory

Let the position of the object be $P(x, y)$ after a time interval t from the start at A.

Considering horizontal motion,

$$x = (u\cos\alpha) \times t$$

or $\quad t = \dfrac{x}{(u\cos\alpha)}$

Considering vertical motion,

$$y = (u\sin\alpha)t - \frac{1}{2}gt^2$$

Substituting $t = \dfrac{x}{u\cos\alpha}$,

$$y = \left[x\tan\alpha - \frac{1}{2}\frac{gx^2}{u^2\cos^2\alpha} \right] \quad \text{...(Eq. 12.9)}$$

This is the equation to the trajectory or path of the projectile, which is obviously a parabola.

Maximum Height

When the object reaches the highest point C, the velocity in the vertical direction is zero.

Thus, for the part of the flight A to C, if the time required is t,

$$0 = u\sin\alpha - gt$$

or $\quad t = \dfrac{u\sin\alpha}{g}$ $\quad\quad$ ...(Eq. 12.10)

This is the time required to reach the highest point.

If H is the maximum height reached,

$$0 - u^2\sin^2\alpha = -2gH$$

or $\quad H = \dfrac{u^2\sin^2\alpha}{2g}$ $\quad\quad$ ...(Eq. 12.11)

Time of Flight

Considering the vertical component of the motion, we have already seen that y is given by

$$y = (u\sin\alpha)t - \frac{1}{2}gt^2$$

$y = 0$ at the end of the flight.

$\therefore\quad 0 = (u\sin\alpha)t - \dfrac{1}{2}gt^2$

$$= t\left(u\sin\alpha - \frac{1}{2}gt \right)$$

This leads to $t = 0$ (for the beginning of the flight)

or $\quad t = \dfrac{2u\sin\alpha}{g}$ (for the end of the flight)

$\therefore$ The time of flight, $T = \dfrac{2u\sin\alpha}{g}$ $\quad$ ...(Eq. 12.12)

This is double the time required to reach the maximum height; in other words, the time for ascent is equal to that for descent.

Range

Range, T, is the horizontal distance covered in the time of flight, T, at the constant velocity $u\cos\alpha$.

$\therefore\quad R = (u\cos\alpha)\cdot T$

$$= u\cos\alpha\frac{2u\sin\alpha}{g}$$

or $\quad R = \dfrac{u^2\cdot\sin 2\alpha}{g}$ $\quad\quad$ ...(Eq. 12.13)

Maximum Range

For R to be maximum, $\sin 2\alpha$ has to be the maximum or unity.

or $\quad 2\alpha = 90°$

or $\quad \alpha = 45°$

For this angle of projection, the maximum range, R_{max}, is given by

$$R_{max} = \frac{u^2}{g} \quad\quad \text{...(Eq. 12.14)}$$

Angle of Projection for a Specified Range

The range, $R = \dfrac{u^2\sin 2\alpha}{g}$

or $\sin 2\alpha = \dfrac{g}{u^2} \cdot R = \sin(180° - 2\alpha)$

There will be two values of 2α—i.e. $2\alpha_1$ and $(180° - 2\alpha_1)$, which satisfy this condition.

The values of α are α_1 and $(90° - \alpha_1)$, their sum being 90°

If $\alpha_1 = 45° + \theta$, say

$\alpha_2 = 45° - \theta$ (45° is the value of α for maximum range)

Thus, the directions of projection for a specified range are equally inclined to the direction of projection for maximum range, as shown in Fig. 12.4.

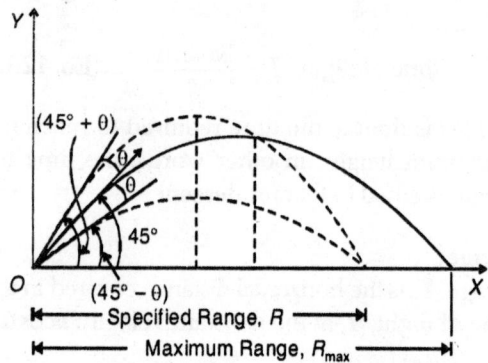

Fig.12.4: Directions of Projection for Specified Range

Since the maximum heights reached are different, although the range is the same, the times of flight would be different for $\alpha_1(= 45° + \theta)$ and $\alpha_2(= 45° - \theta)$.

Velocity at any Point in the Flight
The velocity at any point in the flight can be determined if either the time interval (t) or the height (h) reached is known.

The horizontal component v_h of the velocity is a constant value $u\cos\alpha$; the vertical component v_v of the velocity can be determined either from $v = u\sin\alpha + gt$, or from $v^2 = u^2\sin^2\alpha + 2gh$, respectively depending upon whether t is known or h is known.

Once the components v_h and v_v are known, the resultant velocity v can be determined as

$$v = \sqrt{v_h^2 + v_v^2} \qquad \text{...(Eq. 12.15)}$$

$$\tan\gamma = \dfrac{v_v}{v_h} \qquad \text{...(Eq. 12.16)}$$

γ being the angle made by the direction of the velocity with respect to the horizontal, as marked on the Fig. 12.3.

12.5 INCLINED PROJECTION ON SLOPING GROUND

When the ground is sloping downwards (or upwards), the point of projection and the point at which the object strikes the ground will not be at the same level.

Let the difference in level be H_0 as shown in Fig. 12.5.

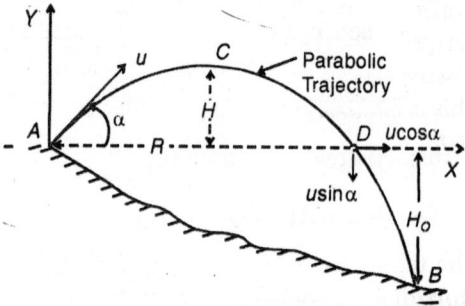

Fig.12.5: Inclined Projection on Sloping Ground

Any one of the approaches given below may be used.

(I) First Approach
For vertical motion

$$y = (u\sin\alpha)t - \dfrac{1}{2}gt^2$$

The time of flight, T, can be solved by putting $y = -H_0$ in the above equation.

Once T is known, the horizontal range R can be got from $R = (u\cos\alpha)t$.

H and the time required to reach the highest point can be obtained from Eqs. 12.11 and 12.10, as usual.

(II) Second Approach
The time of flight, T, can be split up into t_1, the time of travel from A to C and t_2, that from C to B.

$$t_1 = \dfrac{u\sin\alpha}{g} \qquad \text{...(Eq. 12.10)}$$

For the portion of the travel from C to B,

$$y = H + H_0$$

$\therefore (H + H_0) = \frac{1}{2} g t_2^2$, since the velocity at C is zero.

t_2 may be got from this.

The time of flight, $T = t_1 + t_2$

Range, $R = (u \cos \alpha) \cdot T$

(III) Third Approach (Relevant only if B is below A)

If D is at the same level as A on the trajectory, the motion can be spilit up into that form A to D, and the descent from D to B.

For AD:

Horizontal component of velocity is $u \cos \alpha$ at D. Vertical component of velocity at D is $u \sin \alpha$ downwards since D is at the same level as A. This is dealt with in Section 12.4.

For DB.

$$H_0 = (u \sin \alpha) t_2 + \frac{1}{2} g t_2^2$$

The time from D to B, t_2 can be solved.

Horizontal distance from D to B is $(u \cos \alpha) t_2$.

The total values may be obtained by combining the two parts of the motion above.

If B is above the level of A by H_0,

$y = + H_0$ (and not $- H_0$) in the first approach. In the second approach,

$y = (H - H_0)$ for the part of travel from C to B, (The third approach is not applicable for this case). Otherwise, the analysis is the same.

12.6 PROJECTILE ON AN INCLINED PLANE

Let an object be projected up an inclined plane, inclined at B to the horizontal from a point A and let it strike this plane at B (Fig. 12.6).

Let the range AB along the inclined plane be R. Let the velocity of projection be u directed at an angle α with the horizontal.

If AD is the horizontal projection of R,

$$AD = R \cos \beta \text{ and } BD = R \sin \beta$$

The co-ordinates of B with respect to A as origin are $(R \cos \beta, R \sin \beta)$.

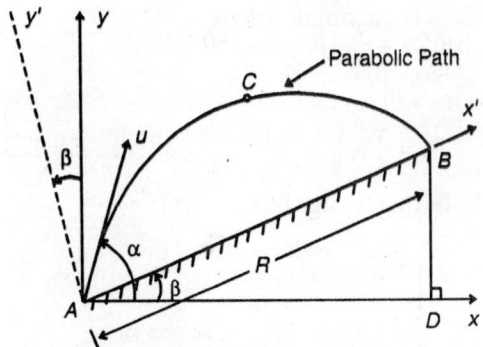

Fig.12.6: Projectile on a Inclined Plane

The equation of the trajectory is given by

$$y = x \tan \alpha - \frac{1}{2} \frac{g}{u^2 \cos^2 \alpha} x^2 \qquad \text{...(Eq. 12.9)}$$

For the point B,

$$R \sin \beta = R \cos \beta \tan \alpha - \frac{g R^2 \cos^2 \beta}{2 u^2 \cos^2 \alpha}$$

By solving for R from this, we have

$$R = \frac{2 u^2 \cos^2 \alpha}{g \cos^2 \beta} (\cos \beta \tan \alpha - \sin \beta) \qquad \text{...(Eq. 12.17)}$$

or

$$R = \frac{2 u^2 \cos \alpha}{g \cos^2 \beta} \sin(\alpha - \beta) \qquad \text{...(Eq. 12.18)}$$

Using $2 \cos A \sin B = \sin(A + B) - \sin(A - B)$

$$R = \frac{u^2}{g \cos^2 \beta} [\sin(2\alpha - \beta) - \sin \beta] \qquad \text{...(Eq. 12.19)}$$

Eq. 12.17 may also be written in the form

$$R = \frac{2 u^2 \cos^2 \alpha}{g \cos \beta} (\tan \alpha - \tan \beta) \qquad \text{...(Eq. 12.20)}$$

Let T be the time of flight.

$$\therefore \quad AD = R \cos B = (u \cos \alpha) \cdot T$$

or

$$T = \frac{R \cos \beta}{u \cos \alpha}$$

$$= \frac{2 u^2 \cos \alpha}{g \cos^2 \beta} \times \sin(\alpha - \beta) \frac{\cos \beta}{u \cos \alpha}$$

or

$$T = \frac{2 u \sin(\alpha - \beta)}{g \cos \beta} \qquad \text{...(Eq. 12.21)}$$

From Eq. 12.19, the range R is a maximum when $\sin(2\alpha - \beta)$ is a maximum (for given values of β and u).

Or R is maximum when

$$(2\alpha - \beta) = \frac{\pi}{2}$$

or $\alpha = \frac{\pi}{4} + \frac{\beta}{2}$ ...(Eq. 12.22)

Referring to Fig. 12.7.

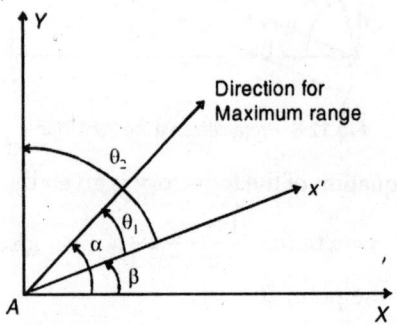

Fig.12.7: Direction for Maximum Range

$\theta_1 = \alpha - \beta$

$\quad = \frac{\pi}{4} + \frac{\beta}{2} - \beta$

$\quad = \frac{\pi}{4} - \frac{\beta}{2}$

$\theta_2 = \frac{\pi}{2} - \beta$

or $\theta_2 = 2\theta_1$

That is, the range on the inclined plane is a maximum, when the direction of projection bisects the angle between the inclined plane and the vertical.

Note: If the projection is down the inclined plane, the entire analysis is applicable if $(-\beta)$ is substituted for β.

Alternative Procedure

With reference to Fig. 12.6, the motion of the projectile may also be considered as the vector sum of its motions along the inclined plane and normal to it.

Motion Normal to the Inclined Plane

Initial velocity $= u\sin(\alpha - \beta)$

Acceleration $= -g\cos\beta$

Let the time of flight be T.

From $s = ut = +\frac{1}{2}at^2$

$$0 = u\sin(\alpha - \beta) - \frac{1}{2}(g\cos\beta)T^2$$

Time of flight,

$T = \dfrac{2u\sin(\alpha - \beta)}{g\cos\beta}$, the same as (Eq. 12.21)

derived earlier.

Motion along the Inclined Plane

Initial velocity $= u\cos(\alpha - \beta)$

Acceleration $= -g\sin\beta$

Using again $s = ut + \frac{1}{2}at^2$

$$R = \{u\cos(\alpha - \beta)\}T - \frac{1}{2} \cdot g\sin\beta \cdot T^2$$

Substituting for T from Eq. 12.23,

$$R = u\cos(\alpha - \beta)\left\{\frac{2u\sin(\alpha - \beta)}{g - \cos\beta}\right.$$

$$\left. -\frac{1}{2}g\sin\beta\frac{2u\sin(\alpha - \beta)}{g\cos\beta}\right\}^2$$

Simplifying and using trigonometrical relationships, we get

$R = \dfrac{2u^2\cos\alpha}{g\cos^2\beta}\sin(\alpha - \beta)$, the same as (Eq. 12.18)

derived earlier.

Equations 12.19 & 12.20 automatically follow.

The maximum range, R_{max}, is got by substituting unity for $\sin(2\alpha - \beta)$ in Eq. 12.19,

$$\therefore \quad R_{max} = \frac{u^2}{g\cos^2\beta}(1 - \sin\beta)$$

or $R_{max} = \dfrac{u^2}{g(1 + \sin\beta)}$ ...(Eq. 12.23)

Note: For projection down the inclined plane, the entire analysis is applicable if $(-\beta)$ is substituted wherever β appears.

All these ideas will be easily understood by going through the relevant problems worked out under the head Illustrative Problems.

ILLUSTRATIVE PROBLEMS

PROBLEM 12.1: An aircraft moving horizontally at a speed of 600 kmph at an altitude of 1500 m towards a target on the ground releases a bomb which hits the target. Find (*i*) time required for the bomb to reach the target on the ground, (*ii*) the horizontal distance of the aircraft from the target when it released the bomb (*iii*) the velocity with which the bomb hits the target, and (*iv*) the direction in which the bomb hits the target.

Solution: Speed of the aircraft in horizontal direction

$$= \frac{600 \times 1000}{60 \times 60} = 166.67 \text{ m/sec}$$

Height, $H = 1500$ m

(*i*) Let $T =$ time required by the bomb to hit the target

Using equation, $H = \frac{1}{2} g T^2$

$$1500 = \frac{1}{2} \times 9.81 \times T^2$$

$$= 17.487 \text{ sec}$$

(*ii*) The horizontal distance of the aircraft from the target when it released the bomb is given by equation

Range, $R = u \times T$

$$= 166.67 \times 17.487$$

$$= 2915 \text{ m}$$

$$= 2.915 \text{ km}$$

(*iii*) Velocity of bomb hitting the target

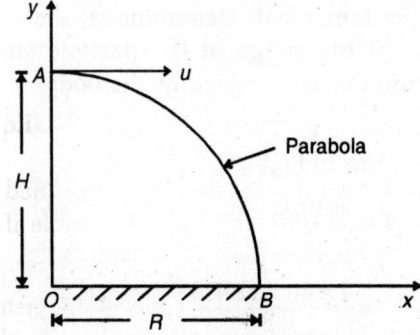

The vertical downward component of velocity at B is given by equation as

$$v = g \times T = 9.81 \times 17.487$$

$$= 171.55 \text{ m/sec}$$

The horizontal component of velocity is constant throughout the motion and is equal to 166.67 m/sec. Hence at target, the horizontal velocity,

$$u = 166.67 \text{ m/sec}$$

$\therefore$ Resultant velocity $= \sqrt{u^2 + v^2}$

$$= \sqrt{166.67^2 + 171.55^2}$$

$$= \sqrt{27778.888 + 29429.402}$$

$$= \sqrt{57208.29}$$

$$= 239.18 \text{ m/sec}$$

(*iv*) The direction in which the bomb hits the target

Let $\gamma =$ Inclination of the resultant velocity with the horizontal at target

$$\tan \gamma = \frac{v}{u}$$

$$= \frac{171.55}{166.67} = 1.0293$$

$$\gamma = \tan^{-1}(1.0293)$$

$$= 45.8272°$$

$$= 45°50'$$

PROBLEM 12.2: A motor cyclist wants to jump over a ditch as shown in Fig. 12.9. Find the necessary minimum velocity u at A of the motor cycle. Also find the direction and magnitude of the velocity of the motor cycle when it clears the ditch.

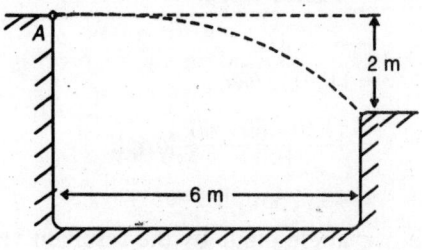

Fig.12.9 (Illustrative Problem 12.2)

Fig.12.8 (Illustrative Problem 12.1)

Solution:

$$H = 2 \text{ m and } R = 6 \text{ m}$$

Let T be the time of flight and u be the minimum horizontal velocity required.

Considering the vertical motion,

$$H = \frac{1}{2} g T^2$$

$$2 = \frac{1}{2} \times 9.81 \times T^2$$

$$T = \sqrt{\frac{2 \times 2}{9.81}}$$

$$= 0.63855 \text{ sec}$$

Considering the horizontal motion of uniform velocity, we get

$$6 = u \times 0.63855$$

$$\therefore \quad u = \frac{6}{0.63855}$$

$$= 9.40 \text{ m/sec}$$

$$= \frac{9.40 \times 60 \times 60}{1000}$$

$$= 33.84 \text{ kmph}$$

Horizontal component of velocity is constant throughout the motion.

Horizontal component of velocity at B

$$= u = 9.40 \text{ m/sec}$$

The vertical downward component of velocity at B is given by equation as

$$v = g \times T = 9.81 \times 0.63855$$

$$= 6.264 \text{ m/sec}$$

Resultant velocity at $B = \sqrt{u^2 + v^2}$

$$= \sqrt{9.40^2 + 6.264^2}$$

$$= \sqrt{88.36 + 39.24}$$

$$= \sqrt{127.6}$$

$$= 11.30 \text{ m/sec}$$

$$= \frac{11.30 \times 60 \times 60}{1000} \text{ kmph}$$

$$= 40.67 \text{ kmph}$$

Let $\gamma =$ angle made by the resultant velocity at B with the horizontal direction

$$\tan \gamma = \frac{v}{u} = \frac{6.264}{9.40}$$

$$= 0.6664$$

$$\gamma = \tan^{-1}(0.6664)$$

$$= 33.68°$$

$$= 33°41'$$

PROBLEM 12.3: A body is projected at such an angle that the horizontal range is $4\sqrt{3}$ times the maximum height. Find the angle of projection.

Solution:

Maximum height of the projectile

$$= \frac{u^2 \sin^2 \alpha}{2g}$$

Horizontal Range

$$= \frac{u^2 \sin 2\alpha}{g}$$

Given:

Horizontal Range $= 4\sqrt{3} \times$ Maximum height reached

$$\text{i.e. } \frac{u^2 \sin 2\alpha}{g} = 4\sqrt{3} \times \frac{u^2 \sin^2 \alpha}{2g}$$

$$\sin 2\alpha = 2\sqrt{3} \sin^2 \alpha$$

$$2 \sin \alpha \cos \alpha = 2\sqrt{3} \sin^2 \alpha$$

$$\tan \alpha = \frac{1}{\sqrt{3}}$$

$$\alpha = \tan^{-1}(1/\sqrt{3}) = 30°$$

PROBLEM 12.4: A particle is projected upwards with a velocity of 40 m/sec at an angle of 45° with the horizontal. Determine (*i*) the time of flight (*ii*) the range of the particle (*iii*) the maximum height attained by the body.

Solution:

(*i*) Time of flight,

$$t = \frac{2u \sin \alpha}{g}$$

$$= \frac{2 \times 40 \times \sin 45°}{9.81}$$

$$= 5.766 \text{ sec}$$

(*ii*) The range of the particle, R

$$= \frac{u^2 \sin 2\alpha}{g}$$

$$= \frac{40^2 \times \sin(2 \times 45°)}{9.81}$$

$$= 163.10 \text{ m}$$

(*iii*) Maximum height attained, h

$$= \frac{u^2 \sin^2 \alpha}{2g}$$

$$= \frac{40^2 \sin^2 45°}{2 \times 9.81}$$

$$= 40.77 \text{ m}$$

PROBLEM 12.5: A particle is projected with a velocity of 30 m/sec in air at an angle 'α' with the horizontal. The x and y coordinates of a point lying on the trajectory of the particle with respect to point of projection are 20 m and 10 m respectively. Find the angle of projection of the particle.

Solution:

$$y = x \tan\alpha - \frac{gx^2}{2u^2 \cos^2\alpha}$$

$$10 = 20 \tan\alpha - \frac{9.81 \times 20^2}{2 \times 30^2} \sec^2\alpha$$

$$10 = 20 \tan\alpha - 2.18(1 + \tan^2\alpha)$$

$$10 = 20 \tan\alpha - 2.18 - 2.18 \tan^2\alpha$$

$$2.18 \tan^2\alpha - 20 \tan\alpha + 12.18 = 0$$

$$\tan^2\alpha - 9.1743 \tan\alpha + 5.5872 = 0$$

$$\tan\alpha = \frac{9.1743 \pm \sqrt{9.1743^2 - 4 \times 5.5872 \times 1}}{2}$$

$$= \frac{9.1743 \pm 7.8625}{2}$$

$$= 8.5184, \text{ and } 0.6559$$

$$= 83°18'16'' \text{ and } 33°15'39''$$

PROBLEM 12.6: A projectile is aimed at a target which lies in the horizontal plane through the point of projection. It falls 'S_1' metres short of the target when the angle of projection is $\alpha°$ and goes 'S_2' metres too far off when the angle of projection is $\beta°$ (Fig. 12.10). Show that the angle of projection $\gamma°$ to hit the target is

$$\frac{1}{2}\sin^{-1}\left(\frac{S_1 \sin 2\beta + S_2 \sin 2\alpha}{S_1 + S_2}\right).$$

Solution: Let S be the distance of the target from the point of projection and u be the velocity of projection.

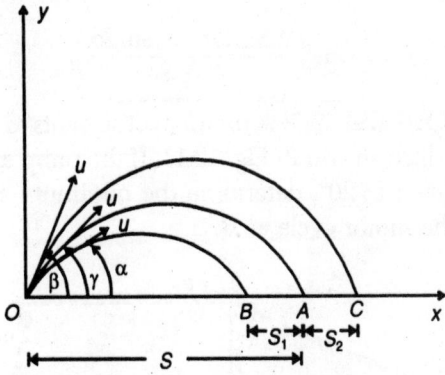

Fig. 12.10 (Illustrative Problem 12.6)

$$S - S_1 = \frac{u^2 \sin 2\alpha}{g} \qquad \ldots(i)$$

$$S + S_2 = \frac{u^2 \sin 2\beta}{g} \qquad \ldots(ii)$$

$$S = \frac{u^2 \sin 2\gamma}{g} \qquad \ldots(iii)$$

(*ii*) – (*i*) gives

$$S_1 + S_2 = \frac{u^2}{g}(\sin 2\beta - \sin 2\alpha) \qquad \ldots(iv)$$

(*ii*) – (*iii*) gives

$$S_2 = \frac{u^2}{g} = (\sin 2\beta - \sin 2\gamma) \qquad \ldots(v)$$

From (*iv*) $\dfrac{u^2}{g} = \dfrac{S_1 + S_2}{\sin 2\beta - \sin 2\alpha} \qquad \ldots(vi)$

From (*v*) $\dfrac{u^2}{g} = \dfrac{S_2}{\sin 2\beta - \sin 2\gamma} \qquad \ldots(vii)$

Equating (*vi*) & (*vii*)

$$\frac{S_1 + S_2}{\sin 2\beta - \sin 2\alpha} = \frac{S_2}{\sin 2\beta - \sin 2\gamma}$$

$(S_1 + S_2)(\sin 2\beta - \sin 2\gamma) = S_2(\sin 2\beta - \sin 2\alpha)$

$(S_1 + S_2)\sin 2\beta - (S_1 + S_2)\sin 2\gamma = S_2\sin 2\beta - S_2\sin 2\alpha$

$(S_1 + S_2 - S_2)\sin 2\beta - (S_1 + S_2)\sin 2\gamma = -S_2\sin 2\alpha$

or $\quad (S_1 + S_2)\sin 2\gamma = S_1\sin 2\beta + S_2\sin 2\alpha$

$\sin 2\gamma = \dfrac{S_1 \sin 2\beta + S_2 \sin 2\alpha}{S_1 + S_2}$

or $\quad 2\gamma = \sin^{-1}\left(\dfrac{S_1 \sin 2\beta + S_2 \sin 2\alpha}{S_1 + S_2}\right)$

or $\quad \gamma = \dfrac{1}{2}\sin^{-1}\left(\dfrac{S_1 \sin 2\beta + S_2 \sin 2\alpha}{S_1 + S_2}\right)$

PROBLEM 12.7: A motor cyclist wants to clear the ditch shown in Fig. 12.11. If the ramp at A is inclined at 20°, determine the minimum speed of the motor cycle at B.

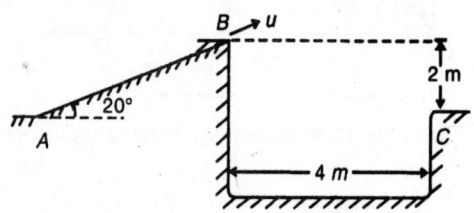

Fig.12.11 (Illustrative Problem 12.7)

Solution:

Angle of ramp, $\alpha = 20°$

Horizontal distance between B and $C = 4$ m

Vertical distance between B and $C = 2$ m

Let $u = $ minimum speed of motor cycle in the direction of AB at B.

The path travelled by the motor cycle from B to C will be parabolic.

The equation of this path is given by

$$y = x\tan\alpha - \dfrac{gx^2}{2u^2\cos^2\alpha}$$

where x and y are the co-ordinates of the point C with respect to point B.

$x = $ Horizontal distance $= 4$ m

$y = $ Vertical distance $= -2$ m

(−ve sign is taken as point 'C' is in the downward direction with respect to B).

Substituting these values in the above equation

$$-2 = 4\tan 20° - \dfrac{9.81 \times 4^2}{2u^2\cos^2 20°}$$

or $\quad \dfrac{9.81 \times 4^2}{2u^2\cos^2 20°} = 2 + 1.4559 = 3.4559$

$$u^2 = \dfrac{9.81 \times 16}{2 \times 0.883 \times 3.4559} = 25.718$$

$$u = \sqrt{25.718} = 5.071 \text{ m/sec}$$

PROBLEM 12.8: A cricket ball thrown from a height of 1.8 m at an angle of 30° with the horizontal with a speed of 18 m/sec is caught by another field man at a height of 0.60 m from the ground. How far apart are the two men?

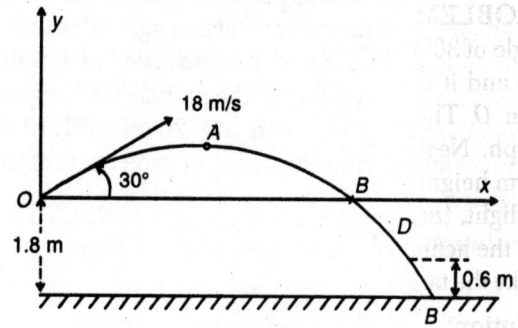

Fig.12.12 (Illustrative Problem 12.8)

Solution:

$$Y = (u\sin\alpha)t - \dfrac{1}{2}gt^2$$

$$Y = -(1.80 - 0.60)$$

$$= -1.29 \text{ m}$$

$$u = 18 \text{ m/sec}$$

$$\alpha = 30°$$

$\therefore \ -1.20 = (18\sin 30°)t - \dfrac{1}{2} \times 9.81 \times t^2$

$$-1.20 = 9t - 4.905t^2$$

$$4.905t^2 - 9t - 1.20 = 0$$

$$t = \dfrac{9 \pm \sqrt{(-9)^2 - 4(4.905)(-1.20)}}{2 \times 4.905}$$

$$= \dfrac{9 \pm \sqrt{81 + 23.544}}{9.81}$$

$$= \frac{9 \pm \sqrt{104.544}}{9.81}$$

$$= \frac{9 \pm 10.2247}{9.81}$$

$$= \frac{9 + 10.2247}{9.81}$$

$$= \frac{19.2247}{9.81}$$

$$= 1.96 \text{ sec}$$

(The negative value is ignored.)

$$\text{Horizontal range} = (u\cos\alpha) \times t$$
$$= (18\cos 30°) \times 1.96$$
$$= 30.55 \text{ metres}$$

The two men are 30.55 metres apart.

PROBLEM 12.9: A bullet is fired upward at an angle of 30° to the horizontal from a point *O* on a hill and it strikes a target which is 100 m lower than *O*. The initial velocity of the bullet is 360 kmph. Neglecting air resistance, find (*i*) maximum height reached by the bullet, (*ii*) total time of flight, (*iii*) horizontal range of the bullet, and (*iv*) the actual velocity with which the bullet will strike the target.

Solution:

(*i*) Maximum height reached by the bullet

$$H = \frac{u^2 \sin^2 \alpha}{2g}$$

$$u = 360 \text{ kmph}$$

$$= \frac{360 \times 1000}{60 \times 60}$$

$$= 100 \text{ m/sec}$$

$$\alpha = 30° \ \& \ g = 9.81 \text{ m/sec}^2$$

$$\therefore \quad H = \frac{100^2 \sin^2 30°}{2 \times 9.81} = 127.42 \text{ m above point } O$$

$$127.42 + 100 = 227.42 \text{ m above the target.}$$

(*ii*) Total time of flight

Method 1

Let t_1 be the time taken by the bullet to reach maximum height

$$t_1 = \frac{u \sin \alpha}{g} = \frac{100 \times \sin 30°}{9.81}$$

$$t_1 = 5.097 \text{ sec}$$

Let t_2 be the time taken by the bullet to reach the target from the maximum height.

The distance traversed during downward motion

$$= 127.42 + 100 = 227.42 \text{ m}$$

Initial velocity $= 0$

$$227.42 = 0 + \frac{1}{2} \times 9.81 \times t_2$$

$$t_2 = \sqrt{\frac{227.42 \times 2}{9.81}} = 6.809 \text{ sec}$$

∴ Total time required for the flight of the bullet,

$$t = t_1 + t_1$$
$$= 5.097 + 6.809$$
$$= 11.906 \text{ sec}$$

Method II

Let time required to travel by the bullet from *O* to *B* be t_1

$$t_1 = \frac{2u \sin \alpha}{g} = \frac{2 \times 100 \times \sin 30°}{9.81}$$

$$= 10.194 \text{ sec}$$

Let t_2 be the time required to travel by the bullet from *B* to *C* (target)

Initial velocity at $B = 100\sin 30°$

$$= 50 \text{ m/sec}$$

$$g = 9.91 \text{ m/sec}^2$$

Distance travelled $= 100$ m

$$\therefore \ 100 = 50t^2 + \frac{1}{2} \times 9.81 \times t_2^2$$

$$t_2^2 + 10.1937 \, t_2 - 20.3874 = 0$$

$$t_2 = \frac{-10.1937 \pm \sqrt{10.1937^2 + 4 \times 20.3874 \times 1}}{2}$$

$$= \frac{-10.1937 \pm 13.6184}{2}$$

$$= 1.712 \text{ sec}$$

∴ Time of flight, *t*

$$= t_1 + t_2$$
$$= 10.194 + 1.712$$
$$= 11.906 \text{ sec}$$

Method III

Let t be the total time of flight of the bullet

$$Y = (u \sin \alpha)t - \frac{1}{2}gt^2$$

$$-100 = (100 \sin 30°)t - \frac{9.81}{2}t^2$$

$$t^2 - 10.1937t - 20.3874 = 0$$

$$i = \frac{10.1937 \pm \sqrt{(-10.1937)^2 + 4 \times 20.3874}}{2}$$

$$= \frac{10.1937 \pm 13.6184}{2}$$

$$= 11.906 \text{ sec (ignoring negative value.)}$$

(*iii*) Horizontal range of the bullet

$$= (u \cos \alpha)t.$$
$$= (100 \cos 30°) \times 11.906$$
$$= 100 \times 0.866 \times 11.906$$
$$= 1031.06 \text{ m}$$
$$= 1.031 \text{ km}$$

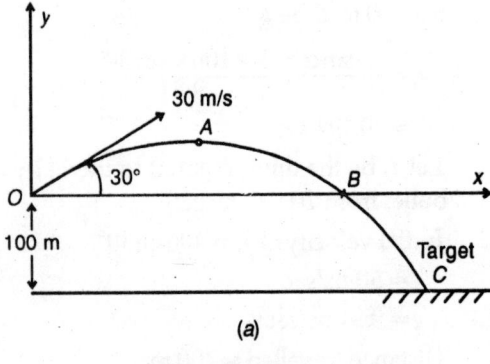

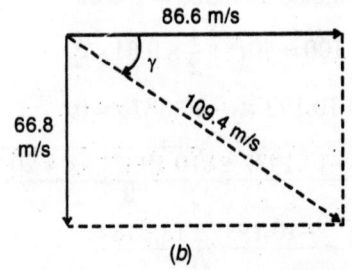

(*b*)

Fig.12.13 (Illustrative Problem 12.9)

(*iv*) Actual velocity of the bullet just before striking the target

Horizontal component of velocity

$$= 100 \cos \alpha$$

$$= 100 \cos 30°$$

$$= 86.6 \text{ m/sec}$$

Vertical component of velocity

$$= 0 + 9.81 \times 6.809$$

$$= 66.796 \text{ m/sec}$$

Actual velocity with which the bullet will strike the target

$$= \sqrt{(86.6)^2 + (66.796)^2}$$

$$= 109.37 \text{ m/sec}$$

$$\gamma = \tan^{-1}\left(\frac{66.796}{86.600}\right)$$

$$\gamma = \tan^{-1}(0.7713)$$

$$= 37°39' \text{ as shown in Fig. 12.13 } (b).$$

PROBLEM 12.10: Two guns are pointed at each other, one upward at an angle of 30°, and the other at the same angle of depression the muzzles being 25 m apart. If the guns are shot with velocities of 1080 kmph upwards and 900 kmph downwards respectively, find when and where they will meet?

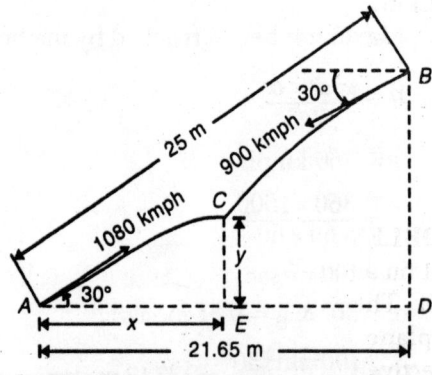

Fig.12.14 (Illustrative Problem 12.10)

Solution: Let the two shots meet at C as shown in Fig. 12.14.

Let x = Horizontal distance between A & C

y = Vertical distance between A & C

Horizontal distance between A & B

$$= 25 \cos 30°$$

$$= 21.65 \text{ m}$$

Distance covered by the shot A in t seconds

$$x = \left(\frac{1080 \times 1000}{60 \times 60}\right)(\cos 30°)(t)$$

$$x = 300 \times 0.866 \times t$$

$$x = 259.8t \qquad \ldots(i)$$

Similarly distance covered by the shot B in t seconds

$$(21.65 - x) = \left(\frac{900 \times 1000}{60 \times 60}\right)(\cos 30°)(t)$$

$$21.65 - x = 250 \times 0.866 \times t$$

$$21.65 - x = 216.5t \qquad \ldots(ii)$$

Adding (i) & (ii)

$$21.65 = t(259.8 + 216.5) = 476.3t$$

$$t = \frac{21.65}{476.3} = 0.0455 \text{ sec}$$

$$\therefore \quad x = 259.8 \times 0.0455$$

$$= 11.821 \text{ m}$$

$$y = (\text{Vertical component of shot } A)\, t - \frac{1}{2}gt^2$$

Vertical component of shot $A = 300\sin 30°$

$$= 300 \times 0.5$$

$$= 150 \text{ m/sec}$$

$$\therefore \quad Y = 150 \times 0.0455 - \frac{1}{2} \times 9.81 \times 0.0455^2$$

$$= 6.825 - 0.010$$

$$= 6.815 \text{ m}$$

PROBLEM 12.11: A particle is projected from a point on an inclined plane with a velocity of 30 m/sec. The angle of projection and the angle of the plane are 50° and 20° to the horizontal respectively. If the motion of the particle is up the plane, determine:

(i) Time of flight
(ii) Range of projectile up the plane
(iii) Angle of projection for maximum range up the plane
(iv) Maximum range up the plane.

Solution:

$$u = 30 \text{ m/sec}$$

$$\alpha = 50°$$

$$\beta = 20°$$

(i) Time of flight

$$T = \frac{2u\sin(\alpha - \beta)}{g\cos\beta}$$

$$= \frac{2 \times 30\sin(50° - 20°)}{9.81 \times \cos 20°}$$

$$= \frac{2 \times 30 \times 0.5}{9.81 \times 0.9397}$$

$$= 3.254 \text{ sec}$$

(ii) Range of projectile up the plane

$$R = \frac{u^2}{g\cos^2\beta}\left[\sin(2\alpha - \beta) - \sin\beta\right]$$

$$= \frac{30^2}{9.81\cos^2 20°}\left[\sin(2 \times 50° - 20°) - \sin 20°\right]$$

$$= 103.90\left[\sin 80° - \sin 20°\right]$$

$$= 103.90\,(0.9848 - 0.3420)$$

$$= 103.90 \times 0.6428$$

$$= 66.79 \text{ m}$$

(iii) Let $\alpha^* = $ Angle of projection for maximum range up the plane

$$2\alpha^* - \beta = 90°$$

$$2\alpha^* = 90 + B = 90° + 20° = 110°$$

$$\therefore \quad \alpha^* = 55°$$

(iv) Maximum range up the plane

$$R_{max} = \frac{u^2}{g(1 + \sin\beta)} = \frac{30^2}{9.81(1 + \sin 20°)}$$

$$= 68.36 \text{ m}$$

PROBLEM 12.12: Show that the greatest range up an inclined plane through the point of projection is equal to the distance through which a particle could fall freely during the time of flight.

Solution: Let u be the velocity of projection.

Let the plane be inclined at an angle β to the horizontal.

For maximum range up the plane, the angle of projection

$$\alpha = \frac{\pi}{4} + \frac{\beta}{2}$$

$T =$ Time of flight up the plane

$$= \frac{2u \sin(\alpha - \beta)}{g \cos \beta}$$

$$= \frac{2u \sin\left(\dfrac{\pi}{4} + \dfrac{\beta}{2} - \beta\right)}{g \cos \beta}$$

$$= \frac{2u \sin\left(\dfrac{\pi}{4} + \dfrac{\beta}{2}\right)}{g \cos \beta}$$

Distance through which the particle would fall freely during the time of flight up the plane

$$= \frac{1}{2} g T^2$$

$$= \frac{1}{2} g \frac{4u^2 \sin^2\left(\dfrac{\pi}{4} - \dfrac{\beta}{2}\right)}{g^2 \cos^2 \beta}$$

$$= \frac{2u^2}{g \cos^2 \beta} \left[\sin \frac{\pi}{4} \cos \frac{\beta}{2} - \cos \frac{\pi}{4} \sin \frac{\beta}{2} \right]^2$$

$$= \frac{2u^2}{g \cos^2 \beta} \left[\frac{1}{\sqrt{2}} \left(\cos \frac{\beta}{2} - \sin \frac{\beta}{2} \right) \right]^2$$

$$= \frac{u^2}{g \cos^2 \beta} \left[\cos^2 \frac{\beta}{2} + \sin^2 \frac{\beta}{2} - 2 \cos \frac{\beta}{2} \sin \frac{\beta}{2} \right]$$

$$= \frac{u^2}{g \cos^2 \beta} [1 - \sin \beta]$$

$$= \frac{u^2 (1 - \sin \beta)}{g(1 - \sin^2 \beta)} = \frac{u^2}{g(1 + \sin \beta)}$$

= maximum range up the plane.

PROBLEM 12.13: What velocity must be given to a projectile at the surface of earth so that it may rise to an infinite height? The radius of the earth may be assumed to be 'R'.

Solution:

Let 'm' be the mass of the projectile

$M =$ Mass of earth

$R =$ Radius of earth

The force with which the body is pulled towards the centre of the earth is equal to $\dfrac{GMm}{R^2}$.

where $G =$ constant of proportionality known as universal constant of gravitation or gravitational constant.

Work done by the body against the gravitational force, W

= force × distance

$$= \frac{GMm}{R^2} \times R = \frac{GMm}{R}$$

which is stored as potential energy in the body on the surface of the earth. When the body is projected with escape velocity this potential energy is converted into kinetic energy

$$\therefore \quad \frac{1}{2} m v_{esc}^2 = \frac{GMm}{R}$$

$$v_{esc}^2 = \frac{2GM}{R}$$

But $\quad g = \dfrac{GM}{R^2}$

$$\therefore \quad v_{esc} = \sqrt{\frac{2 \times gR^2 M}{MR}} = \sqrt{2 \times g \times R}$$

$$= \sqrt{\frac{2 \times 9.81 \times 6.38 \times 10^6}{1000 \times 1000}}$$

$$= 11.2 \text{ km/sec}$$

($\because g = 9.81 \text{ m/sec}^2$ & $R = 6.38 \times 10^6 \text{ m}$).

Escape velocity is independent of mass of the of the projected body. As the escape velocity depends on the constants of g and R, it changes from planet to planet.

PRACTICE PROBLEMS

12.1 An aircarft moving horizontally at a speed of 720 kmph at a height of 1500 m towards a target on the ground, releases a bomb which hits the target. Find (*i*) time required for the bomb to reach the target on the ground and (*ii*) the horizontal distance of the aircraft from the target when it released the bomb.

12.2 A person wants to ju np over a ditch as shown in Fig. 12.15. Find the minimum velocity with which he should jump.

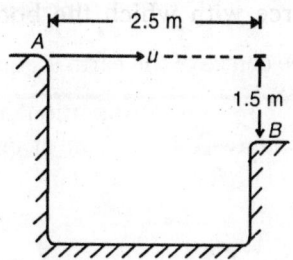

Fig. 12.15 (Practice Problem 12.2)

12.3   A body is projected at such an angle that the horizontal range is 3 times the maximum height. Find the angle of projection.

12.4   If a particle is projected inside a horizontal tunnel which is 6 m high with a velocity of 50 m/sec, find the angle of projection and the greatest possible range.

12.5   A boy throws a ball so that it may just clear a wall 3 m high. The boy is at a distance of 4 m from the wall. The ball was found to hit the ground at a distance of 3 m from the wall (Fig. 12.16). Find the least velocity with which the ball can be thrown.

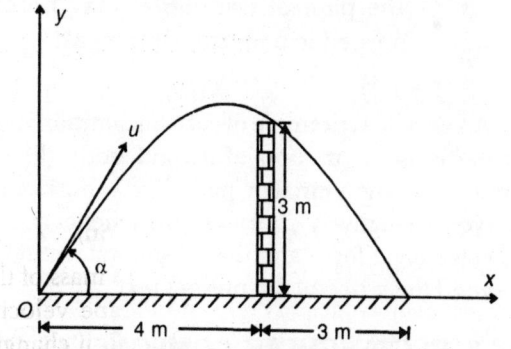

Fig. 12.16 (Practice Problem 12.5)

12.6   A projectile is aimed at a target on the horizontal plane and falls 20 m short when the angle of projection is 20°, while it overshoots by 10 m when the angle is 50°. Find the angle of projection to hit the target.

12.7   A ball is thrown from the top of a hillock 60 metres high with a velocity of 24 m/sec at an elevation of 30° above the horizontal. Find the horizontal distance from the foot of the hillock to the point, where it hits the ground.

12.8   A bullet is fired from the edge of a 120 m high cliff with an initial velocity of 540 kmph at an angle of elevation of 30° with the horizontal. Neglecting air resistance, find (*i*) maximum height reached by the bullet and (*ii*) horizontal range of the bullet.

12.9   A soldier fires a bullet with a velocity of 30 m/sec at an angle α upwards from the horizontal from his position on a hill to strike a target which is 120 m away and 60 m below his position. Find the angle of projection α. Find also the velocity with which the bullet strikes the object.

12.10 A particle has a rise of 2 in 5. A shot is projected with a velocity of 100 m/sec at an elevation of 35°. Find the range of the plane, if (*a*) the shot is fired up the plane, (*b*) the shot is fired down the plane.

12.11 If the greatest range down an inclined plane be three times the greatest range up the plane, show that the plane is inclined at 30° to the horizon.

12.12 A particle is projected from a point on an inclined plane with a velocity of 40 m/sec. The angle of projection and the angle of plane are 55° and 20° to the horizontal respectively. Show that the range up the plane is the maximum for the given plane. Find this range and time of flight.

Chapter 13

Curvilinear Translation

13.1 INTRODUCTION

When a particle in motion traces a curved path, it is said to have a *curvilinear translation;* if the path lies entirely in a single plane, it is *plane* curvilinear translation.

In the case of a body undergoing curvilinear translation, all the particles of the body describe parallel curvilinear paths during planar curvilinear translation (Fig. 13.1)

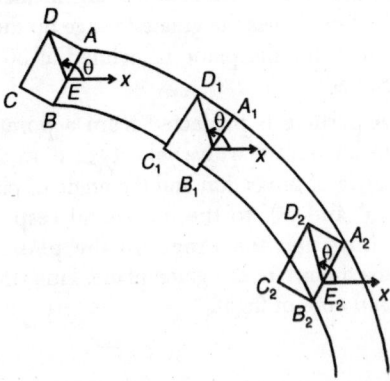

Fig. 13.1: Curvilinear Translation of a Rigid Body

At any given instant all the particles have the same velocity and acceleration, which may change with time. The angle made by any line with a reference axis remains constant, which indicates total absence of rotation.

The motion of a coupling-rod of a locomotive moving on a straight level railway track is an example of planar curvilinear translation.

Curvilinear translation may be viewed as the vector sum of two orthogonal components of rectilinear translations:

If the motion is expressed as:

$s = f(t)$, the components are

$x = f_1(t)$ and $y = f_2(t)$, and the equation of the path of translation may be expressed in a functional form as:

$y = f(x)$

An excellent example of curvilinear translation is the uniform motion of a rigid body (like a vehicle) along a circular path (like a horizontal curve on a highway); however, the kinetics aspect of centrifugal force and that of superelevation is beyond the scope of the present text.

13.2 RECTANGULAR COMPONENTS OF MOTION

Since the direction of the velocity of a particle changes continuously in curvilinear motion, it is convenient to deal with the rectangular components of the velocity (Fig. 13.2).

Velocity v is the vector sum of its rectangular components v_x and v_y, parallel to the x- and y-directions, respectively.

$$v = \sqrt{v_x^2 + v_y^2} \qquad \text{... (Eq. 13.1)}$$

$$\theta = \tan^{-1}\left(\frac{v_y}{v_x}\right) \qquad \text{...(Eq. 13.2)}$$

(θ is the angle made by the direction of the velocity with the *x*-axis).

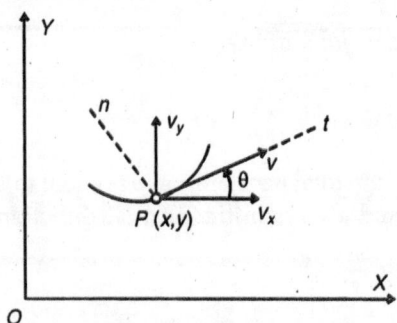

Fig. 13.2: Rectangular Components of Velocity

The velocity is directed along the tangent to the path at the point concerned; it has obviously, no component in the normal direction at the point.

A similar treatment is possible with respect to the rectangular components of acceleration.

$$a = \sqrt{a_x^2 + a_y^2} \qquad \text{...(Eq. 13.3)}$$

$$\theta = \tan^{-1}\left(\frac{a_y}{a_x}\right) \qquad \text{...(Eq. 13.4)}$$

13.3 NORMAL AND TANGENTIAL COMPONENTS OF MOTION

It is often more convenient to obtain the components in the direction of the tangent and normal to the path at any particular point. The velocity vector changes both in magnitude and in direction. We have already seen that, since the velocity *v* is always directed along the tangent to the path at that point, it has no component in the normal direction. However, the change in magnitude of the velocity is related to the acceleration a_t in the tangential direction, while the change in its direction is related to the acceleration a_n in the normal direction.

Thus $v_t = v$ and $v_n = 0$; and a_t and a_n are to be established.

Referring to Fig. 13.3, let *P* and *P'* be two points on the path separated by an infinitesimally small distance δs; let the tangential directions at *P* and *P'* be *Pt* and *P't'* and let the normals at *P* and *P'* meet at *O'*. Let the radius at *P* be *r* and let δs subtends angle δθ at *O'*.

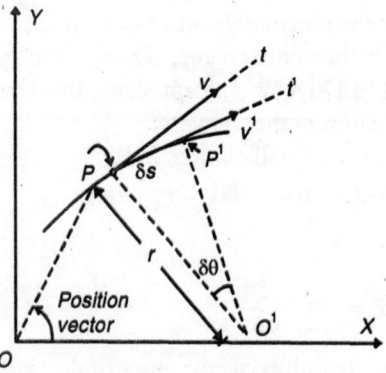

(a) Curvilinear Translation

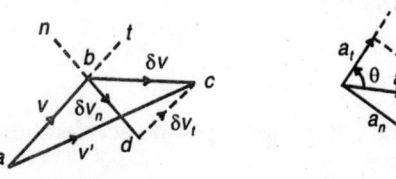

(b) Change in Velocity Vector   (c) Acceleration Components in Tangential and Normal Directions

Fig.13.3: Tangential and Normal Components of Motion

In (*b*), the change in the velocity vector δ*v* is shown as the vector difference (*v'* – *v*). If its tangential and normal components, δv_t and δv_n are summed up vectorially we get δ*v*.

Acceleration *a* is given by

$$a = \operatorname*{Lt}_{\delta t \to 0} \frac{\delta v}{\delta t}$$

$$= \operatorname*{Lt}_{\delta t \to 0} \frac{\delta v_t}{\delta t} + \operatorname*{Lt}_{\delta t \to 0} \frac{\delta v_n}{\delta t}$$

$$= a_t + a_n$$

or $\qquad a = a_t + a_n$, vectorially $\qquad$ ...(Eq. 13.5)

As δ*t* → 0, the point *P'* coincides with the point *P*, and the direction of a_t and a_n coincide with tangential and normal directions to the path at *P*. Change in the magnitude of the velocity is the vector *dc*.

∴ Tangential acceleration a_t is given by

$$a_t = \operatorname*{Lt}_{\delta t \to 0} \frac{\delta v}{\delta t} = \frac{dv}{dt}$$

or $\qquad a_t = \dfrac{dv}{dt} \qquad$ ...(Eq. 13.6)

Thus the tangential component of acceleration is equal to the rate of change of speed of the particle.

Also, the vector bd represents the change in the direction of the velocity.

For a very small change in θ,

Vector $bd = v \cdot \delta\theta$, very nearly

or $\quad \delta v_n \approx v \cdot \delta\theta$

$\therefore \quad a_n = \underset{\delta t \to 0}{\text{Lt}} \dfrac{v \cdot \delta\theta}{\delta t}$

If r is the radius of curvature of the path at the point P;

$$\delta s = r \cdot \delta\theta$$

$$\delta\theta = \frac{\delta s}{r}$$

The normal component a_n of the acceleration is

$$a_n = \underset{\delta t \to 0}{\text{Lt}} \frac{v \cdot \delta\theta}{\delta t} = \underset{\delta t \to 0}{\text{Lt}} \frac{v}{r} \cdot \frac{\delta s}{dt}$$

But $\quad \underset{\delta t \to 0}{\text{Lt}} \dfrac{\delta s}{\delta t} = \dfrac{ds}{dt} = v$

$\therefore \quad a_n = \dfrac{v^2}{r}$ ...(Eq. 13.7)

In words, the normal component of acceleration is equal to the square of the speed divided by the radius of curvature of the path at the point. The direction is always towards the centre of curvature at the point; it is known as *centripetal* (centre-seeking) acceleration.

The acceleration a is a vector—it may be the result of change in magnitude, or change in direction, or both, of the velocity.

Magnitude of $a = \sqrt{a_t^2 + a_n^2}$ ...(Eq. 13.8)

Direction $\theta = \tan^{-1}\left(\dfrac{a_n}{a_t}\right)$ ...(Eq. 13.9)

To understand the physical significance of these concepts, let us consider the motion of a particle along a circular path of radius r with a constant speed v. This is called "uniform circular motion"

$$a_t = \frac{dv}{dt} = 0 \qquad \text{(since } v \text{ is constant)}$$

$$a_n = \frac{v^2}{r}$$

$$a = \sqrt{a_t^2 + a_n^2} = a_n$$

$$\theta = \tan^{-1}\left(\frac{a_n}{0}\right) = 90°$$

Thus the total acceleration a is equal to a_n and is directed always in the normal direction; it is given by $\dfrac{v^2}{r}$.

13.4 RADIAL AND TRANSVERSE COMPONENTS OF MOTION

The position of a particle is sometimes more conveniently expressed in terms of its polar co-ordinates, when the particle undergoes curvilinear translation (Fig. 13.4).

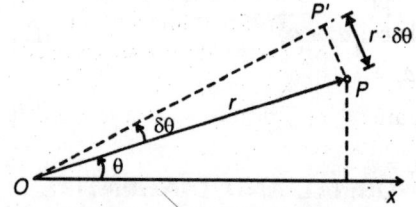

Fig. 13.4: Polar Co-ordinates of a Moving Particle

In such a case it is much simpler to resolve the velocity and acceleration of the particle into components parallel and perpendicular to the position vector of the particle $\overline{OP}$. These are called the 'radial' and 'transverse' components.

By definition, the radial component of velocity, v_r, is

$$v_r = \frac{dr}{dt} = \dot{r} \text{ (directed along the position vector } \vec{r} \text{)}$$

Transverse component of the velocity, v_θ, is

$$v_\theta = r \cdot \frac{d\theta}{dt} = r \cdot \theta \text{ (directed along the normal to the position vector } \vec{r} \text{)}$$

$\because$ Referring to Fig. **13.4**, $P'P = r \cdot \delta\theta$.)

Referring to **Fig. 13.5**,

Total velocity

$v = v_r + v_\theta$, vectorially ...(Eq. 13.10)

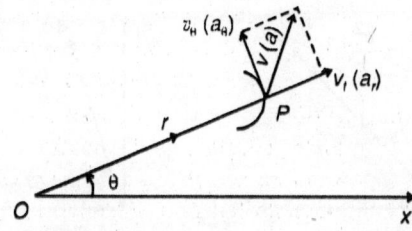

Fig.13.5: Radial and Transverse Components of Velocity/
Acceleration

Again referring to Fig. 13.5 regarding acceleration

The radial component of acceleration, a_r, is

$$a_\theta = r\ddot\theta + 2\dot r\dot\theta \qquad \ldots(\text{Eq. 13.12})$$

(directed along the position vector $\vec r$)
The transverse component of acceleration, a_θ, is

$$a_\theta = r\ddot\theta + 2\dot r\dot\theta \qquad \ldots(\text{Eq. 13.13})$$

(directed along the perpendicular to the
position vector $\vec r$)
Total acceleration

$$a_r = \dot v_r - v_\theta.\theta \qquad \ldots(\text{Eq. 13.14})$$

a_r and a_θ may also be written as

$$a_r = \dot v_r - v_\theta \cdot \dot\theta \qquad \ldots(\text{Eq. 13.14})$$

$$a_\theta = v_\theta + v_r \cdot \dot\theta \qquad \ldots(\text{Eq. 13.15})$$

(It is observed that a_r is not merely $\dot v_r$ and a_θ is not merely $\dot v_\theta$).◄

Also the radial components of velocity and acceleration are taken to be positive in the same sense as the position vector $\vec r$.

Transverse components of velocity and acceleration are taken to be positive in the direction of increasing values of θ.

The physical significance of these concepts may be understood if the following special situations are considered:

(1) Say r is constant and θ varies. The motion
is then rotation along a circular path.
$$\dot r = \ddot r = 0$$
$$v_r = 0 \quad v_\theta = r \cdot \dot\theta$$
$$a_r = -r(\dot\theta)^2 \quad a_\theta = r \cdot \ddot\theta$$
(negative sign indicates that a_r is directed opposite to the position vector $\vec r$, or towards O).

(2) Say θ is constant and r varies. The motion is then a rectilinear motion along a fixed direction.
$$\dot\theta = \ddot\theta = 0$$
$$v_r = \dot r \quad v_\theta = 0$$
$$a_r = \ddot r \quad a_\theta = 0.$$

ILLUSTRATIVE PROBLEMS

PROBLEM 13.1: The motion of a particle is described by the following equations.

$$x = t^2 + 4t + 2$$

$$y = t^3 + 3t^2 + 4t + 2$$

Determine (a) the initial velocity and initial acceleration of the particle (b) velocity and acceleration of the particle at $t = 1$ sec.

Solution:

$$x = t^2 + 4t + 2$$

$$v_x = \frac{dx}{dt} = 2t + 4$$

$$a_x = \frac{d^2x}{dt^2} = 2$$

$$y = t^3 + 3t^2 + 4t + 2$$

$$v_y = \frac{dy}{dt} = 3t^2 + 6t + 4$$

$$a_y = \frac{d^2y}{dt^2} = 6t + 6$$

(a) Initial Velocity of the particle
When $t = 0$

$$v_x = 2 \times 0 + 4 = 4 \text{ m/sec}$$

$$v_y = 3 \times 0^2 + 6 \times 0 + 4 = 4 \text{ m/sec}$$

$$v = \sqrt{(v_x)^2 + (v_y)^2}$$

$$= \sqrt{4^2 + 4^2}$$

$$= \sqrt{16 + 16}$$

$$= \sqrt{32}$$

$$= 5.657 \text{ m/sec}$$

$$\theta = \tan^{-1}(v_y / v_x) = \tan^{-1}(4 / 4)$$

$$= \tan^{-1}(1)$$

$$= 45°$$

Initial acceleration of the particle

When $t = 0$

$$a_x = 2 \text{ m/sec}^2$$

$$a_y = 6 \times 0 + 6 = 6 \text{ m /sec}^2$$

$$a = \sqrt{(a_x)^2 + (a_y)^2} = \sqrt{2^2 + 6^2}$$

$$= \sqrt{40} = 6.325 \text{ m/sec}^2$$

$$\theta = \tan^{-1}\left(\frac{a_y}{a_x}\right)$$

$$\theta = \tan^{-1}\left(\frac{6}{2}\right) = \tan^{-1}(3) = 71.57°$$

(b) Velocity of the particle at $t = 1$ sec

$$v_x = 2 \times 1 + 4 = 6 \text{ m/sec}$$

$$v_y = 3 \times 1^2 + 6 \times 1 + 4 = 13 \text{ m/sec}$$

$$\therefore v = \sqrt{(v_x)^2 + (v_y)^2} = \sqrt{6^2 + 13^2}$$

$$= \sqrt{36 + 169} = \sqrt{205}$$

$$= 14.318 \text{ m/sec}$$

$$\theta = \tan^{-1}\left(\frac{v_y}{v_x}\right) = \tan^{-1}(13 / 6) = \tan^{-1}(2.167)$$

$$= 65.23°$$

Acceleration of the particle at $t = 1$ sec

$$a_x = 2 \text{ m/sec}^2$$

$$a_y = 6 \times 1 + 6 = 12 \text{ m/sec}^2$$

$$a = \sqrt{a_x^2 + a_y^2} = \sqrt{2^2 + 12^2}$$

$$= \sqrt{4 + 144} = \sqrt{148}$$

$$= 12.166 \text{ m/sec}^2$$

$$\theta = \tan^{-1}\left(\frac{a_y}{a_x}\right) = \tan^{-1}\left(\frac{12}{2}\right)$$

$$= \tan^{-1}(6)$$

$$= 80.54°$$

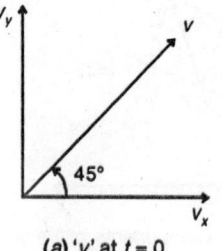

(a) 'v' at $t = 0$

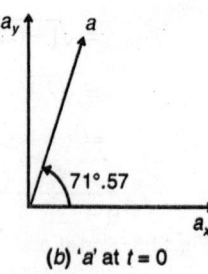

(b) 'a' at $t = 0$

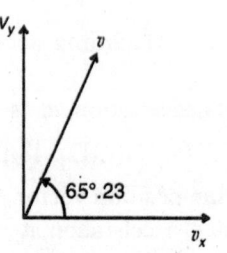

(c) 'v' at $t = 1$ sec

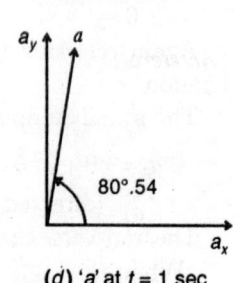

(d) 'a' at $t = 1$ sec

Fig. 13.6 (Illustrative Problem 13.1)

PROBLEM 13.2: The coordinates of a particle which moves in the *x-y* plane are given by $x = 4t - (t^2/2)$ and $y = 1 + 2t - (t^3/6)$ where *x* and *y* are in metres and *t* is in seconds. Determine the velocity and acceleration of the particle when $t = 3$ seconds.

Solution:

Velocity

$$x = 4t - \frac{t^2}{2}$$

$$v_x = \frac{dx}{dt} = 4 - \frac{2t}{2} = 4 - t$$

$$y = 1 + 2t - \frac{t^3}{6}$$

$$v_y = \frac{dy}{dt} = 2 - \frac{3t^2}{2} = 2 - \frac{t^2}{2}$$

When $t = 3$ seconds

$$v_x = 4 - 3 = +1 \text{ m/sec}$$

$$v_y = 2 - \frac{3^2}{2} = 2 - 4.5 = -2.5 \text{ m/sec}$$

(–ve sign shows that the velocity is downward)

$$v = \sqrt{(v_x)^2 + (v_y)^2}$$

$$= \sqrt{(1)^2 + (-2.5)^2}$$

$$= \sqrt{1 + 6.25}$$

$$= \sqrt{7.25}$$

$$= 2.69 \text{ m/sec}$$

$$\theta = \tan^{-1}\left(\frac{v_y}{v_x}\right) = \tan^{-1}\left(\frac{2.5}{1}\right)$$

$$\theta = 68.2°$$

Acceleration

$$a_x = \frac{dv_x}{dt} = -1$$

$$a_y = \frac{dv_y}{dt} = -\frac{2 \times t}{2} = -t$$

When $t = 3$ seconds

$$a_x = -1 \text{ m/sec}^2$$

(–ve sign shows that the acceleration is left)

$$a_y = -3 \text{ m/sec}^2$$

(–ve sign shows that the acceleration is downward)

$$a = \sqrt{(a_x)^2 + (a_y)^2}$$

$$a = \sqrt{(-1)^2 + (-3)^2}$$

$$= \sqrt{1 + 9}$$

$$= \sqrt{10}$$

$$= 3.16 \text{ m/sec}^2$$

$$\theta = \tan^{-1}\left(\frac{a_y}{a_x}\right) = \tan^{-1}\left(\frac{3}{1}\right)$$

$$= \tan^{-1}(3)$$

$$\theta = 71.57°$$

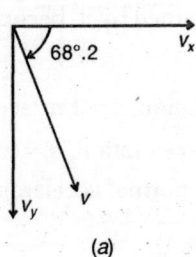

(a)

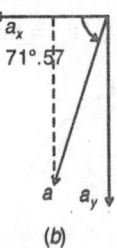

(b)

Fig.13.7 (Illustrative Problem 13.2)

PROBLEM 13.3: The distance s travelled by a particle moving along a circular path of radius r is given by the equation

$$s = Ct^2, \text{ where } C \text{ is a constant.}$$

If the particle starts from rest find,

(a) the tangential velocity and acceleration,

(b) the normal velocity and acceleration of the particle.

Solution:

(a) Tangential velocity

$$v = \frac{ds}{dt} = 2Ct$$

Tangential acceleration

$$a_t = \frac{dv}{dt} = 2C$$

(b) Normal velocity

$$v_n = 0$$

Normal acceleration

$$a_n = \frac{v^2}{\rho} = \left(\frac{2Ct^2}{r}\right) \quad (\because \rho = r)$$

$$a_n = \frac{4C^2t^2}{r^2}$$

PROBLEM 13.4: A car enters a curved section of the road of length equal to the quarter of a circle of radius 200 m at 27 kmph and leaves at 54 kmph (Fig. 13.8). If the car is travelling with a constant tangential acceleration, find the magnitude and direction of acceleration (a) when it enters the curve, (b) when it leaves the curve.

Solution:

$$S = \frac{2\pi r}{4} = \frac{2 \times \pi \times 200}{4} = 314.16 \text{ m}$$

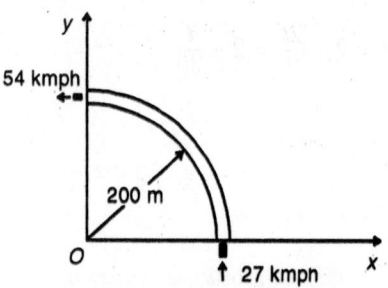

Fig. 13.8 (Illustrative Problem 13.4)

$$u = 27 \,\text{kmph} = \frac{27 \times 1000}{60 \times 60} = 7.5 \ \text{m/sec}$$

$$v = 54 \,\text{kmph} = \frac{57 \times 1000}{60 \times 60} = 15 \ \text{m/sec}$$

$$v^2 - u^2 = 2 \times a_t \times S$$

$$\therefore \quad a_t = \frac{v^2 - u^2}{2 \times S} = \frac{15^2 - 7.5^2}{2 \times 314.16}$$

$$= 0.2686 \ \text{m/sec}^2$$

(*a*) When the car enters the curve

$$a_n = \frac{u^2}{r} = \frac{7.5^2}{200} = 0.28125 \ \text{m/sec}^2$$

$$a_t = 0.2686 \ \text{m/sec}^2$$

$$\therefore \ a = \sqrt{a_n^2 + a_t^2}$$

$$= \sqrt{(0.28125)^2 + (0.2686)^2}$$

$$= 0.389 \ \text{m/sec}^2$$

$$\theta = \tan^{-1}\!\left(\frac{a_n}{a_t}\right) = \tan^{-1}\!\left(\frac{0.28125}{0.2686}\right)$$

$$\theta = 46.32°$$

(*b*) When the car leaves the curve

$$a_n = \frac{v^2}{r} = \frac{15^2}{200} = 1.125 \ \text{m/sec}^2$$

$$a_t = 0.2686 \ \text{m/sec}^2$$

$$\therefore \ a = \sqrt{a_n^2 + a_t^2}$$

$$= \sqrt{(1.125)^2 + (0.2686)^2}$$

$$a = 1.157 \ \text{m/sec}^2$$

$$\theta = \tan^{-1}\!\left(\frac{a_n}{a_t}\right) = \tan^{-1}\!\left(\frac{1.125}{0.2686}\right)$$

$$\theta = 76.57°$$

PROBLEM 13.5: A train is travelling on a curved track of radius of 600 m at the speed of 135 kmph. Find the normal and tangential components of acceleration.

The brakes are suddenly applied, causing the train to slow down at uniform rate, after 10 seconds, the speed has been reduced to 54 kmph. Determine the normal and the tangential components of deceleration immediately after the brakes are applied.

Solution: When travelling at a constant speed of 135 kmph.

$$v = \frac{135 \times 1000}{60 \times 60} = 37.5 \ \text{m/sec}$$

$$a_n = \frac{v^2}{r} = \frac{(37.5)^2}{600} = 2.34 \ \text{m/sec}$$

$$a_t = \frac{dv}{dt} = 0$$

At the instant when brakes are just applied, the train has a tangential speed of 135 kmph (37.5 m/sec) and in addition, experiences a tangential deceleration

$$54 \,\text{kmph} = \frac{54 \times 1000}{60 \times 60} = 15 \,\text{m/sec}$$

$$a_t = \frac{dv}{dt} = \frac{15 - 37.5}{10} = -2.25 \,\text{m/sec}$$

$$(\because \Delta t = 10 \text{ sec})$$

Negative sign shows there is deceleration.

$$a_n = \frac{v^2}{r} = \frac{(37.5)^2}{600} = 2.34 \ \text{m/sec}^2$$

PROBLEM 13.6: A car starts from rest on a curved road of 200 m radius and accelerates at a constant tangential acceleration of 0.8 m/sec². Determine the distance and the time for which the car will travel before the magnitude of the total acceleration attained by it, becomes 1 m/sec².

Solution:

Total acceleration, $a = 1 \ \text{m/sec}^2$

Tangential acceleration, $a_t = 0.8 \ \text{m/sec}^2$

Let a_n be the normal acceleration

$$a = \sqrt{a_t^2 + a_n^2}$$

$1 = \sqrt{0.8^2 + a_n^2}$

or $\quad 0.8^2 + a_n^2 = 1$

$\quad a_n^2 = 1 - 0.64 = 0.36$

$\therefore \quad a_n = \sqrt{0.36} = 0.60$ m/sec^2

But $a_n = \dfrac{v^2}{r} = 0.60$ m/sec^2

$\quad v^2 = 0.60 \times 200$

$\quad v = \sqrt{0.60 \times 200}$

$\quad v = \sqrt{120}$

$\quad\quad = 10.954$ m/sec

$v = u + a_t \cdot t$

$u = 0; \ v = 10.954$ m/sec; $a_t = 0.8$ m/sec^2

$\therefore \quad 10.954 = 0 + 0.8 \times t$

$t = \dfrac{10.954}{0.8} = 13.6925$ m/sec

$v^2 - u^2 = 2a_t S$

$10.954^2 - 0^2 = 2 \times 0.8 \times S$

$S = \dfrac{10.954 \times 10.954}{2 \times 0.8}$

$S = 74.994$ m

Or

$S = u \cdot t + \dfrac{1}{2} a_t \cdot t^2$

$S = 0 + \dfrac{1}{2} \times 0.8 \times 13.6925^2$

$\quad\quad = 74.994$ m

PROBLEM 13.7: A missile is launched from a point O and moves in a vertical plane. The radar placed at O tracks the movement of the missile as co-ordinates r and θ as function of time and it was found that $r = 2t - \dfrac{t^2}{40}$ and $\theta = 2t$, where r is in kilometres and θ is in degrees, t is time in seconds. Find the velocity and acceleration of the missile at $t = 30$ seconds.

Solution:

$\quad r = 2t - \dfrac{t^2}{40} \quad$ and $\theta = 2t$

$\dot{r} = 2 - \dfrac{2t}{40} = 2 - \dfrac{t}{20}$

$\ddot{r} = -\dfrac{1}{20}$

$\dot{\theta} = 2$

$\ddot{\theta} = 0$

When $t = 30$ seconds

$r = 2 \times 30 - \dfrac{30^2}{40} = 60 - 22.5 = 37.5$ km

$\dot{r} = 2 - \dfrac{30}{20} = 0.5$ km/sec

$\ddot{r} = -\dfrac{1}{20}$ km/sec^2

$\theta = 2 \times 30 = 60°$ or $\pi/3$ radians.

$\dot{\theta} = 2$ degrees/sec $= \dfrac{2 \times \pi}{180}$ rad/sec

$\ddot{\theta} = 0$

Velocity

$v_r = \dot{r} = 0.5$ km/sec

$v_\theta = r\dot{\theta} = 37.5 \times \left(\dfrac{2 \times \pi}{180}\right) = 1.309$ km/sec

$v = \sqrt{(v_r)^2 + (v_\theta)^2} = 1.40$ km/sec

$\alpha = \tan^{-1}(v_\theta / v_r)$

$\quad = \tan^{-1}(1.309 / 0.5) = \tan^{-1}(2.618)$

$\quad = 69°$

Acceleration

$a_r = \ddot{r} - r(\dot{\theta})^2$

$\quad = -\dfrac{1}{20} - 37.5\left(2 \times \dfrac{\pi}{180}\right)^2$

$\quad = -0.05 - 0.0457$

$\quad = -0.0957$ km/sec^2

$a_\theta = r\ddot{\theta} + 2\dot{r}\dot{\theta}$

$\quad = 37.5 \times 0 + 2 \times 0.5 \times \left(\dfrac{2 \times \pi}{180}\right)$

$\quad = +0.0349$ km/sec^2

$a = \sqrt{(a_r)^2 + (a_\theta)^2}$

$$= \sqrt{(-0.0957)^2 + (0.0349)^2}$$

$$= 0.102 \ km/sec^2$$

$$\beta = \tan^{-1}\left(\frac{0.0349}{-0.0957}\right) = -20°$$

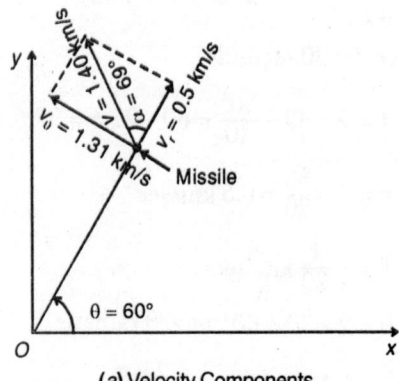

(a) Velocity Components

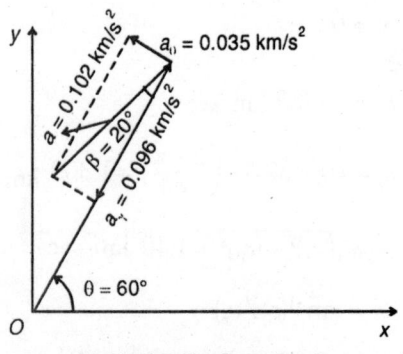

(b) Acceleration Components

Fig. 13.9 (Illustrative Problem 13.7)

PROBLEM 13.8: A particle moves along the spiral shown [Fig. 13.10 (a)]. The motion of the particle is defined by relations.

$$r = 10t \ \text{and} \ \theta = 2\pi t$$

where r is expressed in metres, θ in radians and t in seconds. Determine the velocity and acceleration of the particle when $t = 0.2$ second.

Solution:

Equations of motion are

$$r = 10t \ \text{and} \ \theta = 2\pi t$$

$$\dot{r} = 10 \qquad \dot{\theta} = 2\pi$$

$$\ddot{r} = 0 \qquad \ddot{\theta} = 0$$

When $t = 0.2$ sec

$$r = 10 \times 0.2 = 2 \ m/sec$$

$$\dot{r} = 10 \ m/sec$$

$$\ddot{r} = 0$$

$$\theta = 2\pi \times 0.2 = 0.4 \ \text{radians}$$

$$= \frac{0.4\pi \times 180}{\pi} = 72°$$

$$\dot{\theta} = 2\pi \ \text{radians}$$

$$\ddot{\theta} = 0$$

(*i*) Velocity

$$v_r = \dot{r} = 10 \ m/sec$$

$$v_\theta = r\dot{\theta} = 2 \times 2\pi = 12.566 \ m/sec$$

$$v = \sqrt{(v_r)^2 + (v_\theta)^2}$$

$$= \sqrt{(10)^2 + (12.566)^2}$$

$$= \sqrt{257.904}$$

$$= 16.06 \ m/sec$$

$$\alpha = \tan^{-1}(v_\theta / v_r) = \tan^{-1}(12.566 / 10)$$

$$= 51.49°$$

(*ii*) Acceleration

$$a_r = \ddot{r} - r(\dot{\theta})^2$$

$$= 0 - 2 \times (2\pi)^2$$

$$= -78.957 \ m/sec^2$$

$$a_\theta = r\ddot{\theta} + 2 \cdot \dot{r} \cdot \dot{\theta} = 2 \times 0 + 2 \times 10 \times 2\pi$$

$$= 125.664 \ m/sec^2$$

$$a = \sqrt{(a_r)^2 + (a_\theta)^2}$$

$$= \sqrt{(-78.957)^2 + (125.664)^2}$$

$$= \sqrt{22025.64875}$$

$$= 148.41 \ m/sec^2$$

$$\beta = \tan^{-1}\left(\frac{a_\theta}{a_r}\right) = \tan^{-1}\left(\frac{125.664}{-78.957}\right)$$

$$= \tan^{-1}(-1.59155) = -57.86°$$

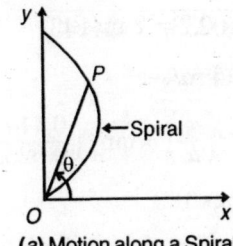

(a) Motion along a Spiral

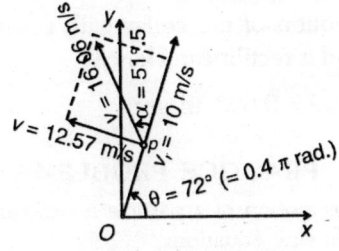

(b) Velocity Components

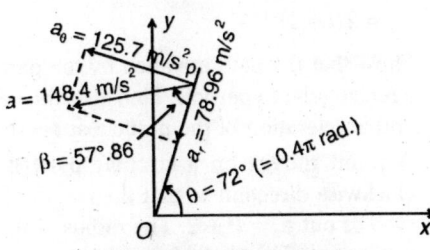

(c) Acceleration Components

Fig. 13.10 (Illustrative Problem 13.8)

PROBLEM 13.9: The rotation of rod OA is defined by the relation $\theta = 0.35t^2$. A collar P slides along this rod in such a way that its distance from O is given by $r = \dfrac{t^2}{4 + 3t}$ [Fig. 13.11(a)]. In these relations θ is expressed in radians, r in meters and t in seconds.

Determine (i) the velocity of the collar, (ii) the total acceleration of the collar, (iii) the acceleration of the collar relative to the rod when $t = 1$ sec.

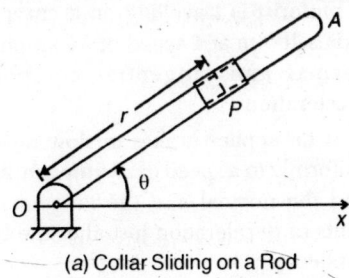

(a) Collar Sliding on a Rod

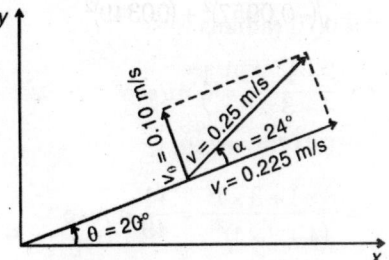

(b) Velocity Components of Collar

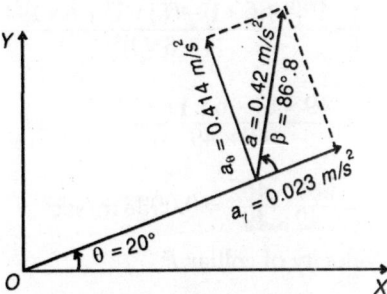

(c) Acceleration Components of Collar

Fig. 13.11 (Illustrative Problem 13.9)

Solution:

Equations of motion are

$$\theta = 0.35t^2$$

and $$r = \frac{t^2}{4 + 3t}$$

$$\dot\theta = 0.35 \times 2 \times t = 0.70t$$

$$\ddot\theta = 0.70 \text{ radians}$$

$$\dot r = \frac{(4 + 3t)2t - t^2(3)}{(4 + 3t)^2}$$

$$\dot r = \frac{8t + 6t^2 - 3t^2}{(4 + 3t)^2}$$

$$\dot r = \frac{8t + 3t^2}{(4 + 3t)^2}$$

$$\ddot r = \frac{(4 + 3t)^2(8 + 6t) - (3t^2 + 8t)2(4 + 3t)3}{(4 + 3t)^4}$$

When $t = 1$ sec

$$\theta = 0.35 \text{ radians}$$

$$\dot\theta = 0.70 \times 1 = 0.70 \text{ radians}$$

$\dot{\theta} = 0.70$ radians

$r = \dfrac{1}{4 + 3 \times 1} = \dfrac{1}{7}$ m/sec

$\quad = 0.14286$ m/sec

$\dot{r} = \dfrac{8 \times 1 + 3 \times 1^2}{(4 + 3 \times 1)^2} = \dfrac{11}{49}$ m/sec

$\quad = 0.22449$ m/sec

$\ddot{r} = \dfrac{(4 + 3 \times 1)^2 (8 + 6 \times 1) - 6(3 \times 1^2 + 8 \times 1)(4 + 3 \times 1)}{(4 + 3 \times 1)^4}$

$\ddot{r} = \dfrac{49 \times 14 - 6 \times 11 \times 7}{49 \times 49}$

$\quad = \dfrac{686 - 462}{49 \times 49} = 0.0933$ m/sec^2

(*i*) Velocity of colllar P

$v_r = \dot{r} = 0.22449$ m/sec

$v_\theta = r \cdot \dot{\theta} = 0.14286 \times 0.70$

$\quad = 0.10$ m/sec

$v = \sqrt{v_r^2 + v_\theta^2}$

$\quad = \sqrt{(0.22449)^2 + (0.10)^2}$

$\quad = 0.2458$ m/sec

$\tan \alpha = \dfrac{v_\theta}{v_r} = \dfrac{0.10}{0.22449} = 0.44545$

$\alpha = 24°$

$\theta = 0.35$ radians

$\quad = \dfrac{0.35 \times 180}{\pi}$

$\quad = 20°$

(*ii*) Acceleration of the collar P

$a_r = \ddot{r} - r(\dot{\theta})^2 = 0.0933 - (0.14286)(0.70)^2$

$\quad = 0.0933 - 0.07 = 0.0233$ m/sec^2

$a_\theta = r\ddot{\theta} + 2\dot{r}\dot{\theta}$

$\quad = 0.14286 \times 0.70 + 2 \times 0.22449 \times 0.70$

$\quad = 0.10 + 0.3143 = 0.4143$ m/sec^2

$a = \sqrt{a_r^2 + a_\theta^2}$

$\quad = \sqrt{(0.0233)^2 + (0.4143)^2}$

$\quad = 0.415$ m/sec^2

$\beta = \tan^{-1}\left(\dfrac{a_\theta}{a_r}\right) = \tan^{-1}\left(\dfrac{0.4143}{0.0233}\right)$

$\quad = 86.78°$

(*iii*) Acceleration of the collar P relative to the rod

Motion of the collar with respect to the rod is rectilinear. Hence

$a_r = \ddot{r} = 0.093$ m/sec^2

PRACTICE PROBLEMS

13.1 The motion of a particle is described by the following equations,

$x = 2(t + 1)^3$

$y = 2(t + 1)^{-3}$

Show that the path travelled by the particle is a rectangular hyperbola. Find also, the velocity and acceleration of the particle at $t = 0$.

13.2 A point moves on a circular path in anti-clockwise direction so that the length of arc it sweeps out $S = t^2 + 2$. The radius of the path is 8 metres. The units of S and t are metre and second respectively. Determine the rectangular components of velocity and acceleration when $t = 2$ sec.

13.3 The displacement-time equation for the oscillation of a simple pendulum is given by

$$s = S \cos\left(\sqrt{\dfrac{g}{l}}\right) t$$

where, S is the maximum displacement of oscillations.

Find (*a*) the maximum velocity, (*b*) the maximum tangential and normal accelerations of the bob.

13.4 A motorist is travelling on a curved road of radius 100 m at a speed of 54 kmph. Find the normal and tangential components of acceleration.

If he applies brakes to slow down his car uniformly to a speed of 27 kmph in 10 seconds, find the normal and the tangential components of deceleration just after the brakes are applied.

13.5 A car starts from rest on a curved road of radius 200 m and attains a speed of 27 kmph at the end of 100 seconds while travelling with a constant tangential acceleration. Find the tangential and normal accelerations of the car 30 seconds after it started.

13.6 A motorist starts from rest on a curved road of 100 m radius and accelerates at a constant tangential acceleration of 0.60 m/sec². Determine the distance and the time for which the car will travel before the magnitude of the total acceleration attained by it becomes 0.75 m/sec².

13.7 A particle moves along the path whose equation is $r = 2\theta$ metre. If the angle θ is related by the equation $\theta = t^2$ radians, where t is in seconds, determine the velocity and acceleration of particle P when $\theta = 90°$.

13.8 A particle moves along the spiral shown (Fig. 13.12). The motion of the particle is defined by the relations:

$r = 10t$ and $\theta = 2\pi t$

where r is expressed in metres, θ in radians and t in seconds. Determine the velocity and acceleration of the particle when $t = 0$.

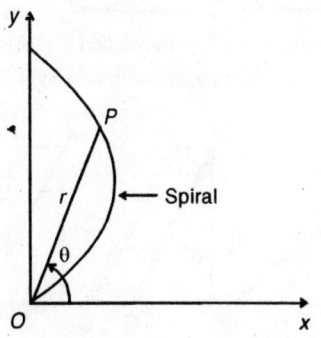

Fig. 13.12 (Practice Problem 13.8)

13.9 The rotation of rod OA is defined by the relation:

$$\theta = \frac{\pi}{3}\left(3t - 2t^2\right)$$

A collar P slides along this rod in such a way that its distance from O is given by $r = t^2 - 0.75t^3$ [Fig. 13.13 (a)]. In these relations θ is expressed in radians, r in metres and t in seconds.

Determine (*i*) the velocity of the collar, (*ii*) the total acceleration of the collar and (*iii*) the acceleration of the collar relative to the rod when $t = 1$ sec.

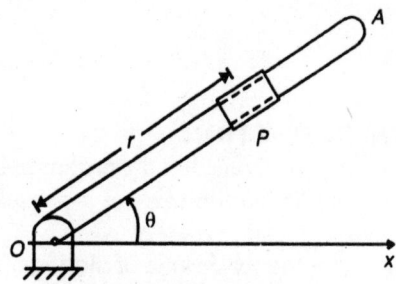

Fig. 13.13 (Practice Problem 13.9)

Chapter 14

Rotation and General Plane Motion of a Rigid Body

14.1 INTRODUCTION

Hitherto we have considered translation—rectilinear as well as curvilinear—of a particle or a rigid body.

There is another basic kind of motion of a rigid body—that is, rotation about a fixed axis; in this case, all the particles of which the rigid body is composed describe concentric circles. The fixed axis, or the centre of all the concentric circles may be within the rigid body or outside it (Fig. 14.1).

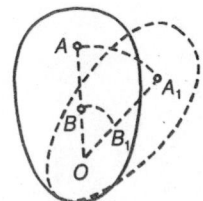

(a) Centre within the Body

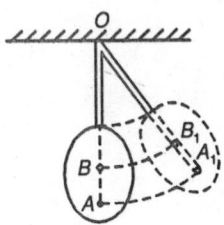

(b) Centre outside the Body

Fig.14.1: Rotation of a Rigid Body about a Fixed Axis

Just as translation involves linear motion, rotation involves angular motion.

General plane motion of a rigid body is a combination of translation and rotation. Two common examples are a rotating wheel or a cylinder and a rod or a ladder sliding against a wall at one end and the floor at the other end (Fig. 14.2).

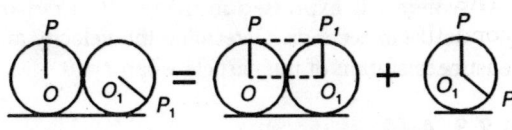

Plane motion = Translation + Rotation
(a) Rotating Wheel or a Cylinder

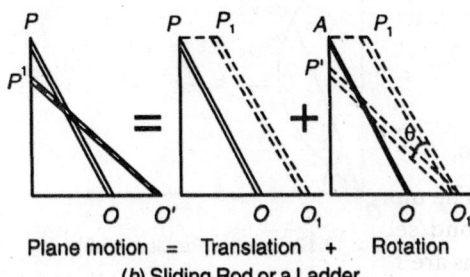

Plane motion = Translation + Rotation
(b) Sliding Rod or a Ladder

Fig.14.2: General Plane Motion of a Rigid Body

Thus, the general plane motion of a rigid body may be analysed by splitting it into translation and rotation.

14.2 DEFINITIONS OF TERMS USED IN ANGULAR MOTION

14.2.1 Angular Displacement

The displacement of a particle in a rigid body undergoing rotation is measured in angular measure. The angular displacement of a particle at A when it moves from A to A' is, say, θ; it is a vector quantity since it has a sense of rotation—clockwise or counter-clockwise, as the case be (Fig. 14.3).

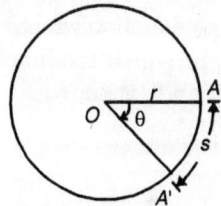

Fig.14.3: Rotation—Terms Used

14.2.2 Angular Velocity

This is the rate of change of angular displacement with time.

Angular Velocity,

$$\omega = \frac{d\theta}{dt} = \dot\theta \qquad \text{...(Eq. 14.1)}$$

The units of angular velocity are radians/second. (Dimensions are s^{-1}, since angle in radian measure is dimensionless).

14.2.3 Angular Acceleration

This is the rate of change of angular velocity with time.

∴ Angular acceleration,

$$\alpha = \frac{d\omega}{dt} = \frac{d^2\theta}{dt^2} = \dot\omega = \ddot\theta \qquad \text{...(Eq. 14.2)}$$

Also, $\alpha = \dfrac{d\omega}{dt} = \dfrac{d\omega}{d\theta}\cdot\dfrac{d\theta}{dt} = \omega\cdot\dfrac{d\omega}{d\theta}$ ...(Eq. 14.3)

The units of angular acceleration are radians/second/second, or radians/second^2 (the dimensions are s^{-2}).

14.3 RELATION BETWEEN ANGULAR MOTION AND LINEAR MOTION

It is easy to relate angular motion to linear motion if the radius of the path of rotation is r and linear (or arc) distance through which the particle moves in time t is s (Fig. 14.3).

$$s = r\theta \ (\theta \text{ in radian measure}) \text{ ...(Eq. 14.4)}$$

The linear velocity is the tangential velocity of the particle, say, v.

Then $v = \dfrac{ds}{dt} = r\cdot\dfrac{d\theta}{dt} = r\cdot\dot\theta = r\cdot\omega$ ...(Eq. 14.5)

Thus, while ω is independent of the location of the point in the rigid body with respect to the centre/axis of rotation, the linear velocity v is dependent on it.

The tangential component a_t, of the linear acceleration is given by

$$a_t = \frac{dv}{dt} = r\cdot\frac{d^2\theta}{dt^2} = r\cdot\ddot\theta = r\cdot\frac{d\omega}{dt} = r\cdot\alpha \text{ ...(Eq. 14.6)}$$

The normal component, a_n of the linear acceleration is given by

$$a_n = \frac{v^2}{r} = \frac{r^2\omega^2}{r} = r\cdot\omega^2 \qquad \text{...(Eq. 14.7)}$$

14.4 UNIFORM ANGULAR MOTION

Rotation of a rigid body with a constant angular velocity (that is, with zero angular acceleration) is called "uniform angular motion".

Since ω is constant and $\alpha = 0$,

$$\theta = \omega t \ (\text{radians}) \qquad \text{...(Eq. 14.8)}$$

Sometimes the angular velocity is given in terms of revolutions per minute (rpm), of course, when it is constant; it is then designated by the letter symbol, n.

Since one revolution is equivalent to an angular displacement of 2π radians, the angular velocity, ω, is given by

$$\omega = \frac{2\pi n}{60}(\text{radians / second}) \quad \text{...(Eq. 14.9)}$$

The time taken for one revolution (or period, T) is given by

$$T = \frac{2\pi}{\omega} \qquad \text{...(Eq. 14.10)}$$

14.5 ROTATION UNDER CONSTANT ANGULAR ACCELERATION

Let us consider the rotation of a rigid body under constant angular acceleration.

This means that

$$\frac{d\omega}{dt} = \alpha, \text{ a constant.}$$

Integrating,

$$\omega = \alpha t + C_1,$$

C_1 being the constant of integration

Let the initial angular velocity be ω_0 (*at t* = 0) then

$$C_1 = \omega_0$$

$$\therefore \quad \omega = (\omega_0 + \alpha t) \qquad \text{...(Eq. 14.11)}$$

Again

$$\frac{d\theta}{dt} = \omega = \omega_0 + \alpha t$$

Integrating,

$$\theta = \omega_0 t + \frac{1}{2}\alpha t^2 + C_2$$

C_2 being the constant of integration.

When $t = 0$, $\theta = 0$, and so $C_2 = 0$.

$$\therefore \quad \theta = \omega_0 t + \frac{1}{2}\alpha t^2 \qquad \text{...(Eq. 14.12)}$$

By Eq. 14.3,

$$\alpha = \omega \cdot \frac{d\omega}{d\theta}$$

$$\alpha \cdot d\theta = \omega \cdot d\omega$$

Integrating both sides,

$$\alpha\theta = \frac{1}{2}\omega^2 + C_3$$

C_3 being the constant of integration.

When $t = 0$, $\theta = 0$, and $\omega = \omega_0$

This leads to

$$C_3 = -\frac{1}{2}\omega_0^2$$

$$\therefore \quad \alpha\theta = \frac{1}{2}(\omega^2 - \omega_0^2)$$

or $\quad (\omega^2 - \omega_0^2) = 2\alpha\theta \qquad \text{...(Eq. 14.13)}$

The similarity between Eqs. 14.11 to 14.13 and the corresponding equations of rectilinear motion with constant acceleration,

$$v = (u + at),$$

$$s = ut + \frac{1}{2}at^2$$

and $(v^2 - u^2) = 2as$

is strikingly evident.

If it is a case of uniform angular retardation instead of acceleration, $(-\alpha)$ may be substituted for (α) in the above equations, and the problem may be solved.

14.6 ROTATION UNDER VARYING ANGULAR ACCELERATION

Problems involving rotation of a rigid body under angular acceleration which varies with time can be solved, provided the functional relationship between α and *t* is known:

$$\alpha = f(t), \text{ say}$$

Then

$$\alpha = \frac{d\omega}{dt} = f(t)$$

By integrating, ω can be solved.

$$\omega = f_1(t) + C_1$$

Then

$$\omega = \frac{d\theta}{dt}$$

By integrating, $\theta = f_2(t) + C_1 t + C_2$

The constants of integration C_1 and C_2 may be determined from known initial and boundary conditions of the problem.

14.7 GENERAL PLANE MOTION OF A RIGID BODY

As has been defined in the introduction to this Chapter, a body is said to have general plane motion if it undergoes translation and rotation simultaneously. Both these components shall be in a single plane, the overall motion being *planar*, any particle of the rigid body would remain in the same fixed plane, and all other particles would lie in planes parallel to this during the motion. This would facilitate the study of this motion— say in one plane—and conclude that the motion is the same in all other parallel planes. As an example, the motion of a cylinder rolling, without slipping, in one plane, can be studied by considering the motion of one of the cross-sections, a circle on a straight line lying in the plane. Thus, rolling bodies constitute an excellent example of general plane motion.

The examples shown in Fig. 14.2 help one to understand the nature of general plane motion as the vector sum of its components—pure translation and pure rotation; the analysis of general plane motion, therefore, is based on this concept.

14.8 ABSOLUTE AND RELATIVE VELOCITY IN GENERAL PLANE MOTION

Let us consider a rigid body in general plane motion with respect to a fixed set of axes OX and OY. This motion can be fully represented by the motion of a plane section which represents the motion of one of its parallel planes (Fig. 14.4).

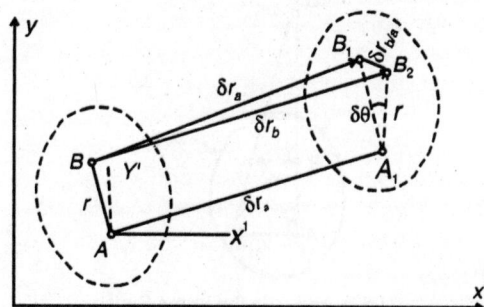

Fig. 14.4: General Plane Motion of a Rigid Body—Velocities

Let a line AB in the body be displaced to $A_1 B_2$ in a small interval of time δt. This displacement can be treated to be the sum of a translation from AB to $A_1 B_1$ and a rotation about A_1 from $A_1 B_1$ to $A_1 B_2$. A reference set of axes AX' and AY' to A such that this set translates but remains parallel to the fixed set OX and OY during the plane motion.

Total displacement of AB

= Translation of AB + Displacement due to the rotation of B about A

or $\delta r_b = (\delta r_a + \delta r_{b/a})$, vectorially, from the triangle law.

Dividing both sides by δt and taking the limit as $\delta t \to 0$, we have,

$$v_b = (v_a + v_{b/a}) \qquad \text{...(Eq. 14.14)}$$

here, v_b is the absolute velocity of B with respect to the fixed set of axes OX and OY,

v_a is the absolute velocity of A and corresponds to the translation with respect to the fixed set of axes.

$v_{b/a}$ is the relative velocity of B with respect to A associated with its rotation about A

[In fact, $v_{b/a} = r \cdot \omega$, r being the fixed distance of B from A and ω the angular velocity of motion of B about A].

A is called the 'pole'.

The results embodied in Eq. 14.14 may be generalised as follows:

"The velocity of any point in the rigid body is given by the vector sum of the velocity of the pole and the relative velocity of the point with respect to the pole".

A similar statement can be shown to hold good even for acceleration—by substituting the term acceleration for velocity and relative acceleration for relative velocity in the above.

14.9 INSTANTANEOUS CENTRE OF ROTATION

In the previous section, we have seen that the plane motion of a rigid body can be viewed as a combination of translation of a reference point called "pole" and a rotation about this plane.

When a rigid body undergoes a general plane motion, it is possible to locate a point in the plane which appears to be instantaneously at rest, and hence the motion of other points appears to be pure rotation about this point. Such a point is called the "instantaneous centre of rotation", the velocity of which is zero. The axis through this centre and normal to the plane of motion is called the "Instantaneous axis of rotation".

It should be understood that the instantaneous centre has zero velocity only at a particular instant of time; thus, the instantaneous centre is not a fixed point, but appears to vary with time.

Let us consider a rigid body in plane motion, as shown in Fig. 14.5.

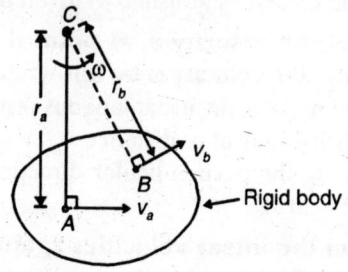

Fig. 14.5: Instantaneous Centre of Rotation—Concept

Let A be a point with a velocity v_a at any instant; let us choose a point C at a perpendicular distance r_a to the direction of this velocity.

From the concept of plane motion, a

$$v_a = v_c + r_a \cdot \omega \qquad \text{...(from Eq. 14.14)}$$

If we choose $r_a = \dfrac{v_a}{\omega}$, we get

$$v_a = v_c + \frac{v_a}{\omega} \cdot \omega$$

or $\qquad v_c = 0$

That is to say, C has zero velocity, or it is the instantaneous centre, the plane motion of A reduced to pure rotation about this centre.

If B is any other point at a distance r_b from C, its velocity r_b is given by

$$v_b = v_c + r_b \cdot \omega$$

$$= r_b \cdot \omega, \text{ since } v_c = 0$$

v_b is at right angles to CB.

Thus, if the instantaneous centre is located, the motion of all other points at that instant can be determined as per rotation about C.

Since the velocity of the instantaneous centre is zero only at a particular instant of time and the velocity at any later instant is non-zero, the acceleration is not zero at this centre. Thus, the general plane motion can be treated as pure rotation about the instantaneous centre—only for the purpose of velocity calculation and not for acceleration.

14.10 LOCATION OF INSTANTANEOUS CENTRE OF ROTATION

Depending upon the known quantities, the procedure for locating the instantaneous centre of rotation can be established as given below:

(i) **Let the velocity v_a at point A and the angular velocity ω be known:** Referring to Fig. 14.5, the instantaneous centre C can be located at a distance r_a/ω measured along the perpendicular direction to that of v_a at A.

(ii) **Let the linear velocities v_a and v_b at A and B be known:** Again referring to Fig. 14.5, the instantaneous centre C is located

by obtaining the intersection of the perpendicular to the directions of v_a and v_b at A and B, respectively.

$$\left(\omega \text{ is given by } \frac{v_a}{r_a} \text{ or } \frac{v_b}{r_b} \right).$$

(iii) **Let the velocities v_a and v_b at A and B be known to be parallel:** Referring to Fig. 14.6, let v_a and v_b at A and B be known to be parallel.

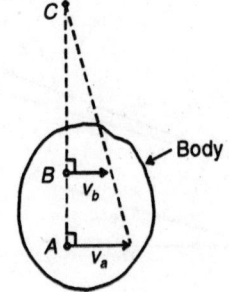

(a) v_a and v_b in the Same Sense

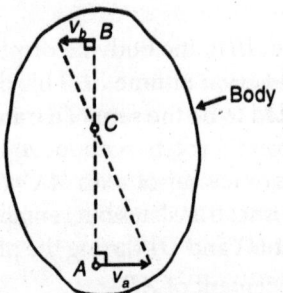

(b) v_a and v_b in Opposite Senses

Fig. 14.6: Location of Instantaneous Centre—the Velocities at two Points Known to be Parallel

In case (a) v_a and v_b are in the same sense, *but unequal in magnitude*. The instantaneous centre is located by joining the ends of the vectors v_a and v_b to meet AB produced in C.

In case (b), v_a and v_b are in opposite senses. The instantaneous centre is again located by joining the ends of the vectors v_a and v_b to meet AB in C. (It may be noted that the stipulation that v_a and v_b should be unequal in magnitude does not apply here).

(iv) **Let the velocities v_a and v_b at A and B be known to be equal and parallel:** Referring to Fig. 14.7, Let v_a and v_b, the velocities at two points A and B, be equal and parallel as shown. In this case the intersection of the two lines AB and the line joining the ends of the velocity vectors will be at infinity; thus the instantaneous centre for this special case lies at infinity.

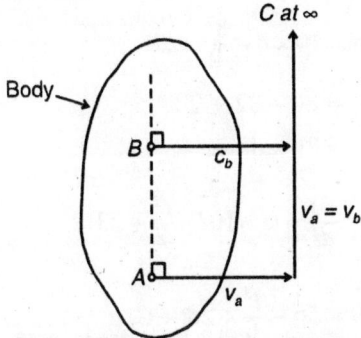

Fig. 14.7: Location of Instantaneous Centre— Velocities at two Points Known to be Equal and Parallel

ILLUSTRATIVE PROBLEMS

PROBLEM 14.1: The angle of rotation of a body is given by the equation $\theta = 3t^3 - 5t^2 + 7t + 9$ where θ is expressed in radians and t in seconds, find *(a)* angular velocity and *(b)* angular acceleration of the body when $t = 0$, and $t = 3$ secs.

Solution:

$$\theta = 3t^3\ 5t^2 + 7t + 9$$

Differentiating with respect to 't'

$$\omega = \frac{d\theta}{dt} = 9t^2 - 10t + 7$$

When $t = 0$

$$\omega = 9 \times 0^2 - 10 \times 0 + 7 = 7 \text{ rad/sec}$$

When $t = 3$ secs

$$\omega = 9 \times 3^2 - 10 \times 3 + 7$$

$$= 81 - 30 + 7$$

$$= 58 \text{ rad/sec}$$

$$\omega = 9t^2 - 10t + 7$$

Differentiating once again with respect to 't'

$$\alpha = \frac{d\omega}{dt} = \frac{d^2\theta}{dt^2} = 18t - 10$$

When $t = 0$

$$\alpha = -10 \text{ rad/sec}^2$$

When $t = 3$ sec

$$\alpha = 18 \times 3 - 10 = 44 \text{ rad/sec}^2$$

PROBLEM 14.2: The angle of rotation of a body is given as a function of time by the equation

$$\theta = \theta_0 + at + bt^2$$

where θ_0 is initial angular displacement, a and b are constants.

Obtain general expression for *(a)* the angular velocity and *(b)* angular acceleration of body. If the initial angular velocity be π radians/sec and after 1.5 seconds the angular velocity is 7π radians/sec, find the constants a and b.

Solution:

$$\theta = \theta_0 + at + bt^2$$

$$\omega = \frac{d\theta}{dt} = a + 2bt$$

$$\omega = \pi \text{ rad/sec when } t = 0$$

$\therefore$ $$\pi = a + 2 \times b \times 0$$

or $$a = \pi$$

$$\omega = 7\pi \text{ rad/sec when } t = 1.5 \text{ sec}$$

$$7\pi = \pi + 2(b)(1.5)$$

$$3b = 7\pi - \pi = 6\pi$$

or $$b = 2\pi$$

General expression for the angular velocity

$$\omega = \pi + 2(2\pi)t$$

$$\omega = \pi + 4\pi t$$

General expression for the angular acceleration

$$\alpha = \frac{d\omega}{dt} = \frac{d}{dt}(\pi + 4\pi t)$$

or $$\alpha = 4\pi$$

$$a = \pi;\ b = 2\pi$$

$$\omega = \pi + 4\pi t\ \&\ \alpha = 4\pi$$

PROBLEM 14.3: A car is moving at 60 kmph, if the wheels are 750 mm diameter, find the angular velocity of the tyre about its axis. If the car comes to rest in a distance of 15 metres, under a uniform retardation, find the angular retardation of the wheels.

$$\omega = \frac{v}{r}$$

$$v = 60 \text{ kmph} = \frac{60 \times 1000}{60 \times 60} = \frac{50}{3} \text{ m/sec}$$

$$d = 750 \text{ mm} = 0.75 \text{ m}$$

$$\therefore \quad r = \frac{d}{2} = \frac{0.75}{2} = 0.375 \text{ m}$$

$$\therefore \quad \omega = \frac{(50/3)}{(0.375)} = 44.44 \text{ rad/sec}$$

Angular velocity of the tyre = 44.44 rad/sec

$$v^2 = u^2 + 2as$$

$$v = 0$$

$$u = 50/3 \text{ m/sec}$$

$$s = 15 \text{ m}$$

$$\therefore \quad a = \frac{0^2 - \left(\frac{50}{3}\right)^2}{2 \times 15}$$

$$= -9.26 \text{ m/sec}^2$$

(Minus sign indicates retardation)

Angular retardation of the wheel,

$$\alpha = \frac{a}{r} = -\frac{9.26}{0.375} = -24.69 \text{ rad/sec}^2$$

(Minus sign indicates retardation)

PROBLEM 14.4: The angular acceleration of a fly wheel is given by $\alpha = 10 - t$, where, α is in rad/sec^2 and t is in seconds. If the angular velocity of the fly wheel is 65 rad/sec at the end of 6th second, determine the angular velocity at the end of 8th second. How many revolutions take place in these 8 seconds?

Solution:

$$\alpha = 10 - t$$

i.e. $$\frac{d\omega}{dt} = 10 - t$$

$$\omega = 10t - \frac{t^2}{2} + C_1$$

where, C_1 is constant of integration.

When $t = 6$ sec, $\omega = 65$ rad/sec

$$65 = 10 \times 6 - \frac{6^2}{2} + C_1$$

$$65 = 60 - 18 + C_1$$

$$C_1 = 23$$

$$\therefore \quad \omega = 10t - \frac{t^2}{2} + 23$$

When $t = 8$ sec

$$\omega = 10 \times 8 - \frac{8^2}{2} + 23$$

$$= 80 - 32 + 23$$

$$= 71 \text{ rad/sec}$$

Now $$\frac{d\theta}{dt} = \omega = 10t - \frac{t^2}{2} + 23$$

$$\therefore \quad \theta = 5t^2 - \frac{t^3}{6} + 23t + C_2$$

where, C_2 is constant of integration.

When $t = 0$, $\theta_o = C_2$

When $t = 8$ sec

$$\theta_8 = 5 \times 8^2 - \frac{8^2}{6} + 23 \times 8 + C_2$$

$$\theta_8 = 320 - 85.33 + 184 + C_2$$

$$\theta_8 = 418.67 + C_2$$

$$\theta_8 = 418.67 + \theta_o$$

$$\theta_8 - \theta_o = 418.67 \text{ rad}$$

$\therefore$ No. of revolutions

$$= \frac{418.67}{2\pi} = 66.63$$

PROBLEM 14.5: The step pulley shown in Fig 14.8 starts from rest and accelerates at 3.5 rad/sec^2. What time is required for block A to move 30 m? Find also the velocity of A and B at that time.

Solution:

When A moves by 30 m, the angular displacement of pulley, θ is given by

$$r\theta = s$$

$$0.75 \times \theta = 30$$

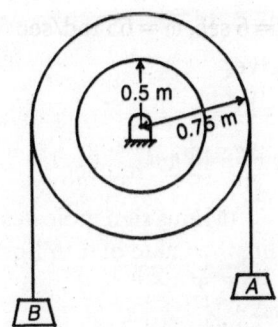

Fig. 14.8 (Illustrative Problem 14.5)

$\theta = \dfrac{30}{0.75} = 40$ radians

$\alpha = 3.5$ rad/sec^2 and $\omega_o = 0$

$\theta = \omega_o t + \dfrac{1}{2}\alpha t^2$

$40 = 0 + \dfrac{1}{2} \times 3.5 \times t^2$

$t = \sqrt{\dfrac{40 \times 2}{3.5}} = 4.78$ sec

Angular velocity of pulley at this time

$\omega = \omega_o + \alpha t$

$\omega = 0 + 3.5 \times 4.78 = 16.73$ rad/sec

Velocity of A, $v_A = 0.75 \times 16.73$

$\qquad = 12.55$ m/sec

and

Velocity of B, $v_B = 0.5 \times 16.73$

$\qquad = 8.365$ m/sec

PROBLEM 14.6: A fly wheel starts rotating from rest and is given an acceleration of 3 rad/sec^2. Find the angular velocity and speed in r.p.m. after 5 minutes.

If the flywheel is brought to rest with a uniform angular retardation of 2 rad/sec^2, find the time taken by the flywheel to come to rest.

Solution:

Angular initial velocity, $\omega_o = 0$

Angular acceleration, $\alpha = 3$ rad/sec^2

Time, $t = 5$ minutes $= 5 \times 60 = 300$ sec

Let the final angular velocity be ω.

$\omega = \omega_o + \alpha t$

$= 0 + 3 \times 300$

$= 900$ rad/sec

Speed in r.p.m., $N = \dfrac{60\omega}{2\pi}$

$= \dfrac{60 \times 900}{2\pi}$

$= 8594.367$ r.p.m.

Let angular initial velocity, $\omega_o = 900$ rad/sec

Final angular velocity, $\omega = 0$

Angular retardation 2 rad/sec^2

Angular acceleration, $\alpha = -2$ rad/sec^2

Let the time taken by the flywheel in coming to rest $= t$

$\omega = \omega_o + \alpha t$

$0 = 900 - 2 \times t$

or $\qquad t = 450$ sec

or $\qquad$ 7 minutes 30 seconds

PROBLEM 14.7: A wheel rotating about a fixed axis at 30 rpm is uniformly accelerated for 60 seconds, during which time it makes 90 revolutions. Find (*a*) the angular velocity at the end of this interval, and (*b*) the time required for the velocity to reach 180 r.p.m.

Solution:

Initial r.p.m. of the wheel, $N_o = 30$

$\therefore$ Angular initial velocity, $\omega_o = \dfrac{2\pi N_o}{60}$

$\omega_o = \dfrac{2\pi \times 30}{60}$

$= \pi$ radians

Angular displacement, $\theta = 90$ revolutions

$= 90 \times 2\pi = 180\pi$ rad

Let $\alpha = $ Angular acceleration

$t = 60$ sec

$\theta = \omega_o t + \dfrac{1}{2}\alpha t^2$

$180\pi = \pi \times 60 + \dfrac{1}{2}\alpha \times 60^2$

$\dfrac{1}{2} \times \alpha \times 60^2 = 120\pi$

$$\alpha = \frac{120 \times \pi \times 2}{6 \times 60} = 0.2094 \text{ rad/sec}^2$$

(a) Angular velocity at the end of 60 sec interval

$$\omega = \omega_0 + \alpha t$$

$$\omega = \pi + 0.2094 \times 60$$

$$= 15.7056 \text{ rad/sec}$$

(b) Let time required for the velocity to reach 180 rpm be 't' sec.

Now final speed, $N = 180$ rpm

$\therefore$ Final angular velocity, $\omega = \dfrac{2\pi N}{60}$

$$= \frac{2 \times \pi \times 180}{60}$$

$$= 6\pi \text{ rad/sec}$$

$$\omega = \omega_0 + \alpha t$$

$$6\pi = \pi + 0.2094 \times t$$

$$t = \frac{\pi}{0.2094} = 75 \text{ sec}$$

or   1 minute 15 seconds

PROBLEM 14.8: A wheel rotates for 6 sec with a constant angular acceleration and describes during this time 120 radians. It then rotates with a constant angular velocity and during the next 6 seconds describes 90 radians. Find the initial angular velocity and the angular acceleration.

Solution: Considering the angular motion of the wheel with constant acceleration for 6 seconds.

$$\theta = \omega_0 t + \frac{1}{2}\alpha t^2$$

$$120 = \omega_0 \times 6 + \frac{1}{2} \times \alpha \times 6^2$$

$$120 = 6\omega_0 + 18\alpha$$

$$20 = \omega_0 + 3\alpha \qquad \qquad \dots(1)$$

$$\dot{\omega} = \omega_0 + \alpha t$$

$$\omega = \omega_0 + \alpha \times 6$$

$$\omega = \omega_0 + 6\alpha$$

Considering the angular motion of the wheel with a constant angular velocity of $(\omega_0 + 6)$ for 6 sec and describing 90 radians.

$$90 = 6 \times (\omega_0 + 6\alpha)$$

or $\omega_0 + 6\alpha = 15$ ...(2)

(2) – (1) gives

$$3\alpha = -15$$

or $\alpha = -\dfrac{5}{3} = -1.67 \text{ rad/sec}^2$

(Minus sign indicates retardation)

Substituting the value of α in Eqn. (1)

$$20 = \omega_0 + 3\left(-\frac{5}{3}\right)$$

$\therefore$ $\omega_0 = 25$ rad/sec.

PROBLEM 14.9: A flywheel is making 240 r.p.m. and after 30 sec it is running at 150 r.p.m. How many revolutions will it make, and what time will elapse before it stops, if the retardation is uniform?

Solution:

Revolutions of the wheel before it stops:

$$\omega = \omega_0 + \alpha t$$

where α is uniform acceleration

$$\omega_0 = \frac{2\pi \times 240}{60} = 8\pi \text{ rad/sec}$$

$$\omega = \frac{2\pi \times 150}{60} = 5\pi \text{ rad/sec}$$

$$t = 30 \text{ sec}$$

$\therefore$ $\alpha = \dfrac{\omega - \omega_0}{t}$

$$= \frac{5\pi - 8\pi}{30} = -0.1\pi \text{ rad/sec}^2$$

(Minus sign indicates retardation)

Considering angular motion of the flywheel from 240 r.p.m. to 0 r.p.m. with a constant retardation of 0.1π rad/sec^2.

$$\omega^2 = \omega_0^2 + 2\alpha\theta$$

$$0 = (8\pi)^2 + 2(-0.1\pi)\theta$$

$$\theta = \frac{64\pi^2}{2 \times 0.1 \times \pi} = 320\pi \text{ rad}$$

$$= \frac{320\pi}{2\pi} = 160 \text{ revolutions}$$

Time in which the flywheel will come to rest:

Let t be time in which the wheel comes to rest in sec.

$$\omega = \omega_0 + \alpha t$$

$$0 = 8\pi + (-0.1\pi)t$$

$$t = \frac{8\pi}{0.1\pi} = 80 \text{ sec}$$

$$= 1 \text{ minute } 20 \text{ secs.}$$

PROBLEM 14.10: A shaft is uniformly accelerated from 12 rev/sec to 24 rev/sec in 6 seconds. The shaft continues to accelerate at this rate for the next 12 seconds. Thereafter, the shaft rotates with a uniform angular speed. Find the total time to complete 780 revolutions.

Solution: Let α be the angular acceleration in rad/sec^2 considering the motion of the shaft in first 6 seconds.

$$\omega = \omega_0 + \alpha t$$

$$\omega_0 = (2\pi)(12) = 24\pi \text{ rad/sec}$$

$$\omega = (2\pi)(24) = 48\pi \text{ rad/sec}$$

$$t = 6 \text{ sec}$$

$$\therefore \quad \alpha = \frac{\omega - \omega_0}{t} = \frac{(48 - 24)}{6} = 4\pi \text{ rad/sec}^2$$

Angular displacement during these 6 seconds

$$\theta_1 = \omega_0 t + \frac{1}{2}\alpha t^2$$

$$= (24\pi)(6) + \frac{1}{2}(4\pi)(6^2)$$

$$= (144 + 72)\pi$$

$$= 216\pi \text{ rad}$$

or $\quad \dfrac{216\pi}{2\pi} = 108 \text{ rev}$

Considering the shaft for the next 12 seconds with an angular acceleration of 4π rad/sec^2

Final angular velocity of the shaft,

$$\omega = \omega_0 + \alpha \cdot t$$

$$= 48\pi + (4\pi)(12)$$

$$= 96\pi \text{ rad/sec}$$

or $\quad \dfrac{96\pi}{2\pi} = 48 \text{ rev/sec}$

Angular displacement,

$$\theta_2 = \omega_0 \times t + \frac{1}{2} \times \alpha \times t^2$$

$$= 48\pi \times 12 + \frac{1}{2} \times 4\pi \times 12^2$$

$$= 576\pi + 288\pi = 864\pi \text{ rad}$$

or $\quad 864\pi/2\pi = 432 \text{ rev}$

Total number of revolutions completed by the shaft at a speed of 48 rev/sec,

$$\theta_3 = \theta - (\theta_1 + \theta_2)$$

$$= 780 - (108 + 432) = 780 - 540 = 240 \text{ rev.}$$

Time taken by the shaft to complete 240 revolutions

$$= \frac{240}{48} = 5 \text{ sec.}$$

Total time to complete 780 revolutions

$$= 6 + 12 + 5$$

$$= 23 \text{ seconds}$$

PROBLEM 14.11: A wheel of radius 1.2 m rolls freely with an angular velocity of 6 rad/sec clockwise as shown in Fig. 14.9. Determine the velocities of points B & D by instantaneous centre method.

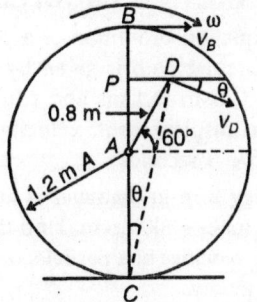

Fig. 14.9 (Illustrative Problem 14.11)

Solution:

Angular velocity, $\omega = 6$ rad/sec

Wheel radius, $r = 1.2$ m

$\therefore$ Velocity of geometric centre, A, $v_A = r\omega$

$$= 1.2 \times 6 = 7.2 \text{ m/sec}$$

It is horizontal towards the right.

Hence instantaneous centre is in the vertical downward direction at a distance

$$= \frac{v_A}{\omega} = \frac{7.2}{6} = 1.2 \text{ m, } i.e., \text{ Point } C,$$

which is in contact with floor (Fig. 14.9)

$$\therefore \quad v_B = CB \times \omega$$

$$= 2.4 \times 6 = 14.4 \text{ m/sec}$$

$$v_D = CD \times \omega$$

$$CD = \sqrt{CP^2 + PD^2}$$

$CP = CA + AP$

$\quad = 1.2 + 0.8 \times \sin 60°$

$\quad = 1.8928$ m

$PD = 0.8 \cos 60° = 0.4$ m

$\therefore \quad CD = \sqrt{(1.8928)^2 + (0.4)^2}$

$\quad = 1.9346$ m

$\therefore \quad v_D = 1.9346 \times 6 = 11.608$ m/sec

Its direction is at right angle to *CD*. Its inclination to horizontal (θ) is given by

$$\theta = \angle PCD = \tan^{-1}\left(\frac{0.4}{1.8928}\right)$$

$$= \tan^{-1}(0.2113)$$

$$= 11.93° = 11°56'$$

PRACTICE PROBLEMS

14.1 The equation of motion of a particle moving on a circular path is given by $\theta = 3t^2 + 2.0$ where θ is in radians and t in seconds. Find angular displacement, velocity and acceleration after 3 seconds.

14.2 A pulley 2 m in diameter is keyed to a shaft which makes 300 r.p.m. Find the angular and linear velocities of a particle, on the periphery of the pulley.

14.3 The rotation of the flywheel is governed by the equation $\omega = 6t^2 - 4t + 2$ where ω is in radians per second and t is in seconds. After one second from start the angular displacement was 5 radians. Determine the angular displacement, angular velocity and angular acceleration when $t = 2$ seconds.

14.4 The equation of motion of a particle moving on a circular path (radius 200 m) is given by: $S = 12t + 2t^2 - t^3$ where S is the total distance covered from the starting point in metres at the end of t seconds. Find (a) the velocity and acceleration at the start, (b) the time when the particle reaches its maximum speed, and (c) the maximum speed of the particle.

14.5 A body is rotating with an angular velocity of 10 radians/sec. After 5 seconds the angular velocity of the body becomes 30 radians per second. Determine the angular acceleration of the body.

14.6 A wheel is rotating about its axis with a constant angular acceleration of 1.5 rad/sec². If the initial and final angular velocities are 10 rad/sec and 20 rad/sec, find the total angle turned through during the time interval this change of angular velocity took place.

14.7 A wheel rotating about a fixed axis at 30 revolutions per minute is uniformly accelerated for 90 seconds during which time it makes 60 revolutions. Find (a) the angular velocity at the end of this interval and (b) the time required for the velocity to reach 120 revolutions per minute.

14.8 A pulley starting from rest is given an acceleration of 0.6 rad/sec². What will be its speed in r.p.m. at the end of 3 minutes? If it is uniformly retarded at the rate of 0.3 rad/sec², in how many minutes the pulley will come to rest?

14.9 A swing bridge turns through 90° in 150 seconds. The bridge is uniformly accelerated from rest for the first 50 seconds. Subsequently, it turns with a uniform angular velocity for the next 75 seconds. Now the motion of the bridge is uniformly retarded for the last 25 seconds. Find (a) angular acceleration, (b) maximum angular velocity, (c) angular retardation of the bridge.

14.10 A slender beam *AB* of length 3.5 m which remains always in the same vertical plane has its ends *A* and *B* constrained to remain in contact with a horizontal floor and a vertical wall respectively as shown in Fig. 14.10. Determine the velocity of the end *B* at the position shown in Fig. 14.10, if point *A* has a velocity of 2.5 m/sec leftward by instantaneous centre method.

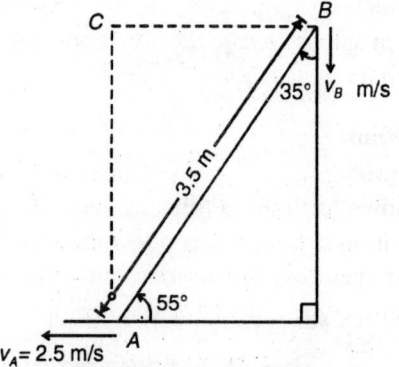

Fig. 14.10 (Practice Problem 14.10)

Chapter 15

Laws of Motion

15.1 INTRODUCTION

We have seen the kinematics of different kinds of motion in the previous four chapters. The aspects of kinetics of motion can be considered only after understanding the "Laws of Motion", enunciated by Sir Isaac Newton.

In this chapter, therefore, the laws of motion and the application of these to different practical situations will be treated in some detail. Finally, what is known as the "D'Alembert's Principle", which is very popularly used by engineers in solving problems involving the kinetics of motion, will be explained.

15.2 DEFINITIONS

Definitions of certain terms relevant to this chapter are given below:

(i) Inertia

The property of a mass or a body by virtue of which it resists motion is called its "inertia".

(ii) Momentum

The product of the mass of a body and its velocity is known as the 'momentum' of the body. Since velocity is a vector quantity, although 'mass' is a scalar quantity, the momentum of a body is a vector quantity. Thus, the change in the direction, or in the magnitude, or in both, of the velocity will result in a change in momentum of a body. The units are kg-m/s (kilogram-metre per second).

(iii) Weight

The force of gravitational attraction of the earth on any mass/body is known as its weight. It is always directed towards the center of the earth; in other words, it always acts vertically. The units of weight are kgf (kilogram-force).

15.3 NEWTON'S LAWS OF MOTION

The reasons for some bodies to be at rest and some to be in motion had engaged the attention of scientists and engineers for several centuries; Sir Isaac Newton, a British Physicist, succeeded in understanding these reasons, embodied in his "Laws of Motion", which laid the foundation for Newtonian Mechanics.

Newton's laws of motion are explained in the following sub-sections.

15.3.1 Newton's First Law of Motion

The statement of this law is:

> *"Every body continues to be in a state of rest or of uniform motion in a straight line unless it is compelled to change that state by a force or a force-system acting on the body".*

This law, therefore, helps us to define the force —"a force is that which changes or tends to change the state of rest or of steady or uniform rectilinear motion of a body".

15.3.2 Newton's Second Law of Motion

The statement of this law is given below.

"The rate of change of momentum of a body is directly proportional to the force acting on it and it takes place in the direction of the force".

This laws helps us to define a force in a quantitative manner.

∴ Force ∝ rate of change of momentum.

or Force ∝ rate of change of (mass × velocity)

or Force ∝ mass × (rate of change of velocity)

    (since the mass of a given body is constant)

or Force ∝ mass × acceleration

That is, $F \propto m \cdot a$ ...(Eq. 15.1)

where F is the force acting on a body,

    m is the mass of the body, and

    a is the acceleration of the body produced by the force

Then $F = k \cdot m \cdot a$ ... (Eq. 15.2)

where k is a constant of proportionality.

If a *unit force* is defined as that which produces a *unit acceleration* while acting on a body of *unit mass,* the constant of proportionality automatically becomes unity. The system of units are appropriately chosen to satisfy this.

Then $F = m \cdot a$ ...(Eq. 15.3)

15.3.3 Newton's Third Law of Motion

The statement of this law is:

"There is always an equal and opposite reaction to every action; in other words, the forces of action and reaction between any two bodies are equal in magnitude and opposite in direction".

Since the action and reaction constitute a two-force system, this law is the direct consequence of the static equilibrium condition for such a system. This law has been frequently used in problems of statics for drawing free-body diagrams and will also be found useful in solving problems of kinetics.

15.3.4 Newton's Law of Universal Gravitation

The statement of this law is:

"Two bodies attract each other along the line joining their centres of mass with a force of magnitude directly proportional to the product of their masses and inversely proportional to the square of the distance between their centres of mass".

or   Force, $F \propto \dfrac{m_1 m_2}{r^2}$ ...(Eq. 15.4)

where m_1 and m_2 are the masses of the two bodies and r, the distance between them.

or $F = G \cdot \dfrac{m_1 m_2}{r^2}$ ...(Eq. 15.5)

where G is the constant of proportionality, called the "Universal Constant of Gravitation".

The weight W of a body of mass m being the force of attraction of earth's gravity on the body directed towards the centre of the earth (from Eq. 15.5),

$$F = W = \frac{G \cdot M \cdot m}{R^2} \quad \text{...(Eq. 15.6)}$$

where M and m are the masses of the earth and of the body respectively and R is the mean radius of the earth.

Since $F = W = m \cdot g$ (from Eq. 15.3), being the acceleration due to gravity.

$$mg = \frac{G \cdot M \cdot m}{R^2}$$

or $g = \dfrac{G \cdot M}{R^2}$ ...(Eq. 15.7)

This gives us a way to obtain the value of G:

$$G = \frac{gR^2}{M} \quad \text{...(Eq. 15.8)}$$

since g, R and M for the earth have been determined experimentally by scientists and engineers.

According to Cavendish, a British scientist,

$$G = 6.673 \times 10^{-11} \ \text{Nm}^2/\text{Kg}^2.$$

Since the earth is not exactly a sphere, R varies slightly from place to place, and only an average value of G is obtained and used. Thus, the weight of a body is considered to be a constant value for a given mass, although it varies slightly from place to place.

This assumption does not lead to any significant error in engineering computations involving gravity.

15.4 UNITS OF FORCE

The units of force depend upon the system of units. These may be categorised as the "Absolute

Units" and the "Gravitational Units". In the former the fundamental quantities are length (L), mass (M) and time (T), and in the latter these are length (L), force (F), and time (T). These are explained in the following sub-sections.

15.4.1 Absolute Units of Force

In the CGS-system of units, the units of length, mass and time are centimetre, gram and second, respectively.

In this system, the fundamental unit of force, a derived quantity, is the "dyne", which is defined as the force that produces an acceleration of 1 cm/s² acting on a mass of 1 gram.

In the MKS and the S.I. systems of units, the units of length, mass and time are metre, kilogram and second, respectively. The primary difference between these two systems is in the units of force.

In the S.I. system, the unit of force is the "newton (N)", in honour of Sir Isaac Newton.

A newton is the amount of force which causes a mass of one kilogram to move with an acceleration of 1 m/s².

Both the dyne and the newton are absolute units of force which do not depend upon the location of the mass on the earth's surface.

For scientific purposes, therefore, absolute units are preferred.

15.4.2 Gravitational Units of Force

In the CGS system of units, a force which causes a mass of one gram to move with an acceleration equal to that due to gravity is called one gram-weight (1 g.wt.).

Since $g = 981$ cm/s²,

1 g.wt. = 981 dynes.

In the MKS system of units, a force which causes a mass of one kilogram to move with an acceleration equal to that due to gravity is called one kilogram-weight (1 kg.wt.).

Since $g = 9.81$ m/s²,

1 kg.wt = 9.81 newtons

In practice, kg.wt. is simply called kg, but should be differentiated from the corresponding unit of mass; this is best done by referring the weight/force as kgf (kg-force) and to the mass as kgm (kg-mass). (However, kgm should not be confused with the unit of moment of a force !).

Obviously, the gravitational units are dependent upon the numerical value of the acceleration due to gravity, and hence these are not absolutely constant, and are subjected to the variations in g, although the error involved is insignificant for engineering purposes. Weight of a body is more easily felt and understood rather than its mass owing to the gravitational force on it; this is the reason for engineers to prefer gravitational units to absolute units of force.

15.5 d'ALAMBERT'S PRINCIPLE

d'Alembert's principle is a re-statement of Newton's second law of motion, viewing the latter from a different angle. Jean le Rond d'Alembert, a French mathematician, enunciated his principle in 1743 in the form of an equation of equilibrium, in fact, of *dynamic* equilibrium.

Eq. 15.3 may be rewritten in the following form:

$$F + (-ma) = 0 \qquad \text{...(Eq. 15.9)}$$

F may be a single force or the resultant of a force system, whichever is acting on the particle/body in motion.

The term in brackets may be looked upon as an imaginary force equal and opposite to the applied force F, and is called the *inertia force* of the body. ('Inertia' has already been defined in Sec. 15.2). This inertia force is the resistance offered by the body to the change in its state of rest or of uniform motion; this is its physical interpretation.

Eq. 15.9 may be considered to be an equation of equilibrium of the body in motion, and hence an equation of *dynamic equilibrium*.

Now d'Alembert's principle may be stated in words as:

"The force or force-system acting on a body in motion is in dynamic equilibrium with the inertia force of the body"

d'Alembert's principle achieves the following:

(*i*) Application of Newton's second law of motion to a system of particles or a rigid body;

(*ii*) Introduction of the concept of inertia force; and,

(*iii*) Conversion of a problem in dynamics to one of statics by the equation of dynamic equilibrium, enabling one to apply all the techniques of solving static problems.

Although scientists have been critical of d'Alembert for clubbing external forces with the internal resistance of a body, engineers have always preferred to use the principle as a convenient tool for solving kinetics problems in dynamics. Its applications are mentioned in the next section.

15.6 APPLICATION OF THE LAWS OF MOTION

The application of the laws of motion, or alternatively, of the d'Alembert's principle, to a few types of problems in kinetics/dynamics will be seen now.

15.6.1 Motion of Two Bodies Connected by a String in the Vertical Plane

In the case of a system of connected bodies, each of which has a rectilinear motion, their individual motions are interrelated, the relationships depending upon the nature of the physical connections; the establishment of these relations will be the kinematics part of the problem. Later, the kinetics part of the problem can be solved by using Newton's laws of motion, especially the second one. Alternatively, the equations of dynamic equilibrium may be written down using d'Alembert's principle, and the methods of statics including the principle of virtual work can be used; this eliminates the need for consideration of the internal forces of the system, which results in substantial reduction in labour.

The dynamic system involved is assumed to be an ideal one, unless otherwise stated—that is, (*i*) friction is absent at all surfaces of contact, (*ii*) wires, ropes or strings, if any, are thin, weightless, and inextensible; and (*iii*) pulleys, if any, are smooth and of negligible weight.

The above discussion is applicable to the problems considered in the remaining subsections also.

Let two bodies of weights W_1 and W_2 (say $W_2 > W_1$) be hung to the ends of a rope passing over an ideal pulley (Fig. 15.1(*a*)).

Let the acceleration of the heavier body be a downwards. The free-body diagrams of the two bodies is shown in (*b*), the tension in the string being T.

For the body of weight W_1:

$$\Sigma V = 0$$

$$T - \frac{W_1}{g}a - W_1 = 0 \qquad \text{...(Eq. 15.10)}$$

For the body of weight W_2:

$$\Sigma V = 0$$

$$T + \frac{W_2}{g}a - W_2 = 0 \qquad \text{...(Eq. 15.11)}$$

The two unknowns T and a can be solved from these two equations.

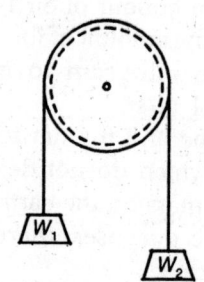

(*a*) System of two bodies

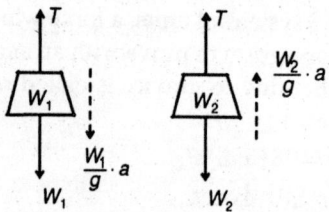

(*b*) Free-body diagrams of the two bodies.

Fig. 15.1: Two Bodies Connected by a String Passing over a Smooth Pulley

If the two bodies are connected by a two-pulley system as shown in Fig. 15.2 (*a*), it will be easily understood that if W_2 moves downward by a distance *y*, W_1 moves up by 2*y*. Therefore, the acceleration of W_1 will be 2*a* if that of W_2 is *a*. The free-body diagrams of W_1 and W_2 will be as shown in (*b*), using d'Alembert's principle.

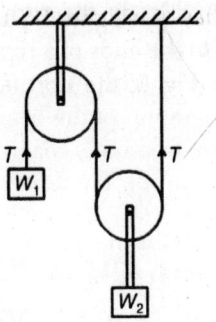

(a) System of Bodies Connected by Two Pulleys

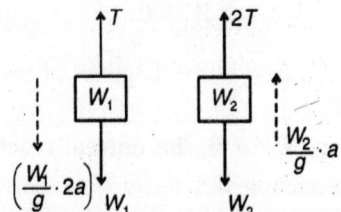

(b) Free-body Diagrams of the Two Bodies

Fig. 15.2: Two Body System Connected through a Two Pulley System

For the body of weight W_1:

$$T - \frac{W_1}{g}(2a) - W_1 = 0 \qquad \text{...(Eq. 15.12)}$$

For the body of weight W_2:

$$2T + \frac{W_2}{g}(a) - W_2 = 0 \qquad \text{...(Eq. 15.13)}$$

The tension in the string, T, and the acceleration of the heavier body, a, may be solved from these equations.

15.6.2 Motion of Bodies on a Horizontal Plane

If a single body of weight W is made to move on a smooth horizontal plane with an acceleration a, the force required to achieve this is obviously $\frac{W}{g} \cdot a$; if the force is known, the acceleration may be determined.

(i) Smooth Horizontal Plane

If two bodies of weights W_1 and W_2 are connected by a string and move along a *smooth horizontal plane* under the action of a force P as shown in Fig. 15.3 (a). The acceleration of the bodies and

the tension in the string can be determined with the aid of d'Alembert's principle as follows:

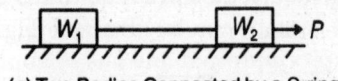

(a) Two Bodies Connected by a String

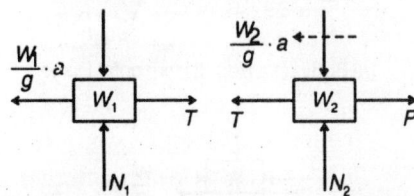

(b) Free Body Diagrams of the Two Bodies

Fig. 15.3: Two body System Moving on a Smooth Horizontal Plane

From the free-body diagram of body of weight W_1 [Fig. (*b*)]:

Applying $\Sigma V = 0$, the reaction N_1 from the horizontal plane is given by

$$N_1 = W_1$$

Applying $\Sigma H = 0$,

$$T - \frac{W_1}{g} \cdot a = 0 \qquad \text{...(Eq. 15.14)}$$

From the free-body diagram of body of weight W_2 [Fig. (*b*)] :

Applying $\Sigma V = 0$, the reaction N_2 from the horizontal plane is given by

$$N_2 = W_2$$

Applying $\Sigma H = 0$,

$$P - T - \frac{W_2}{g} \cdot a = 0 \qquad \text{...(Eq. 15.15)}$$

From these two equations 15.14 and 15.15, the acceleration a of the system, and the tension T in the string are solved.

(ii) Rough Horizontal Plane

If the horizontal plane is rough, the frictional resistance offered to the motion has also to be considered as an additional force in the equations of dynamic equilibrium. This is illustrated below.

If a body of weight W is acted on by a horizontal force P on a rough horizontal plane with a coefficient of friction μ between the body and the plane, the acceleration produced may be determined; alternatively, the force required to produce a given acceleration may be obtained [Fig. 15.4 (a)].

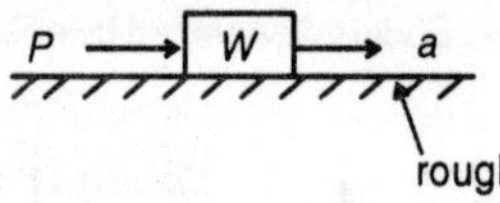

$P \longrightarrow$ | W | $\longrightarrow a$

rough

(a) Body on a Rough Horizontal Plane

$P \longrightarrow$ | W | $\longleftarrow--- \dfrac{W}{g} \cdot a$

$F \longleftarrow$

N

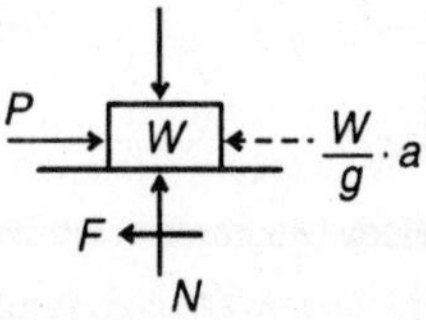

(b) Free Body Diagram of the Body

Fig. 15.4: Body in Motion on a Rough Horizontal Plane

From the free body diagram [Fig. (b)]:

Applying $\Sigma V = 0$, the normal reaction N from the horizontal plane is given by

$N = W$

Applying $\Sigma H = 0$,

$$P - F - \dfrac{W}{g} \cdot a = 0 \qquad \text{...(Eq. 15.16)}$$

$$F = \mu \cdot N = \mu \cdot W \qquad \text{...(Eq. 15.17)}$$

From these two equations, if P is known a may be got, and vice-versa.

For the two-body system already shown in Fig. 15.3, if the horizontal plane is rough, with a co-efficient of friction μ between the bodies and the plane, the analysis is as follows (Fig. 15.5):

| W_1 | ━━━━━ | W_2 | $\rightarrow P$

(a) Two Body System under a Force

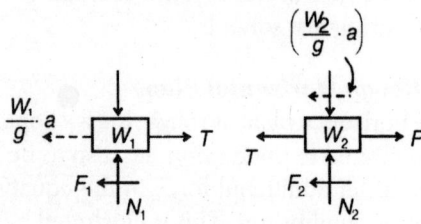

(b) Free Body Diagrams of the Bodies

Fig. 15.5: Two Body System Moving on a Rough Horizontal Plane

From the free body diagram of the body of weight W_1 [Fig. (b)]:

Applying $\Sigma V = 0$, the normal reaction N_1 is given by

$N_1 = W_1$

Applying $\Sigma H = 0$,

$$T - \dfrac{W_1}{g} \cdot a - F_1 = 0$$

But $F_1 = \mu \cdot N_1 = \mu \cdot W_1$ from the laws of friction.

$$\therefore \quad T - \dfrac{W_1}{g} \cdot a - \mu \cdot W_1 = 0 \qquad \text{...(Eq. 15.18)}$$

From the free body diagram of the body of weight W_1:

Applying $\Sigma V = 0$, the normal reaction N_2 is given by

$N_2 = W_2$

Applying $\Sigma H = 0$,

$$T + \dfrac{W_2}{g} \cdot a + F_2 - P = 0$$

But $F_2 = \mu \cdot N_2 = \mu \cdot W_2$

$$\therefore \quad T + \dfrac{W_2}{g} \cdot a - \mu \cdot W_2 - P = 0 \qquad \text{...(Eq. 15.19)}$$

From the two equations 15.18 and 15.19, the tension in the string and the acceleration a of the system may be solved if P, W_1 and W_2 are known.

In fact, of the six quantities—W_1, W_2, P, T, a and μ, any two may be obtained if the other four are known.

15.6.3 Motion of a Two Body System with One Body on an Inclined Plane and the other hanging in free in

Referring to Fig. 15.69 (a), let a body of weight W_1 be pulled up a rough inclined plane (coefficient of friction: μ) by means of a light flexible rope passing over a light frictionless pulley at the vertex of the inclined plane, a body of weight W_2 hanging vertically at the other end.

From the free body diagram of weight W_1:
Forces normal to the inclined plane:

$$N - W_1 \cos \beta = 0$$

$$\therefore \quad N = W_1 \cos\beta \qquad \ldots \text{(Eq. 15.20)}$$

Also,

$$F = \mu N = \mu W_1 \cos\beta \qquad \ldots \text{(Eq. 15.21)}$$

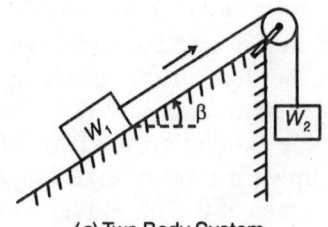

(a) Two Body System

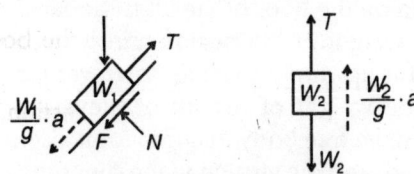

(b) Free Body Diagrams of the Bodies

Fig. 15.6: Two Body System on an Inclined Plane

Forces parallel to the inclined pla

$$T - F - \frac{W_1}{g} a - W_1 \sin\beta = 0$$

or $\quad T - \mu W_1 \cos\beta - \frac{W_1}{g} \cdot a - W_1 \sin\beta = 0$

$$\ldots \text{(Eq. 15.22)}$$

From the free body diagram of W_2:

Applying $\Sigma V = 0$,

$$T + \frac{W_2}{g} \cdot a - W_2 = 0 \qquad \ldots \text{(Eq. 15.23)}$$

The tension T in the string and the acceleration a of the system may be determined by solving Eqs. 15.22 and 15.23 simultaneously.

Note: If the inclined plane were *smooth*, the frictional force $F(= \mu W_1 \cos\beta)$ will vanish; otherwise, the equations and the solution continue to be applicable.

15.6.4 Motion of a Two-Body System with the Bodies Lying on Different Inclined Planes

Referring to Fig. 15.7, let a body of weight W_1 lie on an inclined plane of angle β_1 with horizontal

and another of weight W_2 on an inclined plane of angle β_2, and let them be connected by a light string passing over a pulley at the apex.

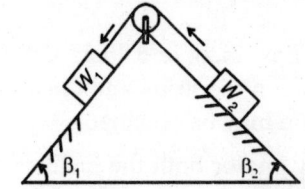

(a) Two Bodies on Two Inclined Planes

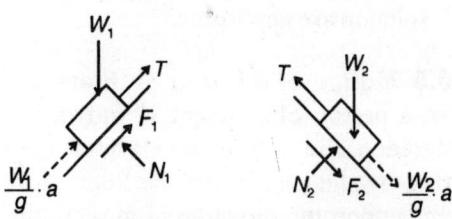

(b) Free Body Diagrams of the Bodies

Fig.15.7: Two Body System on Two Inclinded Planes

Let the two inclined planes be rough, with coefficients of friction μ_1 and μ_2, respectively. Let the assembly move down the inclined plane of $\beta_1 [W_1 > W_2$ and $\beta_1 > \beta_2]$.

Then $F_1 = \mu_1 N_1$ and $F_2 = \mu_2 N_2$

Forces normal to the plane β_1:

$$N_1 = W_1 \cos\beta_1$$

$$\therefore \quad F_1 = \mu_1 W_1 \cos\beta_1$$

Forces normal to the plane β_2:

$$N_2 = W_2 \cos\beta_2$$

$$\therefore \quad F_2 = \mu_2 W_2 \cos\beta_2$$

Forces parallel to the plane β_1:

$$T + \frac{W_1}{g} \cdot a + F_1 - W_1 \sin\beta_1 = 0$$

Substituting for F_1 we get

$$T + \frac{W_1}{g} \cdot a + \mu_1 W_1 \cos\beta_1 - W_1 \sin\beta_1 = 0$$

$$\ldots \text{(Eq. 15.24)}$$

Forces parallel to the plane β_2:

$$T - \frac{W_2}{g} \cdot a - F_2 - W_2 \sin\beta_2 = 0$$

Substituting for F_2 we get

$$T - \frac{W_2}{g} \cdot a - \mu_2 W_2 \cos\beta_2 - W_2 \sin\beta_2 = 0$$

...(Eq. 15.25)

Solving Eqs. 15.24 and 15.25 simultaneously, the tension T in the string and the acceleration a of the system may be obtained.

Note: If any one or both the inclined planes are smooth, the corresponding frictional forces vanish; otherwise, the equations and the solution are applicable.

15.6.5 Motion of a Lift or an Elevator

When a person of a weight W moves with an acceleration a in a lift or an elevator, the force exerted by the person on the floor of the lift depends upon the direction of motion, the free body diagrams of the person are shown in Fig. 15.8.

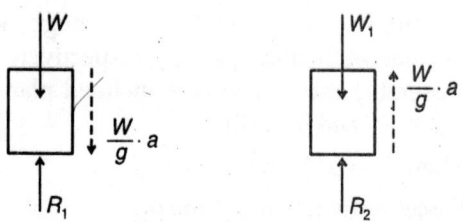

(a) Upward Motion   (b) Downward Motion

Fig. 15.8: Motion of a Person in a Lift or an Elevator

For upward motion, the inertia force of the person, $(W/g) \cdot a$, should be applied downward as shown in a:

From the free body diagram (a):

Applying $\Sigma V = 0$,

$$R_1 - W - \frac{W}{g} \cdot a = 0$$

or $\quad R_1 = W\left(1 + \frac{a}{g}\right)$ $\qquad$ ...(Eq. 15.26)

For downward motion, the inertia force of the person, $\left(\frac{W}{g}\right) \cdot a$, should be applied upward as shown in (b).

From the free body diagram (b):

Applying $\Sigma V = 0$,

$$R_2 - W + \frac{W}{g} \cdot a = 0$$

or $\quad R_2 = W\left(1 - \frac{a}{g}\right)$ $\qquad$ ...(Eq. 15.27)

Thus, R_1 and R_2 give the forces exerted by the person on the floor of the lift for upward and downward motions, respectively. It may be noted that the force exerted on the floor of the lift is more for upward motion than for downward motion. Further, if the lift moves with uniform velocity, the acceleration a is zero; then the force exerted on the floor of the lift is the same as that of the weight of the person and it remains the same for upward as well as downward motions.

If the weight of the lift or elevator cage is known, the free body diagram for it also may be sketched, and the tension in the supporting ropes may be determined if the acceleration is known; the inertia force must simply be shown in a direction opposite to that of the motion.

The versatility of the d'Alembert's principle in solving different kinds of problems in dynamics is illustrated by the examples in the next section.

ILLUSTRATIVE PROBLEMS

PROBLEM 15.1: A force of 300 N acts on a body of mass $m = 150$ kg. Find the acceleration of the body.

Solution:

Method I

Equation of Motion:

Let the acceleration of the body in the direction of the force be a.

Writing the equation of motion of the body

$$\Sigma F = ma$$

$300 = 150 \times 1$ (only one force is acting on the body)

$$a = \frac{300}{150} = 2 \text{ m/sec}^2$$

Method II

Equation of equilibrium:

Apply a fictitious force equal to $m\,a$ to the body, in a direction opposite to the direction of acceleration of the body.

Writing the equation of dynamic equilibrium of the body

$$\Sigma F - ma = 0$$

$$300 - 150 \times a = 0$$

or $a = 2 \text{ m/sec}^2$

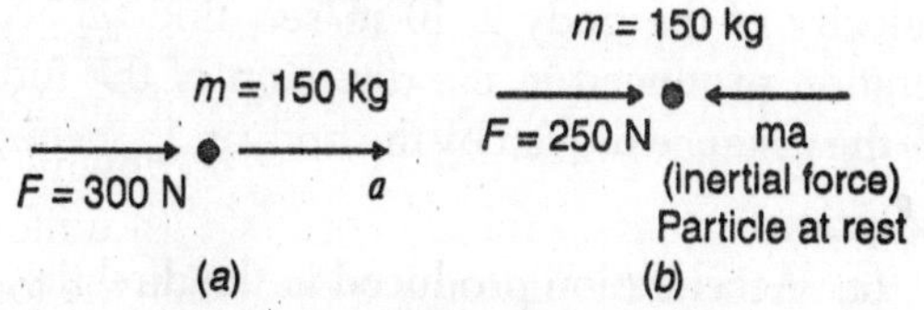

m = 150 kg

m = 150 kg
F = 300 N   a

F = 250 N   ma
(inertial force)
Particle at rest

(a)   (b)

Fig. 15.9 (Illustrative Problem 15.1)

PROBLEM 15.2: A man pulls a cart weighing 1500 N and produces an acceleration of 1.5 m/sec^2. Find the force exerted by the man.

Solution:

$$F = m \times a$$

$$m = \frac{1500}{9.81} \text{ kg}$$

$$a = 1.5 \text{ m/sec}^2$$

∴ $$F = \frac{1500}{9.81} \times 1.5$$

$$F = 229.358 \text{ N}$$

PROBLEM 15.3: A car weighing 30 kN moves on a level road under the 1.5 kN propelling force. Find the time taken by the car to increase its velocity from 27 kmph to 54 kmph.

Solution:

$$F = m \times a$$

$$F = 1.5 \text{ kN}$$

$$m = \frac{W}{g} = \frac{30}{9.81} \text{ kg}$$

$$1.5 = \frac{30}{9.81} \times a$$

$$a = \frac{9.81 \times 1.5}{30}$$

$$= 0.4905 \text{ m/sec}^2$$

$$v = 54 \text{ kmph} = \frac{54 \times 1000}{60 \times 60} = 15 \text{ m/sec}$$

$$u = 27 \text{ kmph} = \frac{27 \times 1000}{60 \times 60} = 7.5 \text{ m/sec}$$

$$v = u + at$$

or $$t = \frac{v - u}{a} = \frac{15 - 7.5}{0.4905}$$

$$= \frac{7.5}{0.4905} = 15.29 \text{ sec}$$

PROBLEM 15.4: A ball of mass 0.225 kg falls freely under gravity through a distance of 7.5 m. In the process of catching, a man allows his hand to drop a distance of 0.30 m. Find the average pressure on his hand during the catch.

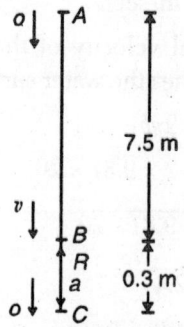

Fig. 15.10 (Illustrative Problem 15.4)

Solution: For the motion of the ball from A to B,

$$v^2 - u^2 = 2gh$$

$$v^2 = 2 \times 9.81 \times 7.5$$

$$v = \sqrt{2 \times 9.81 \times 7.5}$$

$$= 12.13 \text{ m/sec}^2$$

For the motion of ball from B to C

$$v^2 - u^2 = 2as$$

$$0 - (12.13)^2 = 2 \times a \times 0.30$$

$$a = -\frac{12.13 \times 12.13}{2 \times 0.30}$$

$$= -245.23 \text{ m/sec}^2$$

(–ve sign shows that this is retardation)

Equation of motion of the ball is

$$mg - R = ma$$

$$0.225 \times 9.81 - R = 0.225 \times 245.23$$

$$R = 0.225(245.23 + 9.81)$$

$$= 0.225 \times 255.04$$

$$= 57.384 \text{ N}$$

PROBLEM 15.5: A man weighing 650 N dives vertically downward into a swimming pool from a tower of height 20 m. He was found to go down in water by 2.043 m and then started rising. Calculate the average resistance of water. Neglect the resistance of air.

Solution: Considering the motion of the man from the top of tower to water surface.

$$h = 20 \text{ m}$$

$$u = 0$$

$$g = 9.81 \text{ m/sec}^2$$

Let $v =$ Final velocity of the man when he reaches the water surface

$$v^2 = u^2 + 2gh$$

$$v^2 = 0^2 + 2 \times 9.81 \times 20$$

$$v = \sqrt{2 \times 9.81 \times 20}$$

$$= \sqrt{392.4}$$

$$= 19.809 \text{ m/sec}$$

Considering the motion of man from water surface upto the point in water from where he started rising.

$$u = 19.809 \text{ m/sec}$$

$$v = 0$$

$$h = 2.043 \text{ m}$$

$$v^2 = u^2 + 2as$$

$$0^2 = 19.809^2 + 2 \times a \times 2.043$$

$$a = -\frac{19.809^2}{2 \times 2.043}$$

$$= -96.0344 \text{ m/sec}^2$$

(–ve sign indicates that this is retardation)

This retardation is due to water resistance.
Let R be the average resistance of water
Weight of man $= 650$ N

Net force acting on man in the upward direction $= (R - 650)$ Newtons

$$\therefore \qquad R - 650 = \frac{650}{9.81} \times 96.0344 \quad (\because F = m \cdot a)$$

$$R - 650 = 6363.13$$

$$R = 6363.13 + 650$$

$$R = 7013.13 \text{ N}$$

PROBLEM 15.6: A force of 300 N acts on a body having a mass of 10 kg for 15 seconds. If the initial velocity of the body is 10 m/sec, find (a) acceleration produced in the direction of the force, (b) the distance moved by the body in 15 seconds.

Solution:

(a) Acceleration produced in the direction of the force:

$$F = m \times a$$

$$F = 300 \text{ N}$$

$$m = 10 \text{ kg}$$

$$\therefore a = \frac{F}{m} = \frac{300}{10} = 30 \text{ m/sec}^2$$

(b) Distance moved by the body:

$$s = ut + \frac{1}{2}at^2$$

$$u = 10 \text{ m/sec}$$

$$t = 15 \text{ sec}$$

$$a = 30 \text{ m/sec}^2$$

$$s = 10 \times 15 + \frac{1}{2} \times 30 \times 15^2$$

$$s = 150 + 3375$$

$$s = 3525 \text{ m}$$

PROBLEM 15.7: A man weighs 600 N on the earth. Find his weight (a) on the moon, where the gravitational acceleration is 1.635 m/sec^2, and (b) on the sun, where the gravitational acceleration is 270 m/sec^2.

Solution:

$$F = m \times a$$

or weight = mass × gravitational acceleration

$$600 = m \times 9.81$$

$$m = \frac{600}{9.81} = 61.16 \text{ kg}$$

(a) Weight of the man on the moon

$$W = 61.16 \times 1.635$$

$W = 99.9966$

≈ 100 Newtons

(b) Weight of the man on the sun

$W = 61.16 \times 270$

$W = 1651.32$ Newtons

or $\quad = 1.65132$ kN

PROBLEM 15.8: A mass of 0.10 kg is rolled on grass with a velocity of 1 m/sec. If the resistance be 1/10th of the weight, how far will the body move.

Solution:

Mass $= 0.10$ kg

Weight $= 0.10 \times 9.81$

$\quad = 0.981$ Newton

$\therefore$ Resistance $= \dfrac{1}{10}$th of the weight

$\quad = \dfrac{1}{10} \times 0.981$

$\quad = 0.0981$ Newton

$R = m \times a$

$0.0981 = 0.10 \times a$

Retardation, $a = \dfrac{0.0981}{0.10} = 0.981$ m/sec^2

$\therefore$ Acceleration $a = -0.981$ m/sec^2

$v^2 - u^2 = 2as$

$0 - 1 \times 1 = 2 \times (-0.981) \times s$

$s = \dfrac{1 \times 1}{2 \times 0.981} = 0.5097$ m

or $\quad = 509.7$ mm

PROBLEM 15.9: Two bodies weighing 500 N and 300 N are connected to the two ends of a light inextensible string. The string is passing over a smooth pulley. Find (a) acceleration of the system, (b) tension in the string.

Solution:

Let $\quad a =$ Acceleration of the bodies in m/sec^2

$T =$ Tension in both the strings in Newtons

$$T - \frac{W_1}{g} a - W_1 = 0$$

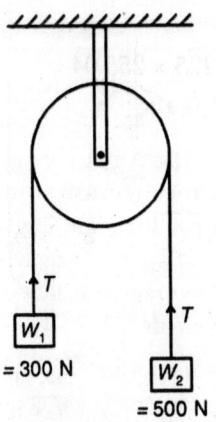

Fig. 15.11 (Illustrative Problem 15.9)

$$T - \frac{300a}{9.81} - 300 = 0$$

$$T - 30.281a = 300 \qquad \ldots(i)$$

$$T + \frac{W_2}{g} a - W_2 = 0$$

$$T + \frac{500}{9.81} a - 500 = 0$$

$$T + 50.968 a = 500 \qquad \ldots(ii)$$

$(ii) - (i)$ gives

$(50.968 + 30.581)a = (500 - 300) = 200$

$$a = \frac{200}{81.549} = 2.4525 \text{ m/sec}^2$$

$\therefore \quad T = 300 + 30.581 \times 2.4525$

$\quad = 300 + 75$

$\quad = 375$ N

or

$T = 500 - 50.968 \times 2.4525$

$\quad = 500 - 125$

$\quad = 375$ N

PROBLEM 15.10: Find the tension in the string and accelerations of blocks A and B weighing 500 N and 2000 N connected by a light inextensible string, and a frictionless and weightless pulley as shown in Fig. 15.12.

Solution:

Let $\quad T =$ Tension in the string

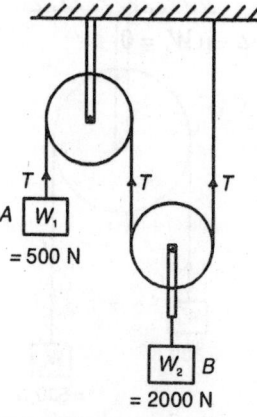

Fig. 15.12 (Illustrative Problem 15.10)

a = acceleration of W_2

$\therefore$ the acceleration of W_1 will be $2a$

$$T - \frac{W_1}{g}(2a) - W_1 = 0$$

$$T - \frac{500}{g}(2a) - 500 = 0$$

$$T - \frac{1000}{g}a = 500 \qquad \ldots(i)$$

$$2T + \frac{W_2}{g}a - W_2 = 0$$

$$2T + \frac{2000}{g}a - 2000 = 0$$

$$2T + \frac{2000}{g}a = 2000 \qquad \ldots(ii)$$

From equation (i), $T = 500 + (1000/g)\,a$
Substituting the value of T in equation (ii)

$$2\left(500 + \frac{1000}{g}a\right) + \frac{2000}{g}a = 2000$$

$$\frac{4000}{g}a = 1000$$

$$a = \frac{1000}{4000} \times 9$$

$$a = \frac{g}{4} = 2.4525 \text{ m/sec}^2$$

$$2a = 4.905 \text{ m/sec}^2$$

Acceleration of weight 2000 N
      = 2.4525 m/sec^2

Acceleration of weight 500N = 4.905 m/sec^2
Tension in the string,

$$T = 500 + \frac{500}{g} \times \frac{g}{4}$$

$$T = 750 \text{ N}$$

or

$$2T = 2000 - \frac{2000}{g}a$$

$$2T = 2000 - \frac{2000}{g} \times \frac{g}{4}$$

$$2T = 1500$$

$$T = 750\text{N}$$

PROBLEM 15.11: 1000 N block rests on a horizontal plane. A force of 500 N is applied to the block as shown in Fig. 15.13 (a). Find the acceleration of the block. The coefficient of friction between the block and the plane is 0.20.

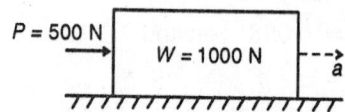

(a) Block Resting on a Rough Horizontal Plane

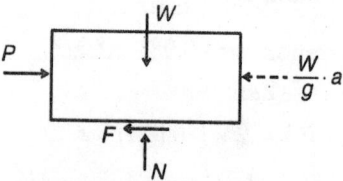

(b) Free Body Diagram of the Block

Fig. 15.13 (Illustrative Problem 15.11)

Solution: From the free body diagram of weight W:

Applying $\Sigma V = 0$, the normal reaction N from the plane is $N = W$

Applying $\Sigma H = 0$

$$P - F - \frac{W}{g}a = 0$$

But $F = \mu N = \mu W$

$\therefore$ $P - \mu W - \dfrac{W}{g}a = 0$

$$500 - 0.2 \times 1000 - \frac{1000}{9.81} \times a = 0$$

$$a = \frac{(500 - 200) \times 9.81}{1000} = 2.943 \text{ m/sec}^2$$

ROBLEM 15.12: Two bodies weighing 1000 N and 250 N are connected by a thread and move along a rough horizontal plane under the action of a force 500 N applied to the first body of weight 1000 N as shown in Fig. 15.14 (a). The coefficient of friction between the sliding surfaces of the bodies and the plane is 0.2. Determine the acceleration of the two bodies and the tension in the thread.

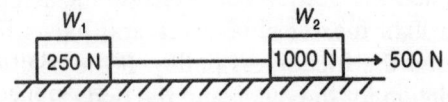

(a) Two Bodies Resting on a Rough Horizontal Plane

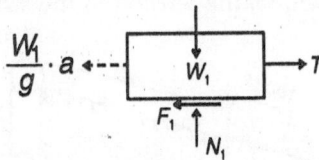

(b) Free Body Diagram of Weight W_1

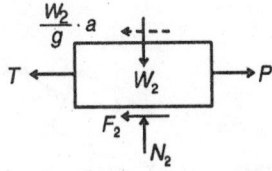

(c) Free Body Diagram of Weight W_2

Fig. 15.14 (Illstrative Problem 15.12)

Solution:

Let a = acceleration of the bodies
T = tension in the thread

From the free body diagram of the body of weight W_1:

Applying $\Sigma V = 0$, the normal reaction N_1 is given by

$$N_1 = W_1$$

Applying $\Sigma H = 0$

$$T - \frac{W_1}{g}a - F_1 = 0 \text{ but } F_1 = \mu N_1 = \mu W_1$$

$\therefore \quad T - \dfrac{W_1}{g}a - \mu W_1 = 0$

$$T - \frac{250}{9.81}a - 0.2 \times 250 = 0$$

$$T - 25.4842a = 50 \qquad \qquad ...(i)$$

From the free body diagram of the body of weight W_2:

Applying $\Sigma V = 0$, $N_2 = W_2$

Applying $\Sigma H = 0$

$$T + \frac{W_2}{g}a + F_2 - P = 0 \text{ But } F_2 = \mu N_2 = \mu W_2$$

$$T + \frac{W_2}{g} + \mu W_2 - P = 0$$

$$T + \frac{1000}{9.81}a + 0.2 \times 1000 - 500 = 0$$

$$T + 101.9368a = 500 - 200 = 300 \qquad ...(ii)$$

$(-) \; T - 25.4842a = 50 \qquad \qquad \qquad ...(i)$

$$\overline{127.421 \times a = 250}$$

$$a = \frac{250}{127.421} = 1.962 \text{ m/sec}^2$$

$\therefore \quad T = 50 + 25.4842 \times 1.962$

$$T = 50 + 50$$

$$T = 100 \text{ N}$$

or

$$T = 300 - 101.9368 \times 1.962$$

$$T = 300 - 200$$

$$T = 100 \text{ N}$$

PROBLEM 15.13: Two blocks of weight 20 N and 10 N are connected to the two ends of a light inextensible string, passing over a smooth pulley. The weight of 20 N is placed on a smooth horizontal surface while the weight 10 N is hanging free in air. Find (a) the acceleration of the system and (b) the tension in the string.

Solution: Let the weight W move down with an acceleration of a m/sec². The acceleration of W_1 is same as that of W_2.

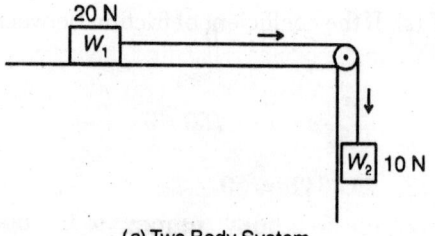

(a) Two Body System

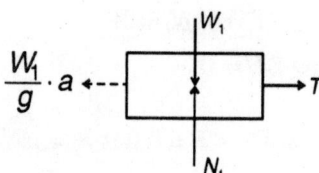

(b) Free Body Diagram of Weight W_1

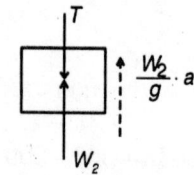

(c) Free Body Diagram of Weight W_2

Fig. 15.15 (Illustrative Problem 15.13)

Let T be the tension in the string.
From the free body diagram of weight W_1
Applying $\Sigma V = 0$; $N_1 = W_1$
Applying $\Sigma H = 0$

$$T - \frac{W_1}{g} \times a = 0$$

$$T - \frac{20}{g} \times a = 0 \qquad \dots(i)$$

From the free body diagram of weight W_2
Applying $\Sigma V = 0$

$$T - W_2 + \frac{W_2}{g} a = 0$$

$$T - 10 + \frac{10}{g} a = 0 \qquad \dots(ii)$$

$(ii) - (i)$ gives

$$-10 + \frac{10}{g} a + \frac{20}{g} a = 0$$

$$\frac{30}{g} \times a = 10$$

$$a = \frac{10}{30} \times g = \frac{10}{30} \times 9.81$$

$$a = 3.27 \text{ m/sec}^2$$

$$\therefore \qquad T = \frac{20}{g} a$$

$$= \frac{20}{9.81} \times 3.27$$

$$T = 6.67 \text{ N}$$

PROBLEM 15.14: A body of weight 100 N rests on a rough planed inclined at 15° to the horizontal. It is being pulled by a body of weight 75 N. The 75 N body is connected to the first body by a light inextensible string and hangs freely beyond the frictionless pulley. If the coefficient of friction for the plane and the body is 0.2, find (*a*) the acceleration with which the body will come down, (*b*) the tension in the string.

(a) Two Body System

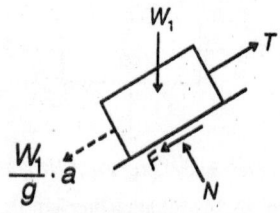

(b) Free Body Diagram of Weight W_1

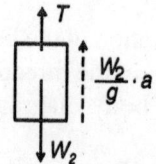

(c) Free Body Diagram of Weight W_2

Fig. 15.16 (Illustrative Problem 15.14)

Solution: From the free body diagram of weight W_1:
Forces normal to the inclined plane:

$$N - W_1 \cos\beta = 0$$

$\therefore \quad N = W_1 \cos \beta$

$\qquad N = 100 \times \cos 15°$

Also, $\quad F = \mu N = 0.2 \times 100 \cos 15°$

$\qquad F = 20 \cos 15°$

Forces parallel to the inclined plane:

$$T - F - \frac{W_1}{g} a - W_1 \sin \beta = 0$$

$$T - 20 \cos 15° - \frac{100}{g} a - 100 \sin 15° = 0$$

$$T - \frac{100}{g} a = 100 \sin 15° + 20 \cos 15°$$

$$T - \frac{100}{g} a = 25.8819 + 19.3185$$

$$T - \frac{100}{g} a = 45.20 \qquad \qquad ...(i)$$

From the free body diagram of W_2:

Applying $\Sigma V = 0$

$$T + \frac{W_2}{g} a - W_2 = 0$$

$$T + \frac{75}{g} a = 75 \qquad \qquad ...(ii)$$

$(ii) - (i)$ gives

$$\frac{175 \times a}{g} = 75 - 45.20 = 29.80$$

$$a = \frac{29.80 \times 9.81}{175}$$

$$a = 1.6705 \text{ m/sec}^2$$

Substituting the value of 'a' in Eqn. (i)

$$T = 45.20 + \frac{100}{9.81} \times 1.6705$$

$$= 45.20 + 17.03$$

$$T = 62.23 \text{ N}$$

PROBLEM 15.15: Two rough inclined planes whose inclination with the horizontal are 60° and 30°, are placed back to back. Two bodies of weight 150 N and 50 N are placed on them and are connected by a light inextensible string passing over a smooth pulley as shown in Fig.

15.17 (*a*). If the coefficient of friction between the blocks and the planes is 0.3, find the acceleration of the system and the tension in the string.

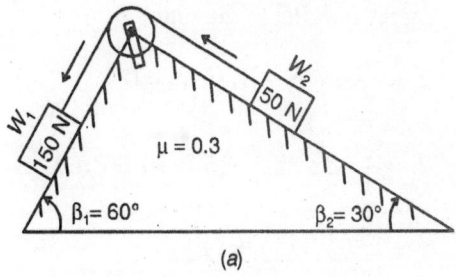

(a)

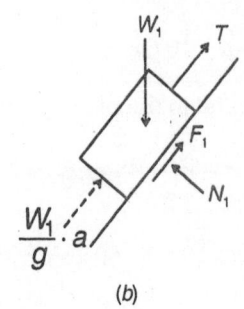

(b)

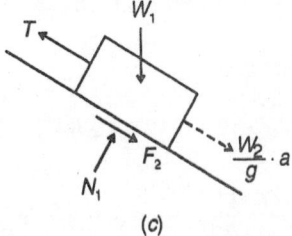

(c)

Fig. 15.17 (Illustrative Problem 15.15)

Solution:

$$F_1 = \mu_1 N_1$$

$$F_2 = \mu_2 N_2$$

$$\mu_1 = \mu_2 = 0.3$$

$\therefore \quad F_1 = 0.3 N_1 \ \& \ F_1 = 0.3 N_2$

Forces normal to the plane β_1 :

$$N_1 = W_1 \cos \beta_1$$

$$F_1 = \mu_1 W_1 \cos \beta_1$$

$$= 0.3 \times 150 \times \cos 60°$$

$$= 22.5 \text{ N}$$

Forces normal to the plane β_2:

$$N_2 = W_2 \cos \beta_2$$

$$F_2 = \mu_2 W_2 \cos\beta_2$$

$$= 0.3 \times 50 \times \cos 30°$$

$$= 12.99 \text{ N}$$

Forces parallel to the plane β_1:

$$T + \frac{W_1}{g}a + F_1 - W_1\sin\beta_1 = 0$$

$$T + \frac{150}{g}a + 22.5 - 150 \times \sin 60° = 0$$

$$T + \frac{150}{g}a = 129.9 - 22.5$$

$$T + \frac{150}{g}a = 107.4 \qquad ...(i)$$

Forces parallel to the plane β_2:

$$T - \frac{W_2}{g}a - F_2 - W_2\sin\beta_2 = 0$$

$$T - \frac{50}{g}a - 12.99 - 50 \times \sin 30° = 0$$

$$T - \frac{50}{g}a = 12.99 + 25 = 37.99 \qquad ...(ii)$$

$(i) - (ii)$ gives

$$\frac{200}{g}a = 107.4 - 37.99 = 69.41$$

$$a = \frac{69.41 \times 9.81}{200} = 3.4046 \text{ m/sec}^2$$

$$T = 107.4 - \frac{150}{9.81} \times 3.4046$$

$$= 107.4 - 52.0581$$

$$T = 55.3419 \text{ N}$$

Or

$$T = 37.99 + \frac{50}{9.81} \times 3.4046$$

$$= 37.99 + 17.3527$$

$$= 55.3427 \text{ N}$$

$$a = 3.405 \text{ m/sec}^2$$

And

$$T = 53.342 \text{ N}$$

PROBLEM 15.16: A lift carries a weight of 1000 N and is moving with a uniform acceleration of 1.962 m/sec^2. Calculate the tension in the cables supporting the lift, when (a) lift is moving upwards and (b) lift is moving downward:

Take $g = 9.81$ m/sec^2

$$W = 1000 \text{ N}$$

$$a = 1.962 \text{ m/sec}^2$$

$$g = 9.81 \text{ m/sec}^2$$

Soution:

(a) When the lift moves up

$$R_1 = W\left(1 + \frac{a}{g}\right)$$

$$= 1000\left(1 + \frac{1.962}{9.81}\right)$$

$$= 1000(1 + 0.20)$$

$$= 1200 \text{ N}$$

(b) When the lift moves down

$$R_2 = W\left(1 - \frac{a}{g}\right)$$

$$= 1000\left(1 - \frac{1.962}{9.81}\right)$$

$$= 1000(1 - 0.2)$$

$$= 1000 \times 0.8$$

$$= 800 \text{ N}$$

PROBLEM 15.17: The elevator in a mine vertical shaft weighs 10,000 N and requires 20 sec to descent 93 m, from rest until it stops at the bottom. The velocity of the elevator is 6 m/sec, except during the starting and stopping periods. If the tension in the cable, supporting the elevator, is 9000 N during the starting period, what is its value during the stopping period, assuming constant deceleration?

Solution: Weight of the elevator, $W = 10,000$ N

Let *OABC* represent the velocity-time graph of the elevator, in which *OA* represents the period of acceleration, *AB* the period of uniform velocity and *BC* the period of retardation as shown in Fig. 15.18. First of all consider the motion of the elevator from *O* to *A*.

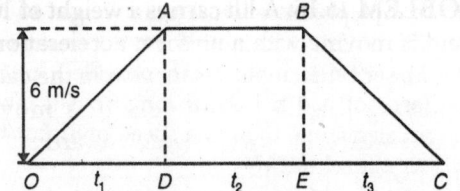

Fig. 15.18: Velocity-time Graph (Illustrative Problem 15.17)

Tension in cable, $R_1 = 9000$ N

Let $a =$ Acceleration of the cage

$$W = 10,000 \text{ N}$$

$$R_1 = W\left(1 - \frac{a}{g}\right) \quad (\because \text{descending})$$

$$9000 = 10000\left(1 - \frac{a}{9.81}\right)$$

$$1 - \frac{a}{9.81} = 0.9$$

or $a = 0.981 \text{ m/sec}^2$

In triangle OAD,

$$a = \frac{AD}{OD}$$

$$0.981 = \frac{6}{t_1}$$

or $t_1 = \dfrac{6}{0.981} = 6.1162 \text{ sec}$

$OC = 20 \text{ sec}$

$t_2 =$ time required to travel from A to B

Area of the figure $OABC = 93$

$$93 = \left(\frac{AB + OC}{2}\right) AD$$

$$93 = \left(\frac{t_2 + 20}{2}\right) \times 6$$

$$t_2 = \frac{93 \times 2}{6} - 20$$

$$t_2 = 31 - 20$$

$$t_2 = 11 \text{ sec}$$

∴ Time required to travel from B to C,

$$t_3 = 20 - t_1 - t_2$$

$$= 20 - 6.1162 - 11$$

$$= 2.8838 \text{ sec}$$

Now consider the motion of the elevator from B to C

$$u = 6 \text{ m/sec}$$

$$v = 0$$

$$v = u + at$$

$$0 = 6 + a \times 2.8838$$

$$\therefore \quad a = -\frac{6}{2.8838} = -2.0806 \text{ m/sec}^2$$

Let $R_2 =$ Tension in the cable, during stopping period

$$R_2 = W\left(1 - \frac{a}{g}\right)$$

$$= 10000\left(1 + \frac{2.0806}{9.81}\right)$$

$$= 12120.9 \text{ N}$$

PROBLEM 15.18: An elevator weighs 3000 N and is moving vertically downwards with a constant acceleration. Starting from rest it travels a distance of 45 metres during an interval of 12 seconds. Find the cable tension during this time. Neglect all other resistances to motion. What are the limits of cable tension?

Solution:

Distance travelled by the elevator, $s = 45$ m

$$u = 0$$

$$t = 12 \text{ sec}$$

$$a = ?$$

$$s = ut + \frac{1}{2}at^2$$

$$45 = 0 \times 12 + \frac{1}{2}(a)(12)^2$$

or $a = \dfrac{45 \times 2}{144} = 0.625 \text{ m/sec}^2$

Elevator is moving vertically downwards

$$\therefore \quad R_2 = W\left(1 - \frac{a}{g}\right)$$

$$= 3000\left(1 - \frac{0.625}{9.81}\right)$$

$$= 2808.87 \text{ N}$$

Limits of cable tension: The cable tension will have two limits i.e., when acceleration is zero and when acceleration is maximum (i.e. 9.81 m/sec²).

∴ Cable tension when acceleration is zero.

$$R_2 = W\left(1 - \frac{a}{g}\right)$$

$$R_2 = 3000\left(1 - \frac{0}{g}\right) = 3000 \text{ N}$$

Cable tension when acceleration is maximum (i.e. 9.81 m/sec²)

$$R_2 = W\left(1 - \frac{a}{g}\right)$$

$$= 3000\left(1 - \frac{9.81}{9.81}\right) = 0$$

PRACTICE PROBLEMS

15.1   A vehicle of mass 300 kg can accelerate at an acceleration of 8 m/sec². Find the tractive force developed by the car.

15.2   A body weighing 100 N is moving over a smooth surface, whose equation of motion is given by the relation

   $$s = 4t + 3t^2$$

   where s is in metres and t is in seconds. Find the magnitude of force responsible for the motion.

15.3   A force of 15 N acts on a body for 10 seconds, and causes it to move 20 m in this time. Find the mass of the body.

15.4   A mass of 5 kg falls from rest vertically a distance 10 m and is brought to rest by penetrating a distance of 1.0 m into sand. Find the average pressure exerted by sand.

15.5   A railway car weighing 75 kN is moving on a level track. The tractive force exerted by the railway car is 2500 N. If the frictional resistance is 5 N per kN of the railway car's weight, find the acceleration of the railway car.

15.6   A vehicle of mass 600 kg, is moving with a velocity of 30 m/sec. A force of 300 N acts on it for 100 seconds. Find the final velocity of the vehicle: (a) when the force acts in the direction of the motion, and (b) when the force acts in the opposite direction of the motion.

15.7   A multiple-unit electric train weighs 10 MN. The resistance to motion is 1% of train weight. The electric motors can provide a tractive force of 250 kN. How long does it take to accelerate the train to a speed of 81 km/hour on a level track?

15.8   A pulley whose axis passes through the centre O carries load as shown in Fig. 15.19. Neglecting the inertia of the pulley and assuming that the cord is inextensible, find the acceleration of the block B. How much weight should be added to or taken away from the block B, if the acceleration of the block B is required to be $g/3$ downwards ?

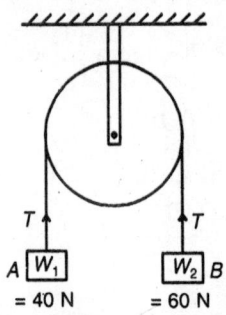

Fig. 15.19 (Practice Problem 15.8)

15.9   A system of weights connected by string passing over pulleys A and B is shown in Fig. 15.20. Find acceleration of the three weights, assuming weightless strings and ideal conditions for pulleys.

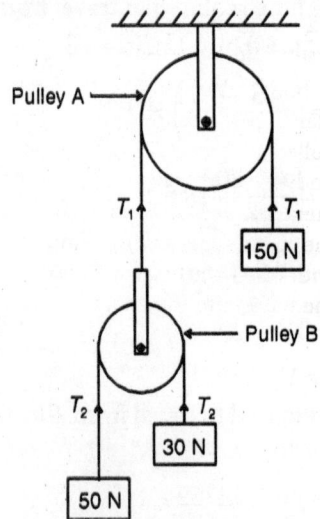

Fig. 15.20 (Practice Problem 15.9)

15.10 1000 N block rets on a horizontal plane as shown in Fig. 15.21. Find the magnitude of the force *P* required to give the block an acceleration of 2 m/sec^2 to the right. The coefficient of friction between the block and the plane is 0.30.

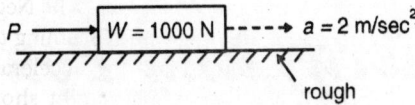

Fig. 15.21 (Practice Problem 15.10)

15.11 Two bodies weighing 1000 N and 250 N are connected by a thread and move along a smooth horizontal plane under the action of a force 250 N applied to the first body of weight 1000 N as shown in Fig. 15.22. Determine the acceleration of the two bodies and the tension in the thread.

Fig. 15.22 (Practice Problem 15.11)

15.12 Two blocks of weight 40 N and 20 N are connected to the two ends of a light inextensible string, passing over a smooth pulley. The weight of 40 N is placed on a rough horizontal surface while the weight 20 N is hanging free in air. The coefficient of friction between the weight 40 N and the plane surface equal to 0.2. Determine (*a*) the acceleration of the system (*b*) the tension in the string.

15.13 A body of weight 100 N, lying on a smooth plane inclined at 30° to the horizontal, is being pulled by a body of weight 75 N. The 75 N body is connected to the first body by a light inextensible string and hangs freely beyond the frictionless pulley. Find the acceleration with which the body will come down. Also find the tension in the string.

15.14 Two smooth inclined planes whose inclinations with the horizontal are 30° and 20°, are placed back to back. Two bodies of weight 50 N and 30 N are placed on them and are connected by a light inextensible string passing over a smooth pulley as shown in Fig. 15.23. Calculate the acceleration of the system and the tension in the string.

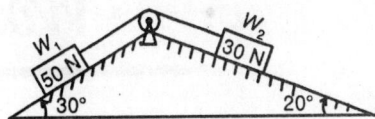

Fig. 15.23 (Practice Problem 15.14)

15.15 An elevator is required to lift a body of mass 60 kg. Find the acceleration of the elevator, which would cause a pressure of 750 N on the floor.

15.16 A lift has an upward acceleration of 1 m/sec^2. What pressure will a man weighing 600 N will exert on the floor of this lift? What pressure would he exert, if the lift had an acceleration of 1 m/sec^2 downward?

Also, find the upward acceleration, which would cause the man to exert a pressure of 722.324 N on the floor.

15.17 An elevator of total weight 5000 N starts to move upwards with a constant acceleration and acquires a velocity of 3 m/sec after travelling a distance of 3 m. Find the tensile force in the cable during the accelerated motion.

The above elevator while moving up with a velocity of 3 m/sec is uniformly decelerated to stop in 3 seconds. Find the pressure at the floor of the elevator under the effect of a man weighing 600 N riding in the elevator.

15.18 An elevator weighing 3000 N is ascending with an acceleration of 2 m/sec^2. During this ascent its operator whose weight is 600 N is standing on the scale placed on the floor. What is the scale reading? What will be the total tension in the cables of the elevator during this motion?

Chapter 16

Work, Energy and Power

16.1 INTRODUCTION

The approaches of Newton's laws of motion and d'Alembert's principle for solving kinetics problems have been dealt with in the previous chapter. In this chapter an alternative approach—the work-energy method—is considered. Work and energy relate the mass, force, velocity, and displacement of a particle/body, or a system of particles/bodies.

The advantage of the work-energy method is that the velocity can be obtained directly without the need to obtain the acceleration; further, work and energy are scalar quantities.

Therefore the concepts of work, energy and power, and the related ideas, including applications to dynamics problems, are set out in the following sections.

16.2 WORK

The 'work' done by a force on a particle/body is defined as the product of the force and the displacement of the particle/body in the direction of the force.

If the displacement occurs in a direction inclined at an angle α to the direction of the force P, the work done U by the force is:

$$U = P(s \cos \alpha) \qquad \text{...(Eq. 16.1)}$$

This may also be written as

$$U = (P \cos \alpha)s \qquad \text{...(Eq. 16.2)}$$

From this equation, the work done by a force may be defined as *the product of the component of the force in the direction of motion and the distance moved.*

The units of work are obviously Newton-metres (N.m); one N.m is called one 'Joule' (J), in honour of the scientist by that name. Thus one Joule is the amount of work done by a force of one Newton acting on a particle/body when the latter moves through one metre. The other related units are kN.m (kJ) and N.mm (mJ).

If a force is not constant but varies during the motion of a particle/body on which it acts in a specific and known manner, the work done may be represented graphically as the area under the P-s diagram as shown in Fig. 16.1.

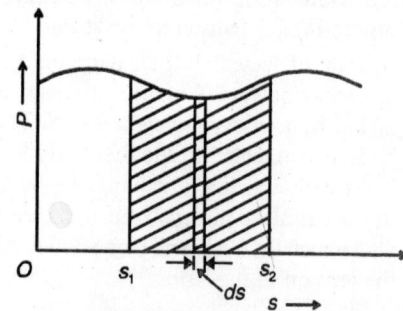

Fig. 16.1: Work Done by a Varying Force

Under an infinitesimal displacement ds, the small amount of work done dU is $dU = P \cdot ds$

The work done U under a finite displacement s—say form s_1 to s_2 is:

$$U = \int_{s_1}^{s_2} P \cdot ds \qquad \text{...(Eq. 16.3)}$$

The following points may be noted about the work done by a force:

(*i*) Work done by a force is zero if either the displacement is zero or if the force acts in a direction normal to that of motion. For example, weight of a body or gravity force does not do any work when the body moves horizontally.

(*ii*) Work done by a force is *positive* if the displacement occurs in the direction of the force, and *negative* if it occurs opposite to the direction of the force. For example, the work done *by* the weight of a body or by the gravity force is *positive* if the body moves downward, and work is done *against* the force of gravity if the body moves upward.

(*iii*) Work is a scalar quantity; it has magnitude and sign, but no direction.

(*iv*) Work done by a force does not depend upon the path between specified points in the case of forces of gravity, spring force and elastic force, which are called "conservative forces". However, it does depend upon the path of motion in the case of forces like friction, which are called "non-conservative forces".

More about this differentiation will be seen in a later section.

16.3 ENERGY

'Energy' is defined as 'the capacity to do work'. Energy, therefore, has the same units as work, and also is a scalar quantity.

Energy can manifest itself in different forms—mechanical energy, electrical energy, thermal (heat) energy, chemical energy, Nuclear energy, and so on.

Mechanical energy can be further classified as 'Potential energy' and 'Kinetic energy'.

In mechanics, we are concerned with mechanical energy in its different forms; these are considered in the next section.

16.4 FORMS OF MECHANICAL ENERGY

The two forms of mechanical energy are (*i*) Potential energy and (*ii*) Kinetic energy.

16.4.1 Potential Energy

Potential energy is that which a body possesses by virtue of its *position*.

A body of weight W held at a height h above an arbitrary datum possesses a potential energy $W \cdot h$.

16.4.2 Kinetic Energy

Kinetic energy is that which a body possesses by virtue of its velocity of motion.

To obtain a general expression for the kinetic energy of a body in motion in terms of its mass and velocity, we can proceed as follows:

Let a car of weight W be moving with a velocity v. If the engine is cut off, it still moves forward, doing work against frictional resistance, and comes to a stop after covering a certain distance s from the point where the engine was cut off (Fig. 16.2).

From the kinematics of rectilinear motion, we get

$$0 - v^2 = 2as$$

or the acceleration $= -\dfrac{v^2}{2s}$ (or the retardation is $= \dfrac{v^2}{2s}$).

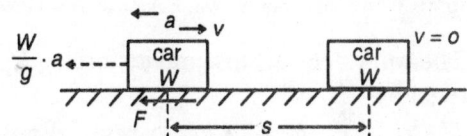

Fig. 16.2: Concept of Kinetic Energy of a Car Doing Work

Applying d'Alembert's principle,

$$\Sigma H = 0$$

$$F + \frac{W}{g} \cdot a = 0$$

Substituting for a,

$$F - \frac{W}{g} \cdot \frac{v^2}{2s} = 0$$

or $\quad F = \dfrac{1}{2} \cdot \dfrac{W}{gs} \cdot v^2$

Work done against frictional resistance F is given by

$$F \times s = \frac{1}{2} \cdot \frac{W}{g} \cdot v^2$$

This work done was because of the kinetic energy possessed by the body while moving at the velocity v just before the engine was cut off.

∴ Kinetic energy,

$$E_k = \frac{1}{2} \cdot \frac{W}{g} \cdot v^2 = \frac{1}{2} m \cdot v^2 \qquad ...(Eq. 16.4)$$

where m is the mass of the body.

16.5 POTENTIAL ENERGY AND CONSERVATIVE FORCES

The potential energy of a body is the energy it possesses by virtue of its position.

Let us consider a body of mass m—moving from position 1 to position 2 along a certain path-I; let the elevation of positions 1 and 2 be h_1 and h_2 above an arbitrary datum as shown in Fig. 16.3.

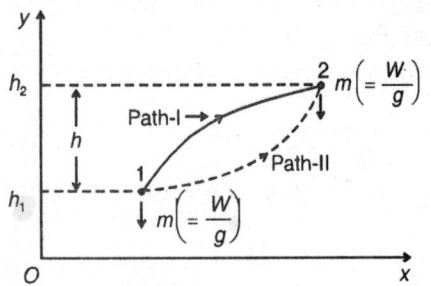

Fig. 16.3: Potential Energy—Work Done against Gravity

The work done against gravity force—U_{1-2}—is given by

$$U_{1-2} = W(h_1 - h_2) = -Wh = -mgh \quad ...(Eq. 16.5)$$

(negative since the body is moved against gravity)

The increase in potential energy, E_p, from position 1 to position 2 is given by,

$$E_p = W(h_2 - h_1) = Wh = mgh \quad ...(Eq. 16.6)$$

Thus it may be seen that the work done against gravity is stored in the form of increase in potential energy of the body.

Even if the body moves along an alternative path-II between the same positions 1 and 2, the difference in elevation remains the same, and hence the work done against gravity remains the same, as given by Eq. 16.5.

When the work done by a force in moving a body from one position to another is independent of the path followed, such a force is called a *conservative* force. Thus, gravity force is obviously a conservative force. In addition, spring force and elastic force are also conservative forces.

However, frictional forces are of the non-conservative type since the work done in this case certainly depends on the length of the path followed by the moving body in resisting the friction.

16.6 WORK-ENERGY EQUATION FOR TRANSLATION

Let us consider a body weight W, acted on by a force P(or the resultant P of a system of forces) in the x-direction, making the body move with an acceleration a in this direction (Fig. 16.4).

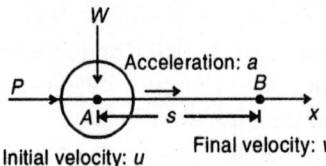

Initial velocity: u Final velocity: v

Fig. 16.4: Work-Energy Concept in Translation

Let the initial velocity of the body be u at A and its final velocity v at B, AB being s.

From Newton's second law of motion,

$$P = \frac{W}{g} \cdot a$$

Multiplying both sides by an infinitesimal distance ds,

$$P \cdot ds = \frac{W}{g} \cdot a \cdot ds$$

$$= \frac{W}{g} \cdot \frac{dv}{dt} \cdot ds$$

$$= \frac{W}{g} \cdot \frac{dv}{ds} \cdot \frac{ds}{dt} \cdot ds$$

$$= \frac{W}{g} \cdot v \cdot \frac{dv}{ds} \cdot ds$$

or $P \cdot ds = \dfrac{W}{g} \cdot v \cdot dv$

Integrating both sides, and applying the limits applicable at A and B,

$$\int_0^s P \cdot ds = \int_u^v \frac{W}{g} \cdot v \cdot dv$$

or $\quad P \cdot s = \dfrac{1}{2} \cdot \dfrac{W}{g} \cdot (v^2 - u^2)$ $\qquad$ ...(Eq. 16.7)

The left-hand side of this equation gives the work done by the force/force-system on the body; on multiplying out the right-hand side, the first term gives the final kinetic energy and the second gives the initial kinetic energy of the body as it moves through the distance s from A to B.

Thus, it is seen that the work done in rectilinear motion/translation is equal to the change in kinetic energy.

Put in symbols:

$\qquad U = E_k(\text{final}) - E_k(\text{initial})$ $\qquad$ ...(Eq. 16.8)

This equation is called the "Work-Energy equation".

(This is applicable even for curvilinear translation, if tangential components are considered).

The work-energy principle may now be stated:

"The work done by a system of forces on a body during a certain displacement equals the change in its kinetic energy during the same displacement."

Note: If the motion is a general plane motion involving translation and rotation, the work done shall include that in rotation also ($M \cdot \theta$, if M is the moment and θ is the angular displacement), and the kinetic energy also shall include that in rotation $(1/2)I \cdot \omega^2$, I being the moment of inertia of the body about the axis of rotation and ω the angular velocity). Then the principle can be applied.

The work done by internal forces like the tension in a string has also to be included unless these appear as action and reaction moving through equal distances. This principle is found to be very useful in solving certain problems in kinetics when acceleration is not of interest.

16.7 LAW OF CONSERVATION OF ENERGY

This is a phenomenon of nature and hence may be called a 'law' rather than a 'principle'. The law may be stated:

"The sum of the potential energy and kinetic energy of a particle/body in motion under the action of conservative forces remains constant."

The proof of this statement is rather simple, given the work-energy equation of the previous sections.

$\quad U_{1\text{-}2} = E_k(\text{final}) - E_k(\text{initial})$ $\quad$ ...(Eq. 16.8)

$\qquad = E_{k2} - E_{k1}$

$\quad U_{1\text{-}2} = W(h_1 - h_2) = -Wh$ $\qquad$ ...(Eq. 16.6)

$\qquad = -(E_{p2} - E_{p1})$

(negative change in potential energy)

$\therefore \quad E_{k2} - E_{k1} = -(E_{p2} - E_{p1})$

or $\quad (E_{k1} + E_{p1}) = (E_{k2} + E_{p2})$ $\quad$ ...(Eq. 16.9)

which proves the statement of the law.

It may be noted that this is applicable only for conservative forces, and not applicable when frictional forces, which are non-conservative are involved.

For example, the law is valid in the case of a simple pendulum if the friction at the support and the resistance of the air during the motion of the bob are considered negligible.

Note: In the more general case of the law of conservation of energy, even the work done by non-conservative forces may be included, but this manifests itself in other forms of energy like thermal energy.

Energy can neither be created nor destroyed in nature, which is a fact of nature.

16.8 POTENTIAL ENERGY AND EQUILIBRIUM

The potential energy of a system depends upon the position of the system, which may be defined by one or more independent variables. This depends upon the degrees of freedom of the system, which was defined in an earlier chapter. The degree of freedom depends upon the Kinematic constraints (restrictions to motion) imposed on the system.

In Section 16.5, it has been shown that work done *against* conservative forces would result in a corresponding increase in the potential energy. The principle of virtual work explained in Chapter 10 may be restated in terms of potential energy of a system with a single degree of freedom as mentioned below:

"For a system in equilibrium the derivative of its potential energy is zero".

or $\dfrac{dE_p}{dx} = 0$

if x is the direction in which the degree of freedom exists.

16.9 KINDS OF EQUILIBRIUM

When a small displacement is given to a system in equilibrium one of three things can happen when the disturbing force is removed.

(i) The system may return to its original position. This kind of equilibrium is called "Stable equilibrium".

(ii) The system may move farther and farther away from the original or equilibrium position. This kind of equilibrium is known as "Unstable equilibrium".

(iii) The system may continue to remain in the new position without returning to its original position or moving away from it. This type is known as "neutral equilibrium".

These are shown in Fig. 16.5.

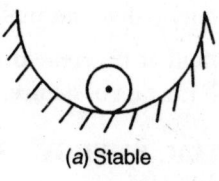

(a) Stable

(b) Unstable

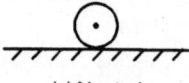

(c) Neutral

Fig. 16.5: Kinds of Equilibrium

The nature of the equilibrium is dependent upon the potential energy of the system with respect to a convenient reference.

A system will be in stable equilibrium when the potential energy of the system is a *minimum*; it will be in unstable equilibrium when the potential energy is a maximum; and it will be in neutral equilibrium when the potential energy is a constant.

Mathematically speaking, the first derivative of the potential energy with respect to the degree of freedom shall be zero; and the second derivative shall be respectively positive, negative, or zero for the three conditions stated above :

$\dfrac{dE_p}{dx} = 0$ ...(Eq. 16.10)

$\dfrac{d^2 E_p}{dx^2} > 0$ for stable equilibrium ...(Eq. 16.11)

$\dfrac{d^2 E_p}{dx^2} < 0$ for unstable equilibrium ...(Eq. 16.12)

$\dfrac{d^2 E_p}{dx^2} = 0$ for neutral equilibrium ...(Eq. 16.13)

Note:

(1) A body with more than a single degree of freedom may be stable for some motion and unstable for others.

(2) The potential energy of a system with multi-degree freedom depends upon several variables, thus making the analysis more complex.

(3) The positions of equilibrium of the system may be determined by using $\dfrac{dE_p}{dx} = 0$.

(4) The nature of the equilibrium for any position may be determined by checking whether $\dfrac{d^2 E_p}{dx^2}$ is positive, negatve or zero.

Thus the prerequisite for determining the equilibrium positions and the nature of the equilibrium at each of these is the determination of the potential energy of the system.

16.10 WORK AND ENERGY OF A RIGID BODY IN PLANE MOTION

Work of a Rigid Body

When a rigid body is in plane motion, the work done on it is the sum of the work done in translation and that done in rotation.

Let us consider a rigid body of mass m rotating with an angular velocity ω about its centre of mass G under the influence of a moment M, and acted on by a resultant force P. Let it undergo a displacement s in the direction of P and an angular displacement θ as shown in Fig. 16.6.

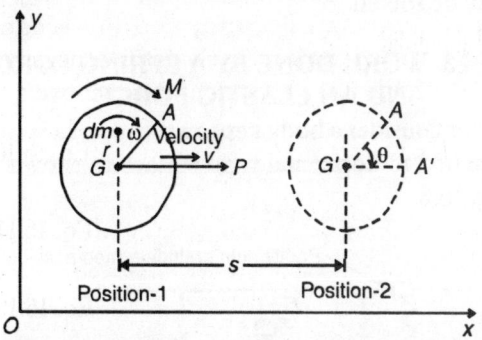

Fig. 16.6: Work Done on a Rigid Body in Plane Motion

Work done in translation

$$U_t = P \cdot s$$

Work done in rotation

$$U_r = M \cdot \theta$$

Total work done in plane motion from position 1 to position 2

$$U_{1\text{-}2} = P \cdot s + M \cdot \theta \qquad \text{...(Eq. 16.14)}$$

Note:

(1) The reaction of a pin about which the body rotates does not do any work since the pin does not translate.

(2) The normal reaction from a smooth surface on which a body is in motion does not do any work.

(3) When a body rolls on a fixed surface without sliding or slipping, the friction force, if any, between the body and the surface does not do any work.

Energy of a Rigid Body

The potential energy E_P of a rigid body continues to be mgh or Wh (W being its weight), h being its elevation above an arbitrary but convenient datum.

The kinetic energy of a rigid body in plane motion is the sum of its kinetic energy in translation and that in rotation.

Kinetic energy of the body in translation

$$E_k = \frac{1}{2} m \cdot v^2 = \frac{1}{2} \cdot \frac{W}{g} \cdot v^2 \qquad \text{...(Eq. 16.4)}$$

Kinetic energy of the body in rotation may be got as follows:

Let us consider a particle of mass dm located at a distance r from the axis of rotation (Fig. 16.6).

Tangential velocity v_t of the particle $= \omega\, r$

Kinetic energy of the particle in rotation

$$dE_k = \frac{1}{2} \cdot dm.(\omega r)^2$$

$$= \frac{1}{2}\omega^2 \cdot (dm \cdot r^2)$$

$\therefore$ Kinetic energy of the body in rotation

$$E_k = \int \frac{1}{2}\omega^2 (r^2 \cdot dm) = \frac{1}{2}\omega^2 \int r^2 \cdot dm$$

$\int r^2 \cdot dm$ represents the mass moment of inertia I of the body about the axis of rotation.

$$\therefore \quad E_k (\text{in rotation}) = \frac{1}{2} I\omega^2 \qquad \text{...(Eq. 16.15)}$$

as indicated in the note at the end of Section 16.6.

The total energy $E_{1\text{-}2}$ of the body *in plane motion*, therefore, is given by

$$E = E_{P_{1\text{-}2}} + E_{k_{1\text{-}2}}$$

or
$$E = mgh + \frac{Wv^2}{2g} + \frac{1}{2} I\omega^2 \qquad \text{...(Eq. 16.16)}$$

16.11 WORK-ENERGY PRINCIPLE FOR A RIGID BODY IN PLANE MOTION

In Section 16.6, we have seen the work-energy principle in translation. This principle may be extended for a rigid body in plane motion, provided the work done and kinetic energy in rotation of the body are included along with those in translation in the respective sides of Eq. 16.8.

Similarly, the principle or law of conservation of energy can also be extended to this case provided the total energy in translation and rotation is considered in Eq. 16.9.

16.12 MOTION OF CONNECTED BODIES

The work-energy principle may be applied to systems of connected bodies also. In this case the internal forces in the connecting strings need not be considered.

The work done by each of the forces has to be obtained by multiplying its component in the direction of motion by the displacement in this direction. The summation of these has to be equated to the change in kinetic energy of the bodies in the system from the initial position to the final.

A simple example (Fig. 16.7) will serve to illustrate this.

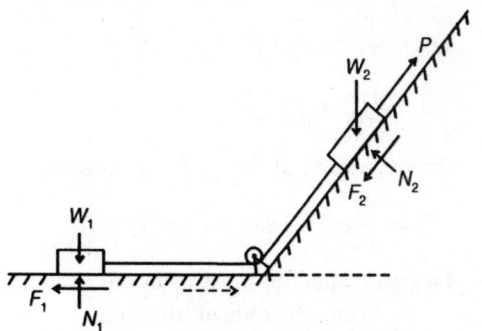

Fig. 16.7: Two Connected Bodies in Translation in Different Planes

Let two bodies W_1 and W_2 move on rough planes—W_1 on the horizontal plane and W_2 on a plane inclined at θ to the horizontal; let the two bodies be connected by a string passing over a pulley arranged at the junction of the planes, and let the coefficients of friction be μ_1 and μ_2 respectively. The normal reactions on these planes are $N_1 (= W_1)$ and $N_2 (= W_2\cos\theta)$, and the frictional resistance F_1 and F_2 opposing motion are given by

$$F_1 = \mu_1 N_1 \text{ and } F_2 = \mu_2 N_2,$$

when a force P acts parallel to the inclined plane on the body of weight W_2, thus tending to pull the system up the plane as shown.

On the first body, only F_1 does work.

On the second body, P, F_2, and $W_2\sin\theta$ will do work.

If the initial and final velocities of the system are u and v in a displacement s parallel to the planes, which are the same for both the bodies, the work-energy equation may be written down as follows:

$$-F_1 s + (P - F_2 - W_2\sin\theta)s = \frac{(W_1 + W_2)}{2g}(v^2 - u^2)$$

or $(P - F_1 - F_2 - W_2\sin\theta)s = \dfrac{(W_1 + W_2)}{2g}(v^2 - u^2)$

$$...\text{(Eq. 16.17)}$$

Making use of this idea, problems involving the motion of any system of connected bodies may be solved.

16.13 WORK DONE BY A SPRING FORCE AND AN ELASTIC FORCE

Let us consider a body kept on the top of a spring attached to horizontal rigid surface as shown in Fig. 16.8.

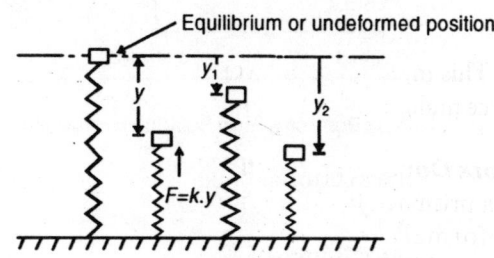

Fig. 16.8: Work Done by a Spring Force

Let the stiffness of the spring be k (force per unit deflection or displacement). In the undeformed position there is no force in the spring. If the spring is pushed down a compressive force develops, and if it is pulled up a tensile force develops; in other words, the spring force is always directed towards its equilibrium position. Since the force is directly proportional to the displacement.

$$F \propto y$$

or $F = k \cdot y$,

k being called the spring-constant, or its stiffness. k is obviously the force required to produce unit displacement, or its 'stiffness'.

For any small displacement, dy, the work done dU is:

$$dU = -F \cdot dy = -ky \cdot dy$$

(since the force is in the opposite direction to the displacement, negative sign is used).

The work done, U_{1-2}, in displacing the spring from y_1 to y_2 is:

$$U_{1-2} = \int_{y1}^{y2}(-ky)dy = -\frac{1}{2}k(y_2^2 - y_1^2)...\text{(Eq. 16.18)}$$

In general, the work done U in displacing the spring by y from the equilibrium position, is given by

$$U = -\frac{1}{2}k \cdot y^2 \qquad \text{...(Eq. 16.19)}$$

If the deformed spring is allowed to move towards its equilibrium position, positive work will be done by the spring. In other words, the work done on the spring earlier is stored by it, and is restored as it does work when called for.

Equation 16.19 may also be expressed as

$$U_{1-2} = -\frac{1}{2}k(y_2 + y_1)(y_2 - y_1)$$

$$= -\frac{1}{2}(F_2 + F_1)(y_2 - y_1) \qquad \text{...(Eq. 16.20)}$$

This means that the work done is the average force multiplied by the displacement, numerically.

Work Done by an Elastic Force

If a prismatic bar of sectional area A, length l, is deformed by an external force P, the deformation dl is given by $\dfrac{Pl}{AE}$, E being the modulus of elasticity of the material (Fig. 16.9).

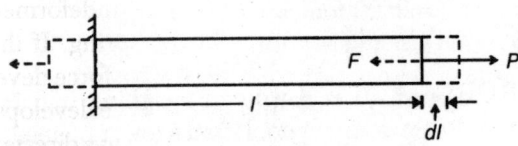

Fig. 16.9: Work Done by Elastic Force

The work done by the force may be got analogus to a spring of stiffness k, which is given by

$$k = \frac{AE}{l} \qquad \text{...(Eq. 16.21)}$$

(k is the force required per unit deformation)

16.14 POWER

'Power' is the time-rate of doing work or work done per unit of time. For turning out a given work, a machine with a smaller power will require more time. Thus, in certain instances, power becomes more important than the work done.

Average power during a small time δt, when the work done is δU, is given by $\dfrac{\delta U}{\delta t}$.

Instantaneous power at any particular instant is therefore given by $\dfrac{\delta U}{\delta t}$.

The unit of power is the "Watt" (in honour of the scientist James Watt), and is given by

1 Watt (W) = 1 Joule/Second = 1 N.m/s.

In practice, one kilowatt (kW) is used, and in measuring large amounts of power, generated by generating stations, one megawatt (MW) is used.

In the intermediate ranges, the horse-power (H.P.) is used.

In the MKS system, 1 metric H.P. = 735.75 Watts.

However, 1 British H.P. = 745.80 Watts.

ILLUSTRATIVE PROBLEMS

PROBLEM 16.1: Calculate the work done by the force on a body weighing 1000 N when (*a*) it moves 5 m along a horizontal surface under the action of horizontal force 300 N, (*b*) it moves 5 m along a horizontal surface under the action of a force 250 N inclined at 30° with the horizontal.

Solution:

(*a*) $P = 300$ N

$s = 5$ m

$\alpha = 0$

Work done, $U = (P\cos\alpha) \cdot s$

$= (300 \times \cos 0) \times 5$

$= 1500$ N.m

$= 1500$ Joules

(*b*) $P = 250$ N

$s = 5$ m

$\alpha = 30°$

Work done, $U = (P\cos\alpha) \cdot s$

$= (250 \times \cos 30°) \times 5$

$= 1082.53$ N.m

or $= 1082.53$ Joules

PROBLEM 16.2: A spring is stretched by 60 mm by the application of a force. Find the work done, if the force required to stretch 1 mm of the spring is 12 N.

Solution: $s = 60$ mm

Maximum force required to stretch the spring by 60 mm

$$= 12 \times 60$$
$$= 720 \text{ N}$$

$\therefore$ Average force, $P = \dfrac{0 + 720}{2} = 360 \text{ N}$

Work done, $U =$ Average force $\times$ Distance
$$= P \times s$$
$$= 360 \times 60$$
$$= 21600 \text{ N.mm}$$
$$= 21.6 \text{ N.m}$$
or
$$= 21.6 \text{ Joules}$$

PROBLEM 16.3: Calculate the work done in pulling up a block of wood weighing 3 kN for a length of 12 m on a smooth plane inclined at an angle of 18° with the horizontal.

Solution:

Weight of the block = 3 kN

Angle made by the inclined plane with the horizontal = 18°

$\therefore$ Resistance due to inclination = $3 \times \sin 18^{\circ}$
$$= 3 \times 0.3090$$
$$= 0.927 \text{ kN}$$

Distance travelled = 12 m

$\therefore$ Work done = Resistance $\times$ Distance
$$= 0.927 \times 12 \text{ kN.m}$$
$$= 11.12 \text{ kJ}$$

Note: The work done is negative, since the resistance developed is in opposite direction.

PROBLEM 16.4: Calculate the work done by electric motor in winding up a uniform cable which hangs from a hoisting drum if its free length is 10 m and weight 500 N. The drum is rotated by the motor.

Solution:

Free length of cable = 10 m

Weight of the cable = 500 N

Weight of the cable per unit length
$$= \dfrac{500}{10} = 50 \text{ N/m}$$

Considering an element dx at distance x,

measured from the bottom end of the cable, as shown in Fig. 16.10.

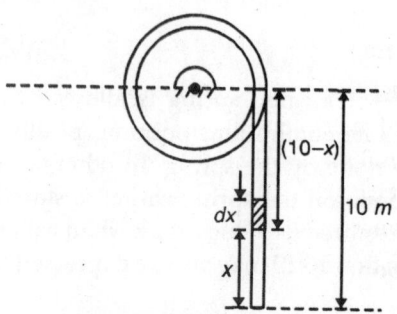

Fig. 16.10 (Illustrative Problem 16.4)

Weight of element = $50 \times dx$ N

Distance moved by element = $(10 - x)$ metres

Work done on element = $50 \, dx \times (10 - x)$

$\therefore$ Total work done = $\int_0^{10} 50(10 - x) \, dx$

$$= 50 \left[10x - \dfrac{x^2}{2} \right]_0^{10}$$
$$= 50 \, [100 - 50]$$
$$= 2500 \text{ N.m}$$
$$= 2500 \text{ Joules}$$
or
$$= 2.5 \text{ kJ}$$

PROBLEM 16.5: *A* woman weighing 650 N, jumps in a swimming pool from a tower of height 20 m. She was found to go down in water by 2.5 m and then started rising. Find the average resistance of the water.

Solution:

Let $P =$ Average resistance of the water in newtons

Potential energy of the woman before jumping
$$= m \cdot g \cdot h$$
$$= \dfrac{650}{9.81} \times 9.81 \times 20$$
$$= 13{,}000 \text{ N.m}$$

Work done by the average resistance of water
$$= P \times 2.5 \text{ N.m}$$

The entire potential energy of the woman is used in the work done by the water.

$\therefore \quad P \times 2.5 = 13000$

or $\quad P = \dfrac{13000}{2.5}$

$\qquad = 5200\ N$

Average resistance of water $= 5200\ N$

PROBLEM 16.6: A bullet moving with a velocity of 300 m/sec is fired into thick target and penetrates upto 750 mm. It is fired into a 375 mm thick target, find the velocity of emergence. Take resistance to be uniform in both the cases.

Solution:

Let m be the mass of the bullet in kg and

$\quad R =$ Force of resistance in newtons (N,

Velocity of bullet, $v = 300$ m/sec

Penetration of bullet $= 750$ mm $= 0.75$ m

Work done by the force of resistance

$\qquad = R \times 0.75$ N.m

K. E. of Bullet $= \dfrac{1}{2} mv^2$

$\qquad = \dfrac{1}{2} m 300^2$

$\qquad = (45000)\ (m)$ N.m

The entire *K.E.* of bullet is used in 750 mm penetration.

$\quad R \times 0.75 = (45000)\ (m)$

$\quad R = \dfrac{(45000)(m)}{0.75}$ newtons

$\qquad = \dfrac{(45000)(m)}{0.75 \times 1000}$

$\qquad = (60)\ (m)$ kN

Let velocity of bullet after 375 mm penetration is v_1 m/sec.

Work done by force of resistance

$\qquad = (60000\ m) \times 0.375$ N.m

$\qquad = (22500\ m)$ N.m

K.E. of bullet after penetration $= \dfrac{1}{2} mv_1^2$

K.E. of bullet used in 750 mm penetration
$\qquad =$ Work done by force of resistance for 375 mm penetration + *K.E.* of bullet after a penetration of 375 mm

$\quad (45000m) = (22500m) + \dfrac{1}{2} mv_1^2$

$\quad \dfrac{1}{2} mv_1^2 = 22500\ m$

$\quad v_1 = \sqrt{45000}$

$\qquad = 212.13$ m/sec

PROBLEM 16.7: A body weighing 400 N is pushed up at 35° plane by a 500 N force acting parallel to the plane. If the initial velocity of the body is 2 m/sec and coefficient of friction is 0.3, find the velocity of the body after moving 5 m.

Solution:

Weight of the body, $W = 400$ N

The component of weight acting normal to the plane

$\quad N = W \cos \alpha$

$\qquad = 400 \times \cos 35° = 327.67$ N

$\therefore$ The frictional force, $F = \mu \cdot N = 0.3 \times 327.67$

$\qquad = 98.30$ N

The component of weight acting parallel to the plane $= W \sin \alpha$

$\qquad = 400 \times \sin 35° = 229.43$ N

$\therefore$ Work done by the forces

$\qquad = (500 - F - W \sin \alpha)\ s$

$\qquad = (500 - 98.30 - 229.43) \times 5$

$\qquad = 172.27 \times 5 = 861.35$ N.m

Change in kinetic energy

$\qquad = \dfrac{1}{2} m \left(v^2 - u^2\right)$

$\qquad = \dfrac{1}{2} \times \dfrac{400}{9.81} \left(v^2 - 2^2\right)$ N.m

According to work-energy equation

$\quad$ Work done = Change in *K.E.*

$\quad 861.35 = \dfrac{1}{2} \times \dfrac{400}{9.81} (v^2 - 2^2)$

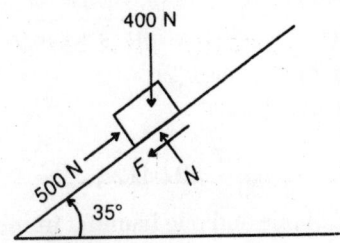

Fig. 16.11 (Illustrative Problem 16.7)

$$v^2 - 4 = \frac{861.35 \times 2 \times 9.81}{400} = 42.25$$

$$v^2 = 42.25 + 4 = 46.25$$

$$v = \sqrt{46.25} = 7.09 \text{ m/sec}$$

Velocity of the body after moving 5 m
= 7.09 m/sec

PROBLEM 16.8: A hammer of mass 1200 kg drops from a height of 0.80 m on a pile of 600 kg. Find (*a*) the common velocity pile and pile hammer after impact, (*b*) the average resistance of the ground if the pile comes to rest after penetrating 60 mm into the ground.

Solution:

(*a*) Mass of the pile hammer, $m = 1200$ kg

Mass of the pile, $M = 600$ kg

Height through which the pile hammer falls before striking the pile, $h = 0.80$ m

Distance through which the pile is driven into the ground, $x = 60$ mm.

Velocity of the hammer just before impact,

$$v = \sqrt{2gh}$$

$$= \sqrt{2 \times 9.81 \times 0.80}$$

$$= \sqrt{15.696}$$

$$= 3.9618 \text{ m/sec}$$

Let $V =$ Common velocity of the pile and pile hammer after impact and

$R =$ Average resistance of the soil

Momentum of the pile hammer and pile just before impact $= mv + M \times 0$

$$= 1200 \times 3.9618$$

$$= 4754.16 \text{ kg.m/sec} \qquad ...(1)$$

Momentum of the pile hammer and pile just after impact $= (m + M)V$

$$= (1200 + 600)V = 1800V \text{ kg.m/sec} \quad ...(2)$$

Equating (1) and (2)

$$4754.16 = 1800V$$

or $V = \dfrac{4754.16}{1800} = 2.6412$ m/sec

(*b*) K.E. of pile and pile hammer immediately after impact,

$$= \frac{1}{2}(m + M)V^2$$

$$= \frac{1}{2}(1200 + 600)(2.6412)^2$$

$$= 6278.34 \text{ N.m}$$

P.E. of pile and pile hammer immediately after impact

$$= (m + M) \cdot g \cdot x$$

$$= (1200 + 600) \times 9.81 \times 0.06$$

$$= 1059.48 \text{ N.m}$$

$\therefore$ Total energy $= 6278.34 + 1059.48$

$$= 7337.82 \text{ N.m} \qquad ...(3)$$

Work done by the soil resistance
= Average resistance of the soil × penetration

$$= R \cdot x$$

$$= R \times 0.06 \text{ N.m} \qquad ...(4)$$

Since the total energy of the hammer and pile is used in the work done by the soil resistance, therefore equating (3) & (4)

$$73337.82 = R \times 0.06$$

or $R = \dfrac{7337.82}{0.06}$

$$= 122297 \text{ N}$$

or $= 122.3$ kN

Check

$$R = \frac{m^2 gh}{x(m + M)} + (m + M) \cdot g$$

$$= \frac{1200^2 \times 9.81 \times 0.8}{0.6 \times 1800} + 1800 \times 9.81$$

$$= 104640 + 17658$$

$$= 122298 \text{ N}$$

$$= 122.3 \text{ kN}$$

PROBLEM 16.9: Two blocks of weights 500 N and 600 N are connected by inextensible wire running around a smooth pulley as shown in Fig. 16.12 (*a*) starting from rest, find what will be the velocity of the system if distance moved by the blocks is 3.5 m. Also find the tension developed in the wire.

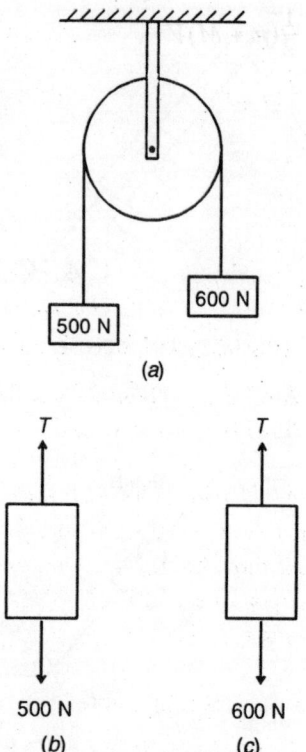

(a)

(b)   (c)

Fig. 16.12 (Illustrative Problem 16.9)

Solution: 600 N block moves down and 500 N block moves up. By observing the pulley system it may be concluded that both the blocks will move by the same distance and will be having same velocity.

Let 'v' be the final velocity of the system writing work-energy equation for the system, we get,

$$(600s - 500s) = \left(\frac{600}{2 \times 9.81}\right)(v^2 - u^2)$$

$$+ \left(\frac{500}{2 \times 9.81}\right)(v^2 - u^2)$$

$$100 \times 3.5 = \left(\frac{600}{2 \times 9.81}\right)(v^2 - 0^2)$$

$$+ \left(\frac{500}{2 \times 9.81}\right)(v^2 - 0^2)$$

or $\quad \left(\dfrac{1100}{2 \times 9.81}\right)v^2 = 350$

$$v = \sqrt{\frac{350 \times 2 \times 9.81}{1100}}$$

$$= \sqrt{6.2427}$$

$$= 2.49855 \approx 2.5 \text{ m/sec}$$

Let T be the tension in the string. Considering work energy equation for the 600 N block.

$$(600 \times s - T \times s) = \frac{600}{2 \times 9.81}(v^2 - u^2)$$

$$(600 - T) \times 3.5 = \frac{600}{2 \times 9.81}(2.49858^2)$$

$$(600 - T) \times 3.5 = 190.91$$

$$600 - T = \frac{190.91}{3.5} = 54.55$$

$$T = 600 - 54.55 = 545.45 \text{ N}$$

$$\text{Or}$$

$$(T - 500)s = \frac{500}{2 \times 9.81}(2.49855^2 - 0^2)$$

$$(T - 500) \times 3.5 = 159.09$$

$$(T - 500) = \frac{150.09}{3.5} = 45.45$$

or $\quad T = 500 + 45.45$

$$T = 545.45 \text{ N}$$

PROBLEM 16.10: Two blocks are connected by inextensible wires as shown in Fig. 16.13. Find how much distance block 300 N will move in increasing its velocity to 4 m/sec from 2 m/sec. Assume pulleys are frictionless and weightness.

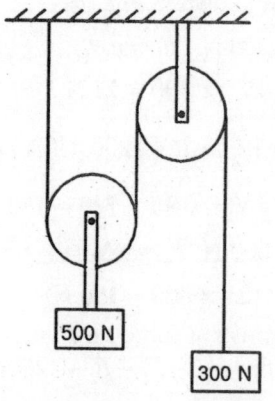

Fig. 16.13 (Illustrative Problem 16.10)

Solution: 300 N block moves down and 500 N block moves up. By observing the pulley system, it may be concluded that if 300 N body moves a distance 's', 500 N body moves a distance of 0.5 s, and if velocity of 300 N block is v that of 500 N block will be $0.5v$.

Writing work-energy equation for the system, we get:

$$(300 \times s - 500 \times 0.5 \times s) = \left(\frac{300}{2 \times 9.81}\right)(v^2 - u^2)$$

$$+ \left(\frac{500}{2 \times 9.81}\right)\left[(0.5v)^2 - (0.5u)^2\right]$$

$$50 \times s = \frac{300}{2 \times 9.81}(4^2 - 2^2) + \frac{500}{2 \times 9.81}$$

$$\left[(0.5 \times 4)^2 - (0.5 \times 2)^2\right]$$

$50 \times s = 183.486 + 76.453$

$50 \times s = 259.939$

or $\quad s = \dfrac{259.939}{50} = 5.199$ m ≈ 5.20 m

PROBLEM 16.11: Determine the constant force P that will give the system of bodies shown in Fig. 16.14 (a) a velocity of 3.5 m/sec after moving 5 m from rest. Coefficient of friction between the blocks and the plane is 0.25. Pulleys are frictionless and weightless.

Solution: The system of forces acting on connecting bodies is shown in Fig. (b).

$N_1 = 300$ N; $\quad F_1 = \mu N_1$

$\qquad = 0.25 \times 300 = 75$ N

$N_2 = W_2 \cos\alpha = 1200 \times \dfrac{3}{5} = 720$ N

$\therefore \quad F_2 = \mu N_2 = 0.25 \times 720 = 180$ N

$N_3 = 600$ N; $\quad F_3 = \mu N_3$

$\qquad = 0.25 \times 600 = 150$ N

Let the constant force be P newtons.

Work done $= (P - F_1 - F_2 - 1200\sin\alpha - F_3) \times s$

$\qquad = \left(P - 75 - 180 - 1200 \times \dfrac{4}{5} - 150\right) \times 5$

$\qquad = (P - 1365) \times 5$ N.m

$u = 0$ and $v = 3.5$ m/sec

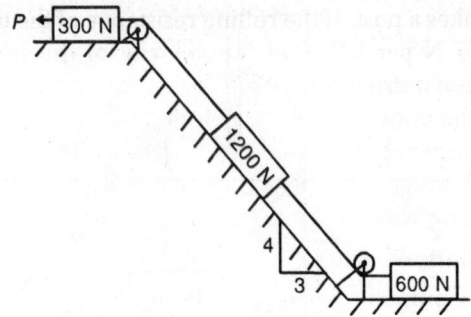

(a) System of Bodies

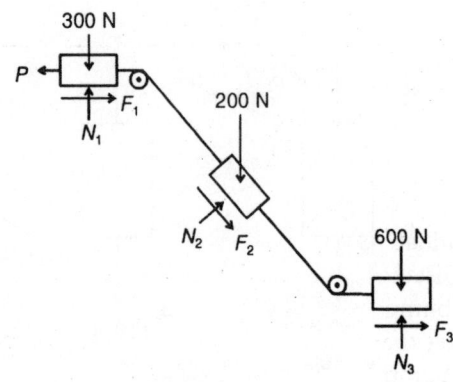

(b) Free Body Diagrams

Fig. 16.14 (Illustrative Problem 16.11)

Initial $K.E. = 0$

Final $K.E. = \dfrac{1}{2}mv^2$

$\qquad = \dfrac{1}{2}\left(\dfrac{300 + 1200 + 600}{9.81}\right) \times 3.5^2$

$\qquad = 1311.162$ N.m

$\therefore \quad$ Change in $K.E.$ = Final $K.E.$ – Initial $K.E.$

$\qquad = 1311.162 - 0$

$\qquad = 1311.162$ N.m

According to work-energy equation,
Work done = Change in $K.E.$

$(P - 1365) \times 5 = 1311.162$

$P - 1365 = \dfrac{1311.162}{5} = 262.23$

or $\quad P = 262.23 + 1365.00$

$\qquad P = 1627.23$ N

PROBLEM 16.12: A wagon weighing 400 kN starts from rest, runs 30 m down a 1% grade and strikes a post. If the rolling resistance of the track is 5 N per kN, find the velocity of the wagon when it strikes the post.

If the impact is to be cushioned by means of one bumper spring, which compresses 1 mm per 20 kN weight, determine how much the bumper spring will be compressed.

Solution:

Component of weight down the plane
$$= W\sin\theta$$
$$= 400 \times \frac{1}{100}$$
$$= 4 \text{ kN}$$

Track resistance, $F = 5$ N per kN
$$= 5 \times 400$$
$$= 2000 \text{ N}$$
$$= 2 \text{ kN}$$

Initial velocity of the wagon, $u = 0$

Distance moved by the wagon before striking the bumper, $s = 20$ m

Let the velocity of the wagon while striking the bumper be v m/sec.

Work done $= (W\sin\theta - F) s$
$$= (4 - 2) \times 20 = 40 \text{ kN.m}$$

Initial kinetic energy $= 0$

Final kinetic energy $= \frac{1}{2} mv^2$
$$= \frac{1}{2} \times \frac{400}{9.81} v^2$$
$$= 20.3874 v^2 \text{ kN.m}$$

∴ Change in kinetic energy

= Final kinetic energy – Initial kinetic energy
$$= 20.3874 v^2 - 0$$
$$= 20.3874 v^2$$

According to work energy equation,
Work done = Change in kinetic energy
$$40 = 20.3874 v^2$$
∴
$$v = \sqrt{40 / 20.3874}$$
$$v = \sqrt{1.962}$$
$$v = 1.40 \text{ m/sec}$$

Work done in compressing the spring
= Average load × displacement

Let x = Compression of the spring in mm

Compression of the spring 1 mm per 20 kN weight

∴ Compressive load $= \dfrac{x}{(1 / 20)} = 20x$ kN

∴ Work done in compressing the spring
$$= \frac{1}{2} \times 20x \times x$$
$$= 10x^2 \text{ kN.mm}$$

Since the entire kinetic energy of the wagon is used to compress the spring, therefore,

$$10x^2 = \frac{1}{2} mv^2$$
$$10x^2 = \frac{1}{2} \times \frac{400}{9.81} (1.40)^2$$
$$x^2 = \frac{400 \times 1.40 \times 1.40 \times 1000}{2 \times 9.81 \times 10}$$
$$= 3995.92 \text{ mm}^2$$
$$x = \sqrt{3995.92}$$
$$x = 63.21 \text{ mm}$$

PROBLEM 16.13: A locomotive draws a train of weight 2000 kN, including its own weight, on a level ground with a uniform acceleration, until it acquires a velocity of 60 kmph in 6 minutes.

If the frictional resistance is 10 N per kN of weight and the air resistance varies with the square of the velocity, find the power of the engine. Take air resistance as 500 N at 20 kmph.

Solution:

Frictional resistance = 10 N/kN of weight
Weight of locomotive & train = 2000 kN

∴ Frictional resistance $= 10 \times 2000$
$$= 20000 \text{ N}$$
or
$$= 20 \text{ kN}$$

Velocity of locomotive after $6 \times 60 = 360$ sec
$$= 60 \text{ kmph}$$
$$= \frac{60 \times 1000}{60 \times 60} = \frac{50}{3} \text{ m/sec}$$

Let f = acceleration
$$v = u + f \cdot t$$
$$\frac{50}{3} = 0 + f \times 360$$

$$\therefore \quad f = \frac{50}{3 \times 360} = \frac{5}{108} \text{ m/sec}^2$$

Force required for this acceleration
$$= m \cdot f$$
$$= \frac{2000}{9.81} \times \frac{5}{108}$$
$$= 9.44 \text{ kN}$$

Air resistance at 60 kmph
$$= 5000 \left(\frac{60}{20}\right)^2$$
$$= 4500 \text{ N}$$
$$= 4.5 \text{ kN}$$

Total resistance $= 20 + 9.44 + 4.50$
$$= 33.94 \text{ kN}$$

$\therefore$ Power $=$ Total resistance $\times$ velocity
$$= 33.94 \times (50/3) \text{ kN.m/sec}$$
$$= 565.67 \text{ kN.m/sec}$$
$$= 565.67 \text{ kJ/sec}$$
or $\quad = 565.67 \text{ kW}$

PROBLEM 16.14: A locomotive of weight 500 kN pulls a train of weight of 2500 kN. The tractive resistance, due to friction is 10 N per kN. The train can go with a maximum speed of 27 kmph on a grade of 1 in 100. Determine (*a*) the power of the locomotive, and (*b*) the maximum speed it can attain on a straight level track with the tractive resistance remaining same.

Solution:

(*a*) Power of locomotive:

Weight of locomotive $= 500$ kN
Weight of train $= 2500$ kN
The weight of train including locomotive
$$= 3000 \text{ kN}$$
Tractive resistance $= 10$ N per kN
Tractive resistance for 3000 kN
$$= \frac{10 \times 3000}{1}$$
$$= 30000 \text{ N} = 30 \text{ kN}$$
Resistance due to inclination
$$= 3000 \times \frac{1}{100}$$
$$= 30 \text{ kN}$$

$\therefore$ Total resistance $= 30 + 30 = 60$ kN
Velocity of the train $= 27$ kmph
$$= \frac{27 \times 1000}{60 \times 60} = 7.5 \text{ m/sec}$$

$\therefore$ Power of the locomotive $=$ Total resistance $\times$ velocity
$$= 60 \times 7.5 \text{ kN.m/sec}$$
$$= 450 \text{ kW}$$

(*b*) **Maximum speed the locomotive can attain on straight level track.**

The locomotive has to overcome the tractive resistance of 30 kN only.
$$\therefore \quad 30 \times v = 450$$
$$v = 450/30 = 15 \text{ m/sec}$$
or $\quad \dfrac{15 \times 60 \times 60}{1000} = 54 \text{ kmph}$

The maximum speed the locomotive can attain $= 54$ kmph

PROBLEM 16.15: A pump lifts 50 m³ of water to a height of 60 m and delivers it with a velocity of 6 m/sec. Find the amount of energy spent during this process. If the job is done in 45 minutes, what is the input power of the pump. Take the efficiency of the pump as 80%.

Solution:

Weight of 50 m³ of water $= 50 \times 9.81$ kN
$$(\because 1 \text{ m}^3 \text{ of water weighs } 9.81 \text{ kN})$$
Work done in lifting 50 m³ of water to a height of 60 m $= Wh$
$$= 50 \times 9.81 \times 60$$
$$= 29430 \text{ kN.m}$$
or $\quad = 29430 \text{ kJ}$

Kinetic energy at delivery $= \dfrac{1}{2}mv^2$
$$= \frac{1}{2}\frac{W}{g}v^2$$
$$= \frac{1}{2} \times \frac{50 \times 9.81}{9.81} \times (6)^2$$
$$= 900 \text{ kJ}$$

Total energy spent $= 29430 + 900$
$$= 30330 \text{ kJ}$$

This energy is spent by the pump in 45 minutes.

$$\text{Output power of pump} = \frac{30330}{5 \times 60} \frac{\text{kJ}}{\text{sec}}$$

$$= 11.2333 \text{ kW}$$

Efficiency of the pump $= 80\% = 0.80$

$$\text{Input power} = \frac{\text{Output}}{\text{Efficiency}}$$

$$= \frac{11.230}{0.80} = 14.0416 \text{ kW}$$

PROBLEM 16.16: A tangential force of 3000 N is acting on a shaft of diameter 12 mm. The shaft is rotating at 300 r.p.m. Find the power of the shaft.

Solution:

Tangential force, $P = 3000$ N

Diameter of shaft, $d = 12$ mm $= 0.012$ m

Radius of shaft, $r = \dfrac{0.012}{2} = 0.006$ m

∴ Torque, $T = P \cdot r = 3000 \times 0.006 = 18$ N.m

Angle turned in 1 revolution $= 2\pi$

Number of revolutions performed in 1 minute $= 300$

∴ Angle turned in one minute $= 300 \times 2\pi$

Angle turned in one second $= \dfrac{300 \times 2\pi}{60}$

$= 10\pi$ radians

Power = Work done per second

= Torque × Angle turned in one second

$= 18 \times 10\pi$

$= 565.49$ N.m/sec $= 565.49$ Watts

PRACTICE PROBLEMS

16.1 A horse pulling a cart exerts a steady horizontal pull of 250 N and walks at the rate of 4 kmph. How much work is done by the horse in 6 minutes?

16.2 A trolley of weight 3000 N moves on a level track for a distance of 400 m. The resistance of the road is 12 N per 1000 N weight of the trolley. Determine the work done in moving the trolley.

16.3 A trolley of weight 2000 N is placed on a horizontal surface. An initial force of 500 N applied horizontally is sufficient just to move trolley. The applied force varies uniformly with the distance moved by the trolley. If the trolley moves a distance of 30 m and force applied is given by the relation $P = P_0 + 10x$, determine the work done by the applied force. P_0 is the initial force in Newtons and x is distance moved in metres.

16.4 A body of weight 15 N is moving with a velocity of 60 m/sec. What will be the kinetic energy of the body?

16.5 A truck of weight 20 kN is travelling at a speed of 36 kmph on a level road. It is brought to rest in 16 metres. Find the average force of resistance acting on the truck.

16.6 A bullet of mass 0.10 kg and moving with a velocity of 300 m/sec is fired into a block of wood and it penetrates 100 mm. If the bullet moving with the same velocity, were fired into a similar piece of wood 75 mm thick, with what velocity would it emerge? Determine the force of resistance, assuming it to be uniform.

16.7 An automobile is moving with 54 kmph when the driver applies his brakes and goes into a skid in the direction of motion. If the car weighs 10 kN, and coefficient of friction between the road and tyres be 0.6. How far will the car move before stopping?

16.8 A hammer of mass 0.6 kg hits a nail of 0.03 kg with a velocity of 6 m/sec and drives it into a fixed wooden block by 30 mm. Find the resistance offered by the wooden block.

16.9 Two bodies weighing 300 N and 500 N are hung to the ends of a rope passing over an ideal

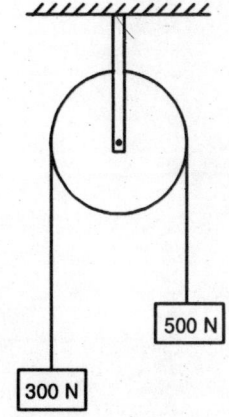

Fig. 16.15 (Practice Problem 16.9)

pulley as shown in Fig. 16.15. How much distance the blocks will move in increasing the velocity of system from 2 m/sec to 5 m/sec. How much is the tension in the string?

16.10   In what distance will body *A* shown in Fig. 16.16 attain a velocity of 2.5 m/sec starting from rest ? Take coefficient of friction between the blocks and the plane as 0.2. Assume the pulley is frictionless and weightless.

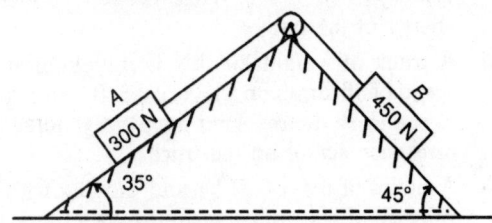

Fig. 16.16 (Practice Problem 16.10)

16.11   A truck of weight 200 kN travelling at 9 kmph impacts with a buffer spring compresses 1.5 mm per kN. Find the maximum compression of the spring.

16.12   A train of weight 2000 kN is pulled by an engine on a level track at a constant speed of 60 kmph. The resistance due to friction is 1% of the train. Determine the power of the engine.

16.13   A train of weight 2000 kN moves at a uniform speed of 36 kmph. The tractive resistance on the level being 5 N/kN, what power will be required to pull the train?

   (*a*)   On level surface
   (*b*)   Up an incline of 1 in 400   and
   (*c*)   Down an inclination of 1 in 400.

16.14   A car of weight 12 kN accelerates from rest to a speed of 36 kmph in a distance of 40 m against a resistance of 120 N. Find the average driving force acting on the car and the power developed by the engine.

16.15   A tangential force of 3000 N is acting on a shaft of diameter 12 mm. Find the work done by the force for one revolution of the shaft.

Chapter 17

Impulse and Momentum

17.1 INTRODUCTION

In the previous chapters two basic methods (including a variation of one of them) for solving kinetics problems have been discussed— Newton's second law of motion (including d'Alembert's principle, which is merely a varaition of this) and work-energy method. In this Chapter, the third basic method for this purpose—'impulse' and 'momentum' approach is given. The principle relates force, mass, velocity, and time; it is particularly suitable when large magnitudes of forces act for a small interval of time.

17.2 IMPULSE

An 'impulse' is defined as a large force acting over a short period of time; quantitatively speaking, an impulse is given by the product of the force and time:

Impulse, $i = P \cdot t$ ...(Eq. 17.1)

P being the force acting over time-interval t.

The units of impulse are obviously newton-seconds (N.s); it is a vector quantity, and is called "Linear Impulse".

Graphically, impulse is represented by the area under the curve representing the variation of force with time. It can also be put down as follows for a finite time interval from t_1 to t_2:

$$i = \int_{t_1}^{t_2} P \cdot dt$$ ...(Eq. 17.2)

17.3 MOMENTUM

"Momentum" of a body is defined as the product of mass and its velocity; it is called "Linear Momentum" in the case of rectilinear motion.

Momentum, M_m of a body of mass m, moving with a velocity v, is given by

$M_m = m \cdot v$ ...(Eq. 17.3)

The units of momentum may be established from fundamentals:

$$m \cdot v = \frac{W}{g} \cdot v$$

∴ The unit of momentum are

$$\frac{N}{(m/s^2)} \times \frac{m}{s} = N.s$$

Thus the units of momentum are the same as those for impulse. Since velocity is a vector, momentum also is a vector quantity.

17.4 IMPULSE-MOMENTUM EQUATION

If a force P acts on a body of mass m, producing an acceleration a,

$P = m \cdot a$

$$= m \cdot \frac{dv}{dt}$$

or $P \cdot dt = m \cdot dv$

Integrating both sides between appropriate limits,

$$\int_o^t P \cdot dt = \int_u^v m \cdot dv$$

u and v being the initial velocity and final velocity after time t, respectively.

or $P \cdot t = (m \cdot v - m \cdot u)$ ...(Eq. 17.4)

The left hand side represents the impulse and the right-hand side the change in momentum in the corresponding time interval.

Eq. 17.4 is called the "impulse-momentum equation" or "impulse-momentum principle", which may be stated:

"The linear impulse in any direction is equal to the change in linear momentum in that direction over the chosen interval of time."

Using this principle, certain problems in kinetics involving force, velocity and time may be solved easily.

17.5 CONSERVATION OF MOMENTUM

An inspection of Eq. 17.4 reveals that when the resultant force is zero in a system, the impulse is zero, and the final momentum is the same as the initial momentum; in other words, the momentum is *conserved* in such a system.

This is known as the principle of conservation of momentum, which is stated:

"When the resultant force of a system is zero, the vector sum of the impulses of all external forces is zero, and the momentum of the system remains constant or conserved".

When a force system consists of only action and reaction, such a situation arises.

Two examples of such problems are the recoil of a gun firing a bullet, and a man jumping off a boat. In both cases, the momentum of the system is conserved and problems involving these may be solved using the principle of conservation of momentum, impulse-momentum equation, and work-energy equation, according to the need. This is illustrated by the worked examples at the end of the chapter.

17.6 ANGULAR MOMENTUM

Let us consider a body of mass m moving along a curvilinear path (Fig. 17.1); let its velocity at a particular point A be v in the tangential direction as shown.

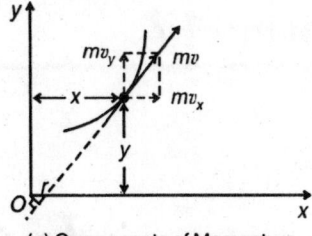

(a) Components of Momentum

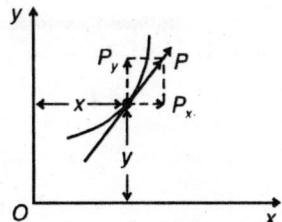

(b) Components of Resultant Force

Fig. 17.1: Principle of Angular Momentum

Its momentum at A is mv; its rectangular components are mv_x and mv_y [Fig. 17.1(a)].

The "*angular momentum*" with respect to the origin O is defined as the product of the momentum and the perpendicular distance from the origin on to the line of action of the momentum. This is also known as the "moment of momentum".

Angular momentum,

$M_{am} = M_m \cdot r = m \cdot v \cdot r$ ...(Eq. 17.5)

It is often convenient to consider the rectangular components even for angular momentum.

$M_{am} = (mv_y)x - (mv_x)y$ ...(Eq. 17.6)

(Taking counter-clockwise moment as positive)

The rate of change of angular momentum is given by:

$$\frac{dM_{am}}{dt} = m \cdot \frac{d}{dt}\left(x \cdot v_y - y v_x\right)$$

$$= m\left(v_y \frac{dx}{dt} + x \frac{dv_y}{dt} - v_x \cdot \frac{dy}{dt} - y \frac{dv_x}{dt}\right)$$

Recognising that

$$\frac{dx}{dt} = v_x, \quad \frac{dy}{dt} = v_y, \quad \frac{dv_x}{dt} = a_x, \text{ and } \frac{dv_y}{dt} = a_y,$$

We get

$$\frac{dM_{am}}{dt} = m\left(v_x v_y + x a_y - v_y v_x - y a_x\right)$$

$$= m(x a_y - y a_x). \qquad ...(Eq.\ 17.7)$$

Applying Newton's second law, $ma_x = P_x$ and $ma_y = P_y$ [Fig. 17.1(b)].

$$\therefore \quad \frac{dM_{am}}{dt} = (xP_y - yP_x) \qquad ...(Eq.\ 17.8)$$

here P_x and P_y are the rectangular components of the resultant force P on the system.

Since the right-hand side of this equation represents the moment M of the resultant force P about the origin O, Eq. 17.8 may also be written as:

$$\frac{dM_{am}}{dt} = \left(x \cdot P_y - yP_x\right) = M_o \qquad ...(Eq.\ 17.8)$$

Put in words, this means that the rate of change of angular momentum or that of moment of momentum equals the moment of the resultant force on the system about the moment centre or the origin.

17.7 CONSERVATION OF ANGULAR MOMENTUM

From Eq. 17.8, if M_o is zero,

$$\frac{dM_{am}}{dt} = 0$$

Integrating with resepct to time,

$$M_{am} = k,\ \text{a constant} \qquad ...(Eq.\ 17.9)$$

Stated in words :

"The angular momentum of a particle/body about a point is constant (or is conserved) if the moment of the resultant force acting on the particle about the same point is zero".

This is known as the "Moment of Momentum Principle". An important corollary of this is:

"The moment of momentum about a point shall be conserved if the resultant force passes through that point"

In the case of '*central force motion*', the only force acting on the body is directed towards or away from the centre of rotation; automatically thus, the moment of momentum is conserved in this case.

It will be found that this principle is very useful in solving certain kinds of kinetics problems.

17.8 APPLICATIONS OF IMPULSE— MOMENTUM PRINCIPLE

There are several problems of kinetics for the solution of which the impulse-momentum principle and conservation of momentum may be conveniently applied.

Among them are problems involving

(i) Motion of connected bodies,
(ii) Force of jet on a vane,
(iii) The kinetics of pile-driving, and
(iv) The recoil of a gun.

These are given in the following sub-sections.

17.8.1 Motion of Connected Bodies

If the connected bodies are such that the displacement of each body is the same in a given time, the impulse in the internal forces in the connecting chords will be eliminated as they will constitute action and reaction at either end.

Hence the free body diagaram of the system as a whole may be considered, and the impulse-momentum equation may be applied in the direction of motion of the connected bodies.

Alternatively, free body diagrams of each body in the system may be drawn separately, and impulse-momentum equation for each body written in its direction of motion. These equations are solved to obtain the unknown values.

17.8.2 Forces of Jet on a Vane

Hydraulic turbines are used in hydroelectric power stations to generate power. A jet of water is made to impinge on the vanes of the turbine, and gets deflected through a predetermined angle. In this process, a force is exerted by the deflecting jet on the vane, which causes the turbine to rotate. This mechanical energy is further converted to electrical energy.

The force exerted by the jet on the vane, which may be stationary or moving, can be determined by applying the impulse-momentum equation and conservation of momentum.

17.8.3 Pile and Pile Hammer

A pile is a member driven into the ground to improve the safe bearing capacity or load-carrying capacity of the foundation soil. It may be of reinforced concrete or steel. Piles are commonly used in groups with a concrete pile cap at the top, the superstructure being built over it.

A pile-hammer is used to drive the piles into the ground. It consists of a falling weight, called the 'monkey', raised to a convenient height h and dropped freely on to the head of the pile through guides (Fig. 17.2).

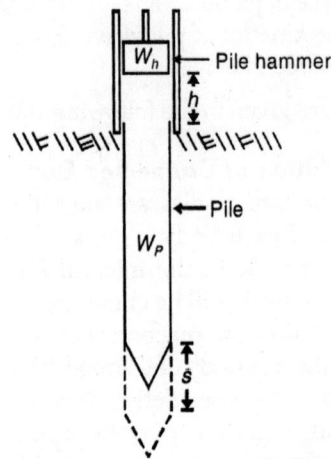

Fig. 17.2: Pile Driving by a Pile Hammer

After the hammer strikes the pile, both move downward together. The kinetic energy is made to do work against the resistance of the soil, making the pile to sink through a ceratin depth s. The pile is driven to the desired depth by giving blows through the hammer in succession. If the set s per blow is known, the resistance of the soil can be calculated.

Let the weights of pile and hammer be W_p and W_h, respectively.

Hammer: Initial velocity is zero.

Distance moved before striking the pile-head
$$= h$$

Velocity of the hammer at the instant of striking the pile, is, say, v.
$$v^2 - u^2 = 2gh$$
or $\quad v = \sqrt{2gh}$ ...(Eq. 17.10)

After the hammer strikes the pile, both move together; the momentum is conserved.

Applying the principle of conservation of momentum to the system,

$$\frac{W_h}{g} \cdot v = \frac{(W_h + W_p)}{g} \cdot V \qquad ...(Eq. 17.11)$$

V being the velocity of the hammer and the pile, immediately after the impact.

From this,

$$V = \frac{W_h}{(W_h + W_p)} \cdot v \qquad ...(Eq. 17.12)$$

With this as the initial velocity, the pile and hammer start moving downward and will stop after the pile set s.

Applying the work-energy equation, we get,

$$(W_h + W_p)s - R \times s = \frac{1}{2}\frac{(W_h + W_p)}{g} \cdot V^2$$

R being the average resistance of the soil.

Substituting for V and v from Eqs. 17.11 and 17.12, and simplifying, we get,

$$R = (W_h + W_p) + \frac{W_h^2}{(W_h + W_p)} \cdot \frac{h}{s} \qquad ...(Eq. 17.13)$$

Loss of kinetic energy during the impact is

$$\frac{W_h}{2g} \cdot v^2 - \frac{(W_h + W_p)}{2g} \cdot V^2$$

Using again Eqs. 17.11 and 17.12, and simplying,

Loss of kinetic energy during impact,

$$E_{k_n} = \frac{W_h \cdot W_p}{(W_h + W_p)} \cdot h \qquad ...(Eq. 17.14)$$

The time during which the pile is in motion may be got applying the impulse-momentum principle.

$$[(W_h + W_p) - R] \cdot t = \frac{(W_h + W_p)}{g}(-V)$$

Substituting for V and R from Eqs. 17.11, 17.12 and 17.13, and simplyfing, we have,

$$t = \frac{(W_h + W_p)}{W_h} \cdot s \cdot \sqrt{\frac{2}{gh}} \qquad ...(Eq. 17.15)$$

All these applications will be illustrated through numerical examples.

17.8.4 Recoil of a Gun

When a gun is fixed, it recoils, or it is thrown backwards; this is simply a corollary of the principle of conservation of momentum. Let M_g and m_s be the masses, and v_g and v_s the velocities of the gun and shell, respectively, just after firing. Let the gun be backed up by springs to take the recoil (Fig. 17.3).

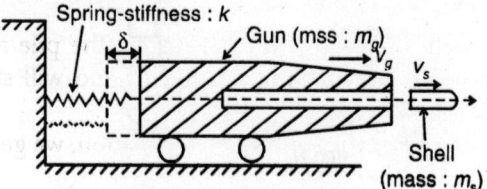

Fig. 17.3: Recoil of a Gun

Since there is no external force on the system during the extremely small interval of time of passage of the shell in the barrel, the momentum of the system is conserved.

Initial momentum is $(m_g \cdot v_g + m_s \cdot v_s)$

$$\therefore \quad m_g \cdot v_g + m_s \cdot v_s = 0 \qquad \text{...(Eq. 17.16)}$$

or $\quad v_g = -\left(\dfrac{m_s}{m_g}\right) \cdot v_s \qquad \text{...(Eq. 17.17)}$

The negative sign indicates that v_g is opposite in direction to v_s. Kinetic energy acquired by the gun,

$$E_k = \frac{1}{2} m_g \cdot v_g^2$$

If the spring force is P, the elastic energy stored in the spring,

$$= \frac{1}{2} P \cdot \delta$$
$$= \frac{1}{2} k \cdot \delta^2$$

k being the stiffness of the spring.
δ is the maximum recoil of the gun.

If there is no loss of energy during recoil,

$$= \frac{1}{2} k \cdot \delta^2 = \frac{1}{2} m_g v_g^2$$

or $\quad \delta = \sqrt{\dfrac{m_g}{k}} \cdot v_g \qquad \text{...(Eq. 17.18)}$

Maximum force generated by the recoil is

$$P = k \cdot \delta = m_g \cdot k \cdot v_g \qquad \text{...(Eq. 17.19)}$$

The kinematic aspects of the system may also be solved from the usual principles of kinematics.

ILLUSTRATIVE PROBLEMS

PROBLEM 17.1: A cricket ball of mass 100 gms moving horizontally at 25 m per sec was hit straight back with a speed of 15 m per sec. If the contact lasted 1/20 second, find the average force exerted by the bat.

Solution:

Change of momentum produced

$$= \frac{100}{1000}[25 - (-15)]$$
$$= 4 \text{ kg.m/sec}$$

Change of momentum = Force × time

$$4 = \text{Force} \times \frac{1}{20}$$

Force = 80 kg.m/sec²

Force = 80 N

PROBLEM 17.2: A ball of mass 100 gms, falls from a height of 7.5 m and rebounds to a height of 5 m. Find the blow on the ball and the average force between the floor and the ball if the duration of contact is 1/20 sec.

Solution:

$$v^2 - 0^2 = 2 \times 9.81 \times 7.5$$
$$v = \sqrt{2 \times 9.81 \times 7.5}$$
$$= 12.13 \text{ m/sec.}$$
$$0^2 - u^2 = 2 \times -9.81 \times 5$$
$$u = \sqrt{2 \times 9.81 \times 5}$$
$$= 9.9 \text{ m/sec}$$

Change in momentum

$$= \frac{100}{1000}[9.9 - (-12.13)]$$
$$= 2.203 \text{ kg.m/sec}$$

Change in momentum = Force × time

$$2.203 = \text{Force} \times \frac{1}{20}$$

Force = 2.203 × 20

$$= 44.060 \text{ N}$$

PROBLEM 17.3: A automobile weighing 20 kN is moving at a speed of 60 kmph when the brakes are fully applied causing all four wheels to skid. Determine the time required to stop automobile.

   (*a*) on concrete road for which $\mu = 0.70$,

   (*b*) on ice for which $\mu = 0.08$

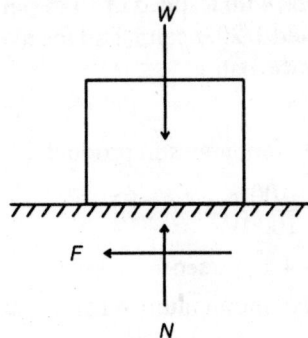

Fig. 17.4 (Illustrative Problem 17.3)

Solution: Initial velocity of the vehicle,

$$u = \frac{60 \times 1000}{60 \times 60}$$

$$= 16.67 \text{ m/sec}$$

Final velocity $= 0$

$$F = N\mu = W\mu$$

$$F = 20\,\mu$$

Applying impulse momentum equation,

$$-F \cdot t = \frac{W}{g}(v - u)$$

$$-20\mu t = \frac{20}{9.81}(0 - 16.67)$$

$$\therefore \qquad t = \frac{1.699}{\mu}$$

   (*a*) on concrete road, $\mu = 0.70$

$$t = \frac{1.699}{0.70} = 2.427 \text{ sec}$$

   (*b*) on ice, $\mu = 0.08$

$$\therefore\ t = \frac{1.699}{0.08} = 21.234 \text{ sec}$$

PROBLEM 17.4: The initial velocity of 600 N block is 5 m/sec leftward. At this stage a weight of 300 N is applied as shown in Fig. 17.5 (*a*).

Determine the time at which the block has (*a*) no velocity, (*b*) a velocity of 3 m/sec to the right. Take $\mu = 0.2$.

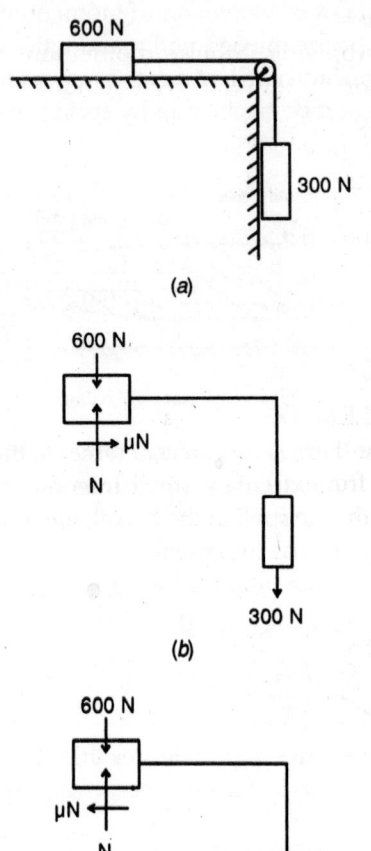

Fig. 17.5 (Illustrative Problem 17.4)

Solution:

   (*a*) No velocity $(v = 0)$; $u = 5$ m/sec
      Applying impulse momentum equation for the motion upto stationary condition,

$$(-\mu N - 300)t_1 = \left(\frac{500 + 300}{9.81}\right)(0 - 5)$$

$$(-0.2 \times 600 - 300)t_1 = -\frac{800}{9.81} \times 5$$

$$(-120 - 300)t_1 = -\frac{800}{9.81} \times 5$$

$$t_1 = \frac{800 \times 5}{9.81 \times 420} = 0.97 \text{ sec}$$

(b) Rightward velocity of 3 m/sec ($v = 3$ m/sec)
$u = 0$

Applying impulse momentum equation for the motion from stationary position to rightward motion.

$$(300 - \mu N)(t_2 - 0.97) = \left(\frac{500 + 300}{9.81}\right)(3 - 0)$$

$$(300 - 0.2 \times 600)(t_2 - 0.97) = 244.648$$

$$t_2 - 0.97 = \frac{244.648}{180} = 1.359$$

$$t_2 = 1.359 + 0.97 = 2.33 \text{ sec}$$

PROBLEM 17.5: In what distance will body A shown in Fig. 17.6 (*a*) attain a velocity of 3 m/sec starting from rest? Take $\mu = 0.2$. Assume the pully is smooth.

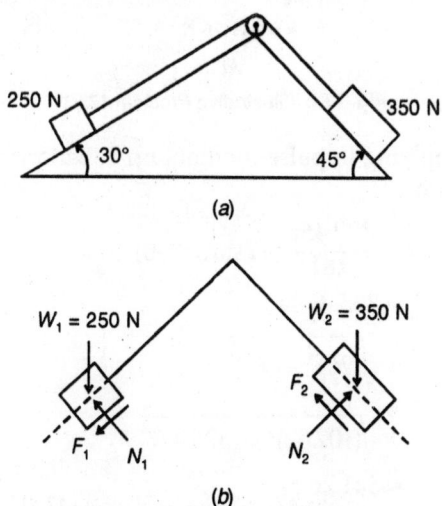

Fig. 17.6 (Illustrative Problem 17.5)

Solution:

$$N_1 = 250 \times \cos 30° = 216.506 \text{ N}$$
$$F_1 = \mu N_1 = 0.2 \times 250 \times \cos 30° = 43.30 \text{ N}$$
$$N_2 = 350 \times \cos 45° = 247.49 \text{ N}$$
$$F_2 = \mu N_2 = 0.2 \times 350 \times \cos 45° = 49.5 \text{ N}$$
$$\mu = 0; \; v = 3 \text{ m/sec}$$

Writing impulse mometum equation in the direction of motion.

$$(350 \sin 45° - F_2 - 250 \sin 30° - F_1)t$$

$$= \left(\frac{250 + 350}{9.81}\right)(3 - 0)$$

$$(247.49 - 49.5 - 125 - 43.30)t = 183.486$$

$$29.69 \times t = 183.486$$

$$t = 6.18 \text{ sec}$$

Distance, $s = \left(\frac{u + v}{2}\right)t$

$$s = \left(\frac{0 + 3}{2}\right)(6.18)$$

$$s = 9.27 \text{ m}$$

PROBLEM 17.6: Determine the tension in the strings and the velocity of 1200 N block shown in Fig. 17.7(*a*) 3 seconds after starting with a downward velocity of 2 m/sec.

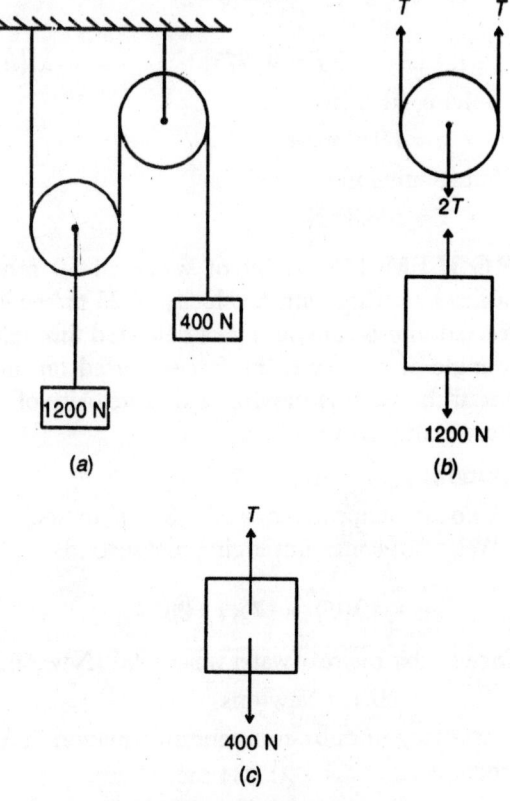

Fig. 17.7 (Illustrative Problem 17.6)

Solution: When 1200 N block moves a distance *s*, in the same time 400 N block moves a distance

2 s. Hence if velocity of 1200 N block is v m/sec that of 400 N block will be $2v$ m/sec. Let T be the tension in the string connecting 400 N block. Hence the tension in the string connecting 1200 N block will be $2T$ (Fig. 17.7(b)).

Initial velocity $u = 2$ m/sec, $t = 3$ sec.

Applying impulse-momentum equation for 400 N block.

$$(T - 400)t = \frac{400}{9.81}(2v - u)$$

$$(T - 400)3 = \frac{400}{9.81}(2v - 2)$$

or $\quad T - 372.817 = 27.183v \qquad …(i)$

Applying impulse-momentum equation to 1200 N block,

$$(1200 - 2T)t = \frac{1200}{9.81}(v - u)$$

$$(1200 - 2T)3 = \frac{1200}{9.81}(v - 2)$$

or $\quad 1281.549 - 2T = 40.775v \qquad …(ii)$

Solving (i) & (ii)

$$v = 4.929 \text{ m/sec}$$

Substituting in equation (i)

$$T = 506.80 \text{ N}$$

PROBLEM 17.7: A jet of water of 50 mm diameter moving with a velocity of 25 m/sec is directed into a vane and gets deflected through an angle of 35°. Find the force exerted on the vane if the vane is moving with a velocity of 8 m/sec to the right.

Solution:

Velocity of approach = 25 – 8 = 17 m/sec

Weight of water impinging in t seconds

$$= \frac{\pi}{4}(0.05)^2 \times 17 \times t \times 9810$$

(Since 1 cubic metre of water weighs 9810 Newtons)

$$= 327.45t \text{ Newtons}$$

Applying impulse-momentum equation in X-direction

$$-P_x \cdot t = \frac{327.45t}{9.81}(17\cos 35° - 17)$$

$$P_x = 102.62 \text{ N}$$

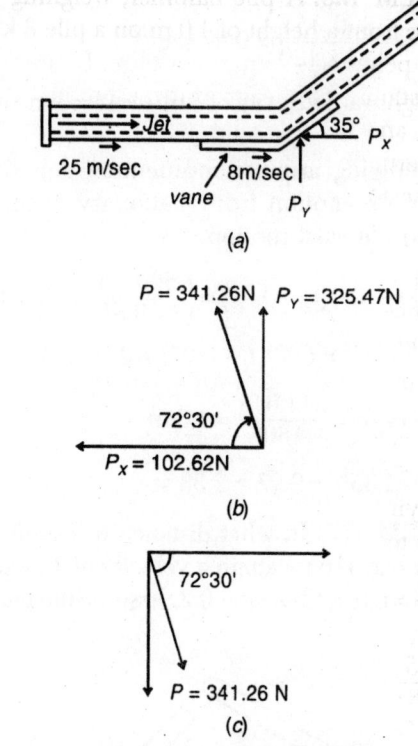

Fig. 17.8 (Illustrative Problem 17.7)

Applying impulse momentum equation in Y-direction

$$P_y \cdot t = \frac{327.45t}{9.81}(17\sin 35° - 0)$$

$$P_y = 325.47 \text{ N}$$

$\therefore \quad P = \sqrt{\left(P_x\right)^2 + \left(P_y\right)^2}$

$$= \sqrt{(102.62)^2 + (325.47)^2}$$

$$= 341.26 \text{ N}$$

Its inclination with horizontal

$$0 = \tan^{-1}\left[\frac{P_y}{P_x}\right]$$

$$= \tan^{-1}\left(\frac{325.47}{102.62}\right)$$

$$= 72°30'$$

Reaction of vane P is shown in Fig. 17.8 (b). The force of jet is equal and opposite to the reaction of the vane as shown in Fig. 17.8 (c).

PROBLEM 17.8: A pile hammer, weighing 15 kN drops from a height of 1.0 m on a pile 8 kN. The pile penetrates 75 mm per blow. Determine the resistance of the ground assumed to be constant and the pile supposed to be inelastic. Also determine the loss of kinetic energy during the impact.

Solution:

Initial velocity of hammer $u = 0$
Distance moved $h = 1.0$ m
Acceleration due to gravity $g = 9.81$ m/sec^2
$\therefore$ Velocity at the time of striking, $v = \sqrt{2gh}$

$$= \sqrt{2 \times 9.81 \times 1}$$
$$= 4.43 \text{ m/sec}$$

Applying principle of conservation of momentum to the system of pile and hammer.

$$\frac{wv}{g} = \left(\frac{w+W}{g}\right)V$$

$$\frac{15}{9.81} \times 4.43 = \frac{15+8}{9.81}V$$

where $V =$ Velocity of the pile and hammer

$$\therefore \quad V = \frac{15 \times 4.43 \times 9.81}{9.81 \times 23} = 2.89 \text{ m/sec}$$

Applying work-energy equation to the motion of hammer and pile

$$(w+W)s - R \times s = \frac{1}{2}\left(\frac{w+W}{g}\right)(0-V^2)$$

$$(15+8-R)s = \left(\frac{15+8}{2 \times g}\right)(0-V^2)$$

where $R =$ Resistance of the ground in kN
$s = 75$ mm $= 0.075$ m

$$\therefore \quad (15+8-R)0.075 = \left(\frac{23}{2 \times 9.81}\right)(-(2.89)^2)$$

$$23 - R = -\frac{23(8.3521)}{2 \times 9.81 \times 0.075} = -130.5$$

or $\quad R = 130.5 + 23$

or $\quad R = 153.5$ kN

Loss of kinetic energy

$$= \frac{wWh}{w+W}$$

$w = 15$ kN
$W = 8$ kN
$h = 1$ m

$\therefore$ Loss of kinetic energy during the impact

$$= \frac{15 \times 8 \times 1}{15 + 8}$$

$$= 5.217 \text{ kN-m}$$

$$= 5.217 \text{ kJ}$$

PROBLEM 17.9: A pile hammer weighing 20 kN drops from a height of 600 mm on a pile of 10 kN. Find the penetration of the pile for each blow of the hammer, if the resistance to penetration of the ground is 150 kN. Assume that the resistance of the ground is constant.

Solution:

Initial velocity of hammer $u = 0$
Distance moved $h = 600$ mm

$$= 0.60 \text{ m}$$

Acceleration due to gravity $g = 9.81$ m/sec^2
$\therefore$ Velocity at the time of striking $v = \sqrt{2gh}$

$$= \sqrt{2 \times 9.81 \times 0.60}$$

$$= 3.431 \text{ m/sec}$$

Let V be the velocity of the pile and hammer immediately after impact.

Applying principle of conservation of momentum to the system of pile and hammer.

$$\frac{wv}{g} = \left(\frac{w+W}{g}\right)V$$

$$\frac{20 \times 3.431}{9.81} = \left(\frac{20+10}{9.81}\right)V$$

or $\quad V = \frac{20 \times 3.431}{30} = 2.287$ m/sec

Applying work energy equation to the motion of hammer and pile,

$$(w+W)s - R \times s = \frac{1}{2}\left(\frac{w+W}{g}\right)(0-V^2)$$

$$(20+10-150)s = \frac{1}{2}\left(\frac{20+10}{9.81}\right)(-(2.287)^2)$$

$$S = \frac{1 \times 30 \times 5.23}{2 \times 9.81 \times 120} = 0.0666 \text{ m}$$

$$= 66.6 \text{ mm}$$

PROBLEM 17.10: A gun of mass 4000 kg fires horizontally a shell of mass 75 kg with a velocity of 400 m/sec. With what velocity will the gun recoil ? Also determine the uniform force required to stop the gun in 0.65 m. In how much time will it stop?

Solution:

Initial velocity of the gun = 0

Initial velocity of the shell = 0

Final velocity of the gun = V_g

Final velocity of the shell = V_s = 400 m/sec

Mass of gun = 4000 kg

Mass of shell = 75 kg

Applying principle of conservation of momentum to the system of gun and the shell,

$$0 = m_g\, v_g + m_s \cdot v_s$$

$$0 = 4000 \times v_g + 75 \times 400$$

$$v_g = \frac{-75 \times 400}{4000} = -7.5 \text{ m/sec}$$

i.e., gun will have a velocity of 7.5 m/sec in the direction opposite to that of shell.

$$v = 0$$

$$u = 7.5 \text{ m/sec}$$

$$s = 0.65 \text{ m}$$

$$\therefore \quad v^2 - u^2 = 2as$$

$$0 - (7.5)^2 = 2 \times a \times 0.65$$

$$a = -\frac{(7.5)^2}{2 \times 0.65}$$

$$= -43.27 \text{ m/sec}^2$$

Force required to stop the gun

$$= \text{Mass} \times \text{Acceleration}$$

$$= 4000 \times 43.27$$

$$= 173080 \text{ kN}$$

$$= 173 \text{ kN}$$

Time required to stop the gun $v = u + at$

$$t = \frac{7.5}{43.27} = 0.173 \text{ sec}$$

PRACTICE PROBLEMS

17.1   A 1 N ball is bowled towards a batsman. The velocity of the ball was 25 m/sec horizontally just before batsman hit it. After hitting it went away with a velocity of 50 m/sec at an inclination of 36° to the horizontal as shown in Fig. 17.9. Find the average force exerted on the ball by the bat if the impact lasts (1/40)th of a second.

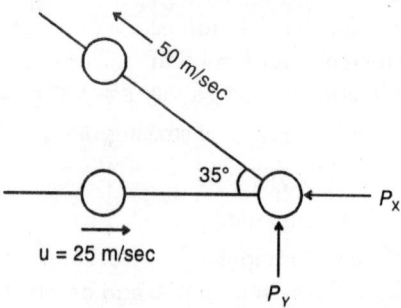

Fig. 17.9 (Practice Problem 17.1)

17.2   A glass marble, whose weight is 0.3 N, falls from a height of 12 m and rebounds to a height of 9 m. Find the impulse and the average force between the marble and floor, if the duration of contact is (1/20)th of a second.

17.3   A car weighing 12 kN and running at 12 m/sec holds three men each weighing 600 N. The men jump off from the back end gaining a relative velocity of 6 m/sec with the car. Find the speed of the car if the three men jump off (*a*) in succession, (*b*) all together.

17.4   A block of wood *A* of weight 200 N is held on a rough horizontal table. A light inextensible string connected to the block passes over a smooth pulley at the end of the table. Another wooden block *B* of weight 100 N is connected to the other end of the string. Find the acceleration of the system and tension in the string.

17.5   Determine the time required for the weights shown in Fig. 17.10 to attain a velocity of 6 m/sec.

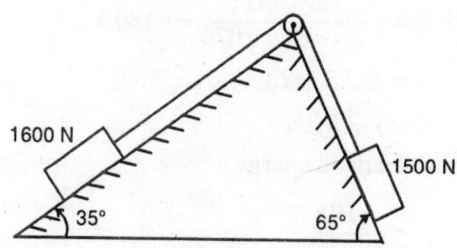

Fig. 17.10 (Practice Problem 17.5)

Also determine the tension in the string. Take $\mu = 0.2$ for both planes. Assume the pulleys as frictionless.

17.6 Determine the tension in the strings and the velocity of 1200 N block shown in Fig. 17.11 three seconds after starting from rest. Assume pulleys as weightless and frictionless.

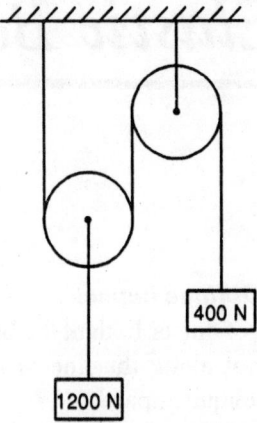

Fig. 17.11 (Practice Problem 17.6)

17.7 A jet of water of 50 mm diameter moving with a velocity of 25 m/sec is directed into a vane and get deflected through an angle of 35°. Find the force exerted on the vane.

17.8 A pile hammer, weighing 10 kN drops from a height of 500 mm on a pile of 5 kN. The pile penetrates 50 mm per each blow. Find the resistance to penetration of the ground. Assume that the resistance of the ground is constant. Also find the time occupied by the movement of the pile.

17.9 A gun of mass 4000 kg fires horizontally a shell of mass 75 kg with a velocity of 400 m/sec. With what velocity will the gun recoil? If the recoil is overcome by an average force of 175 kN, how far will the gun travel? How long will it take?

Chapter 18

Impact and Collision of Elastic Bodies

18.1 INTRODUCTION

An 'impact' is defined as the collision between two bodies for a very short interval of time, during which the bodies exert very large force on each other.

During the phenomenon of impact, the bodies deform first and then recover due to their elastic properties, and later move with different velocities. The analysis of the velocity and other characteristics during the short period of impact is complex and is not of interest in the present treatment. Only the conditions before and after impact will be analysed, and the loss of kinetic energy during the impact will be determined.

There are several factors such as the velocity of approach, elastic properties, size and shape of the bodies, besides the line of impact, which in-fluence the conditions after impact.

18.2 DEFINITIONS

Definitions of common terms relevant to this topic are given below:

Line of Impact

The common normal to the surfaces of the two bodies in contact during the impact is called the 'line of impact'.

Direct Impact

If the motion of the two bodies before collision is along the line of impact, it is called 'direct impact'.

Indirect or Oblique Impact

If the motion of one or both of the bodies before collision is not along the line of impact, it is 'indirect or oblique impact'.

Central Impact

If the centres of mass of the colliding bodies lie on the line of impact, it is called 'central impact'.

Non-central or Eccentric Impact

Even if the centre of mass of one of the colliding bodies, does not lie on the line of impact, it is 'non-central' or 'eccentric impact'.

Period of Deformation

The short time interval during which the colliding bodies undergo elastic deformation is the 'period of deformation'.

Period of Restitution

The short interval of time during which the colliding bodies recover from the elastic deformation and just get separated is the 'period of restitution (regain or recovery)'.

Period of Collision or Impact

The sum of the periods of deformation and restitution is the total 'period of collision or impact'. Some of these terms are illustrated in Fig. 18.1.

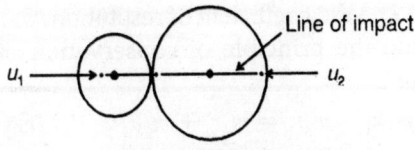

(a) Direct Central Impact

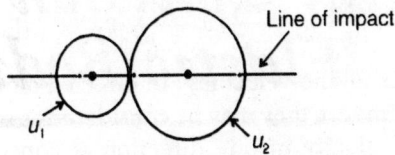

(b) Oblique (indirect) Central Impact

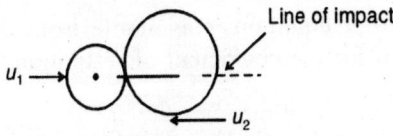

(c) Direct Eccentric Impact

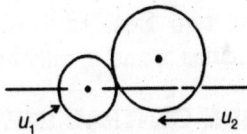

(d) Oblique Eccentric (non–central) Impact

Fig. 18.1: Kinds of Impact

18.3 COEFFICIENT OF RESTITUTION

As stated in the introduction to this chapter, the phenomenon of impact consists of two important phases–the period of deformation and that of restitution (Fig. 18.2).

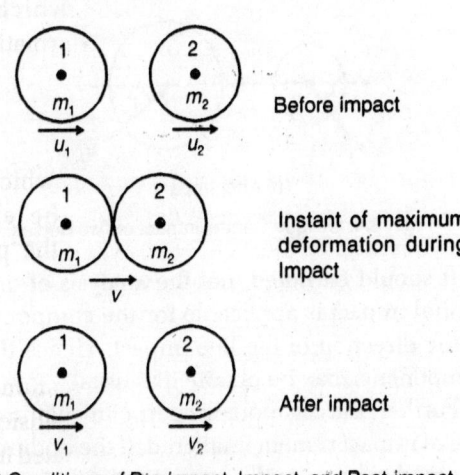

(a) Conditions of Pre-impact, Impact, and Post-impact

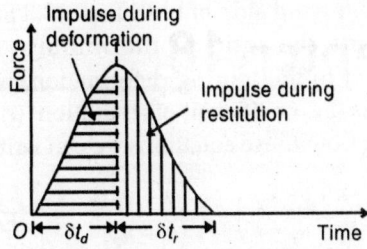

(b) Graphical Representation of Impulses during Impact

Fig. 18.2: Phenomenon of Impact

Let the periods of deformation and of restitution be δt_d and δt_r respectively. Impulse during deformation, $I_d = \int F_d \cdot dt$, F_d being the impulsive force during the period of deformation δt_d of the integral being taken over this period.

Similarly, impulse during restitution, $I_r = \int F_r \cdot dt$, F_r being the impulsive force during the period of restitution, δt_r the integral being taken over this period.

Let the two bodies of mass m_1 and m_2, move with initial velocities u_1 and u_2, move together with a common velocity v at the instant of maximum deformation, and get separated after restitution, and move with final velocities v_1 and v_2 as shown.

Applying the impulse-memontum principle for the first body–during deformation,

$$\int_{t_d} F_d \cdot dt = m_1 v - m_1 u_1 = m_1(v - u_1) \quad \text{...(Eq. 18.1)}$$

during restitution,

$$\int_{t_r} F_r \cdot dt = m_1 v_1 - m_1 v = m_1(v_1 - v) \quad \text{...(Eq. 18.2)}$$

Dividing Eq. 18.2 by Eq. 18.1,

$$\frac{\int_{t_r} F_r \cdot dt}{\int_{t_d} F_r \cdot dt} = \left(\frac{v_1 - v}{v - u_1}\right) \quad \text{...(Eq. 18.3)}$$

By a similar analysis for the second body, we obtain

$$\frac{\int_{t_r} F_r \cdot dt}{\int_{t_d} F_r \cdot dt} = \left(\frac{v - v_2}{u_2 - v}\right) \quad \text{...(Eq. 18.4)}$$

The left hand side of both Eqs. 18.3 and 18.4 represents the ratio of the impulse during restitution to that during deformation, which is known as the 'coefficient of restitution' (e).

Using both these equations we can write,

$$e = \left(\frac{v_1 - v + v - v_2}{v - u_1 + u_2 - v} \right) = -\frac{v_1 - v_2}{(u_1 - u_2)} \quad \text{...(Eq. 18.5)}$$

We may also state that

$$e = -\left(\frac{\text{Relative velocity of separation}}{\text{Relative velocity of approach}} \right)$$

$$\text{...(Eq. 18.6)}$$

(The relative velocities are to be obtained along the line of impact and the signs of the velocities are to be carefully considered).

The coefficient of restitution is a parameter which is a measure of the energy loss during impact and can be determined experimentally; in fact, Newton found that it is a constant for any two given bodies.

For perfectly elastic bodies, e will be 1; for perfectly plastic bodies, e will be zero, the restitution being zero. e will always lie between 0 and 1.

18.4 DIRECT CENTRAL IMPACT

Let us consider the direct central impact of two bodies of masses m_1 and m_2, the velocities before impact and after impact being u_1, u_2 and v_1, v_2, respectively (Fig. 18.3).

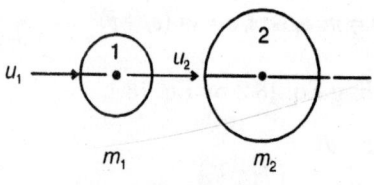

(a) Before Impact

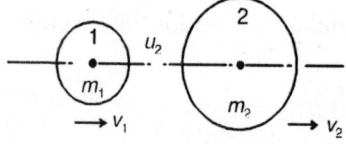

(b) After Impact

Fig. 18.3: Direct Central Impact of Two Bodies

Let e be the coefficient of restitution.

From the principle of conservation of momentum,

$$m_1 u_1 + m_2 u_2 = m_1 v_1 + m_2 v_2 \quad \text{...(Eq. 18.7)}$$

In terms of weights W_1 and W_2 of the bodies,

$$\frac{W_1}{g} \cdot u_1 + \frac{W_2}{g} \cdot u_2 = \frac{W_1}{g} \cdot v_1 + \frac{W_2}{g} \cdot v_2 \quad \text{...(Eq.18.8)}$$

Since all the velocities are directed along the line of impact, they may be considered as scalars. If the velocity in one direction is considered positive, that in the opposite direction is taken to be negative.

Another equation is available from the expression for the coefficient of restitution, e (Eq. 18.5).

$$e = -\left(\frac{v_1 - v_2}{u_1 - u_2} \right) \quad \text{...(Eq. 18.5)}$$

From these two Eqs. 18.8 and 18.5, the unknown velocities v_1 and v_2 may be solved.

18.5 OBLIQUE CENTRAL IMPACT

Let us consider the oblique central impact of two bodies of weights W_1 and W_2 as shown in Fig. 18.4.

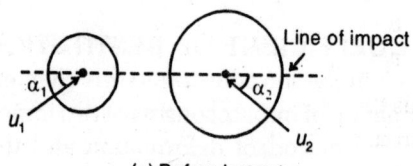

(a) Before Impact

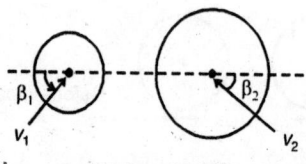

(b) After Impact

Fig. 18.4: Oblique Central Impact of Two Bodies

It should be noted that the analysis of direct central impact is applicable for the components in the direction of the line impact. Hence these components may be obtained as usual.

Further, the components perpendicular to the line of impact remain unaltered, if the bodies are assumed to be smooth and frictionless.

The final velocities after impact—v_1 and v_2—may now be got by compounding of their components parallel and perpendicular to the line of impact.

18.6 SPECIAL CASES OF ELASTIC IMPACT

A few particular and special cases of elastic impact are considered in the following sub-sections.

18.6.1 Perfectly Elastic Impact of Two Equal Masses

When the two masses are equal in a direct central impact,

$$m_1 = m_2 = m$$

Conservation of momentum gives

$$(u_1 + u_2) = (v_1 + v_2) \qquad \text{...(Eq. 18.9)}$$

Coefficient of restitution, e, is given by

$$e = -\left(\frac{v_1 - v_2}{u_1 - u_2}\right) = 1$$

$$(u_1 - u_2) = (v_2 - v_1) \qquad \text{...(Eq. 18.10)}$$

Solving Eqs. 18.9 and 18.10 simultaneously, we get,

$$\left. \begin{array}{l} v_2 = u_1 \\ v_1 = u_2 \end{array} \right\} \qquad \text{...(Eq. 18.11)}$$

Thus the two masses exchange their velocities of approach.

18.6.2 Perfectly Elastic Impact of Two Equal Masses when One is at Rest

Say $m_1 = m_2 = m$, and $u_2 = 0$

Conservation of momentum gives

$$u_1 = (v_1 + v_2) \text{ or } v_2 + v_1 = u_1 \qquad \text{...(Eq. 18.12)}$$

Coefficient of restitution, e, is given by

$$e = -\left(\frac{v_1 - v_2}{u_1}\right) = 1 \text{ or } v_2 - v_1 = u_1 \qquad \text{...(Eq. 18.13)}$$

Solving Eqs. 18.12 and 18.13 simultaneously, we get

$$v_1 = 0 \text{ and } v_2 = u_1 \qquad \text{...(Eq. 18.14)}$$

That is, the striking mass stops after imparting its velocity to that originally at rest.

This can be viewed as a corollary of the case given in the previous subsection.

18.6.3 Impact of a Body on a Rigid Plane

Let a body of mass of m_1 impinge a rigid plane normally with a velocity u_1 as shown in Fig. 18.5.

The mass of the rigid plane is considered to be infinite.

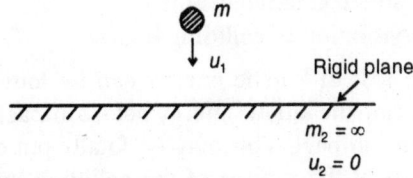

Fig. 18.5: Direct Impact of a Body on a Rigid Plane

Conservation of momentum gives,

$$m_1 u_1 + m_2 u_2 = m_1 v_1 + m_2 v_2$$

Dividing by m_2,

$$\frac{m_1}{m_2} \cdot u_1 + u_2 = \frac{m_1}{m_2} \cdot v_1 + v_2$$

When m_2 is very very large and $u_2 = 0$

$$v_2 = 0$$

Coefficient of restitution, e, gives

$$e = 1 = -\left(\frac{v_1 - v_2}{u_1 - u_2}\right)$$

$$v_1 = u_1$$

In other words, the body will rebound with the same velocity with which it strikes the rigid plane.

In the case of oblique impact, the above analysis is applicable for the normal component of the velocity; the component parallel to the rigid plane remains unaltered.

The resultant final velocity will be numerically the same as the striking velocity but will be deflected as shown in Fig. 18.6.

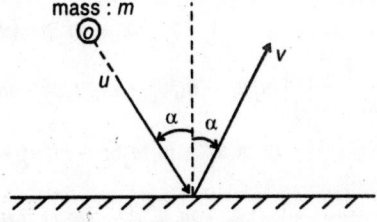

Fig. 18.6: Oblique Impact of a Mass on a Rigid Plane

18.7 LOSS OF KINETIC ENERGY DURING IMPACT

During collison of two bodies, energy is lost on account of the following:

   (*i*) imperfect elastic action;
   (*ii*) heat generated;
   (*iii*) sound generated; and
   (*iv*) vibration of colliding bodies.

The loss of kinetic energy can be found by comparing the kinetic energy before impact with that after impact. This may be finally put down in terms of the masses of the colliding bodies, their velocities of approach, and the coefficient of restitution.

Let the masses be m_1 and m_2, their approach velocities u_1 and u_2, their final velocities v_1 and v_2, and the coefficient of restitution e.

Kinetic energy before impact,

$$= \frac{1}{2}\left(m_1 u_1^2 + m_2 u_2^2\right)$$

Kinetic energy after impact

$$= \frac{1}{2}\left(m_1 v_1^2 + m_2 v_2^2\right)$$

$\therefore$ Loss of kinetic energy during impact,

$$E_k' = \frac{1}{2}\left[\left(m_1 u_1^2 + m_2 u_2^2\right) - \left(m_1 v_1^2 + m_2 v_2^2\right)\right] \ldots\text{(Eq. 18.15)}$$

or $E_k = \frac{1}{2}\left[m_1(u_1^2 - v_1^2) + m_2(u_2^2 - v_2^2)\right] \ldots$(Eq. 18.16)

Conservation of momentum gives

$$m_1 u_1 + m_2 u_2 = m_1 v_1 + m_2 v_2 \quad \ldots\text{(Eq. 18.17)}$$

From the coefficient of restitution principle,

$$(v_1 - v_2) = -e(u_1 - u_2) \qquad \ldots\text{(Eq. 18.18)}$$

Multiplying the numerator and denominator of the R.H.S. of Eq. 18.16 by $(m_1 + m_2)$, we have

$$E_k' = \frac{1}{2(m_1 + m_2)}\Big[(m_1 + m_2)\,m_1(u_1^2 - v_1^2)$$
$$+ (m_1 + m_2)m_2(u_2^2 - v_2^2)\Big]$$

$$= \frac{1}{2(m_1 + m_2)}\Big[m_1^2 u_1^2 - m_1^2 v_1^2 + m_1 m_2 u_2^2 - m_1 m_2 v_2^2$$
$$+ m_1 m_2 u_1^2 - m_1 m_2 v_1^2 + m_2^2 u_2^2 - m_2^2 v_2^2\Big]$$

$$= \frac{1}{2(m_1 + m_2)}\Big[(m_1 u_1 + m_2 u_2)^2 - 2m_1 m_2 u_1 u_2$$

$$-(m_1 v_1 + m_2 v_2)^2 + 2m_1 m_2 v_1 v_2 + m_1 m_2 (u_1^2 + u_2^2)$$
$$-m_1 m_2(v_1^2 + v_2^2)\Big]$$

Using Eq. 18.17,

$$(m_1 u_1 + m_2 u_2)^2 = (m_1 v_1 + m_2 v_2)^2$$

$\therefore \quad E_k' = \dfrac{m_1 m_2}{2(m_1 + m_2)}\Big[u_1^2 + u_2^2 - 2u_1 u_2\Big)$

$$-(v_1^2 + v_2^2 - 2v_1 v_2)\Big]$$

$$= \frac{m_1 m_2}{2(m_1 + m_2)}\Big[(u_1 - u_2)^2 - (v_1 - v_2)^2\Big]$$

Using Eq. 18.18,

$$E_k' = \frac{m_1 m_2}{2(m_1 + m_2)}\Big[(u_1 - u_2)^2 - e^2(u_1 - u_2)^2\Big]$$

or $\quad E_k' = \dfrac{m_1 m_2}{2(m_1 + m_2)}\Big[(u_1 - u_2)^2 \cdot (1 - e^2)\Big]$

$$\ldots\text{(Eq. 18.19)}$$

Thus, the loss of kinetic energy is expressed in terms of the masses m_1 & m_2, the approach velocities u_1 & u_2, and the coefficient of restitution, e.

ILLUSTRATIVE PROBLEMS

PROBLEM 18.1: A ball of mass 4 kg impinges directly with a ball of mass 2 kg, which is at rest. If the velocity of the smaller mass after impact is the same as that of the first ball before impact, find the coefficient of restitution.

Solution:

$$m_1 = 4 \text{ kg}, \ u_1, \ v_1 = ?$$
$$m_2 = 2 \text{ kg}, \ u_2 = 0 \ v_2 = u_1$$

Let e be the coefficient of restitution.

By principle of conservation of linear momentum

$$m_1 \cdot u_1 + m_2 \cdot u_2 = m_1 \cdot v_1 + m_2 \cdot v_2$$
$$4 \times u_1 + 2 \times 0 = 4 \times v_1 + 2 \times u_1$$
$$2u_1 = 4v_1$$
$$u_1 = 2v_1$$
$$u_1 = 2v_1; \ v_1 = v_1$$
$$u_2 = 0; \ v_2 = 2v_1$$

$\therefore \qquad e = -\dfrac{(v_1 - v_2)}{(u_1 - u_2)}$

$$e = -\frac{(v_1 - 2v_1)}{(2v_1 - 0)}$$

$$e = -\frac{(-v_1)}{2v_1} = \frac{v_1}{2v_1} = \frac{1}{2}$$

$$e = \frac{1}{2}$$

PROBLEM 18.2: A ball of mass 1.5 kg moving with a velocity of 2 m/sec impinges directly on a ball of mass 3.0 kg at rest. The first ball, after impinging, comes to rest. Find the velocity of the second ball after the impact and the coefficient of restitution.

Solution:

$$m_1 = 1.5 \text{ kg}$$
$$u_1 = 2 \text{ m/sec}$$
$$m_2 = 3.0 \text{ kg}$$
$$u_2 = 0$$
$$v_1 = 0$$
$$v_2 = ?$$
$$e = ?$$

According to the law of conversation of momentum

Total initial momentum = Total final momentum

$$m_1 u_1 + m_2 u_2 = m_1 v_1 + m_2 v_2$$
$$(1.5 \times 2) + (3.0 \times 0) = (1.5 \times 0) + (3.0 \times v_2)$$
$$3 = 3v_2$$
$$\therefore \quad v_2 = \frac{3}{3} = 1 \text{ m/sec}$$

$$e = \frac{v_2 - v_1}{u_1 - u_2} = \frac{1 - 0}{2 - 0} = 0.5$$

PROBLEM 18.3: A bullet of mass 50 gm is fired into a freely suspended target of mass 6 kg. On impact, the target moves with a velocity of 10 m/sec along with the bullet in the direction of firing. Find the velocity of bullet.

$$m_1 = 50 \text{ gm}$$
$$u_1 = ?$$
$$m_2 = 6 \text{ kg} = 6000 \text{ gms}$$

$$u_2 = 0 \text{ m/sec}$$
$$v_1 = v_2 = v = 10 \text{ m/sec}$$

According to law of conservation of momentum

$$m_1 u_1 + m_2 u_2 = m_1 v_1 + m_2 v_2$$
$$m_1 u_1 + m_1 u_2 = (m_1 + m_2)v$$
$$50 \times u_1 + 6000 \times 0 = (50 + 6000) \times 10$$
$$50 u_1 = 6050 \times 10$$

$$u_1 = \frac{6050 \times 10}{50} = 1 \text{ m/sec}$$

PROBLEM 18.4: Three perfectly elastic balls A, B and C of masses m, $2m$ and $3m$ are placed in a straight line. The first impinges directly on the second with a velocity u and then the second impinges on the third. Find the velocity of the third ball after impact.

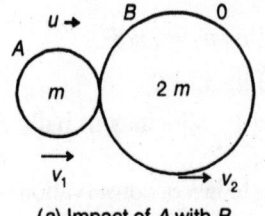

(a) Impact of A with B

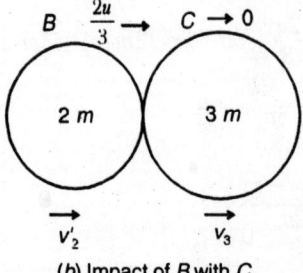

(b) Impact of B with C

Fig. 18.7 (Illustrative Problem 18.4)

Solution: Here $e = 1$

$$m_1 = m$$
$$m_2 = 2m$$
$$u_1 = u$$
$$u_2 = 0$$

Considering impact of A & B

Let v_1 and v_2 velocities of balls A and B after impact.

According to law of conservation of momentum

$$m_1 \cdot u_1 + m_2 \cdot u_2 = m_1 \cdot v_1 + m_2 \cdot v_2$$

$$m \cdot u + 2m \cdot 0 = m \cdot v_1 + 2m \cdot v_2$$

$$v_1 + 2v_2 = u \qquad \qquad \ldots(i)$$

Also $\quad e = \dfrac{v_2 - v_1}{u_1 - u_2} = 1$

$$v_2 - v_1 = u \qquad \qquad \ldots(ii)$$

Solving (i) & (ii)

$$v_2 = \frac{2u}{3}$$

$$v_1 = u - 2\left(\frac{2u}{3}\right)$$

$$v_1 = -\frac{u}{3}$$

Considering impact of B and C

The second ball now moves with a velocity $\dfrac{2u}{3}$ and strikes the third ball

$$m_2 = 2m; \; u_2' = 2u/3$$

$$m_3 = 3m; \; u_3 = 0$$

Let v_2' and v_3 velocities of balls B and C after impact.

According to law of conservation of momentum

$$m_2 u_2' + m_3 u_3 = m_2 v_2' + m_3 v_3$$

$$2m \cdot \frac{2u}{3} + 3m \cdot 0 = 2m \cdot v_2' + 3m \cdot v_3$$

$$2v_2' + 3v_3 = \frac{4u}{3} \qquad \qquad \ldots(iii)$$

Also $\quad e = \dfrac{v_3 - v_2'}{\dfrac{2u}{3} - 0}$

$$v_3 - v_2' = \frac{2u}{3} \qquad \qquad \ldots(iv)$$

$(iii) \quad = 2v_2' + 3v_3 = \dfrac{4u}{3}$

$(iv) \quad \times 2 = \dfrac{2v_3 - 2v_2' = 4(u/3)}{5v_3 = \dfrac{8u}{3}}$

or $\quad v_3 = \dfrac{8u}{15}$

$$v_2' = v_3 - \frac{2u}{3} = \frac{8u}{15} - \frac{2u}{3}$$

$$= \frac{8u - 10u}{15}$$

$$v_2' = -\frac{2}{15}u$$

$$v_1 = -u/3; \; v_2 = 2u/3; \; v_2' = -\frac{2}{15}u; \; v_3 = 8(u/15)$$

Velocity of the third ball after impact $= \dfrac{8u}{15}$

PROBLEM 18.5: A heavy elastic ball drops from the ceiling of a room, and after rebounding twice from the floor, reaches a height equal to half that of ceiling, show that the coefficient of restitution is $\dfrac{1}{\sqrt[4]{2}}$.

Solution:

Let the height of ceiling $= h$

Height ascended after second rebound $= \dfrac{h}{2}$

Velocity on reaching the floor for the first rebound $= \sqrt{2gh}$

Let e be the coefficient of restitution

Velocity after first rebound $= e \cdot \sqrt{2gh}$

When the ball strikes the floor a second time, its velocity $= e \cdot \sqrt{2gh}$

Velocity after second rebound $= e \cdot e \cdot \sqrt{2gh}$

$$= e^2 \cdot \sqrt{2gh}$$

Now, for motion after second rebound

Initial velocity $= e^2 \cdot \sqrt{2gh}$

Final velocity $= 0$

Height attained $= \dfrac{h}{2}$

Using, $v^2 - u^2 = 2gh$, we have

$$0^2 - \left(e^2\sqrt{2gh}\right)^2 = 2(-g)\left(\frac{h}{2}\right)$$

$$-e^4 \, 2gh = -gh$$

$$e^4 = \frac{1}{2}$$

Hence $e = \dfrac{1}{\sqrt[4]{2}}$

PROBLEM 18.6: A vehicle of 500 kg moving with a velocity of 10 m/sec strikes another vehicle of mass 300 kg moving at 6 m/sec in the same direction. Both the vehicles got coupled together due to impact. Find the common velocity with which the two vehicles will move. Also find the loss of K.E. due to impact.

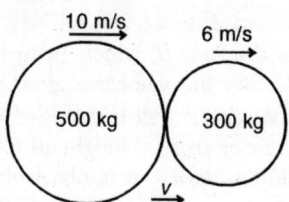

Fig. 18.8 (Illustrative Problem 18.6)

Solution:

$$m_1 = 500 \text{ kg}$$

$$u_1 = 10 \text{ m/sec}$$

$$m_2 = 300 \text{ kg}$$

$$u_2 = 6 \text{ m/sec}$$

Let v be the common velocity

By the principle of conservation of linear momentum,

$$m_1 \cdot u_1 + m_2 \cdot u_2 = m_1 \cdot v_1 + m_2 \cdot v_2$$

$$500 \times 10 + 300 \times 6 = (500 + 300)v$$

$$800v = 5000 + 1800 = 6800$$

$$v = \frac{6800}{800} = 8.5 \text{ m/sec}$$

Initial K.E. $= \frac{1}{2} m_1 u_1^2 + \frac{1}{2} m_2 u_2^2$

$$= \frac{1}{2} \times 500 \times 10^2 + \frac{1}{2} \times 300 \times 6^2$$

$$= 25000 + 5400$$

$$= 30400 \text{ Joules}$$

Final K.E. $= \frac{1}{2}(500 + 300)(8.5)^2$

$$= 28900 \text{ Joules}$$

Loss of K.E. = Initial K.E. – Final K.E.

$$= 30400 - 28900$$

$$= 1500 \text{ Joules}$$

PROBLEM 18.7: A ball impinges directly upon another ball at rest and is itself reduced to rest by the impact. If half of the initial kinetic energy is destroyed in the collision, find the coefficient of restitution.

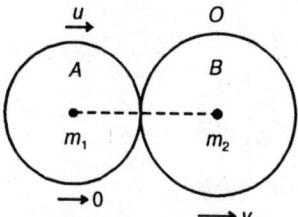

Fig. 18.9 (Illustrative Problem 18.7)

Solution: Let u be the velocity of the first ball before impact and v be the velocity of the second ball after impact.

Let m_1 & m_2 be their masses.

Let e be the coefficient of restitution.

By principle of conservation of linear momentum,

$$m_1 \times v_1 + m_2 \times v_2 = m_1 \times u_1 + m_2 \times u_2$$

$$m_2 \times v + m_1 \times 0 = m_1 \times u + m_2 \times 0$$

$$m_2 v = m_1 u \qquad \qquad \ldots(1)$$

$$e = -\left(\frac{v_1 - v_2}{u_1 - u_2}\right)$$

$$e = -\left(\frac{0 - v}{u - 0}\right) = \frac{v}{u}$$

or $\qquad v = e \cdot u \qquad \qquad \ldots(2)$

Initial K.E. $= \frac{1}{2} m_1 u^2 + \frac{1}{2} m_2 \times 0^2$

$$= \frac{1}{2} m_1 u^2$$

Final K.E. $= \frac{1}{2} m_1 \times 0^2 + \frac{1}{2} m_2 \times v^2$

$$= \frac{1}{2} m_2 v^2$$

$$= \frac{1}{2} m_2 e^2 u^2$$

Loss of K.E. $= \frac{1}{2} m_1 u^2 - \frac{1}{2} m_2 e^2 u^2$

Half of the initial K.E. is destroyed in the collision.

$$\frac{1}{2}m_1u^2 - \frac{1}{2}m_2e^2u^2 = \frac{1}{2} \cdot \frac{1}{2}m_1u^2$$

$$\frac{m_1u^2}{4} = \frac{1}{2}m_2e^2u^2$$

$$m_1 = 2m_2e^2$$

or $\dfrac{m_1}{m_2} = 2 \cdot e^2$ ...(3)

From (1) $\dfrac{m_1}{m_2} = \dfrac{v}{u}$

From (2) $v = e \cdot u$

$\therefore$ $\dfrac{m_1}{m_2} = \dfrac{e \cdot u}{u} = e$ ...(4)

Equating equations (3) & (4)

$$2e^2 = e$$

or $e = \dfrac{1}{2}$

PRACTICE PROBLEMS

18.1 A body of mass 120 kg, moving with a velocity of 6 m/sec, collides with a stationary body of mass 60 kg. If the two bodies become coupled so that they move on together after the impact, what is the common velocity?

18.2 Two balls of masses 1.5 kg and 3.0 kg are moving with velocities 2 m/sec and 3 m/sec towards each other. If the coefficient of resistution is 0.5, find the velocity of two balls after impact.

18.3 Three perfectly elastic balls A, B and C of masses 2 kg, 6 kg and 10 kg are moving in the same direction with velocities 10 m/sec, 6 m/sec and 2 m/sec respectively. If the ball A strikes the ball B, which in turn, strikes the ball C, find the velocities after impact.

18.4 A heavy elastic ball is dropped upon a horizontal floor from a height of 6 m and after rebounding twice, it is observed to attain a height of 3 m. Find the coefficient of restitution.

18.5 Two spheres A and B of masses 4 kg and 8 kgs moving with velocities 9 m/sec and 3 m/sec in opposite directions collide. If A rebounds with a velocity of 1 m/sec, find the velocity of B after impact, the coefficient of restitution and the loss of kinetic energy.

18.6 A sphere impinges directly on an equal sphere which is at rest. Show that a fraction $\frac{1}{2}(1-e^2)$ of original kinetic energy is lost during the impact.

Chapter 19

Transmission of Power

19.1 INTRODUCTION

The universal use of machines led to the need for transmission of power from one machine to another through rotating shafts, connected by running over pulleys, fixed to the shafts. Flexible belts and ropes are used. Of course, in the modern era, electrical transmission through motors is used in place of mechanical transmission through rope or belt drives. In view of the flexibility of belts and ropes, there is a possibility of "slip"; so these are called "non-positive" drives.

The simplest arrangement to transmit power from one shaft to another is shown in Fig. 19.1

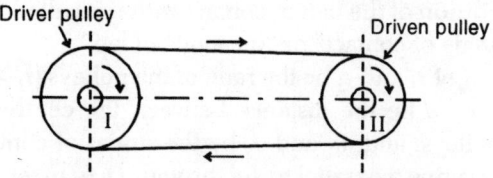

Fig. 19.1: Transmission of Power through Belts

To shaft-I is fixed the driver pulley from which power is transmitted to shaft-II through the driven pulley, the pulleys being fixed to the shafts through keys.

The driver pulley rotates and carries with it the belt by virtue of the friction between the rim of the pulley and the belt, which in turn, rotates the driven pulley and with it the other shaft.

19.2 TYPES OF BELTS

Belts are classified, based on their cross-section,

as (*i*) Flat belt, (*ii*) V-belt, and (*iii*) Circular belt; the last one is also called a 'rope'.

Flat belt has rectangular-shaped section.

V-belt has a cross section of the shape of the letter, "V".

Circular belt has a cross-section of circular shape, as the name indicates. These are shown in Fig. 19.2.

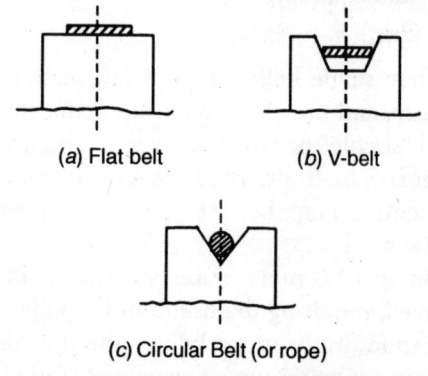

(*a*) Flat belt   (*b*) V-belt

(*c*) Circular Belt (or rope)

Fig. 19.2: Types of Belts

19.3 VELOCITY RATIO OF BELT DRIVE

Let the speed of the driver pulley be n_1 rpm and its radius r_1; the speed of the driven pulley is, say, n_2 rpm and its radius r_2.

The angular velocity ω_1 of the driver pulley is

$$\omega_1 = \frac{2\pi n_1}{60}$$

Similarly, the angular velocity of the driven pulley is,

$$\omega_2 = \frac{2\pi n_2}{60}$$

The velocity of the belt, v, is,

$$v = \omega_1 r_1$$

provided there is no slip between the pulley and the belt.

Considering the driven pulley, the velocity of the belt is,

$$v = \omega_2 r_2$$

Since the same belt runs over both the pulleys, the two values of v shall be equal:

$$\omega_1 r_1 = \omega_2 r_2$$

$$\frac{2\pi n_1 r_1}{60} = \frac{2\pi n_2}{60} \cdot r_2$$

or $n_1 r_1 = n_2 r_2$

or $\dfrac{n_1}{n_2} = \dfrac{r_2}{r_1}$ ...(Eq. 19.1)

That is, the rotational speeds of the pulleys are inversely proportional to their radii.

Note: The same result could have been obtained by considering the length of the belt that runs over both the pulleys in one minute and equating,

$$(2\pi r_1) \cdot n_1 = (2\pi r_2) \cdot n_2$$

When single belt or rope is adequate for the purpose, and only two shafts are connected it is called "simple" belt drive. However, when a large power is to be transmitted, single belt will not be sufficient; so a number of belts or ropes in parallel may be used.

Sometimes, more than two shafts may be involved, requiring the use of more pulleys and correspondingly more belts. This is called a "compound belt drive". The velocity ratio has to be obtained in stages till the velocities of the first and last shaft are related.

19.4 TYPES OF BELT DRIVES

The belt over two pulleys can be laid in two different arrangements:

(*i*) Open-belt drive and (*ii*) Cross-belt drive.

These are shown in Fig. 19.3.

In the first, both pulleys rotate in the same direction, and in the second, they rotate in opposite directions.

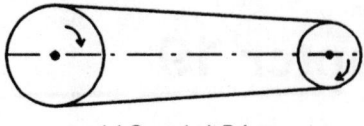

(*a*) Open-belt Drive

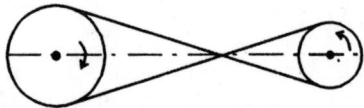

(*b*) Cross-belt Drive

Fig. 19.3: Types of Belt Drives

Length of the belt required may be determined in each case from the geometry of the figure.

19.4.1 Open-Belt Drive

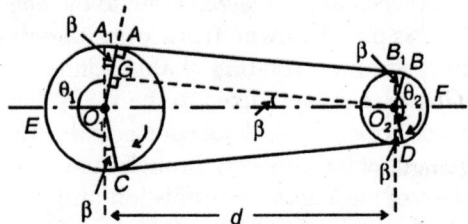

Fig. 19.4: Length of Belt for an Open-Belt Drive

With reference to Fig. 19.4, only a part of the belt will be in contact with each pulley. The angle subtended at the centre of the pulley by the portion of the belt in contact with it is called "the angle of contact" or "the angle of lap".

Let r_1 and r_2 be the radii of the pulleys ($r_1 > r_2$, say); d be the distance between the centres of pulleys; and, θ_1 and θ_2 be the angles of contact. Imagine a parallel to BA through O_2 to meet O_1A in G, and let $G\hat{O}_2 O_1$ be a small angle β. (A, C and B, D are the points of contact of the belt with the two pulleys, and so $O_1\hat{A}B = O_2\hat{B}A = 90°$).

O_1A_1 and O_2B_2 are perpendicular to O_1O_2.

Since $G\hat{O}_2 O_1 = \beta$, $A_1\hat{O}_1 A = B_1\hat{O}_2 B = \beta$

Also $AB = GO_2$, by the construction.

Length of the belt, L, is given by

$$L = 2(\text{Arc}EA + \text{Arc}BF + AB)$$
$$= 2(\text{Arc}EA + \text{Arc}BF + GO_2)$$

Arc $EA = \left(\dfrac{\pi}{2} + \beta\right) r_1$

Arc $BF = \left(\dfrac{\pi}{2} - \beta\right) r_2$

From $\Delta\, O_1\, G\, O_2$,

$$GO_2 = \sqrt{d^2 - (r_1 - r_2)^2}$$

$$= d\sqrt{1 - \left(\dfrac{r_1 - r_2}{d}\right)^2}$$

$$= d\left[1 - \left(\dfrac{r_1 - r_2}{d}\right)^2\right]^{1/2}$$

$$= d\left[1 - \dfrac{1}{2}\left(\dfrac{r_1 - r_2}{d}\right)^2\right], \text{ using binomial theorem}$$

or $GO_2 = \left[d - \left(\dfrac{(r_1 - r_2)^2}{2d}\right)\right]$

Length of the belt,

$$L = 2\left[\left(\dfrac{\pi}{2} + \beta\right) r_1 + \left(\dfrac{\pi}{2} - \beta\right) r_2 + \left\{d - \dfrac{(r_1 - r_2)^2}{2d}\right\}\right]$$

$$= 2\left[\pi(r_1 + r_2) + 2\beta(r_1 - r_2) + 2d - \dfrac{(r_1 - r_2)^2}{d}\right]$$

From $\Delta\, O_1 G O_2$, $\quad \sin\beta = \dfrac{(r_1 - r_2)}{d}$

Since β is very small in practice, $\sin\beta = \beta$ (rad.)

$\therefore \qquad \beta \approx \left(\dfrac{r_1 - r_2}{d}\right)$

Now $L = \left[\pi(r_1 + r_2) + 2d + \dfrac{(r_1 - r_2)^2}{d}\right]$ ...(Eq. 19.2)

Angle of lap for the larger pulley
$$\theta_1 = (\pi + 2\beta) \qquad\qquad \text{...(Eq. 19.3)}$$
Angle of lap for the smaller pulley
$$\theta_2 = (\pi - 2\beta) \qquad\qquad \text{...(Eq.19.4)}$$

Obviously $\theta_1 > \theta_2$, or the angle of lap for the bigger pulley is greater than that for the smaller one.

19.4.2 Cross-Belt Drive

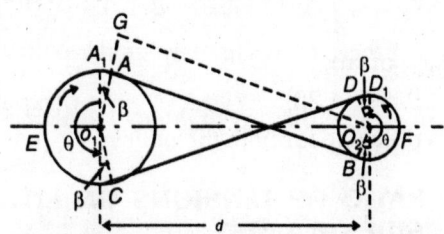

Fig. 19.5: Length of Belt for a Cross-Belt Drive

In this, the two pulleys rotate in opposite directions; however, both the pulleys have the same angle of lap. Usng the same notation of Fig. 19.4,

Length of the belt, L, is,

$$L = 2(\text{Arc } EA + \text{Arc} FD + AB)$$

$O_2 G$ is drawn parallel to AB to meet OA produced in G.

$AB = GO_2$

From $\Delta O_1 G O_2$

$$GO_2 = \sqrt{d^2 - (r_1 + r_2)^2} = d\left[1 - \left(\dfrac{r_1 + r_2}{d}\right)^2\right]^{1/2}$$

$$= d\left[1 - \dfrac{1}{2}\left(\dfrac{r_1 + r_2}{d}\right)^2\right], \text{ using Binomial theorem.}$$

or $\quad AB = GO_2 = \left[d - \left(\dfrac{(r_1 + r_2)^2}{2d}\right)\right]$

$\therefore \quad L = 2\left[r_1\left(\dfrac{\pi}{2} + \beta\right) + r_2\left(\dfrac{\pi}{2} + \beta\right) + \left\{d - \dfrac{(r_1 + r_2)^2}{2d}\right\}\right]$

$$= \left[\pi(r_1 + r_2) + 2\beta(r_1 + r_2) + 2d - \dfrac{(r_1 + r_2)^2}{d}\right]$$

But $\sin\beta = \left(\dfrac{r_1 + r_2}{d}\right)$.

For small values of β,

$$\sin\beta \approx (\text{rad.}) = \left(\dfrac{r_1 + r_2}{d}\right)$$

Substituting for β, we obtain,

$$L = \left[\pi(r_1 + r_2) + 2\left(\dfrac{r_1 + r_2}{d}\right)(r_1 + r_2) + 2d - \dfrac{(r_1 + r_2)^2}{d}\right]$$

or $$L = \left[\pi(r_1 + r_2) + 2d + \frac{(r_1 + r_2)^2}{d}\right] \quad \text{...(Eq. 19.5)}$$

Angle of lap,
$$\theta = (\pi + 2\beta) \qquad \text{...(Eq. 19.6)}$$

It is the same for both the pulleys.

19.5 RATIO OF TENSIONS ON EITHER SIDE OF A BELT

Power tranmission by a belt is possible only if friction is present between it and the surface of the pulley with which it is in contact. Otherwise, the driver pulley rotates, but the belt remains stationary; consequently the driven pulley also does not rotate, the purpose of power transmission not being served.

Friction can be developed by the belt wrapping the pulley tightly, with same initial tension in it with a tendency for the pulleys to be pulled towards each other. (This initial tension required will be analysed in a later section).

As the driver pulley starts rotating reaching its designed speed, the initial tension increases on the other side; the former is called the 'tight' side and the latter 'slack' side.

19.5.1 Ratio of Tension for a Flat Belt

The ratio of the tensions will be dependent upon the angle of contact or lap, as shown in Fig. 19.6.

Referring to Fig. 19.6,

Let T_1 be the tension in the belt on the tight side (on the side of the belt leaving the pulley);

T_2 be the tension in the belt on the slack side (on the side of the belt approaching the pulley);

and θ the angle of contact or lap

Let us consider the equilibrium of an element ab of the belt, located at α from OB and with a central angle $d\alpha$ (Fig. 19.6 (b)).

Let the tension at a be T and that at b be $T + dT$, when the belt is just about to slip; also let the normal reaction be dN and the friction force be dF.

$$dF = \mu \cdot dN,$$

μ being the coefficient of friction between the pulley and the belt.

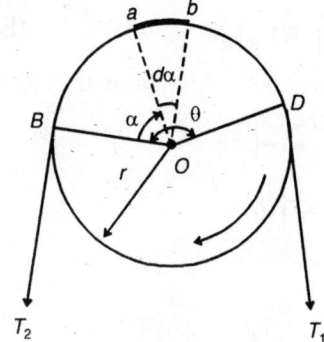

(a) Belt Passing over a Pulley

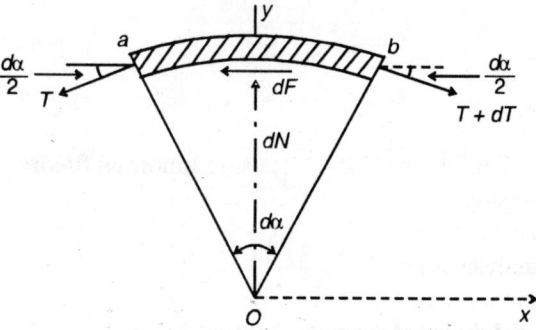

(b) Equilibrium of Element ab

Fig. 19.6: Ratio of Tensions on either Side of a Belt

Applying the static equilibrium condition:

$$\Sigma H = 0 : T\cos\frac{d\alpha}{2} - (T + dT)\cos\frac{d\alpha}{2} + dF = 0$$
$$\text{...(Eq. 19.7)}$$

$$\Sigma V = 0 : dN - T \cdot \sin\frac{d\alpha}{2} - (T + dT)\sin\frac{d\alpha}{2} = 0$$
$$\text{...(Eq. 19.8)}$$

Since $d\alpha$ is very small, we can use,

$$\cos\frac{d\alpha}{2} \approx 1 \text{ and } \sin\frac{d\alpha}{2} \approx \frac{d\alpha}{2}(\text{rad.})$$

Also using $dF = \mu \cdot dN$,
$$T + \mu \cdot dN - (T + dT) = 0$$

or $\quad dT - \mu dN = 0 \qquad \text{...(Eq. 19.9)}$

And $\quad dN - T\dfrac{d\alpha}{2} - (T + dT)\dfrac{d\alpha}{2} = 0$

or $\quad dN - T \cdot d\alpha = 0 \qquad \text{...(Eq. 19.10)}$

Eliminating dN from Eqs. 19.9 and 19.10,
$$dT = \mu \cdot T \cdot d\alpha$$

or $\quad \dfrac{dT}{T} = \mu \cdot d\alpha$ $\qquad$ ...(Eq. 19.11)

Integrating both sides of equation 19.11 within the appropriate limits, we obtain,

$$\int_{T_2}^{T_1} \frac{dT}{T} = \int_0^\theta \mu \cdot d\alpha$$

$$\left[\text{Log } T \right]_{T_2}^{T_1} = \mu \cdot \theta$$

$$\log_e \frac{T_1}{T_2} = \mu\theta$$

or $\quad \dfrac{T_1}{T_2} = e^{\mu\theta}$ $\qquad$ ...(Eq. 19.12)

This is applicable in the case of a flat belt passing over a pulley, a band brake, or a rope *wound round a cylindrical surface*. However, the analysis is slightly different in the case of V-belt, and a rope going over the rim of a pulley appropriately grooved to accommodate the V-belt or the rope. This is given in the following subsection.

19.5.2 Ratio of Tensions for a V-belt and Rope in a Groove

Fig. 19.7 shows a V-belt section [(a)], a rope section [(b)], and the free body of an element of the belt [(c)].

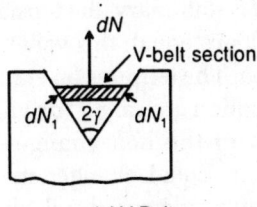

(a) V-Belt

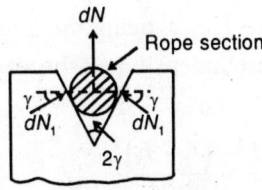

(b) Rope in a Groove

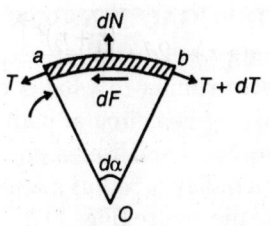

(c) Free Body of Element *ab* of the Belt

Fig. 19.7: Ratio of Tensions for a V-belt and a Rope in a Groove

Let 2γ be the angle of the groove.

The other notation is the same as that for a flat belt. The normal reaction and friction forces will be as shown in (a) and (b) for a V-belt and a rope in a groove, respectively. dN is the resultant of the normal reactions dN_1, on either side.

$$dN = 2 \cdot dN_1 \cdot \sin\gamma$$

or $\quad dN_1 = \dfrac{dN}{2\sin\gamma}$

Friction force,

$$dF = 2\mu \cdot dN_1 = \frac{\mu \cdot dN}{\sin\gamma} = \left(\mu \cdot \text{cosec}\,\gamma\right)dN$$

Equilibrium conditions yield the same equations 19.7 and 19.8. But, for dF, we have to substitute $(\mu \cdot \text{cosec}\,\gamma) \cdot dN$, and the rest of the analysis is applicable.

Thus the following result will be obtained in this case:

$$\frac{T_1}{T_2} = e^{\left(\mu\,\text{cosec}\,\gamma\right)\theta}$$ $\qquad$ ...(Eq. 19.13)

[γ is the half–angle of the groove which accommodates the V-belt or the rope].

19.5.3 Advantages of Rope Drive

Compared to a flat-belt drive, the following are the advantages of a rope drive and a V-belt drive:

(1) Friction is more than that for flat-belt drive.
(2) Particularly suitable for greater distance between the pulleys (and shafts).
(3) More power can be transmitted than flat-belt drive.
    (This will become obvious at the end of Section 19.9)

19.6 CENTRIFUGAL TENSION

A belt running over a pulley experiences a centrifugal force similar to what is experienced by a body moving in a circular path.

Let us consider a small element ab of a belt running over a pulley of radius r, and subtending an angle $d\alpha$ at the centre (Fig. 19.8).

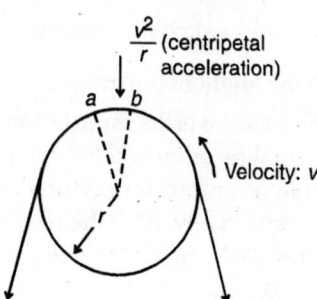

(a) Belt Running with no Load

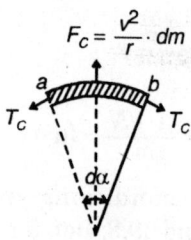

(b) Equilibrium of an Element of Belt

Fig. 19.8: Centrifugal Tension in Belts

Let there be no load on the pulley; hence it takes no power, resulting in equal tension on both sides of the belt. The small element ab of the belt is in equilibrium under the inertial force $\left(\dfrac{v^2}{r}\right) \cdot dm$, $\left(\dfrac{v^2}{r}\right)$ being the normal component of the acceleration (centripetal value), directed towards the centre, dm being the mass of the element, and the tensions T_C at the ends in the tangential directions. The inertial force will be directed opposite to this, i.e., away from the centre, as shown in Fig. 19.8 (b). This force is called the "centrifugal force", and the tension T_C is called the "centrifugal tension". (Weight of the element is considered negligible).

Centrifugal force, $F_C = \dfrac{v^2}{r} \cdot dm$

$dm = \dfrac{w}{g}\left(r d\alpha\right)$, w being the weight per unit length of the belt.

$\therefore$ Centrifugal force, $F_C = \dfrac{wv^2}{g} \cdot d\alpha$

Resolving in the horizontal and vertical directions,

$$\Sigma H = 0 : T_C \cos\frac{d\alpha}{2} - T_C \cos\frac{d\alpha}{2} = 0$$

This does not help

$$\Sigma V = 0 : F_C - 2T_C \cdot \sin\frac{d\alpha}{2} = 0$$

or $\quad \dfrac{wv^2}{g} \cdot d\alpha - 2T_C \sin\dfrac{d\alpha}{2} = 0$

Since $d\alpha$ is small, $\sin\dfrac{d\alpha}{2} \approx \dfrac{d\alpha}{2}$ (rad.)

$\therefore \qquad T_C = \dfrac{w}{g} \cdot v^2 \qquad\qquad$ ...(Eq. 19.14)

Note:

(1) This does not depend on the radius of the pulley, but only on the weight per unit length of the belt and the speed. So T_C is the same for a flat-belt, V-belt, and a rope.

(2) This is additional tension and so gets added to the tensions T_1 and T_2 on either side when the belt is in motion.

19.7 INITIAL TENSION IN THE BELT

When the belt is stationary, the tensions on either side of the belt are equal; this value is called the "initial" tension. The tensions on either side when the belt is in motion are called "driving tensions".

The length of the belt changes due to the applied tension. Equal changes in tension will occur on the tight side and the slack side.

The increase in tension on the tight side

$\qquad = (T_1 - T_i)$, T_i being the initial tension.

The decrease in tension on the slack side,

$\qquad = (T_i - T_2)$

$$\frac{(T_1 - T_i)L}{AE} = \frac{(T_i - T_2)L}{AE}$$

(Since net change in length is zero).

or $\quad (T_1 - T_i) = (T_i - T_2)$

or $\quad T_i = \left(\dfrac{T_1 + T_2}{2}\right)$ $\qquad$ ...(Eq. 19.15)

Thus, the initial tension is merely the average of the tensions on the tight side and the slack side.

19.8 MAXIMUM TENSION IN THE BELT

The maximum tension in the belt is that on the tight side, which is supposed to include the centrifugal tension also.

Thus $T_{max} = T_1 + T_C$

This is supposed to be limited by the maximum permissible tension in the belt material, based on the tensile strength and the cross-section of the belt.

This is to be designed accordingly.

19.9 POWER TRANSMITTED BY A BELT

The primary purpose of a belt drive is to transmit power from one shaft to another.

An expression for power can be derived in terms of the driving tensions and the velocity of the belt; further, the conditions for maximum power to be transmitted by a given system may also be established. These two aspects are covered in the following sub-sections.

19.9.1 Expression for Power Transmitted

Let the pulley I be driving pulley and II be the driven pulley of radii r_1 and r_2, respectively, running with angular velocities ω_1 and ω_2 (Fig. 19.9).

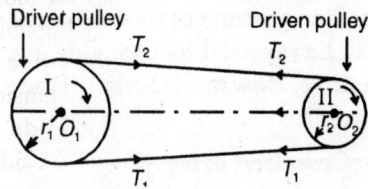

Driver pulley $\qquad$ Driven pulley

Fig. 19.9: Power Transmitted by a Belt

Work done per second $= (T_1 - T_2)v$; here v is the velocity of the belt.

Note: The centrifugal tension, T_C, does not affect the power transmitted as it equally affects T_1 and T_2.

Power transmitted, $P = (T_1 - T_2)v$ ...(Eq. 19.16)

19.9.2 Condition for Maximum Power to be Transmitted

The following are the limitations:

(i) The magnitude of tight-side tension, T_1, is limited by the strength or rather the safe permissible value of tension in the material of the belt.

(ii) The centrifugal tension, T_C, is equal on both sides, so that this part of the tension T_1 does not contribute to power transmission.

(iii) The centrifugal tension, T_C, increases proportionate to the square of the velocity v.

(iv) The tensions are further controlled by friction as the belt must not slip over the pulley.

With these limitations, the power transmitted for a given system has a maximum value at a chosen speed; this will be derived now:

Power transmitted by the belt, P, is given by,

$$P = (T_1 - T_2)v$$

Let $\dfrac{T_1}{T_2} = k$

Then $= T_1\left(1 - \dfrac{1}{k}\right) \cdot v$

Maximum permissible tension in the belt,

$$T_{max} = T_1 + T_C$$

where $T_C = \dfrac{w}{g} \cdot v^2$ $\qquad$ ...(Eq. 19.14)

(centrifugal tension)

$$T_1 = T_{max} - T_C$$

Hence $P = \left(T_{max} - T_C\right)\left(1 - \dfrac{1}{k}\right) \cdot v$

or $\quad P = \left(T_{max} - \dfrac{wv^2}{g}\right)\left(1 - \dfrac{1}{k}\right) \cdot v$

For P to be maximum,

$\dfrac{dP}{dv}$ should be equal to zero.

(v is the only variable in the system)

$$\frac{dP}{dv} = \left(1 - \frac{1}{k}\right)\left(T_{max} - \frac{3wv^2}{g}\right) = 0$$

or $T_{max} = \frac{3wv^2}{g} = 3T_C$

or $T_C = \frac{1}{3}T_{max}$ ...(Eq. 19.17)

Hence, the power transmitted is maximum when the centrifugal tension is one-third the maximum permissible tension in the belt.

Further, $T_{max} = T_1 + T_C$

or $T_1 = T_{max} - T_C = T_{max} - \frac{1}{3}T_{max} = \frac{2}{3}T_{max}$

Maximum power that can be transmitted, P_{max}, is given by

$$P_{max} = T_1\left(1 - \frac{1}{k}\right) \cdot v$$

or $P_{max} = \frac{2}{3}T_{max}\left(1 - \frac{1}{k}\right) \cdot v$ ...(Eq. 19.18)

T_{max} depends upon the tensile strength or the maximum permissible tensile stress, τ_{max} for the meterial of the belt.

If the cross-sectional area of the belt is a,

$T_{max} = \tau_{max} \cdot a$ ...(Eq. 19.19)

ILLUSTRATIVE PROBLEMS

PROBLEM 19.1: Find the speed of a shaft which is driven with the help of a belt by an engine running at 360 revolutions per minute. The diameter of the engine pulley is 600 mm and that of the shaft is 400 mm. The thickness of the belt is 10 mm. The total slip is 2%.

Solution:

$$\frac{n_2}{n_1} = \frac{(d_1 + t)}{(d_2 + t)}\left(1 - \frac{S}{100}\right)$$

$n_1 = 360$ r.p.m.

$d_1 = 600$ mm

$d_2 = 400$ mm

$t = 10$ mm

$S = 2\%$

$\therefore \quad n_2 = n_1 \times \frac{(d_1 + t)}{(d_2 + t)}\left(1 - \frac{S}{100}\right)$

$= \frac{360 \times (600 + 10)}{(400 + 10)}\left(1 - \frac{2}{100}\right)$

$= 524.90$ r.p.m.

PROBLEM 19.2: In a work-shop, an engine drives a shaft by a belt. The diameter of the engine pulley and the shaft pulley are 550 mm and 275 mm respectively. Another pulley of 750 mm diameter on the same shaft drives a pulley 300 mm in diameter on a motor shaft. If the engine runs at 210 r.p.m, find the speed of the motor shaft.

Solution:

$$n_4 = n_1 \times \frac{d_1 \times d_3}{d_2 \times d_4}$$

$d_1 = 550$ mm

$d_2 = 275$ mm

$d_3 = 750$ mm

$d_4 = 300$ mm

$n_1 = 210$ r.p.m.

$\therefore \quad n_4 = 210 \times \frac{550 \times 750}{275 \times 300}$

$= 1050$ r.p.m.

PROBLEM 19.3: Two parallel shafts 9 m apart are provided with 1000 mm and 500 mm diameter pulleys and are connected by means of a cross belt. The direction of rotation of the driven pulley is to be reversed by changing over to an open belt drive. How much length of the belt has to be reduced?

Solution: Cross belt drive

$$L = \pi(r_1 + r_2) + 2d + \frac{(r_1 + r_2)^2}{d}$$

$r_1 = \frac{1000}{2}$ mm $= 500$ mm $= 0.50$ m

$r_2 = \frac{500}{2}$ mm $= 250$ mm $= 0.25$ m

$$d = 9 \text{ m}$$

$$\therefore \quad L = \pi(0.50 + 0.25) + 2 \times 9 + \frac{(0.50 + 0.25)^2}{9}$$

$$= 2.3562 + 18 + 0.0625$$

$$= 20.4187 \text{ m}$$

Open Belt Drive

$$L = \pi(r_1 + r_2) + 2d + \frac{(r_1 - r_2)^2}{d}$$

$$= \pi(0.50 + 0.25) + 2 \times 9 + \frac{(0.50 - 0.25)^2}{9}$$

$$= 2.3562 + 18 + 0.0069$$

$$= 20.3631 \text{ m}$$

$\therefore$ The length of the belt to be reduced

$$= 20.4187 - 20.3631$$

$$= 0.0556 \text{ m}$$

$$= 55.60 \text{ mm}$$

PROBLEM 19.4: A pulley 1.2 m in diameter is mounted on the shaft. It is used to transmit the power. Determine what maximum torque can be transmitted if the coefficient of friction between the belt and pulley is 0.25 and the angle of lap is 160°. The maximum tension permitted in the belt is 2.5 kN.

Solution:

$$\frac{T_1}{T_2} = e^{\mu\theta}$$

$$\theta = 160° = \frac{160 \times \pi}{180} = 2.793 \text{ radians}$$

$$\frac{2.5}{T_2} = e^{0.25 \times 2.793} = 2.01$$

$$\therefore \quad T_2 = \frac{2.5}{2.01} = 1.244 \text{ kN}$$

Maximum torque,

$$(T_1 - T_2)r = (2.5 - 1.244) \times \frac{1.2}{2}$$

$$= (2.5 - 1.244) \times 0.6$$

$$= 1.256 \times 0.6$$

$$= 0.7536 \text{ kN.m}$$

or $$= 753.6 \text{ N.m}$$

PROBLEM 19.5: Find the power transmitted by a belt running over a pulley of 800 mm diameter at 300 r.p.m. The coefficient of friction between the pulley and belt is 0.20, angle of lap 180° and maximum tension in the belt is 2 kN.

Solution:

$$v = \frac{\pi d n}{60}$$

$$d = 800 \text{ mm} = 0.8 \text{ m}$$

$$n = 300 \text{ r.p.m.}$$

$$\therefore \quad v = \frac{\pi \times 0.8 \times 300}{60}$$

$$= 4\pi \text{ m/sec}$$

$$\frac{T_1}{T_2} = e^{\mu\theta}$$

$$T_1 = 2 \text{ kN}$$

$$\theta = 180° = \frac{180 \times \pi}{180} = \pi \text{ radians}$$

$$\mu = 0.20$$

$$\therefore \quad \frac{2}{T_2} = e^{0.20 \times \pi}$$

$$\frac{2}{T_2} = 1.8745$$

or $$T_2 = 1.067 \text{ kN}$$

$$P = (T_1 - T_2)v$$

$$= (2 - 1.067) \times 4\pi \text{ kN.m/sec}$$

$$= 11.72 \text{ kW}$$

PROBLEM 19.6: Find the power transmitted by a rope drive, from the following data:

Angle of contact = 160°

Angle of the groove = 45°

Coefficient of friction = 0.25

Maximum tension in the belt = 2.5 kN

Mass of rope = 0.5 kg/metre length

Velocity of rope = 10 m/sec

Solution: When the centrifugal tension is neglected:

$$T = T_1 = 2.5 \text{ kN}$$

$$\frac{T_1}{T_2} = e^{(\mu\,\mathrm{cosec}\,\gamma)\theta}$$

$$\mu = 0.25$$

$$\theta = 160° = \frac{160 \times \pi}{180} = 2.793 \text{ radians}$$

$$2\gamma = 45°$$

or $\gamma = 22.5°$

$$\therefore \quad \frac{2.5}{T_2} = e^{(0.25 \times \mathrm{cosec}\,22.5°)2.793}$$

$$\frac{2.5}{T_2} = e^{1.8246} = 6.2$$

$$T_2 = \frac{2.5}{6.2} = 0.4032 \text{ kN}$$

$$P = (T_1 - T_2)v$$

$$T_1 = 2.5 \text{ kN}; \ T_2 = 0.4032 \text{ kN}; \ v = 10 \text{ m/sec}$$

$$P = (2.5 - 0.4032) \times 10 \text{ kN.m/sec}$$

$$= 20.968 \text{ kW}$$

When the centrifugal tension is considered:

$$T_C = m \cdot v^2$$

$$m = 0.5 \text{ kg/m}$$

$$v = 10 \text{ m/sec}$$

$$\therefore \quad T_C = 0.5 \times (10)^2 = 50 \text{ N} = 0.05 \text{ kN}$$

$$T_1 = T - T_C = 2.50 - 0.05$$

$$= 2.45 \text{ kN}$$

$$\frac{T_1}{T_2} = e^{(\mu\,\mathrm{cosec}\,\gamma)\theta}$$

$$\frac{2.45}{T_2} = e^{(0.25 \times \mathrm{cosec}\,22.5°)2.793}$$

$$\frac{2.45}{T_2} = e^{1.8246} = 6.2$$

$$T_2 = \frac{2.45}{6.2} = 0.3952 \text{ kN}$$

$$P = (T_1 - T_2)v$$

$$= (2.45 - 0.3952) \times 10 \text{ kN.m/sec}$$

$$= 20.548 \text{ kW}$$

PROBLEM 19.7: A belt 8 mm thick and 120 mm wide drives a pulley of 1.0 m diameter at 210 r.p.m. The angle of lap is 180° and mass of the belt material is 1000 kg/m³. If the stress in the belt is not to exceed 1.5 N/mm² and the co-efficient of friction between the belt and the pulley is 0.25, determine the power transmitted when the centrifugal tension is (*i*) considered and (*ii*) neglected.

Solution:

(*i*) Power transmitted when the centrifugal tension is considered:

$$v = \frac{\pi d n}{60} = \frac{\pi \times 1.0 \times 210}{60} = 11 \text{ m/sec}$$

$$T = f \times b \times t = 1.5 \times 120 \times 8 = 1440 \text{ N}$$

$$m = \text{Area} \times \text{length} \times \text{density}$$

$$= (0.008 \times 0.120) \times 1 \times 1000$$

$$= 0.96 \text{ kg/m}$$

$$T_C = m \cdot v^2 = 0.96 \times 11^2 = 116.16 \text{ N}$$

$$T_1 = T - T_C = 1440 - 116.16$$

$$= 1323.84 \text{ N}$$

$$\frac{T_1}{T_2} = e^{\mu\theta}; \ \ \theta = \frac{180 \times \pi}{180} = \pi \text{ radians}$$

$$\frac{1323.84}{T_2} = e^{0.25 \times \pi} = 2.1933$$

$$\therefore \ T_2 = \frac{1323.84}{2.1933} = 603.58 \text{ N}$$

$$P = (T_1 - T_2)v$$

$$= (1323.84 - 603.58) \times 11 \text{ N.m/sec}$$

$$= 7923 \text{ W}$$

or 7.923 kW

(*ii*) Power transmitted when the centrifugal tension is neglected:

$$T_1 = T = 1440 \text{ N}$$

$$\frac{T_1}{T_2} = e^{\mu\theta}; \ \theta = 180° = \pi \text{ radians}$$

$$\frac{1400}{T_2} = e^{0.25 \times \pi} = 2.1933$$

$$T_2 = \frac{1440}{2.1933} = 656.55 \text{ N}$$

$$= 8618 \text{ W}$$

$$= 8.618 \text{ kW}$$

PROBLEM 19.8: Two parallel shafts whose centre lines are 6 m apart are connected by an open belt drive. The diameter of the larger pulley is 1.8 m and that of the smaller pulley is 1.2 m. The initial tension in the belt, when stationary, is 2.5 kN. The mass of the material is 1.5 kg/m length and the coefficient of friction between the belt and the pulley is 0.3. Find the power transmitted, when the smaller pulley rotates at 420 r.p.m.

Solution:

$$v = \frac{\pi d_2 n_2}{60}$$

$$d_2 = 1.2 \text{ m}; \ n_2 = 420 \text{ r.p.m.}$$

$$\therefore \quad v = \frac{\pi \times 1.2 \times 420}{60} = 26.39 \text{ m/sec}$$

$$T_i = \frac{T_1 + T_2}{2}$$

$$T_i = 2.5 \text{ kN}$$

$$\therefore \quad 2.5 = \frac{T_1 + T_2}{2}$$

or $\quad T_1 + T_2 = 2.5 \times 2 = 5 \text{ kN}$ ...(i)

For open belt drive:

$$\sin\beta = \frac{r_1 - r_2}{d}$$

$$d_1 = 1.8 \text{ m}$$

$$\therefore \quad r_1 = \frac{1.8}{2} = 0.9 \text{ m}$$

$$d_2 = 1.2 \text{ m}$$

$$\therefore \quad r_2 = \frac{1.2}{2} = 0.6 \text{ m}$$

$$d = 6 \text{ m}$$

$$\therefore \quad \sin\beta = \frac{0.9 - 0.6}{6} = 0.05$$

$$\beta = \text{Sin}^{-1} (0.05)$$

$$\beta = 2°52'$$

Angle of lap for the smaller pulley,

$$\theta = 180° - 2\beta$$

$$\theta = 180 - 2 \times (2°52')$$

$$= 174.27° \text{ or } 3.0415 \text{ radians}$$

$$\frac{T_1}{T_2} = e^{\mu\theta}; \ \mu = 0.3$$

$$\frac{T_1}{T_2} = e^{0.3 \times 3.0415} = 2.49$$

or $\quad T_1 = 2.49 T_1$ ...(ii)

Substituting (ii) in (i)

$$2.49 T_2 + T_2 = 5$$

$$3.49 T_2 = 5$$

or $\quad T_2 = \frac{5}{3.49} = 1.433 \text{ kN}$

$\therefore \quad T_1 = 2.49 \times 1.433$

$$= 3.568 \text{ kN}$$

$$P = (T_1 - T_2) v$$

$$P = (3.568 - 1.433) \times 26.39 \text{ kN.m/sec}$$

$$P = 56.34 \text{ kW}$$

Power transmitted = 56.34 kW

PRACTICE PROBLEMS

19.1 Find the speed of a shaft which is driven with the help of a belt by an engine running at 360 revolutions per minute. The diameter of the engine pulley is 600 mm and that of the shaft is 400 mm. Neglect the thickness of the belt.

19.2 Find the speed of a shaft which is driven with the help of a belt by an engine running at 360 revolutions per minute. The diameter of the engine pulley is 600 mm and that of the shaft is 400 mm. The thickness of the belt is 10 mm. Assume that there is a firm grip between the belts and the shafts.

19.3 An engine drives an electric motor across a main and a secondary shaft. The diameter of driving pulleys are 800 mm, 600 mm and 400 mm respectively and those of the driven pulleys are 600 mm, 400 mm and 200 mm. If the 800 mm driver on the engine shaft is rotating at 360 r.p.m, find the speed of the motor shaft to which the 200 mm diameter driven pulley is keyed (*i*) if there is no slip (*ii*) if there is a slip of 3% at each drive.

19.4   Find the length of the belt necessary to drive a pulley of 600 mm diameter running parallel at a distance of 12 metres from the driving pulley of diameter 1600 mm, if it is (*i*) open, and (*ii*) crossed.

19.5   The tensions in the two sides of the belt are 900 and 600 newtons respectively. If the speed of the belt is 60 m/sec, calculate the power transmitted by the belt.

19.6   Find the power transmitted by a rope drive from the following data:

Angle of contact = 160°

Angle of the groove = 60°

Coefficient of friction = 0.30

Maximum tension in the belt = 2 kN

Mass of rope = 1 kg/metre length

Velocity of rope = 20 m/sec

19.7   A flat belt is required to transmit 12 kW power from a pulley of 1.2 m diameter running at 270 r.p.m. The angle of contact is 160° and mass of the belt material is 1000 kg/m^3. If the stress in the belt is not to exceed 1.5 N/mm^2 and the coefficient of friction between the belt and pulley is 0.25, determine the width of the belt considering centrifugal tension into account. Take the thickness of belt as 8 mm.

19.8   A belt 120 mm wide and 10 mm thick is transmitting power at a speed of 20 m/sec. The net driving tension is 1.6 times the tension on the slack side. Calculate the power transmitted at this speed, if the permissible stress in the belt is 2.0 N/mm^2. Mass of the belt material is 1000 kg/m^3.

Also calculate the absolute maximum power that can be transmitted by this belt and the speed at which this can be transmitted.

Chapter 20

Mechanical Vibrations

20.1 INTRODUCTION

The phenomenon of vibrations of bodies is of importance in civil engineering and mechanical engineering. Vibrations may be induced in structures and machines, and may cause damage to them unless preventive measures are undertaken. A monumental example is the collapse of the Tachoma suspension bridge in the USA due to torsional oscillations induced by wind in a moderate gale.

A study of the salient features of different kinds of vibrations will be made in the following sections.

20.2 SIMPLE HARMONIC MOTION

The motion of a particle such that its acceleration is proportional to its displacement from an initial equilibrium position and is directed towards this position is called "Simple hormonic motion".

Let us consider a particle P moving round the circumference of a circle of radius r with a constant angular velocity ω radians per second; let P start from a point A and move in the counterclockwise direction (Fig. 20.1).

Let OP be the position of the radius vector OA making an angle θ with respect to x-axis. Let P' be the projection of P on the x-axis or BOA.

$$\theta = \omega t$$

The displacement x of the poiont P' (at any time t) from the mean position O on the x-axis is,

$$x = r\cos\theta = r\cos\omega t \qquad \text{...(Eq. 20.1)}$$

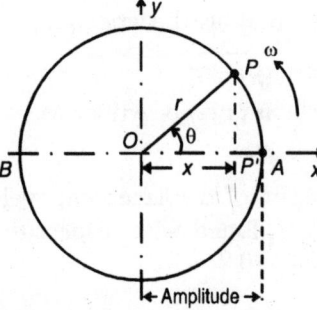

Fig. 20.1: Motion of a Particle Round a Circle with Constant Angular Velocity

This is the displacement vs time relation for the point P'. Differentiating Eq. 20.1 with respect to time,

$$\frac{dx}{dt} = v = \dot{x} = -\omega r\sin\omega t \qquad \text{...(Eq. 20.2)}$$

This is the veloctiy vs time relation. Differentiating again,

$$\frac{d^2x}{dt^2} = a = \ddot{x} = -\omega^2 r\cos\omega t = -\omega^2 x \quad \text{...(Eq. 20.3)}$$

Thus the acceleration is proportional to the displacement and the negative sign indicates that it is always directed towards the mean position.

Obviously the motion of P' is a Simple Harmonic Motion (S.H.M.) as defined already.

This rectilinear motion is 'repetitive' or 'periodic' in nature. In this connection, the following definitions are relevant.

Oscillation

A periodic motion which is repeated in equal intervals of time is called an "oscillation".

Amplitude

The maximum displacement from the mean position is called the "amplitude".

Period

The time required to complete one cycle is called the "period".

$$T = \frac{2\pi}{\omega} \qquad \text{...(Eq. 20.4)}$$

Frequency

The number of oscillations or cycles per second is called the "frequency".

$$f = \frac{1}{T} = \frac{\omega}{2\pi} \qquad \text{...(Eq. 20.5)}$$

Thus $\omega = 2\pi f$ $\qquad$ ...(Eq. 20.6)

Velocity $\dot{x}$ may also be written as,

$$\dot{x} = -\omega\sqrt{r - x^2} \qquad \text{...(Eq. 20.7)}$$

Acceleration $\ddot{x}$ may be written as,

$$\ddot{x} = -\omega^2 x \qquad \text{...(Eq. 20.3)}$$

The graphs of displacement, velocity and acceleration, plotted with respect to time, are shown in Fig. 20.9

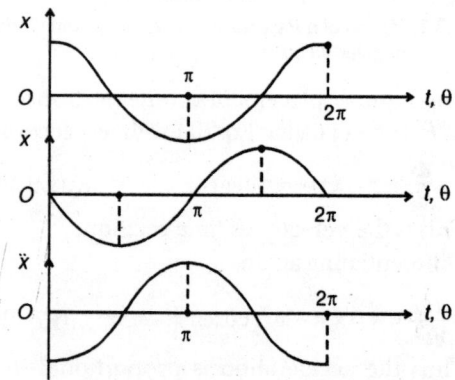

Fig. 20.2: Variation of Displacement, Velocity and Acceleration with Time in S.H.M.

20.3 UNDAMPED FREE VIBRATIONS

A simple example of 'undamped free vibrations' or 'free vibrations without damping' is the oscillation of a weight suspended from a spring and constrained to move in a vertical direction.

The spring gets stretched by the weight of the suspended mass and comes to equilibrium in a certain position, called the equilibrium or mean position. If the body is pulled and then released, it moves up and down about its equilibrium position and tends to execute oscillations indefinitely, if air resistance is negligible. In other words, in the absence of any inhibiting or 'damping force', the oscillations are called un-damped free or natural vibrations.

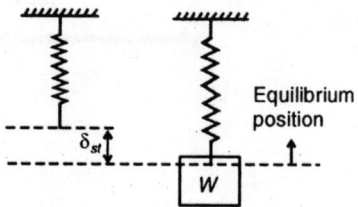

(a) Unstretched $\qquad$ (b) Equilibrium Position with W

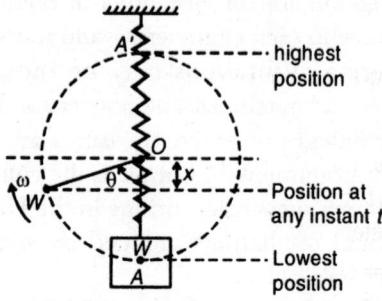

(c) Oscillating Weight about Mean Position

Fig. 20.3: Undamped Free Vibrations Executed by a Spring Supported Weight

With reference to Fig. 20.3:

Let the stiffness of the spring (or spring constant) be k.

$$\delta_{st} = \frac{W}{k} \qquad \text{...(Eq. 20.8)}$$

It can be shown the motion of the body is 'simple harmonic'.

Equation of motion of the body is,

$$\frac{W}{g} \cdot \ddot{x} = W - k(\delta_{st} + x)$$

But $k \cdot \delta_{st} = W$

$$\therefore \quad \frac{W}{g}\ddot{x} = W - W - kx$$

or $\qquad \dfrac{W}{g}\ddot{x} + kx = 0$

$$\ddot{x} + \left(\frac{kg}{W}\right)x = 0$$

Let $\dfrac{kg}{W} = \omega^2$

Then $\ddot{x} + \omega^2 x = 0$

$$\ddot{x} = -\omega^2 x \qquad \text{...(Eq. 20.9)}$$

The acceleration being proportional to the displacement about the mean position and always directed towards it, the motion is obviously simple harmonic.

$$\omega = \sqrt{\frac{kg}{W}}$$

The period $\quad T = \dfrac{2\pi}{\omega} = 2\pi\sqrt{\dfrac{W}{kg}} \qquad \text{...(Eq. 20.10)}$

If the weight W is imagined to be rotating with a constant angular velocity $\omega\left(=\sqrt{\dfrac{kg}{W}}\right)$ in a circle with a radius equal to the amplitude of the vertical vibrations, these oscillations can be linked to the motion of this imaginary rotating weight W on the diameter $A - A'$ as shown in Fig. 20.3 (*c*).

Angle $\theta = \omega t$

Eq. 20.9 gives

$$\frac{d^2 x}{dt^2} + \omega^2 x = 0$$

The general solution of this differential equation is,

$$x = C_1 \cos \omega t + C_2 \sin \omega t \qquad \text{...(Eq. 20.11)}$$

Constants C_1 and C_2 are to be determined from the initial conditions.

Let us say, when $t = 0$, $x = x_0$ and $\dot{x} = 0$.

Solving, we get $C_1 = x_0$ and $C_2 = 0$

The solution is then given by

$$x = x_0 \cos \omega t \qquad \text{...(Eq. 20.12)}$$

20.4 COMPOUND SPRINGS

Two or more springs connected together and made to act together are 'compound springs'.

The arrangement can be either 'parallel' or 'series'. The action of weight which oscillates in each of these arrangements is seen in the following sub-sections.

20.4.1 Parallel Springs

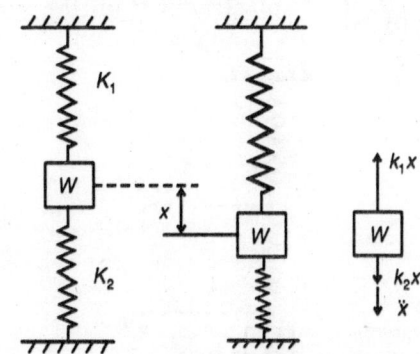

(*a*) Equilibrium Position (*b*) Oscillating Position

Fig. 20.4: Parallel Springs

With reference to Fig. 20.4:

Let the weight W be displaced by x downwards. The top spring extends by x and the bottom one shortens by the same distance.

Tension in the top spring $= k_1 x$

Compresson in the bottom spring $= k_2 x$

Total unbalanced force on W in the direction of motion

$$= -(k_1 + k_2)x$$

The equation of motion is,

$$\frac{W\ddot{x}}{g} = -(k_1 + k_2)x$$

$$\ddot{x} + \frac{(k_1 + k_2)}{W} \cdot g \cdot x = 0$$

or $\quad \ddot{x} + \omega^2 x = 0$

where $\omega = \sqrt{\dfrac{(k_1 + k_2)}{W} \cdot g} \qquad \text{...(Eq. 20.13)}$

Period, $T = \dfrac{2\pi}{\omega} = 2\pi\sqrt{\dfrac{W}{g(k_1 + k_2)}} \qquad \text{...(Eq. 20.14)}$

For an equivalent single spring the spring constant

$$k_{eq} = (k_1 + k_2) \qquad \text{...(Eq. 20.15)}$$

20.4.2 Series Springs

With reference to Fig. 20.5:

Let the two springs be arranged in series. Let the weight W be given a downward displacement *x*. The top spring extends by x_1 and the bottom spring by x_2.

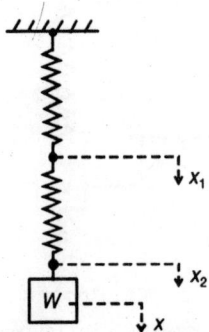

Fig. 20.5: Spring in Series

The tension in both springs is the same, say, F_t

But $x = x_1 + x_2$

or $\qquad x = \dfrac{F_t}{k_1} + \dfrac{F_t}{k_2} = F_t\left(\dfrac{1}{k_1} + \dfrac{1}{k_2}\right)$

$\therefore \qquad F_t = x\left(\dfrac{k_1 k_2}{k_1 + k_2}\right)$

The unbalanced force in the direction of motion is $(-F_t)$.

The equation of motion is,

$$\frac{W}{g}\ddot{x} = -T = -x\left(\frac{k_1 k_2}{k_1 + k_2}\right)$$

$$\frac{W}{g}\ddot{x} + \left(\frac{k_1 k_2}{k_1 + k_2}\right) \cdot x = 0$$

$$\ddot{x} + \frac{g}{W}\left(\frac{k_1 k_2}{k_1 + k_2}\right) \cdot x = 0$$

or $\qquad x + \omega^2 x = 0$

where $\omega^2 = \dfrac{g}{W}\left(\dfrac{k_1 k_2}{k_1 + k_2}\right)$ $\qquad$ ...(Eq. 20.16)

Period, $T = \dfrac{2\pi}{\omega}$

or $\qquad T = 2\pi\sqrt{\dfrac{(k_1 + k_2)W}{k_1 k_2\, g}}$ $\qquad$ ...(Eq. 20.17)

For an equivalent single spring, the spring constant is,

$$k_{eq} = \frac{k_1 k_2}{k_1 + k_2} \qquad \text{...(Eq. 20.18)}$$

20.5 UNDAMPED FORCED VIBRATIONS

In contrast with the free vibrations seen in Section 20.3, vibrations of the spring-mass system, when the suspended weight of Fig. 20.3 is subjected to a periodic external force Q, in the direction of the spring force, through the action of a rotating unbalanced mass attached to the rotor of an electric motor, are called "forced vibrations"; as earlier, if no damping force is present, they are specifically "undamped forced vibrations".

Q is a function of time, $Q = f(t)$, say.

The differential equation of motion is,

$$\frac{W}{g}\ddot{x} = -kx + Q$$

or $\qquad \dfrac{W}{g}\ddot{x} = -kx + f(t)$ $\qquad$ ...(Eq. 20.19)

Q can be expressed as,

$Q = Q_0 \cos\alpha t$ (in the case of the periodic disturbing force, or "exciting force" as it is called) ...(Eq. 20.20)

$\therefore \qquad \ddot{x} = -\dfrac{gk}{W}x + \dfrac{gQ_0}{W}\cos\alpha t$ $\qquad$ ...(Eq. 20.21)

If $\dfrac{Gk}{W} = \omega^2$, where ω is the circular frequency of the natural vibrations, and $\dfrac{gQ_0}{W} = q_0$ (the maximum exciting force per unit mass of the body), we have,

$$\ddot{x} = -\omega^2 x + q_0 \cos\alpha t \qquad \text{...(Eq. 20.22)}$$

The solution of this differential equation may be obtained as the sum of the general solution and the particular integral, obtained without and with the term $q_0\cos\alpha t$, on the right hand side of the Eq. 20.22; this gives an expression for the displacement of the weight from the mean position at any instant.

20.6 DAMPED FREE VIBRATIONS

In nature, all vibrations are damped, or tend to be suppressed by frictional forces acting on the oscillating mass, which oppose the motion. A typical case what is called "viscous damping" in which the frictional force is directly proportional to the velocity of the body on which it acts. An example of this is that of a body moving in a viscous fluid; that is why such a damping is called "viscous damping". The constant of proportionality between the frictional resistance due to viscosity and the velocity of the body moving in the fluid is called "coefficient of viscous damping". (Unit of this coefficient are (N÷m/s) or N.s/m).

Now damped free vibrations can be studied by including a dashpot in the basic spring-mass system (Fig. 20.6).

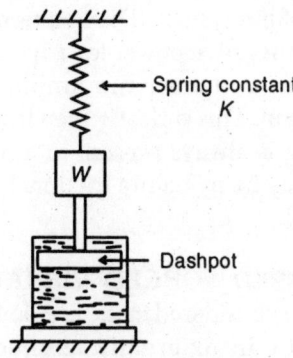

(a) Mass-Spring-Dashpot System

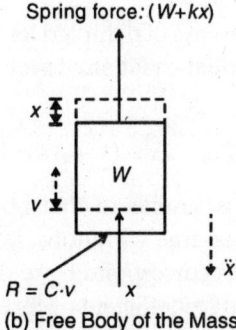

(b) Free Body of the Mass

Fig. 20.6: Mass-Spring-Dashpot System—Damped Free Vibrations

The free body of the mass is shown in (*b*).

The differential equation of motion is obtained as follows:

$$\frac{W}{g}\ddot{x} = -kx + R \qquad \text{...(Eq. 20.23)}$$

But $R = C \cdot v = C \cdot \dfrac{dx}{dt}$ \qquad ...(Eq. 20.24)

$$\therefore \quad \frac{W}{g} \cdot \ddot{x} = -kx + C\frac{dx}{dt} \qquad \text{...(Eq. 20.25)}$$

or $\quad \ddot{x} + \dfrac{gc}{W} \cdot \dot{x} + \dfrac{gk}{W} \cdot x = 0 \qquad$ ...(Eq. 20.26)

The solution of this differential equation is interesting, and may be obtained as follows:

Say $x = e^{\lambda t}$ \qquad ...(Eq. 20.27)

$$\frac{d^2 x}{dt^2} = \lambda^2 \cdot e^{\lambda t} \text{ and } \frac{dx}{dt} = \lambda \cdot e^{\lambda t}$$

Substituting in Eq. 20.26,

$$e^{\lambda t}\left(\lambda^2 + \frac{gC}{W} \cdot \lambda + \frac{gk}{W}\right) = 0$$

or $\quad \lambda^2 + \dfrac{gC}{W} \cdot \lambda + \dfrac{gk}{W} = 0 \qquad$ ...(Eq. 20.28)

($x = e^{\lambda t} = 0$ is the equilibrium position).

From Eq. 20.28,

$$\lambda = -\frac{gC}{2W} \pm \sqrt{\frac{g^2 C^2}{4W^2} - \frac{gk}{W}} \qquad \text{...(Eq. 20.29)}$$

A particular value of C, called the coefficient of critical damping, C_C, is defined as that which makes the redical of Eq. 20.29 zero.

This to say,

$$\frac{g^2 C_C^2}{4W^2} = \frac{gk}{W}$$

or $\quad C_C = 2\sqrt{\dfrac{kW}{g}} \qquad$ ...(Eq. 20.30)

Also $\quad C_C = 2\dfrac{W}{g} \cdot \omega \qquad$ ... (Eq. 20.31)

where ω is the circular frequency of the natural vibrations.

Depending upon the actual value of C, three distinct cases arise as mentioned below:

(*i*) $C > C_C$ (Heavy Damping)

For this case, both roots λ_1 and λ_2 of Eq. 20.28 are real, distinct, and negative. (This is because $\sqrt{\left(\dfrac{gC}{2W}\right)^2 - \dfrac{gk}{W}}$ will be real and less than $\dfrac{gk}{2W}$.

So, with the positive sign of radical also, will be negative.

The general solution of the differential equation (Eq. 20.26) will be,

$$x = A \cdot e^{\lambda_1 t} + B \cdot e^{\lambda_2 t} \qquad \text{...(Eq. 20.32)}$$

$x \rightarrow 0$ as $t \rightarrow \infty$, that is, the system reaches the equilibrium position in infinite time. The system is a periodic, or non-vibrating.

(ii) C = C_C (Critical Damping)
Both roots λ_1 and λ_2 are equal and

$$\lambda_1 = \lambda_2 = \lambda = -\frac{gC_C}{2W} = -\omega$$

The general solution of Eq. 20.26 then is got as,

$$x = (A + Bt)e^{-\omega t} \qquad \text{...(Eq. 20.33)}$$

This motion is also non-vibratory.

(iii) C < C_C (Light Damping)
The quantity under the radical sign in Eq. 20.29 is negative.

Then $\lambda_1 = -\dfrac{gC}{2W} + i\sqrt{\dfrac{gk}{W} - \left(\dfrac{gC}{2W}\right)^2}$

and $\lambda_2 = -\dfrac{gC}{2W} - i\sqrt{\dfrac{gk}{W} - \left(\dfrac{gC}{2W}\right)^2}$ ...(Eq. 20.34)

$$[i = \sqrt{-1}]$$

The roots are complex and conjugate.
The solution then is,

$$x = e^{\left(-\frac{gC}{2W}t\right)}[A\cos qt + B\sin qt] \quad \text{...(Eq. 20.35)}$$

where $q = \sqrt{\dfrac{gk}{W} - \left(\dfrac{gC}{2W}\right)^2}$ ...Eq. 20.36)

But $\dfrac{gk}{W} = \omega^2$ and $C_C = 2m\omega$

$$\therefore q = \sqrt{\omega^2 - \left(\frac{\omega C}{C_C}\right)^2} = \omega\sqrt{1 - \left(\frac{C}{C_C}\right)^2} = \omega\sqrt{1 - d'^2}$$

...(Eq. 20.37)

Here $\left(\dfrac{C}{C_C}\right)$ is called the damping factor 'd'. (Obviously $q < \omega$).
The displacement versus time is depicted in Fig. 20.7.

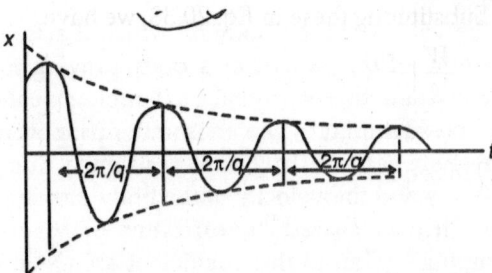

Fig. 20.7: Displacement versus Time for Light Damping

The period of the harmonic motion represented by the term in the square brackets of Eq. 20.35 is greater than that of the undamped natural vibrations of the system. The first term, an exponential with negative power for t, has the effect of continuously reducing the amplitude of the former motion. This is clearly seen from Fig. 20.7.

Damping is always present in any system in nature; it may be by nature frictional, or viscous, or structural.

20.7 DAMPED FORCED VIBRATIONS
If the system considered in the previous section is subjected to an externally applied periodic disturbing force Q defined by

$$Q = Q_0 \sin\alpha t$$

We have the case of damped forced vibrations.

The differential equation of motion is modified as,

$$\frac{W}{g}\ddot{x} + C \cdot \dot{x} + kx = Q_0 \sin\alpha t \quad \text{...(Eq. 20.38)}$$

The solution is similar to that obtained for the case of damped free vibrations; however, all the three variations corresponding to different values of the damping coefficients represent transient motion which is damped out very fast. Only the case of steady-state vibrations represented by the particular solution is of interest.

It can be shown that,

$$x = A\sin(\alpha t - \phi) \qquad \text{...(Eq. 20.39)}$$

is a solution, if A and ϕ are chosen in an appropriate manner.

Then $\dot{x} = \dfrac{dx}{dt} = \omega A(\cos\alpha t - \phi)$

and $\ddot{x} = \dfrac{d^2x}{dt^2} = -\omega^2 A\sin(\alpha t - \phi)$

Substituting these in Eq. 20.38, we have,

$$\frac{W}{g}\left[-\omega^2 A\sin(\alpha t - \phi)\right] + C\omega A\cos(\alpha t - \phi)$$

$$+ k\cdot A\sin(\alpha t - \phi) = Q_0\sin\alpha t \quad \ldots\text{(Eq. 20.40)}$$

This equation shall be valid for all values of t.

$\therefore$ If $t = \dfrac{\phi}{\omega}$ or $(\alpha t - \phi) = 0$

$$C\omega A = Q_0\sin\phi \qquad \ldots\text{(Eq. 20.41)}$$

If $(\omega t - 0) = \pi/2$,

$$-\alpha^2\frac{W}{g}\cdot A + k\cdot A = Q_0\cos\phi$$

or $\quad A\left(k - \dfrac{W}{g}\alpha^2\right) = Q_0\cos\phi \qquad \ldots\text{(Eq. 20.42)}$

Squaring and adding Eqs. 20.41 and Eq. 20.42,

$$A^2\left[\left(k - \frac{W}{g}\alpha^2\right)^2 + (C\alpha)^2\right] = Q_0^2$$

or $\quad A = \dfrac{Q_0}{\sqrt{\left(k - \dfrac{W}{g}\alpha^2\right)^2 + (C\alpha)^2}} \qquad \ldots\text{(Eq. 20.43)}$

Dividing Eq. 20.41 by Eq. 20.42, we have,

$$\tan\phi = \frac{C\alpha}{\left(k - \dfrac{W}{g}\alpha^2\right)} \qquad \ldots\text{(Eq. 20.44)}$$

Therefore, the required solution representing steady-state vibrations will be,

$$x = \frac{Q_0}{\sqrt{\left[\left(k - \dfrac{W}{g}\alpha^2\right)^2 + (C\alpha)^2\right]}}$$

$$\sin\left[\alpha t - \tan^{-1}\left(\frac{C\alpha}{k - \dfrac{W}{g}\alpha^2}\right)\right] \quad \ldots\text{(Eq. 20.45)}$$

This indicates that the motion is a S.H.M. with the same frequency, α, as the exciting frequency. The amplitude, A is given by,

$$A = \frac{Q_0}{\sqrt{\left(k - \dfrac{W}{g}\alpha^2\right)^2 + (C\alpha)^2}} \quad \ldots\text{(Eq. 20.46)}$$

Dividing both the numerator and denominator by k,

$$A = \frac{(Q_0/k)}{\left(\dfrac{1}{k}\right)\sqrt{\left(k - \dfrac{W}{g}\alpha^2\right)^2 + (C\alpha)^2}}$$

If we put $\dfrac{Q_0}{k} = \delta_{st}$, the static deflection for Q_0,

$$A = (\delta_{st})\frac{1}{\sqrt{\left(1 - \dfrac{W\alpha^2}{gk}\right) + \left(\dfrac{C\alpha}{k}\right)^2}}$$

But $\dfrac{gk}{W} = \omega^2$ and $\dfrac{W\alpha^2}{gk} = \left(\dfrac{\alpha}{\omega}\right)^2$

Also $\dfrac{C\alpha}{k} = \dfrac{C}{C_c}\cdot\dfrac{C_c\alpha}{k} = \left(\dfrac{C}{C_C}\right)\dfrac{2m\omega\alpha}{k} = \dfrac{C}{C_C}\cdot\dfrac{2\omega\alpha}{\omega^2}$

or $\quad \dfrac{C\alpha}{k} = 2\left(\dfrac{C}{C_C}\right)\left(\dfrac{\alpha}{\omega}\right)$

$$\therefore A = (\delta_{st})\left[\frac{1}{\sqrt{\left\{1 - \left(\dfrac{\alpha}{\omega}\right)^2\right\}^2 + \left(\dfrac{2C}{C_C}\right)^2\left(\dfrac{\alpha}{\omega}\right)^2}}\right]$$

$$\ldots\text{(Eq. 20.47)}$$

A is written in terms of the magnification factor, M,

$$A = (\delta_{st})\cdot M$$

where M is given by

$$M = \frac{1}{\sqrt{\{(1 - n^2) + (2dn)^2\}}} \qquad \ldots\text{(Eq. 20.48)}$$

where $n = \dfrac{f_e}{f}$, the frequency ratio

$$= \frac{\text{exciting frequency}}{\text{natural frequency}}, \text{ and}$$

$d = \dfrac{C}{C_C}$, the damping factor, as already defined.

The variation of the magnificatin factor M with the frequency ratio, n, for different values of the damping factor, d, are shown in Fig. 20.8.

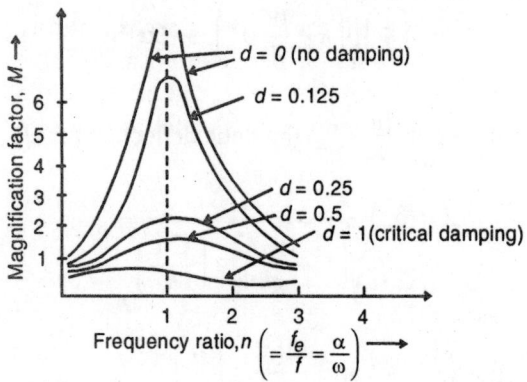

Fig. 20.8: Variation of Magnification Factor with Frequency Ratio

20.8 PENDULUM MOTION

A "pendulum" is a device that executes simple harmonic motion; the period of oscillation can be varied by changing the physical dimensions suitably.

The friction at the pivot support and the air resistance to motion are considered to be negligible in pendulum motion.

The forces involved are: (i) gravity force, (ii) elastic force, and (iii) spring force. These are all conservative forces, and so the principle of conservation of energy is applicable. The equation of motion may be applied using Newton's laws.

The different kinds of pendulums and their motions are considered in the following subsections.

20.8.1 Simple Pendulum

This consists of a belt or a small sphere of mass m tied to the lower end of an inextensible string of negligible weight, the other being tied to a rigid support (Fig. 20.9).

The displacement is measured from the equilibrium position along the circular arc, which is the path of the bob when disturbed from the equilibrium position. The oscillating mass will have tangential as well as normal components of accelerations.

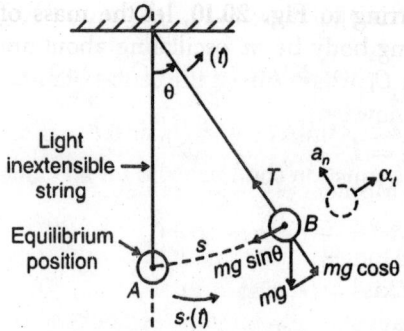

Fig. 20.9: Simple Pendulum

The equation of motion is,
$$ma_t = -mg\sin\theta$$
(tangential)

The negative sign indicates that the force is always directed towards the equilibrium position.
If v is the velocity,

$$a_t = \frac{dv}{dt} = \frac{d^2s}{dt^2}$$

$$\therefore \quad m\frac{d^2s}{dt^2} = -mg\sin\theta$$

If θ is small, $\sin\theta \approx \theta(\text{rad.}) \approx \frac{s}{l}$

$$\therefore \quad m\frac{d^2s}{dt^2} + mg \cdot \frac{s}{l} = 0$$

$$\frac{d^2s}{dt^2} + \frac{g}{l} \cdot s = 0 \qquad \text{...(Eq. 20.49)}$$

This is of the form:

$$\frac{d^2s}{dt^2} + c^2 \cdot s = 0 \qquad \text{...(Eq. 20.50)}$$

where $c^2 = \frac{g}{l}$

Period, $T = \dfrac{2\pi}{c}$

or $\quad T = 2\pi\sqrt{\dfrac{l}{g}} \qquad \text{...(Eq. 20.51)}$

20.8.2 Compound Pendulum

A "compound pendulum" consists of a rigid body oscillating about a fixed point in the body other than its centre of mass.

Referring to **Fig. 20.10**, let the mass of the oscillating body be m, oscillating about an axis through O, distant b from the centre of mass, G.

We know that,

$$M_o = I_o \cdot \alpha \qquad \text{...(Eq. 20.52)}$$

M_o = Moment of the forces acting on the body about O;

I_o = Moment of inertia of the body about the axis of rotation through O;

α = angular acceleration of the body and is equal to $\ddot{\theta}$.

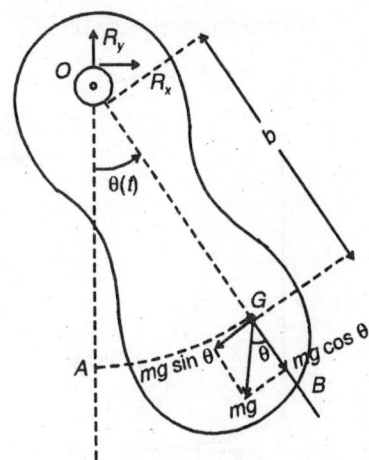

Fig. 20.10: Compound Pendulum

Eq. 20.52 can be written as,

$$-(mg\sin\theta) \cdot b = I_o \cdot \ddot{\theta}$$

For small angular displacements
$\sin\theta \approx \theta (\text{rad.})$

$$\therefore \quad -b\,mg \cdot \theta = I_o \cdot \ddot{\theta}$$

or $\quad \ddot{\theta} + \dfrac{mg \cdot b}{I_o} \cdot \theta = 0 \qquad \text{...(Eq. 20.52)}$

This is of the form:

$$\ddot{\theta} + c^2 \cdot \theta = 0$$

where $c^2 = \dfrac{mg \cdot b}{I_o} \qquad \text{...(Eq. 20.53)}$

Period, $T = \dfrac{2\pi}{c}$

or $\quad T = 2\pi \sqrt{\dfrac{I_o}{b\,mg}} \qquad \text{...(Eq. 20.54)}$

Using $I_o = m \cdot k_o^2$, k_o being the radius of gyration with respect to the axis of rotation through O,

$$T = 2\pi \sqrt{\dfrac{k_o^2}{bg}} \qquad \text{...(Eq. 20.55)}$$

Comparing this with the corresponding expression for a simple pendulum (Eq. 20.51), it can be understood that the period of oscillation of a compound pendulum is equal to that of a simple pendulum of length,

$$l = \left(\dfrac{k_o^2}{b}\right) \qquad \text{... (Eq. 20.56)}$$

If k_g is the radius of gyration of the body with respect to an axis through its centre of mass and parallel to the axis of rotation, we have,

$$k_o^2 = k_g^2 + b^2 \qquad \text{...(Eq. 20.57)}$$

$$\therefore \quad l = \left(\dfrac{k_g^2 + b^2}{b}\right) = \dfrac{k_g^2}{b} + b \qquad \text{...(Eq. 20.58)}$$

This is the equivalent length of the pendulum, the criterion of equivalence being the period, the comparison being with the simple pendulum.

20.8.3 Torsional Pendulum

A "torsional pendulum" is one in which a disc makes oscillatory motion of rotation about an axis under the influence of an applied twisting moment or torque. (Fig. 20.11). The vibrations are called "torsional vibrations".

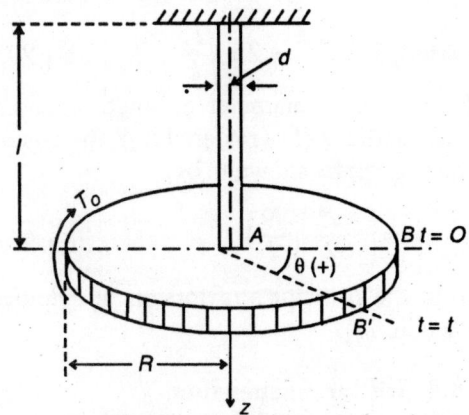

Fig. 20.11: Torsional Pendulum

A circular disc is attached to the lower end of a shaft, the upper end being rigidly fixed. Shaft passes through the centre of the disc and its axis is normal to the plane of the disc.

The bar is assumed to be elastic but of negligible weight.

Let a torque T_o be applied to the disc about the axis of the shaft.

Let θ be the angular displacement after time lapse t.

The equation of motion is,

$$T_o = I_Z \cdot \ddot{\theta} \qquad \ldots\text{(Eq. 20.59)}$$

(I_Z: Moment of inertia of the disc about the axis of the shaft).

The weight of the disc acts along OZ and does not cause any torque, which is only because of the elastic reaction of the shaft on the disc.

When the torque is released, the disc makes torsional oscillations or vibrations since the elastic reaction acts as a torsional spring.

$$\therefore \qquad T_o = -k_t \cdot \theta \qquad \ldots\text{(Eq. 20.60)}$$

(k_t is the torsional stiffness)

$\therefore$ From Eqs. 20.59 & 20.60,

$$I_Z\ddot{\theta} + k_t \cdot \theta = 0 \qquad \ldots\text{(Eq. 20.61)}$$

or

$$\ddot{\theta} + \frac{k_t}{I_Z} \cdot \theta = 0 \qquad \ldots\text{(Eq. 20.62)}$$

This is of the form

$$\ddot{\theta} + c^2\theta = 0$$

where $c^2 = k_t/I_Z$ $\qquad \ldots\text{(Eq. 20.63)}$

Period, $T_o = \dfrac{2\pi}{c} = 2\pi\sqrt{\dfrac{I_Z}{k_t}}$ $\qquad \ldots\text{(Eq. 20.64)}$

If the shaft is diameter d, length l, and the rigidity modulus of its material is N, the torsional stiffness k_t can be shown to be,

$$k_t = \left(\frac{\pi d^4}{32}\right)\left(\frac{N}{l}\right) \qquad \ldots\text{(Eq. 20.65)}$$

from the principles of torsion of cylindrical shafts (Ch. 27).

20.8.4 Trifilar Suspension

A "Trifilar suspension" consists of a disc suspended from three wires of equal length, attached to the disc at its circumference at equidistant points (Fig. 20.12).

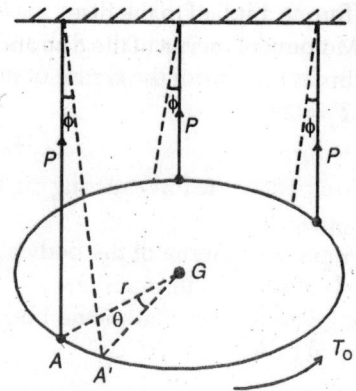

(a) Trifilar Suspension of a Disc

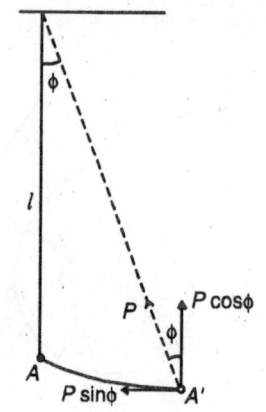

(b) Components of Tension in the Wire

Fig. 20.12: Trifilar Suspension

Let the mass be m and weight W for the disc. Tension in each wire in equilibrium position,

$$P = \frac{W}{3}$$

Let an angle of twist θ be given to the disc.

The corresponding angular displacement of the wire from the vertical, ϕ, is given by,

$$\phi = \frac{r \cdot \theta}{l} \qquad \ldots\text{(Eq. 20.66)}$$

where r is radius of the disc and l is length of the wire.

(For small angular displacements, it may be taken that the tensions in the wires remain unaltered).

The equation of motion is,

$$T_o = I_g \cdot \alpha = I_g \cdot \ddot{\theta} \qquad \ldots\text{(Eq. 20.67)}$$

T_0: Torque applied to the disc

I_g: Moment of inertia of the disc about an axis through G and normal to its plane.

$$T_o = -3(P\sin\phi) \cdot r$$

$$T_o = -3\frac{W}{3}\sin\phi \cdot r = -W\sin\phi \cdot r \quad \text{...(Eq. 20.68)}$$

For small values of ϕ, $\sin\phi \approx \phi$ (rad.)

$$\therefore \quad T_o = -Wr \cdot \phi$$

$$= -W \cdot \frac{r\theta}{l} \cdot r$$

or $$T_o = -\frac{Wr^2\theta}{l} \quad \text{...(Eq. 20.69)}$$

Substituting this in Eq. 20.67, we have,

$$I_g \cdot \ddot{\theta} + \frac{Wr^2\theta}{l} = 0 \quad \text{...(Eq. 20.70)}$$

[In terms of the radius of gyration k_g,

$$I_g = m \cdot k_g^2 = \frac{W}{g}k_g^2]$$

Eq. 20.70 will become,

$$\frac{W}{g} \cdot k_g^2 \cdot \ddot{\theta} + \frac{Wr^2}{l} \cdot \theta = 0 \quad \text{...(Eq. 20.71)}$$

or $$\ddot{\theta} + \frac{gr^2}{k_g^2 \cdot l} \cdot 0 = 0 \quad \text{...(Eq. 20.72)}$$

This is of the form:

$$\ddot{\theta} + c^2\theta = 0$$

where $$c^2 = \frac{gr^2}{k_g^2 \cdot l} \quad \text{...(Eq. 20.73)}$$

Period, $$T = \frac{2\pi}{c} = 2\pi\sqrt{\frac{k_g^2 l}{gr^2}} \quad \text{...(Eq. 20.74)}$$

or $$T = \frac{2\pi k_g}{r}\sqrt{\frac{l}{g}} \quad \text{...(Eq. 20.75)}$$

ILLUSTRATIVE PROBLEMS

PROBLEM 20.1: A harmonic motion has an amplitude of 0.05 m and a frequency of 25 Hz. Find the time period, maximum velocity and maximum acceleration.

Solution:

$$N = 25 \text{ Hz}$$

$$\omega = 2\pi N = 2 \times \pi \times 25 = 50\pi \text{ rad/sec}$$

Time period, $$t = \frac{2\pi}{\omega} = \frac{2\pi}{50\pi} = 0.04 \text{ sec}$$

Maximum velocity $= r \cdot \omega$

$$= 0.05 \times 50\pi$$

$$= 7.854 \text{ m/sec}$$

Maximum acceleration $= r \cdot \omega^2$

$$= 0.05 \times (50\pi)^2$$

$$= 1233.7 \text{ m/sec}^2$$

PROBLEM 20.2: A body, moving with simple harmonic motion, has an amplitude of 0.5 m and a period of oscillation of 1 second. What will be the velocity and acceleration of the body after 0.2 seconds from the extreme position?

Solution:

Amplitude, $r = 0.50$ m

Period of oscillation, $T = 1$ second

$$t = 0.2 \text{ seconds}$$

$$T = \frac{2\pi}{\omega}$$

or $$\omega = \frac{2\pi}{T} = \frac{2\pi}{1} = 2\pi \text{ rad/sec}$$

Let O be the centre, A an extremity of the motion and P the position of the body after 0.2 second from A as shown in Fig. 20.13. Therefore time required by the body to travel from A to $P = 0.2s$.

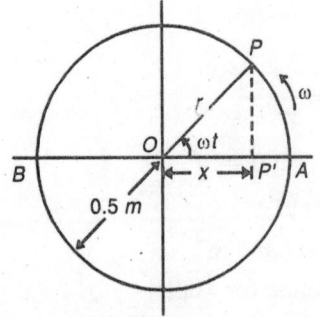

Fig. 20.13 (Illustrative Problem 20.2)

Time required by the body to travel from O to A

$$= \frac{1}{4} \times \text{Time of oscillation}$$

$$= \frac{1}{4} \times 1 = 0.25 s$$

$\therefore$ Time required by the body to travel from O to P',

$$t = 0.25 - 0.2$$
$$= 0.05 s$$
$$v = -\omega \cdot r \cdot \sin\omega\, t$$
$$= -2\pi \times 0.5 \times \sin(2\pi \times 0.05)$$
$$= 0.97 \text{ m/sec}$$
$$a = -\omega^2 \cdot r\cos\omega\, t$$
$$= -(2\pi)^2 \times (0.50) \times \cos(2\pi \times 0.05)$$
$$= -18.773 \text{ m/sec}^2$$

PROBLEM 20.3: A particle is performing a simple harmonic motion. When it is at distances of 1 m and 2 m from the mean position its velocities are 5 m/sec and 3 m/sec respectively. Find (*i*) the amplitude of oscillations, (*ii*) time period of oscillations, (*iii*) its maximum velocity, and (*iv*) its maximum acceleration.

Solution:

$$v = \omega\sqrt{r^2 - x^2}$$

When $x = 1$ m, $v = 5$ m/sec

$\therefore$
$$5 = \omega\sqrt{r^2 - 1^2} \qquad \ldots(i)$$

When $x = 2$ m, $v = 3$ m/sec

$\therefore$
$$3 = \omega\sqrt{r^2 - 2^2} \qquad \ldots(ii)$$

Dividing (*i*) by (*ii*)

$$\frac{5}{3} = \frac{\omega\sqrt{r^2 - 1}}{\omega\sqrt{r^2 - 4}}$$

$$\left(\frac{5}{3}\right)^2 = \frac{r^2 - 1}{r^2 - 4}$$

$$25r^2 - 100 = 9r^2 - 9$$
$$16r^2 = 91$$
$$r = 2.385 \text{ m}$$

Substituting for r in (*i*)

$$5 = \omega\sqrt{(2.385)^2 - 1} = \omega\sqrt{4.688} = 2.165\omega$$
$$\omega = 2.309 \text{ rad/sec}$$
$$T = \frac{2\pi}{\omega} = \frac{2 \times \pi}{2.309} = 2.721 \text{ sec}$$

Maximum velocity $= r \cdot \omega$
$$= 2.385 \times 2.309$$
$$= 5.507 \text{ m/sec}$$

Maximum acceleration $= r \cdot \omega^2$
$$= 2.385 \times (2.309)^2$$
$$= 12.716 \text{ m/sec}^2$$

PROBLEM 20.4: A body is moving with simple harmonic motion. The amplitude of motion is 4 m and period of complete oscillation is 3 seconds. Find the time required by the body in passing between two points which are at distances 3 m and 1 m from the centre and are on the same side of it.

Solution:

$$r = 4 \text{ m}$$
$$T = 3 \text{ sec}$$
$$T = \frac{2\pi}{\omega}$$

or $\quad \omega = \dfrac{2\pi}{T} = \dfrac{2\pi}{3} = 2.0944 \text{ rad/sec}$

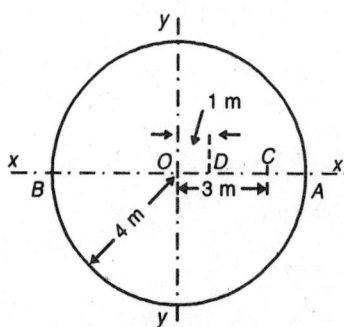

Fig. 20.14 (Illustrative Problem 20.4)

Displacement
$$3 = r\sin(\omega\, t_1)$$
$$3 = 4\sin(\omega\, t_1)$$
$$\omega\, t_1 = \sin^{-1}(3/4) = 0.848$$
$$2.0944 t_1 = 0.848$$

or $\quad t_1 = 0.405 \text{ sec}$

$$1 = r\sin(\omega\, t_2)$$
$$1 = 4\sin(\omega\, t_2)$$
$$\omega\, t_2 = \sin^{-1}(1/4) = 0.2527$$
$$2.0944 t_2 = 0.2527$$

$t_2 = 0.121$ sec

∴ Time required by the body in passing between the two points,

$$t = t_1 - t_2$$
$$= 0.405 - 0.212$$
$$= 0.284 \text{ sec}$$

PROBLEM 20.5: A mass of 25 kg is suspended from a spring as shown in Fig. 20.15. Determine the natural frequency of the system.

$$k_1 = 5 \text{ kN/m and } k_2 = k_3 = 8 \text{ kN/m}$$

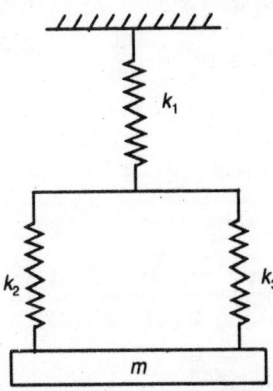

Fig. 20.15: Combined Parallel and Series Springs
(Illustrative Problem 20.5)

Solution:

If k_{e_1} is the effective spring stiffness of the lower two springs in parallel, then

$$k_{e_1} = k_2 + k_3 = 8 + 8 = 16 \text{ kN/m}$$

Now k_1 and k_{e_1} are two springs in series, therefore effective stiffness.

$$\frac{1}{k_e} = \frac{1}{k_1} + \frac{1}{k_{e2}}$$

$$= \frac{1}{5} + \frac{1}{16} = \frac{16 + 5}{80} = \frac{21}{80}$$

or $\quad k_e = \dfrac{88}{21} = 3.81 \text{ kN/m}$

∴ $\quad \omega_n = \sqrt{\dfrac{k_e}{m}}$

$$= \sqrt{\frac{3.81 \times 1000}{25}}$$

$$= 12.345 \text{ rad/sec}$$

or $\quad f_n = \dfrac{\omega_n}{2\pi} = \dfrac{12.345}{2\pi}$

$$= 1.965 \text{ Hz}$$

PROBLEM 20.6: A spring-mass system has spring stiffness of k N/m and a mass of m kg. It has natural frequency of vibration as 12 Hz. An extra 3 kg mass is coupled to m and the natural frequency reduces by 3 Hz. Find k and m.

Solution:

$$\omega_n = \sqrt{\frac{k}{m}}$$

$$2\pi \times 12 = \sqrt{\frac{k}{m}} \qquad \qquad \ldots(i)$$

$$2\pi \times (12 - 3) = \sqrt{\frac{k}{m+3}}$$

$$2\pi \times 9 = \sqrt{\frac{k}{m+3}} \qquad \ldots(ii)$$

Dividing equation (i) by equation (ii)

$$\frac{2\pi \times 12}{2\pi \times 9} = \sqrt{\frac{k}{m} \times \frac{m+3}{k}}$$

$$\frac{4}{3} = \sqrt{\frac{m+3}{m}}$$

or $\quad \dfrac{16}{9} = \dfrac{m+3}{m}$

$$16m = 9m + 27$$
$$7m = 27$$

or $\quad m = 3.857 \text{ kg}$

Substituting the value of m in equation (i)

$$2 \times \pi \times 12 = \sqrt{\frac{k}{3.857}}$$

$$k = (2\pi \times 12)^2 \times 3.857$$
$$= 21927 \text{ N/m}$$
$$= 21.927 \text{ kN/m}$$

PROBLEM 20.7: A mass of 1 kg is to be supported on a spring having a stiffness 10 kN/m. The damping coefficient is 5 N.sec/m. Determine the natural frequency of the system.

Find also the logarithmic decrement and the amplitude after five cycles if the initial amplitude is 50 mm.

Solution:

$$\omega_n = \sqrt{\frac{k}{m}}$$

$$= \sqrt{\frac{10 \times 1000}{1}}$$

$$= 100 \text{ rad/sec}$$

$$f_n = \frac{\omega_n}{2\pi} = \frac{100}{2\pi} = 15.915 \text{ Hz}$$

$$C_C = 2\sqrt{k \cdot m}$$

$$= 2\sqrt{10 \times 1000 \times 1}$$

$$= 200 \text{ N.sec/m}$$

$$d = \frac{C}{C_C} = \frac{5}{200} = 0.025$$

The damped natural frequency of vibration,

$$f_d = \sqrt{1 - d^2} \cdot f_n$$

$$f_d = \sqrt{1 - 0.025^2} \times 15.915$$

$$f_d = 15.910 \text{ Hz}$$

$$\delta = \frac{2\pi d}{\sqrt{1 - d^2}}$$

$$\delta = \frac{2 \times \pi \times 0.025}{\sqrt{1 - 0.025^2}} = 0.1571$$

$$\delta = \frac{1}{\eta} \log_e\left(\frac{x_o}{x_n}\right)$$

$$0.1571 = \frac{1}{5} \log_e\left(\frac{50}{x_5}\right)$$

$$\log_e\left(\frac{50}{x_5}\right) = 0.1571 \times 5$$

$$\log_e\left(\frac{50}{x_5}\right) = 0.7855$$

or $$\frac{50}{x_5} = 2.1935$$

$$\therefore \quad x_5 = \frac{50}{2.1935} = 22.79 \text{ mm}$$

PROBLEM 20.8: The damped natural frequency of a system as obtained from a free vibration test, is 10 Hz. During the forced vibration test, with constant exciting force, on the same system, the peak amplitude of vibration is found to be at 9.8 Hz. Find the damping factor for the system and its natural frequency.

Solution:

$$W_d = 10.0 \times 2\pi \text{ rad/sec}$$

$$W_p = 9.8 \times 2\pi \text{ rad/sec}$$

$$W_d = W_n \sqrt{1 - d^2}$$

$$10.0 \times 2\pi = W_n \sqrt{1 - d^2} \qquad \ldots(i)$$

$$W_p = W_n \sqrt{1 - 2d^2}$$

$$9.8 \times 2\pi = W_n \sqrt{1 - 2d^2} \qquad \ldots(ii)$$

Dividing (ii) by (i)

$$\frac{9.8 \times 2\pi}{10.0 \times 2\pi} = \frac{\omega_n \sqrt{1 - 2d^2}}{\omega_n \sqrt{1 - d^2}}$$

$$0.98 = \sqrt{\frac{1 - 2d^2}{1 - d^2}}$$

or $$\frac{1 - 2d^2}{1 - d^2} = 0.9604$$

$$0.9604 - 0.9604 d^2 = 1 - 2d^2$$

or $$1.0396 d^2 = 0.0396$$

or $$d = \sqrt{\frac{0.0396}{1.0396}} = 0.19517$$

Substituting the value of *d* in equation (i)

$$10.0 \times 2\pi = \omega_n \sqrt{1 - (0.19517)^2}$$

$$\omega_n = \frac{10 \times 2\pi}{0.980769} = 10.2 \times 2\pi \text{ rad/sec}$$

or $$f_n = 10.2 \text{ Hz}$$

PROBLEM 20.9: A single cylinder vertical diesel engine of total mass 300 kg is mounted upon a steel chasis frame and causes a vertical static deflection of 2.06 mm. The reciprocating parts of the engine have a mass of 24 kg and through a vertical stroke of 140 mm with simple harmonic motion. A dashpot is provided, the damping resistance of which is directly proportional to the velocity and amount to 500 N at 0.3 m/sec. Determine (i) the speed of the driving shaft at which resonance will occur, and (ii) the amplitude of steady state forced vibrations when the driving shaft of the engine rotates at 7 Hz.

Solution:

(i) $\omega_n = \sqrt{\dfrac{g}{\delta_{st}}}$

$= \sqrt{\dfrac{9.81}{0.00206}}$

$= 69$ rad/sec

or $f_n = \dfrac{\omega_n}{2\pi} = \dfrac{69}{2\pi} = 10.98$ Hz

Therefore, the speed of the driving shaft at which resonance occurs = 10.98 Hz.

(ii) $\alpha = 2\pi \times 7$ rad/sec

$\omega_n = 2\pi \times 10.98$ rad/sec

$\dfrac{\alpha}{\omega_n} = \dfrac{2\pi \times 7}{2\pi \times 10.98} = 0.6375$

$d = \dfrac{C}{2m\omega_n}$

$= \dfrac{(500/0.3)}{2 \times 300 \times 69}$

$= 0.04026$

$\dfrac{m_o e}{m} = \dfrac{24 \times (0.14/2)}{300}$

$= 0.0056$

$A = \dfrac{\left(\dfrac{m_o e}{m}\right)\left(\dfrac{\alpha}{\omega_n}\right)^2}{\sqrt{\left[1-\left(\dfrac{\alpha}{\omega_n}\right)^2\right]^2 + \left[2 \cdot d \cdot \dfrac{\alpha}{\omega_n}\right]^2}}$

$= \dfrac{0.0056 \times (0.6375)^2}{\sqrt{\left(1-(0.6375)^2\right)^2 + \left(2(0.04026)(0.6375)\right)^2}}$

$= \dfrac{0.0022758}{\sqrt{0.3524 + 0.0026}}$

$= \dfrac{0.0022758}{0.5958}$

$= 0.00382$

or $= 3.82$ mm

PROBLEM 20.10: Determine the length of a second's pendulum.

Solution: A second's pendulum executes one beat per second.

$\dfrac{T}{2} = \pi\sqrt{\dfrac{l}{g}}$

$1 = \pi\sqrt{\dfrac{l}{9.81}}$

$\sqrt{\dfrac{l}{9.81}} = \dfrac{1}{\pi}$

$\dfrac{l}{9.81} = \left(\dfrac{1}{\pi}\right)^2 = 0.10132$

$l = 0.10132 \times 9.81 = 0.9939$ m

or $= 993.9$ mm

PROBLEM 20.11: A pendulum is having a string length of 1.0 m. It gains 4 seconds per day. Determine the change in length of the pendulum to correct the time.

Solution:

$\dfrac{dn}{n} = -\dfrac{dl}{2l}$

$\dfrac{4}{24 \times 60 \times 60} = -\dfrac{dl}{2 \times 1 \times 1000}$

$dl = -\dfrac{4 \times 2 \times 1 \times 1000}{24 \times 60 \times 60}$

$= -0.09259$ mm

Therefore, the length of the pendulum should be increased by 0.09259 mm for correct time.

PROBLEM 20.12: Find how many seconds a clock would lose per day if the length of the pendulum is increased in the ratio 600 : 601.

Solution:

$\dfrac{dn}{n} = -\dfrac{1}{2}\dfrac{dl}{l}$

$$\frac{dn}{24 \times 60 \times 60} = -\frac{1}{2} \times \frac{1}{600}$$

$$dn = -\frac{24 \times 60 \times 60}{2 \times 600}$$

$$= -72 \text{ seconds}$$

Hence, the clock will lose 72 seconds per day.

PROBLEM 20.13: A second's pendulum loses 9 seconds per day at the top of a mountain. Find the height of the mountain. Take the radius of the earth as 6400 km.

Solution:

$$\frac{dn}{n} = -\frac{h}{r}$$

$$\frac{-9}{24 \times 60 \times 60} = -\frac{h}{6400}$$

$$h = \frac{24 \times 60 \times 60}{6400 \times 9}$$

$$= 1.5 \text{ km}$$

$$= 1500 \text{ m}$$

Hence, the height of the mountain is 1500 m.

PROBLEM 20.14: A uniform straight rod of mass 1 kg is 1 m long. The rod is smoothly pivoted about a point, which is 80 mm from one end. Find the frequency of the rod about the pivot, if the rod turns freely in the vertical plane.

Solution:

$$I_g = \frac{ml^2}{12} = \frac{1 \times 1^2}{12}$$

$$= 0.0833 \text{ kg.}m^2$$

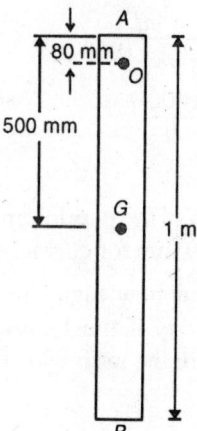

Fig. 20.16 (Illustrative Problem 20.14)

$$b = 500 - 80 = 420 \text{ mm}$$

$$= 0.42 \text{ m}$$

$$I_o = I_g + m \cdot b^2$$

$$= 0.0833 + 1 \times 0.42^2$$

$$= 0.26 \text{ kg.}m^2$$

$$T = 2\pi \sqrt{\frac{I_o}{b \cdot m \cdot g}}$$

$$= 2\pi \sqrt{\frac{0.26}{0.42 \times 1 \times 9.81}}$$

$$= 1.58 \text{ sec}$$

$$\text{Frequency} = \frac{1}{T} = \frac{1}{1.58} = 0.63 \text{ Hz}$$

PROBLEM 20.15: Calculate the natural frequency of vibration of a torsional pendulum with following dimensions:

Length of the shaft = 1.2 m
Diameter of the shaft = 6 mm
Diameter of the disc = 0.25 m
Mass of the disc = 2.5 kg

The modulus of rigidity for the material of the shaft may be assumed to be 0.8×10^{11} N/m².

Solution:

$$I_Z = \frac{m \cdot R^2}{2}$$

$$= \frac{2.5 \times (0.25 / 2)^2}{2}$$

$$= 0.01953 \text{ kg.}m^2$$

$$k_t = \left(\frac{\pi d^4}{32}\right)\left(\frac{N}{l}\right)$$

$$= \frac{\pi (0.006)^4 \times 0.8 \times 10^{11}}{32 \times 1.2}$$

$$= 8.4823 \text{ N.m/rad}$$

$$\omega_n = \sqrt{\frac{k_t}{I_Z}}$$

$$= \sqrt{\frac{8.4823}{0.01953}}$$

$$= 20.84 \text{ rad/sec}$$

$$\text{or} \quad f_n = \frac{\omega_n}{2\pi} = \frac{20.84}{2\pi}$$

$$= 3.32 \text{ Hz}$$

PROBLEM 20.16: A thin circular plate of diameter 1.5 m is suspended from three vertical wires of length 0.50 m each, equally spaced around the perimeter of the plate. Find the period of oscillations when the plate is rotated through a small angle about a vertical axis passing through the mass centre and then released (Fig. 20.17).

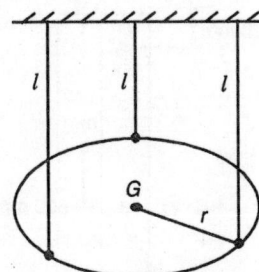

Fig. 20.17 (Illustrative Problem 20.16)

Solution:

$$I_g = \frac{m \cdot r^2}{2}$$

$$= \frac{m(0.75)^2}{2}$$

$$= 0.28125\, m\, \text{kg.}m^2$$

$$k_g = \sqrt{\frac{I_g}{m}}$$

$$= \sqrt{\frac{0.28125\, m}{m}}$$

$$= 0.53033 \text{ metre}$$

$$T = \frac{2\pi k_g}{r}\sqrt{\frac{l}{g}}$$

$$T = \frac{2\times\pi\times 0.53033}{0.75}\sqrt{\frac{0.50}{9.81}}$$

$$T = 1.003 \text{ sec}$$

Or

$$I_g = \frac{m \cdot R^2}{2}$$

$$k_g = \sqrt{\frac{I_g}{m}}$$

$$= \sqrt{\frac{m \cdot r^2}{2 \cdot m}}$$

$$= \sqrt{\frac{r^2}{2}}$$

$$= \frac{r}{\sqrt{2}}$$

$$T = \frac{2\pi k_g}{r}\sqrt{\frac{l}{g}}$$

$$= \frac{2\pi \cdot r}{r \cdot \sqrt{2}}\sqrt{\frac{l}{g}}$$

$$= \sqrt{2}\cdot\pi\times\sqrt{\frac{l}{g}}$$

$$= \sqrt{2}\times\pi\times\sqrt{\frac{0.50}{9.81}}$$

$$T = 1.003 \text{ sec}$$

Note: When the axis of rotation passes through the mass centre of the circular-plate, it can be seen that the period is independent of the radius of the plate.

PRACTICE PROBLEMS

20.1 An instrument has a natural frequency of 10 Hz. It can stand a maximum acceleration of 10 m/sec². Find the amplitude.

20.2 The piston of an engine moves with simple harmonic motion. The crank rotates at 90 r.p.m. and the stroke length is 1.5 metres. Find the velocity and acceleration of the piston, when it is at a distance of 0.50 metre from the centre.

20.3 A body performing simple harmonic motion has a velocity of 16 m/sec when the displacement is 50 mm, and 4 m/sec when the displacement is 100 mm, the displacement being measured from the centre. Determine the frequency and amplitude of the motion. What is the acceleration when the displacement is 75 mm?

20.4 For the system shown in Fig. 20.18 $k_1 = 2$ kN/m, $k_2 = 1.5$ kN/m, $k_3 = 3$ kN/m, $k_4 = k_5 = 0.5$ kN/m and $m = 0.5$ kg. Determine the natural frequency of the system.

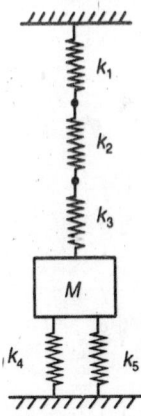

Fig. 20.18: Combined Series and Parallel Springs
(Practice Problem 20.4)

20.5 A spring-mass system has a natural frequency of 9 Hz. The mass of the suspended block is 7 kg. In order to increase the natural frequency of the system to 12 Hz, how much mass is to be removed from the block? What is the spring stiffness?

20.6 The damped vibration record of a spring-mass-dashpot system shows the following data:
Amplitude on second cycle = 12 mm
Amplitude on third cycle = 10.5 mm
Spring stiffness, $k = 8000$ N/m
Mass on the spring, $m = 2$ kg

Determine the damping constant, assuming it to be viscous.

20.7 A pendulum is having a string length of 1.0 m. It loses 5 seconds per day. Determine the change in length of the pendulum in order to correct the time.

20.8 A system of beams supports a motor of mass 1500 kg. The motor has an unbalanced mass of 1.2 kg at 80 mm radius. It is known that the resonance occurs at 40 Hz. What amplitude of vibration can be expected at the motor's operating speed of 24 Hz, if damping factor is assumed to be 0.1?

20.9 The gravity at the poles exceeds the gravity at the equator in the ratio 601 : 600. If a pendulum regulated at poles is taken to the equator, find how many seconds a day will it lose.

20.10 A seconds pendulum gains 13.5 secs per day at the bottom of a mine. Find the depth of the mine. Take the radius of the earth as 6400 km.

20.11 A uniform straight rod of mass 0.600 kg is 600 mm long. The rod is smoothly pivoted about a point, which is 60 mm from one end. Find the frequency of the rod about the pivot, if the rod turns freely in the vertical plane.

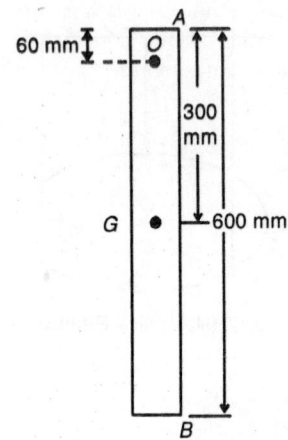

Fig. 20.19 (Practice Problem 20.11)

20.12 A torsion pendulum has a natural frequency of 6 Hz, what length of steel wire of diameter 2 mm should be used for this pendulum? The inertia of the mass fixed at the free end is 0.01 kg.m². Take $N = 0.80 \times 11^{11}$ N/m².

20.13 A horizontal disc of mass 20 kg, 400 mm in diameter and 20 mm thick is attached at its centre to a 1.2 m long shaft of 12 mm diameter.

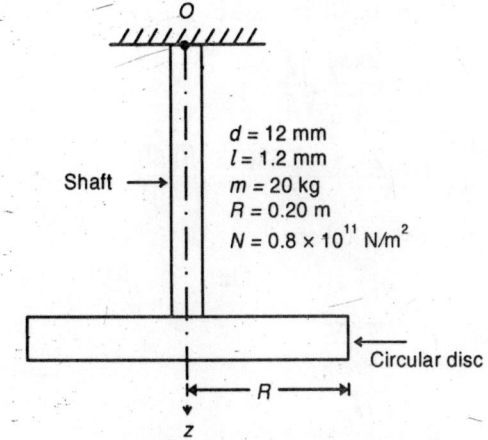

Fig. 20.20 (Practice Problem 20.13)

Calculate the time period of oscillations of the disc if twisted and released (Fig. 20.20). The modulus of rigidity for the material of the shaft may be assumed to be 0.8×10^{11} N/m².

20.14 A thin circular plate of diameter 1.50 m is suspended from three vertical wires of length 1.0 m each, equally spaced around the perimeter of the plate. Find the period of oscillations when the plate is rotated through a small angle about a vertical axis passing through the mass centre and then released (Fig. 20.21).

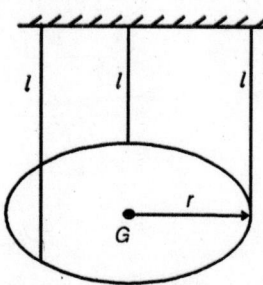

Fig. 20.21 (Practice Problem 20.14)

ELEMENTS OF SOLID MECHANICS AND STRUCTURAL ANALYSIS

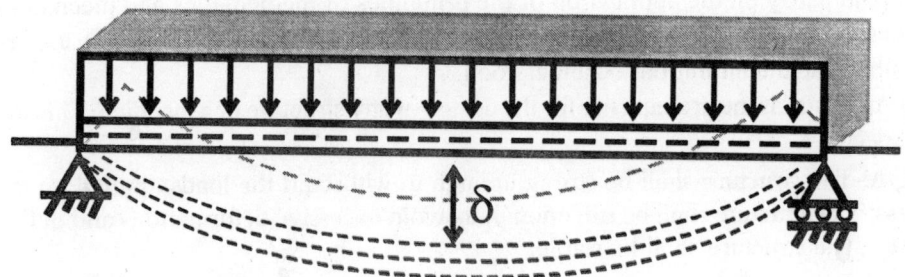

PART - III

Introduction to Solid Mechanics and Structural Analysis

The subject generally called "Strength of Materials", "Mechanics of Solids", or "Statics of Deformable Bodies" includes the study of the distribution of internal forces, the stability and deformation of various elements of structures and machines subjected to externally applied forces. It is based partly on experiment and partly on the application of the principles of mathematics and mechanics. The semi-empirical nature of the subject makes it imperative that its formulae be verified by experiment, wherever possible, and the limitations be understood.

"Structural Analysis" is the prerequisite for the design of any structure or a machine. This involves three major aspects:

(a) **Strength**: The structure shall be strong enough to withstand the loads applied on it.
(b) **Stiffness**: The structure shall be stiff enough to avoid excessive deformations and deflections.
(c) **Stability**: The structure shall be stable.

While engineering mechanics is restricted to the study of external forces on rigid bodies, at rest or in motion, Solid Mechanics or Strength of Materials may be regarded as the statics of deformable or elastic bodies.

Both the strength and stiffness of a member are functions of the size and shape of the member and also of certain physical properties of its material. The study of Strength of Materials aims at predicting how these properties of a member will influence its behaviour under service conditions. The applications of the subject are broad in scope and are found in all branches of engineering.

Certain assumptions are inevitably made in the derivation of any formula; but these are not often realized in practice. The significance of the assumptions should be appreciated in any particular case. Fortunately, the results are applicable to a fair degree of accuracy, despite the assumptions.

'Mechanics of Materials' is a fairly old subject, dating back to Galileo in the seventeenth century; Galileo was the first to explain the behaviour of members under load on a rational and scientific basis. Frenchmen—Coulomb, Poisson, Navier, St. Venant, and Cauchy—made significant contributions in the nineteenth century.

This is one of the most fundamental subjects of an engineering curriculum along with the other basic subjects such as fluid mechanics, thermodynamics, and basic electrical engineering. It is a science blended with experiment and Newtonian postulates of analytical mechanics including statics.

Chapter 21

Simple Stress and Strain

21.1 INTRODUCTION

External forces applied to a body have a tendency to deform the body, which develops internal resistance to counteract the forces. This resistance increases with the increase in the force, but only upto a certain limit, beyond which the forces cause failure of the body. The maximum or ultimate internal resistance offered depends upon the type of deformation and the nature of the material of the body.

In the following sections, 'Stress' and 'Strain' will be defined and simple or basic kinds of stress and strain will be understood. Further concepts relevant to the context will also be dealt with.

21.2 STRESS

When load is applied to a piece of material, internal forces are called into play to resist the applied load. The equal and opposite action and reaction which take place between parts of the same body, transmitting forces constitute a "stress". The material at the interface is said to be stressed or "in a state of stress". The constituent forces, and hence the stress itself, are distributed over the interface either uniformly or in some other manner. The "intensity of stress", generally referred to as merely the "stress", is estimated by the force transmitted per unit area in the case of uniform distribution; this is also some times called "unit stress". If the distribution is not uniform, the stress intensity at a point in the interface must be looked upon as the ratio of force to area when each is decreased indefinitely.

The units of stress are, therefore, N/m^2 (Pascal or Pa), MN/m^2 or N/mm^2, etc.

There are three specially "simple" states of stress which may occur within the material of a body. More complex states of stress may be split into combinations of these simple ones.

21.2.1 Tensile Stress

When two parts of a body pull each other towards itself, 'tensile stress' exists between the two parts. The simplest example is that of a tie-bar sustaining a pull (Fig. 21.1).

Let a bar of uniform cross-section of area A be subjected to a uniaxial tensile force or pull P as shown.

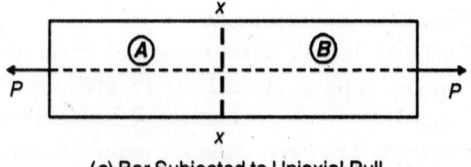

(a) Bar Subjected to Uniaxial Pull

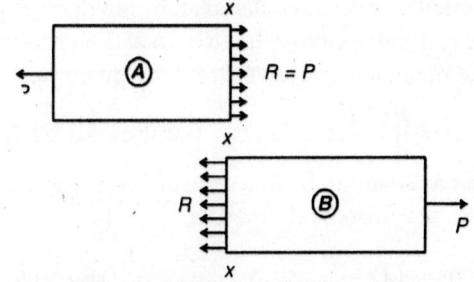

(b) Free Bodies of the Two Parts (A) and (B) of the Bar

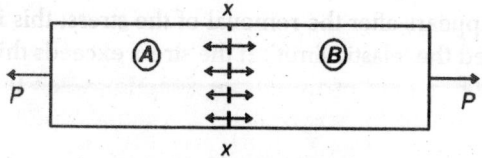

(c) Resistance at the Section xx

Fig. 21.1: Bar under Uniaxial Pull

Let a cross-section *xx* be imagined dividing the bar into two parts (A) and (B) [(a)]. The free bodies of the two parts are shown in (b). Part (B) exerts a pull on (A) or offers a resistance R (equal to P), and vice-versa. Just as the whole bar is in equilibrium, the free-bodies (A) and (B) are also in equilibrium when the internal resisting forces are considered.

The 'free body' concept has already been explained at the end of Chapter 1 and is very valuable in structural analysis.

The internal resistance causes the stress on the section which is taken as the force per unit area.

$$\sigma = \frac{R}{A} = \frac{P}{A} \qquad \ldots(\text{Eq. } 21.1)$$

σ is the average intensity of tensile stress.

The bar is said to be in a state of tension.

21.2.2 Compressive Stress

When two parts of a body push each other from it, there exists a 'Compressive' stress at the interface (Fig. 21.2).

Let a bar of uniform section (area A) be subjected to a uniaxial thrust P as shown.

Let a cross-section *xx* be imagined to divide the bar into two parts (A) and (B). Part (B) exerts a push on (A) or offers a resistance R (equal to P), and vice-versa. The free bodies are in equilibrium under the action of the applied axial thrust P and the internal resistance R offered at the interface (b).

The resistance causes the stress σ at the section and the mean compressive stress σ is given by,

$$\sigma = \frac{R}{A} = \frac{P}{A} \qquad (\text{same as Eq. } 21.1)$$

The bar is said to be in a state of compression. Stress may also be defined as,

$$\sigma = \underset{\Delta A \to 0}{\text{Lt}} \frac{\Delta P}{\Delta A} \qquad \ldots(\text{Eq. } 21.2)$$

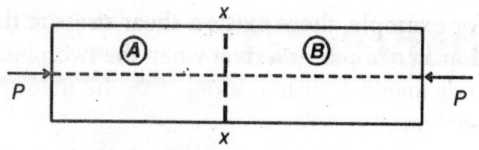

(a) Bar Subjected to Uniaxial Thrust

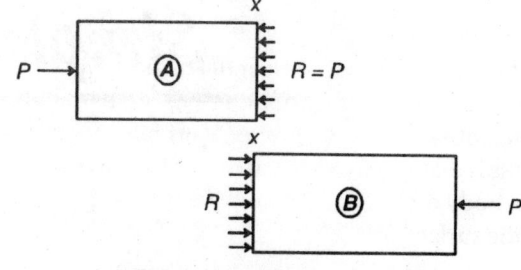

(b) Free Bodies of Points (A) and (B)

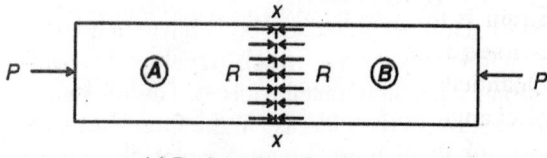

(c) Resistance at Section xx

Fig. 21.2: Bar under Uniaxial Thrust

Tensile and compressive stresses are also called 'normal' stresses or 'direct' stresses on the cross-section, as they occur normal to the area of the section.

21.2.3 Shear Stress

'Shear stress' is said to exist between two parts of a body in contact when the two parts exert equal and opposite forces on each other laterally in a direction tangential to the surface of contact (Fig. 21.3).

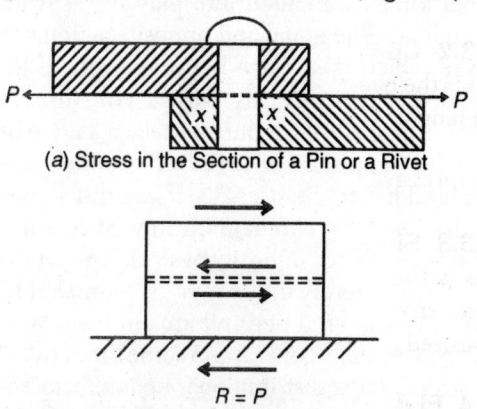

(a) Stress in the Section of a Pin or a Rivet

(b) Body on a Rigid Surface Moved by a Shear Force

Fig. 21.3: Shear Stress

For example, there exists a shear stress at the section *xx* of a pin or a rivet when the two plates it holds together sustain a pull *P* in the plane of the section.

If the area of the section *xx* is *A*, the average intensity of shear stress *q* is given by

$$q = \frac{P}{A} \qquad \ldots(\text{Eq. 21.3})$$

Shear stress occurs tangential to the area of the resisting section and hence is also called 'tangential' stress. [Another example of shear stress is that which occurs at the surface of contact of a body on a rigid surface being pulled parallel to the surface (*b*)].

21.3 STRAIN

Strain is the alteration in shape or dimensions consequent to the application of stress. Quantitatively speaking, it is defined as the ratio of change in length to the initial length in the case of normal stresses, and as the angular displacement produced in the case of shear stress. Thus strain is a measure of the deformation produced by the application of external forces. It is a dimensionless quantity.

21.3.1. Tensile Strain

This is given by the stretch or elongation per unit length.

$$\text{Tensile strain, } \varepsilon_t = \frac{\delta l}{l} \qquad \ldots(\text{Eq. 21.4})$$

δl is the elongation and *l* the initial or original length.

21.3.2 Compressive Strain

This is the contraction or reduction in length per unit length.

$$\text{Compressive strain, } \varepsilon_c = \frac{\delta l}{l} \qquad \ldots(\text{Eq. 21.5})$$

21.3.3 Shear Strain

This is the angular deformation resulting from shear stress. This is designated as ϕ and is measured in radiations.

21.4 ELASTIC LIMIT AND HOOKE'S LAW

For an elastic material, there will be a limit of stress within which the resulting strain completely disappears after the removal of the stress; this is called the 'elastic limit'. If the stress exceeds this value, a part of the strain remains after the removal of the stress; such a residual strain is known as the "permanent set".

In some materials the time allowed for strain to develop or disappear will affect the result obtained for the elastic limit.

The strain produced within the elastic limit is the elastic strain; also the portion of the strain that disappears with the removal of the stress even when the latter exceeds the elastic limit is called so—in the zone of partial elasticity.

"Hooke's Law" (Robert Hooke, 1676) states that, within the elastic limit, the strain produced is proportional to the stress causing it.

[Robert Hooke, an English Physicist, worked with springs, and not with rods. In 1676, he announced an anagram–"Ceiiinosssttuv", which in Latin means "Ut tensio sic Vis" (the force varies as the stretch)].

The law refers to all kinds of stress, and is approximately true for many materials.

Since stress is proportional to load for a given area of cross-section and strain proportional to deformation for a given dimension. Hooke's law also means that, within the elastic limit, deformation produced is proportional to the load producing it.

Strain ∝ stress

$$\text{or} \quad \frac{\sigma}{\varepsilon} = \text{a constant} \qquad \ldots(\text{Eq. 21.6})$$

More will be seen about this constant in the next section.

21.5 ELASTIC MODULI

The constant of proportionality in Hooke's law is called the 'Elastic Modulus' or the stress per unit strain. Three elastic moduli are relevant, the values being dependent on the nature of the material.

21.5.1 Modulus of Elasticity

When a body is subjected to simple tension or compression, the constant of proportionality in Hooke's law is known as the "modulus of Elasticity", it is also called Youngs modulus (after Young, who determined it), and "Direct or Stretch Modulus" (*E*).

It is given by,

$$E = \frac{\sigma}{\varepsilon} = \frac{\sigma_t}{\varepsilon_t} = \frac{\sigma_c}{\varepsilon_c}$$

For most materials *E* is practically the same in tension and in compression.

Since strain is a dimensionless quantity, the units of *E* are the same as those for stress.

21.5.2 Modulus of Rigidity

When a body is subjected to pure shear, the constant of proportionality between shear stress and shear strain is called the "Modulus of Rigidity" or the "Rigidity Modulus'. It is also called the 'Modulus of transverse elasticity' or 'Shear Modulus' (*N*).

It is given by

$$N = \frac{q}{\phi} \qquad \qquad ...(\text{Eq. 21.7})$$

The letter symbols *G* and *C* are also used to denote it.

21.5.3 Bulk Modulus

This is the modulus defined under hydrostatic stress conditions; when a body is subjected to equal direct stresses in three mutually perpendicular directions, the ratio of the direct stress to the volumetric strain (by volumetric strain is meant the ratio of change in volume to the original volume) is known as the 'Bulk Modulus'.

For example, a body immersed in a liquid will be subjected to equal pressure (or compressive stress) in all directions according to Pascal's law; the idealisation of this is the application of equal compressive stresses in three mutually perpendicular directions, as shown in Fig. 21.4. The volumetric strain in this case is negative as there will be slight reduction in volume.

Thus the bulk modulus (*K*) is given by,

$$K = \frac{\sigma}{\varepsilon_v} \qquad \qquad ...(\text{Eq. 21.8})$$

σ = direct stress acting in the three mutually perpendicular directions.

ε_V = Volumetric strain = $\frac{\Delta V}{V}$.

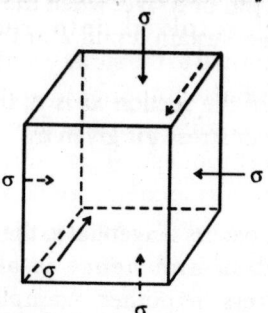

Fig. 21.4: Hydrostatic Stress Condition

Let us consider a cube of side *l*; let the contraction in the side be a small quantity δl.

Change in volume,

$$V = l^3 - (l - \delta l)^3$$

$$= 3l^2 \cdot \delta l - 3l(\delta l)^2 + (\delta l)^3$$

$$\approx 3l^2 \cdot \delta l \quad (\text{neglecting higher powers } \delta l)$$

$\therefore$ Volumetric strain,

$$\varepsilon_V = \frac{\Delta V}{V} = \frac{3l^2 \cdot \delta l}{l^3} = \frac{3 \cdot \delta l}{l} = 3 \cdot \varepsilon \quad ...(\text{Eq. 21.9})$$

ε being the linear strain.

Thus the volumetric strain is *three times* the linear strain, in the case of a cube.

For a rectangular parallelopiped of dimensions *x*, *y* and *z* and if the strains in the respective directions are ε_x, ε_y and ε_z,

Change in volume,

$$\Delta V = \{x(1 + \varepsilon_x) \cdot y(1 + \varepsilon_y) \cdot z(1 + \varepsilon_z)\} - xyz$$

$$= xyz(\varepsilon_x + \varepsilon_y + \varepsilon_z)$$

$\therefore$ Volumetric strain,

$$\varepsilon_V = \frac{\Delta V}{V} = \frac{xyz(\varepsilon_x + \varepsilon_y + \varepsilon_z)}{xyz}$$

$$= \varepsilon_x + \varepsilon_y + \varepsilon_z \qquad ...(\text{Eq. 21.10})$$

Thus the volumetric strain is also the *sum of the linear strains* in the three principal directions.

21.6 STRESSES ON OBLIQUE SECTION OF A BODY UNDER UNIAXIAL STRESS

When a bar is subjected to uniaxial stress any oblique section of the bar will be subjected in general to a stress which is neither normal nor tangential to that section. Such a stress may be conveniently resolved into rectangular components, normal to the surface and tangential to it. The normal components are tensile or compressive according to their directions, and the tangential components are shear stresses.

The method of resolution can be understood from the following simple example of a bar under uniaxial tension (Fig. 21.5).

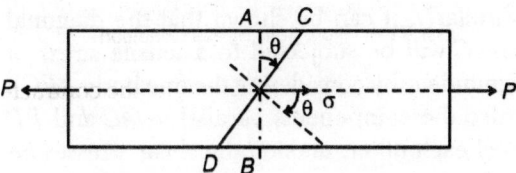

Fig. 21.5: A Bar under Uniaxial Tension

Let a bar of cross-sectional area A be subjected to an axial pull P. The intensity of tensile stress, σ, is given by,

$$\sigma = \frac{P}{A} \text{ (normal to the section } AB) \text{ in the}$$

axial direction

Let the normal and shear stresses on the section CD, which makes an angle θ with the plane of the section AB.

Resolving the force P normal to CD, the component is

$$P_n = P \cdot \cos\theta$$

Area of the oblique section $CD = A \cdot \sec\theta$

∴ The normal component, σ_n, is given by,

$$\sigma_n = \frac{P\cos\theta}{A\sec\theta} = \sigma \cdot \cos^2\theta \qquad \text{...(Eq. 21.11)}$$

(tensile)

Obviously σ_n is zero for $\theta = \pi/2$ or on the horizontal section.

Resolving parallel to CD, the tangential component of the force P is

$$P_i = P \cdot \sin\theta$$

∴ The tangential or shear stress on the oblique section CD is given by,

$$q = \frac{P\sin\theta}{A\sec\theta} = \sigma \cdot \sin\theta \cdot \cos\theta = (1/2)\sigma\sin2\theta$$

$$\text{...(Eq. 21.12)}$$

Obviously P_t reaches a maximum value of $(1/2)$ σ when $\theta = 45°$; that is to say, all sections inclined at 45° to AB as also to the direction of the axis are subject to the maximum shear stress.

The analysis for uniaxial compression is similar except that σ_n is compressive. It can be under-stood that if a material is such that its shear strength is less than half of its tensile strength, then the material will fail in shear when subjected to a tensile force.

21.7 STATE OF SIMPLE SHEAR—COMPLEMENTARY SHEAR STRESSES

A shear stress in a given plane will be automatically accompanied by a balancing shear stress of equal intensity in a plane perpendicular to the former.

This balancing shear stress is known the 'Complementary shear stress'.

Let us consider an infinitesimal rectangular block $ABCD$ under a shear stress of intensity q (Fig. 21.6).

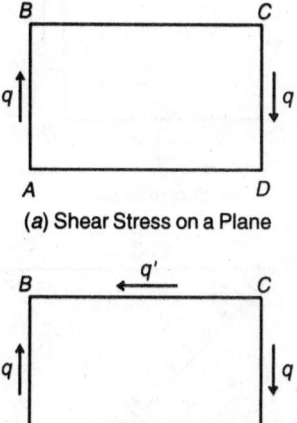

(a) Shear Stress on a Plane

(b) Complementary Shear Stress

Fig. 21.6: Principle of Complementary Shear Stresses

This will cause equal and opposite tangential forces on faces AB and CD, constituting a couple. Hence, for maintaining static equilibrium of the element, an equal and opposite couple must be induced by tangential forces on faces AD and BC as shown in (b). Let the shear stress on these planes be q' and let the thickness of the element perpendicular to the plane of the figure be unity.

For equilibrium, the moments of the couples must be equal,

$$(AB) \cdot 1 \cdot q \, (BC) = (BC) \cdot 1 \cdot q' \, (AB)$$

and this leads to $q = q'$.

Obviously, the intensities of shear stresses on planes at right angles to each other are *equal*; this is true irrespective of the normal stresses which may act on these planes.

The state of stress shown in (b) is called 'Simple Shear', and the stresses are said to be 'Complementary' to each other.

21.7.1 Stresses on Diagonal Planes in Simple Shear

Let us consider a square block *ABCD* in a state of simple shear, q, as shown in Fig. 21.7.

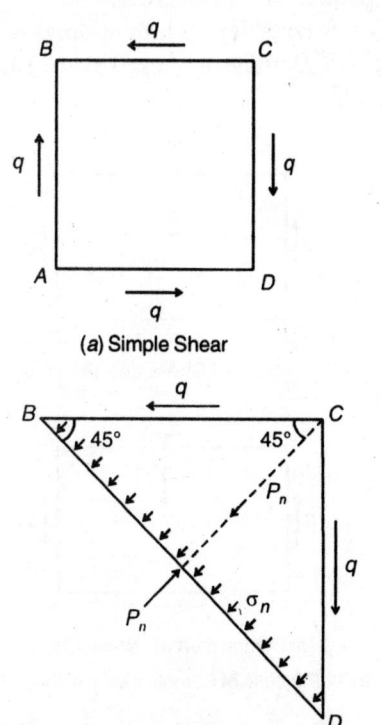

(a) Simple Shear

(b) Normal Stress on Diagonal Plane *BD*

Fig. 21.7: Stresses on Diagonal Planes in Simple Shear

Let the side of the square be a and the thickness of the block be unity perpendicular to the plane of the figure.

Force on face $BC = P_1 = q \cdot a \cdot 1$

Force on face $CD = P_2 = q \cdot a \cdot 1$

Resultant P_n of P_1 and P_2 will be perpendicular to the diagonal plane *BD* and is $P_1 \cos 45° + P_2 \cos 45° = 2P_1 \cos 45° = \sqrt{2} \cdot q \cdot a$

This causes a compressive stress σ_n on the diagonal face *BD*.

$$\sigma_n = \frac{P_n}{\text{Area of face } BD} = \frac{\sqrt{2} \cdot q \cdot a}{\sqrt{2} \cdot q \cdot 1} = q \text{ itself}$$

Similarly, it can be shown that the diagonal face *AC* will be subjected to a tensile stress of magnitude q (by considering the free body *ABC*).

Also the components parallel to *AC* and *BD* cancel each other, making the shear stresses on these faces zero.

Hence a state of pure shear produces pure tensile and compressive stresses across diagonal planes and the intensities of these are numerically equal to the shear stress.

21.8 POISSON'S RATIO

A direct stress produces a strain in its own direction and an opposite kind of strain in a direction perpendicular to its own. Thus, when a bar is subjected to uniaxial pull along its axis, its length increases while its lateral dimensions decrease. Similarly, if a strut is subjected to an axial compression, the length decreases but its lateral dimensions tend to increase.

A strain in the direction of the applied force is known as 'longitudinal' strain (or 'primary' strain), while those in a perpendicular direction to that of the applied force are called 'lateral strains' (or 'secondary' strains).

Within the elastic limit, the ratio of the lateral strain to the longitudinal strain is a constant for any material; it is known as the 'Poisson's Ratio' after Poisson who first defined it.

$$\text{Poisson's ratio, } \mu = \frac{\text{Lateral strain}}{\text{Longitudinal strain}}$$

...(Eq. 21.13)

Sometimes the letter symbols ν or σ are also used to designated it.

It is also designated by $1/m$.

Experiments showed that Poisson's ratio ranges from 0 to 0.5; it is in the range of 0.25 to 0.33 for most metals, and is the same in tension and compression.

Using the concept of Poisson's ratio the changes in dimensions of a bar of known dimensions may be calculated when it is subjected to uniaxial stress.

21.9 RELATION BETWEEN ELASTIC MODULI

The elastic moduli and the Poisson's ratio are interrelated; by eliminating Poisson's ratio, it is possible to establish a relationship between the three elastic moduli.

The following derivation consists of five steps.

(i) Linear Strain of a Diagonal in Simple Shear

Assuming one of the faces to be fixed, we can arrive at the linear strain of a diagonal of a square in simple shear (Fig. 21.8).

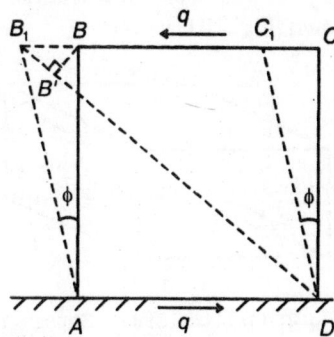

Fig. 21.8: Linear Strain of a Diagonal

The square *ABCD* will get deformed as shown into AB_1C_1D. Let ϕ be the shear strain accompanying the state of simple shear q.

Linear strain of diagonal *BD*,

$$(\varepsilon_d) = \frac{(B_1D - BD)}{BD} \approx \frac{B_1B'}{BD} \approx \frac{BB_1 \cos 45°}{BD} \approx$$

$$\left(\frac{BB_1}{\sqrt{2}}\right)\left(\frac{1}{\sqrt{2} \cdot AB}\right) = \frac{1}{2}\left(\frac{BB_1}{AB}\right) \approx \frac{1}{2}\phi,$$

Since ϕ is small

$$\therefore \quad \varepsilon_d = \frac{1}{2}\phi = \frac{1}{2} \cdot \frac{q}{N} \qquad \ldots\text{(Eq. 21.14)}$$

(ii) Linear Strain of Diagonal from the Stress on Diagonal Planes in Simple Shear

In Section 21.7, it has already been seen that a state of simple shear produces direct tensile stress along one diagonal (*BD* in this case), and direct compressive stress along the other (*AC* in this case), each being numerically equal to the shear stress q.

Hence, the tensile strain in the direction of *BD* due to the tensile stress q along $BD = (q/E)$; and, the tensile strain in the direction of *BD* due to the compressive stress q along $AC = (1/m) \cdot (q/E)$.

∴ Resultant tensile strain along *BD*,

$$(\varepsilon_d) = \frac{q}{E} + \frac{1}{m} \cdot \frac{q}{E}$$

$$= \frac{q}{E}\left(1 + \frac{1}{m}\right) \qquad \ldots\text{(Eq. 21.15)}$$

(iii) Relation between E, N and 1/m

Equating the two expressions for ε_d (Eqs. 21.14 and 21.15),

$$\frac{q}{2N} = \frac{q}{E}\left(1 + \frac{1}{m}\right)$$

or $\quad E = 2N\left(1 + \frac{1}{m}\right) \qquad \ldots\text{(Eq. 21.16)}$

(iv) Relation between E, K and 1/m

Now let us consider the hydrostatic condition with the compressive stress being σ in the three mutually perpendicular directions (Fig. 21.9).

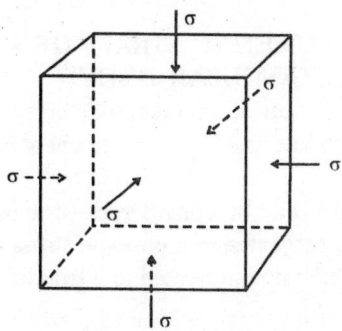

Fig. 21.9: Hydrostatic Condition

Linear strain in any one direction

$$= \frac{\sigma}{E} - \frac{\sigma}{mE} - \frac{\sigma}{mE}$$

$$= \frac{\sigma}{E}\left(1 - \frac{2}{m}\right) \qquad \text{(Compressive)}$$

Volumetric strain $= 3 \times$ (linear strain)

$$= \frac{3\sigma}{E}\left(1 - \frac{2}{m}\right)$$

But, by definition, volumetric strain $=$

here K is bulk modulus.

Equating,

$$\frac{\sigma}{K} = \frac{3\sigma}{E}\left(1 - \frac{2}{m}\right)$$

$$\therefore \qquad E = 3K\left(1 - \frac{2}{m}\right) \qquad \qquad \text{...(Eq. 21.17)}$$

(v) *Relation between the Elastic Moduli*

Now the Poisson's ratio $1/m$ can be eliminated from the equations 21.16 and 21.17 to get a relation between E, N and K. This may be done in more than one way.

$$E = 2N\left(1 + \frac{1}{m}\right) = 3K\left(1 - \frac{2}{m}\right)$$

From this,

$$\frac{1}{m} = \frac{3K - 2N}{6K + 2N} \qquad \qquad \text{...(Eq. 21.18)}$$

Substituting this in Eq. 21.16, and simplifying we obtain,

$$E = \frac{9KN}{(3K + N)} \qquad \qquad \text{...(Eq. 21.19)}$$

21.10 VOLUMETRIC STRAIN OF A RECTANGULAR BODY

Volumetric strain of a rectangular body may be determined making use of the concept of Poisson's ratio.

Two cases may be considered—one in which the body is subjected to a uniaxial stress and the other in which it is subjected to a triaxial state of stress.

21.10.1 Body Subjected to a Uniaxial Stress

Let a rectangular prismatic body of dimensions $l \times b \times h$ be subjected to a uniaxial stress—say, in the direction of l (Fig. 21.10).

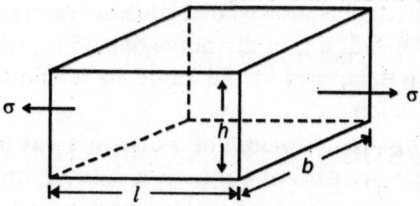

Fig. 21.10: Uniaxial Stress

Strain in the direction of $l = \dfrac{\sigma}{E}$ (tensile)

Strain in the other two directions $= \dfrac{\sigma}{mE}$

(Compressive)

Volumetric strain,

$$\varepsilon_V = \frac{\sigma}{E} - \frac{\sigma}{mE} - \frac{\sigma}{mE} = \frac{\sigma}{E}\left(1 - \frac{2}{m}\right) \quad \text{...(Eq. 21.20)}$$

The changes in the dimensions as well as in the volume may be easily determined.

21.10.2 Body Subjected to a Triaxial State of Stress

Let the body be subjected to a triaxial state of stress as shown (Fig. 21.11).

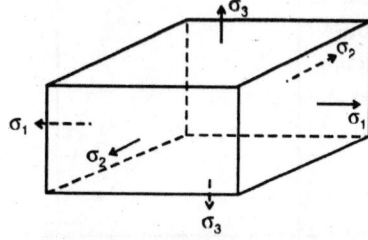

Fig. 21.11: Triaxial State of Stress

$\varepsilon_1 =$ Strain in the direction of σ_1

$$= \frac{\sigma_1}{E} - \frac{\sigma_2}{mE} - \frac{\sigma_3}{mE}$$

$\varepsilon_2 =$ Strain in the direction of σ_2

$$= \frac{\sigma_2}{E} - \frac{\sigma_3}{mE} - \frac{\sigma_1}{mE}$$

$\varepsilon_3 =$ Strain in the direction of σ_3

$$= \frac{\sigma_3}{E} - \frac{\sigma_1}{mE} - \frac{\sigma_2}{mE}$$

Volumetric strain,

$$\varepsilon_V = \varepsilon_1 + \varepsilon_2 + \varepsilon_3 = \frac{1}{E}(\sigma_1 + \sigma_2 + \sigma_3)$$

$$- \frac{2}{mE} \times (\sigma_1 + \sigma_2 + \sigma_3)$$

or $\quad \varepsilon_V = \dfrac{(\sigma_1 + \sigma_2 + \sigma_3)}{E}\left(1 - \dfrac{2}{m}\right) \quad …(\text{Eq. } 21.21)$

21.11 STRESS-STRAIN CURVE FOR MILD STEEL CHARACTERISTICS

Certain properties of metals such as elasticity, ductility, malleability, brittleness, and plasticity are of great importance from an engineering stand-point.

'Ductility' is the property of the material which allows it to be drawn out into a wire by tension. During ductile extension, the material generally shows a certain degree of elasticity, together with considerable plasticity.

"Brittleness" is lack of ductility.

"Malleability" is similar to ductility; it is the property by virtue of which a material can be beaten or rolled into plates.

No material can possess all these properties— for example, mild steel is elastic, copper ductile, wrought iron malleable, cast iron brittle, and lead plastic.

A simple tensile test on a bar of a ductile material like mild steel, carried to destruction, will yield valuable information regarding its strength, its elasticity, and its ductility. Some of these properties are of premier importance in design.

In general, the stress-strain diagram for a ductile material in tension will be as shown in Fig. 21.12: (Stress on *y*-axis and strain on the *x*-axis).

The tensile strain is proportional to the corresponding tensile stress upto a certain point, called the limit of proportionality or 'Proportional Limit' (*LP*); obviously, this is the range of validity of Hooke's law. However, the property of elasticity is exhibited upto a stress a *little higher* than the proportional limit, this value of stress being called the 'Elastic Limit' (*EL*; the portion of the stress-strain diagram between the proportional limit and the elastic limit is slightly curved with convexity upwards. Strains upto this are usually very small and are measured with the aid of extensometers on a small "gauge length", (50 mm to 150 mm). (In the case of a few materials, the elastic limit may coincide with the proportional limit). Any strain that is left after the removal of stress is called the permanent set; it can occur only beyond elastic limit.

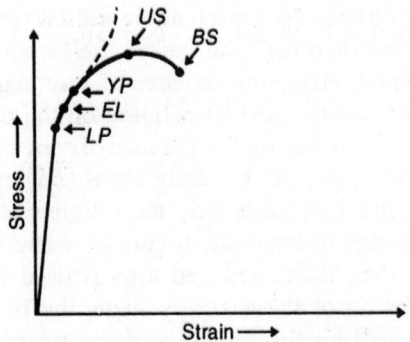

(a) Stress-Strain Curve for Mild Steel in Tension

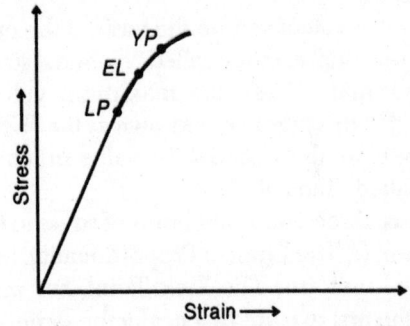

(b) Diagram Plotted to a Larger Scale upto Yield Point

Fig. 21.12: Stress-Strain Diagram for a Ductile Material in Tension

The portion of the diagram upto the elastic limit is shown separately with strain plotted to a larger scale, for the purpose of clarity. The range of stress over which the material shows elasticity is called the 'Elastic range'; 'Plastic range' is also similarly defined.

At a stress slightly greater than the elastic limit, the strain starts to increase more rapidly than the stress; that is, even with a small increase in stress the material "yields" or flows causing very rapid increase in strain. The stress at which this sudden stretch occurs is known as the "Yield Point" of the material. (Some materials exhibits two such points—in this case, these are known as the 'Lower Yield Point' and 'Upper Yield Point'). Soon after, the material enters the plastic stage. If the load is further increased, apparently the maximum value of stress will be reached (ultimate stress *US*); now the material suddenly stretches locally, forming a "neck" or a "waist". In view of this local reduction in area of section, the load required to break the bar at the waist is considerably less than the load before the formation of the waist. Also, the stresses at any stage are conventionally calculated based on the original area of cross–section for the sake of convenience. Consequently, the breaking stress (*BS*) appears to be much smaller than the ultimate stress— some what anomalous. It can be easily understood that, if the reduced area is used for the calculation of stress at any stage, the breaking stress will be the largest and the stress continuously increases as shown by the dotted line in the figure.

Stresses calculated on the basis of the original cross-sectional area are called "Nominal stresses". The ultimate stress, the maximum value calculated in this manner, is known as the "Ultimate Strength" of the material; its value in tension is also called "Tenacity".

Thus, three important limits of stress in tensile tests are; (*i*) The Limit of Proportionality, (*ii*) The Elastic Limit, (*iii*) The Yield Point. For wrought iron, the first two are practically the same.

Mile steel or "Structural Steel" shows two distinct limits at which yield occurs—the upper and the lower yield points, as shown in Fig. 21.13.

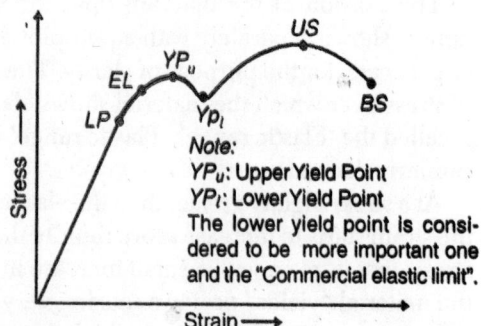

Fig. 21.13: Lower and Upper Yield Points for Mild Steel

21.11.1 Index of Ductility—Percentage Elongation and Percentage Reduction in Area

Before the occurence of local yielding, the elongation of the bar is distributed practically uniformly along the bar; however, after this there will be more elongation on either side of the waist than in the other portion.

'Percentage Elongation' is one of the indices used to measure the ductility of a material; it is the more common one.

$$\% \text{ Elongation} = \left(\frac{l'-l}{l}\right) \times 100 \quad \text{...(Eq. 21.22)}$$

Here l' = final length of the bar at failure.

l = original length of the bar.

This is to some extent dependent upon the initial length of the bar owing to the phenomenon of local yield; the value decreases with increase in length, the effect of the local yield becoming smaller with larger initial lengths of the bar. Thus it is imperative that the length be stated whenever the percentage elongation is given (Common gauge length is 200 mm). There is a further complication that the zone of local yielding is more for thicker sections.

For bars of different cross-sectional areas, Prof. Unwin of U.K. has shown that comparable results may be obtained if the bars are geometrically similar.

The total extension is made up of a uniformly distributed extension and a local extension, the former being proportional to the length of the bar, and the latter almost independent of the length.

If e is the total extension, e_l is the local extension, l the gauge length,

$$e = e_l + e_1 l$$

where $e_1 l$ is the part of extension proportional to the gauge length.

The local extension, e_l, is found to be nearly proportional to the square root of the cross-sectional area of the bar.

$$e_l = e_2 \sqrt{A}$$

e_2 being the constant of proportionality.

$$\therefore \qquad e = e_2 \sqrt{A} + e_1 \cdot l \qquad \text{...(Eq. 21.23)}$$

Thus if e_1 and e_2 are established, a rough idea can be obtained of the elongation of another bar of different dimensions, but of the same material.

'Percentage Reduction in Area of Cross-Section' is another measure of ductility.

It is defined as follows:

$$\% \text{ Reduction in Area} = \left(\frac{A - A'}{A}\right) \times 100\%$$

$$\text{...(Eq. 21.24)}$$

A being the original area, A' the final area of section. For a prismatic bar, nearly uniform contraction of area occurs throughout the length. If the volume of the gauge length of the bar remains constant, as is very nearly true, the percentage reduction in area reckoned on the original area will be same as the percentage elongation reckoned on the final length at the time of measurement.

If l and l' are the initial and final lengths,

$$l \cdot A = l' \cdot A' \quad \text{or} \quad \frac{l}{l'} = \frac{A'}{A}$$

$$\therefore \quad \frac{l}{l'} - 1 = \frac{A'}{A} - 1$$

$$\frac{l - l'}{l'} = \frac{A' - A}{A} \quad \text{or} \quad \left(\frac{l' - l}{l'}\right) = \left(\frac{A - A'}{A}\right)$$

as stated above.

When the metal "flows" near the neck, the extension *at fracture* will be greater than this amount.

21.12 WORKING STRESS AND FACTOR OF SAFETY

In the design of any machine element or a structural component, the maximum stress allowed to be reached under working conditions is always much less than the ultimate strength of the material. This is to ensure that no permanent set occurs and due allowance is made for inaccurate estimation of loads, bad workmanship, and so on. Safety should be ensured consistent with economy.

Such a limit of stress upto which the stress is allowed to be reached is called the "Working Stress" (WS). It is also called the "permissible" stress or "Allowable" stress. This value is invariably less than the elastic limit to avoid functional failure.

The ratio of the ultimate strength to the working stress is known as the "Factor of Safety" (Or is it a "factor of ignorance"!)

This factor varies with the material, its reliability, and the nature of loading, among other factors. For example, a repeated stress requires a larger factor of safety than a constant stress.

Impact and dynamic stresses will require a much larger factor safety,

$$\text{Factor of safety, } \eta = \frac{\text{Ultimate Stress (US)}}{\text{Working Stress (WS)}}$$

$$\text{...(Eq. 21.25)}$$

21.13 ELONGATION OF A BAR

The elongation of a bar under uniaxial stress may be determined from fundamentals for any given case.

The simplest case, of course, is that of a prismatic bar under uniaxial tension.

Other cases of a slightly different and more complex nature may also be solved making use of integration and other simple mathematical tools. A few of them are dealt with in the following subsections.

21.13.1 Elongation of a Prismatic Bar under Uniaxial Tension

A prismatic bar under uniaxial tension is shown in Fig. 20.14.

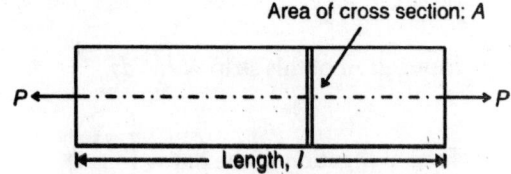

Area of cross section: A

Fig. 21.14: Prismatic Bar under Uniaxial Tension

The tensile stress in every section of the bar is given by,

$$\sigma = \frac{P}{A}$$

The strain is given by,

$$\varepsilon = \frac{\sigma}{E} = \frac{P}{AE}$$

The elongation δl is given by,

$$\delta l = l \cdot \varepsilon = \frac{Pl}{AE} \quad \text{...(Eq. 21.26)}$$

21.13.2 Elongation of a Prismatic Bar Due to its Self-Weight

When a bar of uniform cross-ssection A and length l is suspended as shown in Fig. 21.15, a small elongation occurs due to its self-weight. This may be determined as follows:

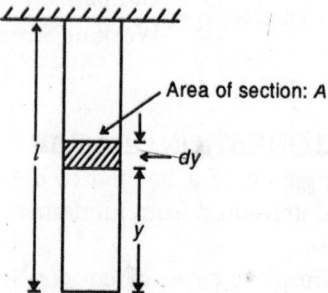

Fig. 21.15: Prismatic Bar Suspended from the Top

Let the unit weight of the material of the bar be γ. Let us consider a thin strip of the bar of thickness dy at a distance y from the bottom.

The strip will be subjected to a pull due to the weight of the bar of length y below it.

Force $= \gamma \cdot A \cdot y$ (pull)

Tensile stress $\sigma = \dfrac{\gamma A y}{A} = \gamma \cdot y$

Strain, $\varepsilon = \dfrac{\gamma y}{E}$

Elongation of this strip $= \dfrac{\gamma y}{E} \cdot dy$

Total elongation of the bar $= \displaystyle\int_0^l \dfrac{\gamma y}{E} \, dy$

$$= \dfrac{\gamma}{E} \int_0^l y \, dy = \dfrac{\gamma}{2E} (y^2)_0^l = \dfrac{\gamma l^2}{2E}$$

$\therefore \qquad \delta l = \dfrac{\gamma l^2}{2E}$ ...(Eq. 21.27)

(due to self-weight)

This may also be written in terms of the weight W of the bar.

$$\delta l = \dfrac{(\gamma A l) \cdot l}{2AE} = \dfrac{Wl}{2AE} \qquad \text{...(Eq. 21.28)}$$

Thus, the elongation of a bar due to its self-weight is half of that due to the action of a uniform pull equal to its weight.

21.13.3 Elongation of a Bar of Varying Section
If the cross-section of a bar is different for different portions of its length, the deformation under an axial load may be determined for each portion separately and added algebrically to obtain the total deformation for the bar.

This is demonstrated for a bar of three sections as shown in Fig. 21.16.

Fig. 21.16: Bar of Varying Section

Let the lengths and areas of section of the three portions of the bar be l_1, A_1, l_2, A_2, and l_3, A_3, respectively, and let the pull be P.

$$\sigma_1 = \dfrac{P}{A_1}; \ \sigma_2 = \dfrac{P}{A_2}; \ \sigma_3 = \dfrac{P}{A_3}$$

$$\varepsilon_1 = \dfrac{\sigma_1}{E}; \ \varepsilon_2 = \dfrac{\sigma_2}{E}; \ \varepsilon_3 = \dfrac{\sigma_3}{E}$$

$$\delta l_1 = \varepsilon_1 l_1; \ \delta l_2 = \varepsilon_2 l_2; \ \delta l_3 = \varepsilon_3 l_3$$

The total elongation of the bar,

$$\delta l = \delta l_1 + \delta l_2 + \delta l_3$$

$$\delta l = \dfrac{P}{E} \left[\dfrac{l_1}{A_1} + \dfrac{l_2}{A_2} + \dfrac{l_3}{A_3} \right] \qquad \text{...(Eq. 21.29)}$$

21.13.4 Elongation of a Bar of Tapering Section
Let us consider a bar of uniformly varying section—diameter d_1 at one end and d_2 at the other $(d_2 > d_1)$, subjected to an axial pull P as shown in Fig. 21.17.

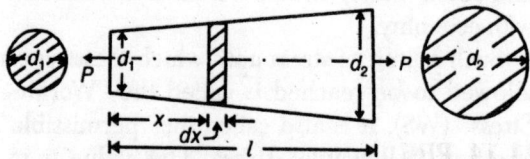

Fig. 21.17: Bar of Tapering Section

Let an elementary strip of thickness dx at a distance x from the smaller side be considered. Since the thickness is infinitesimally small, the portion dx may be taken to be prismatic.

The diameter for this strip

$$= d_1 + \dfrac{(d_2 - d_1)}{l} \cdot x$$

$$= (d_1 + kx),$$

where $k = \left(\dfrac{d_2 - d_1}{l} \right)$

$$\text{Stress} = \dfrac{P}{\left(\dfrac{\pi}{4} (d_1 + kx)^2 \right)} = \dfrac{4P}{\pi (d_1 + kx)^2}$$

$$\text{Strain} = \dfrac{4P}{\pi E (d_1 + kx)^2}$$

Elongation for the length, dx

$$= \dfrac{4P}{\pi E (d_1 + kx)^2} \cdot dx$$

Total elongation of the bar

$$= \dfrac{4P}{\pi E} \int_0^l \dfrac{dx}{(d_1 + kx)^2}$$

$$= -\dfrac{4P}{\pi k E} \left[\dfrac{1}{d_1 + kx} \right]_0^l$$

$$= \dfrac{4P}{\pi k E} \left[\dfrac{1}{d_1} - \dfrac{1}{(d_1 + kl)} \right]$$

$$= \dfrac{4P}{\pi k E} \left[\dfrac{1}{d_1} - \dfrac{1}{d_2} \right]$$

or $\quad \delta l = \dfrac{4Pl}{\pi d_1 d_2 E}$ ...(Eq. 21.30)

If the bar is prismatic, $d_1 = d_2 = d$, and

$$\delta l = \dfrac{4Pl}{\pi d^2 E}$$

$$= \dfrac{Pl}{AE}, \text{ as already known.}$$

21.14 PRINCIPLE OF SUPERPOSITION

For an elastic body acted upon by several forces, the resulting internal stresses and strains are equal to the algebraic sum of the values due to each force acting independently, when the stresses are within the elastic limit for the material.

21.15 STRESSES IN COMPOSITE BARS

A tension or a compression member which consists of more than one bar or tube of different materials is called a 'composite bar'; some times it is also called a 'compound bar'. The bars are so connected as to ensure composite action such that the materials behave like one.

A simple example of a composite section in common practice is one of reinforced concrete, consisting of steel reinforcement and cement concrete.

Let us consider a composite bar of two different materials with moduli of elasticity E_1 and E_2. Let the areas of section be A_1 and A_2. Let the two bars be kept *parallel* and be subjected to an axial load P. Let the stresses induced be σ_1 and σ_2 in the two materials and δl be the total deformation in the length l of the bar.

The equations governing the composite action are:

Strain is the same,

$$\varepsilon = \dfrac{\delta l}{l} = \dfrac{\sigma_1}{E_1} = \dfrac{\sigma_2}{E_2} \qquad \text{...(Eq. 21.31)}$$

Also, the load is shared,

$$P = P_1 + P_2 = \sigma_1 A_1 + \sigma_2 A_2 \quad \text{...(Eq. 21.32)}$$

Equation 21.31 may also be written as:

$$\dfrac{\sigma_1}{\sigma_2} = \dfrac{E_1}{E_2}, \text{ called the 'Modular Ratio'.}$$

If the composite bar is formed by two or more bars of different materials *in series*, the load on each will be the same and the deformation of the composite bar will be the algebraic sum of those of the different bars of which it is made.

$$\sigma_1 = \dfrac{P}{A_1} \qquad \text{...(Eq. 21.33)}$$

$$\sigma_2 = \dfrac{P}{A_2} \qquad \text{...(Eq. 21.34)}$$

$$\varepsilon_1 = \dfrac{\sigma_1}{E_1}$$

$$\varepsilon_2 = \dfrac{\sigma_2}{E_2}$$

$$\delta l = \delta l_1 + \delta l_2 = \varepsilon_1 l_1 + \varepsilon_2 l_2 = \dfrac{\sigma_1}{E_1} l_1 + \dfrac{\sigma_2}{E_2} l_2$$

$$\text{...(Eq. 21.35)}$$

These principles are illustrated in a few of the problems given under the heading Illustrative Problems.

21.16 TEMPERATURE STRESSES

Temperature changes in a structural element or a machine component give rise to stress just as applied loads do. When the material is allowed to undergo natural expansion or contraction as a result of a rise or fall in temperature, no resistance is set up and hence there will be no stresses and strains. Only when such a natural expansion or contraction due to a temperature change is *prevented*, resistance will be set up and stresses and strains will be produced.

21.16.1 Temperature Stresses in Simple Bars

Let us assume that a bar is heated up by a certain temperature above its normal temperature, the ends fixed up by rigid grips, and allowed to cool; the ends are thus prevented from coming back to their position. As a result of this, the bar will be under tension.

On the other hand, if the bar, gripped at the ends, is heated up and is constrained or prevented by the grips from expanding, the bar will be under compression.

The grips exert pushes on the bar in the former case and pulls in the latter.

Different materials expand or contract differently under temperature changes; this is characterised by a property called the "coefficient of linear expansion" (α), which is defined as the strain (or change in length per unit length) per unit or one degree of change of temperature. Its numerical value depends upon the scale of temperature used besides the material.

Let the length under normal temperature be l.

The modified length, l', if the bar were free, would be,

$$l' = l \pm l\alpha t \qquad \text{...(Eq. 21.36)}$$

(Positive sign for rise of temperature and negative sign for fall)

$$\text{Strain}, \varepsilon = \frac{l\alpha t}{l} = \alpha t \qquad \text{...(Eq. 21.37)}$$

$$\text{Stress induced}, \sigma = E \cdot \varepsilon = E \cdot \alpha \cdot t \quad \text{...(Eq. 21.38)}$$

The push or pull exerted on the grips, P, will be,

$$P = \sigma \cdot A = E \cdot A \cdot \alpha \cdot t \qquad \text{...(Eq. 21.39)}$$

If the grips yield by an amount δ, the change in length prevented is $(l\alpha t - \delta)$

$$\text{The strain}, \varepsilon = \left(\frac{l\alpha t - \delta}{l}\right) = \left(\alpha t - \frac{\delta}{l}\right)$$

$$\text{The stress}, \sigma = E \cdot \varepsilon = E\left(\alpha t - \frac{\delta}{l}\right) \quad \text{...(Eq. 21.40)}$$

21.16.2 Temperature Stresses in Composite Bars

The treatment for a simple bar of a single material is seen to be relatively simple. However, if a composite bar of different materials is subjected to a change of temperature, the materials tend to suffer different strains, and due to composite action, a resultant compromise strain is reached. This is the basis for the computation of the stresses and strains in such a case.

The principles are illustrated in the relevant problems in the following section.

ILLUSTRATIVE PROBLEMS

PROBLEM 21.1: A steel rod, 25 mm diameter and 5 m long, is subjected to an axial pull of 60 kN. If $E = 2 \times 10^5$ MN/m^2, determine the stress, strain, and elongation.

Solution:

Diameter, $d = 25$ mm

Length, $l = 5$ m $= 5000$ mm

Pull, $P = 60$ kN

$E = 2 \times 10^5$ MN/m$^2 = 2 \times 10^5$ N/mm^2

$$\text{Stress} \frac{P}{A} = \frac{60 \times 1000}{\left(\frac{\pi}{4} \times 25^2\right)}$$

$$= 122.23 \text{ N/mm}^2 \text{ (tensile)}$$

$$\text{Strain}, e = \frac{122.23}{2 \times 10^5} = 6.112 \times 10^{-4}$$

$$\text{Elongation}, \delta l = 5000 \times 6.112 \times 10^{-4}$$

$$= 3.056 \text{ mm}$$

PROBLEM 21.2: A short hollow cast iron cylinder of wall thickness of 10 mm is to carry a compressive load of 500 kN. Assuming the ultimate compressive strength of the material as 500 MN/m² and a factor of safety of 5, determine the size of the cross-section.

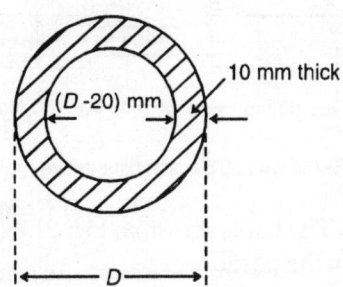

Fig. 21.18: Cross-Section of Hollow Cylinder
(Illustrative Problem 21.2)

Solution:

$$A = \frac{\pi}{4}\left[D^2 - (D-20)^2\right]$$

$$= \frac{\pi}{4} \times 20 \times (2D - 20) = 10\pi(D - 10) \text{ mm}^2$$

Working stress

$$= \frac{500}{5} = 100 \text{ MN/m}^2 \text{ (or N/mm}^2)$$

Working load $= 100 \times 10\pi(D - 10)$

$$= \pi(D - 10) \text{ kN}$$

This is equal to 500 kN

∴ $\pi(D - 10) = 500$

$D = 169.155$ mm, say 170 mm

So, the cylinder should be of 150 mm internal diameter and 170 mm external diameter.

PROBLEM 21.3: A rectangular base plate is fixed at each of its four corners by 20 mm diameter bolts and nuts with washers as shown in Fig. 21.19. The plate rests on washers of 22 mm internal diameter and 48 mm external diameter. Upper washers, 22 mm internal diameter and 44 mm external diameter are placed between the nut and the plate. If the base plate transmits a load of 25 kN to each of the bolts, calculate the stress in the lower washer before the nut is tightened. Also calculate the stress in lower and upper washers when the nuts are so tightened as to produce a tension of 5 kN in each bolt.

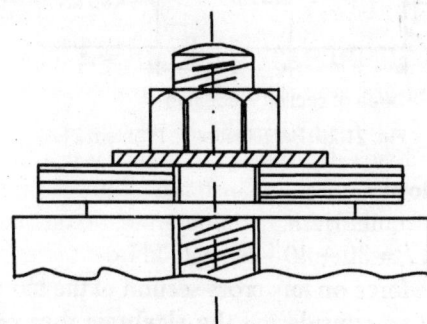

Fig. 21.19: Base Plate, Washers and Nut
(Illustrative Problem 21.3)

Solution: Prior to tightening the nut, only the lower washer will be stressed by the load transmitted from the base plate. After tightening the nut, the tension in the bolt will induce equal compression in the washers. The resultant stress in the lower washer can be found, as also the stress in the upper washer, which is only due to tightening of the nut.

Thus, the compressive stress in the lower washer due to load from the base plate

$$= \frac{P}{A} = \frac{25 \times 1000}{\frac{\pi}{4}(48^2 - 22^2)} = 17.5 \text{ N/mm}^2$$

Compressive stress in the lower washer due to tightening of the nut

$$= \frac{5 \times 1000}{\frac{\pi}{4}(48^2 - 22^2)} = 3.5 \text{ N/mm}^2$$

Total compressive stress in the lower washer
$$= 17.5 + 3.5 = 21 \text{ N/mm}^2$$

Compressive stress in the upper washer due to tightening of the nut

$$= \frac{5 \times 1000}{\frac{\pi}{4}(44^2 - 22^2)} = 4.4 \text{ N/mm}^2$$

PROBLEM 21.4: Determine the force P for equilibrium of the bar and find the change in the length of the bar shown.

Area of cross-section $= 800$ mm²
$E = 2 \times 10^5$ MPa

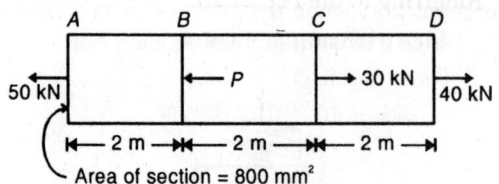

Fig. 21.20: Bar (Illustrative Problem 21.4)

Solution:

For equilibrium,

$$P = 30 + 40 - 50 = 20 \text{ kN} \leftarrow$$

The force on any cross-section of the bar may be got by considering the algebraic sum of the forces acting on to one side—the left or the right of the section. Using this principle, the free body diagrams of the three parts—*AB*, *BC* and *CD* are shown below:

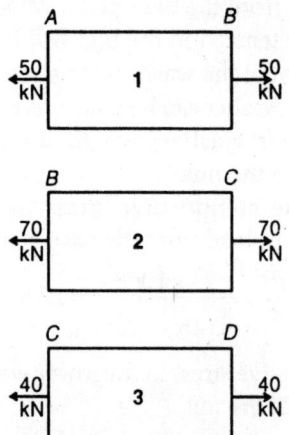

Fig. 21.21: Free Body Diagrams of the Three Parts (Illustrative Problem 21.4)

$$\delta l_1 = \frac{50 \times 1000 \times 2000}{800 \times 2 \times 10^5} = 0.625 \text{ mm}$$

$$\delta l_2 = \frac{70 \times 1000 \times 2000}{800 \times 2 \times 10^5} = 0.875 \text{ mm}$$

$$\delta l_3 = \frac{40 \times 1000 \times 2000}{800 \times 2 \times 10^5} = 0.500 \text{ mm}$$

Total elongation of the bar

$$= \Sigma \delta l = 0.625 + 0.875 + 0.500$$
$$= 2.00 \text{ mm}$$

PROBLEM 21.5: A bar 450 mm long is 50 mm square in section for the first 150 mm of its length,

25 mm diameter for the next 150 mm, and 50 mm diameter for the remaining 150 mm length. Determine the stress in each portion and the total elongation when a pull of 100 kN is applied. $E = 2 \times 10^5$ MPa.

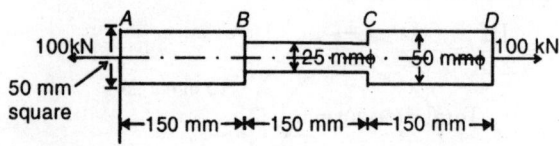

Fig. 21.22: Bar of Varying Section (Illustrative Problem 21.5)

Solution: The bar is shown in Fig. 21.22.

Stress in the portion,

$$AB = \frac{100 \times 1000}{50 \times 50} = 40 \text{ MPa} \quad \text{(Tensile)}$$

Stress in the portion,

$$BC = \frac{100 \times 1000}{\left(\frac{\pi}{4} \times 25^2\right)} = 203.7 \text{ MPa} \quad \text{(Tensile)}$$

Stress in the portion,

$$CD = \frac{100 \times 1000}{\left(\frac{\pi}{4} \times 50^2\right)} = 50.9 \text{ MPa} \quad \text{(Tensile)}$$

$$\text{Total elongation} = \frac{P}{E}\left(\frac{l_1}{A_1} + \frac{l_2}{A_2} + \frac{l_3}{A_3}\right)$$

$$= \left(\frac{100 \times 1000}{2 \times 10^5}\right)\left(\frac{150}{50 \times 50} + \frac{150 \times 4}{\pi \times 25^2} + \frac{150 \times 4}{\pi \times 50^2}\right) \text{mm}$$

$$= 75\left(\frac{1}{2500} + \frac{4}{625\pi} + \frac{4}{2500\pi}\right) \text{mm}$$

$$= 0.221 \text{ mm}$$

PROBLEM 21.6: A uniform rope, 300 metres long and 10 mm diameter, is hung vertically while hoisting a load of 2 kN. (*a*) Determine the elongation of the rope due to the combined effect of the load and the self-weight; (*b*) if the rope hangs under its own weight, determine the elongation of the top 150 m length. The unit weight of steel = 78. kN/m³. $E = 2 \times 10^8$ kN/m².

Solution:

(a) Elongation due to load

$$= \frac{Pl}{AE} = \frac{3 \times 300 \times 1000}{\frac{\pi}{4} \times (0.1)^2 \times 2 \times 10^8} \text{ mm} = 38.20 \text{ mm}$$

Elongation due to self-weight

$$= \frac{Wl}{2AE} = \frac{300 \times A \times 78 \times 300 \times 1000}{2 \times A \times 2 \times 10^8} \text{ mm} = 17.55\text{mm}$$

Total elongation $= (38.20 + 17.55)$ mm

$$= 55.75 \text{ mm}$$

(b) Elongation of the top 150 m length:

This will be due to self-weight of the top 150 m-length of rope plus that due to the weight of the bottom 150 m length taken to act as the force at the lower end of the top 150 m length.

$$\therefore \quad \text{Elongation} = \frac{Wl}{2AE} + \frac{Pl}{AE}$$

$$= \frac{150 \times A \times 78 \times 150 \times 1000}{2A \times 2 \times 10^8}$$

$$+ \frac{150 \times A \times 78 \times 150 \times 1000}{A \times 2 \times 10^8}$$

$$= (4.39 + 8.78) \text{ mm} = 13.17 \text{ mm}$$

PROBLEM 21.7: Determine the force required to punch a 10 mm diameter hole in a mild steel plate 10 mm thick, if the shear strength of mild steel is 360 MPa. Also calculate the compressive stress in the punch.

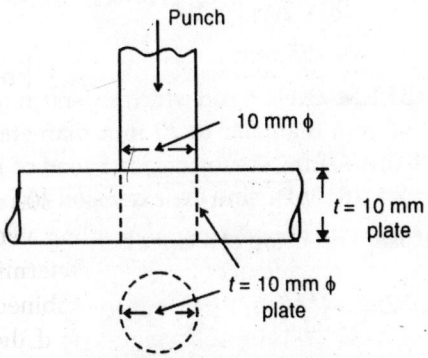

Fig. 21.23: Punching a Hole in a Plate
(Illustrative Problem 21.7)

Solution:

Referring to the Fig. 21.23,

Area of shearing $= \pi \cdot d \cdot t$

Force required $= \pi \cdot d \cdot t \cdot q$

$$= \pi \times 10 \times 10 \times 360 \text{ N}$$

$$= 36\pi \text{ kN}$$

$$= 113.1 \text{ kN}$$

Compressive stress in the punch

$$= \frac{113.1 \times 1000}{\left(\frac{\pi}{4} \times 12^2\right)} \text{ MPa} = 1000 \text{ MPa}$$

PROBLEM 21.8: Two parts of a tie-rod are connected by a bolt as shown. The tie-rod carries a pull of 50 kN. If the allowable shear stress is 90 MPa, determine the necessary diameter of the bolt required.

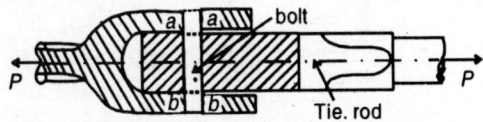

(Shear occurs on the cross-sections *aa* and *bb* of the bolt)

Fig. 21.24: Tie-rod in Tension (Illustrative Problem 21.8)

Solution:

Pull in the tie-rod, $P = 50$ kN

This is resisted by the two cross-sections *aa* and *bb* of the bolt. (The bolt is therefore said to be in double shear).

Shear stress

$$= \frac{P}{\left(2 \times \frac{\pi d^2}{4}\right)} = \frac{50 \times 1000}{\left(2 \times \frac{\pi}{4} d^2\right)}$$

This is to be equated to 90 MPa

$$\therefore \quad \frac{50000}{\left(2 \times \frac{\pi}{4} d^2\right)} = 90$$

Solving, $d = 18.8$ mm

Rounding off appropriately, a bolt of 20 mm diameter may be used to connect the two parts of the tie-rod.

PROBLEM 21.9: A tensile test was conducted on a mild steel bar. The following data were obtained from the test.

(*i*) Diameter of the bar = 30 mm
(*ii*) Gauge length = 200 mm
(*iii*) Load at elastic limit = 240 kN
(*iv*) Maximum load = 360 kN
(*v*) Extension at a load of 150 kN = 0.20 mm
(*vi*) Total extension = 60 mm
(*vii*) Diameter of the rod at failure = 22.5 mm

Determine:
(*a*) The Young's modulus;
(*b*) The stress at elastic limit;
(*c*) Percentage elongation; and
(*d*) Percent reduction in area.

Solution:

(*a*) Stress $= \dfrac{150 \times 1000}{\pi \times 15^2} = 212.2$ MPa

Strain $= \dfrac{\text{Extension}}{\text{Gauge length}} = \dfrac{0.20}{200} = 0.001$

$E = \dfrac{\text{Stress}}{\text{Strain}} = \dfrac{212.2}{0.001} = 2.122 \times 10^5$ MPa

(*b*) Stress at elastic limit

$= \dfrac{240 \times 1000}{\pi \times 15^2} = 339.55$ MPa

(*c*) Percentage elongation

$= \dfrac{60}{200} \times 100 = 30$

(*d*) Percent reduction in area

$= \dfrac{\{(30)^2 - (22.5)^2\}}{30^2} \times 100 = 43.75$

PROBLEM 21.10: A square bar of 20 mm side is held between two rigid plates and is loaded axially with a force of 150 kN as shown in Fig. 21.25. Determine the reactions at the ends *A* and *C* and the extension of the portion *AB*. $E = 2 \times 10^8$ kN/m².

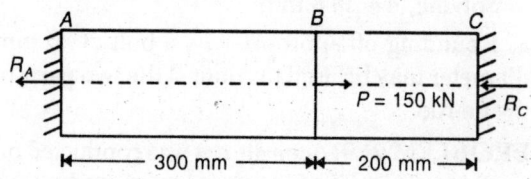

Fig. 21.25: Square Bar between Rigid Plates
(Illustrative Problem 21.10)

Solution: Both reactions R_A and R_B tend to oppose the applied force as shown. As the bar is held between rigid plates, the left portion will be subjected to tension and the right one to compression; further, the elongation of the portion *AB* shall be equal to the contraction of the portion *BC*.

For the equilibrium of the bar,

$$R_A + R_C = P = 150 \qquad \ldots(i)$$

The free bodies of *AB* and *BC* are shown below:

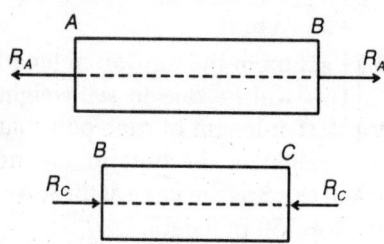

Fig. 21.26: Free Bodies of Portions *AB* and *BC*
(Illustrative Problem 21.10)

$$\delta l_{AB} = \frac{R_A \cdot l_{AB}}{A \cdot E}; \ \delta l_{BC} = \frac{R_C \cdot l_{AC}}{AE}$$

Equating the two, $300 R_A = 200 R_C$

or $R_C = 1.5 R_A$ $\qquad \ldots(ii)$

Solving eqs. (*i*) and (*ii*)

We get $R_A = 60$ kN and $R_C = 90$ kN

$$\delta_{AB} = \frac{R_A \cdot l_{AB}}{A \cdot E} = \frac{60 \times 300}{\left(\dfrac{20 \times 20}{1000 \times 1000}\right) \times 2 \times 10^8}$$

$$= \frac{60 \times 300 \times 10^6}{20 \times 20 \times 2 \times 10^8} \text{ mm}$$

$$= 0.225 \text{ mm}$$

PROBLEM 21.11: A rod which tapers uniformly from 40 mm diameter to 20 mm diameter in a length of 400 mm, is subjected to a load of 10 kN. If $E = 2 \times 10^5$ MPa, find the extension of the rod.

Solutoin: Extension of a taper bar

$$\delta L = \frac{4PL}{\pi E d_1 d_2}$$

where *L* is the length of the bar, tapering from a diameter d_1 to d_2.

$$\therefore \quad \delta L = \frac{4 \times 10 \times 1000 \times 400}{\pi \times 2 \times 10^5 \times 40 \times 20} \text{ mm}$$

$$= 0.032 \text{ mm}$$

PROBLEM 21.12: A bar of steel is of length l and uniform thickness t. The width of the bar varies uniformly from a at one end to b at the other. Find the extension of the rod under an axial pull P. The bar is shown in Fig. 21.27.

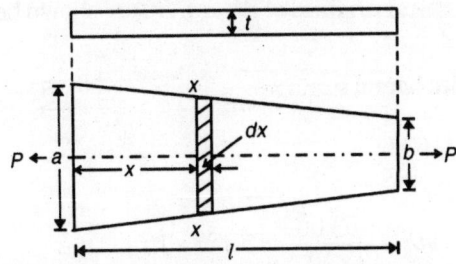

Fig. 21.27: Taper Bar of Uniform Thickness
(Illustrative Problem 21.12)

Solution: Let us consider a strip of elemental thickness dx at a distance x from the end of breadth a.

Width of this strip $= a - \dfrac{(a-b)x}{l}$

$$= (a - kx), \text{ say,}$$

where $k = \dfrac{(a-b)}{l}$

Area of the section at distance $x = t(a - kx)$

Stress on this section $= \dfrac{P}{t(a - kx)}$

Extension of the strip of elemental thickness,

$$dx = \frac{P}{tE(a - kx)} \cdot dx$$

Total extension of the taper bar,

$$\delta l = \frac{P}{tE} \int_0^l \frac{dx}{(a - kx)}$$

$$= -\frac{P}{ktE} \cdot \left[\log(a - kx) \right]_0^l$$

$$= \frac{P}{ktE} \cdot \log\left[\frac{a}{(a - kl)} \right]$$

Substituting back for k,

$$\delta l = \frac{P \cdot l}{tE(a - b)} \log_e \frac{a}{b}$$

PROBLEM 21.13: A steel bar is of 20 mm × 40 mm section and 400 mm long. It is subjected to an axial pull of 200 kN. $E = 2 \times 10^5$ MPa and Poisson's ratio $= 0.3$. Determine the changes in the dimensions of the bar and in the volume.

Solution:

Axial tensile stress in the bar,

$$\sigma = \frac{P}{A} = \frac{200 \times 1000}{20 \times 40} = 250 \text{ MPa}$$

Axial or longitudinal strain,

$$e = \frac{\sigma}{E} = \frac{250}{2 \times 10^5} = 1.25 \times 10^{-3}$$

Elongation or increase in length of the bar,
$$\delta l = 1.25 \times 10^{-3} \times 400$$
$$= 0.5 \text{ mm}$$

Compressive strain in the lateral directions
$$= \mu \cdot e = 0.3 \times 1.25 \times 10^{-3}$$
$$= 3.75 \times 10^{-4}$$

Decrease in breadth
$$= 40 \times 3.75 \times 10^{-4} = 0.015 \text{ mm}$$

Decrease in thickness
$$= 20 \times 3.75 \times 10^{-4} = 0.0075 \text{ mm}$$

Volumetric strain, $e_V = e_l + e_b + e_t$
$$= (12.5 \times 10^{-4} - 3.75 \times 10^{-4} - 3.75 \times 10^{-4})$$
$$= 5 \times 10^{-4}$$

Since this is positive, the volume increases. Increase in volume,
$$\delta V = 5 \times 10^{-4} \times 20 \times 40 \times 4 = 160 \text{ mm}^3$$

PROBLEM 21.14: An axial pull of 100 kN is applied to a steel rod, 30 mm diameter and 500 mm long. Determine the change in volume. $E = 2 \times 10^5$ MPa and Poisson's ratio $= 0.25$.

Solution:

Axial stress,

$$\sigma = \frac{P}{A} = \frac{100 \times 1000}{\pi \times 15^2} = 141.47 \text{ MPa}$$

Axial tensile strain,

$$e_l = \frac{\sigma}{E} = \frac{141.47}{2 \times 10^5} = 7.0735 \times 10^{-4}$$

Compressive strain in the two lateral directions,

$$e_d = 0.25 \times 70.735 \times 10^{-4} = 1.7684 \times 10^{-4}$$

Volumetric strain,

$$e_V = e_l + 2e_d = (7.0735 - 2 \times 1.7684) \times 10^{-4}$$
$$= 3.5367 \times 10^{-4}$$

∴ Increase in volume,

$$\delta V = 3.5367 \times 10^{-4} \times \frac{\pi}{4} \times 30^2 \times 500 \text{ mm}^3$$
$$= 125 \text{ mm}^3$$

PROBLEM 21.15: A bar of 20 mm diameter is subjected to a pull of 27 kN. The measured extension over a gauge length of 200 mm is 0.08 mm and the change in diameter is 0.0025 mm. Determine the Poisson's ratio, and the three elastic moduli.

Solution:

Longitudinal strain $= \dfrac{0.08}{200} = 4 \times 10^{-4}$

Lateral strain $= \dfrac{0.0025}{20} = 1.25 \times 10^{-4}$

Poisson's ratio, $\dfrac{1}{m} = \dfrac{1.25 \times 10^{-4}}{4 \times 10^{-4}} = 0.3125$

Stress, $\sigma = \dfrac{27 \times 1000}{\left(\dfrac{\pi}{4} \times 20^2\right)} = 85.94 \text{ MPa}$

$E = \dfrac{\text{Stress}}{\text{Strain}} = \dfrac{85.94}{4 \times 10^{-4}} = 2.15 \times 10^5 \text{ MPa}$

$E = 2N\left(1 + \dfrac{1}{m}\right)$

∴ $2.15 \times 10^5 = 2N(1 + 0.3125)$

$N = 0.8175 \times 10^5 \text{ MPa}$

$E = 3K\left(1 - \dfrac{2}{m}\right)$

∴ $2.15 \times 10^5 = 3K(1 - 2 \times 0.3125)$

$K = 1.91 \times 10^5 \text{ MPa}$

PROBLEM 21.16: The modulus of rigidity of a material is 4×10^4 MPa. A 10 mm diameter rod of this material is subjected to an axial pull of 5 kN and the change in diameter is observed to be 0.002 mm. Calculate the modulus of elasticity and the Poisson's ratio of this material.

Solution:

$$N = 4 \times 10^4 \text{ MPa}$$

Stress, $\sigma = \dfrac{P}{A} = \dfrac{5 \times 1000 \times 4}{\pi \times 10^2}$ MPa $= 63.66$ MPa

Lateral strain $= \dfrac{0.002}{10} = 2 \times 10^{-4}$

But lateral strain $= \dfrac{\sigma}{mE}$

∴ $\dfrac{63.66}{mE} = 2 \times 10^{-4}$

or $mE = \dfrac{63.66}{2 \times 10^{-4}} = 3.183 \times 10^5$

Also $E = 2N\left(1 + \dfrac{1}{m}\right) = 2N\left(\dfrac{m+1}{m}\right)$

or $mE = 2N(m + 1)$

∴ $3.183 \times 10^5 = 2 \times 4 \times 10^4 (m + 1)$

$m = 2.98$

Poisson ratio, $\dfrac{1}{m} = 0.3356$

Modulus of elasticity,

$$E = \dfrac{3.183 \times 10^5}{2.98} = 1.068 \times 10^5 \text{ MPa}$$

PROBLEM 21.17: A member is formed by connecting a steel bar to an aluminium bar end to end as shown in Fig. 21.28. Calculate the thrust P so that the total decrease in length is 0.25 mm.

$E_{\text{steel}} = 2 \times 10^5$ MPa and

$E_{\text{aluminium}} = 0.7 \times 10^5$ MPa

Solution: Let P be the required force.

$$\delta l = \left(\dfrac{P \times 300}{60 \times 60 \times 2 \times 10^5} + \dfrac{P \times 400}{100 \times 100 \times 0.7 \times 10^5}\right)$$

But $\delta l = 0.25$ mm

∴   Equating δl to 0.25 mm and solving for P, we get,

$$P = 253 \text{ kN}$$

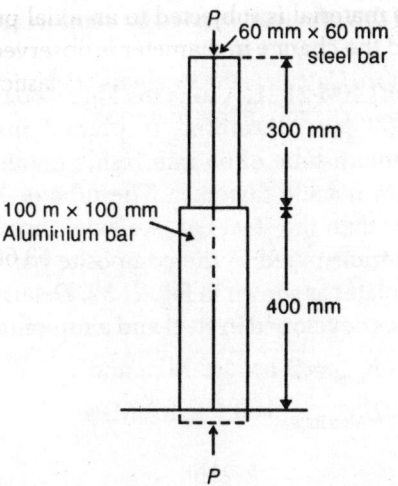

Fig. 21.28: Composite Bar (Illustrative Problem 21.17)

PROBLEM 21.18: A steel cylinder is enclosed in a copper tube as shown in Fig. 21.29. The cylinder and tube are compressed between rigid parallel plates. Find the stresses in steel and copper and the compressive strain under an axial thrust of 420 kN. Internal diameter is 100 mm and external diameter 200 mm.

$E_{steel} = 20 \times 10^5$ MPa and

$E_{copper} = 1.2 \times 10^5$ MPa

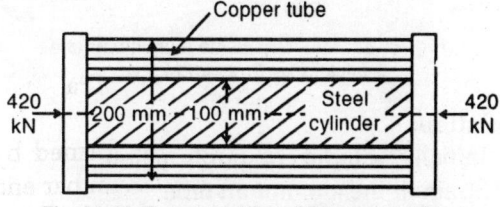

Fig. 21.29: Compound Bar of Copper and Steel
(Illustrative Problem 21.18)

Solution: Let P_1 be the load taken by steel bar and P_2 by copper tube.

$$P_1 + P_2 = 420 \qquad \ldots(i)$$

The strain is the same in both.

$$\frac{P_1}{A_1 E_1} = \frac{P_2}{A_2 E_2}$$

$$A_1 = \frac{\pi}{4} \times 100^2 \ mm^2 = 2500\pi \ mm^2$$

$$A_2 = \frac{\pi}{4}(200^2 - 100^2) = 750\pi \ mm^2$$

$$\frac{P_1}{2500\pi \times 2 \times 10^5} = \frac{P_2}{7500\pi \times 1.2 \times 10^5}$$

$$9P_1 - 5P_2 = 0 \qquad \ldots(ii)$$

Solving eqs. (i) and (ii),

$$P_1 = 150 \ kN; \ P_2 = 270 \ kN$$

Stress in steel

$$= \frac{P_1}{A_1} = \frac{150 \times 1000}{2500\pi} \ MPa = 19.10 \ MPa$$

Stress in copper

$$= \frac{P_2}{A_2} = \frac{270 \times 1000}{7500\pi} \ MPa = 11.46 \ MPa$$

$$\text{Compressive strain} = \frac{19.10}{2 \times 10^5} = 9.55 \times 10^{-5}$$

PROBLEM 21.19: A load of 44 kN hangs from three wires of equal length through a rigid plate as shown in Fig. 21.30. The middle wire is of steel and the two outer wires are of copper. If the sectional area of each wire is 300 mm², determine the load carried by each wire. $E_{steel} = 2 \times 10^5$ MPa; $E_{appear} = 1.2 \times 10^5$ MPa.

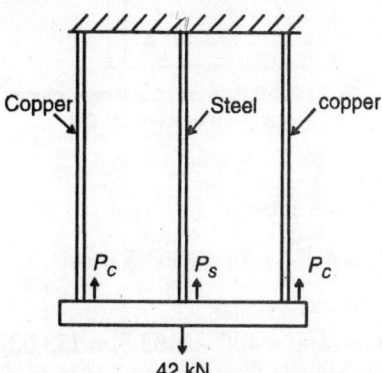

Fig. 21.30: Load Hanging from Three Wires
(Illustrative Problem 21.19)

Solution: Let each of the copper wires carry a load P_C and the steel wire P_S.

$$P_S + 2P_C = 44 \qquad \ldots(i)$$

Strain is the same in each wire.

$$e = \frac{\sigma_C}{E_C} = \frac{\sigma_S}{E_S}$$

Since the area of section of each wire is also the same,

$$\frac{\sigma_C}{\sigma_S} = \frac{P_C}{P_S}$$

$$\therefore \quad \frac{P_C}{P_S} = \frac{E_C}{E_S} = \frac{120}{200} = \frac{3}{5}$$

$$3P_S - 5P_C = 0 \qquad \qquad \dots(ii)$$

Solving Eqs. (*i*) and (*ii*)

$$P_C = 2 \text{ kN and } P_S = 20 \text{ kN}$$

PROBLEM 21.20: A reinforced concrete column of square section, 400 mm side, has four reinforcing steel bars, 25 mm diameter, as shown in Fig. 21.31. Determine the safe axial load ~n the column if the maximum allowable stress in concrete is 5 MPa. What is the corresponding stress in the steel bars and what portion of the load is borne by the reinforcing bars? Modular ratio of steel and concrete is 18.

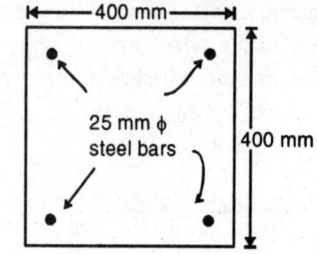

Fig. 21.31: Reinforced Concrete Column
(Illustrative Problem 21.20)

Solution:

Area of steel bars,

$$A_S = 4 \times \frac{\pi}{4} \times 25 = 1963.5 \text{ mm}^2$$

Area of concrete,

$$A_C = (400 \times 400 - 1963.5) = 158{,}037.5 \text{ mm}^2$$

Strain is the same in both steel and concrete.

$$\therefore \quad \frac{\sigma_S}{E_S} = \frac{\sigma_C}{E_C}$$

or $\quad \sigma_S = \sigma_C \cdot \dfrac{E_S}{E_C} = 5 \times 18 = 90 \text{ MPa}$

Safe central load, $P = \sigma_C A_C + \sigma_S A_S$

$$= (5 \times 158{,}037.5 + 90 \times 1963.5) \text{ N}$$

$$= (790.1875 + 176.715) \text{ kN}$$

$$= 966.9 \text{ kN Say } 960 \text{ kN}$$

Of this, the load carried by reinforcing bars

$$= 176.7 \text{ kN}$$

PROBLEM 21.21: A solid steel bar, 480 mm long and 60 mm diameter, is placed inside an aluminium tube of 65 mm inside diameter and 90 mm outside diameter. The tube is 0.12 mm longer than the steel bar. An axial thrust of 500 kN is transmitted to the composite bar through rigid plates as shown in Fig. 21.32. Determine the stresses developed in steel and aluminium.

$$E_{\text{steel}} = 2.1 \times 10^5 \text{ MPa and}$$

$$E_{\text{Aluminium}} = 0.7 \times 10^5 \text{ MPa}$$

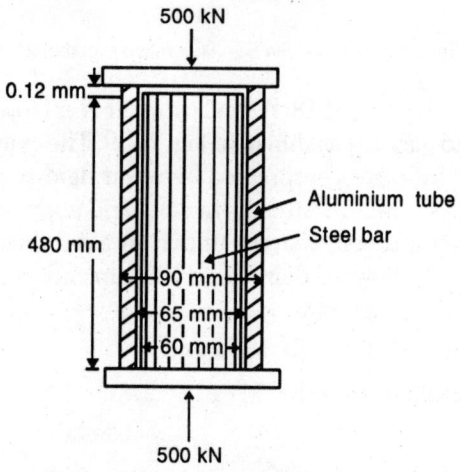

Fig. 21.32: Steel Bar Inside Aluminium Tube
(Illustrative Problem 21.21)

Solution:

Initially, without composite action:
Strain in the aluminium tube

$$= \frac{0.12}{480.12} = 2.5 \times 10^{-4}$$

Stress in the aluminium tube,

$$\sigma_{Al} = 2.5 \times 10^{-4} \times 0.7 \times 10^5 = 17.5 \text{ MPa}$$

Load required to cause this stress $= \sigma_{Al} \cdot A_{Al}$

$$= 17.5 \times \frac{\pi}{4}(90^2 - 65^2) \text{ N}$$

$$= 53.25 \text{ kN}$$

The remaining load to be borne by composite action

$$= 500 - 53.25 = 446.75 \text{ kN}$$

With composite action:

Let P_{Al} and P_S be the loads shared by aluminium and steel during the phase of composite action.

$$P_{Al} + P_S = 446.75 \qquad \ldots(i)$$

Strain is the same in both during composite action.

$$\frac{P_{Al}}{A_{Al} E_{Al}} = \frac{P_S}{A_S E_S}$$

$$P_{Al} : P_S = A_{Al} \cdot E_{Al} : A_S E_S = \frac{\pi}{4}(90^2 - 65^2)$$

$$\times 0.7 \times 10^5 : \frac{\pi}{4} \times 60^2 \times 2.1 \times 10^5$$

$$= 0.3588 : 1$$

$$\therefore \quad P_{Al} = \frac{0.3588}{1.3588} \times 446.75 \approx 118 \text{ kN}$$

and $\quad P_S = (446.75 - 118)$

$$= 328.75 \text{ kN}$$

$$\sigma_{Al} = \frac{P_{Al}}{A_{Al}} = \frac{118 \times 1000 \times 4}{\pi(90^2 - 65^2)} \text{ MPa} = 38.77 \text{ MPa}$$

$$S = \frac{P_S}{A_S} = \frac{328.75 \times 1000 \times 4}{\pi \times 60^2} = 116.27 \text{ MPa}$$

Resultant stresses:

Aluminium: $17.50 + 38.77 = 56.27$ MPa

Steel: 116.27 MPa

PROBLEM 21.22: A railway line is laid so that there is no stress in the rails at 10 °C. Compute (*i*) the stress in the rails at 50 °C if there is no allowance for expansion; (*ii*) the stress in the rails at 50 °C if the expansion allowance is 10 mm; (*iii*) the expansion allowance required if the stress is to be zero at 50 °C, and (*iv*) the maximum temperature if there should be no stress in the rails for an expansion allowance of 15 mm. The rails are 25 m long. The modulus of elasticity is 2 × 10⁵ MPa and the coefficient of linear expansion of rail steel is 12.5 × 10⁻⁶/°C.

Solution:

(*i*) Strain due to temperature change, $e = \alpha \cdot t$

$$= \alpha(t_1 - t_0) = 12.5 \times 10^{-6} (50 - 10) = 5 \times 10^{-4}$$

Stress in the rails with no expansion allowance

$$= E \cdot e = 2 \times 10^5 \times 5 \times 10^{-4} = 100 \text{ MPa}$$

(*ii*) Expansion allowance provided = 10 mm

Free expansion:

$$l \cdot \alpha \cdot t = 25 \times 1000 \times 5 \times 10^{-4} = 12.5 \text{ mm}$$

Expansion prevented $= (12.5 - 10) = 2.5$ mm

Stress in the rails

$$= \frac{2.5}{25 \times 1000} \times 2 \times 10^5 = 20 \text{ MPa}$$

(*iii*) For the stress to be zero at 50 °C, the expansion allowance required equals the free expansion at 50 °C, *i.e.*, 12.5 mm.

(*iv*) If there is to be no stress in the rails for an expansion allowance of 15 mm, the maximum temperature t °C can be got from

$$15 = 25 \times 1000 \times 12.5 \times 10^{-6} \times (t - t_0)$$

$$(t - t_0) = \frac{15 \times 10^6}{25 \times 1000 \times 12.5} = 48°$$

$$\therefore \qquad t = 48° + 10° = 58 \text{ °C}$$

PROBLEM 21.23: A tapered bar, 100 mm diameter at one end and 200 mm diameter at the other, and 1000 mm long, is initially free of stress. If the temperature of the bar drops by 20 °C, determine the maximum stress in the bar.

$E = 2 \times 10^5$ MPa and $\alpha = 12.5 \times 10^{-6}$ °C.

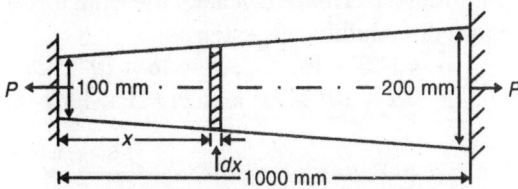

Fig. 21.33: Tapered Bar under Temperature Stress (Illustrative Problem 21.23)

Solution:

Free contraction of the bar $= l \cdot \alpha \cdot t$

$$= 1000 \times 121.5 \times 10^{-6} \times 20 = 0.25 \text{ mm}$$

If P is the pull induced due to the unyielding supports, the free elongation of the bar under this pull should be equated to 0.25 mm.

Let the radius of the bar be r mm at a distance of x mm from the smaller end; consider a small thickness dx at this distance.

$$r = 50 + \frac{x}{1000} \times 50 = \left(50 + \frac{x}{20}\right) \text{ mm}$$

$$\text{Stress} = \frac{P}{\pi\left(50 + \dfrac{x}{20}\right)^2}$$

$$\therefore \quad \text{Elongation in } dx = \frac{P}{E\pi\left(50 + \dfrac{x}{20}\right)^2} \cdot dx$$

$$\therefore \text{Total elongation} = \frac{400P}{\pi E}\int_{0}^{1000}\frac{dx}{(1000+x)^2} = \frac{P}{5\pi E},$$

on evaluation

$$\therefore \quad \frac{P}{5\pi E} = 0.25$$

$$\text{or} \quad P = \frac{0.25 \times 5\pi \times 2 \times 10^5}{10^3}\text{kN} = 785.4\,\text{kN}$$

Maximum stress at the smaller end

$$= \frac{785.4 \times 1000}{\pi \times 50^2}\text{ MPa} = 100\text{ MPa}$$

PROBLEM 21.24: A mild steel bar, 20 mm diameter and 400 mm long, is enclosed in a brass tube of external diameter 30 mm and internal diameter 25 mm. The composite bar is heated through 50 °C. Determine the stresses induced in each metal. Determine also the extension of the composite bar. Hence calculate the axial thrust P required to nullify the extension.

$\alpha_s = 11.2 \times 10^{-6}/°C;\ \alpha_b = 16 \times 10^{-6}/°C;$
$E_s = 2 \times 10^5\text{ MPa; and } E_b = 1 \times 10^5\text{ MPa}$

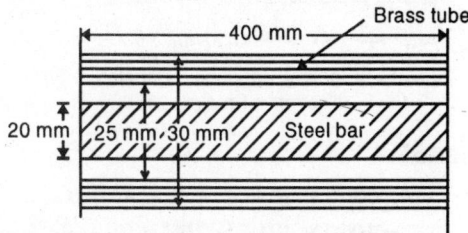

Fig. 21.34: Composite Bar under Temperature Stress (Illustrative Problem 21.24)

Solution:

Area of steel bar, $A_s = \dfrac{\pi}{4} \times 20^2 = 100\pi$ mm^2

Area of brass tube,

$$A_b = \frac{\pi}{4}(30^2 - 25^2) = \frac{275}{4}\pi \text{ mm}^2$$

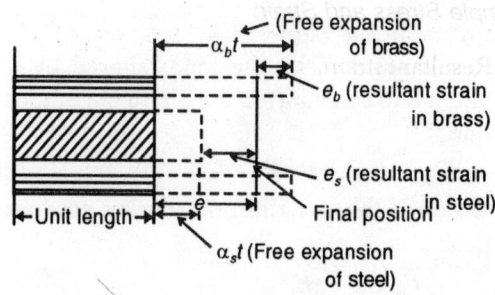

Fig. 21.35: Compromise Position in Composite Bar under Temperature Change (Illustrative Problem 21.24)

Let us consider a unit length of the bar as shown in Fig. 21.35; let the resultant final position of the bar be such that it is some where in between the positions, which the brass tube and steel bar would have occupied had they been free to expand. The steel bar would be pulled by the brass tube and will stretch more than its natural expansion, thus suffering a tensile strain. The brass tube will be restrained by the steel rod, and hence will suffer a compressive strain.

Let e_s be the tensile strain of the steel rod and e_b the compressive strain of the brass tube, the stresses being σ_s and σ_b, respectively.

$$e_s = e - \alpha_s t$$
$$e_b = \alpha_b t - e$$

Adding, $e_s + e_b = t(\alpha_b - \alpha_s)$

$$\text{or} \quad \frac{\sigma_s}{E_s} + \frac{\sigma_b}{E_b} = t(\alpha_b - \alpha_s) \qquad \dots(i)$$

Also, since no external force is there, the pull on steel shall equal the push on brass for equilibrium.

$$\sigma_s A_s = \sigma_b \cdot A_b \qquad \dots(ii)$$

$$\frac{\sigma_s}{2 \times 10^5} + \frac{\sigma_b}{1 \times 10^5} = 50(16 - 11.2) \times 10^{-6}$$

$$\text{or} \quad \sigma_s + 2\sigma_b = 48 \qquad \dots(i)$$

$$100\pi\sigma_s = \frac{275}{4}\pi\sigma_b$$

$$16\sigma_s - 11\sigma_b = 0 \qquad \dots(ii)$$

Solving equations (i) and (ii),

$\sigma_b = 17.86$ MPa; and $\sigma_s = 12.28$ MPa
(compressive)   (tensile)

Resultant strain,

$$e = e_s + \alpha_s t = \frac{\sigma_s}{E_s} + \alpha_s t = \frac{12.28}{2 \times 10^5}$$

$$+ 11.2 \times 10^{-6} \times 50$$

$$= 621.4 \times 10^{-6}$$

Extension of the composite bar

$$= e \cdot l = 400 \times 621.4 \times 10^{-6} \, \text{mm}$$

$$= 0.24856 \, \text{mm}$$

Check:

Resultant strain, $e = \alpha_b \cdot t - e_b$

$$= \alpha_b \cdot t - \frac{\sigma_b}{E_b}$$

$$= 16 \times 10^{-6} \times 50 - \frac{17.86}{1 \times 10^5}$$

$$= 621.4 \times 10^{-6}$$

PRACTICE PROBLEMS

21.1   A rod, 1 m long and of 20 mm square cross-section is subjected to a pull of 12 kN. If the modulus of elasticity is 2×10^5 MPa, determine the elongation of the rod.

21.2   A steel rod of 25 mm diameter and 2.5 m long is subjected to an axial load of 120 kN. Its extension is measured and found to be 3 mm. Determine the modulus of elasticity.

21.3   A rod, 1.50 m long and 20 mm diameter, is subjected to an axial pull of 24 kN. If $E = 2 \times 10^5$ MPa; determine the stress, the strain, and the elongation of the rod.

21.4   A brass bar with a cross-sectional area of 750 mm² is subjected to axial forces as shown in Fig. 21.36. Assuming $E = 1 \times 10^5$ MPa, determine the resultant change in length of the bar.

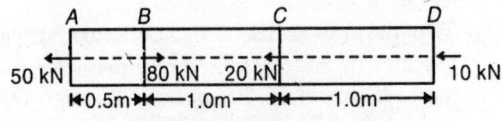

Fig. 21.36: A Brass Bar under Forces
(Practice Problem 21.4)

21.5   A bar *ABCD* as shown, is subjected to a pull of 180 kN. If the tensile stress is limited to 135 MPa, find the diameter of the portion *BC* and its length, if the total elongation of the bar is 0.3 mm. Assume $E = 2.05 \times 10^5$ MPa.

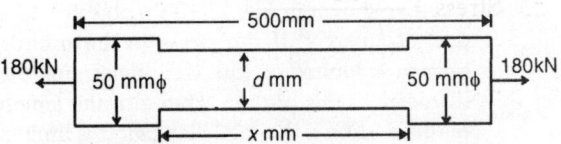

Fig. 21.37: A Bar under Pull (Practice Problem 21.5)

21.6   The maximum crushing stress in a punch is 3 times the maximum shearing stress of the plate. Show that the biggest hole that can be punched in the plate is of diameter equal to $1\frac{1}{3}$ times the plate thickness.
(**Hint:** Equate the thrust on the punch to the shear force needed to punch the hole and solve for the diameter of the hole in terms of the thickness of the plate).

21.7   An axial pull of 36 kN acts on a bar of variable section as shown in Fig. 21.38. Determine the stresses in each portion and the total extension of the bar $E = 2.1 \times 10^5$ MPa.

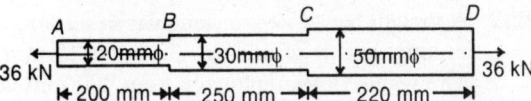

Fig. 21.38: Bar of Varying Section (Practice Problem 21.7)

21.8   A composite bar *ABC*, rigidly fixed at *A*, and 1 mm above the floor is loaded as shown in Fig. 21.39. If the sectional area of the portion *AB* is 100 mm² and that of *BC* is 200 mm², determine the reactions at the ends and the stresses in the two portions. Assume $E = 2 \times 10^5$ MPa.

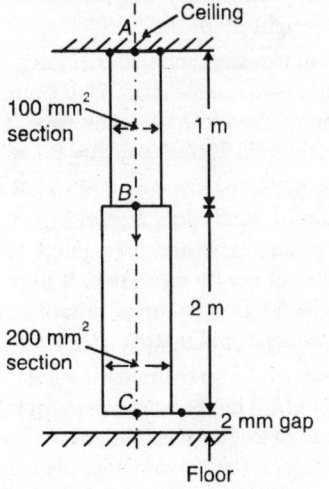

Fig. 21.39: Composite Bar (Practice Problem 21.8)

21.9 The bar shown in Fig. 21.40 is subjected to a pull of 160 kN. If the stress in the middle portion is limited to 150 MPa, determine the diameter of this portion. Find also the length of this portion if the total elongation is limited to 0.2 mm. $E = 2 \times 10^5$ MPa.

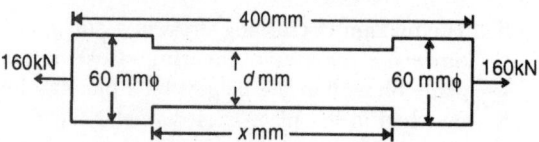

Fig. 21.40 Bar of Varying Section (Practice Problem 21.9)

21.10 A rod, which tapers uniformly from 40 mm diameter to 20 mm diameter in a length of 400 mm, is subjected to an axial load of 5 kN. If $E = 2 \times 10^5$ MPa, find the extension of the rod.

21.11 In a tension test, the bar is found to taper from diameter $(D + a)$ to $(D - a)$, prove that the error in using the mean diameter to calculate the modulus of elasticity is $(10a/D)^2$ per cent.

21.12 A straight bar of steel rectangular in section, 3 m long, and is of uniform thickness 15 mm. The width varies uniformly from 100 mm at one end to 40 mm at the other. If the rod is subjected to an axial pull of 40 kN, find the extension of the bar. Assume $E = 2 \times 10^5$ MPa.

21.13 Determine the elongation of a solid conical bar under its own weight (W). Length is L, diameter at the top is D and unit weight is γ. (**Hint:** Use the principle of integration by considering a thin strip at a distance x from the tip, and finally put the result in terms of the weight W of the bar).

21.14 A uniform steel rope, 250 m long, hangs vertically down a mine shaft. Determine the elongation of the top half of the rope. Unit weight of steel is 75 kN/m³ and $E = 2.1 \times 10^5$ MPa.

21.15 A tie-bar is to carry a pull of 15 kN and is to be made of steel with a working stress of 200 MPa. The extension on a gauge length of 200 mm shall not be more than 0.10 mm. $E = 2 \times 10^5$ MPa. Determine a suitable diameter to the nearest millimetre.

21.16 A bar of 30 mm diameter is subjected to a pull of 80 kN. The measured extension on a gauge length of 200 mm is 0.111 mm. The change in diameter is 0.0050 mm. Calculate the Poisson's ratio and the elastic moduli.

21.17 For a given material, $E = 1 \times 10^5$ MPa and $N = 0.4 \times 10^5$ MPa. Find the Poisson's ratio and bulk modulus. Determine the reduction in diameter of 50 mm dia. bar and 2.5 long when stretched 3.75 mm.

21.18 Determine the stresses that will be produced in a steel bolt and a copper tube by turning the nut half-a-turn. The bolt is 900 mm long, the pitch of the bolt thread = 4 mm, the sectional area of the bolt is 900 mm² and that of the tube 1500 mm². E of steel = 2×10^5 MPa and E of copper = 1.2×10^5 MPa.

21.19 A ball of steel of diameter 120 mm is placed at the bottom of a sea at 1 km depth. If sea water weighs 10 kN/m³, find decrease in diameter and in volume of the ball. $E = 2 \times 10^5$ MPa and $\mu = 1/4$.

21.20 A reinforced concrete column has a square section of side 500 mm, with four steel bars 25 mm diameter embedded in it. The column carries a compressive load of 2.25 MN. Determine the stresses in concrete and steel if the modular ratio is 15.

21.21 A steel rod of diameter 20 mm passes centrally through a hollow copper tube of external diameter 30 mm and internal diameter 20 mm, and is secured by nuts and washers. The nuts are tightened till the tension in the rod is 90 kN. Find the tensile force is to be applied to the steel rod so as to relieve the copper tube of all the compressive stress. $E_S = 2 \times 10^5$ MPa and $E_C = 1 \times 10^5$ MPa.

21.22 A thin steel tyre is to be shrunk on to a rigid wheel of 2 m diameter. If the temperature stress is to be limited to 100 MPa. Calculate the suitable inner diameter of the tyre. What is the minimum temperature necessary to slip the tyre on to the wheel? $E = 2 \times 10^5$ MPa and $\alpha = 11 \times 10^{-6}$ /°C.

21.23 Two parallel walls, 5 m apart, are stayed together by a steel rod, 30 mm diameter, at a temperature of 75 °C, passing through nuts and washers at each end. Calculate the pull exerted by the rod on the wall when it has cooled to 15 °C—(a) if the walls do not yield; and (b) if the total yield at both ends is 1.14 mm. $E = 2 \times 10^5$ MPa and $\alpha = 11 \times 10^{-6}$/°C.

21.24 Determine the stresses in the bars of the system shown, if the sectional area of each wire is 60 mm², the load is 25 kN and the temperature of the system rises by 15 °C. The bar remains horizontal. $\alpha_C = 16 \times 10^{-6}$ /°C; $\alpha_S = 12 \times 10^{-6}$ /°C; $E_C = 1.2 \times 10^5$ MPa; and $E_S = 2 \times 10^5$ MPa.

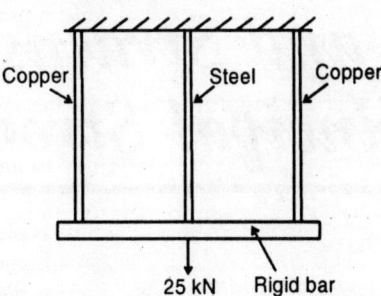

Fig. 21.41: Composite Bar (Practice Problem 21.24)

(**Hint:** Since the bar remains horizontal, the total elongation of each wire will be the same—the sum of elastic elongation and the free expansion due temperature rise; the second equation is obtained from the principle of sharing of the load in composite action. The load carried by each wire is solved first and then the stress.)

Chapter 22

Compound Stresses and Strains— Principal Planes Principal Stresses

22.1 INTRODUCTION

When a body is acted upon by several forces which cause wholly normal or wholly tangential stresses, or a combination of these, across different planes of known orientations, we can determine the state of stress across other planes by resolving the known forces and applying the laws of static equilibrium. The stress will be a complex combination of the three basic or simple kinds of stress.

22.2 PRINCIPAL PLANES AND PRINCIPAL STRESSES

However complex the stress conditions at a point within a body be, there always exist three mutually perpendicular planes on which the resultant stress is wholly normal. Such planes are known as 'Principal Planes' and the normal stresses on these planes are known as 'Principal Stresses". Since the resultant stresses on principal planes are wholly normal, these planes are free of tangential or shear stresses. Further, one of the principal stresses will be the largest and another will be the smallest; the former is called the 'major' principal stress and the latter 'minor' principal stress. The one in between these two is called the 'intermediate' principal stress.

In most practical cases, there is a plane normal to which the stress is negligibly small; that is to say, one of the principal stresses is zero or negligibly small. In such cases, compounding and resolution of stresses reduces·to a two-dimensional problem as in coplanar statics, making the analysis essentially two-dimensional.

The concept of principal planes and principal stresses is an ingenious tool in the hands of an engineer to analyse a problem in which the stresses at any point are a compound or a complex combination of the simple or basic kinds of stress seen in the previous chapter; as such this has great value in the analysis and design of a structural component or machine element.

Now let us investigate a number of cases which involve the process of compounding and resolution of forces/stresses.

22.3 STATE OF STRESS ON AN OBLIQUE SECTION OF A BODY UNDER PURE SHEAR

If a small element $ABCD$ of a material is subjected to pure shear stress q as shown in Fig. 22.1, the state of stress on an oblique section BE inclined at an angle θ with one of the planes can be determined as follows:

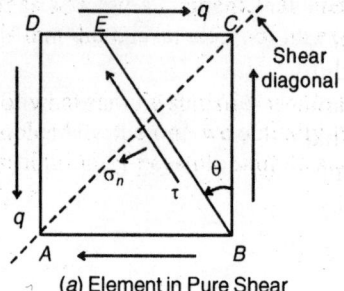

(a) Element in Pure Shear

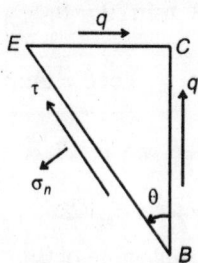

(b) Equilibrium of Wedge *BCE*

Fig. 22.1: State of Stress on an Oblique Section of a Body in Pure Shear

Note: The direction of *q* in the state of pure shear as shown with the shear diagonal in the first and third quadrants is treated as positive.

Let us consider the equilibrium of the wedge *BCE* as shown in (*b*), the normal and tangential components of stresses being σ_n and τ on the plane *BE*.

Resolving perpendicular to *BE*,

$\sigma_n \cdot BE = q \cdot BC \cdot \sin\theta + q \cdot CE \cdot \cos\theta$ (with the direction of σ_n as indicated—tensile on plane *BE*)

$$\sigma_n = q \cdot \frac{BC}{BE} \sin\theta + q \cdot \frac{CE}{BE} \cdot \cos\theta$$

$$= q \cos\theta \cdot \sin\theta + q \sin\theta \cdot \cos\theta = 2q \sin\theta \cdot \cos\theta$$

or $\sigma_n = q \cdot \sin2\theta$ ... (Eq. 21.1)

Resolving parallel to *BE*,

$\tau \cdot BE = -q \cdot BC \cdot \cos\theta + q \cdot CE \cdot \sin\theta$ (with the direction of τ as indicated)

$$\tau = -q \cdot \frac{BC}{BE} \cdot \cos\theta + q \cdot \frac{CE}{BE} \sin\theta$$

$$\tau = -q \cos^2\theta + q \sin^2\theta = q(\sin^2\theta - \cos^2\theta)$$

$$\tau = -q \cdot \cos2\theta$$ ... (Eq. 21.2)

(τ will be negative on plane *BE* if a negative value is obtained from this).

σ_n is obviously tensile for positive θ (counter clockwise) and positive *q* (shear diagonal in the first and third quadrants), further, for varying θ, σ_n varies and reaches its maximum value of *q* for $\theta = 45°$. That is, the 45°—planes will be subjected to maximum tensile and compressive stresses, numerically equal to *q*, and the shear stresses on these planes are zero.

Thus, if the tensile or compressive strength is less than the shear strength of a material, it will fail in tension or compression along 45°-planes of pure shear (similar comments apply in the case of negative pure shear).

22.4 STRESSES ON OBLIQUE SECTION OF A BODY–PRINCIPAL STRESSES KNOWN

Let σ_x and σ_y be the principal stresses acting on the two perpendicular planes, these planes being free of shear as shown in Fig. 22.2.

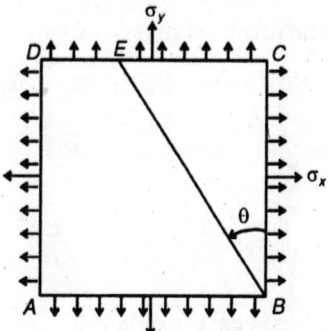

(a) Principal Stresses Known

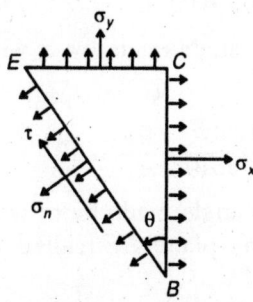

(b) Equilibrium of wedge *BCE*

Fig. 22.2: Stresses or Oblique Section of a Body—Principal Stresses Known

Let *BE* be a plane making an angle θ with respect to the plane on which σ_x acts. Let the thickness of the element be unity perpendicular to the plane of the paper. Let σ_n and τ be the normal and shear stresses on the plane *BE*.

Considering the equilibrium of the wedge *BCE* as shown in (*b*) and resolving all the forces pendicular to *BE*,

$$\sigma_n \cdot BE = \sigma_x \cdot BC \cdot \cos\theta + \sigma_y \cdot CE \cdot \sin\theta$$

$$\sigma_n = \sigma_x \cdot \frac{BC}{BE} \cdot \cos\theta + \sigma_y \cdot \frac{CE}{BE} \sin\theta$$

$$\sigma_n = \sigma_x \cos^2\theta + \sigma_y \sin^2\theta \qquad \text{...(Eq. 22.3)}$$

Resolving all the forces parallel to *BE*,

$$\tau \cdot BE = \sigma_x \cdot BC \cdot \sin\theta - \sigma_y \cdot CE \cdot \cos\theta$$

$$\tau = \sigma_x \cdot \frac{BC}{BE} \cdot \sin\theta - \sigma_y \cdot \frac{CE}{BE} \cdot \cos\theta$$

$$\tau = \sigma_x \sin\theta\cos\theta - \sigma_y \cdot \sin\theta\cos\theta$$

$$\tau = \left(\frac{\sigma_x - \sigma_y}{2}\right) \cdot \sin2\theta \qquad \text{...(Eq. 22.4)}$$

If σ_r is the resultant stress,

$$(\sigma_r \cdot BE)^2 = (\sigma_x \cdot BC)^2 + (\sigma_y \cdot CE)^2$$

$$\sigma_r \cdot BE = \sqrt{(\sigma_x \cdot BC)^2 + (\sigma_y \cdot CE)^2}$$

$$\sigma_r = \sqrt{\left(\sigma_x \cdot \frac{BC}{BE}\right)^2 + \left(\sigma_y \cdot \frac{CE}{BE}\right)^2}$$

or $\quad \sigma_r = \sqrt{\sigma_x^2 \cos^2\theta + \sigma_y^2 \sin^2\theta} \qquad \text{...(Eq. 22.5)}$

Also, $\sigma_r = \sqrt{\sigma_n^2 + \tau^2} \qquad \text{...(Eq. 22.6)}$

If α is the angle made by σ_r with respect to *x*-axis,

$$\tan\alpha = \frac{\sigma_y \sin\theta}{\sigma_x \cos\theta} = \frac{\sigma_y}{\sigma_x} \cdot \tan\theta \qquad \text{...(Eq. 22.7)}$$

If ϕ is the angle made by σ_r with respect to normal to the plane *BE* (called the angle of obliquity),

$$\tan\phi = \frac{\tau}{\sigma_n} = \frac{(\sigma_x - \sigma_y)\sin2\theta}{2(\sigma_x \cos^2\theta + \sigma_y \sin^2\theta)} \quad \text{...(Eq. 22.8)}$$

If ϕ is to be a maximum, $\tan\phi$ should be a maximum, or

$$\frac{d(\tan\phi)}{d\theta} = 0$$

Differentiating with respect to θ and dividing by common factors,

$$(\sigma_x \cos^2\theta + \sigma_y \sin^2\theta)\cos2\theta + (\sigma_x - \sigma_y)$$
$$\sin\theta\cos\theta \cdot \sin2\theta = 0$$

$$\sigma_n \cos2\theta + \tau \sin2\theta = 0$$

$$\tan2\theta = -\frac{\sigma_n}{\tau} = -\cot\phi = \tan(90° + \phi)$$

$\therefore \qquad 2\theta = 90° + \phi \quad$ say $\theta = \theta_1$ & $\phi = \phi_m$ (or ϕ_{max})

$$\theta = \theta_1 = 45° + \phi_m / 2 \qquad \text{...(Eq. 22.9)}$$

Substituting this value of θ,

$$\tan\phi_m = \frac{(\sigma_x - \sigma_y)\cos\phi_m}{\sigma_x(1 - \sin\phi_m) + \sigma_y(1 + \sin\phi_m)}$$

or $\quad \dfrac{\sigma_y}{\sigma_x} = \dfrac{1 - \sin\phi_m}{1 + \sin\phi_m} \qquad \text{...(Eq. 22.10)}$

Also, $\sin\phi_m = \left(\dfrac{\sigma_x - \sigma_y}{\sigma_x + \sigma_y}\right) \qquad \text{...(Eq. 22.11)}$

These are the conditions for maximum obliquity.

Unlike Stresses

If σ_x and σ_y are unlike—say σ_y is compressive,

$$\sigma_n = \sigma_x \cos^2\theta - \sigma_y \sin^2\theta$$

$$\tau = \frac{(\sigma_x + \sigma_y)}{2} \sin2\theta$$

τ_{max} again occurs on 45° planes.

If $\sigma_x = \sigma_y$ numerically, the stresses on these planes,

$$\tau = \left(\frac{\sigma_x + \sigma_y}{2}\right) = \sigma_x = \sigma \text{ and } \sigma_n = 0,$$

which means that it is a state of pure shear.

22.5 STRESSES ON AN OBLIQUE SECTION OF A BODY SUBJECTED TO A GENERAL BIAXIAL STATE OF STRESS

Let a prismatic element of material *ABCD* of a unit thickness be subjected to normal and shear stresses as shown in Fig. 22.3.

Let the normal and shear stresses on planes *AD* and *BC* be σ_1 and *q*, and the normal stress on planes *AB* and *CD* (perpendicular to *AD* and *BC*) be σ_2; the shear stress on the latter planes is automatically q, owing to the principle of complementary shear.

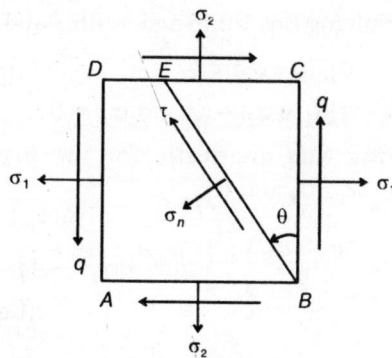

(a) Element in a Biaxial State of Stress

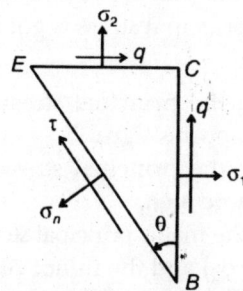

(b) Equilibrium of Wedge *BCE*

Fig. 22.3: Stresses on an Oblique Section of a Body Subjected to a General Biaxial State of Stress

Let it be required to determine the state of stress (the normal and tangential components of stresses) on a particular plane *BE*, inclined at a counterclockwise angle θ with respect to the plane *BC*, on which the normal stress σ_1 acts.

Considering the equilibrium of the wedge *BCE* as in (b) (with unit thickness perpendicular to the plane of the paper), and resolving in the normal direction to the plane *BE*,

$$\sigma_n \cdot BE = \sigma_1 \cdot BC \cdot \cos\theta + \sigma_2 \cdot EC \cdot \sin\theta$$
$$+ q \cdot BC \cdot \sin\theta + q \cdot EC \cdot \cos\theta$$

(with the direction of σ_n as indicated)

$$\sigma_n = \sigma_1 \cdot \frac{BC}{BE} \cdot \cos\theta + \sigma_2 \cdot \frac{EC}{BE} \sin\theta + q \cdot \frac{BC}{BE} \cdot \sin\theta$$

$$+ q \cdot \frac{EC}{BE} \cdot \cos\theta$$

$$= \sigma_1 \cos^2\theta + \sigma_2 \sin^2\theta + q \cdot \cos\theta \cdot \sin\theta + q \sin\theta \cdot \cos\theta$$

$$\sigma_n = \sigma_1 \cos^2\theta + \sigma_1 \sin^2\theta + q \sin 2\theta \quad \text{...(Eq. 22.12)}$$

This can also be written as,

$$\sigma_n = \left(\frac{\sigma_1 + \sigma_2}{2}\right) + \left(\frac{\sigma_1 - \sigma_2}{2}\right) \cdot \cos 2\theta + q \sin 2\theta$$

$$\text{...(Eq. 22.13)}$$

Resolving in the tangential direction to the plane *BE*,

$$\tau \cdot BE = \sigma_1 \cdot BC \cdot \sin\theta - \sigma_1 \cdot EC \cdot \cos\theta$$
$$+ q \cdot EC \cdot \sin\theta - q \cdot BC \cdot \cos\theta$$

(with the direction of τ as indicated).

$$\tau = \sigma_1 \cdot \frac{BC}{BE} \cdot \sin\theta - \sigma_2 \cdot \frac{EC}{BE} \cdot \cos\theta$$

$$+ q \cdot \frac{EC}{BE} \cdot \sin\theta - q \cdot \frac{BC}{BE} \cdot \cos\theta$$

$$= \sigma_1 \cdot \cos\theta \sin\theta - \sigma_2 \cdot \sin\theta \cos\theta + q \sin^2\theta - q \cos^2\theta$$

$$\text{or } \tau = \left(\frac{\sigma_1 - \sigma_2}{2}\right) \times \sin 2\theta - q \cos 2\theta \quad \text{...(Eq. 22.14)}$$

These expressions for σ_n and τ could well have been obtained from the principle of superposition for the biaxial state of stress (Principal stresses known) and the state of pure shear (Eqs. 22.3 and 22.1 & 22.4 and 22.2).

22.6 DETERMINATION OF PRINCIPAL PLANES AND PRINCIPAL STRESSES IN A GENERAL BIAXIAL STATE OF STRESS

In a general biaxial state of stress, if the normal and tangential stresses on any two perpendicular planes are known, the principal planes and principal stresses may be determined as follows (Fig. 22.4).

Let the thickness of the element be unity perpendicular to the plane of paper. Let σ_1 and σ_2 be the normal stresses and q the shear stress on the two perpendicular planes *BC* and *AB*.

Let us assume that the principal planes make an angle θ and $(90° + \theta)$ with the planes on which σ_1 acts. Let the principal plane *BE* make a positive angle θ with respect to *BC*, the plane on which σ_1 acts.

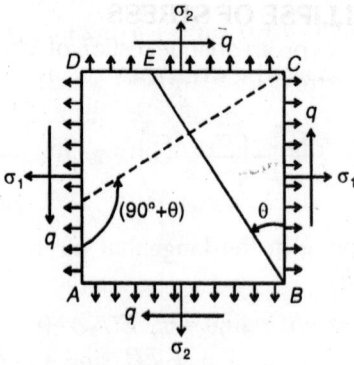

(a) An Element in a General Biaxial State of Stress

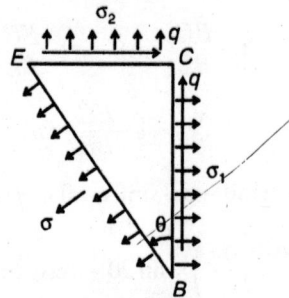

(b) Equilibrium of Wedge *BCE*

Fig. 22.4: Determination of Principal Planes and Principal Stresses in a General Biaxial State of Stress

Considering the equilibrium of the wedge *BCE*, and resolving perpendicular to *BC*

$$\sigma \cdot BE \cdot \cos\theta = \sigma_1 \cdot BC + q \cdot EC$$

(with the direction of σ as indicated)

$$\sigma \cdot \cos\theta = \sigma_1 \cdot \frac{BC}{BE} + q \cdot \frac{EC}{BE} = \sigma_1 \cos\theta + q \cdot \sin\theta$$

$$\therefore \quad (\sigma - \sigma_1) = q \cdot \tan\theta \qquad \text{...(Eq. 22.15)}$$

Resolving parallel to *BC*

$$\sigma \cdot BE \cdot \sin\theta = \sigma_2 \cdot EC + q \cdot BC$$

(with direction of σ as indicated).

$$\sigma \cdot \sin\theta = \sigma_2 \cdot \frac{EC}{BE} + q \cdot \frac{BC}{BE} = \sigma_2 \sin\theta + q \cdot \cos\theta$$

$$\therefore \quad (\sigma - \sigma_2) = q \cdot \cot\theta \qquad \text{...(Eq. 22.16)}$$

Substracting Eq. 21.15 from Eq. 21.16,

$$(\sigma_1 - \sigma_2) = q\,(\cot\theta - \tan\theta) = \frac{2q}{\tan 2\theta}$$

or $\tan 2\theta = \dfrac{2q}{(\sigma_1 - \sigma_2)}$ \qquad ...(Eq. 22.17)

This determines the orientation of the principal planes. The acute angle θ_1 gives the orientation of σ_1 and the obtuse angle $\theta_1 + 90°$ that of σ_2.

Multiplying Eqs. 22.15 and 22.16,

$$(\sigma - \sigma_1)\,(\sigma - \sigma_2) = q^2$$

$$\sigma^2 - \sigma\,(\sigma_1 + \sigma_2) - (q^2 - \sigma_1\sigma_2) = 0$$

Solving this quadratic for the principal stresses— say, σ_x and σ_y,

$$\sigma_{x,\,y} = \left(\frac{\sigma_1 + \sigma_2}{2}\right) \pm \frac{1}{2}\sqrt{(\sigma_1 - \sigma_2)^2 + 4q^2}$$

$$\text{...(Eq. 22.18)}$$

The greatest or the major principal stress is got by taking the positive sign, and the smallest or the minor principal stress is got by taking the negative sign.

If $q^2 > \sigma_1\sigma_2$, the principal stresses will be of opposite signs.

If $q^2 < \sigma_1\sigma_2$, the principal stresses will be of same sign.

If $q^2 = \sigma_1\sigma_2$, the major principal stress will be $(\sigma_1 + \sigma_2)$ and the minor principal stress will be zero. (This is a special case).

Note: If σ_1 or σ_2 is a compressive stress, the modifications are obvious with just a sign change for it. Similarly, if either σ_1 or σ_2 is zero, the modifications are easy and simple.

ALITER

The normal stress, σ_n, on a plane which makes an angle θ with respect to the plane on which σ acts, is:

$$\sigma_n = \left(\frac{\sigma_1 + \sigma_2}{2}\right) + \left(\frac{\sigma_1 - \sigma_2}{2}\right) \cdot \cos 2\theta + q \cdot \sin 2\theta$$

$$\text{...(Eq. 22.13)}$$

For maximum value of σ_n, $\dfrac{d\sigma_n}{d\theta}$ must be zero.

$$\frac{d\sigma_n}{d\theta} = -(\sigma_1 - \sigma_2)\sin 2\theta + 2q\cos 2\theta = 0$$

$$\therefore \quad \tan 2\theta = \frac{2q}{(\sigma_1 - \sigma_2)} \qquad \text{...(Eq. 22.17)}$$

(Also, the shear stress must be zero on the principal planes,

$$\tau = \left(\frac{\sigma_1 - \sigma_2}{2}\right)\sin 2\theta - q\cos 2\theta = 0$$

or $\quad \tan 2\theta = \dfrac{2q}{(\sigma_1 - \sigma_2)}$, again)

Substituting the corresponding values of $\sin 2\theta$ and $\cos 2\theta$—

$$\sin 2\theta = \frac{2q}{\sqrt{(\sigma_1 - \sigma_2)^2 + 4q^2}}$$

$$\cos 2\theta = \frac{(\sigma_1 - \sigma_2)}{\sqrt{(\sigma_1 - \sigma_2)^2 + 4q^2}}$$

in the Eq. 21.13, we get the maximum principal stress, σ_x.

$$\sigma_x = \left(\frac{\sigma_1 + \sigma_2}{2}\right) + \frac{(\sigma_1 - \sigma_2)^2}{2\sqrt{(\sigma_1 - \sigma_2)^2 + 4q^2}}$$

$$+ \frac{4q^2}{2\sqrt{(\sigma_1 - \sigma_2)^2 + 4q^2}}$$

or $\quad \sigma_x = \left(\dfrac{\sigma_1 + \sigma_2}{2}\right) + \dfrac{1}{2}\sqrt{(\sigma_1 - \sigma_2)^2 + 4q^2}$

Same as …(Eq. 22.18)

Similarly, the maximum principal stress may be obtained by substituting $(90° + \theta)$ where θ occurs in the above derivation; it may be shown that

$$\sigma_y = \left(\frac{\sigma_1 + \sigma_2}{2}\right) - \frac{1}{2}\sqrt{(\sigma_1 - \sigma_2)^2 + 4q^2}$$

$\left[\dfrac{d^2\sigma_n}{d\theta^2} = -2(\sigma_1 - \sigma_2)\cos 2\theta - 4q\sin\theta\right.$. For positive values of $\sin 2\theta$ and $\cos 2\theta$, this is negative and indicates a maximum value; while, for negative values of $\sin 2\theta$ and $\cos 2\theta$, i.e., when θ is $(90° + \theta)$, this is positive and indicates a minimum value, and hence the earlier conclusion with regard to the orientation of the major and minor principal planes].

22.7 ELLIPSE OF STRESS

The stresses on an oblique section of a body when principal stresses are known (Sec. 21.4) may also be obtained graphically by the use of what is known as the 'Ellipse of Stress' (Fig. 22.5).

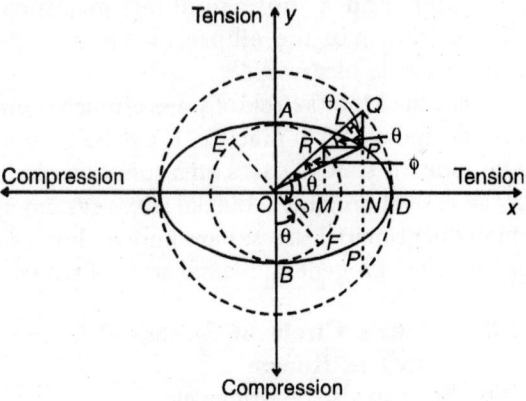

Fig. 22.5: Ellipse of Stress

The major and minor axes shall correspond to the major and minor principal stresses, respectively.

With O as centre two circles are described with radii proportional to σ_x and σ_y. OQ is drawn normal to the plane EF to meet the bigger circle in Q and the smaller in R. QN is dropped perpendicular to OX and RP to OY to meet QN in P. OP represents the resultant stress, σ_r. The locus of P for varying θ is obviously an ellipse:

$$ON = \sigma_x\cos\theta \quad \text{and} \quad PN = \sigma_y\sin\theta$$

Then $\sigma_n = OL \qquad$ and $\tau = LP$

$$L\hat{P}Q = \theta = L\hat{R}P$$

$$PL = PQ\cos\theta = (QN - PN)\cos\theta$$

$$= (\sigma_x - \sigma_y)\sin\theta\cos\theta$$

$$OL = OR + RQ - LQ$$

$$= \sigma_y + (\sigma_x - \sigma_y) - (\sigma_x - \sigma_y)\sin^2\theta$$

$$= \sigma_x - (\sigma_x - \sigma_y)\sin^2\theta = \sigma_x\cos^2\theta + \sigma_y\sin^2\theta$$

Hence $OP = \sqrt{\sigma_n^2 + \tau^2} = \sigma_r$

If σ_x is tensile and σ_y compressive, the reflection P' of P on the compression side σ_y is the required point, OP' is the value of σ_y. Similarly the use of points in the other quadrants can be understood. [If the resultant and normal stresses fall above the plane *EF*, they are tensile, and if below, compressive].

22.8 MOHR'S CIRCLE OF STRESS

A simpler, and a more popular, graphical construction than the ellipse of stress is the "Mohr's Circle of stress'.

In fact, the Mohr's circle of stress is much more versatile and useful in practice. It can be used to determine the state of stress on a specified plane for the case of not only a biaxial stress system in which the principal stresses are known, but also for the case of a general biaxial state of stress.

22.8.1 Mohr's Circle of Stress—Principal Stresses Known

Eqn.21.3 can also be written as:

$$\sigma_n = \left(\frac{\sigma_x + \sigma_y}{2}\right) + \left(\frac{\sigma_x - \sigma_y}{2}\right) \cdot \cos 2\theta \quad \text{...(Eq. 22.19)}$$

$$\tau = \left(\frac{\sigma_x - \sigma_y}{2}\right) \sin 2\theta \quad \text{...(Eq. 22.20)}$$

τ and a part of σ_n which varies with θ may be represented by the two rectangular projections of a radius vector of length $\left(\frac{\sigma_x - \sigma_y}{2}\right)$, making an angle 2θ with *OX*.

This immediately suggests a simple graphical construction for finding stresses on any plane (Fig. 22.6).

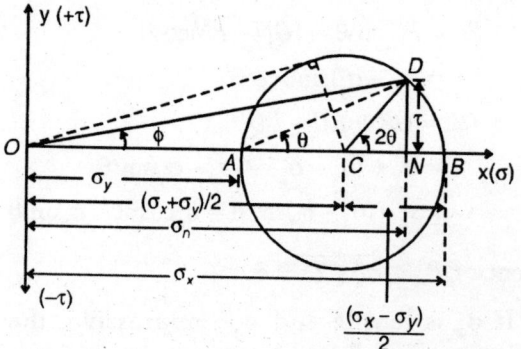

Fig. 22.6: Mohr's Circle of Stress–Principal Stresses Known

Along axis *OX*, σ_x (*OB*) and σ_y (*OA*) are set off to scale. *AB* is bisected in *D*. Then $AC = CB$ $\left(\frac{\sigma_x - \sigma_y}{2}\right)$ and $OC = \left(\frac{\sigma_x + \sigma_y}{2}\right)$. With *C* as centre and *CA* as radius a circle is described. This is the Mohr's circle of stress.

To determine the stress on any plane, *CD* is drawn inclined at 2θ to *OX*. Then

$$ON = OC + CN$$

$$= \left(\frac{\sigma_x + \sigma_y}{2}\right) + \left(\frac{\sigma_x - \sigma_y}{2}\right) \times \cos 2\theta = \sigma_n$$

$$ND = DC \cdot \sin 2\theta = \left(\frac{\sigma_x - \sigma_y}{2}\right) \sin 2\theta = \tau$$

So *OD* represents the resultant stress σ, and $D\hat{O}N$ gives the angle of obliquity ϕ.

Maximum obliquity occurs when *OD* touches the circle, i.e., when,

$$\sin \phi_m = \left(\frac{\sigma_x - \sigma_y}{\sigma_x + \sigma_y}\right) \quad \text{...(Eq. 22.20)}$$

Maximum τ occurs when $2\theta = 90°$ or $270°$ (or $\theta = 45°$ or $135°$).

When $\sigma_y = 0$, *O* and *A* coincide, and when σ_y is of opposite sign to σ_x, *O* falls within the circle.

When the normal and tangential stresses on two planes at right angles are given, it is easy to note that τ will be of the magnitude on both planes, and the sum of the normal stresses on these two planes must be equal to the sum of the principal stresses. The radius vectors form a diameter of the circle.

If the normal and tangential stresses on any two planes are known, we get two points on the circle. By drawing the perpendicular bisector of the line joining these two points to meet the axis, the centre of the circle is got; the circle can be completed and the principal stresses determined. Thus, such problems which may require difficult mathematical manipulation can be easily solved by using the Mohr's circle of stress. Many other useful conclusions can be drawn, once the circle is completed.

22.8.2 More General Case of Mohr's Circle —General Biaxial State of Stress

For the general biaxial state of stress (Sec. 22.5) the state of stress on a specified plane may be determined making use of the Mohr's circle of stress, with a minor adaptation.

This is considered to be a very convenient tool in the solution of problems which, otherwise, is laborious.

For the general biaxial state of stress shown in Fig. 22.3 (Sec. 22.5), the Mohr's circle of stress is given below (Fig. 22.7).

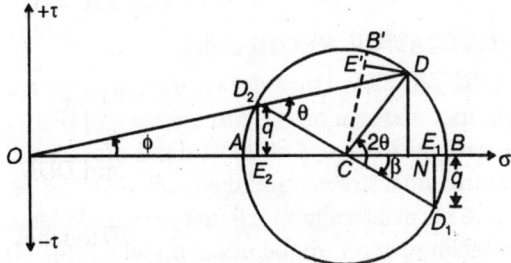

Fig. 22.7: Mohr's Circle of Stress—More General Case for Biaxial State of Stress

From O on the σ-axis, OE_1 and OE_2 are set off to scale to represent the normal stresses σ_1 and σ_2. From E_1 and E_2, E_1D_1, and E_2D_2 are set off perpendicular to the σ-axis to represent the shear stresses q. Then E_1E_2 is bisected in C, OC represents $(1/2)(\sigma_1 + \sigma_2)$ and $CE_1 = CE_2 = (1/2)(\sigma_1 - \sigma_2)$. C also will be the mid point of D_1D_2, which is a straight line.

If a circle be described with C as centre and CD_1 as radius, this is the 'Mohr's Circle of Stress'. To determine the stresses across a plane inclined at an angle θ to the plane on which σ_1 acts, an angle 2θ is set off with respect to D_1C at C_1 to meet the circle in D (Alternatively, an angle θ is set off at D_2 with respect to D_1D_2). The σ- and τ-coordinates of D will then represent the normal stress σ_n and shear stress τ on the specified plane.

If N is the foot of the perpendicular from D on to the σ-axis,

$$ON = OC + CN$$

$$= \frac{1}{2}(\sigma_1 + \sigma_2) + \frac{1}{2}(\sigma_1 - \sigma_2)\cos\theta + q\sin 2\theta$$

(Same as Eq. 22.13 for σ_n)

$$ND = \frac{1}{2}(\sigma_1 - \sigma_2)\sin 2\theta - q\cos 2\theta$$

(Same as Eq. 22.14 for τ)

These two relations can be easily proved by imagining the radius vector CD_2 turned through an angle 2θ carrying with it CB and E_1D_1 to the positions CB' and $E'D$ respectively, and then projecting the length CE' $[= (1/2)(\sigma_1 - \sigma_2)]$ and $ED (= q)$ to the base line and perpendicular it. It will then be clear that ON and ND represent the components σ_n and τ, respectively, and consequently OD represents the resultant stress across the specified plane.

[Alternatively, if R is the radius of the stress circle:

$$CE_1 = R\cos\beta = \frac{1}{2}(\sigma_1 - \sigma_2)$$

$$E_1D_1 = R\sin\beta = q$$

Now from Eq. 22.13,

$$\sigma_n = \frac{1}{2}(\sigma_1 + \sigma_2) + R\cos 2\theta \cdot \cos\beta + R\sin 2\theta \cdot \sin\beta$$

$$= OC + R\cos(2\theta + \beta) = OC + CN = ON.$$

Also from Eq. 22.14,

$$\tau = R\sin 2\theta \cos\beta - R\cos 2\theta \cdot \sin\beta$$

$$= R\sin(2\theta - \beta) = DN]$$

$$R = \frac{1}{2}(\sigma_1 + \sigma_2) + \sqrt{\frac{1}{4}(\sigma_1 - \sigma_2)^2 + q^2}$$

σ_n varies between the upper limit,

$$\frac{1}{2}(\sigma_1 + \sigma_2) + \sqrt{\frac{1}{4}(\sigma_1 - \sigma_2)^2 + q^2}$$

when $2\theta = \beta$ or $\tan 2\theta = \dfrac{2q}{(\sigma_1 - \sigma_2)}$.

This occurs when D falls on B. The lower limit occurs when D falls on A, and is

$$\frac{1}{2}(\sigma_1 + \sigma_2) - \sqrt{\frac{1}{4}(\sigma_1 - \sigma_2)^2 + q^2}$$

when $2\theta = \beta + 180°$ or $\theta = (1/2)\beta + 90°$, i.e., at right angles to the plane for which σ_n reaches its upper limit. These are the principal stresses. For these two values of σ_n, the shear stress τ is zero. The value of τ varies between the limits,

$$\pm\sqrt{\frac{1}{4}(\sigma_1 - \sigma_2)^2 + q^2}$$

when $\theta = \frac{1}{2}\beta + 45°$ and $\theta = \frac{1}{2}\beta + 135°$

A number of other conditions as well as relations may be easily obtained from the Mohr's Circle of Stress.

22.9 PRINCIPAL STRAINS

Principal strains are those occurring in the directions of principal stresses.

If the principal stresses at a point are σ_x, σ_y and σ_z at a point in an isotropic material (with the same elastic properties in all directions), each will independently produce the same strains which it would if acting alone (Fig. 22.8).

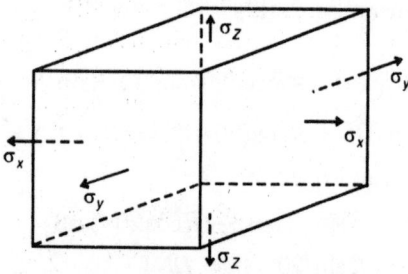

Fig. 22.8: Three Principal Stresses Acting on a Body

$$\varepsilon_x = \frac{\sigma_x}{E} - \frac{(\sigma_y + \sigma_z)}{mE}$$

$$\varepsilon_y = \frac{\sigma_y}{E} - \frac{(\sigma_z + \sigma_x)}{mE}$$

$$\varepsilon_z = \frac{\sigma_z}{E} - \frac{(\sigma_x + \sigma_y)}{mE}$$

If $\sigma_z = 0$, it would be a two-dimensional stress system;

$$\varepsilon_x - \varepsilon_y = \left(\frac{\sigma_x - \sigma_y}{E}\right)\left(1 + \frac{1}{m}\right) = \left(\frac{m+1}{m}\right)\frac{2q_{max}}{E}$$

But shear strain $= \dfrac{2q_{max}}{N}$ and

$$E = 2N\left(1 + \frac{1}{m}\right)$$

$$\therefore \quad \frac{q_{max}}{N} = \frac{q_{max}}{m}\frac{(m+1)2}{E} = (\varepsilon_x - \varepsilon_y)$$

...(Eq. 22.23)

or greatest shear strain = greatest difference of principal strains.

It has already been seen that the volumetric strain, ε_V, is the sum of the principal strains,

$$\varepsilon_V = \varepsilon_x + \varepsilon_y + \varepsilon_z$$

ILLUSTRATIVE PROBLEMS

PROBLEM 22.1: The tensile principal stress at a point in a material on the three principal planes are zero, 30 MPa and 60 MPa. Find the normal and tangential stresses and the resultant stress on a plane perpendicular to the first principal plane, and inclined at 30° to the plane on which the 60 MPa principal stress acts.

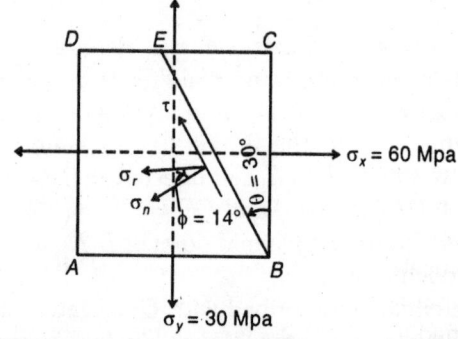

Fig. 22.9: Principal Stresses Known (Illustrative Problem 22.1)

Solution: The data are shown in Fig. 22.9. The plane of the figure is the one on which the stress is zero. The trace of the plane on which the stresses are required is *BE*.

Normal stress on the plane *BE* is

$$\sigma_n = \sigma_x\cos^2\theta + \sigma_y\sin^2\theta$$
$$= 60\cos^2 30° + 30\sin^2 30°$$
$$= 52.5 \text{ MPa (Tensile)}$$

Tangential stress, $\tau = \left(\dfrac{\sigma_x - \sigma_y}{2}\right)\sin 2\theta$

$$= \left(\frac{60-30}{2}\right) \sin 60°$$

$$= 13 \text{ MPa (Left up as shown)}$$

Resultant stresses,

$$\sigma_r = \sqrt{\sigma_n^2 + \tau^2} = \sqrt{(52.5)^2 + 13^2}$$

$$= 54 \text{ MPa (tensile)}$$

$$\tan\phi = \frac{13}{52.5},$$

whence the angle of obliquity, $\phi = 13°54'$

Graphical Solution

The Mohr's circle of stress is shown in Fig. 22.10.

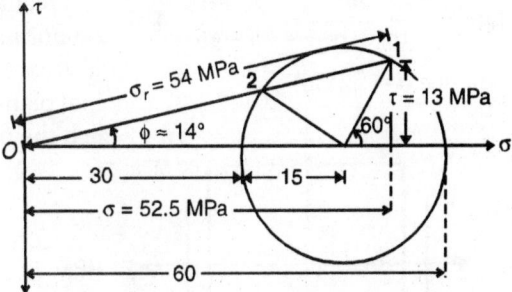

Fig. 22.10: Mohr's Circle (Illustrative Problem 22.1)

The graphical solution is self-explanatory. If it is desired to obtain the resultant stress on the plane on which the angle of obliquity is 14°, the graphical solution is easy. However, there will be two planes on which this condition will be satisfied, the corresponding points on the Mohr circle being (1) and (2). The σ_r—values can be straight away scaled off as 0-1 and 0-2.

PROBLEM 22.2: At a point in a stressed material, the principal stresses are 75 MPa (tensile) and 45 MPa (compressive). Find the resultant stress on a plane inclined at 60° to the axis of the 75 MPa stress, and perpendicular to the plane which has no stress. What is the maximum shear stress?

Solution: The data are shown in Fig. 22.11. Since the plane is inclined at 60° to the direction of the 75 MPa stress the angle which it makes with the plane on which this stress acts, is $\theta = 90° - 60° = 30°$.

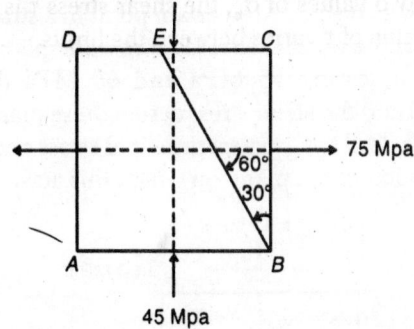

Fig. 22.11: Element (Illustrative Problem 22.2)

Resultant stress, $\sigma_r = \sqrt{\sigma_x^2 \cos^2\theta + +\sigma_y \sin^2\theta}$

$$= \sqrt{75^2 \cos^2 30° + 45^2 \sin^2 30°} = 69 \text{ MPa}$$

$$\tan\phi = \frac{\tau}{\sigma} = \frac{\left[\dfrac{(\sigma_x - \sigma_y)}{2} \cdot \sin 2\theta\right]}{(\sigma_x \cos^2\theta + \sigma_y \sin^2\theta)}$$

$$= \frac{60 \sin 60°}{(75 \cos^2 30° - 45 \sin^2 30°)}$$

$$= 1.15$$

whence the angle of obliquity, $\phi = 49°$

$$\text{Maximum shear stress} = \frac{75 - (-45)}{2} = 60 \text{ MPa}$$

Graphical Solution

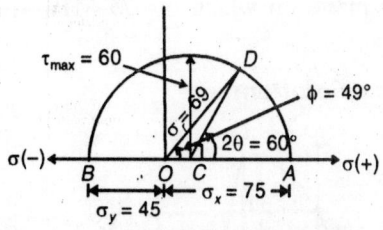

Fig. 22.12: Mohr's Circle (Illustrative Problem 22.2)

The graphical solution by Mohr's circle is obvious. σ_y has to be scaled off onto the negative side as it is compressive. D is fixed such that $D\hat{C}A$ is $2\theta = 60°$. OD is scaled off as $\sigma_r = 69$ MPa. Maximum shear stress is the radius of the Mohr's circle and is scaled off as CP, $P\hat{C}A$ being 90°. It is 60 MPa.

PROBLEM 22.3: At a certain point in a strained material the normal stresses across the planes at right angles are 75 MPa and 45 MPa (both tensile) and the shear stress across these planes is 60 MPa, find the greatest principal stress and the orientation of the plane on which this acts.

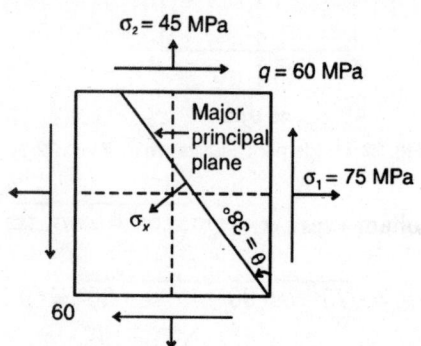

Fig. 22.13: Element with Stresses on Planes at Right Angles (Illustrative Problem 22.3)

Solution: The greatest principal stress

$$\sigma_x = \left(\frac{\sigma_1 + \sigma_2}{2}\right) + \frac{1}{2}\sqrt{(\sigma_1 - \sigma_2)^2 + 4q^2}$$

$$= \left(\frac{75 + 45}{2}\right) + \frac{1}{2}\sqrt{(75 - 45)^2 + 4 \times 60^2}$$

$$= 60 + 62 = 122 \text{ MPa (tensile)}$$

$$\tan 2\theta = \frac{2q}{(\sigma_1 - \sigma_2)} = \frac{2 \times 60}{75 - 45} = 4$$

whence $\theta = 38°$

∴ The major principal plane makes an angle 38° with the plane on which the 75 MPa—normal stress acts.

Graphical Solution:

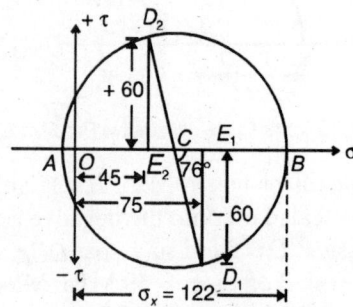

Fig. 22.14: General Case of Mohr's Circle (Illustrative Problem 22.3)

The graphical solution involves the use of the general case of Mohr's circle for a biaxial state of stress. The solution is self-explanatory as shown in Fig. 22.14.

$\overline{OB}$ is scaled off as σ_x and $D_1\hat{C}B$ as 2θ and the results agree with the analytical values.

PROBLEM 22.4: Solve Problem 22.3 if the stress of 45 MPa is compressive.

Major principal stress,

$$\sigma_x = \left(\frac{\sigma_1 + \sigma_2}{2}\right) + \frac{1}{2}\sqrt{(\sigma_1 - \sigma_2)^2 + 4q^2}$$

$$= \left(\frac{75 - 45}{2}\right) + \frac{1}{2}\sqrt{(75 + 45)^2 + 4 \times 60^2}$$

$$= 100 \text{ MPa (Tensile)}$$

$$\tan 2\theta = \frac{2q}{(\sigma_1 - \sigma_2)} = \frac{2 \times 60}{75 + 45} = 1$$

whence $\theta = 22°30'$

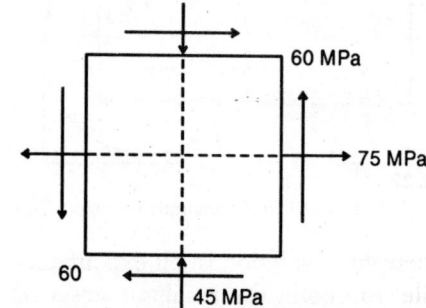

Fig. 22.15: Element under Biaxial State of Stress (Illustrative Problem 22.4)

Hence this principal stress is inclined at 22°30' with the direction of the 75 MPa—stress.

Graphical Solution

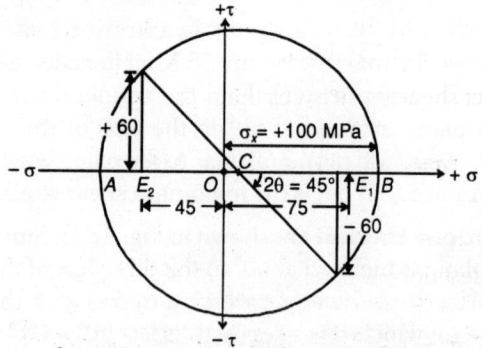

Fig. 22.16: Mohr's Circle (Illustrative Problem 22.4)

The Mohr's circle is constructed with the centre at the mid-point C of E_2E_1, OE_1 being +75 and OE_2 being –45 to the chosen scale. E_1D_1 and E_2D_2 are erected as $-q$ and $+q$ (= 60), respectively. The circle is completed with C as centre and CD_1 as radius, cutting the σ-axis at A and B. The greatest principal stress is scaled off as OB and its orientation θ with respect to the axis of 75 MPa stress is from BCD_1, which is 2θ. The results agree with the analytical values.

PROBLEM 22.5: A bolt of 25 mm diameter is subjected simultaneously to a tensile force of 15 kN and a shear force of 10 kN. Determine the greatest normal stress and its orientation with respect to the longitudinal axis of the bolt.

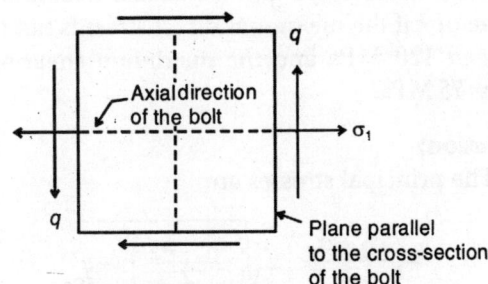

Fig. 22.17: An Element of the Bolt under Stress (Illustrative Problem 22.5)

Diameter of bolt = 25 mm
Tensile force, P = 15 kN
Shear force, F = 10 kN

Axial tensile stress in the bolt $\sigma_1 = \dfrac{P}{A}$

$$= \dfrac{15 \times 1000}{\left(\dfrac{\pi}{4} \times 25^2\right)} \text{ N/mm}^2 \text{ or MPa}$$

$$= 30.56 \text{ MPa}$$

Direct shearing stress in the bolt,

$$q = \dfrac{F}{A} = \dfrac{10 \times 1000}{\left[(\pi / 4) \times 25^2\right]} \text{ MPa}$$

$$= 20.37 \text{ MPa}$$

Greatest normal stress,

$$\sigma_x = \left(\dfrac{\sigma_1 + \sigma_2}{2}\right) + \dfrac{1}{2}\sqrt{(\sigma_1 - \sigma_2)^2 + 4q^2}$$

$$= \dfrac{\sigma_1}{2} + \sqrt{\left(\dfrac{\sigma_1}{2}\right)^2 + q^2} \text{ Since } \sigma_2 = 0$$

$$= \dfrac{30.56}{2} + \sqrt{(15.28)^2 + (20.37)^2}$$

$$= 15.28 + 25.47 = 40.75 \text{ MPa}$$

$$\tan 2\theta = \dfrac{2q}{\sigma_1} = \dfrac{2 \times 20.37}{30.56}$$

whence $2\theta = 53°.12$

or $\qquad \theta = 26°34'$

Graphical Solution:

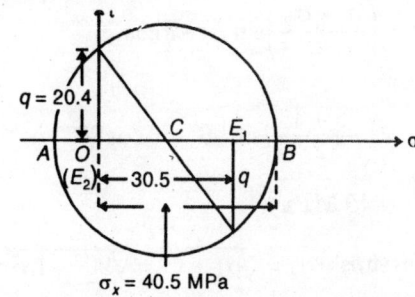

Fig. 22.18: Mohr's Circle (Illustrative Problem 22.5)

The modification in the graphical solution in view of σ_2 being zero is obvious. The results agree with the analytical values.

PROBLEM 22.6: At a point in a strained material the normal stresses on two planes at right angles are 75 MPa and 45 MPa (both tensile) and the shear stresses on these two planes are 30 MPa. Determine the normal stress, the shear stress and the resultant stress on a plane inclined at 45° with the plane on which the 75 MPa, normal stress acts.

Solution:
Normal stress,

$$\sigma_n = \left(\dfrac{\sigma_1 + \sigma_2}{2}\right) + \left(\dfrac{\sigma_1 - \sigma_2}{2}\right)\cos 2\theta + q \sin 2\theta$$

$$\sigma_1 = 75; \ \sigma_2 = 45; \ q = 30; \text{ and } \theta = 45°$$

$$\therefore \ \sigma_n = \left(\dfrac{75 + 45}{2}\right) + \left(\dfrac{75 - 45}{2}\right)\cos 90° + 30\sin 90°$$

$$= 90 \text{ MPa (Tensile)}$$

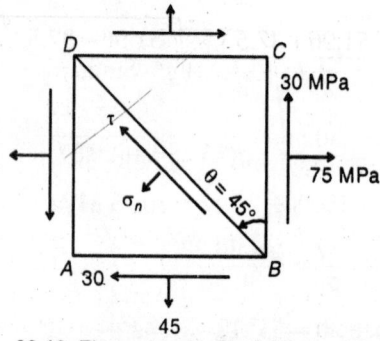

Fig. 22.19: Element under Biaxial State of Stress
(Illustrative Problem 22.6)

Shear stress,

$$\tau = \left(\frac{\sigma_1 - \sigma_2}{2}\right)\sin 2\theta - q\cos 2\theta$$

$$= \left(\frac{75 - 45}{2}\right)\sin 90° - 30\cos 90°$$

$$= 15 \text{ MPa}$$

Resultant stress, $\sigma_r = \sqrt{\sigma_n^2 + \tau^2} = \sqrt{90^2 + 15^2}$

$$= 91.24 \text{ MPa (Tensile)}$$

$$\tan\phi = \frac{\tau}{\sigma_n} = \frac{15}{90} = \frac{1}{6}$$

whence angle of obliquity,
$$= 9°27'36''$$

Graphical Solution:

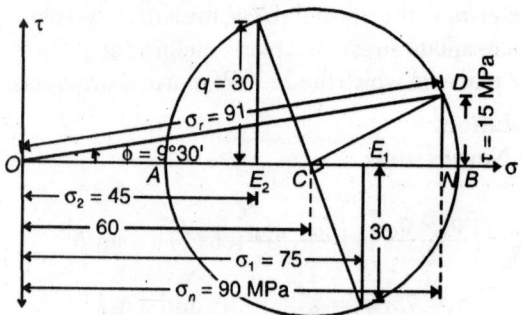

Fig. 22.20: Mohr's Circle (Illustrative Problem 22.6)

The Mohr's circle is completed as usual. D is located on the circle by making $D_1\hat{C}D$ equal to

2θ, or $90°$ in this case. The σ and τ coordinates of the point D are scaled off as σ_n (= 90 MPa-tensile) and τ (= 15 MPa) on the plane inclined at $45°$ with the plane on which the 75 MPa normal stress acts.

OD is scaled off as σ_r (= 91.25 MPa) and $D\hat{O}B$ is measured as the angle of obliquity, ϕ(= 9°30′).

The results agree excellently with those from the analytical solution.

PROBLEM 22.7: At a certain point in a material there are normal stresses of 45 MPa (tension) and 30 MPa (compression) on two planes at right angles to each other, with shearing stress of 22.5 MPa on these planes. If the loading on the material is increased so that the stresses reach values k times those given, find the maximum value of k if the maximum direct stress is not to exceed 120 MPa and the maximum shearing stress 75 MPa.

Solution:
    The principal stresses are

$$\sigma_{x,y} = \left(\frac{\sigma_1 + \sigma_2}{2}\right) \pm \sqrt{\left(\frac{\sigma_1 - \sigma_2}{2}\right)^2 + q^2}$$

$$= \left(\frac{45 - 30}{2}\right) \pm \sqrt{\left(\frac{45 + 30}{2}\right)^2 + (22.5)^2}$$

$$= 7.50 \pm 43.73$$

$\therefore \quad \sigma_x = 51.23$ MPa (tensile)
$\quad \sigma_y = 36.23$ MPa (compression)

Maximum shear stress,

$$\tau_{max} = \left(\frac{\sigma_x - \sigma_y}{2}\right) = 43.73 \text{ MPa}$$

If the given stress are replaced by k times the values,

$$\sigma_k = \frac{k(\sigma_1 + \sigma_2)}{2} + \frac{k}{2}\sqrt{(\sigma_1 - \sigma_2)^2 + 4q^2} = k[\sigma_{x,y}]$$

If σ_k is not to exceed 120 MPa,

$$k = \frac{120}{51.23} = 2.34$$

If τ_{max} is not to exceed 75 MPa,

$$k = \frac{75}{43.73} = 1.71$$

For both these conditions to be simultaneously satisfied, k shall not be greater than 1.71.

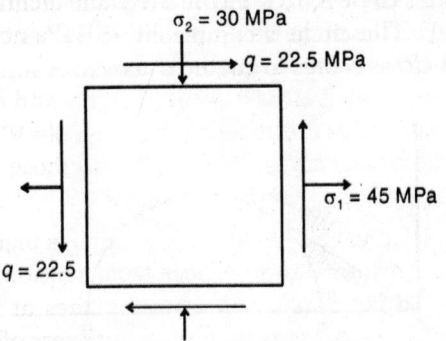

Fig. 22.21: Biaxial State of Stress (Illustrative Problem 22.7)

PROBLEM 22.8: At a point in a material under stress, the intensity of resultant stress on a certain plane is 60 MPa (tensile), inclined at 30° to the normal of that plane. The plane at right angles to this has a normal stress of 37.5 MPa (tensile). Determine the resultant stress on the second plane and the principal stresses and principal planes.

Solution: The shear stress component, q, on the first plane

$$= 60\sin 30° = 30 \text{ MPa}$$

This is also the shear stress on the second plane by the principle of complementary shear.

Resultant stress, σ_r, on the second plane

$$= \sqrt{(37.5)^2 + 30^2} = 48 \text{ MPa (tensile)}$$

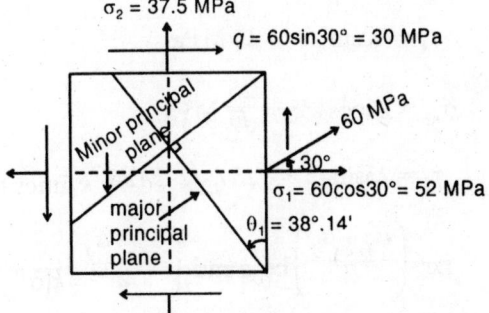

Fig. 22.22: Resultant Stress on Planes at Right Angles (Illustrative Problem 22.8)

Principal stresses are

$$\sigma_{x,\,y} = \left(\frac{\sigma_1 + \sigma_2}{2}\right) \pm \sqrt{\left(\frac{\sigma_1 + \sigma_2}{2}\right)^2 + q^2}$$

$$= \left(\frac{51.96 + 37.5}{2}\right) \pm \sqrt{\left(\frac{51.96 - 37.5}{2}\right)^2 + 30^2}$$

$$= 44.73 \pm 33.32$$

$$= 78.05 \text{ and } 11.41 \text{ MPa (both tensile)}$$

$$\tan 2\theta = \frac{2q}{(\sigma_1 - \sigma_2)}$$

$$= \frac{2 \times 30}{(51.96 - 37.50)} = \frac{60}{14.46} = 4.15$$

whence $2\theta = 76° 27'$ or $\theta_1 = 38°13'\,30''$

$$\theta_2 = 128°13'\,30''$$

The orientation of the principal planes is shown in Fig. 22.22.

PROBLEM 22.9: Direct normal stresses of 100 MPa (tensile) and 80 MPa (compressive) act at a point in an elastic material on planes at right angles to each other, in addition to a shear stress q. If the maximum principal stress developed is 130 MPa (tensile), what is the value of q?

Solution:

$$\sigma_1 = 100 \text{ MPa}$$
$$\sigma_2 = -80 \text{ MPa}$$
$$\sigma_x = 130 \text{ MPa}$$

$$\therefore \quad 130 = \left(\frac{100 - 80}{2}\right) + \sqrt{\left(\frac{100 + 80}{2}\right)^2 + q^2}$$

$$90^2 + q^2 = 120^2$$

$$q = \sqrt{120^2 - 90^2} \text{ MPa} = 79.4 \text{ MPa}$$

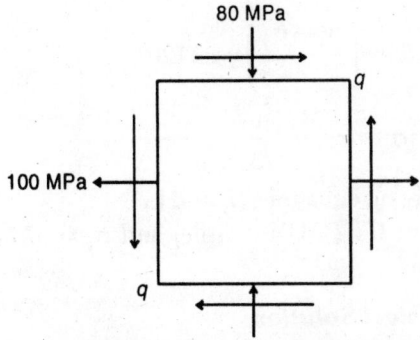

Fig. 22.23: Biaxial State of Stress q is Unknown (Illustrative Problem 22.9)

Graphical Solution:

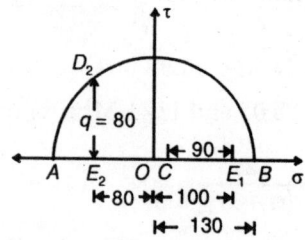

Fig. 22.24: Mohr's Circle (Illustrative Problem 22.9)

With O as centre and $OB (= \sigma_x)$ as radius the Mohr's circle is completed. From E_2 (OE_2 being $\sigma_2 = 80$), the perpendicular $E_2 D_2$, D_2 being the point on the Mohr's circle is scaled off to give the shear stress $q (= 80$ MPa).

PROBLEM 22.10: The maximum angle of obliquity of the resultant stress with reference to the normal to the plane in a stressed member is $30°$ and the magnitude of the resultant stress is 100 MPa (tensile). Determine the principal stresses.

Solution: If ϕ_m is the maximum value of obliquity angle and the angle made by the major principal plane with the plane of maximum obliquity is θ_1,

$$\theta_1 = \frac{\phi_m}{2} + 45° = \frac{30°}{2} + 45° = 60°$$

Also, $\dfrac{\sigma_y}{\sigma_x} = \left(\dfrac{1 - \sin \phi_m}{1 + \sin \phi_m}\right) = \left(\dfrac{1 - \sin 30°}{1 + \sin 30°}\right) = \dfrac{1}{3}$ ...(i)

$$\tau = \sigma_r \sin \phi_m = 100 \sin 30° = 50$$

But $\tau = \left(\dfrac{\sigma_x - \sigma_y}{2}\right) \sin 2\theta$

or $\quad 50 = \left(\dfrac{\sigma_x - \sigma_y}{2}\right) \times \sin 120°$

or $\quad (\sigma_x - \sigma_y) = \dfrac{200}{\sqrt{3}}$ $\qquad$ ...(ii)

Solving equations (i) and (ii),

$\sigma_x = 173.2$ MPa (tensile) and $\sigma_y = 57.7$ MPa (tensile)

Graphical Solution

The graphical solution is relatively simple in this case. OD is scaled off as $\sigma_r = 100$ to the

scale along OD such that DO makes an angle ϕ_m $(= 30°)$ with the σ-axis. The centre C of the Mohr's circle is fixed on the σ-axis such that $O\hat{D}C$ is $90°$. The circle is completed with C as centre and CD as radius to cut the σ-axis in A and B.

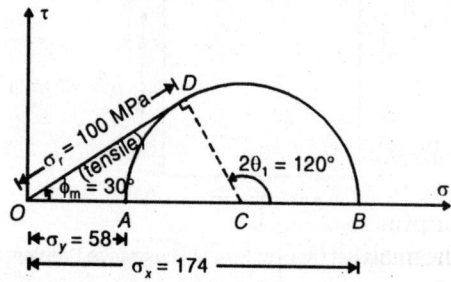

Fig. 22.25: Mohr's Circle (Illustrative Problem 22.10)

OB and OA are scaled off as $\sigma_x (= 174)$ and σ_y $(= 58)$, the principal stresses, to the chosen scale.

The results agree closely with those of the analytical solution.

PROBLEM 22.11: At a point in a material subjected to direct stresses at right angles, the resultant stress on a plane P is 50 MPa (tensile) inclined at $30°$ to the normal, and on a plane Q it is 25 MPa (tensile) inclined at $45°$ to the normal. Determine the principal stresses and principal planes, graphically or otherwise.

Solution:

$$\sigma_{n1} = 50 \cos 30° = 25\sqrt{3} \text{ MPa};$$

$$\tau_1 = 50 \sin 30° = 25 \text{ MPa}$$

$$\sigma_{n2} = 25 \cos 45° = \frac{25}{\sqrt{2}} \text{ MPa};$$

$$\tau_2 = 25 \sin 45° = 25 / \sqrt{2} \text{ MPa}$$

Let $\left(\dfrac{\sigma_x + \sigma_y}{2}\right)$ be a and $\left(\dfrac{\sigma_x - \sigma_y}{2}\right)$ be b.

Then $a + b \cos 2\theta_1 = 25\sqrt{3}$ $\quad$ ⎫ Solve these four

$a + b \cos 2\theta_2 = 25 / \sqrt{2}$ $\qquad$ ⎬ equations simult-

$b \sin 2\theta_1 = 25$ $\qquad\qquad$ ⎪ aneously for a, b, θ_1

$b \sin 2\theta_2 = 25 / \sqrt{2}$ $\qquad$ ⎭ and θ_2.

Note: Take $(180° - 2\theta_2)$ instead of θ_2

Eliminating θ_1 & θ_2,

$$\left.\begin{array}{c} \dfrac{\left(25\sqrt{3}-a\right)^2}{b^2}+\dfrac{25^2}{b^2}=1 \\[3mm] \dfrac{\left(\dfrac{25}{\sqrt{2}}-a\right)^2}{b^2}+\dfrac{\left(25/\sqrt{2}\right)^2}{b^2}=1 \end{array}\right\} \begin{array}{l} \text{Solving} \\ a=36.6 \\ b=25.9 \end{array}$$

$\sigma_x = (a+b) = 62.5\,\text{MPa (Tensile) and}$

$\sigma_y = (a-b) = 10.7\,\text{MPa (Tensile)}$

$\theta_1 = 37°27'$ and $\theta_2 = 68°30'$ with respect to the major principal plane.

The analytical solution, therefore, appears a bit laborious.

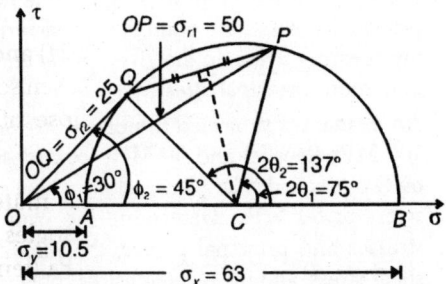

Fig. 22.26: Mohr's Circle (Illustrative Problem 22.11)

The graphical solution is very convenient in this case. OP at $30°$ and OQ at $45°$ with the σ-axis, are scaled off as $\sigma_{r1} = 50$ and $\sigma_{r2} = 25$ to the chosen scale. P and Q are two points on the Mohr's circle. The centre of the circle C is got as the point where the perpendicular bisector of QP intersects the σ-axis. The circle is completed with C as centre CP or CQ as radius. The principal stresses σ_x and σ_y are scaled off as OA and OB to the chosen scale. The angles made by the given planes P and Q with respect to the major principal planes are obtained as half of angles PCB and QCB ($\theta_1 = 37°30'$ and $\theta_2 = 68°30'$). There is excellent agreement of these results with the values from the analytical solution.

PROBLEM 22.12: In a boiler plate, the tensile stress in the circumferential direction is 75 MPa and that in the longitudinal direction is 37.5 MPa. Find what intensity of tensile stress acting alone would produce the same maximum tensile strain if Poisson's ratio is 1/4.

Solution: Maximum tensile strain in the circumferential direction,

$$e_{max} = \frac{75}{E} - \frac{1}{4} \times \frac{37.5}{E}$$

$$= \frac{7}{4E} \times 37.5 = \frac{262.5}{4E}$$

The intensity of tensile stress acting alone required to produce the same maximum tensile strain is,

$$\sigma = E \cdot e_{max} = \frac{262.5}{4E} \times E = 65.63\,\text{MPa}$$

PROBLEM 22.13: Determine the volumetric strain if a bar of steel stretches $\dfrac{1}{1000}$th of its length under simple tension, within the elastic limit. Take Poisson's ratio as 1/4.

Volumetric strain $e_x + e_y + e_z$

$$= \frac{1}{1000} - \frac{1}{4000} - \frac{1}{4000} = \frac{1}{2000}\,\text{(increase)}$$

PROBLEM 22.14: A circle, 300 mm in diameter, is described on a mild steel plate before it is stretched as shown in Fig. 22.27. After stressing, the circle deforms into an ellipse. Calculate the lengths of the major and minor axes of the ellipse and also their directions. $E = 2 \times 10^5$ MPa and $\dfrac{1}{m} = 0.3$.

Solution: The principal stresses are,

$$\sigma_{x,y} = \left(\frac{\sigma_1 + \sigma_2}{2}\right) \pm \sqrt{\left(\frac{\sigma_1 - \sigma_2}{2}\right)^2 + q^2}$$

$$= \left(\frac{70+14}{2}\right) \pm \sqrt{\left(\frac{70-14}{2}\right)^2 + 21^2}$$

$$= 42 \pm \sqrt{28^2 + 21^2}$$

$$= 42 \pm 35$$

$\therefore$ $\sigma_x = 77$ MPa and $\sigma_y = 7$ MPa (both tensile)

Directions of the principal planes are given by,

$$\tan 2\theta = \frac{2q}{(\sigma_1 - \sigma_2)} = \frac{2 \times 21}{(70-14)} = \frac{3}{4} = 0.75$$

whence $\theta_1 = 18°25'$ and $\theta_2 = 108°25'$ with the plane on which the 70 MPa stress acts.

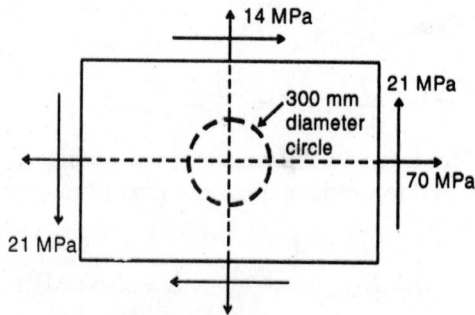

Fig. 22.27: Mild Steel Plate under Stress
(Illustrative Problem 22.14)

From the principal strains, for the ellipse into which the circle gets deformed:

The major axis

$$= \frac{300}{2 \times 10^5}[77 - 0.3 \times 7] + 300 = 300.11235 \text{ mm}$$

The minor axis

$$= 300 + \frac{300}{2 \times 10^5}[7 - 0.3 \times 77]$$

$$= 300 - 0.02415 = 299.97585 \text{ mm}$$

Note: It can be seen that almost any problem may be solved graphically with the aid of the Mohr's circle, and that, in certain cases, the graphical solution proves to be relatively simple.

PRACTICE PROBLEMS

22.1   At a certain point in a strained material, the principal stresses are 100 MPa and 50 MPa (both tensile). Determine the normal, tangential, and resultant stresses on a plane through the point inclined at 45° to the major principal plane.

22.2   At a certain point in a strained material, the principal streses are 100 MPa (tensile) and 50 MPa (compressive). Determine the resultant stress on a plane inclined at 30° to the major principal plane. What is the maximum shear stress at the point?

22.3   In a stressed material the normal stresses on two-planes at right angles to each other are

MPa and 20 MPa (both tensile), with a shear stress of 50 MPa on these planes. Determine the principal stresses and principal planes. What is the maximum shear stress and the normal stress on this plane?

22.4   The principal stresses at a point in a strained material are 40 MPa and 20 MPa (both tensile). Find the normal, shear, and resultant stresses on a plane inclined at 30° with the major principal plane. What is the maximum shear stress? What is the maximum angle of obliquity and what is the inclination of this plane with the major principal plane?

22.5   The normal stresses on two perpendicular planes at a point in a stressed material are 75 MPa and 15 MPa, both tensile. There are also shearing stresses on those planes. If the minor principal stress at the point is zero, determine the shearing stress on the given plane and the maximum principal stress.

22.6   An element is subjected to normal stresses of 100 MPa (tensile) and 50 MPa (compressive) on two perpendicular planes, along with a shear stress of 60 MPa. Determine the principal stresses and principal planes, the maximum shear stress, and the resultant stress on a plane inclined at 30° to the major principal plane.

22.7   There are normal stresses of 100 MPa (tensile) and 80 MPa (compressive) on two perpendicular planes at a point in an elastic material, in addition to a certain shear stress. If the maximum principal stress is 130 MPa, what is the shear stress acting on the perpendicular planes?

22.8   At a point in a stressed material, there are normal stresses of 40 MPa (tensile) and 30 MPa (compressive) on two perpendicular planes along with shear stress of 20 MPa. If these stresses are increased to k times those given, find the maximum value of k if the maximum direct stress and maximum shear stress are not to exceed 120 MPa and 80 MPa respectively.

22.9   In a complex stress system there are two planes on which the resultant stress makes an angle 30° with the normal of the plane in each case. The resultant stress on one plane is 80 MPa (tensile) and on the other it is 40 MPa (tensile). Find the inclination between the two planes, and the principal stresses.

22.10 An element is subjected to principal stresses 100 MPa and 60 MPa (both tensile). Find the normal and shear stresses on a plane inclined at 30° to the plane of major principal stress. Calculate also the maximum principal strain, if Poisson's ratio is 0.25.

22.11 The principal stresses at a point in a stressed material are $\sigma_x = 100$ MPa (T), $\sigma_y = 60$ MPa (C); and $\sigma_z = 40$ MPa (T). $E = 2 \times 10^5$ MPa; and $1/m = 1/4$. Determine the principal strains and the volumetric strain.

22.12 A circle 150 mm in diameter is described on a mild steel plate before it is subjected to normal stresses 70 MPa and 14 MPa (both tensile) on its perpendicular edges with a shear stress of 21 MPa on these edge planes. Determine the major and minor axes of the ellipse into which the circle gets deformed. $E = 2 \times 10^5$ MPa and $1/m = 0.3$.

22.13 The principal stresses at a point in a material are 500 MPa (tensile) and 300 MPa (compressive). $E = 2 \times 10^5$ MPa and $\mu = 1/3$. Determine the major and minor axes of an ellipse into which a circle of 40 mm radius will got deformed upon stressing.

22.14 At a certain point in a steel structural element, the directions of the principal stresses, σ_x and σ_y are known. Measurements by strain gauges show that the strain $e_x = 8 \times 10^{-4}$ (tensile) and $e_x = 5 \times 10^{-4}$ (compressive). Determine the values σ_x and σ_y and the maximum shear stress. $E = 2 \times 10^5$ MPa and $1/m = 4$.

Chapter 23

Strain Energy and Resilience

23.1 INTRODUCTION

Whenever a body undergoes any kind of strain, work is done externally by the applied load. This work is stored as 'strain energy' internally in the material. Neglecting any loss of energy, the internal strain energy stored is equal to the external work done by the load. This is the case with all elastic materials. Work is done under any kind of straining action—tension, compression, or shear, or for that matter under any kind of complicated straining action. Hence the concept of strain energy is applicable to any kind of straining action.

This and other relevant ideas will be dealt with in the sections to follow.

23.2 WORK DONE DURING STRAIN

During the application of, say, a gradually increasing tensile load to a bar, elongation takes place in the direction of the applied load, and work is done. If the load W is considered constant during an infinitesimal extension δx the work done is $W \cdot \delta x$.

During an elongation δl, the work done may be got conveniently by the principle of summation or integration, as

$$\Sigma(W \cdot \delta x) \text{ or } \int_0^{\delta l} W \cdot \delta x$$

This may be interpreted graphically also on a load-extension diagram as shown in Fig. 23.1.

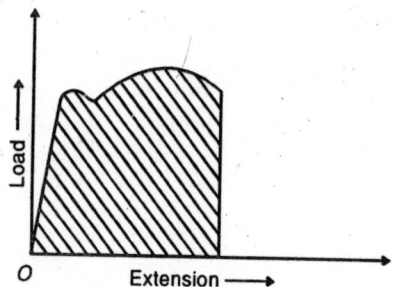

Fig. 23.1: Work Done During Tensile Strain

$\int_0^{\delta l} W \cdot \delta x$ is obviously the area under the curve for the entire extension—for a tensile strain of a ductile material like mild steel.

23.3 ELASTIC STRAIN ENERGY IN TENSION AND COMPRESSION

The energy expended (measured by the work done) in producing an elastic strain is stored as 'strain energy' in the material; this reappears on the removal of the load. However, the work done during non-elastic strain is spent in overcoming the cohesion of the particles, and the energy expended as heat in the strained material.

In materials which follow Hooke's law, the elastic portion of the load-extension diagram is a straight line, and the strain energy stored within the elastic limit during the tensile strain equals half the product of the load and extension (Fig. 23.2).

Similar ideas are applicable in compression also within the elastic limit.

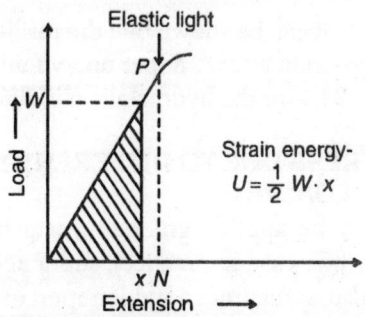

Fig. 23.2: Strain Energy in Tension

23.4 RESILIENCE

Literally speaking, the word 'Resilience' means the power to spring back on the removal of the stress. But, technically the term means the amount of energy restored by the strained body.

Within the elastic limit, this is generally the product of half the load and the extension.

If a body is acted upon by a uniform tensile stress σ, the cross-sectional area and length being A and l, the extension is

$$\delta l = \frac{\sigma}{E} \cdot l = \frac{p \cdot l}{AE}, \; P \text{ being the tensile force.}$$

Hence the strain energy or resilience U is,

$$U = \frac{1}{2} \cdot P \cdot \delta l = \frac{1}{2} \cdot \sigma \cdot A \cdot \frac{\sigma}{E} \cdot l = \frac{1}{2} \frac{\sigma^2}{E} \times$$

$$\text{(Volume of the body)}$$

The strain energy per unit volume of the material, u is

$$u = \frac{1}{2} \frac{\sigma^2}{E} \qquad \qquad ...(\text{Eq. 23.1})$$

(The coefficient 1/2 is applicable only for uniform σ).

23.5 PROOF RESILIENCE AND MODULUS OF RESILIENCE

The maximum strain energy which can be stored in a piece of material without permanent strain is called its 'Proof Resilience".

If the elastic limit or 'proof stress' is σ_e, the proof resilience is given by $(\frac{1}{2})(\sigma_e^2 / E)$ (Volume of the material). This is represented by the area OPN (Fig. 23.2) for a material obeying Hooke's law.

The proof resilience per unit volume of the material $(\frac{1}{2}) \cdot (\sigma_e^2 / E)$ is considered to be a property of the material, and is called the "Modulus of Resilience".

Note. Strain energy beyond the elastic limit cannot be called 'Resilience'!

23.6 STRAIN ENERGY IN SHEAR

When a body suffers shear strain within the elastic limit, elastic strain energy is stored just as in the case of direct stress and strain.

For uniform distribution of shear stress the shearing resilience or elastic shear strain energy is easily calculated (Fig. 23.3).

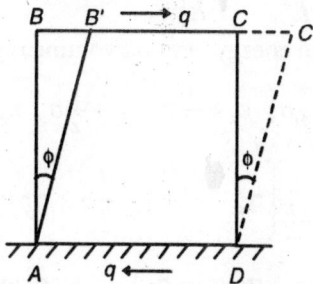

Fig. 23.3: Uniform Shear—Elastic Shear Strain Energy

Let a $ABCD$ be a piece of material, say of unit thickness perpendicular to the plane of the figure, acted on by a shear stress q on the plane BC, and produce a shear strain ϕ and a lateral displacement BB'.

The elastic strain energy U_s is a given by

$$\frac{1}{2} \times (\text{force}) \times (\text{displacement}) = \frac{1}{2} \cdot BC \cdot 1 \cdot q (BB')$$

$$= \frac{1}{2} BC \cdot 1 \cdot q(AB \cdot \phi)$$

$$= \frac{1}{2} BC \cdot 1 \cdot q AB \cdot \left(\frac{q}{N} \right)$$

$$U_S = \frac{1}{2} \cdot \frac{q^2}{N} \text{ (Volume)} \qquad ...(\text{Eq. 23.2})$$

Thus the elastic strain energy in shear per unit volume is given by,

$$u_S = \frac{1}{2} \cdot \frac{q^2}{N} \qquad \qquad ...(\text{Eq. 23.3})$$

N being the modulus of rigidity.

This expression bears similarity with that given by Eq. 23.1.

23.7 STRAIN ENERGY IN THE MORE GENERAL CASE

If the principal stresses at a point in a strained material are σ_x, σ_y and σ_z, respectively:

$$\left.\begin{array}{l} \varepsilon_x = \dfrac{\sigma_x}{E} - \dfrac{(\sigma_y + \sigma_z)}{mE} \\[3mm] \varepsilon_y = \dfrac{\sigma_y}{E} - \dfrac{(\sigma_z + \sigma_x)}{mE} \\[3mm] \varepsilon_z = \dfrac{\sigma_z}{E} - \dfrac{(\sigma_x + \sigma_y)}{mE} \end{array}\right\} \quad \text{...(Eq. 22.22)}$$

The strain energy per unit volume is given by,

$$u = \frac{1}{2}\sigma_x \cdot \varepsilon_x + \frac{1}{2}\sigma_y \cdot \varepsilon_y + \frac{1}{2}\sigma_z \cdot \varepsilon_z$$

$$= \frac{1}{2E}\left[\sigma_x^2 - \frac{\sigma_x(\sigma_y + \sigma_z)}{m} + \sigma_y^2 \right.$$

$$\left. - \frac{\sigma_y(\sigma_z + \sigma_x)}{m} + \sigma_z^2 - \frac{\sigma_z(\sigma_x + \sigma_y)}{m}\right]$$

or

$$u = \frac{1}{2E}\left[(\sigma_x^2 + \sigma_y^2 + \sigma_z^2)\right.$$

$$\left. - \frac{2}{m}(\sigma_x\sigma_y + \sigma_y\sigma_z + \sigma_z\sigma_x)\right] \quad \text{...(Eq. 23.4)}$$

For two-dimensional stress analysis, if $\sigma_z = 0$,

$$u = \frac{1}{2E}\left[(\sigma_x^2 + \sigma_y^2) - \frac{2}{m}\sigma_x\sigma_y\right] \quad \text{...(Eq. 23.5)}$$

For the interesting case of pure shear

$$\sigma_x = q \text{ and } \sigma_y = -q, \ \sigma_z = 0$$

the elastic strain energy per unit volume, u_s, reduces to

$$u_s = \frac{1}{2E}\left(2q^2 + \frac{2}{m}q^2\right) = \frac{q^2}{E}\left(1 + \frac{1}{m}\right)$$

In view of Eq. 21.16 $\left[E = 2N\left(1 + \dfrac{1}{m}\right)\right]$, this reduces to

$$u_s = \frac{1}{2}\cdot\frac{q^2}{N} \quad \text{...(same as Eq. 23.3)}$$

Similarly it can be shown that the resilience in volumetric strain is $(\sigma^2/2K)$ per unit volume. (From Eq. 23.4 for the hydrostatic condition!)

23.8 STRESS DUE TO DIFFERENT TYPE OF LOADING

Loads may be applied gradually (which is the most common case in practice), suddenly, or by impact. Also, sometimes a sudden spurt of energy is supplied to a material (manually as in the case of blasting operations, or by nature through seismic activity or earthquakes); in this case, the material is said to experience "shock loading" or merely "shock".

The effect or the stress induced in each case will be different—this will be seen in the following subsections.

23.8.1 Gradually Applied Load

When a load P is gradually applied, the load is increased from zero to the finite value P in a gradual manner. The work done or the internal strain energy is given by $(1/2)\cdot(P)\cdot(x)$ or the shaded area as shown in Fig. 23.2.

The stress caused is merely P/A, A being the area of cross-section over which P acts. This is the most common situation in practice.

23.8.2 Suddenly Applied Load

Let an axial tensile load P be applied *suddenly* to a bar of cross-sectional area A, causing an *instantaneous* elongation of δl in a length l. Let σ be the instantaneous stress caused, which is, say, within the elastic limit. The bar then behaves like any other perfect spring, and makes oscillations, the amplitude being equal to the extension which would be produced by the same load gradually applied. Hence the maximum instantaneous strain produced is *double* that which would be produced by the same load applied gradually.

This may be proved as follows:

The stress is zero when the bar is unstretched and increases with increase in stretch. The stress is finally the maximum when the bar stretches fully by δl. Thus the average stress in the bar is $(1/2)\,\sigma$.

As the load is P throughout the deformation, the work done on the bar is $P \cdot \delta l$, which is the strain energy, U.

But $U = \left(\frac{1}{2} \sigma \times A\right) \cdot \delta l$

$\therefore \quad \frac{1}{2} \sigma \cdot A \cdot \delta l = P \cdot \delta l$

or $\quad \sigma = \dfrac{2P}{A}$ ...(Eq. 23.6)

Thus the instantaneous stress developed in a bar subjected to a suddenly applied load is *twice* the stress produced by the same load applied gradually.

$$\delta l = \frac{\sigma}{E} \cdot l = \frac{2Pl}{A \cdot E} \qquad \text{...(Eq. 23.7)}$$

Hence, the instantaneous elongation produced in a bar subjected to a suddenly applied load is *twice* that produced by the same load applied gradually.

23.8.3 Impact Load

Let a load W be allowed to fall freely from a height h on to a collar fitted at the lower end of a bar of length l and cross-sectional area A (Fig. 23.4), rigidly fixed at the upper end. It then undergoes more extension than when the same load is applied gradually.

If no failure occurs due to impact, the bar oscillates about its equilibrium position. Let δl be the maximum instantaneous elongation, σ be the maximum instantaneous stress, and let P be the equivalent gradually applied load which would produce the same elongation δl,

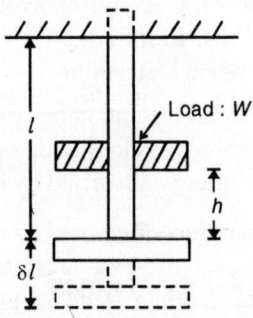

Fig. 23.4: Impact Loading on a Bar

Work done is equal to the strain energy stored, neglecting losses, if any.

Work done by the falling weight $= W(h + \delta l)$

Strain energy stored $= \frac{1}{2} \cdot P \cdot \delta l$

But $\delta l = \dfrac{Pl}{AE}$

$\therefore \quad W\left(h + \dfrac{Pl}{AE}\right) = \dfrac{1}{2} P \times \dfrac{Pl}{AE}$

$Wh + \dfrac{WPl}{AE} = \dfrac{1}{2} \dfrac{P^2 l}{AE}$

$P^2 - 2WP - \dfrac{2AEWh}{l} = 0$

$\therefore \quad P = \dfrac{2W + \sqrt{4W^2 + \dfrac{8AEWh}{l}}}{2}$

(negative value is ignored).

or $\quad P = W\left[1 + \sqrt{\left(1 + \dfrac{2AEh}{Wl}\right)}\right]$ ...(Eq. 23.8)

Then $\sigma = \dfrac{P}{A} = \dfrac{W}{A}\left[1 + \sqrt{\left(1 + \dfrac{2AEh}{Wl}\right)}\right]$ ...(Eq. 23.9)

However, if δl is very small relative to h,

$Wh = \dfrac{1}{2} \dfrac{P^2 l}{AE}$

$\therefore \quad P^2 = \dfrac{2WAEh}{l}$

$\therefore \quad \dfrac{P^2}{A^2} = \dfrac{2WEh}{Al}$

$\therefore \quad \sigma = \dfrac{P}{A} \approx \sqrt{\dfrac{2WEh}{Al}}$

After knowing P, the value of $\delta l \left(= \dfrac{Pl}{AE}\right)$ and $\sigma \left(= \dfrac{P}{A}\right)$ can be found out.

In the specific case when $h = 0$; $P = 2W$; and $\sigma = \dfrac{2W}{A}$.

Thus, the stress produced by a suddenly applied load is twice the one produced by the same load when gradually applied.

23.8.4 Shock Loading

If U is the amount of energy transmitted by shock, it will be absorbed by the bar and stored, provided the elastic limit is not exceeded. Equating it to the energy stored by the bar:

$$U = \frac{1}{2}\sigma \cdot A \cdot \frac{\sigma l}{E} = \frac{1}{2}\frac{\sigma^2}{E}(Al)$$

We solve for σ, which is the maximum instantaneous stress induced in the bar by the shock.

The corresponding deformation, $\delta l = \dfrac{\sigma}{E} \cdot l$

The idea will be understood from the illustrative problems solved in the following Section.

ILLUSTRATIVE PROBEMS

PROBLEM 23.1: Calculate the strain energy in a bar 2 m long and the 50 mm diameter, when it is subjected to a pull of 100 kN. What is the resilience per unit volume? $E = 2 \times 10^5$ MN/m^2.

Solution:
Sectional area of bar,

$$A = \frac{\pi}{4} \times 50^2 = 1963.5 \text{ mm}^2$$

Volume of the bar,

$$V = 1963.5 \times 2000 = 3.927 \times 10^6 \text{ mm}^3$$

Stress,

$$\sigma = \frac{P}{A} = \frac{100 \times 1000}{1963.5} = 50.93 \text{ N/mm}^2$$

Strain energy of the bar,

$$U = \frac{\sigma^2}{2E} \times V = \frac{(50.93)^2}{2 \times 2 \times 10^5} \times 3.927 \times 10^6 \text{ N.mm}$$

$$= 25465.27 \text{ N.mm}$$

Resilence per unit volume

$$= \frac{\sigma^2}{2E} = 6.48467 \times 10^{-3} \text{ N.mm/mm}^3$$

PROBLEM 23.2: A steel bar, 40 mm × 40 mm × 3 m long, is subjected to an axial pull of 120 kN. Calculate the energy stored in the bar during the extension. $E = 2 \times 10^5$ N/mm^2.

Solution: Sectional area, A, of the bar

$$= 40 \times 40 \text{ mm}^2 = 1600 \text{ mm}^2$$

Stress,

$$\sigma = \frac{P}{A} = \frac{120 \times 1000}{1600} = 75 \text{ MPa}$$

Energy stored in the bar,

$$U = \frac{\sigma^2}{2E} \times V$$

$$= \frac{75^2}{2 \times 2 \times 10^5} \times 1600 \times 3000 \text{ N.mm}$$

$$= 67,500 \text{ N.mm}$$

PROBLEM 23.3: A pull of 75 kN is gradually applied to a bar, 40 mm diameter and 5 m long. If $E = 2 \times 10^5$ MPa, determine the strain energy absorbed by the rod.

Solution:
Gradually applied load, $P = 75000$ N

Area of section of the bar

$$= \frac{\pi}{4} \times 40^2 = 400\pi \text{ mm}^2$$

Volume of rod,

$$V = 400\pi \times 5000 \text{ mm}^2 = 2\pi \times 10^6 \text{ mm}^3$$

Stress,

$$\sigma = \frac{P}{A} = \frac{75000}{400\pi} = 59.68 \text{ N/mm}^2$$

Strain energy absorbed,

$$U = \frac{\sigma^2}{2E} \times V = \frac{(59.68)^2}{2 \times 2 \times 10^5} \times 2\pi \times 10^6$$

$$= 55,947 \text{ N.mm}$$

PROBLEM 23.4: A pull of 75 kN is suddenly applied to a bar, 40 mm diameter and 5 m long. $E = 2 \times 10^5$ MPa. Determine:

(*i*) maximum instantaneous stress induced,
(*ii*) instantaneous elongation of the bar, and
(*iii*) strain energy absorbed by the bar.

(*i*) Stress induced,

$$\sigma = \frac{2P}{A} = \frac{2 \times 75 \times 1000}{\left(\frac{\pi}{4} \times 40^2\right)} = 119.36 \text{ MPa}$$

(ii) Instantaneous elongation,

$$\delta l = \frac{\sigma}{E} \cdot l = \frac{119.36}{2 \times 10^5} \times 5000 \text{ mm}$$

$$= 2.984 \text{ mm}$$

(iii) Strain energy absorbed,

$$U = \frac{\sigma^2}{2E} \times V$$

$$= \frac{(119.36)^2}{2 \times 2 \times 10^5} \times \frac{\pi}{4} \times 40^2 \times 5000 \text{ N.mm}$$

$$= 2,23,788 \text{ N.mm}$$

PROBLEM 23.5: An axial pull of 60 kN is suddenly applied to a steel bar, 2 m long and 1000 mm^2 cross-section. Calculate the strain energy absorbed by the bar. $E = 2 \times 10^5$ MPa.

Solution:

Pull, $P = 60,000$ N

Area $A = 1000$ mm^2

Stress, $\sigma = \frac{2P}{A} = \frac{2 \times 60,000}{1000} = 120$ MPa

Strain energy,

$$U = \frac{\sigma^2}{2E} \times V = \frac{120 \times 120 \times 1000 \times 2000}{2 \times 2 \times 10^5} \text{ N.mm}$$

$$= 72,000 \text{ N.mm}$$

PROBLEM 23.6: Water under a pressure of 10 MPa is suddenly admitted on a plunger of 100 mm diameter, attached to a rod 25 mm diameter and 10 m long. Find the maximum instantaneous stress and deformation of the rod. $E = 2 \times 10^5$ MPa.

Solution:

Force, $P = 10 \times \frac{\pi}{4} \times 100^2$ N $= 7.854 \times 10^4$ N

Stress,

$$\sigma = \frac{2P}{A} = \frac{2 \times 7.85 \times 10^4}{\left(\frac{\pi}{4} \times 25^2\right)} = 320 \text{ MPa}$$

Elongation of the rod,

$$\delta l = \frac{320}{2 \times 10^5} \times 10 \times 1000 \text{ mm} = 16 \text{ mm}$$

PROBLEM 23.7: A wagon of weight 50 kN is attached to a wire rope, 50 mm diameter, and moves down in slope at a speed of 3 kmph when the rope jams and the wagon is brought to rest. If the length of rope is 50 m at the time of sudden stoppage, calculate the instantaneous maximum stress and elongation. $E = 2 \times 10^5$ N/mm^2.

Solution:

Speed, $v = \frac{3 \times 1000}{60 \times 60} = \frac{5}{6}$ m/s

Kinetic energy of the wagon,

$$E_k = \frac{Wv^2}{2g} = \frac{50 \times 1000 \times 25 \times 1000}{36 \times 2 \times 9.81} \text{ N.mm}$$

$$= 1.77 \times 10^6 \text{ N.mm}$$

This energy is transmitted to the rope. Strain energy of the rope,

$$U = \frac{\sigma^2}{2E} \times (\text{Volume}) \text{ N.mm},$$

σ being the stress in N/mm^2, E the Young's Modulus of the material of the rope, and volume of the rope being expressed in mm^3.

$$\therefore U = \frac{\sigma^2}{2 \times 2 \times 10^5} \times 50 \times 1000 \times \frac{\pi}{4} \times 50^2 \text{ N.mm}$$

Equating E_k and U,

$$\frac{\sigma^2}{2 \times 2 \times 10^2} \times 50 \times 1000 \times \frac{\pi}{4} \times 50^2$$

$$= \frac{50 \times 1000 \times 25 \times 1000}{36 \times 2 \times 9.81}$$

$$\therefore \sigma = \sqrt{\frac{2 \times 2 \times 10^5 \times 4^2 \times 50 \times 1000 \times 25 \times 1000}{50 \times 1000 \times \pi \times 50^2 \times 36 \times 2 \times 9.81}}$$

$$\text{N/mm}^2$$

$$= \sqrt{\frac{10^6}{88.29\pi}} = \frac{1000}{\sqrt{88.26\pi}} \text{ N/mm}^2 = 60 \text{ MPa}$$

$\therefore$ The instantaneous maximum stress $= 60$ MPa

Instantaneous elongation,

$$\delta l = \frac{\sigma}{E} \cdot l = \frac{60}{2 \times 10^5} \times 50 \times 10^3 \text{ mm}$$

$$= 15 \text{ mm}$$

PROBLEM 23.8: A bar of 12 mm diameter gets elongated by 3.18 mm under a steady load of 8 kN. What stress would be produced in the bar when a weight of 1 kN falls through a height of 80 mm, if the bar is initially unstressed. $E = 2 \times 10^5$ MPa.

Solution:

Area of section of the bar,

$$A = \frac{\pi}{4} \times 12^2 = 36\pi \text{ mm}^2$$

$$\delta l = \frac{Pl}{AE} \rightarrow 3.18 = \frac{8000 \times l}{36\pi \times 2 \times 10^5}$$

∴ Length, l, of the bar

$$= \frac{3.18 \times 36\pi \times 2 \times 10^5}{8000} \text{ mm}$$

$$= \frac{3.18 \times 36\pi \times 2}{80} \text{ m}$$

$$= 9 \text{ m}$$

Under the impact of load

$$\sigma = \frac{W}{A}\left[1 + \sqrt{\left(1 + \frac{2AEh}{Wl}\right)}\right]$$

$$= \frac{1000}{36\pi}\left[1 + \sqrt{\left(1 + \frac{2 \times 36\pi \times 2 \times 10^5 \times 80}{1000 \times 9 \times 1000}\right)}\right]$$

$$= 186.4 \text{ MPa}$$

PROBLEM 23.9: Determine the largest weight that can be dropped from a height of 250 mm on to a collar at the lower end of a bar, 40 mm diameter and 2.4 m long, without exceeding the elastic limit of 330 MPa. $E = 2 \times 10^5$ MPa.

Solution:

Let the largest weight be W newtons.

Work done by the falling load

$$= W(250 + \delta l) \text{ N.mm}$$

$$\delta l = \frac{\sigma}{E} \times l = \frac{330}{2 \times 10^5} \times 2400 \text{ mm} \approx 4 \text{ mm}$$

Strain energy,

$$U = \frac{\sigma^2}{2E} \times (\text{Volume})$$

$$= \frac{(330)^2}{2 \times 2 \times 10^5} \times \frac{\pi}{4} \times 40^2 \times 2400 \text{ N.mm}$$

Equating the work done by the falling load to the strain energy stored (neglecting lossses),

$$254W = \frac{(330)^2 \times \pi \times (40)^2 \times 2400}{16 \times 10^5} \text{ N}$$

or $$W = \frac{330 \times 330 \times \pi \times 1600 \times 2400}{16 \times 10^5 \times 254 \times 1000} \text{ kN}$$

$$= 3.23 \text{ kN}$$

(Direct application of the formula and solving for W is also possible).

PROBLEM 23.10: A vertical tie-rod, rigidly fixed at the upper end, consists of a steel rod 2.5 m long and 20 mm diameter, encased in a brass tube, 20 mm internal diameter and 30 mm external diameter. The rod and casing are fixed at the lower end. The composite rod is subjected to a load of 10 kN falling freely through 5 mm before being arrested by the tie. Calculate the stresses in steel and brass. $E_s = 2 \times 10^5$ MPa and $E_b = 1 \times 10^5$ MPa.

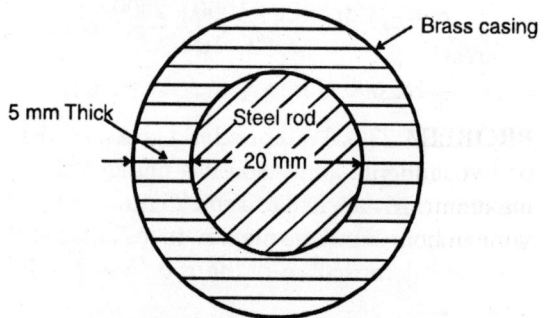

Fig. 23.5: Composite Rod under Impact Load
(Illustrative Problem 23.10)

Solution: Let the strain in the composite bar be e.

Strain in the steel rod, $\sigma_s = E_s \cdot e = 2 \times 10^5 \cdot e$

Stress in the brass rod, $\sigma_b = E_b \cdot e = 1 \times 10^5 \cdot e$

Sectional area of steel rod,

$$A_s = \frac{\pi}{4} \times 20^2 = 100\pi \text{ mm}^2$$

Sectional area of brass tube,

$$A_b = \frac{\pi}{4}(30^2 - 20^2) = 125\pi \text{ mm}^2$$

Strain energy in the steel rod,

$$U_s = \frac{4 \times 10^{10} \times e^2}{2 \times 2 \times 10^5} \times 100\pi \times 2500$$

$$= 25\pi e^2 \cdot 10^9$$

Strain energy in the brass rod,

$$U_b = \frac{1 \times 10^{10} \times e^2}{2 \times 1 \times 10^5} \times 125\pi \times 2500$$

$$= 15.625\pi\, e^2 \cdot 10^9$$

Total energy stored, $U = 40.625\pi \cdot e^2 \cdot 10^9$

$$\delta l = e \cdot l = 2500 \cdot e$$

Work done by the load $= 10000\,(5 + 2500 \cdot e)$

Assuming no loss of energy,

$$50000(1 + 500 \cdot e) = 40.625\pi \cdot 10^9 \cdot e^2$$

$$1 + 500e = \frac{40.625\,\pi \cdot 10^5 \cdot 10^4\, e^2}{50000} = 81.25 \times 10^4\,\pi e^2$$

$$= 25,52,544 e^2$$

$$e = \frac{+500 \pm \sqrt{250000 + 4 \times 2552544}}{51,05,088} = 7.3 \times 10^{-4}$$

∴ Stress in steel,

$$\sigma_s = 2 \times 10^5 \times 7.3 \times 10^{-4} = 146 \text{ MPa}$$

∴ Stress in brass,

$$\sigma_b = 1 \times 10^5 \times 7.3 \times 10^{-4} = 73 \text{ MPa}$$

PROBLEM 23.11: Compare the strain energies of two steel bars shown in Fig. 23.6. The maximum tensile stress is 150 MPa and is the same in both the bars. $E = 2 \times 10^5$ MPa.

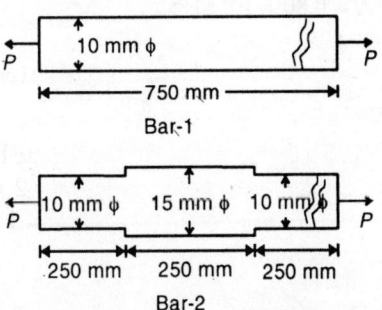

Fig. 23.6: Strain Energies of Two Bars (Illustrative Problem 23.11)

Solution: Since the minimum cross-sectional areas and the maximum stress in both the bars are the same, the tensile forces also must be equal.

Minimum sectional cross,

$$A = \frac{\pi}{4} \times 10^2 = 25\,\pi \text{ mm}^2$$

Tensile force, $P = \sigma \cdot A = 150 \times 25 = 11,780$ N

Strain energy of the first bar,

$$U_1 = \frac{P^2}{2AE} \cdot l = \frac{P^2 \times 750}{2 \times 2 \times 10^5 \times 25\,\pi} \text{ N.mm}$$

Strain energy of the first bar,

$$U_2 = \frac{P^2}{2AE} \cdot l = \frac{P^2}{2 \times 2 \times 10^5} \times \left[\frac{500}{25\pi} + \frac{250}{56.25\pi} \right] \text{ N.mm}$$

$$\therefore \quad \frac{U_1}{U_2} = \frac{3}{\{2 + (1/2.25)\}} = \frac{3}{2.44} = 1.23$$

(cancelling all the common factors in both U_1 and U_2)

PROBLEM 23.12: The shear stress in a material at a point is 50 MPa. Determine the strain energy per unit volume due to shear stored in the material. Take $N = 8 \times 10^4$ MPa.

Solution: Strain energy per unit volume in shear,

$$u_s = \frac{q^2}{2N} = \frac{(50)^2}{2 \times 2 \times 10^4}$$

$$= 1.5626 \times 10^{-2} \text{ N.mm/mm}^3$$

PRACTICE PROBLEMS

23.1 Calculate the strain energy in a bar, 50 mm diameter and 1.5 m long, when it is subjected to a pull of 90 kN. What is the resilience per unit volume?

23.2 A steel bar, 30 mm square in section and 2 m long, is subjected to an axial pull of 60 kN. Calculate the strain energy stored in the bar.

23.3 A pull of 60 kN is applied on to a bar, 30 mm diameter and 3 m long. $E = 2 \times 10^5$ MPa. What is the strain energy absorbed by the rod?

23.4 A pull of 30 kN is suddenly applied to a bar of square section, 30 mm size, and 3 m long. $E = 2 \times 10^5$ MPa. Determine (*i*) the maximum instantaneous stress induced; (*ii*) instantaneous elongation of the bar; and (*iii*) the strain energy absorbed by the bar.

23.5 An axial pull of 50 kN is suddenly applied on to a steel bar, 1 m long and 1000 mm² in cross–section. What is the strain energy absorbed? $E = 2 \times 10^5$ MPa.

23.6 A weight of 1500 N falls through 100 mm on to a collar attached to a steel bar, 25 mm diameter and 2.5 m long. Determine the maximum instataneous stress and elongation. $E = 2 \times 10^5$ MPa.

23.7 An unknown weight falls through 90 mm on to a collar rigidly attached to the lower end of a vertical bar, 3 m long, and 600 mm² in section. If the maximum instantaneous elongation was found to be 3 mm, what is the corresponding stress and the unknown falling weight? Assume $E = 2 \times 10^5$ MPa.

23.8 Two bars A and B of the same material are each 500 mm long. For bar A, the area of section for the first 200 mm length is 600 mm² and for the remaining 300 mm length it is 900 mm². For bar B, it is 600 mm² throughout. If bar B receives a blow inducing a stress of 90 MPa, find the maximum stress produced by the same blow in bar A.

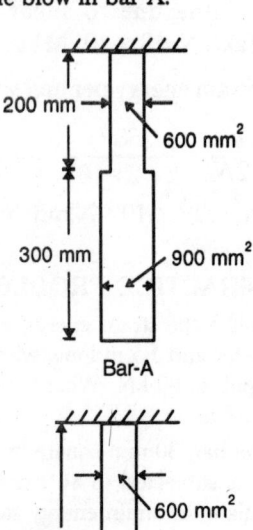

Bar-A

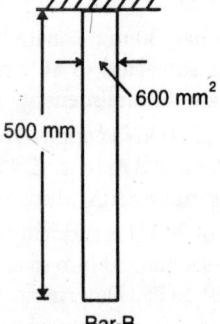

Bar-B

Fig. 23.7: Two Bars Absorbing Energy from a Blow (Practice Problem 23.8)

23.9 Two round bars A and B are each 600 mm long as shown in Fig. 23.8. The bar A receives an axial blow, which produces a maximum stress of 200 MPa. Find the maximum stress produced by the same blow on bar B. Determine also the ratio of the energies stored in bars A and B if the bar also is stressed to a maximum of 200 MPa.

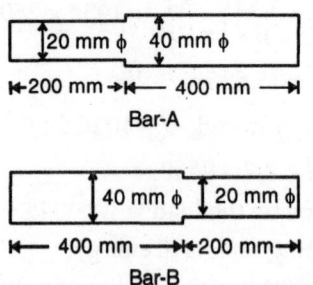

Fig. 23.8: Two Bars Absorbing Energy (Practice Problem 23.9)

23.10 A bar, 2 m long, is subjected to an axial pull such that the maximum stress is 150 MPa. The cross-section is 200 mm² for length of 1.90 m, while it is only 100 mm² for the middle 0.10 m length. $E = 2 \times 10^5$ MPa. Calculate the strain energy stored in the bar.

23.11 Compare the abilities of the two axially loaded bars to take impact load from a height of 40 mm, if one bar is 400 mm long and 100 mm² in section, while the other is 150 mm long and 125 mm² in section. The maximum allowable stress as 150 MPa. $E = 2 \times 10^5$ MPa.

23.12 A rectangular bar, 100 mm × 50 mm × 1000 mm long is subjected to a shear stress of 40 MPa. Determine the strain energy stored, if $N = 0.80 \times 10^5$ MPa.

Chapter 24

Shearing Force and Bending Moment in Beams

24.1 INTRODUCTION

As already defined in Chapter 3 (Section 3.9), a "beam" is a structural member acted on by transverse forces and/or couples that lie in a plane containing the longitudinal axis. Beams are usually straight and horizontal and the external forces are commonly vertical. If the loads are inclined to the longitudinal axis, the effects of the transverse components and the axial components are to be determined separately.

Beams shall be supported appropriately to provide stability and the capability to carry/transmit external loads. The different kinds of supports, of structures, and of loads have already been dealt with in Chapter 3 (Section 3.9). In this chapter only the effects of transverse components of the loads are treated; the nature of the straining action caused in beams is primarily "bending".

24.2 STRAINING ACTIONS ON BEAMS

Let us consider a beam carrying a number of transverse loads as shown in Fig. 24.1.

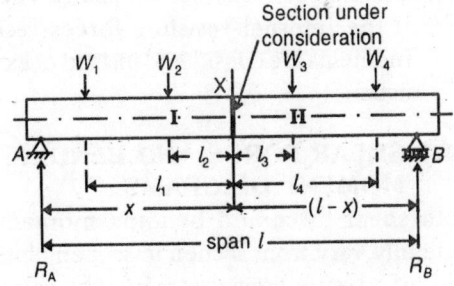

Fig. 24.1: Loaded Simple Beam

The entire beam is in equilibrium under the combined action of the external loads W_1, W_2, W_3, W_4 and the support reactions at A and B. Also, if we imagine the beam to be divided into two portions I and II, each part is in equilibrium, according to the free body concept. The system which keeps portion I in equilibrium consists of the force W_1, W_2 and R_A, together with the forces exerted on I by II across the section X by virtue of the stress in the beam. We may conveniently consider these latter forces by estimating their components and their moments. The reactions are such that $R_A + R_B = W_1 + W_2 + W_3 + W_4$, numerically,

Applying static equilibrium conditions, we have:

(i) the total horizontal force across the section X is zero since there are no horizontal forces acting on portion I;

(ii) since the algebraic sum of the vertical upward forces on I is:

$$(R_A - W_1 - W_2);$$

the total or resultant *downward* vertical force exerted by II on I is $(R_A - W_1 - W_2)$, which is also equal to a downward force, $(W_3 + W_4 - R_B)$. If the former is treated as the external action, the latter is the internal resistance or reaction.

The action of I on II is equal and opposite.

The resultant effect on the section is to cause tangential force or "shear".

(*iii*) The moment of the external forces on I about any axis perpendicular to the figure and in the section X is,

$$M = R_A x - W_1 l_1 - W_2 l_2$$

which is also equal to $W_3 l_3 + W_4 l_4 - R_B$ $(l - x)$, and is of clockwise sense if these expressions yield positive values. The moment exerted by II on I must balance the above value and is therefore of equal magnitude. The former is the external moment, and the latter is the internal resistance.

The resultant effect on the section is to cause "bending" or "flexure".

24.3 SHEARING FORCE

"Shearing Force" at any section of a loaded beam is defined as the algebraic sum of all the forces, including support reactions, to one side of the section, either to the left or to the right.

This causes tangential or shear stresses on the section, as will be seen in Chapter 26.

24.4 BENDING MOMENT

"Bending Moment" at any section of a loaded beam is defined as the algebraic sum of the moments of all the forces, including support reactions, to one side of the section, either to the left or to the right.

This causes bending stresses or normal stresses on the section, as will be seen in Chapter 25.

24.5 SIGN CONVENTIONS

It is obvious that a sign convention is necessarily to be adopted for both the shearing force and the bending moment.

The sign convention adopted herein for shearing force (S.F.) is as follows (Fig. 24.2):

Upward force to the *left/ Downward* force to the *right—Positive.*
Downward force to the *left/ Upward* force to the *right—Negative.*

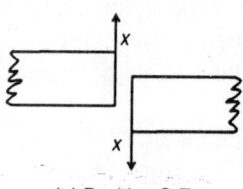

(*a*) Positive S.F.

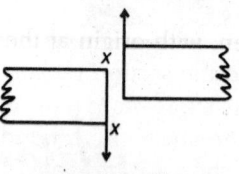

(*b*) Negative S.F.

Fig. 24.2: Sign Convention for Shearing Force

The sign convention adopted herein for bending moment (B.M.) is as follows (Fig. 24.3):

A bending moment tending to bend the beam with concavity *upward* is considered *positive.* (This corresponds to clockwise moment to the left or counterclockwise moment to the right).

This means the beam tends to assume a cup-shape as though to hold water—also said to be "sagging".

A bending moment tending to bend the beam with *convexity upwards* is considered *negative.*

(This corresponds to counterclockwise moment to the left or clockwise moment to the right).

This means the beam tends to assume an inverted cup shape as though to shed or drain away water–also said to be "hogging".

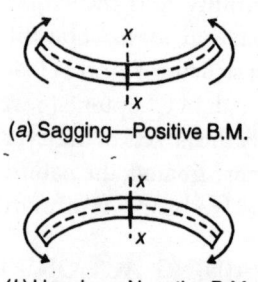

(*a*) Sagging—Positive B.M.

(*b*) Hogging—Negative B.M.

Fig. 24.3: Sign Convention for Bending Moment

Note: The sign conventions will just get reversed if the internal resisting forces/resisting moments are considered instead of external actions.

24.6 SHEAR FORCE AND BENDING MOMENT DIAGRAMS

Both shear force and bending moment will generally vary from section to section along the span of a beam; their values may be calculated arithmetically, or general algebraic expressions may be written, with origin at the left end, and the distance to the section reckoned positive to the right. These expressions indicate the nature of variation of the shear force/bending moment, which may be shown as ordinates from section to section. These diagrams with the span or length of the beam as the base and the ordinates as shearing force/bending moment are known as the shearing force/bending moment diagrams (S.F. and B.M. diagrams). From these diagrams, it is easy to find the salient values, like zero values and maximum values, and other interesting information.

These diagrams may also be drawn to scale graphically by a certain procedure; however, it is ordinarily sufficient to sketch the S.F. and B.M. diagrams by the analytical method.

24.7 RELATION BETWEEN SHEAR FORCE AND BENDING MOMENT

There is a distinct relationship between the rate of loading, shearing force, and bending moment in a loaded beam, which may be shown as follows (Fig. 24.4).

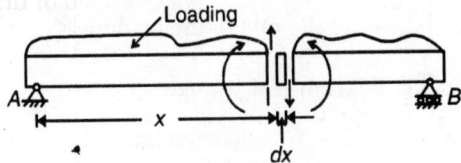

Loading

A — x — B

dx

(a) Conditions at a Section of a Loaded Beam

Intensity of loading: (w)
(assumed to be constant over dx)

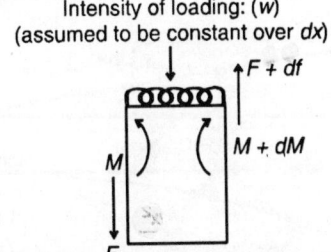

$F + df$

M $M + dM$

F

(b) Free Body of Slice of Thickness dx

Fig. 24.4: Relation between Shear Force and Bending Moment

Let a beam be loaded by a distributed load (not necessarily uniform) as shown; let us consider a thin slice distant x from the left end and of infinitesimal thickness dx. The loading may be assumed to be uniform over the infinitesimal length dx, and let it be w. Let the shearing force be F and bending moment be M at the left end

of the strip at distance x, and $(F + dF)$ and $(M + dM)$ at the right end of it at a distance $(x + dx)$, as shown in (b). (Positive directions are shown for S.F. and B.M. for convenience).

Applying the static equilibrium condition $\Sigma V = 0$ for the freebody of the slice,

$$F + dF = F + w \cdot dx$$

This leads to the relation

$$\frac{dF}{dx} = w \text{ (numerically)} \qquad ...(Eq.\ 24.1)$$

(It may be seen that as we go from left to right, the shear force decreases due to downward load, and hence dF is actually negative here).

This means that the rate of change of shearing force with respect to the span at any point gives the rate of loading at the point.

Integrating both sides between the limits x_1 and x_2 (say),

$$\int_{x_1}^{x_2} dF = \int_{x_1}^{x_2} w \cdot dx \qquad ...(Eq.\ 24.2)$$

The left-hand side represents the change in the shearing force from x_1 to x_2, while the right-hand side represents the area of the loading diagram within the same limits. The equality of these two is thus established.

Taking moments about D and applying $\Sigma M = 0$ to the free body of the slice,

$$M + dM = M + F \cdot dx + w \cdot dx (1/2)(dx)$$

Neglecting second order of infinitesimals,

$$M + dM = M + F \cdot dx$$

or $$\frac{dM}{dx} = F \qquad ...(Eq.\ 24.3)$$

This means that the rate of change of bending moment with respect to the span at any point equals the shearing force at the point.

Integrating both sides between x_1 and x_2,

$$\int_{x_1}^{x_2} dM = \int_{x_1}^{x_2} F \cdot dx$$

This means that the change in bending moment between any two sections is equal to the area of the shearing force diagram between these two sections.

(In the case of concentrated loads, the average rate of loading in a portion of the span is to be considered as it is a discrete variation). For the maximum value of M to occur, $\dfrac{dM}{dx}$ should be zero, or the shear force should be zero; alternatively F must change sign, as it then passes through zero at that section.

Note: For a cantilever, these criteria may not apply as such, as will be seen later.

The first step is to determine the reactions using static equilibrium conditions for the beam as a whole; the next is to determine the shear force diagram, and then the B.M. diagram.

For a cantilever, the reactions at the fixed end need not be determined, since the SF and BM can be got by considering the portion towards the free end.

24.8 S.F. AND B.M. DIAGRAMS FOR SOME TYPICAL CASES

Now the S.F. and B.M. diagrams will be obtained for some typical cases of cantilevers, simply supported beams, and overhanging beams, with different kinds of loading.

24.8.1 Cantilevers

(1) *Cantilever with concentrated load at the free end:* With reference to Fig. 24.5–

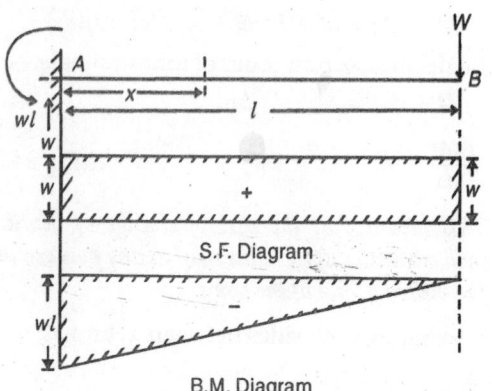

Fig. 24.5: Cantilever with Concentrated Load at the Free End

S.F.

Considering a section distant x from the fixed end A,

$F_x = W(+ve)$

It is a constant from A to the free end, B.

Even at B, at an infinitesimally small distance from B, the S.F. is $+W$ (considering the free end side).

The S.F. diagram is a rectangle.

B.M.

Bending moment at this section is

$M_x = W(l - x)\ (-ve)$

This is the maximum of Wl at the fixed end and zero at the free end (at $x = l$). The variation is linear and the B.M. diagram is a triangle.

[The reactions at the fixed end are–

$V_A = W\uparrow$

$M_A = Wl\ (\circlearrowleft)$].

(2) *Cantilever with uniformly distributed load throughout the length:* Referring to Fig. 24.6 and taking the free end as the origin–

S.F.

$F_x = wx(+ve)$

Linear variation,

$$\left[\dfrac{dF}{dx} = w,\ \text{the rate of the loading}\right]$$

S.F. dagram is a triangle.

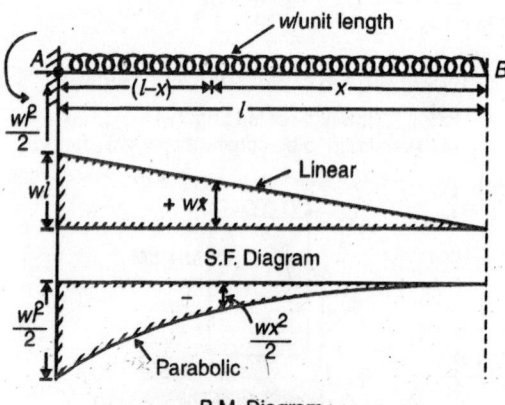

Fig. 24.6: Cantilever with Uniform Loading

B.M.

$$M_x = wx \times \dfrac{x}{2} = \dfrac{wx^2}{2}(-ve)$$

$$\left[\dfrac{dM}{dx} = wx = F,\ \text{numerically.}\right]$$

Parabolic variation.

$x = 0,\ M = 0$

$x = l,\ M = \dfrac{wl^2}{2}$

Mathematically, convexity may be shown to be towards the base as shown in the B.M. diagram.

(3) *Cantilever with a concentrated load at an intermediate point:* Referring to Fig. 24.7—this is similar to case 1, except that the portion *CB* between the load and the free end will be free of both shearing force and bending moment. This portion is therefore, unstressed.

Thus the cantilever acts as one of length *a* instead of length *l*.

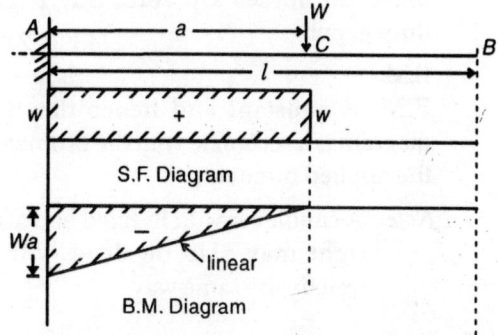

Fig. 24.7: Cantilever with Point Load at an Intermediate Section

In all kinds of loading, this is applicable—*any portion towards the free end which is not loaded upto the free end can be ignored.*

(4) *Cantilever with more than one concentrated load:* Referring to Fig. 24.8—let two concentrated loads W_1 and W_2 act at the free end and at a distance *a* from the free

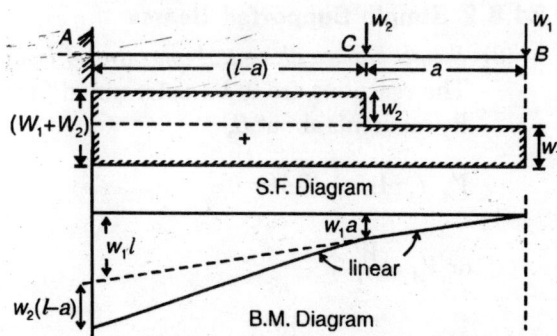

Fig. 24.8: Cantilever with more than One Point Load

end. The S.F. and B.M. diagrams can be sketched from fundamentals as shown in the figure. However, by the principle of superposition, these may also be got by adding the diagrams for the independent action of W_1 and W_2 on the cantilever, the portion of the diagrams for the two loads being shown separated by the dashed lines.

(5) *Cantilever loaded uniformly for a portion of its length from the free end:* Referring to Fig. 24.9—S.F. diagram will be a straight line from *B* to *C*—zero at *B* and + *wa* at *C*. From *C* to *A* the S.F. is a constant value of *w a* as there is no load in this portion.

The B.M. diagram is parabolic from *B* to *C*; the ordinate at *B* is zero and that at *C* is $\dfrac{wa^2}{2}$. The variation is linear between *C* and *A*.

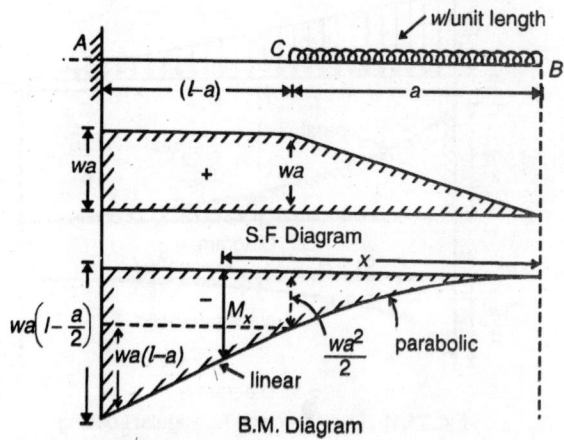

Fig. 24.9: Cantilever with Uniform Load for a Portion of the Length from the Free End

The general expression for *M* between *A* & *C* at a section distant *x* from the free end is,

$M_x = wa\,(x - a/2)$

The maximum value occurs at the fixed end and is $wa(l - a/2)$.

Note: The general approach is the same when only a portion of the cantilever from the free end is loaded, the other portion being free from loading.

(6) *Cantilever with a uniformly varying load (triangular loading) (with maximum intensity at fixed end):* Referring to Fig. 24.10 –

S.F.

$$F_x = \frac{1}{2} x \times \frac{wx}{l} = \frac{wx^2}{2l}(+ve)$$

Parabolic variation,

At $x = l$, $F = \frac{wl}{2}$ (at the fixed end)

B.M.

$$M_x = \frac{wx^2}{2l} \times \frac{x}{3} = \frac{wx^2}{6}(-ve)$$

Cubic parabola,

At $x = l$, $M = \frac{wl^2}{6}$ (at fixed end)

The diagrams is shown below.

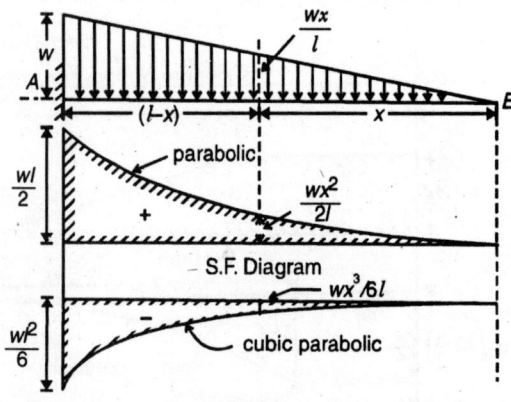

B.M. Diagram

Fig. 24.10: Cantilever with Traiangular Loading

Aliter:

$$\frac{dF}{dx} = \text{rate of loading}, \ w_x = \frac{wx}{l}$$

Integrating, $F = \frac{wx^2}{2l} + C_1$

$F = 0$ when $x = 0$ $C_1 = 0$.

But $\frac{dM}{dx} = F = \frac{wx^2}{2l}$

Integrating again, $M = \frac{wx^3}{6l} + C_2$

$M = 0$ when $x = 0$ $\therefore C_2 = 0$

Hence the results already derived.

Note: If x is chosen from the fixed end, and $\frac{d^2M}{dx^2}$ or w as negative, the S.F. and B.M. values will be obtained along with their appropriate signs.

If the maximum intensity of loading occurs at the free end, the maximum S.F. will remain the same, but the maximum B.M. will get *doubled*; this should appear natural. However, this case is uncommon in practice, owing to its severity.

(7) *Cantilever with a couple at the free end:* Referring to Fig. 24.11–

S.F.

Since the forces are zero, S.F. is zero throughout.

B.M.

B.M. is constant and hence the B.M. diagram is a rectangle with the ordinate as the applied moment.

Note: A cantilever with its fixed end at the right may also be dealt with in exactly the same way.

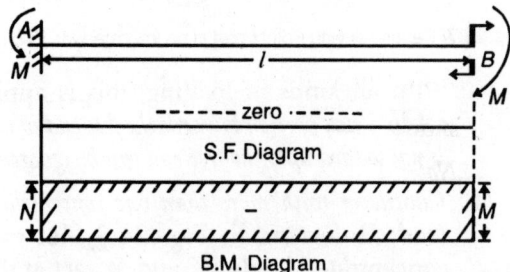

B.M. Diagram

Fig. 24.11: Cantilever with a Couple at the Free End

24.8.2 Simply Supported Beams

(8) *Simple beam with central concentrated load:* The reactions are to be determined first. By moments about B,

$$V_A \cdot l = W \cdot \frac{l}{2} \ (\Sigma M = 0)$$

or $V_A = \frac{W}{2} \uparrow$

$$V_B = W - V_A = \frac{W}{2} \uparrow (\Sigma v = 0)$$

However, these values could have been obtained owing to the symmetry about the middle of the beam and of the loading.

(Since $V_A + V_B = W$, $V_A = V_B = \dfrac{W}{2}\uparrow$)

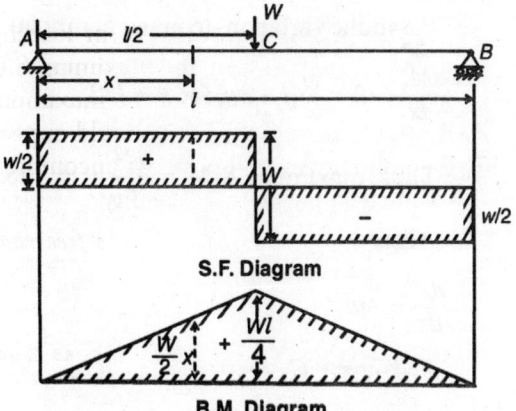

S.F. Diagram

B.M. Diagram

Fig. 24.12: Simple Beam with Central Point Load

S.F.
Between A & C,

$$F_x = \frac{W}{2}(+\text{ve})$$

Between C & B,

$$F_x = \frac{W}{2} \quad \cdots(?)$$

The S.F. appears to be changing sign suddenly at C, under the point load.

Note: The S.F. at C, directly under the point load, is not defined. A little to the left of C, it is $+W/2$ and a little to the right of C it is $-W/2$. Thus, right under a point load, the S.F. is strictly speaking, not defined.

B.M.

Portion AC: LHS $M_x = \dfrac{W}{2}x(+\text{ve})\curvearrowright$

(From the RHS)

$$\frac{W}{2}(l-x) - W\left(\frac{l}{2}-x\right) = \frac{Wx}{2} \text{ i.e., } +\text{ve}$$

Linear variation, $x = 0$, $M = 0$

At $x = l/2$, $M_{max} = \dfrac{W}{2} \cdot \dfrac{l}{2} = \dfrac{Wl}{4}(+\text{ve})$

Portion CB: RHS

$$M_x = \frac{W}{2}x \text{ with } x \text{ form } B$$

LHS

$x > l/2$

$$M_x = \frac{Wx}{2} - W(x - l/2)$$

$$= \frac{W}{2}(l-x)(+\text{ve})$$

At $x = l/2$, $M_{max} = \dfrac{Wl}{4}$

where $x = l$, $M = 0$

At the ends of a simple beam, the bending moment is obviously zero.

(9) *Simple beam with non-central concentrated load*: Referring to Fig. 24.13.

Let the reactions be V_A and V_B
By moments about B,

$$V_A l = W \cdot b$$

or $V_A = \dfrac{Wb}{l}\uparrow$

$$V_B = W - \frac{Wb}{l} = \frac{W(l-b)}{l} = \frac{Wa}{l}\uparrow$$

S.F.
Region AC,

$$F_x = V_A = \frac{Wb}{l}(+\text{ve})$$

Region CB,

$$F_x = V_B = \frac{Wa}{l}(-\text{ve})$$

B.M.
Region AC,

$$M_x = V_A \cdot x = \frac{Wb}{l} \cdot x(+\text{ve})$$

Linear variation,

Maximum *BM* at $C = \dfrac{Wab}{l}$

The diagrams are shown below.

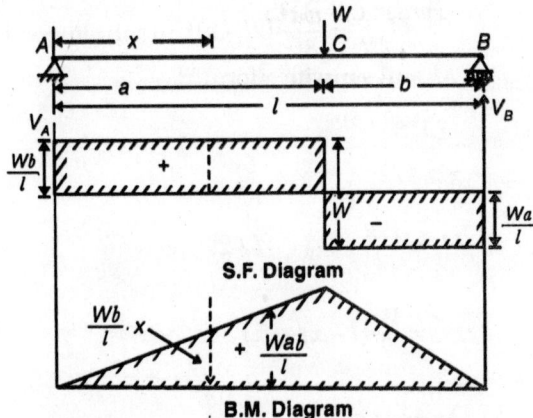

Fig. 24.13: Simple Beam with Non-Central Point Load

(10) *Simple beam with uniformly distributed load throughout the span*: Referring to Fig. 24.14—

$$V_A = V_B = \frac{wl}{2} \text{ by symmetry.}$$

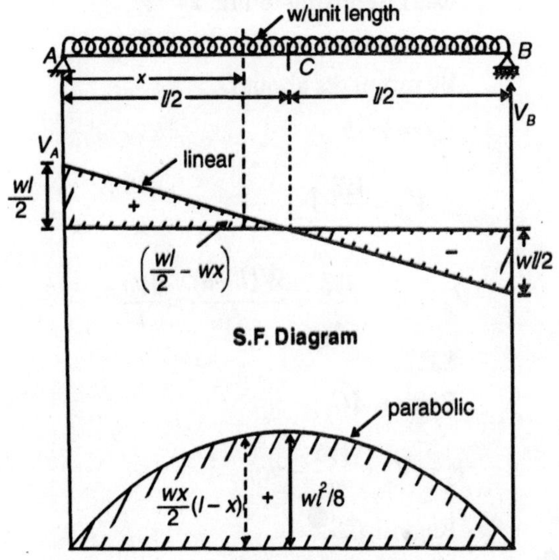

Fig. 24.14: Simple Beam with Uniform Loading

S.F.

$$F_x = \frac{wl}{2} - wx$$

Linear variation

F_x is maximum at $x = 0$ and $x = l$

Maximum S.F. $= \pm \dfrac{wl}{2}$

S.F. is obviously zero at $x = l/2$.

B.M.

$$M_x = \frac{wlx}{2} - \frac{wx^2}{2}$$

$$= \frac{wx}{2}(l - x)$$

Parabolic variation–(convex parabola)

$$\frac{dM_x}{dx} = 0 = \frac{w}{2}(l - 2x) \text{ for } x = l/2$$

Thus, maximum B.M. $= \dfrac{wl^2}{8}$ at mid span.

Aliter

$$\frac{dF}{dx} = -w$$

$$F = -wx + C_1$$

$$F = 0 \text{ at } x = \frac{l}{2} \therefore C_1 = \frac{wl}{2}$$

$$F = -wx + \frac{wl}{2}; F \text{ at } x = 0 \text{ is } +\frac{wl}{2}$$

$$F \text{ at } x = l \text{ is } -\frac{wl}{2} \text{ (Linear variation)}$$

$$\text{or } \frac{dM}{dx} = F = -wx + \frac{wl}{2}$$

$$M = -\frac{wx^2}{2} + \frac{wlx}{2} + C_2$$

$$M = 0 \text{ at } x = 0 \therefore C_2 = 0$$

$$= -\frac{wx^2}{2} + \frac{wlx}{2} \text{ Parabolic variation.}$$

M_{max} occurs at $\dfrac{l}{2}$ by symmetry, and is $\dfrac{wl^2}{8}$

(11) *Simple beam loaded uniformly for a portion of the span*: Referring to Fig. 24.15– Reactions,

$$V_B l = \frac{wa^2}{2}$$

$$\therefore V_B = \frac{wa^2}{2l} \uparrow$$

$$V_A = wa - \frac{wa^2}{2l} = \frac{wa}{2l}(2 \cdot l - a) \uparrow$$

S.F.

Region AC,

$$F_x = V_A - wx$$

$$= \frac{wa}{2l}(2l - a) - wx$$

$$F_x = 0 \text{ at } x = \frac{a}{2l}(2l - a)$$

$F_x = +V_A$ at A and $-V_B$ at B

$$F_c = wa\left\{\left(\frac{2l - a}{2l}\right) - 1\right\} = -\frac{wa^2}{2l} = -V_B$$

Region CB,

F_x is constant value $-V_B$

B.M.

Region CB,

$M_x (x \text{ from } B) = V_B \cdot x$ (linear)

At $x = (l - a)$ or C

$$M_C = \frac{wa^2}{2l}(l - a)(+\text{ve})$$

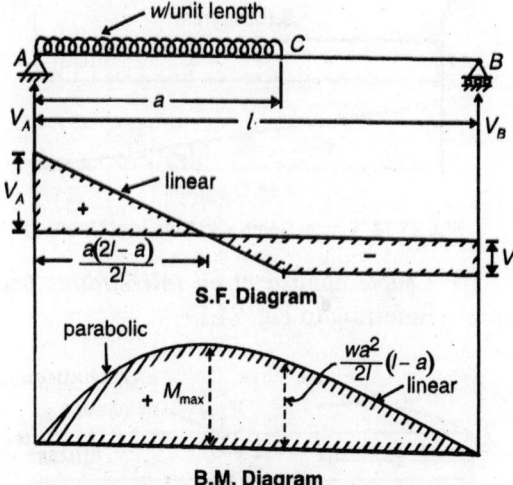

S.F. Diagram

B.M. Diagram

Fig. 24.15: Simple Beam with UDL for a Part of the Span

Region AC,

x from A

$$M_x = \frac{wa}{2l}(2l - a)x - \frac{wx^2}{2} \text{ (parabolic)}$$

M_{max} occurs at the section where S.F. is

zero, or at $x = \frac{a(2l - a)}{2l}$

$\therefore M_{\text{max}} = \frac{wa^2}{8l^2}(2l - a)^2$ on substitution for

x and simplification.

Special case :

When $a = l/2$, $V_A = \frac{3wl}{8}$, $V_B = \frac{wl}{8}$

M_{max} at $\frac{3l}{8}$ from $A = \frac{9\,wl^2}{128}$ (Verify!)

(12) *Simple beam with symmetrical triangular loading.* Referring to Fig. 24.16–

Reactions,

$$V_A = \frac{1}{2} \times \frac{wl}{2} = \frac{wl}{2} = V_B, \text{ by symmetry.}$$

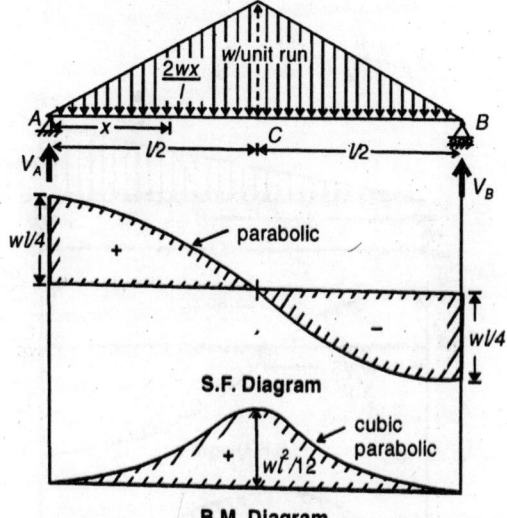

S.F. Diagram

B.M. Diagram

Fig. 24.16: Simple Beam with Symmetrical Triangular Loading

S.F.

$$F_x = \frac{wl}{4} - \frac{1}{2} \times x \times \frac{2wx}{l}$$

$$= \left(\frac{wl}{4} - \frac{wx^2}{l}\right)$$

Parabolic variation,

S.F. is zero at $x = l/2$

S.F. is maximum near supports and is equal to the reaction.

B.M.

$$M_x = \frac{wl}{4}x - \frac{wx^2}{l} \times \frac{x}{3}$$

$$= \frac{wlx}{4} - \frac{wx^3}{3l}$$

Cubic parabolic variation.

M_{max} (at $x = l/2$)

$$= \frac{wl}{4} \cdot \frac{l}{2} - \frac{w}{3l} \times \frac{l^3}{8} = \frac{wl^2}{12}$$

The S.F. and B.M. diagrams are shown in Fig. 24.16.

(13) *Simple beam with triangular loading*: Referring to Fig. 24.17—

Reactions:

$$V_A = \frac{1}{3} \times \frac{wl}{2} = \frac{wl}{6} \uparrow$$

$$V_B = \frac{wl}{2} - \frac{wl}{6} = \frac{wl}{3} \uparrow$$

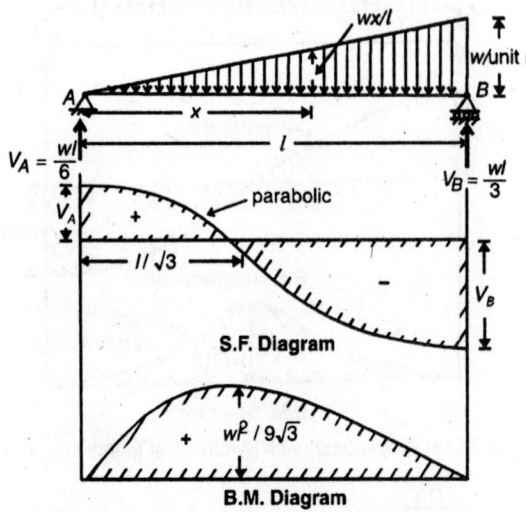

Fig. 24.17: Simple Beam with Triangular Loading

S.F.

x from A

$$F_x = \frac{wl}{6} - \frac{wx^2}{2l}$$

Parabolic variation,

$$F_x = 0 \text{ when } x = \frac{l}{\sqrt{3}} \text{ from } A.$$

B.M.

$$M_x = \frac{wl}{6} \cdot x - \frac{wx^3}{6l}$$

Cubic parabolic variation,

M_{max} at $x = l/\sqrt{3}$ is

$$\frac{wl}{6} \times \frac{l}{\sqrt{3}} - \frac{w}{6l} \cdot \frac{l^2}{3} \cdot \frac{l}{\sqrt{3}} = \frac{wl^2}{9\sqrt{3}}$$

The S.F. and B.M. diagrams are shown in Fig. 24.17.

(14) *Simple beam with a couple*

(a) *Couple applied at one end*: Referring to] 24.18—

Reactions:

$$V_A = \frac{M}{l} \downarrow \quad V_B = \frac{M}{l} \uparrow$$

S.F. diagrams is a negative rectangle.
B.M. diagram is a triangle with maximum value M at A, and zero at B.

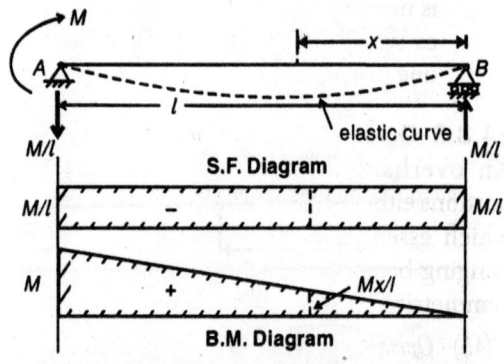

Fig. 24.18: Simple Beam with Couple at One End

(b) *Couple applied at an intermediate point*: Referring to Fig. 24.19—

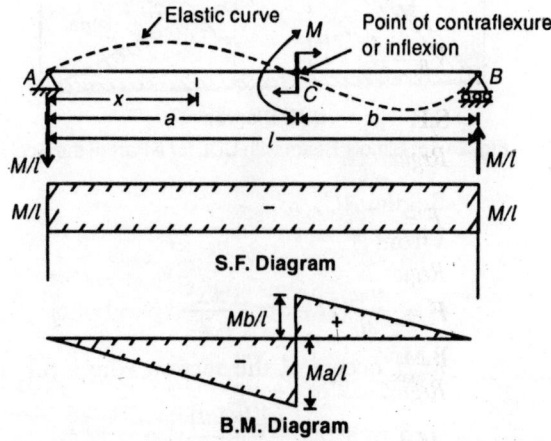

Fig. 24.19: Simple Beam with Couple at an Intermediate Section

Reactions:

$$V_A = \frac{M}{l} \downarrow \quad V_B = \frac{M}{l} \uparrow$$

S.F.

Negative rectangle as in case (*a*).

B.M.

At *x* from *A*

$$M_x = \frac{M}{l} \cdot x(-\text{ve})$$

Linear variation

Max. B.M. at *C* is $\dfrac{Ma}{l}$

B.M. suddenly changes at *C* by *M* as shown.

Special Case: When the couple acts at mid span section, *a* = *b* = *l*/2. The maximum is numerically *M*/2, and suddenly changes sign from −ve to +ve at mid-span, as one proceeds from left to right.

24.8.3 Overhanging Beams

An overhanging beam will have overhang portions either on to one side or on to both sides, which essentially act as cantilevers. An overhanging beam and loading may or may not be symmetrical. A few typical cases are dealt with.

(15) *Overhanging beam with single overhang—point load at the free end*:

Reactions:

$$V_A = \frac{Wa}{l} \downarrow$$

$$V_B = W\left(1 + \frac{a}{l}\right) \uparrow$$

S.F.

Region AB

$$F = \frac{Wa}{l}(-\text{ve})$$

Region BC

$$F = W(+\text{ve})$$

B.M.

Region AB

$$M_x = V_A \cdot x = \frac{Wa}{l} \cdot x(-ve)$$

Linear variation—maximum at *x* = *l* is *Wa.*

Region BC

$$\frac{Wa}{l} \cdot x - W\left(1 + \frac{a}{l}\right)(x - l)$$

or $W(l + a - x)$

Linear variation

At *x* = *l*, $M_{\text{max}} = Wa.$

The S.F. and B.M. diagrams are shown in Fig. 24.20.

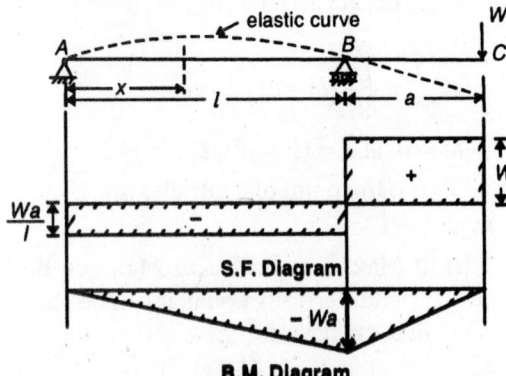

Fig. 24.20: Overhanging Beam with Single Overhang-Point Load at the Free End

(16) *Overhanging beam with single overhang with uniform load throughout the length*: Referring to Fig. 24.21–

Reactions:

$$V_B \cdot l = \frac{w(l + a)^2}{2}$$

$$\therefore V_B = \frac{w}{2l}(l + a)^2 \uparrow$$

$$V_A = w(l + a) - \frac{w}{2l}(l + a)^2$$

$$= \frac{w}{2l}\left(l^2 - a^2\right) \uparrow$$

S.F.

At *A*, it is $V_A.$

Linear variation,

$$F = 0 \quad at \; x = \left(\frac{l^2 - a^2}{2l}\right)$$

At *C*, the free end, *F* = 0.

Zone BC

Triangle with wa at B.

B.M.

At C, $M = 0$

Zone BC Parabolic,

At $B = M = \dfrac{wa^2}{2}$ (−ve)

At A, $M = 0$

Zone AB Parabolic

M_{max} occurs at $x = \left(\dfrac{l^2 - a^2}{2l}\right)$

$M_{max} = \dfrac{w}{8l^2}\left(l^2 - a^2\right)^2$ (+ve)

$M = 0$ at $x = (l^2 - a^2)/l$

This is the point of contraflexure.

Note:

  (i) If Max. +ve B.M. and Max −ve B.M. are numerically equal, it can be shown that $a = 0.414l$.

  (ii) If $M = 0$ at $x = \dfrac{(l + a)}{2}$, it can be shown that $a = \dfrac{l}{2}$;

  (iii) If $M = 0$ at $x = l/2$, it can be shown that $a = \dfrac{l}{\sqrt{2}} = 0.707l$

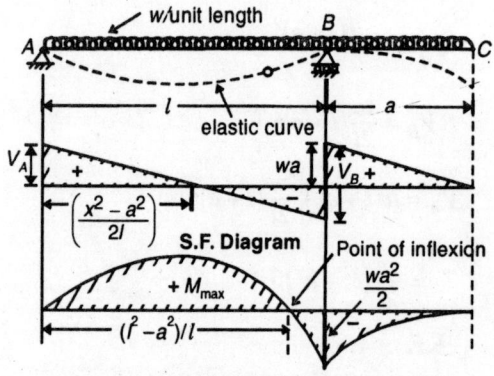

Fig. 24.21: Overhanging Beam with Single Overhang-Uniform Loading

(17) *Overhanging beam with double overhangs with point loads at the free ends:* Referring to Fig. 24.22–

Reactions:

$V_A = V_B = W$ by symmetry

S.F.

Zone AC: $-W$

Zone BC: zero.

Zone BD: $+W$

B.M.

CA & BD

Linear variation

At A: $-Wa$

At B: $-Wa$

In AB: Rectangle with Wa as the ordinate.

Note: Zone AB is obviously in a state of what is called "Simple or pure bending" as it is free of shear force and the bending moment is constant—this will be seen in the next chapter.

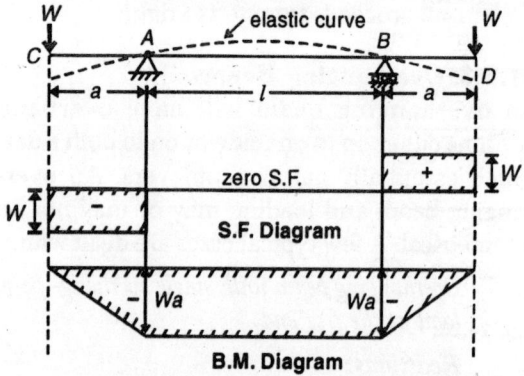

Fig. 24.22: Overhanging Beam with Double Overhangs-Point Loads at Free Ends

(18) *Overhanging beam with double overhangs-loaded uniformly throughout the length:* Referring to Fig. 4.23–

Reactions:

$V_A = V_B = w(a + l/2)$, by symmetry

S.F.

CA & BD

Linear $-wa$ and $+wa$ at A and at B

AB

$+\dfrac{wl}{2}$ and $-\dfrac{wl}{2}$ at A and at B.

Linear variation

B.M.

CA & *BD* parabolic

$-\dfrac{wa^2}{2}$ at *A* & *B*

AB parabolic.

$$M_E = w(a + l/2)\dfrac{l}{2} - \dfrac{w}{2}\left(a + \dfrac{l}{2}\right)^2$$

$$= \dfrac{w}{2}\left(\dfrac{l^2}{4} - a^2\right)$$

This is +ve, if $a < l/2$;

Zero, if $a = l/2$; and

Negative, if $a > l/2$.

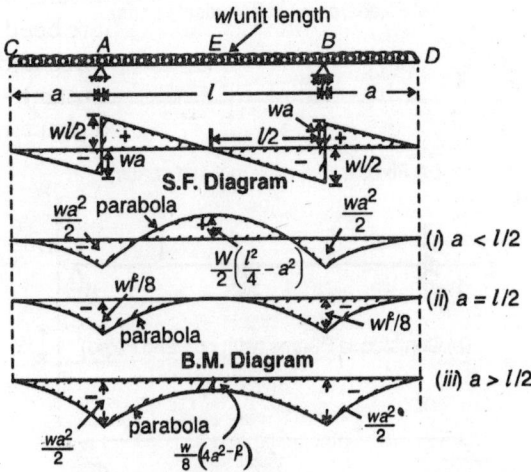

Fig. 24.23: Overhanging Beam with Double Overhangs— Uniform Loading

Say $a < l/2$

For maximum to be as small as possible,

max. +ve BM = max. $-ve$ B.M. numerically. This leads to the condition that,

$$a = l / 2\sqrt{2} = 0.3535l.$$

For this condition, $M_{max} = \dfrac{wl^2}{16}$

This is the best design condition.

Points of contraflexure:

If the distance is *x* from *A*.

$$\dfrac{w(a + x^2)}{2} = \dfrac{w}{2}(l + 2a)x$$

or $x^2 - lx + a^2 = 0$

or $x = \dfrac{l}{2}\left[1 \pm \sqrt{1 - \dfrac{4a^2}{l^2}}\right]$ or

or at a distance of $\dfrac{l}{2} \times \sqrt{1 - \dfrac{4a^2}{l^2}}$ from the centre *E*, on either side.

With +ve sign we get one point of inflexion, and with −ve sign, we get the other.

Note. Interestingly, in the case of best design, this is $l / 2\sqrt{2}$ or $0.3535l$ from the centre on either side.

24.9 BEAMS SUBJECTED TO INCLINED LOADS

When a beam is subjected to loads inclined at an angle with vertical (but contained in the longitudinal plane of bending), the loads are resolved into vertical and horizontal components, and treated separately. The vertical components are transverse loads causing shearing force and bending moment in the beam; this is already seen in the previous sections.

The horizontal components are axial forces, which tend to cause tensile or compressive stresses in the cross-sections of the beam depending upon the directions of these and the corresponding support reactions.

One or two of the illustrative problems will be dealt with later.

24.10 LOADING AND BENDING MOMENT DIAGRAMS FROM SHEAR FORCE DIAGRAM

When the S.F. diagram is given, the loading diagram and the B.M. diagram may be obtained in that order.

In this connection, the following points are to be remembered:

(*i*) SF diagram consists of horizontal lines between point loads and the corresponding portion of the BM diagram will be an inclined straight line. At a concentrated load, the SF changes suddenly by an amount equal to the magnitude of the load, the SF changes suddenly by an amount equal to the magnitude of the load.

(*ii*) SF diagram consists of inclined straight lines in the case of uniformly distributed load, and the corresponding portions of the B.M. diagram will be parabolic curves.

(iii) S.F. diagram consists of parabolic curves for a uniformly varying load, and the corresponding portions of the B.M. diagram will be cubic parabolic curves.

Note:

(1) The sign convention for the SF and BM should be specified. Some times, a unique solution may not be there, with more than one possible solution.

(2) After the loading diagram is found, statics must be checked for the loading and the reactions. Only when this is satisfied, should the BM diagram be determined.

One or two illustrative problems will help the student understand these ideas.

24.11 STATICALLY DETERMINATE AND INDETERMINATE BEAMS

All the beams for which the S.F. and B.M. diagrams have been determined in Section 24.8 could be fully solved with the aid of the laws of static equilibrium alone. Such beams for which the support reactions could be solved, and the shearing force and bending moment diagrams sketched, thus completing the analysis, making use of the laws of static equilibrium alone are known as "Statically Determinate Beams". Examples of this are cantilevers, simply supported beams, and overhanging beams with not more than two supports.

However, in practice many types of beams cannot be solved, or analysed fully (in other words, the S.F. and B.M. diagrams cannot be determined) purely with the help of the static equilibrium conditions alone. Such beams are known as "Statically Indeterminate Beams"; we will need to devise one or more additional conditions/equations for solving all the reactions, and complete the analysis by determining the S.F. and B.M. diagrams, which are necessary for designing the beams.

A few examples of these are shown in Fig. 24.24.

The number of additional conditions needed to solve the problem is known as the "Degree of Static Indeterminacy" of the beam. This will equal

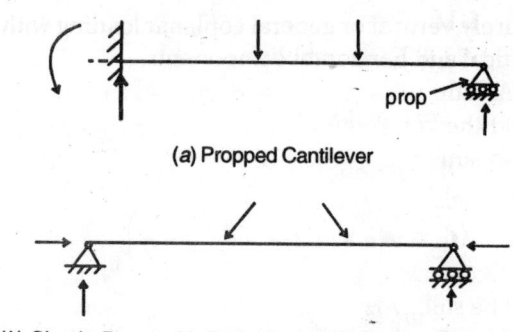

(a) Propped Cantilever

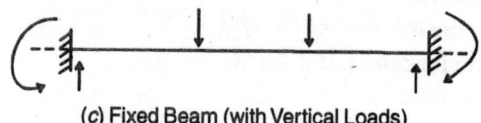

(b) Simple Beam with Both Hinged Supports (with Inclined Loads)

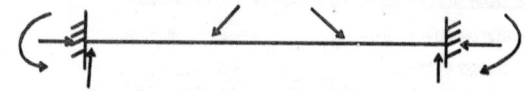

(c) Fixed Beam (with Vertical Loads)

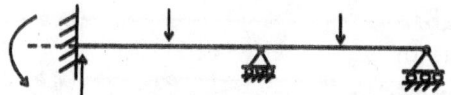

(d) Fixed Beam (with Inclined Loads)

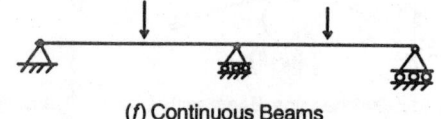

(e) Continuous Beams (with one End Fixed)

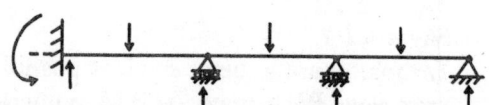

(f) Continuous Beams

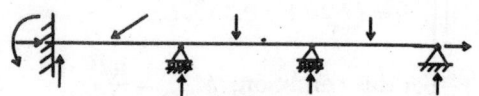

(g) Continuous Beam (with Four Supports and Vertical Loading)

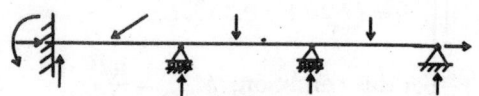

(h) Continuous Beam (with Four Supports and General Loading)

Fig. 24.24: Some Statically Indeterminate Beams

the number additional reaction components to be solved over and above the number of useful laws of static equilibrium available; the former will depend upon the number and nature of supports, while the latter on the nature of loading

–purely vertical or general coplanar loading with vertical and horizontal components.

A little inspection based on these ideas will yield the "Degree of Indeterminacy" in each of the beams in Fig. 24.24 as follows:

(a) 1   (b) 1   (c) 2   (d) 3
(e) 2   (f) 1   (g) 3   (h) 4

A comparison of (c) and (d), as also (g) and (h), will be sufficient to know that for the same beam the degree of indeterminacy may depend upon the nature of loading. Hence, for the degree of indeterminacy to be ascertained in any case, not only the nature of the structure, but also the nature of the loading must be known.

Several different methods have been devised and used for the analysis of statically indeterminate structures, which are preferred in practice in spite of the relative complexity in analysis, because of the economy involved in their design.

ILLUSTRATIVE PROBLEMS

PROBLEM 24.1: Sketch the S.F. and B.M. diagrams for the cantilever shown in Fig. 24.25.

Solution: With reference to Fig. 24.25–

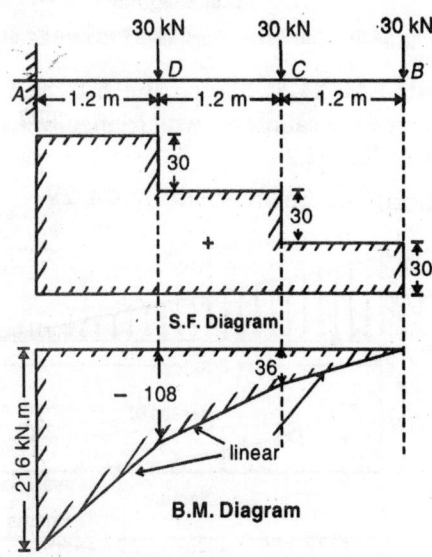

Fig. 24.25: Cantilever (Illustrative Problem 24.1)

S.F.
From the free-end side, from B to C,
$$F = 30 \text{ kN } (+)$$

From C to D,
$$F = 30 + 30 = 60 \text{ kN } (+)$$
From D to A
$$F_{\max} = 30 + 30 + 30 = 90 \text{ kN } (+)$$

B.M.
From the free-end side,
$$M_C = 30 \times 1.2 = 36 \text{ kN.m } (-)$$
$$M_D = 30 \times 2.4 \times 30 \times 1.2 = 108 \text{ kN.m } (-)$$
$$M_A = 30 \times 3.6 + 30 \times 2.4 + 30 \times 1.2, \text{ or}$$
$$M_{\max} = 216 \text{ kN.m } (-)$$

PROBLEM 24.2: Sketch the S.F. and B.M. Diagrams for the cantilever shown in Fig. 24.26.

Solution: With reference to Fig. 24.26–

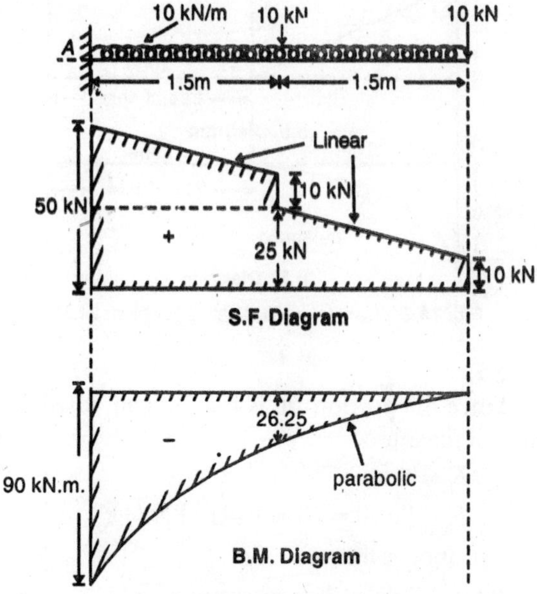

Fig. 24.26: Cantilever (Illustrative Problem 24.2)

S.F.
$$F_B = 10 \text{ kN } (+)$$
$$F_C = 10 + 10 \times 1.5 = 25 \text{ kN } (+)$$

F_C suddenly increases by the value of point load at C as one goes from free-end side to fixed end side.

$$F_{\max} = 10 + 10 + 10 \times 3 = 50 \text{ kN } (+)$$
Variation is linear.

B.M.
$$M_B = 0$$
$$M_C = 10 \times 1.5 + 10 \times 1.5 \times 0.75$$

$$= 26.25 \text{ kN.m } (-)$$
$$M_A = 10 \times 3 + 10 \times 1.5 + 10 \times 3 \times 1.5$$
or $M_{max} = 90$ kN.m $(-)$

Variation is parabolic in view of the uniform loading.

PROBLEM 24.3: Sketch the S.F. and B.M. diagrams for a cantilever, 2.00 m long, carrying a uniformly distributed load of 10 kN/m over a length of 1 m from the fixed end.

Solution: With reference to Fig. 24.27–

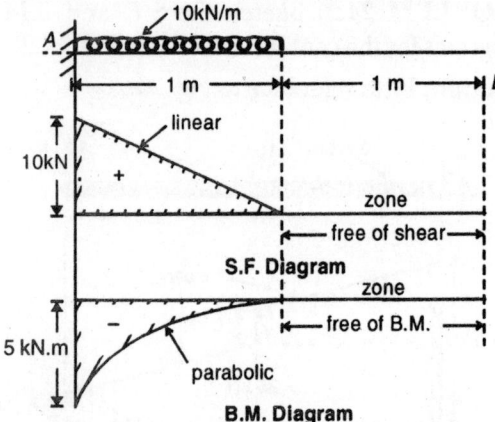

Fig. 24.27: Cantilever (Illustrative Problem 24.3)

S.F.
Since the portion *BC* is free of load, it will be free of shear too.
$$F_C = 0$$
$$F_{max} \text{ (at } A) = 10 \times 1 = 10 \text{ kN } (+)$$
Variation is linear.

B.M.
Since the portion *BC* is free of load, it is free of B.M. also,
$$M_C = 0$$
$$M_{max}(\text{at } A) = 10 \times 1 \times 1/2 = 5 \text{ kN.m } (-)$$
Variation is parabolic.

PROBLEM 24.4: Sketch the S.F. and B.M. diagrams for a cantilever, 2 m long, carrying a uniformly distributed load of 10 kN/m over a length of 1 m from the free end.

Solution: With reference to Fig. 24.28–

S.F.
$$F_B = 0$$

$$F_C = 10 \times 1 = 10 \text{ kN } (+)$$
Since there is no load between *C* and *A*, S.F. continues to be constant at 10 kN.
$$F_{max} = 10 \text{ kN } (+)$$
B.M.
$$M_B = 0$$
$$M_C = 10 \times 1 \times 1/2 = 5 \text{ kN.m } (-)$$
Variation from *B* to *C* is parabolic.
$$M_{max} \text{ (at } A) = 10 \times 1 \times 1.5 = 15 \text{ kN.m } (-)$$
Variation from *C* to *A* is linear since there is no load between *C* and *A*.

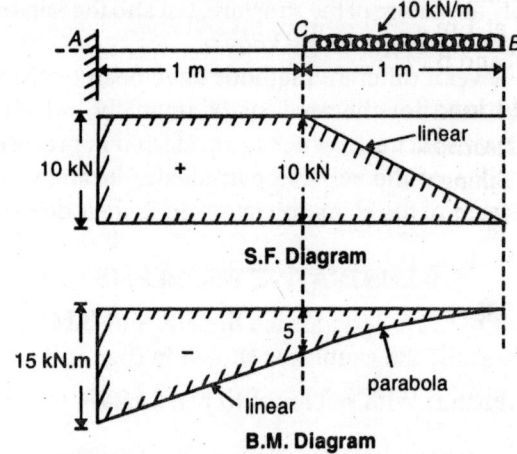

Fig. 24.28: Cantilever (Illustrative Problem 24.4)

PROBLEM 24.5: Sketch the S.F. and B.F. diagrams for a cantilever with triangular load as shown in Fig. 24.29.

Solution: With reference to Fig. 24. 29–

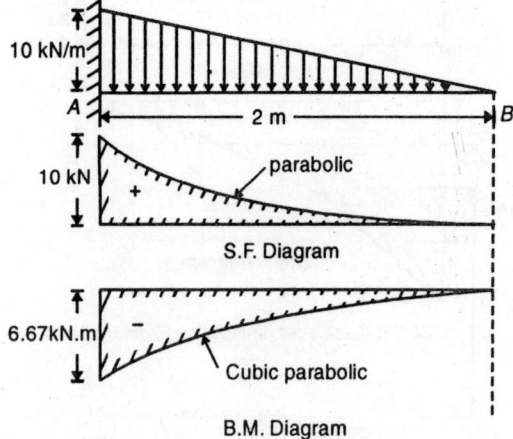

Fig. 24.29: Cantilever (Illustrative Problem 24.5)

S.F.

$F_B = 0$

Variation is parabolic.

F_{max} (at A) $= 10 \times 2 \times 1/2 = 10$ kN (+)

B.M.

$M_B = 0$

Variation is cubic parabolic.

M_{max} (at A) $= 10 \times 2 \times 1/2 \times 2/3$
$= 6.67$ kN.m (–)

PROBLEM 24.6: A simply supported beam of 3 m span is loaded with point loads of 5 kN and 20 kN at 1 m and 2 m from the left end. Sketch the S.F. and B.M. diagrams.

Solution: Referring to Fig. 24.30–

Reactions:

Taking moments about B,

$V_A \cdot 3 = 5 \times 2 + 20 \times 1 = 30$

$V_A = 10$ kN↑ $V_B = (5 + 20 - 10) = 15$ kN↑

S.F.

$F_A = V_A = 10$ kN (+)

$F_C =$ (a little to the right)
$= (10 - 5) = 5$ kN (+)

$F_D =$ (a little to the right)
$= 10 - 5 - 20 = 15$ kN (–)

$F_B = V_B = 15$ kN (–) (F_{max})

B.M.

$M_A = M_B = 0$

$M_C = 10 \times 1 = 10$ kN.m (+)

$M_D = 15 \times 1 = 15$ kN.m (+) M_{max}

Variation is linear.

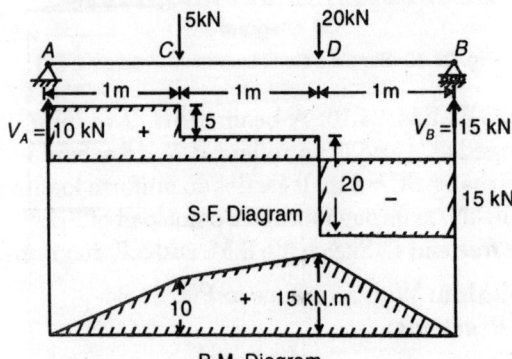

B.M. Diagram

Fig. 24.30: Simple Beam (Illustrative Problem 24.6)

PROBLEM 24.7: A simply supported beam of 3 m span is loaded with point loads of 10 kN/m over the right half of its span of 4 m. Sketch the S.F. and B.M. diagrams.

Solution:

Reactions: Referring to Fig. 24.31-
Taking moments about B

$V_A \cdot 4 = 10 \times 2 \times 1 = 20$

$V_A = 5$ kN↑

$V_B = (10 \times 2 - 5) = 15$ kN ↑

S.F.

$F_A = V_A = 5$ kN (+)

$F_C = 5$ kN (+) (since there is no load between A & C)

Between C and B, the S.F. varies from +5 kN to –15 kN linearly, in view of the uniform loading in this portion.

F will be zero at 1.5 m from B.

B.M.

$M_A = M_B = 0$

$M_C = 5 \times 2 = 10$ kN.m (+)

Between A and C, the variation is linear. Between C and B, the variation is parabolic.

M_{max} occurs at the section where S.F. is zero.

$\therefore M_{max} = (15 \times 1.5 - 10 \times 1.5 \times 0.75)$
$= (22.50 - 11.25) = 11.25$ kN.m (+)

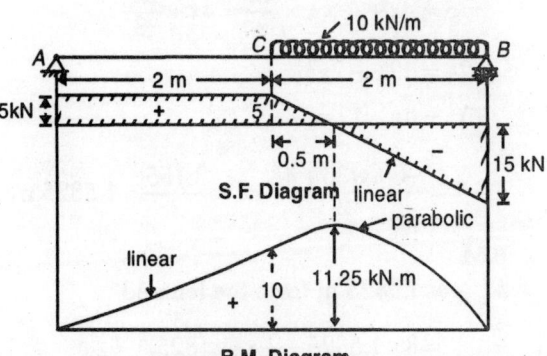

B.M. Diagram

Fig. 24.31: Simple Beam (Illustrative Problem 24.7)

PROBLEM 24.8: A simple beam of 3 m span carries a trapezoidal loading with 4 kN/m at left end and 8 kN/m at the right end. Determine the position and magnitude of the maximum bending moment.

Solution: With reference to Fig. 24.32–

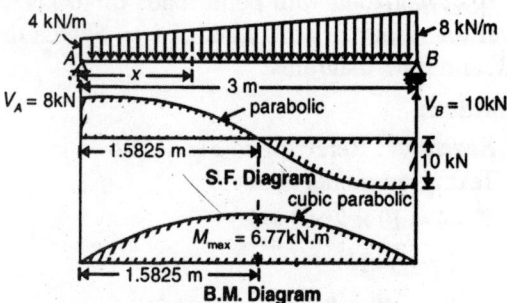

Fig. 24.32: Simple Beam with Trapezoidal Loading
(Illustrative Problem 24.8)

Reactions:

Moments about B,

$$3V_A = 4 \times 3 \times 3/2 + 1/2 \times 4 \times 3 \times 3/3$$

$$= 18 + 6 = 24$$

$$V_A = 8 \text{ kN}\uparrow$$

$$V_B = \left(\frac{4+8}{2}\right) \times 3 - 8 = 10 \text{kN}\uparrow$$

S.F.

$$F_A = 8 \text{ kN } (+)$$

$$F_B = 10 \text{ kN } (-)$$

Variation is parabolic.

Let the S.F. be zero at a distance x m from the left end.

$$8 = 4x + \frac{1}{2} \times x \times \frac{4x}{3} = 4x + \frac{2x^2}{3}$$

$$24 = 12x + 2x^2$$

$$x^2 + 6x - 12 = 0$$

$$x = \frac{-6 + \sqrt{36 + 48}}{2} \text{ m} = \frac{3.165}{2} = 1.5825 \text{ m}$$

B.M.

M_{max} at 1.5825 m from the left end

$$= 8 \times 1.5825 - \frac{4}{2} \times (1.5825)^2$$

$$- \frac{1}{2}(1.5825)^3 \times \frac{4}{3} \times \frac{1}{3}$$

$$= 12.66 - 2 \times (1.5825)^2 \left[1 + \frac{1.5825}{9} \right]$$

$$= (12.66 - 5.89) = 6.77 \text{ kN.m}$$

PROBLEM 24.9: Sketch the S.F. and B.M. diagrams for the simple beam loaded symmetrically as shown in Fig. 24.33–

Solution:

Reactions:

By symmetry,

$$V_A = V_B = \frac{1}{2}(10 + 20 + 10 + 6 \times 10)$$

$$= 50 \text{ kN}$$

S.F.

$$F_A = V_A + 50 \text{ kN}$$

$$F_B = V_B - 50 \text{ kN}$$

The SFD is as shown.

The variation in the segments is linear.

B.M.

$$M_D = M_E = 50 \times 1.5 - 10 \times 1.5 \times 0.75$$

$$= 75 - 11.25 = 63.75 \text{ kN.m } (+)$$

$$M_C \text{ (or } M_{max}) = 50 \times 3 - 10 \times 1.5 - 10 \times 3 \times 1.5$$

$$= 150 - 15 - 45 = 90 \text{ kN.m } (+)$$

The S.F. and B.M. diagrams are as shown below.

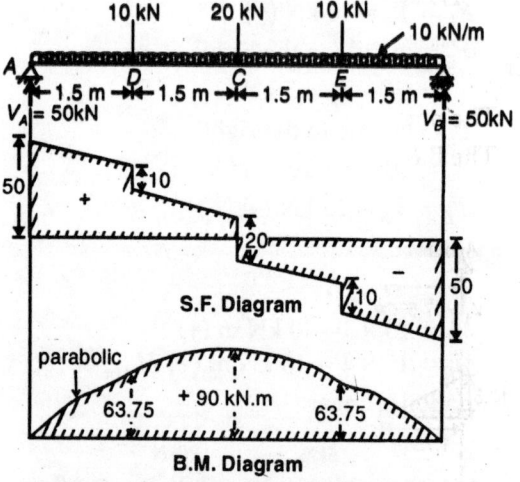

Fig. 24.33: Simple Beam (Illustrative Problem 24.9)

PROBLEM 24.10: A beam ABC, 9 m long, is hinged at A and is on rollers at B. $AB = 6$ m and overhang $BC = 3$ m. It carries an uniform loading of 10 kN/m throughout and a point load of 10 kN at the free end C. Sketch the B.M. and S.F. diagrams.

Solution: With reference to Fig. 24.34-

Reactions:

$$6V_B = 10 \times 9 + 10 \times 9 \times \frac{9}{2}$$

$\therefore \quad V_B = 82.5$ kN↑

$V_A = (10 \times 9 + 10 - 82.5) = 17.5$ kN↑

S.F.

$F_A = V_A = 17.5$ kN (+)

$F_B = 17.5 - 10 \times 6 = 42.5$ kN (–)

$F_C = 17.5 - 10 \times 6 + 82.5 - 30 = 10$ kN (+)

S.F. is zero at 1.75 m from A.

The S.F. diagram is shown.

Variation is linear.

B.M.

M_{max} (+) at 1.75 m from A

$$= 17.5 \times 1.75 - \frac{10 \times (1.75)^2}{2} = 15.3 \text{ kN.m}$$

$M_C = 0$

$$M_B = 10 \times 3 + 10 \times 3 \times \frac{3}{2} = 75 \text{ kN.m}(-)$$

Variation is parabolic.

If M is zero at x m from A,

$$17.5x - 10 \times \frac{x^2}{2} = 0$$

or $\quad x = 17.55 = 3.5$ m from A

The point of contraflexure is at 3.5 from A.

The B.M. diagrams is shown below.

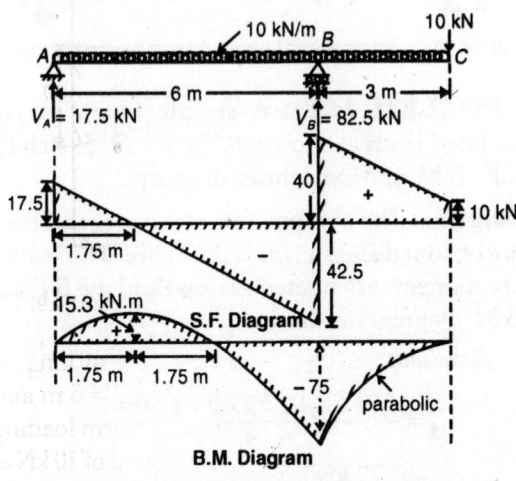

B.M. Diagram

Fig. 24.34: Overhang Beam (Illustrative Problem 24.10)

PROBLEM 24.11: An overhang beam with overhangs on both sides is shown in Fig. 24.35 along with the loading. Sketch the S.F. and B.M. diagrams.

Solution:

Reactions:

$3V_A = 20 \times 4.5 + 30 \times 3 \times 1.5 - 30 \times 1.5$

$V_A = 30 + 45 - 15 = 60$ kN↑

$V_B = 90 + 20 + 30 - 60 = 80$ kN↑

S.F.

With reference to Fig. 24.35–

$F_C = 20$ kN (–)

$F_D = 30$ kN (+)

S.F is zero at $1\frac{1}{3}$ m from A.

The S.F. diagram is shown.

B.M.

$M_A = 20 \times 1.5 = 30$ kN.m (–)

$M_B = 30 \times 1.5 = 45$ kN.m (–)

M_{max} at $1\frac{1}{3}$ from A.

$$= 60 \times \frac{4}{3} - 20 \times \frac{17}{6} - 30 \times \frac{4}{3} \times \frac{2}{3}$$

$$= 80 - 56.7 - 26.7 = 3.4 \text{ kN.m } (-ve)$$

The variation is parabolic.

The B.M. diagram is shown; it is observed that the B.M. is throughout negative.

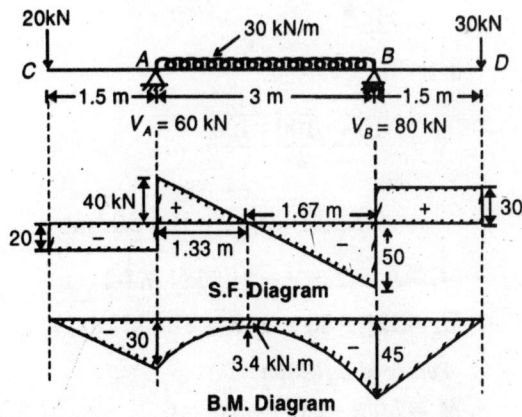

B.M. Diagram

Fig. 24.35: Overhanging Beam (Illustrative Problem 24.11)

PROBLEM 24.12: A beam, 10 m long, is supported symmetrically with equal overhangs on either side, and is uniformly loaded throughout with 10 kN/m. Determine the length of the overhang for the maximum B.M. to be the minimum.

Solution: With reference to 24. 36–

For the maximum B.M. to be the minimum, the maximum negative B.M. at the supports must equal the maximum positive B.M. at mid-span

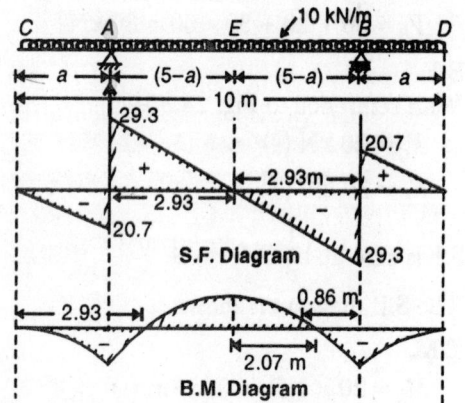

Fig. 24.36: Overhanging Beam (Illustrative Problem 24.12)

Reactions:

$V_A = V_B = 50$ kN (by symmetry)

$M_A = 10a \times \dfrac{a}{2} = 5a^2 \;(-)$

$M_E = 50\,(5 - a) - 10 \times 5 \times 2.5$

$= 250 - 50a - 125$

$= 125 - 50a \;(+)$

Equating the numerical values

$5a^2 = 125 - 50a$

$a^2 + 10a - 25 = 0$

$a = \dfrac{-10 + \sqrt{100 + 100}}{2}$

$= -5 + 5\sqrt{2} = 5(\sqrt{2} - 1) = 2.07\,\text{m}$

$\therefore \quad M_A = 5 \times (2.07)^2 = 21.5$ kN.m $(-)$

$M_E = 125 - 50 \times 2.07 = 21.5$ kN.m $(+)$

Points of inflexion:

$M_x = 50(x - 2.07) - 5x^2 = 0$

$x^2 - 10x + 20.7 = 0$

$x = \dfrac{10 - \sqrt{100 - 82.8}}{2}$

$= \left(\dfrac{10 - 4.15}{2}\right) = \dfrac{5.85}{2}\,\text{m} = 2.93$ m from the ends.

These are 2.07 m from the mid span on either side.

The S.F. and B.M. diagrams are easily constructed.

PROBLEM 24.13: The S.F. diagram for a beam is as shown in Fig. 24.37. Determine the beam and loading.

Solution:

S.F. is +ve when it is upward force to the left

$V_A = 10.25$ kN↑ $V_B = 9.75$ kN↑

Point load at *D* is 5 kN and that at *E* is 4 kN.

Point load *C* is 3 kN.

Portion *AD* and *EB* carry uniform loading of 2 kN/m, obviously.

The reasoning is very simple.

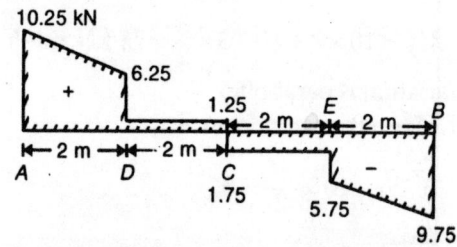

Fig. 24.37: S.F. Diagram (Illustrative Problem 24.13)

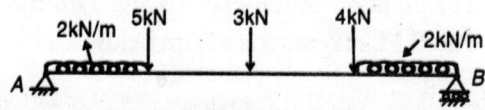

Fig. 24.38: Beam alongwith Loading (Illustrative Problem 24.13)

PROBLEM 24.14: A simple beam carries inclined loads as shown in Fig. 24.39. Sketch the S.F., B.M. and axial thrust diagrams.

Solution: The components of the inclined loads are obtained and shown in the figure. The vertical components are treated as usual and the S.F. and B.M. diagrams as shown.

Reactions:

$V_A = \dfrac{0.71 \times 3 + 1.73 \times 2 + 1.50 \times 1}{4}$

$= 1.77$ kN↑

$V_B = 2.17$ kN↑

Since *B* is a roller support, and *A* is a hinge

Axial thrust in the zone

DA is $0.71 + 1 + 2.60 = 4.31$ kN.

In the zone CD, it is $1 + 2.60 = 3.60$ kN

In the zone EC, it is 2.60 kN.

In the zone BE, it is zero.

The thrust diagram is shown.

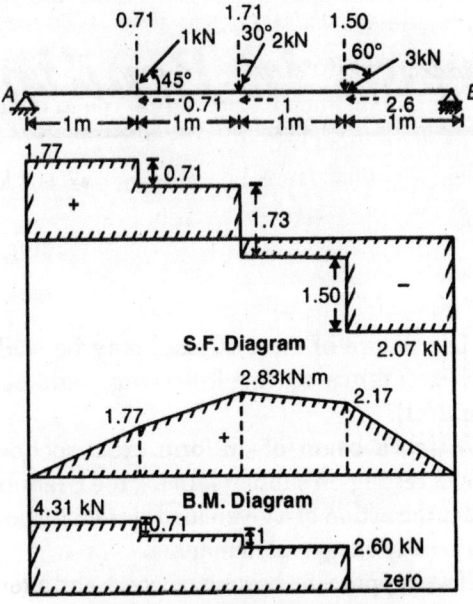

Fig. 24.39: Simple Beam with Inclined Loads
(Illustrative Problem 24.14)

PRACTICE PROBLEMS

24.1 A cantilever 6 m long, is loaded uniformly with 5 kN/m; in addition, there are point loads of 10 kN at the free end and 20 kN at the middle of its length. Sketch the S.F. and B.M. diagrams.

24.2 A cantilever, 6 m long, is loaded uniformly with 5 kN/m; in addition, there are point loads of 10 kN each at the free end and at 3 m from the free end. Sketch the S.F. and B.M. diagrams.

24.3 A simple beam carries a trapezoidal loading with intensity of loading 10 kN/m at the left end and 20 kN/m at the right end on a span of 6 m. Sketch the S.F. and B.M. diagrams.

24.4 A cantilever, 2 m long, carries a uniform load of 10 kN/m for the half towards the free end. Sketch the S.F. and B.M. diagrams.

24.5 A simple beam, 3 m span, carries a symmetrical triangular load with intensity 4 kN/m at the middle. Sketch the S.F. and B.M. diagrams.

24.6 A simple beam 4 m span, a clockwise couple of 16 kN/m acts at mid-span. Sketch the S.F. and B.M. diagrams.

24.7 A simple beam of 5 m span overhangs the roller supports by 1 m. It is loaded at 20 kN/m for 4 m from the hinge on the left. Determine the point load required at the free end such that the reactions are equal. Sketch also the S.F. and B.M. diagrams.

24.8 A simple beam of 6 m span overhangs the roller support by 3 m. It is loaded uniformly with 20 kN/m throughout and also a point load of 50 kN at the free end of the overhang. Sketch the S.F. and B.M. diagrams.

24.9 A simple beam of span 4 m has overhangs of 1 m on either side. It is loaded uniformly with 10 kN/m throughout its length, along with 10 kN point loads at the free ends. Sketch the S.F. and B.M. diagrams.

24.10 A beam ABC is 9 m long and is hinged at A and simply supported at B; $AB = 6$ m; overhang $BC = 3$ m. It carries a uniform load of 20 kN/m over the whole length and a point load of 50 kN at the free end C. Sketch the S.F. and B.M. diagrams.

24.11 The S.F. diagram for a beam is given in Fig. 24.40. Determine the beam and load diagram.

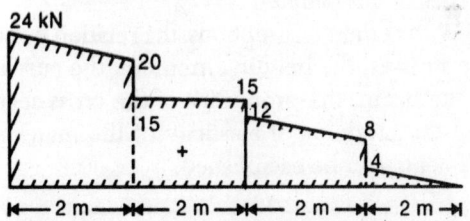

Fig. 24.40: S.F. Diagram for a Beam (Practice Problem 24.11)

24.12 A simple beam AB of span 3 m is hinged at A and is on rollers at B. At the mid-span point C, a point load of 25 kN acts inclined towards A at an angle $(\tan^{-1}4/3)$ with the horizontal. Sketch the S.F., B.M. and axial thrust diagrams.

Chapter 25

Normal Stresses in Bending

25.1 INTRODUCTION

It has been seen that the effect of applied loads on a beam is to cause shearing force and bending stress at any cross-section. The beam will undergo deformation due to these and the material resists these, setting up stresses in the process.

Shearing stresses will be set up to resist the shearing force, and bending stresses will be set up to resist the bending moment—in the cross-section (The ultimate nature of the bending stresses is only normal stresses on the cross-section, as will be seen later on). The former will be seen in the next chapter, while the latter is the subject matter of this chapter.

With certain assumptions, the relation between the stresses, the bending moment, the curvature of the beam, the properties of the cross-section, and the modulus of elasticity of the material of the beam can be established.

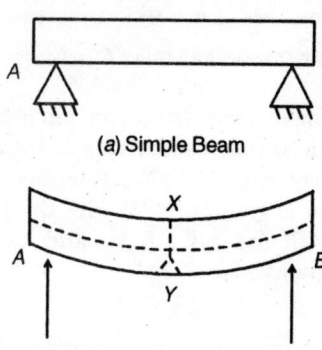

(a) Simple Beam

(b) Bent Shape of the Beam

Fig. 25.1: Nature of Stress in a Bent Beam

The nature of these stresses may be studied with reference to the following bent beam (Fig. 25.1).

In (a), a beam of uniform cross-section is shown resting on supports. In (b), the beam bent under the action of a given load system is shown, in a greatly exaggerated manner.

If we suppose the beam is of wood, and attempt to make a saw-cut in the direction X to Y, it is a matter of common experience that the beam will tend to close in on the saw, indicating a state of compression on one side of the beam. If the saw-cut is made in the reverse direction, i.e., Y to X, it is also well known that the cut opens out rapidly, thus indicating that the other side of the beam is in a state of tension. The fibres on the compression side tend to get shortened and those on the tension side get elongated. Since the transition from compression to tension, or vice-versa, has to be gradual, there must be a layer of fibres which suffer no stress, and consequently remain their original length. This layer is known as the 'Neutral Surface', and the trace of this surface on the cross-section is called the 'Neutral Axis' (N.A.).

25.2 THEORY OF SIMPLE BENDING

When a beam is bent by equal and opposite couples applied at its ends, it has uniform or constant bending moment throughout its length; further, if there is no shearing force, it is said to be in a state of 'pure' or 'simple bending'.

In other words, it is flexure by pure couples without shear force.

A beam under the action of loads may also suffer simple bending for a portion of its length—for example, a simple beam with two equal concentrated loads, placed symmetrically with respect to the ends, between the two loads; or, a symmertrical double overhang beam with equal point loads placed symmetrically in the overhang portions, within the zone between the supports (See Sec. 24.8).

The relations existing between the straining action, the dimensions, the stresses, the strains, the elastic properties of the material, and the curvature of a beam in simple bending, may be easily established under-certain simplifying assumptions. These simple relations may be generally used as close approximations in cases of flexure which are not "simple", but which are far more common in practice, the strains produced by the shear force being negligible. The justification for this is the fact that the results agree with those from experiments, and even with those from the more rigorous approach of the theory of elasticity (Refer "Theory of Elasticity by Timoshenko and Goodier), within reasonable and acceptable limits.

Assumptions in the Theory of Simple Bending

The following are the important assumptions in the theory of simple bending:

1. The material is homogeneous, isotropic, and obeys Hooke's law, and the limits of elasticity are not exceeded.
2. The beam is originally straight and is of constant cross-section.
3. Plane transverse sections of the beam remain plane and normal after bending (This is called "Bernoulli's assumption").
4. The modulus of elasticity has the same value in compression, as in tension.
5. Every layer of the material is free to expand or contract longitudinally and laterally under stress, as if it is separate from other layers.

6. The transverse section of the beam is symmetrical about an axis passing through the centroid of the section, and contained by the plane of bending.
7. The radius of curvature of the beam after bending is very large in comparison with the transverse dimensions of the beam.

25.3 SIMPLE BENDING FORMULA

A portion of a prismatic beam subjected to simple or pure bending is shown in Fig. 25.2.

Let GH be a small portion of a fibre, at a distance y from the neutral surface, the fibre being bounded by two parallel planes CD and EF, very close together. After bending, the planes assume the position shown in (d), being inclined at angle θ and intersecting at O. Let R be the radius of the neutral surface, the radius of the curved fibre G_1H_1 is $(R + y)$. Its length is altered to G_1H_1 from the original value of GH, since it is away from the neutral surface.

Now $\dfrac{G_1H_1}{JK} = \dfrac{(R+y)\theta}{R\theta} = \left(\dfrac{R+y}{R}\right)$

Also the strain in the fibre is given by

$\left(\dfrac{G_1H_1 - GH}{GH}\right) = \dfrac{G_1H_1 - JK}{JK}$

$= \dfrac{G_1H_1}{JK} - 1 = \left(1 + \dfrac{y}{R}\right) - 1 = \dfrac{y}{R}$

If f is the intensity of the stress in the fibre,

$\dfrac{f}{(y / R)} = E$, by definition.

or $\quad \dfrac{f}{y} = \dfrac{E}{R}$ \qquad ...(Eq. 25.1)

This may also be put in the form

$f = \dfrac{E}{R} \cdot y$, and

We can put $f = k \cdot y$, k being a constant, for a given beam under a given bending moment. This means that the stress is directly proportional to the distance of the fibre from the neutral surface.

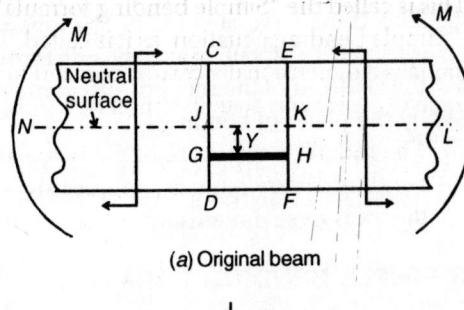

(a) Original beam

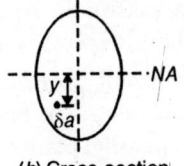

(b) Cross-section

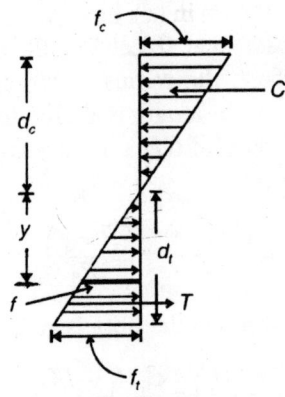

(c) Stress Diagram

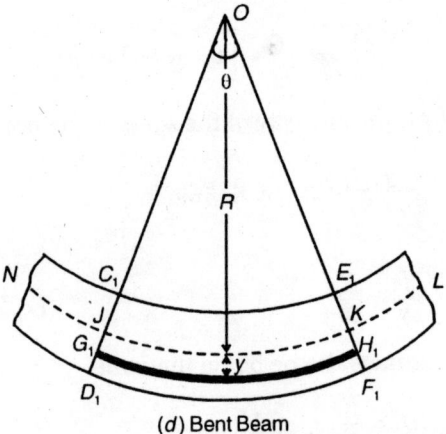

(d) Bent Beam

Fig. 25.2: Beam under Simple Bending

From equation 25.1, it is observed that, since the beam is of uniform section, R is constant, and so NL will be an arc of a circle.

The variation of the normal stress on the plane of the cross-section (b) (the stress being parallel to the longitudinal axis) is shown in (c). The direction of the arrow-heads show the nature of the stress. The total force on the compression side, C, must be equal to the force on the tension side, T, for equilibrium. These forces cause a couple which balances the external one in the plane of bending.

Position of the Neutral Surface

Let us consider an elemental area δa at a distance y from the neutral axis, the position of which is to be ascertained [cross-section shown in (b)].

Total force on the element $= f \cdot \delta a$

But $\dfrac{f}{y} = \dfrac{f_t}{y_t}$ or $f = \dfrac{f_t}{y_t} \cdot y$

Total force on the element $= f \cdot \delta a = \dfrac{f_t}{y_t} \cdot y \cdot \delta a$

Hence the total force on the traverse section below neutral axis

$$= \sum \frac{f_t}{y_t} \cdot y \delta a = \frac{f_t}{y_t} \Sigma y \cdot \delta a$$

Similarly, it can be shown that the total force on the section above neutral axis

$$= \frac{f_c}{y_c} \Sigma y \cdot \delta a$$

The vertical shearing force being nil, the internal forces exerted across the section are wholly horizontal (or longitudinal), and since they form a couple, the total tensile force must balance the compressive one; that is to say, the algebraic sum of the horizontal internal forces must, like the external ones, be zero. Putting this in symbols, we can find the position of the neutral axis.

We know, $\dfrac{f_t}{y_t} = \dfrac{f_c}{y_c} = \dfrac{E}{R}$

(assuming E to be the same in tension as in compression)

Also $(f_t/y_t)\Sigma y \cdot \delta a$ is the total tensile force on the section (for the area below the neutral axis), and $(f_c/y_c)\Sigma y \cdot \delta a$ is the total compressive force on the section (for the area above the neutral axis). Since the resultant has to be zero,

$\Sigma y \cdot \delta a$ for the area below N.A = $\Sigma y \cdot \delta a$ for the area above N.A., numerically, with opposite signs.

Or $\Sigma y \cdot \delta a = 0$ for the whole section above the neutral axis. From the properties of the section, the first moment of the area about an axis is zero only if the axis passes through the centroid of the section.

Thus, we conclude that, in the case of simple bending, the neutral axis passes through the *centroid* of the section.

If the section is symmetrical about the neutral axis, $y_t = y_c$, and the maximum values of the tensile and compressive stresses are equal. Otherwise, they are unequal (i.e., if $y_t \neq y_c$).

Moment of Resistance

The value of the couple formed by the internal resisting forces set up due to the normal stresses on the cross-section is called as the 'Moment of Resistance', which is equal and opposite to the applied bending moment.

Considering the elemental area δa again, the moment of the force on it about the neutral axis is given by

$$\frac{f_t}{y_t} \cdot \delta a \cdot y^2$$

The total moment of all such forces in the cross-section, or the moment of resistance, M_r, is given by,

$$M_r = \frac{f_t}{y_t} \Sigma y^2 \cdot \delta a = M \qquad \text{...(Eq. 25.2)}$$

But $\Sigma y^2 \cdot \delta a$ is the second moment of area (or the moment of inertia as it is loosely referred to), I, of the cross-section about the neutral axis.

$$\therefore M = \frac{f_t}{y_t} \cdot I = \frac{f_c}{y_c} \cdot I = \frac{f}{y} I, \text{ generally speaking} \qquad \text{...(Eq. 25.3)}$$

$$\therefore \quad \frac{M}{I} = \frac{f}{y} \qquad \text{...(Eq. 25.4)}$$

Combining Eqs. 25.1 and 25.4, we get

$$\frac{M}{I} = \frac{f}{y} = \frac{E}{R} \qquad \text{...(Eq. 25.5)}$$

This is called the "Simple bending formula" or the "Simple bending equation' as it is called. This equation is a dimensionally correct one and hence the units of the various quantities must be consistent.

25.4 A CRITICAL APPRAISAL OF THE ASSUMPTIONS IN THE THEORY OF SIMPLE BENDING

Assumption 1 regarding homogeneity, isotropy, obedience of Hooke's law, and the limits of elasticity not being exceeded is imperative in the validity of Eq. 25.1

$$\frac{f}{y} = \frac{E}{R}$$

Assumption 2 is necessary because the consideration of initially curved bars, or beams of non-prismatic sections, requires more complex treatment. Prismatic section is implied in the validity of Eq. 25.1, R being constant in pure bending.

Assumption 3 appears reasonable since the straining action is the same on every section; this has been used in establishing the strain in a longitudinal fibre, and also, the resultant force on the section taken to be zero in obtaining the moment of resistance.

Assumption 4 is imperative in using Eq. 25.1— $\frac{f}{y} = \frac{E}{R}$ — on both the tension and compression sides of the cross-section, without differentiating between the values of the modulus of elasticity E on both the sides.

Assumption 5 is also necessary for the validity of Eq. 25.1— $\frac{f}{y} = \frac{E}{R}$ —since the strain in a particular layer is considered as if it is free to expand or to contract longitudinally and laterally under stress without interference from the adjacent fibres, or as if it is separate from others – (Otherwise E in Eq. 25.1 would not be the modulus of elasticity in the ordinary sense, but some modified value!).

Assumption 6 is necessary because the consideration of bending of a section without a vertical axis of symmetry in the plane of bending is more complex (Satisfaction of this assumption ensures the deflection of the beam to be purely vertical under the action of the bending moment).

Assumption 7 implies that the strain, y/R, in a longitudinal fibre is small, so that the satisfaction of Hooke's law is automatically ensured.

25.5 ORDINARY BENDING—APPLICATION OF SIMPLE BENDING EQUATION TO PRACTICAL CASES

Cases of simple bending are not common; in practice, it is usually found that the bending moment in a beam varies from section to section along the span, and it is also accompanied by a shearing force. Hence it would appear that the simple bending equation is not strictly applicable to such cases. The existence of shearing force leads to the introduction of shearing or tangential stresses in the plane of the cross-section, in addition to the normal stresses due to bending. This may lead to plane sections not remaining plane after bending, thus violating one of the most important assumptions in the theory of simple bending.

However, it is found that in most practical cases, the bending moment is maximum where the shearing force is zero, and where the shearing force is maximum, the bending moment is zero or negligible. Thus, at the section of maximum bending moment, the conditions of simple bending are approximately satisfied, and hence it appears justifiable to apply the simple bending equation for this section. The stresses are the most important at the section of the maximum bending moment, since the strength of the beam will be more than sufficient when the beam is designed for the maximum bending moment.

In view of these comments, and also as experimental results in bending suggest close agreement with those from simple bending equation, it is considered that the application of simple bending theory for most practical cases of ordinary bending is justifiable.

25.6 STRENGTH IN BENDING— EFFICIENT SHAPE OF SECTION

Strength of a beam section in bending is defined as the "capacity to resist moment" without the maximum allowable stress being exceeded for the material.

Equation 24.3 may be written also as,

$$M = f_t \cdot \frac{I}{y_t} = f_c \cdot \frac{I}{y_c}$$

$\dfrac{I}{y_t}$ and $\dfrac{I}{y_c}$ are called the tension and compression "section moduli", respectively, and are denoted by Z_t and Z_c.

$$\therefore \qquad M = f_t \cdot Z_t = f_c \cdot Z_c \qquad \text{...(Eq. 25.6)}$$

For symmetrical sections, $y_t = y_c$, and so $Z_t = Z_c = Z$, say.

Z is now called the "Modulus of Section" for the cross-section.

$$\text{or} \qquad M = f \cdot Z \qquad \text{...(Eq. 25.7)}$$

since $f_t = f_c = f$, for symmetrical sections.

Z is also written as $Z = \dfrac{I}{c}$ $\qquad$ ...(Eq. 25.8)

where c is the extreme fibre distance from the neutral axis.

Since the material in proximity to the neutral zone offers very little resistance, a beam section should be such that the maximum area is disposed or arranged as far away from the neutral axis as possible.

Such a section is an "efficient shape" in bending; this is exemplified in the well-known sections adopted in engineering practice, such as I, channel, and T-sections.

25.7 BEAM OF UNIFORM STRESS

The bending moment generally varies along the span of a beam in some manner depending upon the loading. The beam is usually designed for the maximum bending moment; so the section will be larger than necessary at all other sections, if the cross-section is constant throughout the length of the beam.

A beam with its cross-section proportioned such that its modulus of section is proportional to the bending moment at each section is called a "beam of *uniform stress*" (loosely, but inappropriately, called a "beam of *uniform strength*"). This results in the most economical design and efficient use of the strength of material. This idea, with certain limitations, is attempted in built-up or compound girders of various types. In other cases, there is not much of a practical advantage in adopting an exactly proportional variable cross-section, although variable sections are common, for example, tapered masts, carriage springs, and certain cantilevers.

We know $M = f \cdot Z$

or $f = \dfrac{M}{Z}$ must be a constant—for a beam of uniform stress.

If a rectangular beam section is used,

$$Z = \frac{I}{c} = \frac{bd^3 \times 2}{12 \times d} = \frac{1}{6}bd^2 \qquad \ldots \text{(Eq. 25.9)}$$

In this, either b or d (or both) may be varied so that bd^2 is proportional to M.

If b is constant, d has to be proportional to $\sqrt{M}$; if d is constant, b has to be proportional to M.

If a circular section is used,

$$Z = \frac{\pi d^4 \times 2}{64 \times d} = \frac{\pi d^3}{32} \qquad \ldots \text{(Eq. 25.10)}$$

In this, the diameter should be proportional to cube root of the bending moment.

25.8 STRAIN ENERGY IN BENDING

We have seen in Chapter 22 that the strain energy due to normal stress is given by $\dfrac{\sigma^2}{2E}$ per unit volume.

When a beam undergoes bending within elastic limits, the material is subjected to different and varying values of tensile and compressive stresses due to bending moments, and hence possesses elastic strain energy. The total flexural strain energy may be obtained in various ways—it may be conveniently expressed in the form:

$$k \cdot \frac{f^2}{E} \cdot \text{(Volume of the material in the beam)}$$

where f is the maximum bending stress and k is a coefficient depending upon the manner in which the beam is supported and loaded—but this is always less than 1/2, the value for uniform normal stress. [It may be noted that the capacity to absorb energy varies not only upon the safe limit of stress,

f, but is also inversely upon the value of the modulus of elasticity, E. In this respect, wood and such other light materials may compare favourably with metals like steel].

Let us consider a small length dx of a beam between two sections, and let dA be a small/elemental area over which the stress is f. (Fig. 25.3).

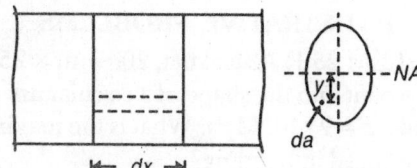

Fig. 25.3: Strain Energy in Bending

Strain energy in the area dA over the small length dx

$$= \frac{f^2}{2E} \cdot dx \vdots dA$$

But $f = \dfrac{M}{I} \cdot y$,

where y is the distance of dA from the neutral axis.

Thus the strain energy in the small piece

$$= \frac{M^2 y^2}{2I^2 E} \cdot dx \cdot dA$$

The strain energy in the length dx of the beam

$$= \sum \frac{M^2 y^2}{2I^2 E} \cdot dx \cdot dA = \frac{M^2 dx}{2I^2 E} \Sigma y^2 dA$$

But since $\Sigma y^2 \cdot dA = I$, this comes to

$$= \frac{M^2}{2EI} \cdot dx.$$

Total strain energy of the beam in bending,

$$U_b = \int_0^l \frac{M^2}{2EI} \cdot dx \qquad \ldots \text{(Eq. 25.11)}$$

or $U_b = \dfrac{1}{2EI} \displaystyle\int_0^l M^2 \cdot dx \qquad \ldots \text{(Eq. 25.12)}$

(for a prismatic cross-section)

[This can be applied to non-prismatic sections by putting $1/I$ within the integral sign].

Thus, the strain energy depends on the span, the loading, and the properties of the section.

For a constant bending moment,

$$U_b = \frac{1}{2EI} \cdot M^2 l.$$

For other cases, it can be derived from fundamentals by applying Eq. 25.12.

ILLUSTRATIVE PROBLEMS

PROBLEM 25.1: A steel flat, 200 mm × 25 mm thick, is bent into the shape of a circular arc of 25 m radius. $E = 2 \times 10^5$ MPa. What is the maximum stress induced in bending?

Solution: Extreme fibre distance,

$$c = \frac{1}{2}t = \frac{1}{2} \times 25 = 12.5 \text{ mm}$$

Radius of the circular arc,

$$R = 25 \text{ m} = 25 \times 10^3 \text{ mm}$$

$$\frac{f}{c} = \frac{E}{R} \text{ or } f = \frac{E}{R} \cdot c$$

$$= \frac{2 \times 10^5}{25 \times 10^3} \times 12.5 = 100 \text{ MPa}$$

PROBLEM 25.2: A timber joist, 300 mm × 600 mm deep, simply supported over a span of 3 m, carries a uniform load of 3 kN/m. Calculate the maximum bending stress.

Solution:

$$M = \frac{wl^2}{8} = \frac{3 \times 10^3 \times 3 \times 3 \times 10^3}{8}$$

$$= 3.375 \times 10^6 \text{ N.mm}$$

$$\frac{M}{I} = \frac{f}{c} \text{ or } f = \frac{M \cdot c}{I}$$

$$I = \frac{1}{12} \times bd^3 = \frac{1}{12} \times 300 \times (600)^3 \text{mm}^4$$

$$c = \frac{1}{2}d = \frac{1}{2} \times 600 = 300 \text{ mm}$$

$$\therefore \quad f = \frac{3.375 \times 10^6 \times 300 \times 12}{300 \times 600 \times 600 \times 600} \text{ MPa}$$

$$= 0.2 \text{ MPa}$$

PROBLEM 25.3: The moment of inertia of a beam section, 400 mm deep, is 8×10^8 mm⁴. Determine the maximum span over which this beam can carry a uniform load of 300 kN/m. The maximum stress in the material should not exceed 120 MPa.

Solution:

$I = 8 \times 10^8$ mm⁴ Extreme fibre distance,

$$c = \frac{400}{2} = 200 \text{ mm}$$

$$f = 120 \text{ MPa}$$

$$M_r = f \cdot Z = f \cdot \frac{I}{c} = \frac{120 \times 8 \times 10^8 \times 2}{400}$$

$$= 480 \times 10^6 \text{ N.mm} = 480 \text{ kN.m}$$

$$M_r = M = \frac{wl^2}{8}$$

$$\therefore \quad l = \sqrt{\frac{8M}{w}} = \sqrt{\frac{8 \times 480}{300}}$$

or $l = 3.58$ m,   say 3.50 m

PROBLEM 25.4: A beam of symmetrical section of depth 200 mm with a moment of inertia 1×10^8 mm⁴ units is simply supported over a span of 4 m. What uniformly distributed load can it carry if the maximum bending stress is not to exceed 120 N/mm²? With the same permissible stress, what concentrated load may be carried by the beam at mid-span?

$$I = 10^8 \text{ mm}^4 \qquad f = 120 \text{ MPa}$$

$$c = (1/2) \times 200 = 100 \text{ mm}$$

$$M_r = f \cdot Z = 120 \times \frac{10^8}{100} = 1.2 \times 10^8 \text{ N.mm}$$

$$M_r = M = 120 \text{ kN.m}$$

If w is the intensity of load in kN/m, l the span in metres,

$$M = \frac{wl^2}{8} \text{ kN.m}$$

$$\therefore \quad 120 = \frac{w \times 4^2}{8}$$

$$\therefore \quad w = \frac{8 \times 120}{16} \text{ kN/m} = 60 \text{ kN/m}$$

If W is the central concentrated load in kN,

$$M = \frac{Wl}{4}$$

$$\therefore \quad W = \frac{4M}{l} = \frac{4}{4} \times 120$$

$$= 120 \text{ kN}$$

PROBLEM 25.5: Find the dimensions of the cross-section of a timber joist of span 4 m to carry a maximum bending moment of 20 kN.m. The maximum bending stress is to be limited to 10 MPa. The depth of the joist is to be twice the breadth.

Solution:

$$M = M_y = 20 \text{ kN.m} = 2 \times 10^7 \text{ N.mm}$$

$$\text{depth, } d = 2b \ (b : \text{breadth})$$

$$\therefore \quad b = \frac{d}{2}$$

Modulus of Section,

$$Z = \frac{1}{6} bd^2 = \frac{1}{6} \times \frac{d}{2} \times d^2 = \frac{d^3}{12}$$

$$M_r = f \cdot Z = 10 \times \frac{d^3}{12} = 2 \times 10^7$$

$$\therefore \quad d^3 = \frac{2 \times 10^7 \times 12}{10} = 24 \times 10^6$$

or $\quad d = \sqrt[3]{24} \times 10^2 \text{ mm}$

$$= 288.4 \text{ mm, Say } 300 \text{ mm}$$

A rectangular section 150 mm × 300 mm may be used.

PROBLEM 25.6: A C.I. water pipe, 500 mm internal diameter and 25 mm thick, is supported on a span of 6 m. Determine the maximum bending stress in the material when the pipe is running full. Unit weight of cast-iron is 70 kN/m³ and that of water 10 kN/m³.

Solution:

External diameter
$$= (500 + 2 \times 25) = 550 \text{ mm}$$

Sectional area, A of the material

$$= \frac{\pi}{4} (0.55^2 - 0.50^2)$$

$$= 4.12 \times 10^{-2} \text{ m}^2$$

Self-weight of the pipe for the span of 6 m
$$= 4.12 \times 10^{-2} \times 6 \times 70 \text{ kN}$$

$$= 34.64 \text{ kN}$$

Weight of running water

$$= \frac{\pi}{4} \times (0.50)^2 \times 6 \times 10 \text{ kN} = 23.56 \text{ kN}$$

Total weight, uniformly distributed over the span,

$$wl = (34.64 + 23.56) = 58.20 \text{ kN}$$

Maximum B.M,

$$M = wl \times \frac{l}{8} = \frac{58.20 \times 6}{8} \text{ kN.m}$$

$$= 43.65 \text{ kN.m}$$

I_{NA} of the pipe section

$$= \frac{\pi}{64} (550^4 - 500^4) \text{ mm}^4$$

$$= 1.424 \times 10^9 \text{ mm}^4$$

Maximum bending stress,

$$f = \frac{Mc}{I} = \frac{43.65 \times 10^6 \times (550/2)}{1.424 \times 10^9}$$

$$= 8.43 \text{ MPa}$$

PROBLEM 25.7: Two 120 mm × 30 mm timber sections are glued together to form a T-section as shown in Fig. 25.4. If a sagging B.M. of 4 kN.m is applied to this beam about the horizontal axis through the centroid, determine the stresses at the extreme fibres and the total tension and total compression.

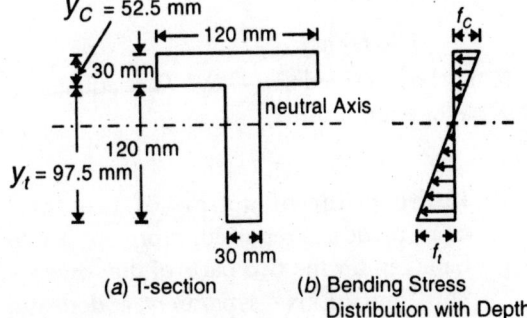

(a) T-section   (b) Bending Stress Distribution with Depth

Fig. 25.4: Beam of T-section (Illustrative Problem 25.7)

Solution:

Area of section, $A = 2 \times 120 \times 30 = 7\,200 \text{ mm}^2$

By moments about the top,

$$\bar{y} = \left(\frac{120 \times 30 \times 15 + 120 \times 30 \times 90}{7200}\right) \text{ mm}$$

$$= 52.2 \text{ mm}$$

$$I_{NA} = \left[\frac{120 \times 30^3}{12} + 120 \times 30 \times (37.5)^2\right.$$

$$\left. + \frac{30 \times 120^3}{12} + 30 \times 120 \times (37.5)^2\right] \text{ mm}^4$$

$$= 14.715 \times 10^6 \text{ mm}^4$$

Maximum compressive stress, f_c, occurs at the top fibre and the maximum tensile stress, f_t, occurs at the bottom fibre.

$$f_c = \frac{4 \times 10^6 \times 5.25}{14.715 \times 10^6} \text{ N/mm}^2 = 14.27 \text{ MPa}$$

$$f_t = \frac{4 \times 10^6 \times 97.5}{14.715 \times 10^6} \text{ N/mm}^2 = 26.53 \text{ MPa}$$

Stress at the junction of the flange and the web

$$= \frac{22.5}{52.5} \times 14.27$$

$$= 6.12 \text{ MPa (compressive)}$$

Total compression,

$$C = \left[\frac{120 \times 30 \times (14.27 + 6.12)}{2} + \frac{6.12}{2} \times 30 \times 22.5\right]$$

$$= (36.70 + 2.07) \approx 38.80 \text{ kN}$$

Total tension,

$$T = 30 \times 97.5 \times \frac{26.53}{2} \text{ N}$$

$$= 38.80 \text{ kN}$$

It may be observed that C and T are equal, as it shall be for equilibrium.

Note:

(1) If the point of action of the total compression is required, moments are to be taken for the two parts of the forces—36.70 and 2.1 kN—separately, added, and divided by the total compression.

(2) Similarly, the point of action of the total tension, is at the centroid of the stress distribution diagram on the tension side.

(3) The product of C or T with the lever arm will be equal to the applied moment on the section.

PROBLEM 25.8: A rectangular section is to be cut from a circular log of wood of diameter D. Find the dimensions of the strongest section in bending.

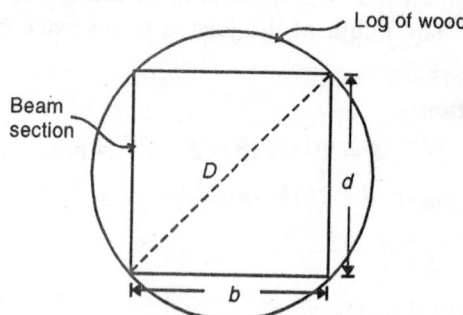

Fig. 25.5: Strongest Section from a Log of Wood (Illustrative Problem 25.8)

Solution: Let the breadth be b and depth d.

For the strongest section in bending, the section modulus, Z, must be maximum.

$$Z = \frac{1}{6}bd^2$$

$$b^2 + d^2 = D^2$$

$$d^2 = D^2 - b^2$$

$$\therefore \quad Z = \frac{1}{6}b(D^2 - b^2)$$

$$\frac{dZ}{db} = \frac{1}{6}(D^2 - 3b^2) = 0$$

or $\quad b = D/\sqrt{3}$

$$d^2 = D^2 - \frac{D^2}{3} = \frac{2D^2}{3}$$

or $\quad d = \frac{\sqrt{2}}{\sqrt{3}}D$

Also, $\dfrac{d}{b} = \sqrt{2}$

PROBLEM 25.9: Two beams of circular section, one solid and the other hollow are of the same material and are equally strong in bending If the inside diameter of the hollow section is three-fifths of the external diameter, compare the areas of section of the two beams.

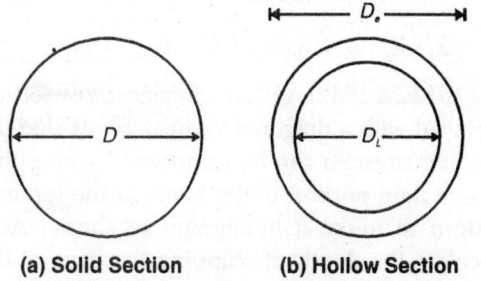

(a) Solid Section   **(b) Hollow Section**

Fig. 25.6: Comparison of Areas of Section of Equally Strong Beams in Bending (Illustrative Problem 25.9)

Solution: Let the diameter of solid section be D, and the external and internal diameters of the hollow section be D_e and D_i, respectively.

Area of solid section,

$$A_s = \frac{\pi}{4} D^2 = 0.25\pi D^2$$

Area of hollow section,

$$A_h = \frac{\pi}{4}(D_e^2 - D_i^2)$$

$$D_i = 0.6\, D_e$$

$$\therefore \quad A_h = \frac{\pi}{4}\{D_e^2 - (0.6D_e)^2\}$$

$$= 0.16\pi D_e^2$$

Modulus of section, Z_s, for the solid section, is given by

$$Z_s = \frac{\pi D^4 \times 2}{64 \times D} = \frac{\pi D^3}{32}$$

Z_h for the hollow section is

$$Z_h = \frac{\pi}{32D_e}(D_e^4 - D_i^4)$$

$$= \frac{\pi}{32}(0.8704 D_e^3)$$

Since the beams are equally strong in bending,
$$Z_s = Z_h.$$

$$\therefore \quad D^3 = 0.8704\, D_e^3 \quad \text{or} \quad D = 0.955 D_e$$

The ratio of the areas of section of solid to hollow section is

$$\frac{A_s}{A_h} = \frac{0.25D^2}{0.16D_e^2} = \frac{0.25 \times (0.955)^2 D_e^2}{0.16D_e^2} = 1.425$$

Thus the solid section will be 42.5% heavier than the hollow section specified.

PROBLEM 25.10: For a given allowable stress, compare the moments of resistance of a beam of square section placed (i) with a side horizontal and (ii) with a diagonal horizontal.

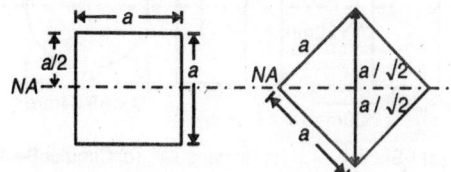

(a) Square Section   **(b) Diamond Section**

Fig. 25.7: Square Section and Diamond Section (Illustrative Problem 25.10)

Solution:

For the square section I_1 (about the *NA*)

$$= \frac{a^4}{12}$$

For the diamond section I_2 (about the *NA*)

$$= 2\left(\frac{\sqrt{2}a}{12} \times \frac{a^3}{2\sqrt{2}}\right) = \frac{a^4}{12}$$

So $I_1 = I_2$, interestingly.

Extreme fibre distances are:

$$c_1 = \frac{a}{2} \qquad \text{and} \qquad c_2 = \frac{a}{\sqrt{2}}$$

Moduli of section are:

$$Z_1 = \frac{I_1}{c_1} = \frac{a^4}{12} \times \frac{2}{a} = \frac{a^3}{6}$$

$$Z_2 = \frac{I_2}{c_2} = \frac{a^4}{12} \times \frac{\sqrt{2}}{a} = \frac{a^3}{6\sqrt{2}}$$

Since $M_r = f \cdot Z$, f being the same, to compare M_r, it is sufficient to compare Z-values.

$$\therefore \quad Z_1 / Z_2 = \sqrt{2} = 1.414$$

That is, the square section is 41.4% stronger than the diamond section of the same size, in bending.

PROBLEM 25.11: Compare the bending strength of an I-section, a rectangular section with depth 1.5 times the breadth, and a circular section, all of the same area as shown in Fig. 25.8.

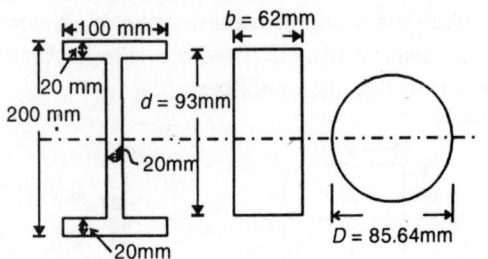

(a) I-Section   (b) Rectangular   (c) Circular Section
               Section

Fig. 25.8: Comparison of Bending Strengths
(Illustrative Problem 25.11)

Solution:

$$\text{Area, } A_i = 160 \times 20 + 2 \times 100 \times 20$$

$$= 5760 \text{ mm}^2$$

$$A_r = b \times d = b \times 1.5b = 5760$$

or $b = 62$ mm

      $d = 93$ mm

$$A_c = \frac{\pi}{4} D^2 = 5760$$

$\therefore$ $D = 85.64$ mm

$$I_i = \left[\frac{20 \times 160^3}{12} + 2\left(\frac{100 \times 20^3}{12} + 100 \times 20 \times 90^2 \right) \right]$$

$$= 39.36 \times 10^6 \text{ mm}^4$$

$$I_r = \frac{62 \times (93)^3}{12} = 4.156 \times 10^6 \text{ mm}^4$$

$$I_c = \frac{\pi}{64} \times (85.64)^4 = 2.64 \times 10^6 \text{ mm}^4$$

$$Z_i = \frac{I_i}{c_i} = \frac{39.36 \times 10^6}{100} = 0.3936 \times 10^6 \text{ mm}^3$$

$$Z_r = \frac{I_r}{c_r} = \frac{4.156 \times 10^6}{46.5} = 0.0894 \times 10^6 \text{ mm}^3$$

$$Z_c = \frac{I_c}{c_i} = \frac{2.648 \times 10^6}{42.82} = 0.0617 \times 10^6 \text{ mm}^3$$

$$Z_i : Z_r : Z_c = 6.38 : 1.45 : 1$$

PROBLEM 25.12: A beam of square cross-section is placed with a diagonal vertical. Show that the bending strength can be improved by chipping off a certain portion of the beam at the top and bottom to make it hexagonal in shape. Also calculate the depth of chipping in terms of the original depth for maximum strength. What is the percent gain in strength for this condition?

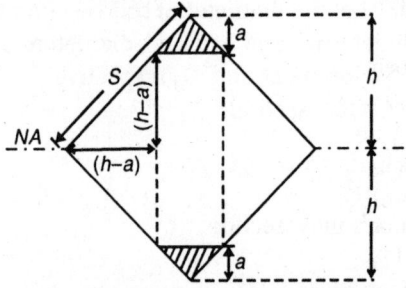

Fig. 25.9: Chamfered Section of a Diamond
(Illustrative Problem 25.12)

Solution: With reference to Fig. 25.9.

Original Section

$$I_{NA} = \frac{s^4}{12} = \frac{h^4}{3}$$

$$Z_1 = \frac{I_1}{c_1} = \frac{h^3}{3}$$

Chamfered Section

$$I_{NA} = \frac{(h-a)^4}{12} \times 4 + \frac{2a \times 8(h-a)^3}{12}$$

$$= \frac{(h-a)^3}{3}(h+3a)$$

$\therefore$ $Z_2 = \frac{I_2}{c_2} = \frac{(h-a)^2}{3}(h+3a)$

$$Z_2 > Z_1$$

∴ The chamfered section is stronger than the original diamond section.

For Z_2 to be maximum, $\dfrac{dZ_2}{da} = 0$.

This equation leads to $a = h/9$.

Depth of cut is one-ninth of the original depth, distributed equally at the top and bottom, for maximum bending strength; substituting $a = h/9$,

$$Z_2 = 1.0535(h^3/3) = 1.0535Z_1$$

Hence, the maximum gain in strength is 5.35%.

PRACTICE PROBLEMS

25.1 What is the maximum bending stress induced in a steel flat 100 mm × 10 mm thick, if it is bent into a circular arc of 10 m radius? $E = 2 \times 10^5$ MPa.

25.2 A timber joist, 300 mm × 600 mm, is simply supported over a span of 3.6 m, and carries a uniform load of 4.5 kN/m. What is the maximum bending stress in the beam?

25.3 The moment of inertia of a beam section, 500 mm deep, is 5×10^8 mm⁴ units. What is the maximum permissible span for a simple beam with a uniform load of 200 kN/m so that the maximum permissible stress does not exceed 100 MPa?

25.4 A beam of symmetrical section of depth 300 mm has a moment of inertia of the section 9×10^7 mm⁴, and is simply supported over a span of 6 m. If the bending stress is not to exceed 120 MPa, what is the maximum uniform intensity of loading it can carry?

25.5 A beam of symmetrical section, 200 mm deep, has a moment of inertia of the section 1×10^8 mm⁴ units, and is simply supported over a span of 4 m. If the permissible stress is 180 MPa, what is the maximum central concentrated load it can carry?

25.6 An equal T-section, 150 mm × 150 mm × 15 mm thick, is used as a simple beam with a maximum bending moment of 6 kN.m. Determine the maximum tensile and compressive stresses induced in the beam section.

25.7 A steel joist of span 6 m carries a maximum bending moment of 300 kN.m. The maximum bending stress is to be limited to 100 N/mm². Find the dimensions of a suitable rectangular section with depth equal to 1.5 times the breadth.

25.8 Compare the weights of two equally strong beams in bending—one of solid circular section and the other of hollow circular section with internal diameter half the external, the material being the same for both.

25.9 Compare the flexural strengths of a circular section with that of a regular hexagonal section, kept with a side horizontal. The side of the hexagonal section equals the radius of the circular section.

25.10 Compare the bending strengths of a circular section with that of a regular hexagonal section, the radius of the former and the side of the latter being equal. The hexagonal section is kept with a side vertical.

25.11 Compare the flexural strengths of an I-section, a rectangular section with depth twice the width (shown in Fig. 25.10), and a circular section, all of the same area.

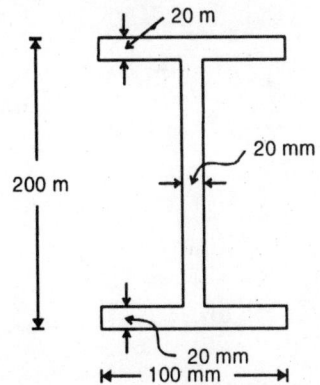

Fig. 25.10: I-Section (Practice Problem 25.11)

25.12 An I-beam with unequal flanges is supported over a span 6 m. If the permissible stresses are 90 MPa in compression and 30 MPa in tension, what uniformly distributed load can it carry safely? $I_{NA} = 2.5 \times 10^8$ mm⁴, $y_c = 200$ mm; and $y_t = 100$ mm (Fig. 25.11).

[**Hint:** If the stress on the compression side reaches 90 MPa at the top fibre, that on the

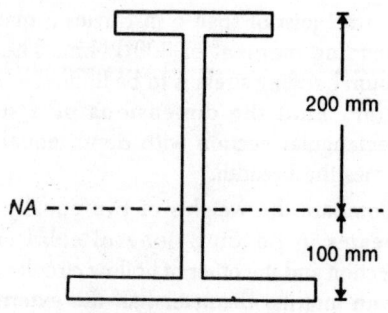

Fig. 25.11: Unsymmetrical I-Section
(Practice Problem 25.12)

200 mm

100 mm

NA

tension side will be half of this, or 45 MPa, which is more than the permissible value of 30 MPa. Hence assume the maximum tensile stress reaches 30 MPa, and the corresponding value on the compression side will be double this, or 60 MPa. Calculate the M_r value accordingly].

Chapter 26

Shear Stresses in Bending

26.1 INTRODUCTION

It has been seen in the previous chapter that the effect of the bending moment upon the cross-section of a beam is to cause bending stresses, which are, in fact, normal stresses on the plane of the cross-section. The variation of these with the depth of the section has been found to be linear.

Similarly, the effect of the vertical shearing force upon the cross-section is to cause vertical shearing stress at any point, it is natural to assume that the shearing stress acts parallel to the shearing force; this is automatically accompanied by horizontal shearing stress (parallel to the longitudinal axis of the beam) of equal magnitude, owing to the principle of complementary shear.

In this chapter, the determination of the shear stress at any point in the cross-section of a beam in bending, and its distribution with depth of the section will be considered.

26.2 DISTRIBUTION OF SHEAR STRESS IN BEAM SECTION—SHEAR STRESS FORMULA

The distribution of the shear stress is taken to be uniform across the width of the cross-section, for all practical purposes, although it is not so, strictly speaking. Also, due to the principle of complementary shear, there will be induced a horizontal shear stress in fibres parallel to the neutral plane, the value of which is numerically equal to the vertical shearing stress at the point (Fig. 26.1).

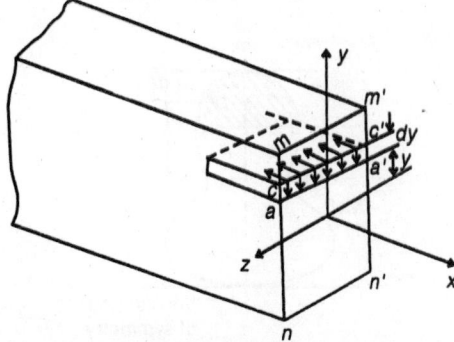

Fig. 26.1: Vertical and Horizontal Shear Stresses in a Beam Section—Pictorial View

The horizontal shear stress intensity is obtained as follows, and the same value is taken to be the vertical shearing stress, owing to the principle of complementary shear (Fig. 26.2).

Let us consider the general case of a varying bending moment M at a cross-section mn and $(M + dM)$ at a cross-section, $m_1 n_1$, at an infinitesimally small distance dx from it.

In the cross-section [Fig. (*b*)], let us consider the normal force acting on an elemental area dA of the side mp of the element $mm_1 p_1 p$ as shown.

The normal force will be,

$$f \cdot dA = \frac{M \cdot y}{I} \cdot dA$$

I being the second moment of the sectional area about the neutral axis.

(since the bending stress $f = \dfrac{M \cdot y}{I}$)

The sum of all such forces distributed over the side (mp) of the element is,

$$F_1 = \int_{y_1}^{c} \frac{M \cdot y}{I} \cdot dA \qquad \text{...(Eq. 26.1)}$$

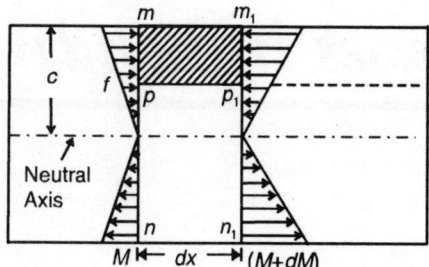

(a) A Small Length of Beam

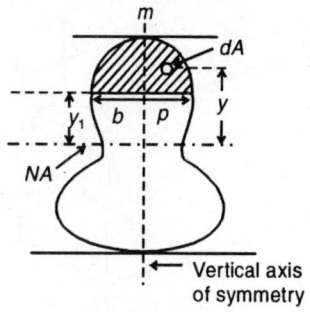

(b) Cross-section

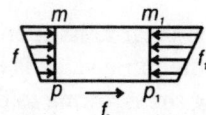

(c) Equilibrium of Element $mm_1\,p_1p$

Fig. 26.2: Determination of Horizontal Shear Stress in Beam Section

Similarly, the sum of the normal forces acting on the side m_1p_1 is

$$F_2 = \int_{y_1}^{c} \left(\frac{M + dM}{I} \right) y \cdot dA \qquad \text{...(Eq. 26.2)}$$

The force due to the horizontal shear stress f_s acting on the face pp_1 of the element is

$$F_3 = f_s \cdot b \cdot dx \qquad \text{...(Eq. 26.3)}$$

The forces F_1, F_2 and F_3 must keep the element mm_1p_1p in equilibrium.

$$\therefore \quad \int_{y_1}^{c} \frac{My}{I} \cdot dA + f_s \cdot b \cdot dx = \int_{y_1}^{c} \frac{(M + dM)}{I} \cdot y \cdot dA$$

or $$f_s = \frac{1}{I \cdot b} \frac{dM}{dx} \cdot \int_{y_1}^{c} y \cdot dA$$

Since $\dfrac{dM}{dx} = V$, the vertical shear force at the section,

$$f_s = \frac{V}{I \cdot b} \cdot Q \qquad \text{...(Eq. 26.4)}$$

where Q is the moment of the cross-section above the fibre about the neutral axis (above/below the layer concerned !). This is the same as the vertical shear stress. This is the famous "shear stress formula".

26.3 SHEAR STRESS DISTRIBUTION FOR SOME STANDARD SECTIONS

Now the shear stress distribution will be determined for some standard sections used for beams, based on the fundamental shear stress formula (Eq. 26.4).

26.3.1 Rectangular Section

The simplest and the most common standard beam section is rectangular in shape (Fig. 26.3).

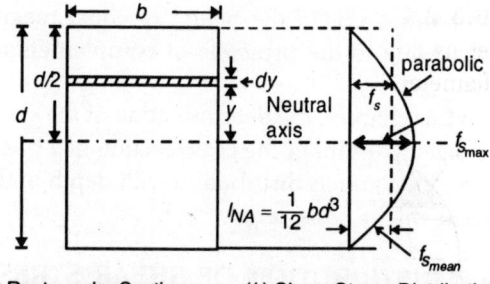

(a) Rectangular Section   (b) Shear Stress Distribution with Depth

Fig. 26.3: Shear Stress Distribution for a Rectangular Section

Let us consider a thin strip of thickness dy at a distance y from the neutral axis. The shear stress, f_s, in this strip may be assumed to be uniform. From Eq. 26.4,

$$f_s = \frac{V}{I \cdot b} \cdot Q = \frac{V \times 12}{bd^3 \times b} \int_{y}^{d/2} yb \cdot dy$$

or $\quad f_s = \dfrac{3}{2} \cdot \dfrac{V}{bd^3}(d^2 - 4y^2)$ ...(Eq. 26.5)

The distribution of f_s with depth is obviously parabolic.

At $y = \pm d/2$, $f_s = 0$, *i.e.*, at the extreme fibres, the shear stress is zero.

At $y = 0$, $f_{s_{max}} = \dfrac{3}{2} \cdot \dfrac{V}{bd} = \dfrac{3}{2} \cdot \dfrac{V}{A} = \dfrac{3}{2} \cdot f_{s_{mean}}$

...(Eq. 26.6)

This is to say, the maximum shear stress occurs at the neutral axis, and is equal to $1\frac{1}{2}$ times the mean or average shear stress in the section.

(The ratio of the maximum to mean shear stress is $1\frac{1}{2}$)

Area of the shear stress distribution diagram

$= \dfrac{2}{3} \times d \times \dfrac{3}{2} \times \dfrac{V}{bd} = \dfrac{V}{b}$

Thus the area of the shear stress distribution diagram multiplied by the width gives the shear force at the section.

In other words, the shear stress distribution diagram integrated over the width and the depth should yield shearing force at the section (This statement is always true for any section !).

26.3.2 Circular Section

Let us consider a circular section of radius, R (diameter, D), as shown in Fig. 26.4.

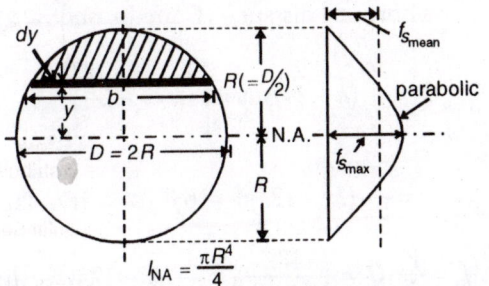

$I_{NA} = \dfrac{\pi R^4}{4}$

(a) Circular Section   (b) Shear Stress Distribution with Depth

Fig. 26.4: Shear Stress Distribution for a Circular Section

Let us consider a fibre distant y from the neutral axis and a strip of infinitesimal thickness dy. Let

the shear stress in the fibre be f_s, considered to be constant for the thin strip.

$$f_s = \dfrac{V}{I \cdot b} \cdot Q$$

V being the vertical shear force acting at the section.

Also let the radii to the ends of the fibre subtend an angle θ with the vertical axis of symmetry.

From the geometry of the figure,

$y = R\cos\theta$; Breadth of the strip, $b = 2R\sin\theta$

$dy = -R \cdot \sin\theta \cdot d\theta$

$\therefore Q = \int_y^R by \cdot dy = \int_\theta^0 (2R\sin\theta)(R\cos\theta)(-R\sin\theta \cdot \cos\theta)$

$\qquad = 2R^3 \int_0^\theta \sin^2\theta \cdot \cos\theta \cdot d\theta = \dfrac{2R^3}{3}(\sin^3\theta)_0^\theta$

$\qquad = \dfrac{2}{3}R^3 \sin^3\theta$

$Q = \dfrac{2}{3}R^3 \sin^3\theta$ ...(Eq. 26.7)

$\therefore f_s = \dfrac{4V}{\pi R^4} \times \dfrac{1}{2R\sin\theta} \times \dfrac{2}{3}R^3\sin^3\theta = \dfrac{4V}{3\pi R^2}\sin^2\theta$

$\qquad = \dfrac{4V}{3\pi R^2}\left(1 - \dfrac{y^2}{R^2}\right)$

or $\quad f_s = \dfrac{4V}{3\pi R^4}(R^2 - y^2)$ ...(Eq. 26.8)

The variation with depth is obviously parabolic.

For $y = \pm R$ (the ends of the vertical diameter),
$f_s = 0$
f_s is obviously maximum at $y = 0$ or at the neutral axis.

$f_{s_{max}} = \dfrac{4V}{3\pi R^2} = \dfrac{4}{3}\left(\dfrac{V}{A}\right) = \dfrac{4}{3} \cdot f_{s_{mean}} = 1\dfrac{1}{3}(f_{s_{mean}})$

...(Eq. 26.9)

or the ratio of the maximum to the mean shear stress is $1\frac{1}{3}$.

Integrating the shear stress distribution diagram over the depth and the breadth of the section, we have

$$2\int_0^R \left\{ \frac{4V}{3\pi R^4}(R^2 - y^2)\right\}(b) \cdot dy$$

$$= 2\int_{\pi/2}^{0} \left(\frac{4V}{3\pi R^2} \cdot \sin^2\theta \right)(2R\sin\theta)(-R\sin\theta)d\theta$$

$$= \frac{16V}{3\pi}\int_0^{\pi/2} \sin^4\theta \cdot d\theta = \frac{16V}{3\pi} \times \frac{3\pi}{16} = V$$

the applied shear force at the section.

Special Case

For the maximum value of shear stress which occurs at the neutral axis, Q will be

$$Q = \int_0^{\pi/2} 2R^3 \cdot \sin^2\theta \cdot \cos\theta\, d\theta$$

$$= \frac{2R^3}{3}\left[\sin^3\theta\right]_0^{\pi/2} = \frac{2R^3}{3}$$

$$\therefore f_{s_{max}} = \frac{4V}{\pi R^4} \times \frac{1}{2R} \times \frac{2R^3}{3}$$

$$= \frac{4V}{3\pi R^2} = \frac{4}{3}\left(\frac{V}{A}\right) = 1\frac{1}{3} f_{s_{mean}}$$

as shown already.

26.3.3 Hollow Circular Section

Let the hollow circular section be of external radius, R, and internal radius, r (Fig. 26.5).

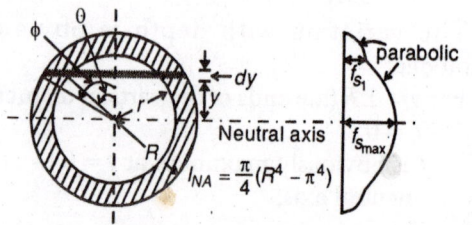

(a) Annular Section   (b) Shear Stress Distribution
Fig. 26.5: Hollow Circular Section

$$Q = \frac{2}{3}(R^3\cos^3\phi - r^3\cos^3\theta) \quad \text{(From Eq. 26.7)}$$

Q (for the semi-circular annular area)

$$= \frac{2}{3}(R^3 - r^3) \qquad \qquad ...(Eq. 26.10)$$

(for f_{max} at N.A.)

$$\therefore f_{s_{max}} = \frac{4V}{2\pi(R^4 - r^4)(R - r)} \times \frac{2}{3}(R^3 - r^3)$$

$$f_{s_{max}} = \frac{4V}{3\pi}\left[\frac{R^2 - Rr + r^2}{(R^4 - r^4)}\right] \qquad ...(Eq. 26.11)$$

This can be easily obtained from fundamentals in a numerical problem. The distribution with depth is shown in (b).

26.3.4 Diamond Section

This is a square section with a diagonal vertical as shown in Fig. 26.6.

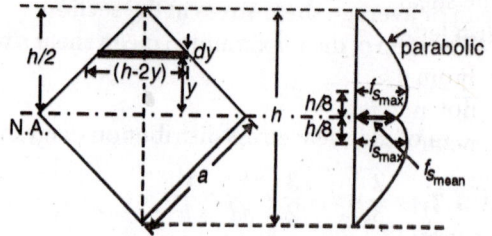

(a) Diamond Section   (b) Shear Stress Distribution with Depth
Fig. 26.6: Shear Stress Distribution for a Diamond Section

If the side is a and diagonal h,

$$h = \sqrt{2}\,a$$

$$I_{NA} = 2\left[\frac{h}{8} \times \frac{h^3}{12}\right] = \frac{h^4}{48} = \frac{a^4}{12} \qquad ...(Eq. 26.12)$$

(The same as that when the square is kept with a side vertical !)

For a fibre at a distance y from the neutral axis.

$$Q = \int_y^{h/2} (h - 2y)y\,dy = \left[\frac{hy^2}{2} - \frac{2y^3}{3}\right]_y^{h/2}$$

$$= \frac{1}{24}(h^3 - 12hy^2 + 16y^3) \qquad ...(Eq. 26.13)$$

$$\therefore f_s = \frac{V}{Ib} \cdot Q = \frac{48V}{h^4(h - 2y)} \times \frac{1}{24}(h^3 - 12hy^2 + 16y^3)$$

$$\text{or} \quad f_s = \frac{2V}{h^4}(h^2 + 2hy - 8y^2) \qquad ...(Eq. 26.14)$$

Variation of the shear stress is parabolic with depth.

$$\frac{df_s}{dy} = \frac{2V}{h^4}(2h - 16y) = 0,$$

for maximum value of f_s.

So, f_s is maximum at $y = h/8$.

$$f_{s_{max}} = \frac{9V}{4h^2} = \frac{9}{8}\frac{V}{a^2} = \frac{9}{8}f_{s_{mean}} = 1\frac{1}{8}f_{s_{mean}}$$

...(Eq. 26.15)

$f_s = 0$ at $y = \pm h/2$

At $y = 0$ at Neutral axis,

$$f_{s_{N.A.}} = \frac{2V}{h^2} = \frac{V}{a^2} = \frac{V}{A} = f_{s_{mean}} \quad \text{...(Eq. 26.16)}$$

The shear stress distribution is in shape of the English letter "*B*".

Note. From this it is clear that the shear stress need not necessarily be the maximum at the neutral axis.

26.3.5 Triangular Section

Let us consider an isosceles triangular section with the base horizontal and the vertex at the top as shown in Fig. 26.7.

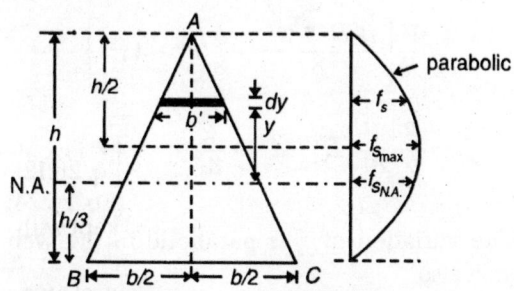

(a) Triangular Section (b) Shear Stress Distribution with Depth

Fig. 26.7: Isosceles Triangular Section—Shear Stress Distribution

Let the base width be b and height be h.

Consider a fibre at a distance y from the neutral axis and a thin strip of thickness dy.

The width b' of the fibre is

$$b' = \left(\frac{\dfrac{2h}{3} - y}{h}\right) \cdot b$$

$$I_{NA} = \frac{b}{h}\int_{-h/3}^{2h/3}\left(\frac{2h}{3} - y\right)y^2 dy = \frac{bh^3}{36} \quad \text{...(Eq. 26.17)}$$

Q for the area above this fibre about the Neutral axis.

$$\int_y^{2h/3}\left(\frac{2h}{3} - y\right)\cdot\frac{b}{h}\cdot y \cdot dy = \frac{b}{h}\int_y^{2h/3}\left(\frac{2h}{3} - y\right)y\, dy$$

$$= \frac{b}{3h}\left[\left(\frac{4h^3}{9} - \frac{8h^3}{27}\right) - (hy^2 - y^3)\right]$$

$$f_s = \frac{V}{Ib}\cdot Q$$

$$= \frac{Vh}{Ib\left(\dfrac{2h}{3} - y\right)}\cdot\frac{b}{3h}\left[\left(\frac{4h^3}{9} - \frac{8h^3}{27}\right) - (hy^2 - y^3)\right]$$

$$= \frac{V}{I(2h - 3y)}\left(\frac{4h^3}{27} - hy^2 + y^3\right)$$

$$= \frac{V(4h^3 - 27hy^2 + 27y^3)}{27I(2h - 3y)}$$

By factors,

$$f_s = \frac{V}{27I}(2h^2 + 3hy - 9y^2) \quad \text{...(Eq. 26.18)}$$

$f_{s_{(2h)/3}} = 0$ and $f_{s_{-h/3}} = 0$, also.

$$f_{s_{NA}} = \frac{2Vh^2}{27I} = \frac{2Vh^2 \times 36}{27bh^3} = \frac{8}{3}\cdot\frac{V}{bh} \quad \text{...(Eq.26.19)}$$

$$\frac{df_s}{dy} = \frac{V}{27I}(3h - 18y) = 0$$

$f_{s_{max}}$ occurs at $y = \dfrac{h}{6}$ above NA or at mid-depth.

$$\therefore f_{s_{max}} = \frac{3V}{bh} = \frac{3}{2}\cdot f_{s_{mean}} = 1\frac{1}{2}\cdot f_{s_{mean}} \quad \text{...(Eq. 26.20)}$$

The distribution is parabolic.

26.36 I-SECTION AND CHANNEL SECTION

A symmetrical I-section, with an overall depth D, flange width B, and a web with a thickness b and a depth d, is shown in Fig. 26.8.

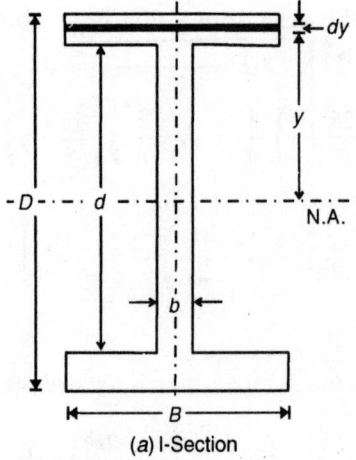

(a) I-Section

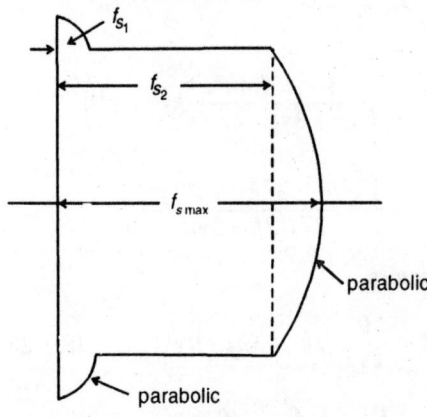

(b) Shear Stress Distribution with Depth

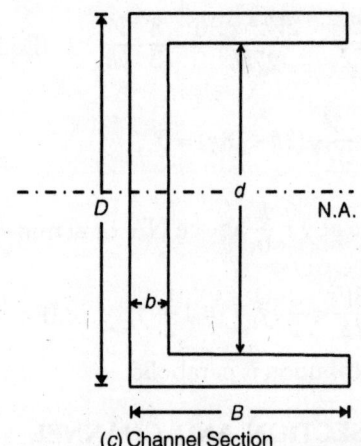

(c) Channel Section

Fig. 26.8: Shear Stress Distribution in I- and Channel Section

Flange:

At a distance y from the neutral axis, width is B.

$$f_s = \frac{V}{IB} \int_y^{D/2} B \cdot y \cdot dy = \frac{V}{8I}(D^2 - 4y^2) \quad \text{...(Eq. 26.21)}$$

The variation of f_s is parabolic.

At the extreme fibre, $y = \dfrac{D}{2}$ and $f_s = 0$.

At the junction of the flange and the web,

$$y = d/2.$$

$$f_{s_1} = \frac{V}{8I}(D^2 - d^2) \qquad \text{... (Eq. 26.22)}$$

Here the width suddenly decreases from B to b.

Hence f_s suddenly increases to a value f_{s_2}

$$f_{s_2} = \frac{B}{b} \cdot \frac{V}{8I}(D^2 - d^2) \qquad \text{...(Eq. 26.23)}$$

Web:

The width of the fibre in the web portion is b.

$$\therefore \quad f_s = \frac{V}{Ib}\left[\frac{B(D^2 - d^2)}{8} + \int_y^{d/2} by\, dy \right]$$

$$= \frac{V}{Ib}\left[\frac{B(D^2 - d^2)}{8} + \frac{b}{8}(d^2 - 4y^2) \right]$$

or $\quad f_s = \dfrac{B}{b} \cdot \dfrac{V}{8I}(D^2 - d^2) + \dfrac{V}{8I}(d^2 - 4y^2)$

$$\text{...(Eq. 26.24)}$$

The variation of f_s is parabolic in the web portion also.

At neutral axis, $y = 0$, and the maximum value of f_s occurs.

$$\therefore\ f_{s_{max}} = \frac{B}{b} \cdot \frac{V}{8I}(D^2 - d^2) + \frac{V}{8I} \cdot d^2 \ \text{...(Eq. 26.25)}$$

The shear stress distribution diagram is shown in Fig. 26.8 (b). From this diagram, it is clear that most of the shear is resisted by the web.

Shear force carried by the two flanges

$$= 2 \times B \times \frac{2}{3} \frac{(D - d)}{2} \cdot \frac{V}{8I}(D^2 - d^2)$$

$$= \frac{VB}{12I}(D - d)(D^2 - d^2)$$

Shear force carried by the web

$$= b \cdot d \frac{B}{b} \cdot \frac{V}{8I}(D^2 - d^2) + b \cdot d \cdot \frac{2}{3} \cdot \frac{V}{8I} \cdot d^2$$

$$= \frac{VBd}{8I}(D^2 - d^2) + \frac{V}{12I} bd^3$$

These two shall sum upto the total shear force V, which can be easily demonstrated in a numerical problem.

Channel Section

For the channel section shown in Fig. 26.8 (*c*), it can be easily guessed that the shear stress distribution diagram will be the same as that for the I-section, as the difference lies only in the lateral shifting of the web, with no change in position relative to the neutral axis.

26.3.7. T-Section and Angle Section

A T-section and an angle section with the same dimensions and the shear stress distribution with depth are shown in Fig. 26.9.

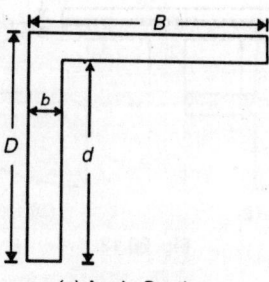

(*c*) Angle Section

Fig. 26.9: Shear Stress Distribution in T- and Angle Section

The shear stress distribution with depth for a T-section and an angle section with similar dimensions is one and the same as shown in Fig. 26.9 (*b*); this distribution may also be guessed easily, as they consist of rectangular components.

26.3.8 Miscellaneous Sections

The shear stress distribution diagrams for a few miscellaneous sections are shown in Figs. 26.10 to 26.13.

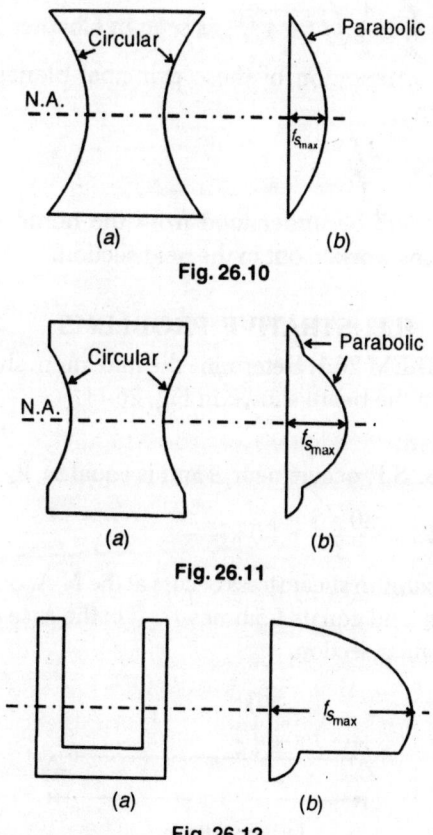

Fig. 26.10

Fig. 26.11

Fig. 26.12

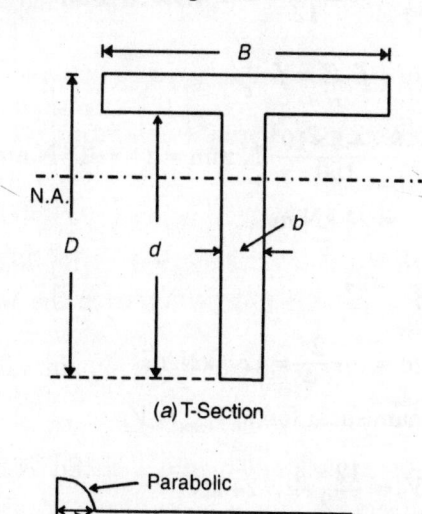

(*a*) T-Section

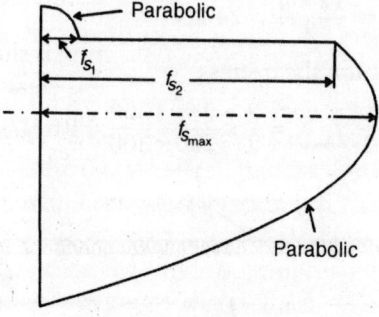

(*b*) Shear Stress Distribution

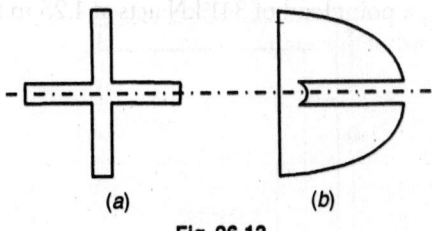

(a)   (b)

Fig. 26.13

26.4 PRINCIPAL STRESSES IN BENDING

The plane of the cross-section of a beam is subjected to normal stresses due to bending moment as seen in Chapter 25. It is also subjected to shearing stresses due to shearing force as has been seen in this chapter.

Hence, at any point in the cross-section any element bounded by its plane and a plane perpendicular to it, there will be shear stresses on both these planes, and normal stresses on the former. If these are f_s and f, respectively, the principal stresses are:

$$f_p = \frac{f}{2} \pm \frac{1}{2}\sqrt{f^2 + 4f_s^2}$$ as seen in Chapter 22.

The orientation of these principal planes is given by,

$$\tan 2\theta = \frac{2f_s}{f}$$

This will be understood from the numerical problems worked out in the next section.

ILLUSTRATIVE PROBLEMS

PROBLEM 26.1: Determine the maximum shear stress in the beam shown in Fig. 26.14.

Solution:

Max. S.F. occurs near A and is equal to V_A

$$V_A = \frac{30 \times 2}{3} = 20 \text{ kN}$$

Maximum shear stress occurs at the N.A. of the section, and equals 1.5 times $f_{s_{mean}}$ in the case of a rectangular section.

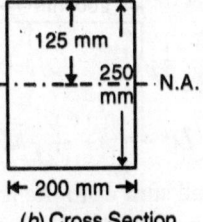

(b).Cross Section

Fig. 26.14: Maximum Shear Stress in a Beam
(Illustrative Problem 26.1)

$$f_{s_{max}} = 1.5 \times \frac{20 \times 1000}{200 \times 250} = 0.6 \text{ N/mm}^2$$

PROBLEM 26.2: A timber beam is simply supported on a span of 4 m and carries a uniform load. What is the maximum shear stress if the maximum normal stress due to flexure is 8 MPa. The cross-section is rectangular—200 mm × 300 mm.

Solution: With reference to Fig. 26.15.

$$I_{NA} = \frac{200 \times (300)^3}{12} = 4.5 \times 10^8 \text{ mm}^4$$

$$M_r = f \cdot Z = f \cdot \frac{I}{c}$$

$$= \frac{8 \times 4.5 \times 10^8}{150} \text{ N.mm} = 24 \times 10^6 \text{ N.mm}$$

$$= 24 \text{ kN.m}$$

$$\therefore \quad \frac{wl^2}{8} = 24$$

$$w = \frac{8 \times 24}{4 \times 4} = 12 \quad \text{kN/m}$$

Maximum shear force, $V_{max} = V_A$

$$or \quad V_B = \frac{12 \times 4}{2} = 24 \text{ kN}$$

Maximum shear stress

$$f_{s_{max}} = 1\frac{1}{2} \cdot f_{s_{mean}} = \frac{3}{2} \times \frac{24 \times 1000}{200 \times 300} = 0.6 \text{ N/mm}^2$$

30 kN

A ⊢— 1m —→| C   B
          ⊢———————— 3 m ————————⊣

(a) Simple Beam

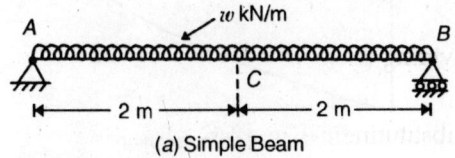

(a) Simple Beam

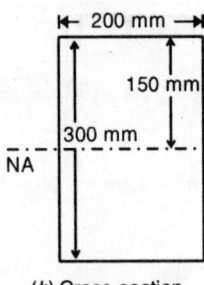

(b) Cross-section

Fig. 26.15: Shear Stress in a Beam
(Illustrative Problem 26.2)

PROBLEM 26.3: A simple beam of span 6 m is of rectangular cross-section, 150 mm × 300 mm deep. It is subjected to a single point load. Determine the position and magnitude of the load, if the maximum values of the bending stress and shear stress in the beam are not to exceed 150 MPa and 9 MPa, respectively.

Solution: With reference to Fig. 26.16:

$$I = \frac{150 \times (300)^3}{12} \text{ mm}^4$$

$$= 3.375 \times 10^8 \text{ mm}^4$$

Max. B.M., $M_{max} = \dfrac{Wa(6-a)}{6}$ kN.m

Max. S.F., $V_{max} = \dfrac{W(6-a)}{6}$ kN

Maximum bending stress,

$$f_{max} = \frac{Mc}{I} = \frac{Wa(6-a) \times 10^6 \times 150}{6 \times 150 \times 2.25 \times 10^6}$$

$$= \frac{Wa(6-a)}{13.5} = 150$$

∴ $Wa(6-a) = 2025$ ...(i)

Maximum shear stress,

$$f_{s_{max}} = \frac{3}{2} \cdot f_{s_{mean}} = \frac{3}{2} \times \frac{W(6-a) \times 1000}{6 \times 150 \times 300} = 9$$

∴ $W(6-a) = 1620$ ...(ii)

Dividing (i) by (ii), $a = \dfrac{2025}{1620} = 1.25$ m

Substituting in (ii), $W = \dfrac{1620}{4.75}$ kN = 341 kN

So, a point load of 341 kN acts at 1.25 m from one end of the span.

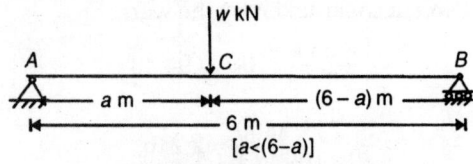

(a) Simple Beam with Non-Central Load

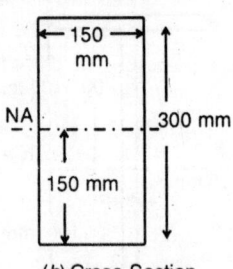

(b) Cross-Section

Fig. 26.16: Normal and Shear Stresses in Bending
(Illustrative Problem 26.3)

PROBLEM 26.4: A beam of I-section, 100 mm × 200 mm deep, and 10 mm thick throughout, carries a shear force of 30 kN at a section. What is the percent shear carried by the web? What is the ratio of the maximum to mean shear stress?

Solution: With reference to Fig. 26.17:
Area, $A = 2 \times 100 \times 10 + 180 \times 10 = 3800$ mm²

$$I_{NA} = 2\left[\frac{100 \times 10^3}{12} + 100 \times 10 \times (95)^2\right] + \frac{10 \times (180)^3}{12}$$

$$= 2.293 \times 10^7 \text{ mm}^4$$

$$f_{s_1} = \frac{30 \times 1000 \times 100 \times 10 \times 95}{2.293 \times 10^7 \times 100}$$

$$= 1.24 \text{ N/mm}^2$$

$$f_{s_2} = \frac{100}{10} \times 1.24 = 12.40 \text{ N/mm}^2$$

$f_{s_{max}}$ at N.A.

$$= \frac{30 \times 1000}{2.293 \times 10^7 \times 10}(100 \times 10 \times 95 + 90 \times 10 \times 45)$$

$$= 17.73 \text{ N/mm}^2$$

S.F. taken by the flanges

$$= 2 \times \frac{2}{3} \times 1.24 \times 10 \times \frac{100}{1000} \text{ kN}$$

$$= 1.656 \text{ kN}$$

S.F. taken by the web

$$= (30 - 1.656) = 28.344 \text{ kN}$$

Percent shear taken by the web

$$= \frac{28.344}{30} \times 100 = 94.48$$

$$\frac{f_{s_{max}}}{f_{s_{mean}}} = \frac{17.73 \times 3800}{30 \times 1000 \times 1} = 2.246$$

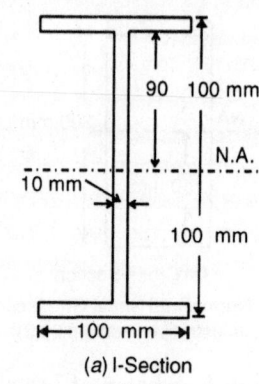

(a) I-Section

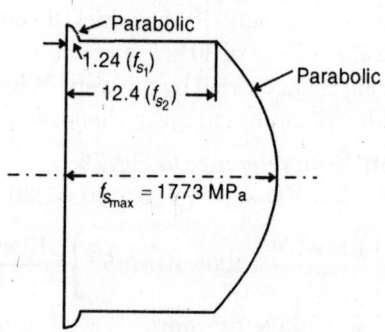

(b) Shear Stress Distribution with Depth

Fig. 26.17: Shear Stress Distribution in I-Section
(Illustrative Problem 26.4)

PROBLEM 26.5: A beam is made of a diamond section with diagonals 100 mm long. It is subjected to a shear force of 50 kN at a section. Determine the shear stress distribution with depth from fundamentals.

Solution: With reference to Fig. 26.18:

$$I_{NA} = \frac{h^4}{48} = \frac{(100)^4}{48} \text{ mm}^4$$
$$= 2.0833 \times 10^6 \text{ mm}^4$$

Let us consider a strip of thickness dy at a distance y from N.A.

Width of strip $= (h - 2y)$

$$Q = \int_y^{h/2} (h - 2y)y \, dy = \left[\frac{hy^2}{2} - \frac{2y^3}{3} \right]_y^{h/2}$$

$$= \frac{1}{24}(h^3 - 12hy^2 + 16y^3)$$

$$f_s = \frac{V}{Ib} \cdot Q$$

$$= \frac{48V}{h^4(h-2y)} \times \frac{(h^3 - 12hy^2 + 16y^3)}{24}$$

$$= \frac{2V}{h^4}(h^2 + 2hy - 8y^2)$$

Variation of shear stress is parabolic with depth.

$$\frac{df_s}{dy} = \frac{2V}{h^4}(2h - 16y) = 0 \text{ for maximum value of } f_s.$$

$\therefore$ f_s is maximum at $y = h/8$ from N.A.

$$\therefore \quad f_{s_{max}} = \frac{9V}{4h^2}$$

$$f_s = 0 \text{ at } y = \pm h/2$$

At N.A., $y = 0$; $f_{s_{NA}} = \frac{2V}{h^2} = \frac{V}{A}$ (Area A being $\frac{h^2}{2}$)

$$= f_{s_{mean}}$$

$$\therefore \quad f_{s_{max}} = \frac{9 \times 50 \times 1000}{4 \times 100 \times 100} \text{ MPa} = 11.25 \text{ MPa}$$

$$f_{s_{NA}} = f_{s_{mean}} = \frac{2 \times 50 \times 1000}{100 \times 100} = 10 \text{ MPa}$$

$f_{s_{mean}}$ occurs at $\pm \frac{100}{8}$ mm or ± 12.5 mm from N.A.

The shear stress distribution diagram resembles the English letter "B" in shape as shown in Fig. 26.18.

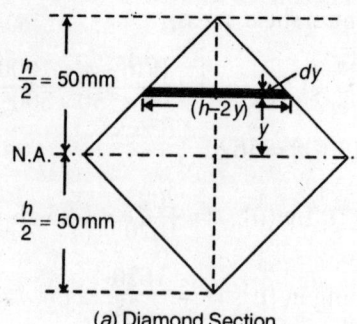

(a) Diamond Section

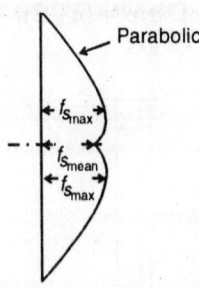

(*b*) Shear Stress Distribution with Depth
Fig. 26.18: Beam of Diamond Section
(Illustrative Problem 26.5)

PROBLEM 26.6: A triangular section of a beam is isosceles with base 180 mm and height of 120 mm. It carries a shear force of 50 kN at a section. Determine the maximum shear stress.

Solution: With reference to Fig. 26.19:

$$I_{NA} = \frac{bh^3}{36} = \frac{180 \times (120)^3}{36} = 8.64 \times 10^6 \text{ mm}^6$$

At a distance y from N.A.

$$Q = \int_y^{80} \left(\frac{80-y}{80}\right) \times 60 \times 2y\,dy$$

$$= \frac{3}{2}\left[40y^2 - \frac{y^3}{3}\right]_0^{80}$$

$$= \frac{1}{2}(120 \times 6400 - 80 \times 6400 - 120y^2 + y^3)$$

$$= \frac{1}{2}(256000 - 120y^2 + y^3)$$

$$f_s = \frac{50 \times 1000 \times 2}{8.64 \times 10^6 \times (80-y) \times 3} \times \frac{1}{2}(256000 - 120y^2 + y^3)$$

$$f_s = \frac{5}{2592}(3200 + 40y - y^2) \rightarrow \text{Parabolic variation}$$

f_s is zero at $y = 80$ mm or -40 mm

$$f_{s_{NA}} = \frac{5}{2592} \times 3200 = 6.17 \text{ MPa}$$

$$\frac{df_s}{dy} = \frac{5}{2592}(40 - 2y) = 0 \text{ for } f_s \text{ to be the maximum.}$$

or   at $y = 20$ mm

That is, f_s will be maximum at mid-height of the section.

$$f_{s_{max}} = \frac{5}{2592}(3200 + 40 \times 20 - 400)$$

$$= \frac{5}{2592} \times 3600 = 6.94 \text{ MPa}$$

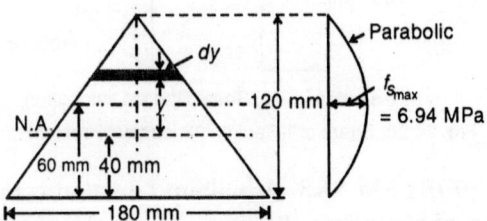

(*a*) Isosceles Triangular Section (*b*) Shear Stress Distribution
Fig. 26.19: Beam of Triangular Section
(Illustrative Problem 26.6)

PROBLEM 26.7: A T-section has a flange 60 mm × 20 mm and web 20 mm × 40 mm. It is used as a beam over a span of 5 m to carry a uniform loading of 10 kN/m. Sketch the shear stress distribution at the section of maximum S.F.

Solution: With reference to Fig. 26.20:

$$V_{max} = V_A \text{ or } V_B = 25 \text{ kN}$$

For the location of NA,

$$\bar{y} \text{ (from top)} = \frac{(60 \times 20 \times 10 + 40 \times 20 \times 40)}{200}$$

$$= 22 \text{ mm}$$

$$I_{NA} = \left(\frac{60 \times 20^3}{12} + 120 \times 12^2 + \frac{20 \times 40^3}{12} \times 80 \times 18^2\right)$$

$$= 5.787 \times 10^7 \text{ mm}^4$$

f_s at the bottom of the flange

$$= \frac{2500 \times 60 \times 20 \times 12}{5.787 \times 10^7 \times 60} = 10.40 \text{ MPa}$$

f_s at the junction of flange and web (in the web)

$$= \frac{10.4 \times 60}{20} = 31.20 \text{ MPa}$$

$f_{s_{max}}$ at the Neutral Axis

$$= \frac{2500(60 \times 20 \times 12 + 20 \times 2 \times 1)}{5.787 \times 10^7 \times 20} = 31.35 \text{ MPa}$$

The shear stress distribution is shown in Fig. 26.20.

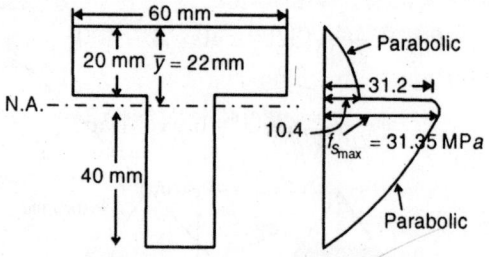

(a) T-Section   (b) Shear Stress Distribution

Fig. 26.20: Beam of T-Section (Illustrative Problem 26.7)

PROBLEM 26.8: A built-up L-section is made up of two plates, 100 m × 10 m each, the total depth of the section being 110 mm. If a shear force of 10 kN acts, determine the shear stress distribution with depth, indicating the salient values.

Solution: With reference to Fig. 26.21:

$V = 10$ kN; Area, $A = 200$ mm²

Let the N.A. be at a height $\bar{y}$ mm from the base.

Taking moments about the base,

$$\bar{y} = \frac{(100 \times 10 \times 60 + 100 \times 10 \times 5)}{2000} = 32.5 \text{ mm}$$

$$I_{NA} = \left[\frac{10 \times (100)^3}{12} + 100 \times 10 \times (27.5)^2 + \frac{100 \times 10^3}{12} \right.$$

$$\left. + 100 \times 10 \times (27.5)^2 \right]$$

$$= 2.354 \times 10^6 \text{ mm}^4$$

f_s at the junction in the flange

$$= \frac{10 \times 1000(100 \times 10 \times 27.5)}{2.354 \times 10^6 \times 100}$$

$$= 1.17 \text{ MPa}$$

f_s at the junction in the web

$$= 1.17 \times \frac{100}{10} = 11.7 \text{ MPa}$$

$f_{s_{max}}$ at the N.A.

$$= \frac{10 \times 1000}{2.354 \times 10^6 \times 10}\left(77.5 \times 10 \times \frac{77.5}{2}\right) = 12.7 \text{ MPa}$$

The shear stress distribution is parabolic and is shown in Fig. 26.21.

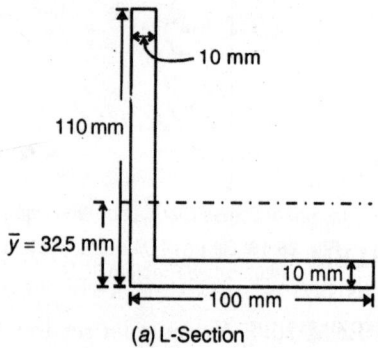

(a) L-Section

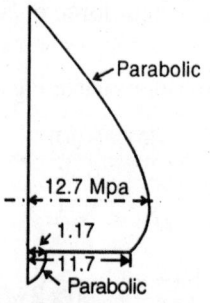

(b) Shear Stress Distribution

Fig. 26.21: Shear Stress Distribution in L-Section (Illustrative Problem 26.8)

PROBLEM 26.9: A hollow circular section, 100 mm internal diameter and 150 mm external diameter, is subjected to a shear force of 30 kN. What is the maximum shear stress?

Solution: With reference to Fig. 26.22:

$$V = 30 \text{ kN}; \ I = \frac{\pi}{64}(150^4 - 100^6) = 2 \times 10^7 \text{ mm}^4$$

Maximum shear stress occurs at the Neutral Axis.

$$Q(\text{at N.A.}) = \frac{2}{3}(R^3 - r^3)$$

$$= \frac{2}{3}(75^3 - 50^3)$$

$$= 1.98 \times 10^5 \text{ mm}^3$$

$$f_{s_{max}} = \frac{30 \times 1000 \times 1.98 \times 10^5}{2 \times 10^7 \times 50} = 5.94 \text{ MPa}$$

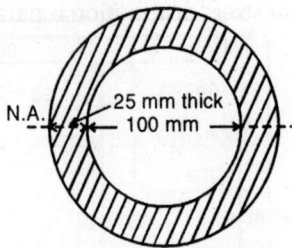

Fig. 26.22: Hollow Circular Section
(Illustrative Problem 26.9)

PROBLEM 26.10: A beam section is in the form of a plus sign as shown in Fig. 26.23. It is subjected to a shear force of 80 kN. Determine the shear stress distribution.

Solution: With reference to Fig. 26.23:

$$V = 80 \text{ kN}$$

$$I_{NA} = \left[\frac{300 \times (100)^3}{12} + 2\left(\frac{100^4}{12} + 1000 \times 100\right)^2\right]$$

$$= 2.417 \times 10^8 \text{ mm}^4$$

f_s at B (with 100 mm width)

$$= \frac{80 \times 1000 \times (100 \times 100 \times 100)}{2.417 \times 10^8 \times 100} = 3.30 \text{ MPa}$$

f_s at B (with 300 mm width)

$$= \frac{3.30 \times 100}{300} = 1.10 \text{ MPa}$$

f_s at N.A. (or C)

$$= \frac{80 \times 1000(100 \times 100 \times 100 + 300 \times 50 \times 25)}{2.417 \times 10^8 \times 300}$$

$$= 1.52 \text{ MPa}$$

The shear stress distribution is shown in Fig. 26.23.

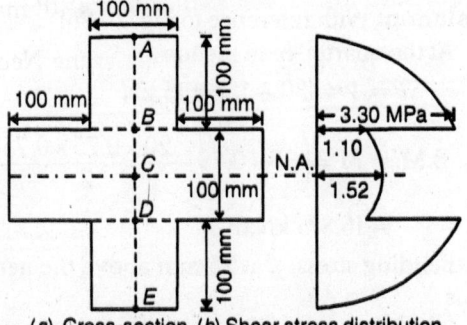

(a) Cross-section (b) Shear stress distribution
Fig. 26.23: Beam with Cross-Section in the Shape of Plus Sign
(Illustrative Problem 26.10)

PROBLEM 26.11: An inverted T-section shown in Fig. 26.24 is used as a beam. It carries a shear force of 100 kN at a section. Determine the shear stress distribution.

Solution: With reference to Fig. 26.24:

$$V = 100 \text{ kN}; \quad A = 50 (100 + 150) = 12500 \text{ mm}^2$$

Let the N.A. be at $\bar{y}$ mm from the base

$$\bar{y} = \frac{(50 \times 100 \times 100 + 150 \times 50 \times 25)}{12500} = 55 \text{ mm}$$

$$I_{NA} = \left[\frac{50 \times 100^3}{12} + 50 \times 100 \times 45^2\right.$$

$$\left. + \frac{150 \times 50^3}{12} + 150 \times 50 \times 30^2\right]$$

$$= 22.604 \times 10^6 \text{ mm}^4$$

$f_{s_{max}}$ at N.A. $= \dfrac{100 \times 1000(50 \times 95 \times 95 / 2)}{22.604 \times 10^6 \times 50}$

$$= 19.96 \text{ MPa}$$

f_s at the junction of rib and flange (in the rib)

$$= \frac{100 \times 1000 \times 50 \times 150 \times 30}{22.604 \times 10^6 \times 50} = 19.91 \text{ MPa}$$

f_s at the junction of rib and flange (in the flange)

$$= \frac{19.91 \times 50}{150} = 6.64 \text{ MPa}$$

The shear stress distribution diagram (with parabolic variation) is shown in Fig. 26.24.

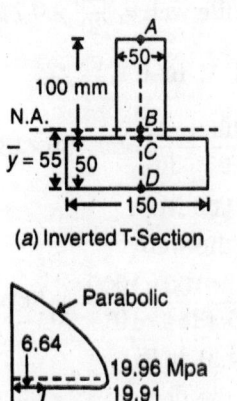

(a) Inverted T-Section

(b) Shear Stress Distribution
Fig. 26.24: Shear Stress Distribution in an Inverted T-Section
(Illustrative Problem 26.11)

PROBLEM 26.12: An I-section has a top flange of 360 mm × 30 mm thick, a bottom flange of 90 mm × 30 mm thick, and a web of 30 mm thickness and 360 mm depth. The overall depth is 420 mm. It has a vertical axis of symmetry. Calculate the maximum shear stress for a shear force of 100 kN.

Solution: With reference to Fig. 26.25:

$$V = 100 \text{ kN}$$

Area, $A = 30(360 + 360 + 90) = 24300 \text{ mm}^2$

Let the N.A. be at $\bar{y}$ mm from the top.

Taking moments about the top, we can obtain $\bar{y}$ as follows:

$$\bar{y} = \frac{(360 \times 30 \times 15 + 360 \times 30 \times 210 + 90 \times 30 \times 405)}{24300}$$

$$= 145 \text{ mm}$$

$$I_{NA} = \left[\frac{360 \times 30^3}{12} + 360 \times 30 \times (130)^2 + \frac{30 \times (360)^3}{12} \right.$$

$$\left. + 30 \times 360 \times 75^2 + \frac{90 \times 30^3}{12} + 90 \times 30 \times (260)^2 \right]$$

$$= 5.4344 \times 10^8 \text{ mm}^4$$

f_s at B in the flange

$$= \frac{100 \times 1000}{5.4344 \times 10^8 \times 360}(360 \times 30 \times 130)$$

$$= 0.72 \text{ MPa}$$

f_s at B in the web $= \dfrac{360}{30} \times 0.72 = 8.64$ MPa

$f_{s_{max}}$ at N.A. (or at C)

$$= \frac{100 \times 1000}{5.4344 \times 10^8 \times 30}\left[360 \times 30 \times 130 + \frac{30}{2} \times (105)^2 \right]$$

$$= 9.63 \text{ MPa}$$

f_s at D in the web

$$= \frac{100 \times 1000}{5.4344 \times 10^8 \times 30}(90 \times 30 \times 260)$$

$$= 4.31 \text{ MPa}$$

f_s at D in the flange

$$= \frac{30}{90} \times 4.31 = 1.44 \text{ MPa}$$

The shear stress distribution diagram is shown in Fig. 26.25.

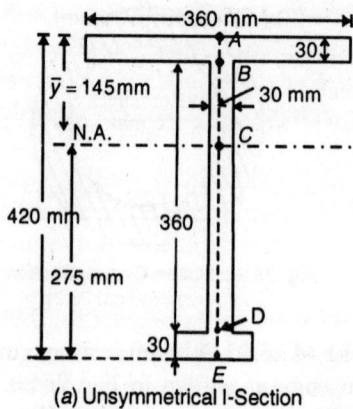

(a) Unsymmetrical I-Section

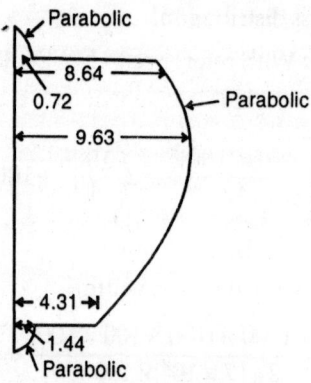

(b) Shear Stress Distribution

Fig. 26.25: Shear Stress Distribution in an Unsymmetrical I-Section (Illustrative Problem 26.12)

PROBLEM 26.13: A simple beam of rectangular section, 750 mm × 150 mm deep, carries a uniform load of 20 kN/m over a span of 3 m. Calculate the principal stress in one of the quarter-span sections at a height 25 mm above the neutral plane.

Solution: With reference to Fig. 26.26:

At the quarter-span section

$$SF = V = (30 - 15) = 15 \text{ kN}$$

$$\text{B.M.} = M = \left(30 \times 0.75 - \frac{20 \times 0.75 \times 0.75}{2} \right)$$

$$= 16.875 \text{ kN.m}$$

Bending stress, f, at 25 mm above the neutral axis

$$= \frac{M \cdot c}{I} = \frac{16.875 \times 10^6 \times 25 \times 12}{75 \times (150)^3}$$

$= 20$ MPa (compressive)

This is the normal stress on the plane of the cross-section at this fibre.

Shear stress, f_s, at 25 mm above the neutral axis $(= q)$

$$= \frac{15 \times 1000 \times 121 \times (75 \times 50 \times 50)}{75 \times (150)^3 \times 75}$$

$$= 1.78 \text{ MPa}$$

An element at this fibre is shown in Fig. 26.27.

$\sigma_1 = 20$ MPa

$q = 1.78$ MPa

$\sigma_2 = 0$

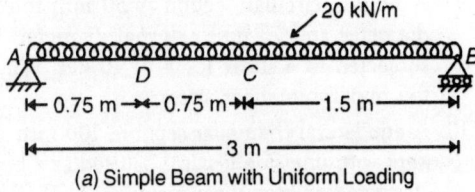

(a) Simple Beam with Uniform Loading

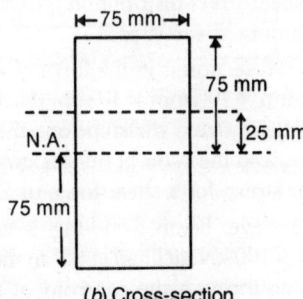

(b) Cross-section

Fig. 26.26: Principal Stresses in Bending
(Illustrative Problem 26.13)

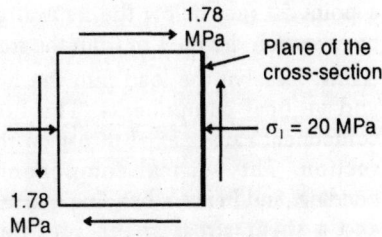

Fig. 26.27: Element at the Specified Fibre in the Cross-Section
(Illustrative Problem 26.13)

The principal stresses are:

$$\sigma_{x, y} = -\frac{\sigma_1}{2} \pm \frac{1}{2}\sqrt{(\sigma_1)^2 + 4q^2}$$

$$= 10 \pm \frac{1}{2}\sqrt{(400) + 4(1.78)^2} = 10 \pm 10.16$$

$\therefore \quad \sigma_x = 20.16$ MPa (Compressive) and

$\sigma_y = 0.16$ MPa (tensile).

PROBLEM 26.14: A uniform rectangular beam, 150 mm × 300 mm cross-section, and 6 m long, is supported at the ends. It is acted on by a uniform load of 30 kN/m over the entire span. Calculate the maximum principal stress at an element situated at a section 2 m from the left end and positioned 60 mm below the neutral axis of the beam.

Solution: With reference to Fig. 26.28:

S.F. at $C = V = (90 - 60)$

$= 30$ kN

B.M. at $C = M = 90 \times 2 - 30 \times 2 \times 1$

$= 120$ kN.m

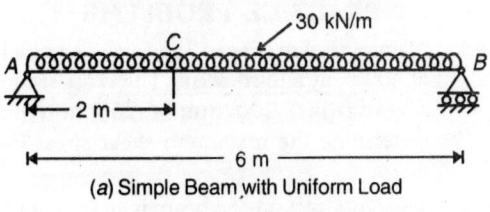

(a) Simple Beam with Uniform Load

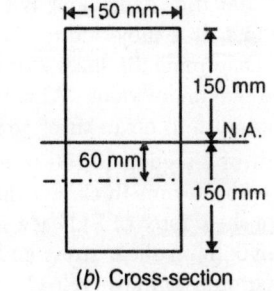

(b) Cross-section

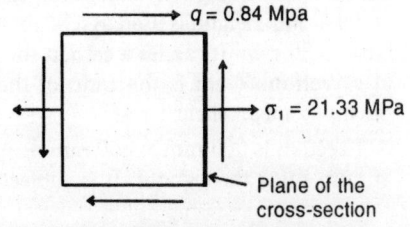

(c) Element at the Specified Fibre in the Cross-Section

Fig. 26.28: Principal Stresses in a Beam
(Illustrative Problem 26.14)

Bending stress,

$$f \text{ (or } \sigma_1) = \frac{Mc}{I} = \frac{120 \times 10^6 \times 150 \times 12}{150 \times (300)^3}$$

$$= 21.33 \text{ MPa (Tensile)}$$

Shear stress, f_s (or q)

$$= \frac{VQ}{I \cdot b} = \frac{30 \times 1000 \times 12 \times (150 \times 90 \times 105)}{150 \times (300)^3 \times 150}$$

$$= 0.84 \text{ MPa}$$

σ_2 being zero,

The maximum principal stress,

$$\sigma_x = \frac{\sigma_1}{2} + \frac{1}{2}\sqrt{\sigma_1^2 + 4q^2}$$

$$= \frac{21.33}{2} + \frac{1}{2}\sqrt{(21.33)^2 + 4 \times (0.84)^2}$$

$$= 21.36 \text{ MPa (Tensile)}$$

PRACTICE PROBLEMS

26.1   A simple beam of span 3 m carries a point load of 60 kN at a third point. The cross-section is rectangular, 200 mm × 250 mm deep. Determine the maximum shear stress in the beam.

26.2   The cross-section of a beam is an I-section, 100 mm × 200 mm deep, and is 20 mm thick throughout. It is subjected to a shear force of 15 kN. Determine the shear stress distribution and the maximum value. What is the ratio of the maximum to mean shear stress?

26.3   A beam of I-section is 100 mm × 200 mm deep, with 10 mm thickness throughout. It carries a shear force of 7 kN at a section. What is the maximum shear stress and the ratio of the maximum to mean value?

26.4   An isosceles triangular section, 80 mm wide and 900 mm height, is used as a beam with the base as 80 mm. It carries a certain shear force at a section. What is the ratio of the maximuzm to mean shear stress?

26.5   A T-section is 600 mm × 600 mm deep and is 20 mm thick throughout. It is subjected to a

shear force of 5 kN. Determine the maximum shear stress.

26.6   An angle section, 110 mm deep and 100 mm wide, is 10 mm thick throughout. If it is subjected to a shear force of 5 kN, what is the maximum shear stress?

26.7   An inverted T-section has a web 50 mm × 100 mm and a flange of 150 mm × 50 mm, the overall depth being 150 mm. What is the maximum shear stress for a shear force of 50 kN?

26.8   The cross-section of a beam is a T-section, 80 mm × 130 mm × 10 mm thick, with the 80 mm side horizontal. Determine the maximum shear stress for a shear force of 10 kN.

26.9   A hollow circular section is 50 mm internal diameter and 75 mm external diameter. It is subjected to a shear force of 10 kN. What is the maximum shear stress?

26.10 A equilateral triangular section, 100 mm side, kept with one side vertical, is used as a beam. It is acted on by a shear force of 10 kN. Sketch the shear stress distribution and determine the maximum shear stress?

26.11 A beam section is in the form of a plus sign— 100 mm × 100 mm × 10 mm thick. Determine the shear stress distribution, the maximum value, and the ratio of the maximum to mean shear stress, for a shear force of 5 kN.

26.12 A cantilever beam, 2 m long, is subjected to a load of 100 kN inclined at 60° to the horizontal, passing through the centroid of the free-end cross-section, and directed towards the fixed end. The cross-section is rectangular, 50 mm × 150 mm. Determine the principal stresses at a point 50 mm below the neutral axis in a cross-section distant 1 m from the free end.

(**Hint**: Resolve the load into the horizontal and vertical components. The horizontal component causes axial thrust on the cross-section. The vertical component causes bending, and hence a bending (normal) stress and a shear stress at the specified fibre. Combine the two normal stresses into one, and determine the principal stresses as usual).

Chapter 27

Deflection of Beams

27.1 INTRODUCTION

It has been pointed out in an earlier chapter that the two important aspects of design of a beam are the 'strength' and 'stiffness'. The former aspect has already been dealt with in the previous Chapters. The latter is the subject matter of this Chapter.

The total "deflection" of a beam at any section is, to a large extent, caused by bending, and, to a much smaller extent, by shear. It is usual to place a limit on the allowable deflection; hence it is important to calculate the deflection of a beam of given section, under the known loading.

Thus the problem of deflection is important with respect to beams for the following reasons:

(1) It is essential to make sure that excessive deflection, which could lead to failure, functional if not structural, does not occur.

(2) Knowledge of deflections helps in solving statically indeterminate structures.

In general, it will be assumed that the deflections are within the elastic limit, and are very small compared to the length of the beam; also, the slope or rotation of the axis at any section is exceedingly small in practice, and hence the radian measure of the angle of slope is approximately equal to the tangent of the angle of slope.

27.2 METHODS OF DETERMINATION OF DEFLECTION

The following methods are available for the determination of deflection:

(a) Successive integration method (double integration of the deflection curve).

(b) Macaulay's approach (modification of double integration)

(c) Moment-area method (Mohr's theorems)

(d) Conjugate beam method

(e) Graphical method

(f) Strain energy method

Of these, the first three will be dealt with in the following sub-sections.

27.2.1 Successive Integration Method

This involves the double integration of the equation to the 'elastic curve' or the bent shape of the axis of the beam in order to obtain the deflection.

Let the curve APB represent the bent shape of the axis of the beam or the elastic curve on loading the beam AB (Fig. 27.1).

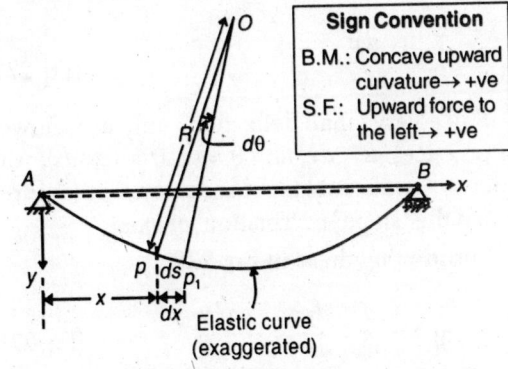

Sign Convention
B.M.: Concave upward curvature→ +ve
S.F.: Upward force to the left→ +ve

Fig. 27.1: Elastic Curve of a Loaded Simple Beam

Bending occurs in the plane of symmetry due to transverse forces acting in that plane. The shape assumed by the longitudinal axis after loading is the 'Elastic curve' or the 'Deflection curve'. Let the x-axis be taken to be positive to the right and the y-axis positive downwards, with respect to the left support A as the origin. Let us assume that the curvature of the deflection curve is influenced only by the bending moment M. In such a case, the principle of pure bending may be applied.

Then $\dfrac{M}{I} = \dfrac{E}{R}$

or $\dfrac{1}{R} = \dfrac{M}{EI}$ ...(Eq. 27.1)

Consider two points p and p_1 adjacent to each other, distance ds apart, on the deflection curve. If the angle between the tangents at p and p_1 is $d\theta$, it will be the same as that between the normals at p and p_1 as shown in Fig. 27.1. The intersection of these normals is the centre of curvature, O, for the elemental curve ds, the radius of curvature, R, is also defined.

Then,

$ds = R \cdot d\theta$, numerically.

Regarding the sign, since M is positive in Eq. 27.1, the curvature $1/R$ is positive, when the centre of curvature is above the curve, as shown. In this case, the angle θ (made by a tangent to the elastic curve at any point with respect to the +ve side of the x-axis) decreases as point p moves along the curve from A to B, a positive increment ds corresponding to a negative increment $d\theta$.

Thus,

$\dfrac{1}{R} = -\dfrac{d\theta}{ds}$ ...(Eq. 27.2)

Since very small deflections only are allowed in practice, $ds \approx dx$, and $\theta \approx \tan\theta = (dy/dx)$, without any significant loss of accuracy. θ or (dy/dx) is called the 'slope' or 'rotation' of axis.

Substituting these in **Eq. 27.2**,

$\dfrac{1}{R} = -\dfrac{d}{dx}\left(\dfrac{dy}{dx}\right) = -\dfrac{d^2y}{dx^2}$ ...(Eq. 27.3)

Eq. 27.1, therefore, becomes,

$$EI \cdot \frac{d^2y}{dx^2} = -M \qquad \text{...(Eq. 27.4)}$$

It should be noted that the sign in this equation depends upon the direction of the coordinate axes. This is called the "Flexure Equation".

(If y is taken positive upwards, $\theta = -\dfrac{dy}{dx}$, and the final sign in the above equation would be positive).

The simplifications used herein do not apply to very slender bars where the deflections are appreciable in magnitude. Then the expression $\theta = \tan^{-1}\left(\dfrac{dy}{dx}\right)$ should be used, leading to

$$\frac{1}{R} = -\frac{d\theta}{ds} = -\frac{d}{dx}\left\{\tan^{-1}\left(\frac{dy}{dx}\right)\right\} \cdot \frac{dx}{ds}$$

$$= \frac{-(d^2y/dx^2)}{\left[1+\left(\dfrac{dy}{dx}\right)^2\right]^{3/2}}$$

In other words, the simplifications used earlier amount to taking $(dy/dx)^2$ is negligible in relation to unity.

By differentiating Eq. 27.4 with respect to x,

$$EI \cdot \frac{d^3y}{dx^3} = -V \text{ (the vertical shear force)}$$
$$\text{...(Eq. 27.5)}$$

$$EI \cdot \frac{d^4y}{dx^4} = w \text{ (rate of loading)} \qquad \text{...(Eq. 27.6)}$$

Eq. 27.6 is sometimes used in considering deflections of beams under distributed loading.

A few standard cases of cantilevers and simple beams will now be considered.

1. Cantilever with Uniformly Distributed Loading

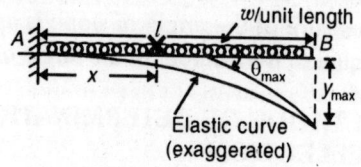

Fig. 27.2: Cantilever with Uniform Loading

Referring to Fig. 27.2, with A, the fixed end, as origin, the B.M., M, at a distance x from A, is

$$M = -\frac{w}{2}(l - x)^2$$

The flexure equation reduces to

$$EI \cdot \frac{d^2y}{dx^2} = \frac{w}{2}(l - x)^2$$

Integrating once with respect to x,

$$EI \cdot \frac{dy}{dx} = -\frac{w}{b}(l - x)^3 + C_1, \; C_1 \text{ being a constant}$$

$$\text{of integration.}$$

At the fixed end, the slope, dy/dx is zero $(x = 0)$

$$\therefore \; C_1 = \frac{wl^3}{6}, \text{Substituting this, we have the general}$$

equation for slope or rotation.

Integrating again,

$$EI \cdot y = \frac{w}{24}(l - x)^4 + \frac{wl^3}{6} \cdot x + C_2$$

$$\text{(after substituting for } C_1\text{).}$$

At the fixed end, $x = 0$, and deflection, y, is zero

$$\therefore \quad C_2 = -\frac{wl^4}{24}$$

The general equation for deflection, therefore, reduces to

$$EI \cdot y = \frac{w}{24}(l - x)^4 + \frac{wl^3}{6} \cdot x - \frac{wl^4}{24}$$

θ_{max} and y_{max} occur at the free end, i.e., at $x = l$.

Substituting $x = l$ in the general equations for slope and deflections, we obtain,

$$\theta_{max} = \frac{wl^3}{6EI} \qquad \text{...(Eq. 27.7)}$$

(+ve sign indicates that the tangent makes a positive or clockwise angle with the +ve direction of the x-axis).

$$y_{max} = \frac{wl^4}{8EI} \qquad \text{...(Eq. 27.8)}$$

(+ve sign indicates downward deflection).

Note: θ & y-values are shown to an exaggerated scale.

2. Cantilever with Point Load at the Free End

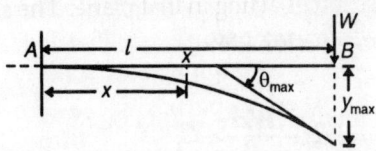

Fig. 27.3: Cantilever with Point Load at the Free End

Referring to Fig. 27.3, M at x, distance x from A, is given by

$$M = -W(l - x)$$

The flexure equation is,

$$EI \cdot \frac{d^2y}{dx^2} = W(l - x)$$

Integrating,

$$EI \cdot \frac{dy}{dx} = -\frac{W}{2}(l - x)^2 + C_1$$

At $x = 0$, $\frac{dy}{dx} = 0$ $\therefore C_1 = \frac{Wl^2}{2}$

$$\therefore \quad EI \cdot \frac{dy}{dx} = -\frac{W}{2}(l - x)^2 + \frac{W}{2}l^2 \rightarrow \text{General}$$

$$\text{equation for slope}$$

Integrating again,

$$EI \cdot y = \frac{W}{6}(l - x)^3 + \frac{W}{2}l^2x + C_2$$

At $x = 0$, $y = 0$ $\therefore C_2 = -\frac{Wl^3}{6}$

$$\therefore EI \cdot y = \frac{W}{6}(l - x)^3 + \frac{W}{2}l^2x - \frac{Wl^3}{6} \rightarrow \text{General}$$

$$\text{equation for deflection.}$$

At $x = l$, the free end section,

$$\theta_{max} = \frac{Wl^2}{2EI} \qquad \text{...(Eq. 27.9)}$$

signs are easily interpreted as earlier.

$$y_{max} = \frac{Wl^3}{3EI} \qquad \text{...(Eq. 27.10)}$$

Aliter: If the two terms are written separately in the expression for M, the constants of integration differ from the above.

$$M = -Wl + Wx$$

$$EI \cdot \frac{d^2y}{dx^2} = Wl - Wx$$

$$EI \cdot \frac{dy}{dx} = Wlx - \frac{Wx^2}{2} + C_1$$

At $x = 0$, $\frac{dy}{dx} = 0$ $\therefore$ $C_1 = 0$

$$EI \cdot y = \frac{Wlx^2}{2} - \frac{Wx^3}{6} + C_2$$

At $x = 0$, $y = 0$ $\therefore$ $C_2 = 0$

$\therefore$ At $x = l$, $\left.\begin{array}{l} \theta_{max} = \dfrac{Wl^2}{2EI} \\[3mm] y_{max} = \dfrac{Wl^3}{3EI} \end{array}\right\}$ as obtained earlier.

3. Cantilever with Triangular Loading

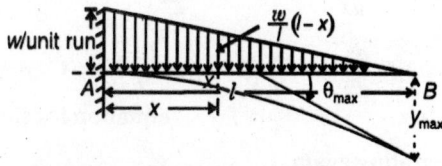

Fig. 27.4: Cantilever with Triangular Loading

Referring to Fig. 27.4, let the cantilever *AB*, fixed at *A* and free at *B*, of length *l*, be loaded with a uniformly varying load with zero at *B* and *w* /unit run at *A* (Triangular loading).

With *A* as origin, *M*, at a distance *x* from *A*, is given by,

$$M = -w\frac{(l-x)}{l}\left(\frac{l-x}{2}\right)\left(\frac{l-x}{3}\right)$$

$$= -\frac{w}{6l}(l-x)^3$$

$\therefore$ $$EI \cdot \frac{d^2y}{dx^2} = \frac{w(l-x)^3}{6l}$$

Integrating,

$$EI \cdot \frac{dy}{dx} = -\frac{w(l-x)^4}{24l} + C_1$$

At $x = 0$, $\frac{dy}{dx} = 0$ $\therefore C_1 = \frac{wl^3}{24}$

$\therefore$ $$EI \cdot \frac{dy}{dx} = -\frac{w(l-x)^4}{24l} + \frac{wl^3}{24}$$

Integrating again,

$$EI \cdot y = \frac{w(l-x)^5}{120l} + \frac{wl^3}{24}x + C_2$$

At $x = 0$, $y = 0$ $\therefore C_2 = -\frac{wl^4}{120}$

$\therefore$ $$EI \cdot y = \frac{w(l-x)^5}{120l} + \frac{wl^3}{24}x - \frac{wl^4}{120}$$

From the general equations for slope and deflection, the maximum values of the slope and deflection, which occur at the free end, are:

$$\theta_{max} = \frac{wl^3}{24EI} \qquad \text{...(Eq. 27.11)}$$

and

$$y_{max} = \frac{wl^4}{30EI} \qquad \text{...(Eq. 27.12)}$$

Aliter: We can start from the general differential equation,

$$EI \cdot \frac{d^4y}{dx^4} = w \text{ (rate of loading)}$$

In this case,

$$EI \cdot \frac{d^4y}{dx^4} = \frac{w}{l}(l-x)$$

Integrating,

$$EI \cdot \frac{d^3y}{dx^3} = -\frac{w(l-x)^2}{2l} + C_1$$

At $x = l$, $EI \cdot \frac{d^3y}{dx^3}$ (or S.F.) $= 0$ $\therefore C_1 = 0$.

$\therefore$ $$EI \cdot \frac{d^3y}{dx^3} = -\frac{w(l-x)^2}{2l}$$

Integrating again,

$$EI \cdot \frac{d^2y}{dx^2} = \frac{w(l-x)^3}{6l} + C_2$$

At $x = l$, $EI \cdot \dfrac{d^2 y}{dx^2}$ (or B.M.) $= 0$ $\therefore$ $C_2 = 0$.

$\therefore$ $EI \cdot \dfrac{d^2 y}{dx^2} = \dfrac{w(l-x)^3}{6l}$, which is the same
equation as got earlier.

Now the work proceeds as usual.

4. Simple Beam with Uniformly Distributed Loading

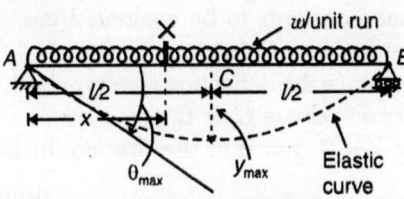

Fig. 27.5: Simple Beam with Uniform Loading

The B.M. at a section ×, distant x from A, is given by,

$$M = \dfrac{wl}{2} x - \dfrac{wx^2}{2}$$

The flexure equation is given by,

$$EI \cdot \dfrac{d^2 y}{dx^2} = -\dfrac{wl}{2} x + \dfrac{wx^2}{2}$$

Integrating,

$$EI \cdot \dfrac{dy}{dx} = -\dfrac{wl}{4} x^2 + \dfrac{wx^3}{6} + C_1$$

Owing to symmetry, the slope at mid-span is zero.

$$\therefore \quad 0 = -\dfrac{wl}{4} \left(\dfrac{l}{2} \right)^2 + \dfrac{w}{6} \left(\dfrac{l}{2} \right)^3 + C_1$$

or $\quad C_1 = \dfrac{wl^3}{24}$

$$\therefore \quad EI \cdot \dfrac{dy}{dx} = -\dfrac{wl}{4} x^2 + \dfrac{wx^3}{6} + \dfrac{wl^3}{24}$$

Integrating again,

$$EI \cdot y = -\dfrac{wl}{12} x^3 + \dfrac{wx^3}{24} + \dfrac{wl^3}{24} x + C_2$$

Since the deflection at the supports is zero, $y = 0$ at $x = 0$

$\therefore \quad C_2 = 0$

or $\quad EI \cdot y = -\dfrac{wl}{12} x^3 + \dfrac{w}{24} x^4 + \dfrac{wl^3}{24} \cdot x$

The maximum slope occurs at the supports.

At $x = 0$, $\theta_A = \dfrac{wl^3}{24 EI}$ $\qquad$ ...(Eq. 27.13)

(+ve value indicates clockwise rotation, or the tangent makes an acute angle with the +ve direction of the X-axis.)

At $x = l$, $\theta_B = -\dfrac{wl^3}{24 EI}$

(−ve value indicates counter clockwise rotation, or the tangent makes an obtuse angle with the +ve direction of the X-axis.)

The maximum deflection occurs at mid-span section, obviously.

$$\therefore \quad EI \cdot y_{max} = \dfrac{5}{384} \cdot wl^4$$

or $\quad y_{max} = \dfrac{5}{384} \cdot \dfrac{wl^4}{EI}$ $\qquad$ ...(Eq. 27.14)

Aliter: We can start from the differential equation involving the rate of loading.

$$EI \cdot \dfrac{d^4 y}{dx^4} = w$$

Integrating,

$$EI \cdot \dfrac{d^3 y}{dx^3} = wx + C_1$$

The S.F. is zero at $x = l/2$,

or $\quad EI \cdot \dfrac{d^3 y}{dx^3} = 0$ at $x = l/2$

$\therefore \quad C_1 = -\dfrac{wl}{2}$

or $\quad EI \cdot \dfrac{d^3 y}{dx^3} = wx - \dfrac{wl}{2}$

Integrating again,

$$EI \cdot \dfrac{d^2 y}{dx^2} = \dfrac{wx^2}{2} - \dfrac{wl}{2} \cdot x + C_2$$

The B.M. is zero at the support,

or $\quad EI \cdot \dfrac{d^2 y}{dx^2} = 0$ at $x = 0$

$$\therefore \quad C_2 = 0$$

or $\quad EI \cdot \dfrac{d^2 y}{dx^2} = \dfrac{wx^2}{2} - \dfrac{wl}{2}x$, which is the same

equation as got above.

Hereafter, the work proceeds as earlier.

5. Simple Beam with Non-Central Point Load

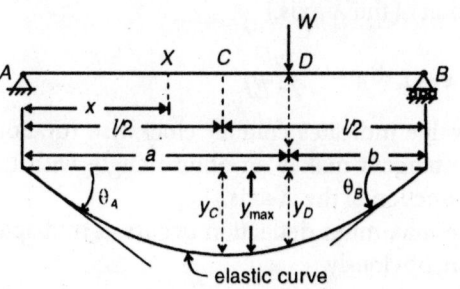

Fig. 27.6: Simple Beam with Non-Central Point Load

Referring to Fig. 27.6, let the point load W act at D, at a distance a from the left support and b from the right, $(a + b)$ being the span l. Let C be the mid-span section, A and B being the supports.

In this case, the bending moment expressions in the two portions AD and DB are different. Hence all the equations have to be written down in pairs as shown below:

$$\left. \begin{array}{l} EI \cdot \dfrac{d^2 y}{dx^2} = -\dfrac{Wb}{l} \cdot x \qquad\qquad x \le a \\[4mm] EI \cdot \dfrac{d^2 y}{dx^2} = -\dfrac{Wb}{l} \cdot x + w(x-a) \; x \ge a \end{array} \right\} \;... \text{(Eq. 27.15)}$$

Integrating,

$$\left. \begin{array}{l} EI \cdot \dfrac{dy}{dx} = -\dfrac{Wb}{2l} \cdot x^2 + C_1 \qquad\qquad ... \; x \le a \\[4mm] EI \cdot \dfrac{dy}{dx} = -\dfrac{Wb}{2l} \cdot x^2 + \dfrac{W(x-a)^2}{2} + C_2 \; ... \; x \ge a \end{array} \right\}$$
$$...\text{(Eq. 27.16)}$$

At $x = a$, $EI \cdot \dfrac{dy}{dx}$ must be the same as obtained from both these.

It is easy to see that $C_1 = C_2$.

Integrating again,

$$\left. \begin{array}{l} EI \cdot y = -\dfrac{Wb}{6l} \cdot x^3 + C_1 x + C_3 \qquad ...x \le a \\[4mm] EI \cdot y = -\dfrac{Wb}{6l} \cdot x^3 + \dfrac{W}{6}(x-a)^3 + C_1 x + C_4 ...x \ge a \end{array} \right\}$$
$$...\text{(Eq. 27.17)}$$

At $x = a$, $EI \cdot y$ must be the same from both these.

This leads to $C_3 = C_4$.

Thus the constants to be evaluated are only two—C_1 and C_3.

Further, since the deflection is zero at both the ends, we can evaluate C_1 & C_3.

Putting $x = 0$, $y = 0$ in the first eq. of 27.17, $C_3 = 0$.

Putting $x = l$, $y = 0$ in the second eq. 27.17,

$$C_1 = \dfrac{-Wb^3}{6l} + \dfrac{Wbl}{6} = \dfrac{Wb}{6l}(l^2 - b^2)$$

Substituting these values of the constants in Eqs. 27.17 and 27.16, we get the general equations for the deflection and the slope.

$$\left. \begin{array}{l} \therefore EI \cdot y = -\dfrac{Wb}{6l} \cdot x^3 + \dfrac{Wb}{6l}(l^2 - b^2)x \quad ... \; x \le a \\[4mm] EI \cdot y = -\dfrac{Wb}{6l} \cdot x^3 + \dfrac{W}{6}(x-a)^3 + \dfrac{Wb}{6l}(l^2 - b^2)x \\[3mm] \hspace{5cm} ... \; x \ge a \end{array} \right\}$$
$$...\text{(Eq. 27.18)}$$

$$\left. \begin{array}{l} EI \cdot \dfrac{dy}{dx} = -\dfrac{Wb}{2l} \cdot x^2 + \dfrac{Wb}{6l}(l^2 - b^2) \quad ... \; x \le a \\[4mm] EI \cdot \dfrac{dy}{dx} = -\dfrac{Wb}{2l} \cdot x^2 + \dfrac{W}{2}(x-a)^2 + \dfrac{Wb}{6l}(l^2 - b^2) \\[3mm] \hspace{5cm} ... \; x \ge a \end{array} \right\}$$
$$...\text{(Eq. 27.19)}$$

The maximum deflection occurs between A and D obviously, $x < a$, when $a > b$.

Equating the first of Eq. 27.19 to zero,

$$\dfrac{Wb}{2l} \cdot x^2 = \dfrac{Wb}{6l}(l^2 - b^2)$$

or $\quad x = \sqrt{\dfrac{(l^2 - b^2)}{3}}$. $\qquad\qquad ...\text{(Eq. 27.20)}$

Substituting this in the first of Eq. 27.18 and simplifying,

$$y_{max} = \frac{Wb(l^2 - b^2)^{3/2}}{9\sqrt{3}\,EIl} \qquad ...\text{(Eq. 27.21)}$$

Deflection under the load,

$$y_D = \frac{Wa^2b^2}{3EIl} \qquad ...\text{(Eq. 27.22)}$$

Substituting $x = l/2$ in the first of Eq. 27.18, and simplifying, the deflection at mid-span, y_C, is given by,

$$y_C = \frac{Wb}{48\,EI}(3l^2 - 4b^2) \qquad ...\text{(Eq. 27.23)}$$

(It is interesting to note the difference between this value and the maximum deflection, in the worst case when b approaches zero is only about 2.5% of the maximum value).

Central Point Laod

When the load is applied at mid-span, y_{max} (which occurs at mid-span itself) is,

$$y_{max} = \frac{Wl/2}{9\sqrt{3}\,EIl}\left\{\frac{3\sqrt{3}\,l^3}{4\times 2}\right\} = \frac{Wl^3}{48EI}$$

or

$$y_D = \frac{Wa^2b^2}{3EIl}, \qquad a = b = l/2,$$

$$y_D = \frac{Wl^2}{48EI}$$

In the case of a single point load, the maximum deflection always occurs nearer the middle of the span.

In the limiting case, when b is very small, x, for maximum deflection, is given by

$$x = \frac{l}{\sqrt{3}}$$

This is only $0.077l$ from the middle.

Thus, the deflection at mid-span is a close approximation to the maximum deflection.

The slopes at the ends A and B may also be obtained from Eq. 27.19 for non-central point load.

$$\theta_A = \frac{Wb}{6EIl}(l^2 - b^2) \qquad ...\text{(Eq. 27.25)}$$

and $\theta_B = -\dfrac{Wab}{6EI\,l}(l+a) = -\dfrac{Wa(l^2 - a^2)}{6EIl}$

$$...\text{(Eq. 27.26)}$$

For a central load,

$$\theta_A = \theta_B = \frac{Wl^2}{16\,EI} \qquad ...\text{(Eq. 27.27)}$$

6. *Simple Beam with a Bending Couple at One End*

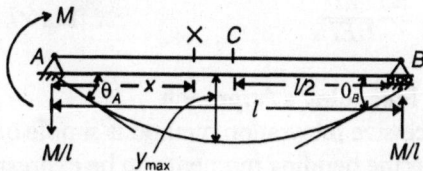

Fig. 27.7: Simple Beam with Couple at one End

Referring to Fig. 27.7, $R_A = R_B = \dfrac{M}{l}$, numerically. R_A is downward and R_B is upward.

$$M_x = M - \frac{M}{l}\cdot x$$

$$EI\cdot\frac{d^2y}{dx^2} = -M + \frac{M}{l}\cdot x$$

$$EI\cdot\frac{dy}{dx} = -Mx + \frac{M}{2l}x^2 + C_1$$

$$EI.y = -\frac{Mx^2}{2} + \frac{M}{6l}x^3 + C_1x + C_2$$

At $x = 0,\ y = 0,\ \therefore C_2 = 0,$

$$x = l,\ y = 0,\ \therefore\ C_1 = \frac{Ml}{3}$$

$$EI\cdot y = -\frac{Mx^2}{2} + \frac{M}{6l}\cdot x^3 + \frac{Ml}{3}x$$

$$= \frac{Mx}{6l}(x^2 - 3l\,x + 2l^2)$$

$$EI\cdot(dy/dx) = -Mx + \frac{Mx^2}{2l} + \frac{Ml}{3}$$

For maximum deflection,

$$\frac{dy}{dx} = 0\ \therefore 0 = -Mx + \frac{Mx^2}{2l} + \frac{Ml}{3}$$

$$3x^2 - 6lx + 2l^2 = 0$$

or $x = (l - l / \sqrt{3})$

$\therefore \quad y_{max} = \dfrac{Ml^2}{9\sqrt{3}EI}$ ...(Eq. 27.28)

$\therefore \quad y_C = \dfrac{Ml^2}{16EI}$ ...(Eq. 27.29)

$\theta_A = \dfrac{Ml}{3EI}$ ...(Eq. 27.30)

$O_B = -\dfrac{Ml}{6EI}$ ...(Eq. 27.31)

27.2.2 Macaulay's Approach

The successive integration method is simple only so long as the bending moment can be expressed in the form of simple expressions. Also if different expressions have to be written for different regions, the method becomes cumbersome.

R. Macaulay (1919) suggested certain modifications, on applying which, the procedure of deriving the deflection equation is simplified.

To understand the method, the case of non-central point load on a simple beam may be considered.

The following points may be observed:

(1) If the expression contributed by the concentrated load, i.e., $W(x - a)$, as it is without expanding the brackets, the constants of integration C_1 and C_2 in the slope equations 27.16 and C_3 and C_4 in the deflection equations 27.17 are one and the same—for the entire span.

(2) Also, if the expression due to concentrated load is in such a way that it can always be recognised and discarded whenever it becomes negative (i.e., for $x < a$), the general expression for bending moment in the farthest zone (taking the left end as the origin) would automatically be applicable for all sections along the span.

Thus, in this case, the expression for B.M. in the farthest zone CB, with A as origin, is

$M = \dfrac{Wbx}{l} \vdots - W(x - a)$

In order to recognise the contribution due to the concentrated load, a dotted line, as shown, is used.

When the B.M. for the zone AC is required $(x < a)$, $(x - a)$ being negative, the term involving this is discarded, and the expression for B.M. would be

$M = \dfrac{Wbx}{l}$

$EI \cdot \dfrac{d^2 y}{dx^2} = -\dfrac{Wbx}{l} \vdots + W(x - a)$ ...(Eq. 27.32)

$EI \cdot \dfrac{dy}{dx} = -\dfrac{Wbx^2}{2l} + C_1 \vdots + \dfrac{W(x - a)^2}{2}$

     ...(Eq. 27.33)

Whenever $x < a$, we have to stop at the dotted line, and when $x > a$, we have to add that term also.

$EI \cdot y = -\dfrac{Wbx^3}{6l} + C_1 x + C_2 \vdots + \dfrac{W(x - a)^3}{6}$

     ...(Eq. 27.34)

Each of these three equations holds good for the entire span, with the rider clause given above. From Eq. 27.33, C_1 and C_2 can be evaluated using the conditions—$x = 0$, $y = 0$ and $x = l$, $y = 0$, thus obtaining the general equations for slope and deflection.

Macaulay's approach is particularly useful when there are several concentrated loads or when part of the span is loaded with uniform intensity of load; this will be understood from the illustrative examples in Section 27.4.

27.2.3 Moment-Area Method

This method is considered suitable when the deflection at a point is required and not the equation to the deflectioin curve. (Otto Mohr, 1868).

The curvature of the deflection curve of a prismatic beam in flexure was found to be

$\dfrac{1}{R} = \dfrac{M}{EI}$ (derived for pure bending)

This equation may also be used for bending of prismatic bars if the effect of shear forces be neglected. Since M varies along the length of the beam, the deflection curve will be of variable curvature.

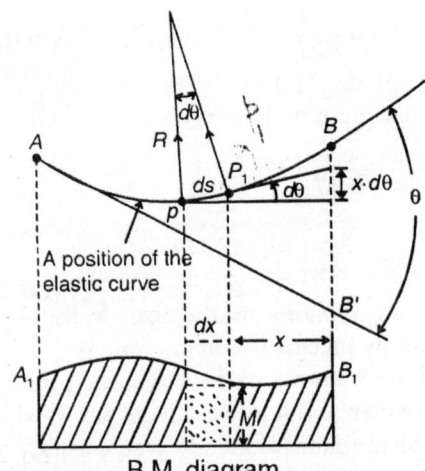

A position of the elastic curve

B.M. diagram

Fig. 27.8: Moment-Area Method (Deflections from B.M. Diagram)

With reference to Fig. 27.8, let AB be a portion of the deflection curve and A_1B_1 be the corresponding portion of the bending moment diagram.

Let pp_1 be an elemental length ds of the elastic curve of a distance x from B and the angle between the tangent at p and p_1 be $d\theta$.

$$d\theta = \frac{1}{R} \cdot ds = \frac{M}{EI} \cdot ds$$

Since the curvature is small, $ds \approx dx$.

Then $d\theta = \frac{1}{EI} \cdot (M \cdot dx)$

Graphically, this means that the elemental angle $d\theta$ between two consecutive radii or two consecutive tangents to the deflection curve equals the shaded elemental area $(M \cdot dx)$ of the bending moment diagram, divided by the flexural rigidity, EI.

This being so for each element, the angle θ between the tangents A and B will be obtained by integrating such elements.

$$\theta = \int_A^B \frac{M \cdot dx}{EI} = \frac{1}{EI} \int_A^B M \cdot dx \quad \ldots(\text{Eq. 27.35})$$

(if EI is constant throughout the span)

Thus the Mohr's (Otto Mohr, 1868) first proposition or theorem is stated:

"The change of slope between any two sections of a loaded beam equals the area of the bending moment diagram between the corresponding verticals, divided by the flexural rigidity, EI (A positive net area denotes that the right-hand tangents makes a counter-clockwise angle with the left-hand tangents)".

The contribution made to the distance or intercept BB' by the bending of the element pp_1 of the bent beam, included between the two consecutive tangents at p and p_1 is

$$x \cdot d\theta = x \cdot \left(\frac{M \cdot dx}{EI} \right)$$

Graphically, this is the moment of the shaded area $(M \cdot dx)$ with respect to the vertical through B, divided by EI.

Integration gives

$$BB' = \int_A^B \frac{M \cdot x \cdot dx}{EI} = \frac{1}{EI} \int_A^B M \cdot x \cdot dx \quad \ldots(\text{Eq. 27.36})$$

(if EI is constant)

Thus the second proposition or theorem of Mohr is stated:

"The intercept of the tangents drawn to the elastic curve between any two points, taken on a vertical line through a moment centre, equals the moment of the bending moment diagram between the verticals through the two points about the moment centre, divided by EI".

By an appropriate choice of the two points and the moment centre, the slope and deflection at any section can be determined.

For convenience in the application of Mohr's propositions, formulae for the areas and the co-ordinates of the centroids of certain regular geometric figures, which could be the shapes of bending moment diagrams, are given in Fig. 27.9.

To demonstrate the use of these principles, a few standard cases of cantilevers and simple beams will now be solved by the moment-area method, which will enable the student to compare the labour involved in this method with that in the successive integration method.

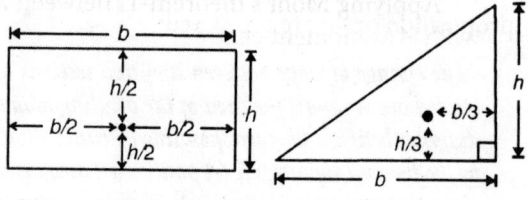

(i) Rectangle $A = bh.$ *(ii)* Right triangle $A = \frac{1}{2}\,bh$

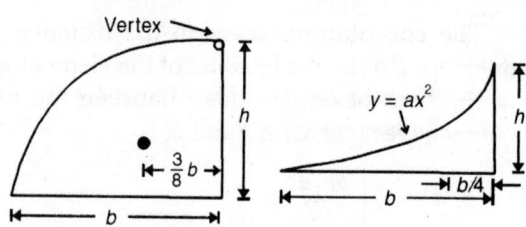

(iii) Convex Parabola $A = \frac{2}{3}\,bh.$ *(iv)* Concave Parabola $A = \frac{1}{3}\,bh$

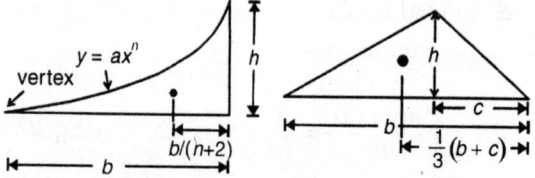

(v) Parabola of nth degree $A = \frac{bh}{(n+1)}.$ *(vi)* Triangle $A = \frac{1}{2}\,bh$

Fig. 27.9: Areas and Co-ordinates of Centroids of Certain Regular Geometric Figures

1. *Cantilever with Uniformly Distributed Loading*

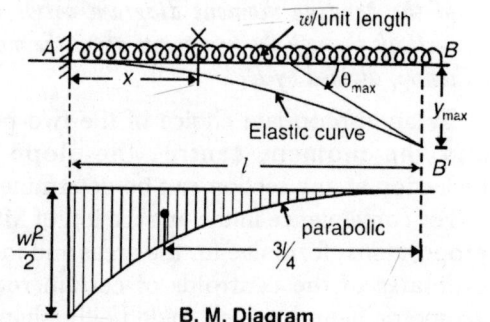

B. M. Diagram

Fig. 27.10: Cantilever with Uniform Loading-Area-Moment Method

Applying Mohr's theorem I between A & B, maximum slope at the free end,

$$\theta_{\max} = \frac{1}{EI} \int_0^l M \cdot dx$$

$$= \frac{1}{EI}\left[\frac{1}{3} \times l \times \frac{wl^2}{2}\right]$$

$$= \frac{wl^3}{6EI}$$

Applying Mohr's theorem II between A & B Maximum deflection at the free end,

$$y_{\max} = \frac{1}{EI} \int M\,x\,dx$$

$$y_{\max} = \frac{1}{EI}\left[\frac{wl^3}{6} \times \frac{3l}{4}\right] = \frac{wl^4}{8EI}$$

these expressions are the same as those derived earlier by successive integration.

If the moment of the B.M. diagram between AX is taken about X, we can get y at X, which also will be the same as that derived earlier.

2. *Cantilever with Point Load at the Free End*

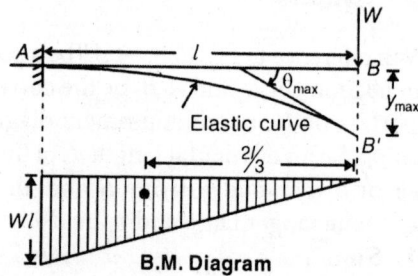

B.M. Diagram

Fig. 27.11: Cantilever with Point Load at the Free End—Area-Moment Method

With reference to Fig. 27.11, applying Mohr's theorem-I between A & B, the maximum slope at the free end,

$$\theta_{\max} = \frac{1}{EI}\left(\frac{1}{2} \cdot l \times Wl\right) = \frac{Wl^2}{2EI}$$

Applying Mohr's theorem-II between A & B, the maximum deflection at the free end,

$$y_{\max} = \frac{1}{EI}\left(\frac{1}{2} \times l \times Wl \times \frac{2l}{3}\right) = \frac{Wl^3}{3EI}$$

These are the same expressions, derived earlier.

3. *Cantilever with Triangular Loading*

With reference to Fig. 27.12,

$$\theta_{\max} = \frac{1}{EI} \int_0^l M \cdot dx = \frac{A}{EI}$$

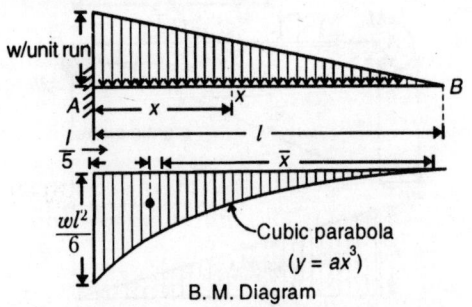

Fig. 27.12: Cantilever with Triangular Loading- Moment- Area Method

$$y_{max} = \frac{1}{EI} \int_0^. M \cdot x \cdot dx = \frac{Ax}{EI}$$

From case (v) of Fig. 27.9

$$A = \frac{bh}{(n+1)} = l \times \frac{wl^2}{6} \times \frac{1}{(3+1)} = \frac{wl^3}{24}$$

$$A\bar{x} = \left(\frac{wl^3}{24}\right)\left(l - \frac{l}{(3+2)}\right) = \frac{wl^3}{24} \times \frac{4l}{5} = \frac{wl^4}{30}$$

$$\left. \begin{array}{l} \therefore \quad \theta_{max} = \dfrac{wl^3}{24\,EI} \\[2ex] \quad y_{max} = \dfrac{wl^4}{30\,EI} \end{array} \right\} \quad \text{which have been derived earlier.}$$

4. Simple Beam with Uniformly Distributed Loading

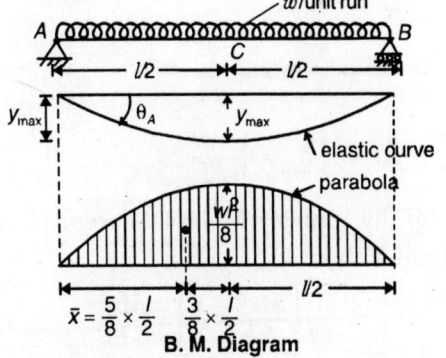

Fig. 27.13: Simple Beam with Uniform Loading-Moment-Area Method

With reference to Fig. 27.13.
Applying Mohr's theorem - I between A & C

$$\theta_{max} = \frac{1}{EI}\left(\frac{1}{2} \times \frac{2}{3} l \times \frac{wl^2}{8}\right) = \frac{wl^3}{24EI}$$

Applying Mohr's theorem-II between A & C, with A as moment-centre

$$y_{max} = \frac{1}{EI}\left(\frac{1}{2} \times \frac{2}{3} \times l \times \frac{wl^2}{8} \times \frac{5}{8} \times \frac{l}{2}\right)$$

$$= \frac{5}{384} \frac{wl^4}{EI}$$

These equations have been derived earlier by successive integration.

5. Simple Beam with a Non-Central Point Load

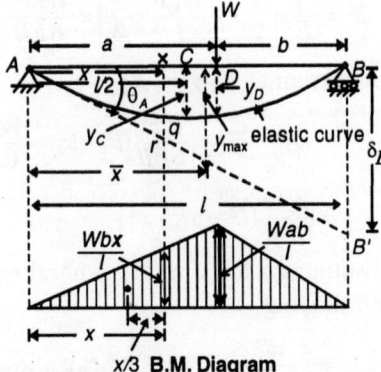

Fig. 27.14: Simple Beam with a Non-Central Point Load-Area-Moment Method

Referring to Fig. 27.14.
Applying Mohr's theorem -I between A & B

$$EI\,\delta_B = \frac{1}{2} \times l \times \frac{Wab}{l} \times \frac{(l+b)}{3} = \frac{Wab(l+b)}{6}$$

$$\delta_B = \frac{Wab(l+b)}{6EI}$$

$$\theta_A = \frac{\delta_B}{l} = \frac{Wab}{6EIl}(l+b)$$

$$= \frac{Wb}{6EIl}(l^2 - b^2)$$

If the $\bar{x}$ is the distance to the section of maximum deflection or zero slope from A, change of slope from A to this section is θ_A.
Applying Mohr's theorem-I, $EI \cdot \theta_A$

$$= \frac{1}{2}\bar{x} \times \frac{\bar{x}}{a} \times \frac{Wab}{l} = \frac{\bar{x}^2}{2l} \cdot Wb$$

$$\therefore \quad \bar{x}^2 = \frac{2l}{Wb} \times \frac{Wb}{6l}(l^2 - b^2) = \left(\frac{l^2 - b^2}{3}\right)$$

$$\therefore \quad \bar{x} = \sqrt{\frac{(l^2 - b^2)}{3}}$$

It is easier to obtain a general expression for y at a distance x from A ($x < a$).

$$\overline{Xr} = \frac{x}{l} \cdot \delta_B = \frac{Wabx}{6EIl}(l + b) = \frac{Wbx(l - b)^2}{6EIl}$$

$\overline{qr}$ (by applying Mohr's theorem-II between A and X and taking moments about X)

$$\overline{qr} = \frac{1}{EI}\left[\frac{1}{2}x \times \frac{Wbx}{l} \times \frac{x}{3}\right] = \frac{Wbx^3}{6EIl}$$

The deflection at X, $y_x = \overline{Xq}$

$$= (\overline{Xr} - \overline{qr}) = \frac{Wbx}{6EIl}(l^2 - b^2) - \frac{Wbx^3}{6EIl}$$

$$= \frac{Wbx}{6EIl}(l^2 - b^2 - x^2)$$

Substituting $\bar{x}$ for x in this general expression and simplifying,

$$y_{max} = \frac{Wb}{9\sqrt{3}\,EIl}(l^2 - b^2)^{3/2}, \text{ as obtained by}$$

successive integration.

Substituting $x = a$, and simplifying,

$$y_D = \frac{Wa^2b^2}{3EIl}$$

Deflection at mid-span, y_C is got by putting $x = l/2$, and simplifying,

$$y_C = \frac{Wb}{48EI}(3l^2 - 4b^2)$$

Similarly, the values for central point load also may be got easily either from this case or from fundamentals.

6. *Simple Beam with a Bending Couple at One End*

Referring to Fig. 27.15.

Applying Mohr's theorem-II between A and B, with moment centre as A,

$$\delta_A = l \cdot \theta_B = \frac{1}{EI}\left[\frac{Ml}{2} \times \frac{l}{3}\right]$$

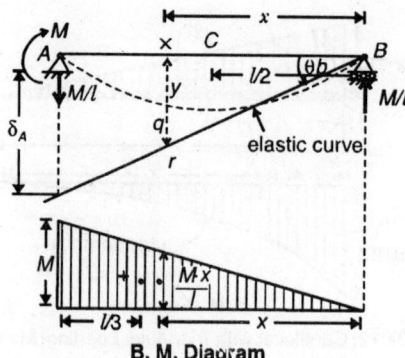

B. M. Diagram

Fig. 27.15: Simple Beam with a Bending Couple at one End-Moment-Area Method

$$= \frac{Ml^2}{6EI}$$

$$\theta_B = \frac{Ml}{6EI}, \text{ numerically.}$$

For maximum deflection, the change in slope between B and X should be θ_B, so that the slope is zero at X.

Applying Mohr's theorem-I between B and X,

$$EI \cdot \theta_B = \frac{1}{2}x \times \frac{Mx}{l} = \frac{Mx^2}{2l}$$

But $\quad EI \cdot \theta_B = \frac{Ml}{6}$

$$\therefore \quad \frac{Mx^2}{2l} = \frac{Ml}{6}$$

$$x^2 = \frac{l^2}{3} \text{ or } x = \frac{l}{\sqrt{3}}$$

$$\overline{Xr} = \frac{x}{l}\delta_A = \frac{x \cdot Ml}{6EI}$$

$\overline{qr}$ (by applying Mohr's theorem-II between B and X about X)

$$= \frac{1}{EI}\left[\frac{Mx^2}{2l} \times \frac{x}{3}\right] = \frac{Mx^3}{6EIl}$$

$$\overline{Xq} = \overline{Xr} - \overline{qr}$$

$$= \frac{Mlx}{6EI} - \frac{Mx^3}{6EIl} = \frac{Mx}{6EIl}(l^2 - x^2)$$

For maximum deflection, we have to substitute

$x = \frac{l}{\sqrt{3}}$ in this general equation.

$$\therefore \quad y_{max} = \frac{M}{6\sqrt{3}\,EI}\left(l^2 - \frac{l^2}{3}\right)$$

$$= \frac{M}{6\sqrt{3}\,EI}\,\frac{2l^2}{3} = \frac{Ml^2}{9\sqrt{3}\,EI}$$

At mid span $(x = l/2)$, y_C may be got by substituting $x = l/2$ in the general equation.

$$\therefore \quad y_c = \frac{M}{12EI}\left(l^2 - \frac{l^2}{4}\right) = \frac{3Ml^2}{12\,EI \times 4} = \frac{Ml^2}{16\,EI}$$

Similarly, the slope at θ_A may be shown to be

$$\frac{Ml}{3EI}.$$

All these expressions have already been derived earlier by successive integration.

Some more interesting cases of beams and loadings will be considered in Section 27.4.

Note: In all the formulae for slope and deflection, the units are to be consistent

W... Load ... N; E ...Youngs modulus ... N/mm^2 (MPa); l... span ... m; l^3 ... l in mm; I... Moment of intertia ... mm^4; w ... intensity of load ... N/m. (W or wl should be applied in Newtons); y in mm; θ in radians.

27.3 LEAF SPRINGS

This is a built-up spring, consisting of a number of curved plates, each of constant width and thickness—designed such that each plate is subjected to the same maximum bending stress at all cross-sections. This is used to support carriages —and so it is called a 'carriage spring' or 'laminated spring'.

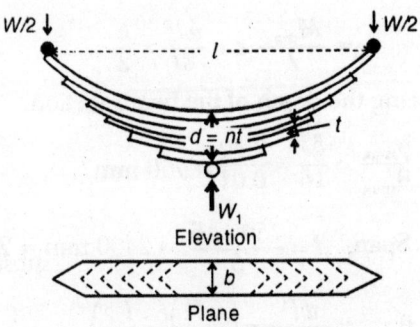

Fig. 27.16: Leaf Spring or Carriage Spring

The load is applied at middle of the span through the wheel of the carriage, and is resisted by the reactions at the pin-supports at the ends. The particular load which causes the plates to straighten is called the "Proof load".

The spring consists of a number of overlapping plates or leaves, each of thickness t (Fig. 27.16), the number decreasing outwards from the centre to the ends.

Every plate has initially the same curvature.

The modulus of section of each plate $(1/6)$ bt^2; the modulus of section of the whole spring at the centre will be $n \cdot (1/6)\,bt^2$ (n being the number of plates), and not $(1/6)\,b\,(nt)^2$, since the strips are separate.

If the proof stress is f in each plate, the bending moment at centre is $Wl/4$.

$$f = \frac{(Wl/4)}{\left(\dfrac{n}{6}bt^2\right)} = \frac{3}{2}\cdot\frac{Wl}{nbt^2} \qquad \text{...(Eq. 27.37)}$$

Since f must be the same at all cross-sections, the modulus of section everywhere must be proportional to the bending moment, from the end to the middle it must be proportional to the distance from the end. Therefore, the number of plates must increase from the ends to the centre, proportional to the distance. The ends of the plates are usually tapered so that the rate of change of resisting moment may be the same — between abrupt changes in the number of plates.

The deflection may be obtained by considering the longest plate since each is an arc of a circle.

$$\delta = \frac{l^2}{8R} \qquad \text{...(Eq. 27.38)}$$

$$\therefore \quad R = \frac{l^2}{8\delta}$$

Also $\quad \delta = \dfrac{Ml^2}{9EI}$ (for a constant M) ...(Eq. 27.39)

$$\delta = \frac{Wl}{4n} \times \frac{l^2 \times 12}{8\,E\,bt^3} = \frac{3}{8}\cdot\frac{Wl^3}{nEbt^3} \quad \text{...(Eq. 27.40)}$$

Resilience, U, can be shown to be

$$U = \frac{1}{6}\cdot\frac{f^2}{E}\ \text{(Volume of the spring)} \quad \text{...(Eq. 27.41)}$$

ILLUSTRATIVE PROBLEMS

PROBLEM 27.1: A cantilever of length l is acted on by a concentrated load W at the free end. Determine the maximum deflection in terms of the working flexural stress f and the depth h of the cross-section, which is rectangular.

Solution: For a cantilever with a point load at the free end, the maximum deflection, y_{max}, at the free end is,

$$y_{max} = \frac{Wl^3}{3EI}$$

$$\frac{M}{I} = \frac{f}{c}$$

c being the extreme fibre distance from the neutral axis.

$$c = \frac{h}{2} \text{ in this case; also } M_{max} = Wl$$

$$\therefore \quad \frac{Wl}{I} = \frac{2f}{h}$$

But $y_{max} = \dfrac{Wl^3}{3EI} = \left(\dfrac{Wl}{I}\right) \times \dfrac{l^2}{3E} = \dfrac{2fl^2}{3Eh}$

PROBLEM 27.2: What is the length of a uniformly loaded cantilever if the deflection and slope at the free end are 25 mm and 0.01 rad., respectively?

Solution: For a cantilever with uniformly distributed load, the maximum values of deflection and slope at free end are,

$$y_{max} = \frac{wl^4}{8EI}; \quad \theta_{max} = \frac{wl^3}{6EI}$$

$$\frac{y_{max}}{\theta_{max}} = \frac{(wl^4/8EI)}{(wl^3/6EI)} = \frac{3l}{4}$$

But, for the given cantilever,

$$\frac{y_{max}}{\theta_{max}} = \frac{25}{0.01} = 2500 \text{ mm}$$

$$\frac{3l}{4} = 2500$$

or $l = \dfrac{10000}{3} = 3{,}333 \text{ mm} = 3\dfrac{1}{3} \text{ m}$

PROBLEM 27.3: A vertical mild steel column, 150 mm diameter, is securely embedded into the

ground, and is 4 m high. The top end is subjected to a horizontal force of 15 kN. Calculate the deflection at the top and also the slope. $E = 2.05 \times 10^5$ MPa.

Solution: This is equivalent to a cantilever with point load at the free end.

At the top,
Deflection,

$$y_{max} = \frac{Wl^3}{3EI} = \frac{15 \times 1000 \times (4000)^3 \times 64}{3 \times 2.05 \times 10^5 \times 11 \times (150)^4} \text{ mm}$$

$$= 62.63 \text{ mm}$$

Slope,

$$\theta_{max} = \frac{Wl^2}{2EI} = \frac{15 \times 1000 \times (4000)^3 \times 64}{3 \times 2.05 \times 10^5 \times \pi \times (150)^4} \text{ rad.}$$

$$= 0.0235 \text{ rad} \approx 1°20'$$

PROBLEM 27.4: A uniformly loaded steel I-beam, simply supported at the ends, has a deflection of 7.5 mm at mid-span and a slope of 0.01 rad. at the ends. Find the depth, h, of the beam, if the maximum bending stress is 1,260 MPa. $E = 2 \times 10^5$ MPa.

Solution: For a uniformly loaded simple beam of span l with intensity of loading w,

$$M_{max} \text{ (at mid - span)} = \frac{wl^2}{8}$$

$$y_{max} \text{ (at mid - span)} = \frac{5}{384} \cdot \frac{wl^4}{EI}$$

$$\theta_{max} \text{ (at the ends)} = \frac{wl^3}{24EI}$$

f_{max} (maximum bending stress)

$$= \frac{M_{max}}{I} \cdot c = \frac{wl^2}{8I} \times \frac{h}{2}$$

h being the depth of the beam section.

$$\frac{y_{max}}{\theta_{max}} = \frac{5l}{16} = \frac{7.50}{0.01} = 750 \text{ mm}$$

$$\therefore \quad \text{Span, } l = \frac{750 \times 16}{5} = 2400 \text{ mm} = 2.4 \text{ m}$$

$$\theta_{max} = \frac{wl^3}{24EI} = \left(\frac{wl^2}{I}\right)\left(\frac{l}{24E}\right)$$

$$= \left(\frac{wl^2}{I}\right) \times \frac{2400}{24 \times 2 \times 10^5} = 0.01$$

or $\left(\dfrac{wl^2}{I}\right) = \dfrac{24 \times 2 \times 10^5 \times 0.01}{2400} = 20$

Since $f_{max} = \left(\dfrac{wl^2}{I}\right)\left(\dfrac{h}{16}\right)$

$$1260 = 20 \times \frac{h}{16}$$

or $h = \dfrac{16 \times 1260}{20} = 1008 \text{ mm} = 1.008 \text{ m}.$

PROBLEM 27.5: A beam of constant I-section is supported freely on a span of 5 m. It carries a point load of 120 kN at 2 m from one end. The moment of inertia of the beam section about the neutral axis is 8×10^7 mm⁴. $E = 2 \times 10^5$ MPa. Find the deflection under the load, at mid-span, and the maximum deflection.

Solution:

$$a = 3 \text{ m} \qquad b = 2 \text{ m}$$

Deflection under the load,

$$\frac{Wa^2b^2}{3EI \cdot l} = \frac{120 \times 10^3 \times (3000)^2 (2000)^2}{3 \times 2 \times 10^5 \times 8 \times 10^7 \times 5000} \text{ mm}$$

$$= 18 \text{ mm}$$

Deflection at mid-span,

$$\frac{Wb}{48EI}(3l^2 - 4b^2)$$

$$= \frac{120 \times 10^3 \times 2000\{3 \times (5000)^2 - 4 \times (2000)^2\}}{48 \times 2 \times 10^5 \times 8 \times 10^7}$$

$$= 18.425 \text{ mm}$$

Maximum deflection,

$$\frac{Wb(l^2 - b^2)^{3/2}}{9\sqrt{3}EIl}$$

$$= \frac{120 \times 10^3 \times 2000\{(5000)^2 - (2000)^2\}^{3/2}}{9\sqrt{3} \times 2 \times 10^5 \times 8 \times 10^7 \times 5000} \text{ mm}$$

$$= 18.52 \text{ mm}$$

Note: It is observed that the difference between mid-span deflection and the maximum value is very small.

PROBLEM 27.6: Determine the maximum deflection at the free end of a cantilever of length *l* if it carries a uniformly distributed load of *w* per unit length for half of its length from the fixed end.

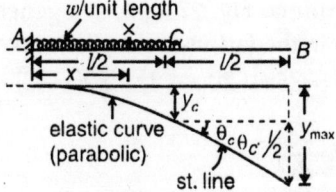

Fig. 27.17: Cantilever with Uniform Loading for Part of the Length (Illustrative Problem 27.6)

Solution: Referring to Fig. 27.17, and applying the standard case of a uniformly loaded cantilever for the slope and deflection at *C*, (substituting *l*/2 for *l*).

$$\theta_c = \frac{w(l/2)^3}{6EI} = \frac{wl^3}{48EI}$$

(This slope continues till the free end, as the right half portion does not carry any load).

$$y_c = \frac{w(l/2)^4}{8EI} = \frac{wl^4}{128EI}$$

y_{max} (at the free end)

$$y_c + \frac{l}{2}\theta_c = \frac{wl^4}{128EI} + \left(\frac{l}{2}\right)\left(\frac{wl^3}{48EI}\right)$$

$$= \frac{wl^4}{EI}\left(\frac{1}{128} + \frac{1}{96}\right)$$

$$= \frac{7wl^4}{384EI} \qquad \qquad \text{...(Eq. 27.42)}$$

This may be remembered and used as standard formula.

PROBLEM 27.7: A simple beam of 4 m span carries point loads of 20 kN each at quarter-span sections and 30 kN at mid-span. Determine the slope at the supports and the deflections under the loads, in terms of the flexural rigidity, *EI*.

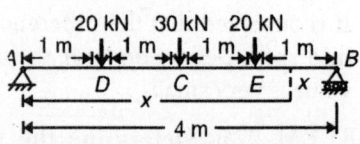

Fig. 27.18: Simple Beam with Point Loads—MaCaulay's Approach (Illustrative Problem 27.7)

Solution: Macaulay's approach will be convenient in this case.

Referring to Fig. 27.18, M at a general section distant x in the farthest zone EB is given by,

$$M = 35x - 20(x-1) - 30(x-2) - 20(x-3)$$

$$\therefore \quad EI \cdot \frac{d^2y}{dx^2}$$

$$= -35x + 20(x-1) + 30(x-2) + 20(x-3)$$

Integrating, $EI \dfrac{dy}{dx}$

$$= -17.5x^2 + 10(x-1)^2 + 15(x-2)^2 + 10(x-3)^2 + C_1$$

Integrating again,

$$EI \cdot y = -\frac{17.5}{3}x^3 + \frac{10}{3}(x-1)^3 + 5(x-2)^3$$

$$+ \frac{10}{3}(x-3)^3 + C_1 x + C_2$$

At $x = 0$, $y = 0$ $\therefore$ $C_2 = 0$ (ignoring all terms involving negative values within the brackets).

At $x = 4$, $y = 0$

$$\therefore 0 = -\frac{17.5}{3}\times 4^3 + \frac{10}{3}\times 3^3 + 5\times 2^3 + \frac{10}{3}\times 1^3 + 4C_1$$

or $C_1 = 60$ (on simplification)

$$\therefore \quad EI \cdot \frac{dy}{dx} = -17.5x^2 + 10(x-1)^2 + 15(x-2)^2$$

$$+ 10(x-3)^2 + 60$$

and $EI \cdot y = -\dfrac{17.5}{3}x^3 + \dfrac{10}{3}(x-1)^3 + 5(x-2)^3$

$$+ \frac{10}{3}(x-3)^3 + 60x$$

These are the general equations for slope and deflection, which are applicable for the entire span, with the rider clause that, whenever any

term within brackets becomes negative, it shall be ignored.

Slope at the support ($x = 0$ in the slope equation) $= \dfrac{60}{EI}$

Deflection under 20 kN loads $= \dfrac{162.5}{3EI}$

Deflection under 30 kN load $= \dfrac{230}{3EI}$

PROBLEM 27.8: A simple beam of span 6 m carries a uniform load of 20 kN/m over the left half of the span. Determine the deflection at midspan and the slope at the left end, by Macaulay's approach.

Solution: Macaulay's approach will be convenient in this case also.

Referring to Fig. 27.19, M at a general section distant x from B in the zone CA is given by,

$$M = 15x - 10(x-3)^2$$

$$\therefore \quad EI \cdot \frac{d^2y}{dx^2} = -15x + 10(x-3)^2$$

Integrating,

$$EI \cdot \frac{dy}{dx} = -7.5x^2 + \frac{10}{3}(x-3)^3 + C_1$$

Integrating again,

$$EI \cdot y = -2.5x^3 + \frac{5}{6}(x-3)^4 + C_1 x + C_2$$

At $x = 0$, $y = 0 \therefore C_2 = 0$

At $x = 6$, $y = 0$. Substituting and simplifying, $C_1 = 78.75$

$$\left. \begin{array}{l} \therefore \quad EI \cdot \dfrac{dy}{dx} = -7.5x^2 + \dfrac{10}{3}(x-3)^3 + 78.75 \\[2ex] \text{and} \quad EI \cdot y = -2.5x^3 + \dfrac{5}{6}(x-3)^3 + 78.75x \end{array} \right\}$$

General equations for slope and deflection.

$$\left. \begin{array}{l} \text{At } x = 6, \quad \theta_A = -0.00724 \text{ rad.} \\[1ex] \text{At } x = 3, \quad y_C = 12.05 \text{ mm} \end{array} \right\}$$

(on substitution in these general equations and simplifying the terms).

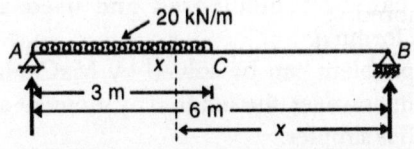

Fig. 27.19: Simple Beam with Half the Span Loaded Uniformly
—MaCaulay's Approach (Illustrative Problem 27.8)

PROBLEM 27.9: A simple beam of span 4 m is loaded uniformly with 40 kN/m for the third quarter of the span from the left end. Determine the maximum deflection. $E = 2 \times 10^5$ MPa and $I = 5 \times 10^7$ mm^4.

Solution: In this case, Macaulay's approach will be found to be very convenient. Referring to Fig. 27.20, imagine downward and upward uniform loading of 40 kN/m as shown in the last quarter (DB) of the span, and choosing a general section, distant x from A, in the zone DB,

$$M = 15x - 20(x-2)^2 + 20(x-3)^2$$

$$\therefore \quad EI \cdot \frac{d^2y}{dx^2} = -15x + 20(x-2)^2 - 20(x-3)^2$$

Integrating,

$$EI \cdot \frac{dy}{dx} = -7.5x^2 + \frac{20}{3}(x-2)^3 - \frac{20}{3}(x-3)^2 + C_1$$

Integrating again,

$$EI \cdot y = -2.5x^3 + \frac{5}{6}(x-2)^4 - \frac{5}{3}(x-3)^4 + C_1x + C_2$$

At $x=0$, $y=0$ $\therefore$ $C_2=0$

At $x=4$, $y=0$

$$\therefore \quad 0 = -2.5 \times 4^3 + \frac{5}{6} \times 2^4 - \frac{5}{3} \times 1^4 + 4C_1$$

$$C_1 = 33.75$$

Substituting,

$$EI\frac{dy}{dx} = -7.5x^2 + \frac{20}{3}(x-2)^3 - \frac{20}{3}(x-3)^2 + 33.75$$

$$EI \cdot y = -2.5x^3 + \frac{5}{3}(x-2)^4 - \frac{5}{3}(x-3)^4 + 33.75x$$

General equations for slope and deflection.

For maximum deflection, the slope is to be equated to zero $2 < x < 3$ (CD), and x is to be solved.

$$-7.5x^2 + \frac{20}{3}(x-2)^3 + 33.75 = 0 \text{ solving by}$$

trial and error, $x = 2.15$ m.

Substituting in the deflection equation,

$$y_{max} = 45.25/EI$$

$$= \frac{45.25 \times 10^6 \times 10^6}{2 \times 10^5 \times 5 \times 10^7} = 4.525 \text{ mm}$$

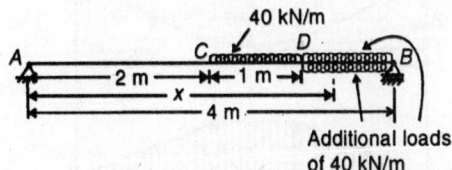

Fig. 27.20: Simple Beam Loaded Uniformly for Part of the Span
—MaCaulay's Approach (Illustrative Problem 27.9)

PROBLEM 27.10: A simple beam of span l carries two point loads W each at a distance a from either support. Determine the maximum deflection and maximum slope by Moment-area method.

Solution: Referring to Fig. 27.21, maximum slope occurs at either support, and by symmetry, zero slope and maximum deflection occur at mid-span.

θ_{max} = Change of slope between A and C.

$$= \frac{1}{EI} \text{ (Area of B.M. diagram between } A \text{ and } C)$$

$$= \frac{1}{EI}\left[\frac{1}{2} \times (a+l-2a) \times Wa\right] = \frac{Wa(l-a)}{2EI}$$

$y_{max} = \frac{1}{EI}$ (Intercept of the tangents to the elastic curve drawn at C and A on a vertical through A)

$$= \frac{1}{EI} \text{ (Moment of B.M. diagram between } A \text{ and } C \text{ about } A)$$

$$= \frac{1}{EI}\left[\frac{Wa^2}{2} \times \frac{2}{3} \times a + wa\left(\frac{l}{2}-a\right)\left(a+\frac{(l-2a)}{4}\right)\right]$$

$$= \frac{1}{EI}\left[\frac{Wa^3}{3} + \frac{Wa(l^2-4a^2)}{8}\right]$$

$$= \frac{Wa}{24EI}(8a^2 + 3l^2 - 12a^2)$$

or　$y_{max} = \dfrac{Wa}{24EI}(3l^2 - 4a^2)$　　...(Eq. 27.43)

This can also be remembered and used as a standard formula.

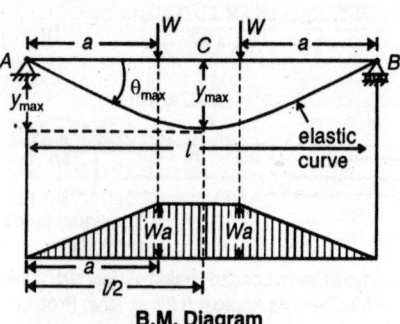

B.M. Diagram

Fig. 27.21: (Simple Beam with Two Symmetrically Placed Point Loads—Moment-Area Method (Ilustrative Problem 27.10)

PROBLEM 27.11: A simple beam of span l carries over the middle-half of the span a uniform load of intensity w. Determine the maximum slope and deflection of the beam by Moment-area method.

Solution: Referring to Fig. 27.22, maximum slope occurs at either support, and by symmetry, zero slope and maximum deflection occur at the mid-span.

$$\theta_{max} = \frac{1}{EI}\,[\text{Area of B.M. diagram between } A \text{ and } C\,]$$

$$= \frac{1}{EI}\left[\frac{1}{2}\times\frac{l}{4}\times\frac{wl^2}{16} + \frac{wl^2}{16}\times\frac{l}{4} + \frac{2}{3}\times\frac{l}{4}\times\frac{wl^2}{32}\right]$$

$$= \frac{11wl^3}{384\,EI}$$

$$y_{max} = \frac{1}{EI}\,[\text{Moment of B.M. diagram between } A \text{ and } C \text{ about } A\,]$$

$$= \frac{1}{EI}\left[\frac{wl^3}{128}\times\frac{2}{3}\times\frac{l}{4} + \frac{wl^2}{16}\times\frac{l}{4}\times\frac{3l}{8}\times\frac{wl^3}{192}\times\frac{13}{8}\times\frac{l}{4}\right]$$

or　$y_{max} = \dfrac{19}{2048}\cdot\dfrac{wl^4}{EI}$　　...(Eq. 27.44)

This can be remembered and used as a standard formula.

This problem can be solved by MaCaulay's approach; however, the solution by moment-area approach is simpler.

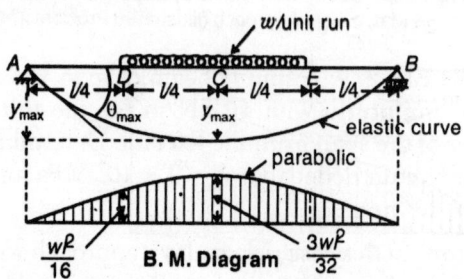

Fig. 27.22: Simple Beam with Middle Half of the Span loaded Uniformly—Moment-Area Method (Illustrative Problem 27.11)

PROBLEM 27.12: Solve Problem 27.7 by Moment-area method.

Solution: Referring to Fig. 27.23, by the Moment-area method, slope at the support,

$$\theta_A = \frac{1}{EI}\,[\text{Area of B.M. diagram between } A \,\&\, C\,]$$

$$= \frac{1}{EI}\left[\frac{1}{2}\times 1\times 35 + 35\times 1 + 15\times 1\times\frac{1}{2}\right]$$

$$= \frac{60}{EI}$$

Deflection at mid-span (under 30 kN load)

$$y_c = \frac{1}{EI}\,[\text{Moment of B.M. diagram between } A \text{ and } C \text{ about } A\,]$$

$$= \frac{1}{EI}\left[\frac{1}{2}\times 35\times\frac{2}{3} + 35\times 1\times 1.5 + 15\times\frac{1}{2}\times\frac{5}{3}\right]$$

$$= \frac{230}{3EI}$$

Deflection under the 20 kN load

$$y_D = \frac{230}{3EI} - \frac{1}{EI}\,[\text{Moment of B.M. diagram between } D \text{ and } C \text{ about } D\,]$$

$$= \frac{230}{3EI} - \frac{1}{EI}\left[35\times 1\times\frac{1}{2} + 15\times\frac{1}{2}\times\frac{2}{3}\right]$$

$$= \frac{230}{3EI} - \frac{22.5}{EI} = \frac{162.5}{3EI}$$

These are the same as those obtained by Macaulay's approach; but the labour involved in the moment-area method is less.

$$\therefore \quad y_c = \overline{mp} = \frac{1}{EI}\left[\frac{wal}{24}(l^2 - a^2) + \frac{wa^4}{8}\right]$$

or

$$y_c = \frac{wa}{24EI}(l^3 - la^2 + 3a^2), \text{ at the free end.}$$

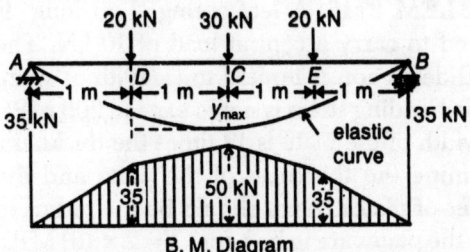

Fig. 27.23: Simple Beam with Point Loads—Moment-Area Method (Illustrative Problem 27.12)

PROBLEM 27.13: An overhanging beam of span l and a single overhang a is loaded uniformly with w/unit length throughout. Determine the deflection at the free end by Moment-area method.

Solution:

$$R_A = \frac{w}{2l}(l^2 - a^2)$$

$$R_A l = \frac{w}{2}(l^2 - a^2)$$

Referring to Fig. 27.24,

$$\overline{mp} = \overline{mn} + \overline{np}$$

$$\overline{mn} = \left(\frac{a}{l}\right)(\overline{qr})$$

$EI \cdot \overline{qr}$ (Moment of BMD between A and B about A)

$$= \frac{1}{2}l \times \frac{w}{2l}(l^2 - a^2) \times l \times \frac{2}{3}l - \frac{1}{3} \times l \times \frac{wl^2}{2} \times \frac{3}{4}l$$

$$= \frac{wl^2}{24}(l^2 - 4a^2)$$

$$\therefore EI \cdot \overline{mn} = \frac{a}{l} \frac{wl^2}{24}(l^2 - 4a^2) = \frac{wal}{24}(l^2 - 4a^2)$$

$EI \cdot \overline{np}$ = (Moment of BMD between B and C about C)

$$= \frac{1}{3} \times a \times \frac{wa^2}{2} \times \frac{3a}{4} = \frac{wa^4}{8}, \text{ numerically.}$$

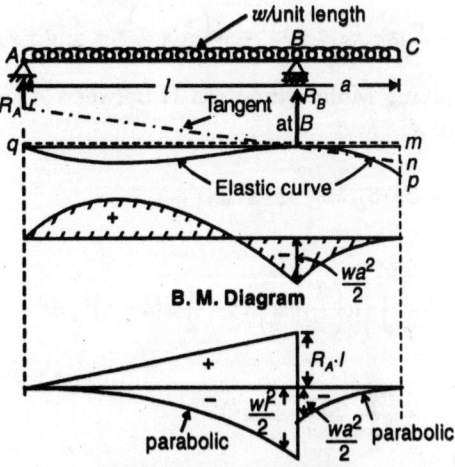

B. M. Diagram (Split-form)

Fig. 27.24: Overhanging Beam with Single Overhang Loaded Uniformly—Moment-Area Method (Illustrative Problem 27.13)

PROBLEM 27.14: An overhanging beam of span l and overhangs a on each side $(a < l/2)$ carries a uniform load of w/unit length throughout. Determine the deflection at mid-span by moment-area method.

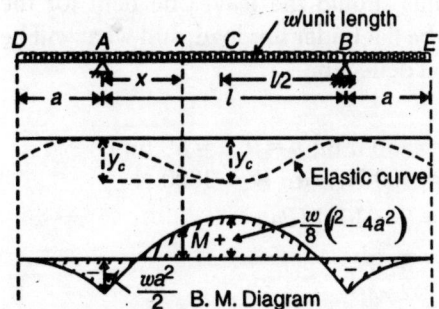

Fig. 27.25: Overhanging Beam with Double Overhangs-Loaded Uniformly—Moment-Area Method (Illustrative Problem 27.14)

Solution: With reference to Fig. 27.25,

$$R_A = R_B = \frac{w(l + 2a)}{2} = w\left(\frac{l}{2} + a\right)$$

$$M_c = R_A \cdot \frac{l}{2} - \frac{w}{2}\left(a + \frac{l}{2}\right)^2 = \frac{w}{8}(l^2 - 4a^2)$$

Since $a < l/2$, M_c is positive.
M at a distance x from A is given by

$$M = R_A x - \frac{w}{2}(a + x)^2 = w\left(\frac{l}{2} + a\right)x - \frac{w}{2}(a + x)^2$$

Applying Mohr's theorem-II between A and C about A,

$$EI \cdot \delta_C = \int_0^{l/2} M \cdot x \cdot dx$$

$$= \int_0^{l/2} \left\{ w\left(\frac{l}{2} + a\right)x^2 - \frac{w}{2}x(a + x)^2 \right\} dx$$

$$= \frac{wl^2}{384}(5l^2 - 24a^2)$$

$$\therefore \quad \delta_C = \frac{wl^2}{384\,EI}(5l^2 - 24a^2)$$

The simplest method in this case is the moment-area approach.

PROBLEM 27.15: A carriage spring, 0.75 m long, is built of 9 leaves of equal thickness and of width 50 mm. The stress is not to exceed 400 MPa under a load of 3 kN at the centre. $E = 2 \times 10^5$ MPa. Determine the thickness of each leaf. To what radius should the leaves be bent for the spring to be flat under this load, and what will be the central deflection?

Solution:
$l = 750$ mm; $n = 9$; $b = 50$ mm;
$f = 400$ N/mm^2 $W = 3000$ N
$E = 2 \times 10^5$ MPa

$$f = \frac{3}{2}\frac{Wl}{nbt^2}$$

$$\therefore \quad 400 = \frac{3}{2} \times \frac{3000 \times 750}{9 \times 50 t^2}, \text{ whence } t = 4.33 \text{ mm}$$

$$R = \frac{EI}{M} = \frac{2 \times 10^5 \times 9 \times 50 \times (4.33)^3}{12 \times 3000 \times (750/4)} \text{ mm} \approx 1.08 \text{ m}$$

$$\delta = \frac{3}{8}\frac{Wl^3}{Ebt^3 n}$$

$$= \frac{3}{8} \times \frac{3000 \times (750)^3}{2 \times 10^5 \times 50 \times (4.33)^3 \times 9} \text{ mm} \approx 65 \text{ mm}$$

Also, $\quad = \frac{l^2}{8R} = \frac{750 \times 750}{8 \times 1080} \approx 65$ mm, as above.

PROBLEM 27.16: A leaf spring, 1 m long, is required to carry a central load of 10 kN. The central deflection is limited to (1/40)th of span, and the bending stress is not to exceed 200 MPa. The width of the plate is 12 times the thickness. Determine the thickness of the plate and the number of plates. What should be the radius to which the plates are to be bent? $E = 2 \times 10^5$ MPa.

Solution:
$l = 1$ m $= 1000$ mm; $W = 10$ kN $= 10000$ N
$f = 200$ N/mm^2; $b = 12t$;
$E = 2 \times 10^5$ N/mm^2

$$f = \frac{3}{2}\frac{Wl}{nbt^2}$$

$$200 = \frac{3}{2} \times \frac{10000 \times 1000}{n \times 12t \times t^2}$$

$$\therefore \quad nt^3 = \frac{3 \times 10000 \times 1000}{200 \times 2 \times 12} = 6250 \qquad \ldots(i)$$

$$\delta = \frac{1 \times 1000}{40} = 25 \text{ mm}$$

$$\delta = \frac{3}{8} \times \frac{Wl^3}{Ebt^3 n}$$

$$\therefore \quad 25 = \frac{3}{8} \times \frac{10000 \times (1000)^3}{2 \times 10^5 \times 12t^4 n}$$

$$\therefore \quad nt^4 = \frac{3 \times 10^4 \times 10^9}{200 \times 24 \times 10^5} = 62500 \qquad \ldots(ii)$$

Dividing (ii) by (i),

$$t = \frac{62500}{6250} = 10 \text{ mm}$$

$$b = 12\ t = 120 \text{ mm}$$

$$n = \frac{6250}{1000} = 6.25, \text{ say } 7.$$

$$\delta = \frac{l^2}{8R}$$

$$\therefore \quad R = \frac{l^2}{8\delta} = \frac{(1000)^2}{8 \times 25} = 5000 \text{ mm} = 5 \text{ m}$$

PRACTICE PROBLEMS

27.1 A vertical mild steel column, 300 mm diameter, is embedded firmly into the ground. The height of the column is 4 m. The top end is subjected to a horizontal load of 20 kN. Calculate the slope and deflection at the top end. $E = 2 \times 10^5$ N/mm^2.

27.2 A cantilever of length l is subjected to a point load W at the middle of its length. Determine the slope and deflection at the free end.

27.3 A beam of constant I-section is simply supported at the ends over a span of 6 m. It carries a point load of 90 kN at 2 m from the left end. Determine the maximum deflection, mid-span deflection, and deflection under the load. $E = 2 \times 10^5$ N/mm^2 and $I = 9 \times 10^7$ mm^4.

27.4 A cantilever of length l is loaded uniformly for half its length, starting from the free end. Show that the maximum values of slope and deflection are $(7/48)\, (wl^3/EI)$ and $(41/384)\, (w\, l^4/EI)$, respectively.

27.5 A cantilever of length l carries a uniform load of intensity w per unit length, spread over the middle half of its length. Show that the maximum deflection is given by $(7/128)\, (wl^4/EI)$.

27.6 Show that the maximum values of slope and deflection for a cantilever of length l with triangular loading with the maximum intensity being w per unit length at the free end are $(w\, l^3/8EI)$ and $(11/120)\, (wl^4/EI)$, respectively.

27.7 A simple beam of span 10 m carries point loads of 100 kN and 60 kN at distances of 2 m and 5 m from the left end. Determine the deflection under each load and the maximum deflection using Macaulay's approach. $E = 2 \times 10^5$ MPa and $I = 18 \times 10^8$ mm^4.

27.8 A simple beam of span l carries point loads $2W$ each at the third-points of the span. Determine, in terms of EI, the maximum slope and deflection.

27.9 A simple beam of span 8 m carries a point load of 80 kN at a quarter-span-point. Determine the slopes at either end, and the deflection under the load, by moment-area method, in terms of the flexural rigidity EI.

27.10 A simple beam of span l carries a uniform load of intensity w per unit length for a length a from one end. Determine the ratio a/l for the condition that the maximum deflection occurs at the end of the load.

[**Hint:** The slope at the section at the end of the load should be equated to zero, and resulting quadratic equation solved].

27.11 An overhanging beam is supported on a span l with equal overhangs a on either side. A point load W acts at the middle of the beam. Show that the ratio a/l should be $1/3$ if the upward deflection at the ends equals the downward deflection at the middle.

27.12 A motor car spring of the laminated type is 1 m long and is made up of plates 80 mm × 10 mm thick. How many plates are required for a central load of 5 kN, if the maximum stress is not to exceed 225 N/mm^2? $E = 2 \times 10^5$ N/mm^2. What is the central deflection?

27.13 A steel carriage spring is 800 mm long and is built up of 9 leaves of equal thickness and 50 mm wide. Find the thickness of leaves for a maximum stress of 380 MPa and a central load of 3.5 kN. To what radius should the leaves be bent initially for the spring to be flat under this load, and what will then be the central deflection? $E = 2 \times 10^5$ N/mm^2.

27.14 A laminated spring, 1 m long, is made up of plates 50 mm × 6 mm thick. If the bending stress is limited to 200 MPa, how many plates are required for a central load of 1 kN? $E = 2 \times 10^5$ MPa. What is the central deflection under this load?

Chapter 28

Torsion of Cylindrical Shafts and Helical Springs

28.1 INTRODUCTION

When a cylindrical rod is twisted by a couple, the axis of which coincides with that of the rod, it is said to be subjected to pure torsion. In other words, the moment of the couple, which is called the "twisting moment" or "torque", acts *in the plane* of the cross-section of the rod (It may be remembered that a moment acting *perpendicular to the plane* of the cross-section causes "bending" of the cross-section, as has already been seen in Chapters 24 and 25).

The stress at any point in the cross-section of the rod is one of pure shear, the two planes being that of the cross-section and the longitudinal plane normal to it. The direction of stress in the plane of the cross-section is every where perpendicular to the radial lines from the axis, or the tangential directions. The principal planes are inclined at 45° to those of the shear stress and the principal stresses, which are of opposite sign, are of the same magnitude as the intensity of shear stress.

The strain is such that any cross-section (normal to the axis of the rod) makes a small rotation about the axis of the rod relative to other similar cross-sections. The rod is also called a "shaft" when used for transmission of power.

28.2 THE TORSION FORMULA

The relation betweeen the twisting couple (or torque), the strain (or angle of twist), the shear

stress, and the geometrical properties of the rod undergoing trosion may be derived in two stages.

The relevant *assumptions* in the following derivation are listed:

(1) The cross-section of the rod perpendicular to the longitudinal axis retains a circular boundary, and remains in a single plane.

(2) The distance between any two cross-sections remains constant after the application of the torque.

(3) Any point within the shaft remains on a circle, perpendicular to the longitudinal axis after loading.

(4) A straight line drawn from any point in the rod to the centre and normal to the longitudinal axis remains straight after loading.

Let a cylindrical rod of length l and radius R be subjected to a twisting couple of magnitude T, applied at one end, the other end being constrained or held by a balancing couple of equal magnitude (Fig. 28.1).

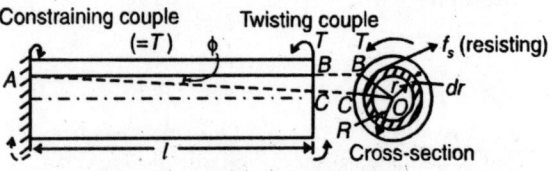

Fig. 28.1: A Cylindrical Rod under Torsion

Relation between Stress, Strain and Angle of Twist

A line AB, on the surface of the rod, which is parallel to the axis before strain, takes the form of a long helix AC after strain; the angle ϕ is the shear strain of the material at the surface.

$BC = l \cdot \phi$ (ϕ is considered to be small)

$\phi = \dfrac{BC}{l} = \dfrac{R \cdot \theta}{l}$ (The angular movement of radius OB due to strain in length l, θ is called the angle of twist).

But $\phi = \dfrac{f_{s_m}}{N}$, ($f_{s_m}$ being the shear stress in the material at the surface of the rod and N the modulus of rigidity of the material).

Equating the two expressions for ϕ,

$$\frac{f_{s_m}}{N} = \frac{R\theta}{l}$$

or $\dfrac{f_{s_m}}{R} = \dfrac{N\theta}{l}$ ...(Eq. 28.1)

For a given twisting moment and length of rod, θ is a constant; hence, (f_{s_m}/R) is a constant.

That is to say, $\dfrac{f_{s_m}}{R} = \dfrac{f_s}{r}$

where f_s is the shear stress at a radial distance r.

Since f_s varies directly with r, obviously the shear stress is the maximum value f_{s_m} at the surface of the rod.

Relation between Twisting Moment and Shear Stress

Considering a thin ring of radius r, in the cross-section at the fixed end A with an infinitesimally small thickness dr, f_s being the shear stress at raduis r,

the total force on the ring $= f_s \times 2\pi r \cdot dr$

Moment of this force about the axis of the rod

$$= f_s \times 2\pi r \cdot dr \times r = \left(\frac{f_s}{r}\right) 2\pi r^3 \cdot dr$$

But $\dfrac{f_s}{r} = \dfrac{f_{s_m}}{R}$,

$\therefore$ The resisting moment of this elemental ring

$$\frac{f_{s_m}}{R} 2\pi r^3 \cdot dr$$

Total resisting moment at the section is obtained by integration, and this shall be numerically equal to the applied twisting moment, T.

$\therefore$
$$T = \frac{2\pi}{R} \cdot f_{s_m} \int_0^R r^3 \cdot dr \qquad \text{...(Eq. 28.2)}$$

$$= \frac{2\pi}{R} \cdot f_{s_m} \left[\frac{r^4}{4}\right]_0^R = \frac{f_{s_m}}{R}\left(\frac{\pi R^4}{2}\right) \quad \text{...(Eq. 28.3)}$$

But $\dfrac{\pi R^4}{2}\left(= \dfrac{\pi D^4}{32}\right)$ is the polar moment of inertia, J, of the circular cross-section of radius R (or diameter D).

$\therefore$
$$T = \frac{f_{s_m}}{R} \cdot J$$

or $\dfrac{T}{J} = \dfrac{f_{s_m}}{R}$ ...(Eq. 28.4)

From Equations 28.1 and 28.4, we can write,

$$\frac{T}{J} = \frac{f_{s_m}}{R} = \frac{N \cdot \theta}{l} \qquad \text{...(Eq. 28.5)}$$

This is the famous "Torsion Formula" for pure torsion of cylindrical rods/bars/shafts.

For a given shaft, $T \left(= (f_{s_m}/R) \cdot J\right)$, f_{s_m} being the permissible shear stress for the material), is called the "Torque-carrying Capacity" or "Strength of the Shaft".

This is a dimensionally correct formula.

Say T is in newton-metres,

J is in (metre)^4 units,

f_{s_m} is in N/m^2 (Pascals),

R is in metres,

N is in N/m^2 (Pascals),

θ is in radian measure,

and l is in metres.

The units of any of these quantities may be changed for convenience.

28.3 TORSIONAL STIFFNESS AND TORSIONAL RIGIDITY

The torque required to cause unit angle of twist is known as the "Torsional Stiffness" of a shaft.

$$T = \frac{N \cdot \theta}{l} \cdot J$$

or $$\frac{T}{\theta} = \frac{NJ}{l}$$ ...(Eq. 28.6)

Torsional stiffness $= \dfrac{NJ}{l}$

This is obviously dependent purely upon the geometrical and material properties of the shaft.

The torsional stiffness for unit length of the shaft is called the "Torsional Rigidity".

Obviously, this is $N \cdot J$—the product of the modulus of rigidity and the polar moment of inertia of the cross-section of the shaft.

28.4 POWER TRANSMITTED BY A SHAFT

When a shaft rotating at a speed of n revolutions per minute transmits power, the mean torque carried by it is related to the power as follows:

$$P\text{(Power in Watts)} = \frac{2\pi n}{60} \times T_{\text{mean}}$$

T_{mean} being in newton-metres,

or $P\text{(Watts)} = \dfrac{2\pi n}{60} \times T_{\text{mean}} \left(\dfrac{\text{N.m}}{\text{sec}}\right)$ ...(Eq. 28.7)

[Actually, 1 N.m/second is equal to 1 Watt]

If T is the torque-carrying capacity (or torsional strength) of a given shaft, the power that can be transmitted by it while running at a given speed is P given by Eq. 28.7, which may be expressed practically in kilowatts (kW).

28.5 HOLLOW CYLINDRICAL SHAFTS

Since the shear stress varies in direct proportion to the radius, the material of the shaft towards the axis will offer very little resistance to the applied torque. This has given rise to the idea that better utilisation of the shear strength of the material is possible by using a hollow shaft instead of a solid shaft, since in the former the material is located away form the axis.

It can be shown that, for the same weight and length (or the same cross-sectional area), a hollow shaft has a greater torque-carrying capacity (or torsional strength) than a solid one. But in a different way, the cross-sectional area required to transmit a given torque, is less for a hollow section than a solid section, other factors remaining the same.

Let the hollow section be of external radius R (diameter D) and internal radius r (diameter d) as shown in Fig. 28.2.

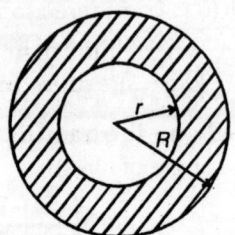

Fig. 28.2: Hollow Circular Section

The value of the resisting moment, which is equal to the applied twisting couple, T, is given by Eq. 28.2, with the exception that limits for the definite integral will be r and R instead of O and R.

$$\therefore \quad T = 2\pi \cdot \frac{f_{s_m}}{R} \int_{r}^{R} r^3 \cdot dr \qquad \text{...(Eq. 28.8)}$$

$$= 2\pi \cdot \frac{f_{s_m}}{R} \left(\frac{r^4}{4}\right)_{r}^{R} = \frac{\pi}{2} \frac{f_{s_m}}{R} \left[R^4 - r^4\right]$$

or $$T = \frac{\pi}{2} \cdot f_{s_m} \left(\frac{R^4 - r^4}{R}\right) \qquad \text{...(Eq. 28.9)}$$

The polar moment of inertia, J, for this shaft is

$$J = \frac{\pi}{2}\left(R^4 - r^4\right) = \frac{\pi}{32}\left(D^4 - d^4\right) \quad \text{...(Eq. 28.10)}$$

The equation $\dfrac{T}{J} = \dfrac{f s}{R}$ (R being the external radius) is still valid.

Let us now compare the torsional strengths of hollow and solid shafts of the same material, weight, and length.

Let R and r be the outer and inner radii of the hollow shaft, and R_1 be the radius of the solid shaft, f_s being the maximum shear stress (allowable value).

$$T_{\text{hollow}} = \frac{\pi}{2} \cdot f_{s_m} \left(\frac{R^4 - r^4}{R}\right) \qquad \text{...(Eq. 28.9)}$$

$$T_{\text{solid}} = \frac{\pi}{2} \cdot f_{s_m} R_1^3 \qquad \text{...(Eq. 28.3)}$$

The maximum shear stress being the same for both shafts,

$$\frac{T_{\text{hollow}}}{T_{\text{solid}}} = \frac{(R^4 - r^4)}{RR_1^3}$$

But since the weight and length (and the unit weight) are the same, the areas of cross-section will be equal.

$$R_1^2 = (R^2 - r^2)$$

$$\frac{T_{\text{hollow}}}{T_{\text{solid}}} = \left(\frac{R^2 + r^2}{RR_1}\right) = \frac{R}{R_1}\left(1 + \frac{1}{k^2}\right)$$

where $k = \dfrac{R}{r}$

$$\frac{T_{\text{hollow}}}{T_{\text{solid}}} = \frac{(k^2 + 1)}{\sqrt{k^2(k^2 - 1)}} = \frac{(k^2 + 1)}{k\sqrt{(k^2 - 1)}} \quad \text{...(Eq. 28.11)}$$

If $k = 2$, as in common practice,

$$\frac{T_{\text{hollow}}}{T_{\text{solid}}} = \frac{5}{2\sqrt{3}} = 1.44$$

28.6 DESIGN OF A SHAFT TO TRANSMIT GIVEN POWER

Let us consider the problem of designing a shaft to transmit a given power of P kW at n r.p.m.

If the mean torque is T_{mean} (N.m),

$$T_{\text{mean}} = \frac{60 \times 1000 P}{2\pi n} \text{N.m}$$

If the torque is variable, the maximum torque, T_{max}, will be a constant times T_{mean}, the constant being greater than unity. For the determination of the maximum shear stress, f_{s_m}, the maximum torque must be used.

If a solid shaft is to be designed, with a radius R,

$$T_{\text{max}} = f_{s_m} \frac{\pi R^3}{2} = \frac{60 \times 1000 P}{2\pi n}$$

(if the ratio $T_{\text{max}}/T_{\text{mean}}$ is unity)

Here $R \propto \sqrt{\dfrac{P}{n}}$ for a given material (since f_{s_m} is the allowable stress).

Similarly, a hollow shaft may also be designed, with an assumed ratio of outer to inner radii.

28.7 TORSION COMBINED WITH BENDING

This case is very common in practice. A propeller shaft, transmitting power, is subjected to bending stresses due to its own weight, or due to the weights of pulleys, or the pull in the belts due to line shafting.

The stresses to be considered are those due to bending, those due to torsion, and those due to shear caused by bending; the last one, however, is considered to be unimportant.

(1) The maximum normal stress f_n, in the cross-section of a solid circular shaft of radius R due to a bending moment, M is

$$f_n = \frac{4M}{\pi R^3} \qquad \text{...(Eq. 28.12)}$$

This occurs at the extreme fibre (or the surface of the shaft.)

(2) The maximum shear stress, f_s, due to a torque, T, is given by

$$f_s = \frac{2T}{\pi R^3}\left(= \frac{16T}{\pi D^3}, \text{where } D \text{ is diameter}\right)$$

$$\text{...(Eq. 28.13)}$$

This also occurs at the extreme fibre (or the surface of the shaft). The principal stresses are given by,

$$f_p = \frac{1}{2}\left[f_n \pm \sqrt{f_n^2 + 4f_s^2}\right] \qquad \text{...(Eq. 28.14)}$$

The normal stress on the other plane being taken as zero.

The maximum value is tensile or compressive according as the bending stress at the point is tensile or compressive.

Substituting for f_n and f_s from Eqs. 26.12 and 26.13,

$$f_p = \frac{1}{2}\left[\frac{4M}{\pi R^3} + \sqrt{\frac{16M^2}{(\pi R^3)^2} + \frac{16T^2}{(\pi R^3)^2}}\right]$$

$$\text{...(Eq. 28.15)}$$

This may be put in the following forms:

$$f_p = \frac{2}{\pi R^3}\left[M + \sqrt{M^2 + T^2}\right] \quad \text{...(Eq. 28.16)}$$

$$f_p = \frac{4}{\pi R^3}\left[\frac{1}{2}\left(M + \sqrt{M^2 + T^2}\right)\right] \quad \text{...(Eq. 28.17)}$$

Since these equations are of the same form as Eqs. 28.13 & 28.12, we may write,

$$f_p = \frac{2T_E}{\pi R^3} \qquad\qquad \text{...(Eq. 28.18)}$$

and $\quad f_p = \dfrac{4M_E}{\pi R^3} \qquad\qquad$ ...(Eq. 28.19)

where $\quad T_E = \left[M + \sqrt{M^2 + T^2}\right] \quad$ ...(Eq. 28.20)

and $\quad M_E = \dfrac{1}{2}\left(M + \sqrt{M^2 + T^2}\right) \quad$ ...(Eq. 28.21)

T_E is the twisting moment, which, unaccompanied by bending, would produce a torsional shearing stress of magnitude f_p, and hence a maximum principal stress of the same value, since a principal stress due to shear stress system has the same magnitude as the shear stress.

This is called the 'Equivalent Twisting Moment' (for principal stress). (It should be remembered that it is not actually a twisting moment).

Similarly M_E is the bending moment, which, unaccompanied by torsion, would produce a direct bending stress of magnitude equal to the principal stress, f_p.

This is called the "Equivalent Bending Moment". The greatest shear stress

$$f_{s_{max}} = \frac{1}{2}(f_{p_1} - f_{p_2})$$

$$= \frac{1}{2}\sqrt{f_n^2 + 4f_s^2} = \frac{1}{2}\left[\sqrt{\frac{16M^2}{(\pi R^3)^2} + \frac{16T^2}{(\pi R^3)^2}}\right]$$

or $\quad f_{s_{max}} = \dfrac{2}{\pi R^3}\left[\sqrt{M^2 + T^2}\right] \qquad$ ...(Eq. 28.22)

This may be written as,

$$f_{s_{max}} = \frac{2}{\pi R^3}\cdot T_{E_s} \qquad\qquad \text{...(Eq. 28.23)}$$

where $T_{E_s} = \sqrt{M^2 + T^2} \qquad\qquad$ ...(Eq. 28.24)

This is called the "Equivalent Twisting Moment for Shear"—which would cause the same maxi-

mum shear stress as that caused by the simultaneous action of a bending moment, M, and a twisting moment, T.

This is used when shear is the design criterion.

28.8 TORSION COMBINED WITH BENDING, SHEAR AND THRUST

Let a solid cylindrical shaft be acted upon by
a bending moment, M,
a twisting moment, T
a shear force due to bending moment, F
and an end thrust, P

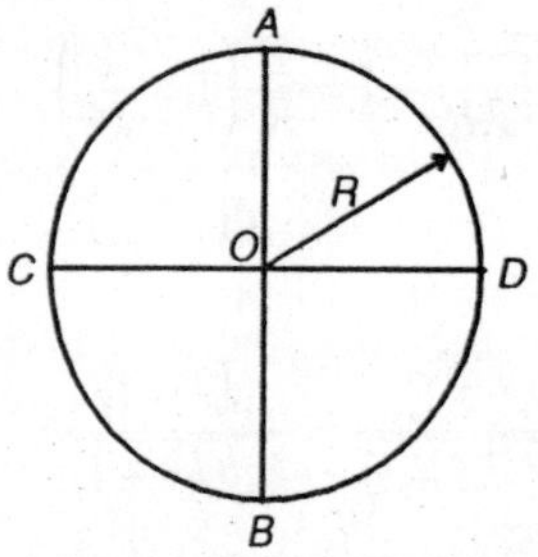

Fig. 28.3: Cross-Section of a Cylindrical Shaft

Let the radius of the shaft be R

(At the ends of a vertical diameter)

Maximum compressive stress,

$$f_{n_{max.}} = \frac{P}{\pi R^2} + \frac{4M}{\pi R^3} \qquad\qquad \text{...(Eq. 28.25)}$$

Minimum compressive stress,

$$f_{n_{min.}} = \frac{P}{\pi R^2} - \frac{4M}{\pi R^3} \qquad\qquad \text{...(Eq. 28.26)}$$

$[f_n$ is zero at the horizontal diameter.]

The shear stress due to F is zero at the top and bottom of the shaft, and maximum at a horizontal diameter.

$$f_{s_{max}} = \frac{4}{3}\frac{F}{\pi R^2} \qquad\qquad \text{...(Eq. 28.27)}$$

Total shear stress at top or bottom $(f_s) = \dfrac{2T}{\pi R^3}$

Total shear stress at each end of a horizontal diameter,

$$f_s = \frac{2T}{\pi R^3} + \frac{4F}{3\pi R^2}.$$

The principal stresses are found as usual from,

$$f_p = \frac{1}{2}\left[f_n \pm \sqrt{f_n^2 + 4f_s^2}\right]$$

The normal and shear stresses are:

For the top or bottom of the shaft (A and B in Fig. 28.3):

$$f_n = \frac{P}{\pi R^2} \pm \frac{4M}{\pi R^3} \text{ and } f_s = \frac{2T}{\pi R^3}$$

For each end of a horizontal diameter (C & D in Fig. 28.3)

$$f_n = \frac{P}{\pi R^2} \text{ and } f_s = \frac{2T}{\pi R^3} + \frac{4F}{3\pi R^2}$$

Thus for any combination of M, T, F and P, we know the values of normal stress, f_n, and shear stress, f_s, at any of the salient points A, B, C or D of the shaft at the surface. This enables us to determine the principal stresses at any of these points.

28.9 RESILIENCE DUE TO TORSION

If a solid cylindrical shaft is subjected to a gradually applied torque T, and θ is the angle of twist produced, the internal energy stored in the shaft is $\frac{1}{2} \cdot T \cdot \theta$.

Resilience, $U = \frac{1}{2} \cdot T \cdot \theta$ ...(Eq. 28.28)

$$= \frac{1}{2} f_s \cdot \frac{J}{R} \times \frac{f_s}{R} \cdot \frac{l}{N}$$

$$= \frac{1}{2} \cdot \frac{f_s^2}{N} \cdot \frac{J}{R^2} \cdot l$$

$$= \frac{1}{2} \cdot \frac{f_s^2}{N} \cdot \frac{\pi R^2 l}{2}$$

or $\quad U = \frac{1}{4} \cdot \frac{f_s^2}{N}$ (Volume of shaft) ...(Eq. 28.29)

For a hollow cylindrical shaft of radii R and r, Resilience,

$$U = \frac{1}{2} \cdot T \cdot \theta$$

$$= \frac{1}{2} \cdot \frac{T^2 l}{NJ}$$

$$= \frac{\frac{1}{2}\left\{\frac{\pi f_s}{2}\left(\frac{R^4 - r^4}{R}\right)\right\}^2}{\frac{\pi}{2}(R^4 - r^4) \cdot N} \cdot l$$

$$= \frac{1}{4} \cdot \frac{f_s^2}{N} \cdot \frac{(R^2 + r^2)}{R^2} \cdot \pi(R^2 - r^2) \cdot l$$

or $\quad U = \frac{1}{4} \cdot \frac{f_s^2}{N} \cdot \frac{(R^2 + r^2)}{R^2}$ (Volume of shaft)

...(Eq. 28.30)

If $\frac{R}{r} = k$, say

$$U = \frac{1}{4} \cdot \frac{f_s^2}{N} \cdot \left(1 + \frac{1}{k^2}\right) \qquad \text{...(Eq. 28.31)}$$

28.10 HELICAL SPRINGS

A 'spring' is a device in which the material is so arranged that it can suffer considerable changes in form without being permanently distorted. The purpose of a spring is to absorb energy and restore it as and when required.

The following are the types of springs used in engineering practice:

1. Helical springs (used in machines, engines, etc.)
2. Flat spiral springs (used in clocks and watches)
3. Leaf springs (Also laminated springs or carriage springs) (used in carriages)

In this section, we shall study helical springs (Leaf springs have been studied in Chapter 27).

A helical spring is one which has coils wound in the shape of a 'helix' which is a curve traced out by a point on the surface of a cylinder, rotating about its longitudinal axis, while the point itself has a velocity parallel to the longitudinal axis.

A schematic diagram of a helical spring is shown in Fig. 28.4.

When the coils are so closely wound that any one coil lies nearly in a plane perpendicular to the axis of the helix, the spring is known as a "close-coiled helical spring".

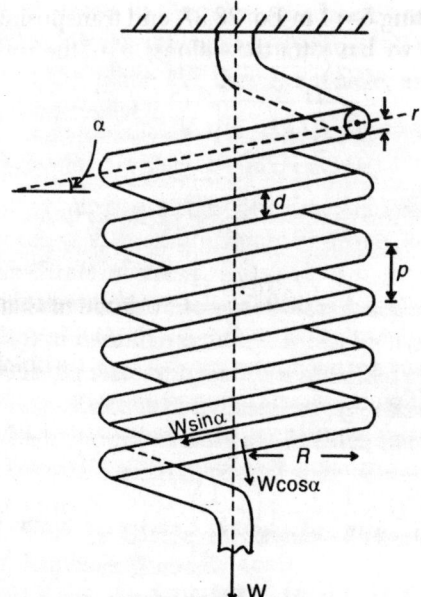

Fig. 28.4: Helical Spring

As against this, a spring in which the coils are so separated that the angle of the helix is a significant value is known as an "open-coiled helical spring".

The structural action in causing stresses in the cross-section of the spring wire will be different in these two cases, as will be brought out in the following sub-sections. Also, a spring may be subjected to an axial force or an axial torque, both of which will be considered.

28.10.1 Close-Coiled Helical Springs

For a helical spring of n coils, the length l of the wire is

$$l = 2\pi R n \text{ (approximately)} \quad \text{...(Eq. 28.32)}$$

R being the mean radius of the coils.

As already stated, a helical spring may be subjected to an axial load/force or an axial torque.

(a) Axial Load/Force

Referring to Fig. 28.4, an axial load/force applied to the helical spring may be resolved into two components:

(i) $W\cos\alpha$—normal to the axis of the spring wire.

(ii) $W\sin\alpha$—parallel to the axis of the spring wire.

The component $W\cos\alpha$ will produce a direct shear stress in the material of the spring wire; further, since it acts at a lever arm R, a twisting moment of $(W\cos\alpha) \cdot R$ is caused. This produces torsional shear stresses in the material which are very large in comparison with the direct shear stresses; hence, invariably, the direct shear stress is considered negligible.

The component $W\sin\alpha$ will produce a direct tensile stress in the material of the spring wire; further, since it acts at a lever arm R, a bending moment of $(W\sin\alpha) \cdot R$ is caused. This produces tensile and compressive stresses in the material, in different zones relative to the neutral axis; these stresses are generally very large in comparison with the direct tensile stress, which may be neglected.

In the case of a close-coiled helical spring, the angle of the plane of the coils with the horizontal, α, is negligibly small. Thus the bending moment $(W\sin\alpha) \cdot R$ is negligible, as also the direct tensile stress. Further, the twisting moment will be very nearly $W \cdot R$.

Let the spring wire be of circular section of diameter d.

The torque, $T = WR$ ...(Eq. 28.33)

If the axial deformation (or extension, in this case) of the spring is δ, and the angle of twist of the wire is θ, the work done by the gradually applied load W will be equal to the resilience in torsion.

$$\frac{1}{2} W \cdot \delta = \frac{1}{2} T \cdot \theta \quad \text{...(Eq. 28.34)}$$

Also, from the torsion formula for circular sections,

$$\frac{T}{J} = \frac{N \cdot \theta}{l} \quad \text{...(Eq. 28.35)}$$

From these three equations, any three quantities can be determined if the rest are known.

$$\theta = \frac{T \cdot l}{N \cdot J} = \frac{(WR)(2\pi Rn)}{N(\pi d^4 / 32)}$$

or $\quad \theta = \dfrac{64WR^2n}{Nd^4}$ $\qquad$ ...(Eq. 28.36)

From Eq. 28.27,

$$\frac{1}{2}W \cdot \delta = \frac{1}{2} \cdot WR \cdot \frac{64WR^2n}{Nd^4}$$

or $\quad \delta = \dfrac{64WR^3n}{Nd^4}$ $\qquad$ ...(Eq. 28.37)

Also, δ could have been obtained from the relation

$$\delta = R \cdot \theta \qquad \text{...(Eq. 28.38)}$$

as can be understood from a comparison of Eqs 28.36 and 28.37. Shear stress, f_s, in the material of the spring wire is given by,

$$f_s = \frac{16WR}{\pi d^3} \text{*} \qquad \text{...(Eq. 28.13)}$$

Resilience (or the internal strain energy) of the spring, U, is given by,

$$U = \frac{1}{2} \cdot T \cdot \theta$$

$$= \frac{1}{2}(WR)\left(\frac{f_s}{r} \cdot \frac{l}{N}\right)$$

$$= \frac{1}{2}\left(\frac{f_s}{r} \cdot \frac{\pi r^4}{2}\right)\left(\frac{f_s}{r} \cdot \frac{l}{N}\right)$$

$$= \frac{1}{4}\frac{f_s^2}{N}(\pi r^2 l)$$

or $\quad U = \dfrac{1}{4}\dfrac{f_s^2}{N}$ (Volume) $\qquad$ ...(Eq. 28.39)

The "stiffness" of a spring is defined as the force required to cause unit deflection, or the torque required to produce unit angle of twist.

---

*Considering the average direct shear stress component also,

$$f_s = \left(\frac{16WR}{\pi d^3}\right)\left(1 + \frac{d}{4R}\right)$$

Even more precisely,

$$f_s = \left(\frac{16WR}{\pi d^3}\right)\left[\frac{4k-1}{4k-4} + \frac{0.615}{k}\right]$$

where $k = \dfrac{2R}{d}$ – Wahl's Eqn. considering the difference in shear strain in the outer and inner sides of the spring wire.

Setting $\delta = 1$ in Eq. 28.37, and transposing the terms, we have, for the stiffness, s, of the spring

$$s = \frac{Nd^4}{64R^3n} \qquad \text{...(Eq. 28.40)}$$

(b) Axial Torque

If an axial torque of magnitude M is applied to a close-coiled helical spring, a constant bending moment of magnitude equal to that of the torque acts on the coils. This bending moment tends to increase or decrease the curvature, of the coils, depending on its direction of action.

If we consider that the wire behaves as a straight beam inspite of its large initial curvature, the strain energy in bending, U, for a constant bending moment M, is given by,

$$U = \frac{1}{2}\frac{M^2 l}{EI}$$

If ϕ is the angle of twist at the free end,

$$\frac{1}{2}M \cdot \phi = \frac{1}{2}\frac{M^2 l}{EI}$$

or $\quad \phi = \dfrac{M \cdot l}{EI}$ $\qquad$ ...(Eq. 28.41)

Also, $\phi = \dfrac{M(2\pi Rn)}{E(\pi d^4 / 64)} = \dfrac{128MRn}{Ed^4}$ $\quad$ ...(Eq. 28.42)

Resilience, $U = \dfrac{1}{2}\dfrac{M^2 l}{EI}$

$$= \frac{1}{2}\left(f \cdot \frac{I}{r}\right)^2 \frac{l}{EI}$$

$$= \frac{1}{2}\frac{f^2}{4E}(\pi r^2 l)$$

or $\quad U = \dfrac{1}{8}\dfrac{f^2}{E}$ (Volume) $\qquad$ ...(Eq. 28.43)

where $f =$ bending stress in the wire due to M.

Maximum bending stress,

$$f_{\max} = \frac{32M}{\pi d^3}$$

Since f_s is zero in this case as there is no torque acting, $f_{\max}$ itself is the major principal stress and the minor principal stress is zero, from the knowledge of biaxial state of stress.

$f_{s_{\text{max}}}$, the greatest shear stress

$$= \frac{1}{2} f_{\text{max}}$$

or $\quad f_{s_{\text{max}}} = \frac{16M}{\pi d^3} \qquad \text{...(Eq. 28.44)}$

28.10.2 Open-Coiled Helical Springs

For an open-coiled helical spring of n coils with a helix angle α, the length of the wire is

$$l = 2\pi \bar{R} n \cdot \sec \alpha \qquad \text{...(Eq. 28.45)}$$

A small length of the spring wire is shown along with the axes in and perpendicular to the plane of the coils in Fig. 28.5.

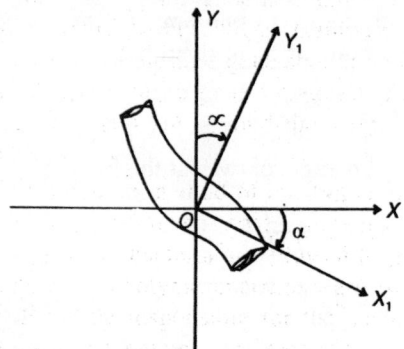

Fig. 28.5: Open-Coiled Helical Spring of a Small Element

(a) Axial Load

Referring to Fig. 28.5, the moment WR about OX, on the normal cross-section at O, may be resolved into two components:

$$T_1 = WR\cos\alpha$$

about OX_1 tangent to the centre line of the wire (i.e., in the plane of the cross-section of the wire).

$$M_1 = WR\sin\alpha$$

about OY_1 perpendicular to the axis of the wire and in the plane of the figure (i.e., normal to the plane of the cross-section of the wire).

If the axial extension, δ, alone is of interest, it may be easily got from strain energy considerations.

$$\frac{1}{2} W \cdot \delta = \frac{1}{2} T_1 \theta_1 + \frac{1}{2} M_1 \cdot \phi_1$$

$$= \frac{1}{2} \frac{T_1^2 l}{N J} + \frac{1}{2} \frac{M_1^2 l}{EI}$$

here θ_1 and ϕ_1 are the angular displacements of the free end about OX_1 and OY_1, respectively.

$$\therefore \quad \delta = WR^2 l \left(\frac{\cos^2\alpha}{NJ} + \frac{\sin^2\alpha}{EI} \right)$$

$$= 2\pi WR^3 n \sec\alpha \left(\frac{\cos^2\alpha}{NJ} + \frac{\sin^2\alpha}{EI} \right)$$

$$= \frac{32 WR^2 l}{\pi d^4} \left(\frac{\cos^2\alpha}{N} + \frac{2\sin^2\alpha}{E} \right)$$

or $\quad \delta = \frac{64 WR^3 n \sec\alpha}{d^4} \left(\frac{\cos^2\alpha}{N} + \frac{2\sin^2\alpha}{E} \right)$

$$\text{...(Eq. 28.46)}$$

This reduces to that for a close-coiled helical spring, if α is set zero.

From the normal stress f_n from bending and the torsional shear stress f_s the major and minor principal stress in the wire section.

f_{p_1} and f_{p_2} can be shown to be,

$$f_{p_{1,2}} = \frac{16WR}{\pi d^3} (\sin\alpha \pm 1)$$

Absolute maximum shear stress,

$f_{s_{\text{max}}} = \left(\frac{f_{p_1} - f_{p_2}}{2} \right) = \frac{16WR}{\pi d^3}$, as in the case of

close-coiled spring.

For $\dfrac{E}{N} = \dfrac{5}{2}$, for $\alpha = 10°$, the effect of obliquity is

only under 1%,

while for $\alpha = 45°$, it is around 10%

(This is the reason for ignoring the effect of α for small values)

Both the bending about OY_1 and twisting about OX_1 cause rotation of the free end of the coil about the axis of the spring.

The components of rotation about OY and OX may be shown to be

$$\phi = 2\pi WR^2 n \sin\alpha \left(\frac{1}{NJ} - \frac{1}{EI} \right) \quad \text{...(Eq. 28.48)}$$

and $\quad \theta = 2\pi WR^2 n \sec\alpha \left(\frac{\cos^2\alpha}{NJ} + \frac{\sin^2\alpha}{EI} \right)$

$$\text{...(Eq. 28.49)}$$

respectively.

$$\delta = R \cdot \theta = 2\pi W R^3 n \sec \alpha \left(\frac{\cos^2 \alpha}{N J} + \frac{\sin^2 \alpha}{EI} \right)$$

as derived earier.

(b) *Axial Torque*

The moment M about OY may be resolved into components $M\cos\alpha$ about OY_1 and $M\sin\alpha$ about OX_1, and the strain energy equation is

$$\frac{1}{2} M \phi = \frac{1}{2} \frac{M^2 \cos^2 \alpha \cdot l}{EI} + \frac{1}{2} \frac{M^2 \sin^2 \alpha \cdot l}{N J}$$

or $\qquad \phi = Ml \left(\dfrac{\cos^2 \alpha}{EI} + \dfrac{\sin^2 \alpha}{N J} \right)$

or $\qquad \phi = 2\pi R n M \sec \alpha \left(\dfrac{\cos^2 \alpha}{EI} + \dfrac{\sin^2 \alpha}{N J} \right)$

...(Eq. 28.50)

The angle of twist, ϕ of the free end is therefore given by,

$$\phi = \frac{64 M R}{d^4} \cdot n \cdot \sec \alpha \left(\frac{2\cos^2 \alpha}{E} + \frac{\sin^2 \alpha}{N} \right)$$

...(Eq. 28.51)

The axial extension caused may be found by resolving the rotations about OX, and mutliplying by R.

The resultant rotation of the free end about OX

$$\theta = Ml \sin \alpha \cos \alpha \left(\frac{1}{N J} - \frac{1}{EI} \right) \text{ ...(Eq. 28.52)}$$

$$\delta = MRl \sin \alpha \cos \alpha \left(\frac{1}{N J} - \frac{1}{EI} \right)$$

$$= \frac{64 M R^2}{d^4} \cdot n \sin \alpha \left(\frac{1}{N} - \frac{2}{E} \right) \text{...(Eq. 28.53)}$$

The principal stresses, $f_{p_{1\&2}}$, are $(16M/pd^3)$ $(\cos \alpha \pm 1)$; the absolute maximum stress, $f_{s_{max}}$, is again $(16M/\pi d^3)$, the same as that for a close-coiled helical spring.

ILLUSTRATIVE PROBLEMS

PROBLEM 28.1: What power can a 75 mm diameter shaft transmit at 100 r.p.m. if the maximum shear stress is not to exceed 75 MPa?

Solution:

$$T = \frac{\pi R^3}{2} \cdot f_{s_m} = \frac{\pi D^3}{16} \cdot f_{s_m} = \frac{\pi}{16} \times (75)^3 \times 75 \text{ N.mm}$$

$$= 6,212.6 \text{ N.m}$$

Power, $P = \dfrac{2\pi n T}{60} = \dfrac{2\pi \times 100 \times 6212.6}{60} \text{ watts}$

$$= 65 \text{ kW}$$

PROBLEM 28.2: A steel shaft of solid circular section has to transmit 300 kW at 200 r.p.m. The maximum shear stress is not to exceed 50 MPa and the angle of twist not more than $1°$ in a length of 3 m. Select a suitable diameter if $N = 8.4 \times 10^4$ N/mm^2.

Solution:

$$P = \frac{2\pi n T}{60}; f_{s_m} = 50 \text{ MPa}$$

$$T = \frac{60 P}{2\pi n} = \frac{60 \times 300 \times 1000}{2\pi \times 200} \text{ N.m}$$

$$= 14,324 \text{ N.m}$$

But $\quad T = \dfrac{\pi D^3}{16} \cdot f_{s_m}$

$\therefore \qquad D = \sqrt[3]{\dfrac{16 \times 14324 \times 1000}{\pi \times 50}} \text{ mm}$

$$= 113.42 \text{ mm}$$

$$\frac{T}{J} = \frac{N\theta}{l}$$

$$\frac{14324 \times 1000}{J} = \frac{8.4 \times 10^4 \times \pi}{180 \times 3000}$$

$$J = \frac{14324 \times 1000 \times 180 \times 3000}{84 \times 1000 \times \pi} = \frac{\pi D^4}{32}$$

$\therefore \quad D = \sqrt[4]{\dfrac{32 \times 14324 \times 1000 \times 180 \times 3000}{84 \times 1000 \times \pi^2}} \text{ mm}$

$$= 131.45 \text{ mm}$$

Select the bigger value, i.e., $D \approx 140$ mm

PROBLEM 28.3: Calculate the diameter of the shaft required to transmit 50 kW of power at 100 r.p.m. for a maximum shear stress of 60 MPa. The maximum torque is likely to exceed the mean by 25%. Calculate the angle of twist for a length of 2 m. $N = 8.4 \times 10^4$ MPa.

Solution:

$$P = \frac{2\pi n T_{mean}}{60}$$

$$\therefore \ T_{mean} = \frac{60P}{2\pi n} = \frac{60 \times 50 \times 1000}{2\pi \times 100} \ \text{N.m}$$

$$= 4775 \ \text{N.m}$$

$$T_{max} = 1.25 \ T_{mean} = 1.25 \times 4775$$

$$= 5970 \ \text{N.m}$$

If D mm is the diameter of the solid shaft required,

$$T_{max} = \frac{\pi D^3}{16} \cdot f_s$$

$$\therefore \ D = \sqrt[3]{\frac{16T}{\pi f_s}} = \sqrt[3]{\frac{16 \times 5970 \times 1000}{\pi \times 60}} \ \text{mm}$$

$$= 79.73 \ \text{mm, say } 80 \ \text{mm}$$

$$\frac{2f_{s_m}}{D} = \frac{N\theta}{l}$$

$$\frac{2 \times 60}{80} = \frac{8.4 \times 10^4 \times \theta}{2000}$$

$$\theta = \frac{120 \times 2000}{80 \times 84 \times 1000} \ \text{rad.} = \frac{1}{28} \text{rad.}$$

$$= 2°.05$$

PROBLEM 28.4: A hollow circular shaft, 200 mm internal diameter and 250 mm external diameter, transmits power at 150 r.p.m. The angle of twist on a length of 3 m was found to be 0°.75. Calculate the power transmitted and the maximum shear stress induced. $N = 9 \times 10^4$ MPa.

Solution:

J for the hollow circular shaft $= \frac{\pi}{2}(R^4 - r^4)$

$$= \frac{\pi}{2}\left(125^4 - 100^4\right) \ \text{mm}^4$$

$$= 2.264 \times 10^8 \ \text{mm}^4$$

$$\frac{T}{J} = \frac{N\theta}{l}$$

$$T = \frac{N\theta}{l} \cdot J$$

$$J = \frac{9 \times 10^4 \times 0.75 \times \pi}{180 \times 3000} \times \frac{2.264 \times 10^8}{10^3} \ \text{N.m}$$

$$= 88,907 \ \text{N.m}$$

Power, $P = \dfrac{2\pi n T}{60} = \dfrac{2\pi \times 150 \times 88907}{60}$ Watts

$$= 5\pi \times 88.907 \ \text{kW} = 1396 \ \text{kW}$$

$$\frac{T}{J} = \frac{2f_{s_m}}{D}$$

$$\therefore \ f_{s_m} = \frac{T}{J} \cdot \frac{D}{2} = \frac{88907 \times 1000}{2.264 \times 10^8} \times \frac{250}{2} \ \text{MPa}$$

$$= 49.1 \ \text{MPa}$$

PROBLEM 28.5: Calculate the dimensions of a hollow shaft to transmit 1500 kW at a speed of 150 r.p.m., the maximum twisting moment being 1.25 times the mean. The internal diameter of the shaft is to be 60% of the outside diameter and the maximum shear stress is limited to 50 MPa.

Solution:

J for the hollow shaft $= \dfrac{\pi}{32}(D^4 - d^4)$

Since $d = 0.6D$,

$$J = \frac{\pi}{32}\left\{D^4 - (0.6)^4 D^4\right\} = 0.8704 \times \frac{\pi D^4}{32}$$

$$P = \frac{2\pi n T_{mean}}{60}$$

$$\therefore \ T_{mean} = \frac{60P}{2\pi n} = \frac{60 \times 1500 \times 1000}{2\pi \times 150} \ \text{N.m}$$

$$= 95,493 \ \text{N.m}$$

$$T_{max} = 1.25 \times 95,493 = 119,366 \ \text{N.m}$$

$$\frac{T_{max}}{J} = \frac{2f_s}{D}$$

$$\frac{119,366 \times 10^3 \times 32}{0.8704 \times \pi D^4} = \frac{2 \times 50}{D}$$

$$= \sqrt[3]{\frac{119366 \times 1000 \times 32}{87.04 \times \pi}} \ \text{mm}$$

$$= 240.84 \ \text{mm say } 250 \ \text{mm}$$

$$d = 0.6 \times 250 = 150 \ \text{mm}$$

$\therefore$ The hollow shaft should be of internal diameter 150 mm and external diameter 250 mm.

PROBLEM 28.6: A vessel with a single propeller shaft, 300 diameter and running at 160 r.p.m., is re-engined with turbines driving two equal propeller shafts at 750 r.p.m and developing 60% more power. If the working stress for the new shafts is 10% more than that of the old shaft, find the diameter of the new shafts.

Solution:

Single shaft

$$P = \frac{2\pi n T}{60 \times 1000}, \ T \text{ being in N.mm}$$

$$T = \frac{\pi}{16} \times (300)^3 \times \underset{(MPa)}{f_s} = \frac{60 \times 1000 \ P}{2\pi \times 160}$$

$$\left(\frac{P}{f_s}\right) = \frac{\pi \times (300)^3 \times 2\pi \times 160}{16 \times 60 \times 1000} \qquad ...(i)$$

Two Shafts

$$1.6P = \frac{2 \times 2\pi \times 750}{60 \times 1000} \times \frac{\pi d^3}{16} \times 1.1 f_s$$

$$\left(\frac{P}{f_s}\right) = \frac{1.1}{1.6} \times \frac{4\pi^2 \times 750 D^3}{60 \times 1000 \times 16} \qquad ...(ii)$$

Equating (*i*) and (*ii*), and solving for *D*,

$$D = 161.2, \text{ say } 165 \text{ mm}$$

∴ The diameter of the new shafts has to be 165 mm

PROBLEM 28.7: What percentage of strength in torsion of a solid circular shaft of 100 mm diameter is lost by boring an axial hole of 50 mm?

$$T_{solid} = \frac{\pi}{16} \times (100)^3 \times f_s = \frac{\pi}{16} f_s \times 10^6$$

$$T_{hollow} = \frac{\pi}{32} \times \frac{2f_s(100^4 - 50^4)}{100}$$

$$= \frac{\pi}{16} f_s \times 0.9375 \times 10^6$$

Loss in strength $= (1 - 0.9375) \times 100\% = 6.25\%$

PROBLEM 28.8: Compare the strength of hollow and solid circular shafts whose material, weight and length are the same. What is the ratio if the external diameter of the hollow shaft is twice its internal diameter?

Solution: Let the diameter of the solid shaft be D_o, and the external and internal diameters of the hollow shaft be *D* and *d*.

$$T_{solid} = \frac{\pi D_o^3}{16} f_s$$

$$T_{hollow} = \frac{\pi}{32} \times \frac{2}{D}(D^4 - d^4) \cdot f_s$$

$$\frac{T_{hollow}}{T_{solid}} = \frac{(D^4 - d^4)}{D \cdot D_o^3}$$

Since the material, weight and length are the same, areas of section are the same.

$$\therefore \quad D_o^2 = (D^2 - d^2)$$

Let $d = \dfrac{D}{k}(k > 1)$

$$D_o^2 = D^2\left(1 - \frac{1}{k^2}\right)$$

$$\frac{D_o^2}{D_o^2} = \frac{k^2}{(k^2 - 1)} \text{ or } \frac{D}{D_o} = \frac{k}{\sqrt{k^2 - 1}}$$

$$\frac{T_{hollow}}{T_{solid}} = \frac{(D^2 + d^2)}{D_o \cdot D} = \frac{D^2\left(1 + \dfrac{1}{k^2}\right)}{D_o \cdot D} = \frac{D(k^2 + 1)}{D_o \cdot k^2}$$

$$\frac{T_{hollow}}{T_{solid}} = \frac{k}{\sqrt{(k^2 - 1)}}\frac{(k^2 + 1)}{k^2}$$

$$= \frac{k^2 + 1}{\sqrt{k^2(k^2 - 1)}} = \frac{k^2 + 1}{k\sqrt{(k^2 - 1)}}$$

If $k = 2$, $\dfrac{T_{hollow}}{T_{solid}} = \dfrac{2^2 + 1}{2 \times \sqrt{2^2 - 1}} = \dfrac{5}{2\sqrt{3}} = 1.44$

PROBLEM 28.9: Compare the weight of a solid shaft with that of a hollow shaft to transmit given power at a given speed with a given maximum shearing stress, the inside diameter of the hollow shaft being 0.6 times the outside diameter.

T_{max} is same for both the solid and the hollow shafts.

Solution: Let the diameter of the solid shaft be D_o, and the external diameter of the hollow shaft be *D*; the internal diameter is therefore 0.6D.

$$\frac{T}{J} = \frac{2f_s}{D}$$

or $\quad \dfrac{T}{f_s} = \dfrac{2J}{D}$

Since f_s is also the same for both, $\left(\dfrac{T}{f_s}\right)$ is the same for both.

$\therefore \quad \dfrac{2J}{D}$ or $\dfrac{J}{D}$ is the same for both.

$$\frac{\pi D_o^4}{32 D_o} = \frac{\pi}{32}\frac{\{D^4 - (0.6)^4 D^4\}}{D}$$

$$D_o^3 = D^3(1 - 0.1296) = 0.8704 D^3$$

$$D = \sqrt[3]{1.149}\,D_o = 1.0474 D_o$$

Area of solid shaft, $A_s = \dfrac{\pi}{4} D_o^2$

Area of hollow shaft, $A_h = \dfrac{\pi}{4}(D^2 - 0.36 D^2)$

$$= 0.64 D^2 \times \frac{\pi}{4}$$

$$= \frac{\pi}{4} \times 0.64 \times 1.097 \times D_o^2$$

$$= \frac{\pi}{4} \times 0.702 D_o^2$$

The weights bear the same ratio as the sectional area

$$W_s : W_h = D_o^2 : 0.702 D_o^2 = 1.4243 : 1$$

PROBLEM 28.10: A 100 mm diameter shaft is subjected to a bending moment M and a twisting moment T. The maximum principal stress due to these 90 MPa. If the maximum bending stress due to M is equal to the maximum shear stress due to T, evaluate M & T.

Solution: Since f_n due to M equals f_s due to T,

$$\frac{4M}{\pi R^3} = \frac{2T}{\pi R^3} \therefore T = 2M \qquad \dots(i)$$

Equivalent bending moment,

$$M_E = \frac{1}{2}\left[M + \sqrt{M^2 + T^2}\right]$$

$$\frac{4 M_E}{\pi \times (50)^3} = 90$$

$$\therefore \quad M_E = \frac{\pi \times (50)^3 \times 90}{4} \text{ N.mm}$$

$$= 8835.73 \text{ N.m}$$

or $\quad M + \sqrt{M^2 + T^2} = 17671.46 \qquad \dots(ii)$

Solving (i) and (ii), we get

$$M = 5,460.8 \text{ N.m}$$

and $\quad T = 10,921.6 \text{ N.m}$

PROBLEM 28.11: A hollow shaft is subjected to a torque of 20 kN.m and a bending moment of 10 kN.m. The internal diameter of the shaft is half the external. If the greatest shear stress is not to exceed 60 MPa, find the size of the shaft.

Solution: If the external diameter is D mm,

$$J = \frac{\pi D^4 (1 - 1/16)}{32} = \frac{\pi}{32} \times \frac{15}{16} D^4$$

Equivalent Twisting Moment for shear,

$$T_{E_s} = \sqrt{M^2 + T^2} = \sqrt{10^2 + 20^2} = 22.36 \text{ kN.m}$$

$$f_{s_{max}} = \frac{D\, T_{E_s}}{2J}$$

$$= \frac{D}{2} \times \frac{22.36 \times 10^6 \times 32 \times 16}{\pi \times 15 D^4} = 60$$

or $\quad D = \sqrt[3]{\dfrac{22.36 \times 256 \times 10^6}{900\pi}}$ mm $= 126.5$ mm,

Say 130 mm (external) and 65 mm (internal)

PROBLEM 28.12: A solid shaft, 400 mm in diameter, is subjected to a bending moment of 360 kN.m and a twisting moment of 150 kN.m. Find the maximum and minimum principal stresses and also the maximum tensile strain. $E = 2 \times 10^5$ MPa and $\mu = 0.3$.

Solution:

$M = 360$ kN.m $= 360 \times 10^6$ N.mm; $T = 150$ kN.m $= 150 \times 10^6$ N.mm

Bending stress, $f_n = \dfrac{4M}{\pi R^3}$

$$= \frac{4 \times 360 \times 10^6}{\pi \times (200)^3} = 57.2 \text{ MPa}$$

Shear stress, $f_s = \dfrac{2T}{\pi R^3}$

$$= \dfrac{2 \times 150 \times 10^6}{\pi (200)^3} = 12 \text{ MPa}$$

Maximum principal stress,

$$\sigma_x = \dfrac{f_n}{2} + \sqrt{\left(\dfrac{f_n}{2}\right)^2 + f_s^2}$$

$$= \dfrac{57.2}{2} + \sqrt{\left(\dfrac{57.2}{2}\right)^2 + 12^2}$$

$$= 59.6 \text{ MPa}$$

This will be tensile or compressive, depending upon the position of the element with respect to the Neutral axis and the sign of the bending moment.

Minimum principal stress,

$$\sigma_y = \dfrac{f_n}{2} - \sqrt{\left(\dfrac{f_n}{2}\right)^2 + f_s^2}$$

$$= \dfrac{57.2}{2} - \sqrt{\left(\dfrac{57.2}{2}\right)^2 + 12^2}$$

$$= 7.4 \text{ MPa (opposite in sign to the maximum principal stress)}$$

Maximum shear stress

$$= \left(\dfrac{59.6 + 7.4}{2}\right) = 33.5 \text{ MPa}$$

Maximum tensile strain (assuming the maximum principal stress to be tensile),

$$e_{max} = \dfrac{1}{E}(\sigma_x - \mu\sigma_y)$$

$$= \dfrac{1}{2 \times 10^5}(59.6 + 0.3 \times 7.4) = 3.091 \times 10^{-4}$$

PROBLEM 28.13: A shaft is required for an engine which indicates a power of 75 kW at 80 r.p.m. The maximum torque exceeds the mean by 80%. The shaft carries a fly wheel weighing 900 kN midway between the bearings, 5 m apart. The steam pressure causes a bending moment equal to 0.8 times the mean torque. If the maximum tensile stress is not to exceed 60 MPa, determine the diameter of the shaft required.

Solution:

$$T_{mean} = \dfrac{60P}{2\pi n} = \dfrac{60 \times 75 \times 1000}{2\pi \times 80} = 8{,}952.5 \text{ N.m}$$

$$T_{max} = 1.8\, T_{mean} = 1.8 \times 8{,}952.5$$

$$= 16{,}114.5 \text{ N.m}$$

B.M. due to steam pressure $= 0.8\, T_{mean}$
$$= 7{,}162 \text{ N.m}$$

B.M. due to fly wheel

$$= \dfrac{Wl}{4} = \dfrac{900 \times 5 \times 1000}{4} \text{ N.m}$$

$$= 11{,}25000 \text{ N.m}$$

Total B.M., $M = 11{,}32{,}162 \text{ N.m}$

$$T = 16{,}114.5 \text{ N.m}$$

Equivalent twisting moment,

$$T_E = M + \sqrt{M^2 + T^2}$$

$$= 11{,}32{,}162 + \sqrt{(1132162)^2 + (16114.5)^2}$$

$$= 11{,}32{,}277 \text{ N.m}$$

$$60 = f_p = \dfrac{2T_E}{\pi R^3} \text{ or } R = \sqrt[3]{\dfrac{2T_E}{\pi f_p}}$$

$$= \sqrt[3]{\dfrac{2 \times 1132277 \times 1000}{\pi \times 60}} \text{mm}$$

$$= 230 \text{ mm}$$

∴ Diameter of the shaft, $D = 460 \text{ mm}$

PROBLEM 28.14: A propeller shaft 200 mm in diameter transmits 2500 kW at 250 r.p.m. The propeller weighs 50 kN and is carried by the shaft overhanging the support by 500 mm. The propeller thrust is 150 kN. Calculate the maximum direct stress.

Solution:

Torque, $T = \dfrac{60P}{2\pi n} = \dfrac{60 \times 2500 \times 1000}{2\pi \times 250}$

$$= 95{,}493 \text{ N.m}$$

Bending Moment, $M = W \cdot l$
(due to cantilever action)

$$= \frac{50 \times 1000 \times 500}{1000} \text{ N.m} = 25,000 \text{ N.m}$$

Propeller thrust, $P = 150 \text{ kN} = 150,000 \text{ N}$

Direct compressive stress due to P,

$$\sigma = \frac{P}{A} = \frac{150,000}{\pi \times (100)^2} = 4.78 \text{ MPa}$$

Bending stress,

$$f_n = \frac{4M}{\pi R^3} = \frac{4 \times 25,000 \times 1000}{\pi \times (100)^3} = 31.83 \text{ MPa}$$

Shear stress,

$$f_s = \frac{2T}{\pi R^3} = \frac{2 \times 95,493 \times 1000}{\pi \times (100)^3} = 60.79 \text{ MPa}$$

Total normal compressive stress,
$$f = 31.83 + 4.79 = 36.61 \text{ MPa}$$

Maximum principal stress (or maximum direct stress),

$$\sigma_x = \frac{36.61}{2} + \sqrt{\left(\frac{36.61}{2}\right)^2 + (60.79)^2}$$

$$= 81.8 \text{ MPa}$$

PROBLEM 28.15: A shaft, 150 mm diameter and 1.5 m long, is subjected to a maximum torsional shearing stress of 50 MPa. Determine the strain energy stored in the shaft. The modulus of rigidity of the material of the shaft is 0.8×10^5 MPa.

Solution:

Diameter, $D = 150$ mm

Length, $l = 1.5$ m $= 1500$ mm

Volume, V of the shaft

$$= \frac{\pi D^2}{4} \times l = \frac{\pi}{4} \times (150)^2 \times 1500 \text{ mm}^3$$

$N = 0.8 \times 10^5$ MPa

Strain energy stored in the shaft,

$$U = \frac{f_s^2}{4N} \times V$$

$$= \frac{50 \times 50}{4 \times 8 \times 10^4} \pi \times (75)^2 \times 1500 \text{ N.mm}$$

$$= \frac{25 \times 100 \times \pi \times 5625 \times 15 \times 100}{32 \times 100 \times 100 \times 1000} \text{ N.m}$$

or $U = 207$ N.m

PROBLEM 28.16: A close-coiled helical spring is made from steel wire 6 mm diameter, and there are ten free coils having a mean diameter of 75 mm. The spring carries an axial load of 100 N. Determine the maximum shear stress, the deflection, and the stiffness. $N = 0.84 \times 10^5$ MPa.

Solution:

$d = 6$ mm; $n = 10$; $R = 37.5$ mm;

$W = 100$ N.

Maximum shear stress,

$$f_s = \frac{16WR}{\pi d^3} = \frac{16 \times 100 \times 37.5}{\pi \times 6^3} = 88.4 \text{ MPa}$$

Axial deflection,

$$\delta = \frac{64WR^3 n}{Nd^4} = \frac{64 \times 100 \times (37.5)^3 \times 10}{0.84 \times 10^5 \times 6^3}$$

$$= 31 \text{ mm}$$

Stiffness of the spring

$$= \frac{W}{\delta} = \frac{100}{31} = 3.24 \text{ N/mm}$$

PROBLEM 28.17: A close-coiled helical spring is made of 6 mm steel wire. Its stiffness is 4 N/mm. If the mean radius of the coils is 37.5 mm, what is the necessary length of wire? $N = 0.84 \times 10^5$ MPa.

Solution:

$d = 6$ mm; $N = 0.84 \times 10^5$ MPa; $\delta = 1$ mm;

$W = 4$ N; $R = 37.5$ mm;

$$1 = \frac{64 \times 4 \times (37.5)^3 \times n}{0.84 \times 10^5 \times 6^4}$$

$$n \approx 8$$

Length of wire $= 2\pi R n = 2 \times \pi \times 37.5 \times 8$

$$= 1890 \text{ mm}$$

PROBLEM 28.18: A close-coiled helical spring with 10 active coils is required to carry a load of 9 kN when compressed axially by 20 mm. Find the mean radius of the coils and the diameter of the spring wire for a maximum shear stress of 420 MPa. $N = 0.8 \times 10^5$ MPa.

Solution:

$n = 10$; $W = 900$ N; $\delta = 20$ mm; $f_s = 420$ MPa

$$900R = \frac{\pi}{16} \cdot d^3 \times 420 \qquad \ldots(i)$$

$$20 = \frac{64 \times 9000 \times R^3 \times 10}{0.8 \times 10^5 d^4} \qquad \ldots(ii)$$

From (i), $\dfrac{R}{d^3} = \dfrac{420\pi}{16 \times 9000}$

$$= 9.163 \times 10^{-3} \therefore \frac{R^3}{d^9} = 7.69 \times 10^{-7}$$

From (ii), $\dfrac{R^3}{d^4} = \dfrac{20 \times 0.8 \times 10^5}{64 \times 9000 \times 10} = 0.278$

$\therefore \quad d^5 = 361219.5$ or $d = 12.92$ mm or 13 mm
and $R = 19.76$ mm or 20 mm

PROBLEM 28.19: Find the necessary weight of a close-coiled helical spring of round wire to withstand a load of 30 kN with a deflection of 25 mm. Maximum shear stress = 385 MPa. $N = 0.8 \times 10^5$ MPa. Unit weight of steel = 7.8×210^{-4} N.mm^3

Solution:

$W = 30,000$ N; $\delta = 25$ mm; $f_s = 385$ MPa; $N = 0.8 \times 10^5$ MPa

$$25 = \frac{64 \times 30000 \times R^3 n}{0.8 \times 10^5 \times d^4} \qquad \ldots(i)$$

$$30000R = \frac{\pi d^3}{16} \times 385 \qquad \ldots(ii)$$

Volume of the wire $= 2\pi R n \cdot \dfrac{\pi d^2}{4}$

From (ii), $R = \dfrac{\pi d^3 \times 385}{30000 \times 16} = 2.52 \times 10^{-3} d^3$

$\therefore$ Volume $= 2\pi n \times \dfrac{\pi d^2}{4} \times \dfrac{\pi d^3 \times 385}{30000 \times 16}$

$$= 0.012435 n d^5$$

Substituting for R in (i)

$$25 = \frac{64 \times 30000}{0.8 \times 10^5} \times \frac{n}{d^4} \times (2.52 \times 10^{-3})^3 \times d^9$$

$$= 3.84 \times 10^{-7} n d^5$$

$\therefore \quad n d^5 = \dfrac{25}{3.84 \times 10^{-7}} = 6.51 \times 10^7$

$\therefore$ Volume $= 0.012435 \times 6.51 \times 10^7$

$$= 8.096 \times 10^5$$

Weight of the spring wire

$$= 8.096 \times 10^5 \times 7.8 \times 10^{-4} \text{ N} = 631.5 \text{ N}$$

PROBLEM 28.20: A close-coiled helical spring, made of 5 mm wire, and having a mean radius of 18.75 mm, joins two shafts. The effective number of coils between the shafts is 15, and 750 Watts of power is transmitted through the spring at 1000 r.p.m. Calculate the relative axial twist (in degrees) between the ends of the spring, and also the bending stress in the material. $E = 2 \times 10^5$ MPa.

Solution:

$d = 5$ mm; $R = 18.75$ mm; $n = 15$;
r.p.m = 1000, $P = 750$ W

$$750 = \frac{2\pi \times 1000 \times T}{60} \therefore T = 7.162 \text{ N.m}$$

$$= 7,162 \text{ N.mm}$$

The axial twisting moment acts as a bending moment, M, on the cross-section of the spring wire.

Angle of twist at the free end,

$$\phi = \frac{Ml}{EI} = \frac{2\pi R n M}{EI}$$

$$= \frac{2\pi \times 18.75 \times 15 \times 7162 \times 64}{2 \times 10^5 \times \pi \times 5^4} \text{ rad}$$

$$\phi = 2.063 \text{ rad}$$

Maximum bending stress in the spring wire—

$$f = \frac{32 \times 7162}{\pi \times 5^3} \text{ N/mm}^2$$

$$= 583.6 \text{ MPa}$$

PROBLEM 28.21: A close-coiled helical spring of a wire of circular section has coils of 37.5 mm mean radius. When loaded axially with 250 N, it is found to extend 135 mm; when subjected to an axial torque of 3500 N.mm, an angular rotation 60°occurs at the free end. Determine the Poisson's ratio of the material.

Solution:

$$\delta = \frac{64WR^3 n}{Nd^4}, \text{ under axial force.}$$

$$\therefore \quad N = \frac{64WR^3 n}{\delta d^4} = \frac{64 \times 250 \times (37.5)^3}{135} \times \frac{n}{d^4}$$

$$= (625 \times 10^4)\left(\frac{n}{d^4}\right)$$

$$\phi = \frac{Ml}{EI}, \text{ under axial torque.}$$

$$\therefore \quad E = \frac{Ml}{\phi I} = \frac{3500 \times 2\pi \times 37.5 \times n \times 180 \times 64}{60 \times \pi \times \pi \times d^4}$$

$$= (1604.3 \times 10^4)\left(\frac{n}{d^4}\right)$$

$$E = 2N(1 + 1/m)$$

$$\frac{E}{2N} = (1 + 1/m) = \frac{1604.3}{1250} = 1.2834$$

$$\therefore \quad \text{Poisson's ratio, } 1/m = 0.2834$$

PROBLEM 28.22: If the stiffness of a close-coiled helical spring is to be 1 N/mm in compression and the maximum shear stress is not to exceed 120 MPa at a maximum axial load of 40 N, find the diameter of the wire, mean coil radius, and the number of coils. The solid length of the spring is 40 mm. $N = 0.8 \times 10^5$ MPa.

Solution:

$$s = 1 \text{ N/mm}; f_s = 120 \text{ MPa}; W = 40 \text{ N}$$

$$\text{Solid length} = 40 \text{ mm} \therefore n = \frac{40}{d} \qquad \ldots(i)$$

$$1 = \frac{64 \times 1 \times R^3 n}{Nd^4}$$

$$= \frac{64 \times 1 \times R^3 \times 40}{0.8 \times 10^5 \times d^5}, \text{ using } (i)$$

$$\frac{R^3}{d^5} = \frac{0.8 \times 10^5}{64 \times 40} = \frac{1000}{32}$$

$$= \frac{62.5}{2} = 31.25 \qquad \ldots(ii)$$

$$40 \times R = \frac{\pi d^3}{16} \times 120$$

$$\frac{R}{d^3} = \frac{\pi \times 120}{16 \times 40} = \frac{3\pi}{16} \qquad \ldots(iii)$$

$$\frac{R^3}{d^9} = \frac{27\pi^3}{256 \times 16} \qquad \ldots(iv)$$

Dividing (ii) by (iv),

$$d^4 = \frac{62.5}{2} \times \frac{16 \times 256}{27 \times \pi^3} = 152.9$$

whence $d = 3.5$ mm; $R = \dfrac{3\pi}{16} \times (3.5)^3 = 25$ mm

$$n = \frac{40}{3.5} = 11.43, \text{ say } 12$$

$\therefore$ Diameter of the wire $= 3.5$ mm;
Mean radius of coils $= 25$ mm;
Number of coils $= 12$

PROBLEM 28.23: Two helical springs of the same material and of equal circular cross-sections and lengths, assembled as shown in Fig. 28.6, are

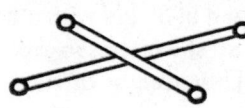

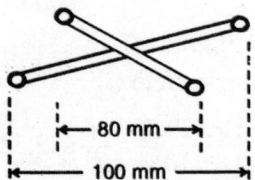

Fig. 28.6: Parallel Springs (Illustrative Problem 28.23)

compressed between parallel planes. Determine the maximum stress in each spring if wire diameter is 10 mm and the axial load is 500 N.

Solution: Referring to Fig. 28.6,

Since $\delta = \dfrac{64WR^3n}{Nd^4}$

W will get distributed in the inverse ratio of the cubes of the mean radii of the coils.

$\therefore\ W_o : W_i = 80^3 : 100^3 = 64 : 125$

Maximum stress in the outer spring,

$$f_{s_o} = \frac{16W_o R_o}{\pi d^3} = 16 \times \frac{500}{\pi} \times \frac{64}{189} \times \frac{50}{10^3}$$

$$= 43 \text{ MPa}$$

Maximum stress in the inner spring,

$$f_{s_i} = \frac{16W_i R_i}{\pi d^3} = 16 \times \frac{500}{\pi} \times \frac{125}{189} \times \frac{40}{10^3}$$

$$= 67.5 \text{ MPa}$$

PROBLEM 28.24: Two springs, one inside the other, not touching each other, have the same overall length and same number of turns. The coil diameter of the outer spring is twice that of the inner. The outer spring is of steel ($N = 0.8 \times 10^5$ MPa) and the inner is of bronze ($N = 0.4 \times 10^4$ MPa). If the two springs are to be designed to share the load equally, what is the ratio of the wire diameters? In this case, what is the ratio of their stresses?

Solution: Since it is the case of parallel springs, δ is same for both; also, since the load is to be shared equally, W is same for both.

$\therefore \qquad \delta = \dfrac{64WR_o^3 n}{N_o d_o^4} = \dfrac{64WR_i^3 n}{N_i d_i^4}$

$$\left(\frac{d_o}{d_i}\right)^4 = \left(\frac{R_o}{R_i}\right)^3 \times \frac{N_i}{N_o} = 2^3 \times \frac{1}{2} = 4$$

$\therefore \qquad \dfrac{d_o}{d_i} = \sqrt{2}$

$$f_{s_o} = \frac{16WR_o}{\pi d_o^3}; \ f_{s_i} = \frac{16WR_i}{\pi d_i^3}$$

$$\frac{f_{s_o}}{f_{s_i}} = \left(\frac{R_o}{R_i}\right)\left(\frac{d_i}{d_o}\right)^3 = 2 \times \frac{1}{2\sqrt{2}} = \frac{1}{\sqrt{2}}$$

PROBLEM 28.25: In an open-coil helical spring of 10 coils, the stresses due to bending and twisting are 100 MPa and 105 MPa respectively when the spring is axially loaded. Assuming the mean radius of the coils to be 4 times the wire diameter, find the maximum permissible axial load and the diameter of the wire for a maximum extension of 20 mm. $E = 2 \times 10^5$ MPa and $N = 0.8 \times 10^5$ MPa.

Solution:

$$n = 10 \qquad R = 4d$$

$$M = f \cdot \frac{I}{c} = \frac{100 \times \pi d^4}{64(d/2)} \text{ N.mm}$$

$$= 3.125\pi d^5 \text{ N.mm}$$

$$T = \frac{f_s}{r} \cdot J = \frac{105 \times \pi d^4}{32(d/2)} = 6.563\pi d^3 \text{ N.mm}$$

$$\tan\alpha = \frac{M}{T} = \frac{3.125}{6.563} = 0.476$$

$$\sin\alpha = 0.430; \cos\alpha = 0.903; \sec\alpha = 1.1076$$

$$\delta = \frac{64WR^3 n}{d^4}\sec\alpha\left(\frac{\cos^2\alpha}{N} + \frac{2\sin^2\alpha}{E}\right)$$

$$20 = \frac{60W \times 64d^3 \times 10}{d^4}$$

$$\times 1.1076\left[\frac{(0.903)^2}{0.8 \times 10^5} + \frac{2 \times (0.430)^2}{2 \times 10^5}\right]$$

$$\left(\frac{W}{d}\right) = \frac{2 \times 10^5}{64 \times 64 \times 1.1076 \times 1.20416} = 36.61 \ \dots(i)$$

$$M = WR\sin\alpha = W \times 4d \times 0.43$$

But $M = 3.125\pi d^3$

$\therefore\ 1.72Wd = 3.125\pi d^3$

$\therefore \qquad \left(\dfrac{W}{d^2}\right) = \dfrac{3.125\pi}{1.72} = 5.708 \qquad \dots(ii)$

Dividing (i) by (ii),

$$d = \frac{36.61}{5.708} = 6.4 \text{ mm}$$

$$W = 36.61 \times 6.4 \text{ N} = 234 \text{ N}$$

PRACTICE PROBLEMS

28.1 Find the power that could be transmitted by a solid shaft, 100 mm diameter, rotating at 200 r.p.m., if the maximum shear stress is limited to 20 N/mm^2.

28.2 Determine the length of a steel shaft of diameter 50 mm which develops a maximum shear stress of 100 MPa for an angle of twist of 6°. Assume $N = 0.8 \times 10^5$ MPa.

28.3 The external and internal diameters of a hollow shaft are 300 mm and 200 mm, respectively. Determine the power that can be transmitted at a speed of 90 r.p.m. if the shear stress is not to exceed 50 N/mm^2.

28.4 Compare the torque transmitted by a solid shaft of diameter D with that transmitted by a hollow shaft of the same material with mean diameter D. The weight per unit length is same for the two shafts.

28.5 A steel shaft has to transmit 75 kW of power at 200 r.p.m. What should be the diameter of the shaft, if the maximum shear stress is not to exceed 80 MPa and the angle of twist is not to exceed 1° over a length of 3 m? Assume $N = 0.8 \times 10^5$ MPa.

28.6 A solid steel shaft, 5 m long, is to transmit 150 kW of power at 160 r.p.m. The angle of twist in this length is not to exceed 2°. The maximum allowable shear stress is 70 MPa and rigidity modulus is 8×10^4 MPa. Find the required size of the shaft if it is to be solid circular in section. If this shaft is replaced with a hollow shaft with external diameter 1.25 times the internal diameter, determine the size of this shaft. What is the saving in weight?

38.7 Compare the torsional strengths of hollow and solid circular shafts of the same material, length, and weight, the external diameter of the hollow shaft being k times its internal diameter. What is this ratio if $k = 1.25$ and $k = 1.50$?

28.8 Compare the weight of a hollow shaft with that of a solid one of the same material to transmit a given power at a given speed with a given maximum shearing stress, the outside diameter of the hollow shaft being k times the internal. What will be this ratio if $k = 1.5$ and $k = 2$?

28.9 A steel shaft is supported on bearings 4 m apart and it carries a pulley weighing 40 kN at mid-span. The shaft is required to transmit 75 kW at 120 r.p.m. Calculate the diameter of the solid shaft if the maximum principal stress is limited to 100 MPa.

28.10 A circular shaft is supported in bearings 3.6 m apart and transmits 90 kW at 150 r.p.m. At the middle of the shaft a pulley exerts a transverse load of 36 kN. Determine a suitable diameter if (*i*) the maximum direct stress is not to exceed 100 MPa, and (*ii*) the maximum shear stress is not to exceed 50 MPa.

28.11 A shaft of 50 mm diameter carries a B.M. of 300 N.m. What torque can be applied if (*i*) the maximum normal stress is not to exceed 110 MPa and (*ii*) the maximum shear stress is not to exceed 80 MPa.

28.12 A solid shaft, 200 mm diameter, transmits 75 kW at 300 r.p.m. It is also subjected to a bending moment of 10 kN.m and an axial thrust of 50 kN. Calculate the maximum principal stress induced.

28.13 A shaft is required for an engine transmitting 450 kW of power running at 100 r.p.m. The maximum torque exceeds the mean by 50%. The main bearings are 5 m apart and the shaft carries a fly wheel of 100 kN mid way between the bearings. Find the diameter of the shaft if the maximum tensile stress is not to exceed 70 MPa.

28.14 A shaft, 100 mm diameter and 4 m long, is freely supported and carries a pulley of weight 2 kN at the middle. It transmits 45 kW of power at 120 r.p.m. Determine the end thrust which can act simultaneously if the maximum shear stress is limited to 30 MPa.

28.15 A shaft, 200 mm diameter and 2 m long, is subjected to a maximum torsional shearing stress of 60 MPa. Determine the strain energy stored in the shaft. $N = 0.84 \times 10^5$ MPa.

28.16 Determine the maximum stress and the extension of a helical spring given the following:

Axial load = 1200 N; Radius of spring = 100 mm; Diameter of spring wire = 20 mm; Number of coils = 20; $N = 0.8 \times 10^5$ MPa.

28.17 A close-coiled helical spring is made of steel wire 5 mm diameter. It has 10 coils of 60 mm mean diameter. Compute the maximum load if the shear stress is limited to 500 MPa, and find the corresponding elongation. $N = 0.8 \times 10^5$ MPa.

28.18 A close-coiled helical spring is made of 6 mm steel wire and is to have a stiffness of 4 N/mm. $N = 0.84 \times 10^5$ MPa. What is the length of wire necessary?

28.19 A close-coiled helical spring has a mean radius of 5 times the wire diameter. Find the wire diameter and number of turns to carry an axial load of 1 kN. The spring constant is 20 N/mm and the maximum shear stress is not to exceed 150 MPa. Also compute the torsional stiffness of the spring. $N = 0.8 \times 10^5$ MPa.

28.20 A close-coiled helical spring is made up of coils of 50 mm mean radius from a wire of 10 mm diameter. The stiffness is 4 N/mm. Compute the number of turns, $N = 0.8 \times 10^5$ MPa.

28.21 A close-coiled helical spring, made of 5 mm diameter wire has coils of mean diameter of 40 mm. The number of coils is 15. The axial torque applied is 7.2 N.m. Determine the bending stress in the wire and the angle of twist in radians.

28.22 A close-coiled spring of circular wire has coils of 75 mm mean diameter. It extends 135 mm under an axial load of 250 N, and the free end rotates through 60° under an axial torque of 3.6 N.m. Determine the Poisson's ratio of the material.

28.23 A composite spring has two close-coiled helical springs, arranged in series, each spring having 12 coils of 30 mm mean diameter. Find the diameter of wire of the second spring if that of the first is 3 mm and the stiffness of the composite spring 720 N/m. Estimate the maximum load that can be carried if the maximum shear stress is 200 MPa and $N = 0.8 \times 10^5$ MPa.

(**Hint:** The load is the same for both springs while the total extension is the sum of those values for the individual springs).

28.24 Two close-coiled springs are placed one inside the other. The mean diameter of the springs are 50 mm and 40 mm, and the wire diameter are 5 mm and 4 mm. The number of coils are 10 and 8. The free lengths are 150 mm and 120 mm. If the springs are subjected to an axial load of 750 N, determine the load carried by each spring and the axial deformation, $N = 0.8 \times 10^5$ MPa.

24.25 An open-coil helical spring has a helix angle of 30°, and is made of steel wire 10 mm diameter. The number of coils is 12 with a mean radius of 50 mm. Determine the axial deflection under an axial force of 500 N. What are the maximum stresses in shear and in bending? $E = 2 \times 10^5$ MPa and $N = 0.8 \times 10^5$ MPa.

Chapter 29

Combined Flexure and Direct Stress and Long Columns

29.1 INTRODUCTION

There are several situations in engineering practice wherein flexural stresses will be accompanied by direct stresses. For example, loads inclined to the axis of the member and an eccentric load acting on a short vertical member do cause a combination of flexural and direct stresses. Analysis and design of several civil engineering structures like masonry walls, chimneys, retaining walls, and dams involve these principles.

An extension of the principle of eccentric loads on short vertical members or 'struts', is the stability aspect of long vertical members or long columns subjected to even axial loads, let alone eccentric loads.

A detailed consideration of all these and related concepts is the subject matter of this chapter.

29.2 COMBINED FLEXURE AND DIRECT STRESS

Let a short prismatic member be vertical and fixed at the bottom, and subjected to an inclined force F, parallel to one of the principal planes of bending (Fig. 29.1).

F may be resolved into the transverse component Q and an axial thrust P, as shown. It is assumed that the member is relatively stiff with a deflection so small that it can be neglected.

The resultant stress at any point is obtained by superimposing the compressive stress due to the axial force P on the bending stress produced by the transverse force Q.

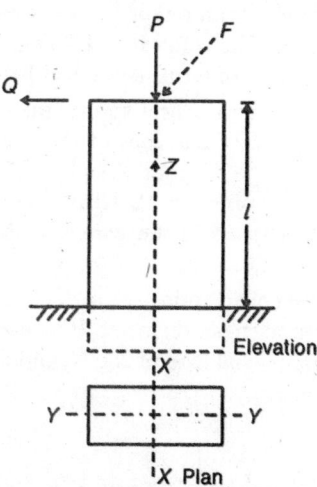

Fig. 29.1: Short Prismatic Member Subjected to an Inclined Force

The compressive stress caused by P is constant for all cross-sections of the member and is equal to P/A, A being the area of cross-section. The bending stress $M \cdot c/I_x$ depends upon the moment, which increases from zero at the top to a maximum of $Q \cdot l$ at the bottom. (I_x is the second moment of the area about XX).

The stress at the built-in end for a point at a distance c from the x-axis (about which bending occurs) is

$$p_z = \frac{P}{A} \pm \frac{Mc}{I_x} \qquad \text{...(Eq. 29.1)}$$

(The negative sign is used if the point lies to the side remote from Q and positive sign is used when it lies to the same side of Q).

The resultant stress is compressive or tensile according as it is positive or negative.

Thus, Resultant stress = Direct stress ± Bending stress

$$(p_z) = (p_d) \pm (p_b) \qquad \text{...(Eq. 29.2)}$$

The same effect may occur due to an eccentric load on a short vertical member or a strut, the length of which is small relative to its lateral dimensions.

Let the strut be subjected to a load, P, parallel to the axis, but at an eccentricity, e (Fig. 29.2).

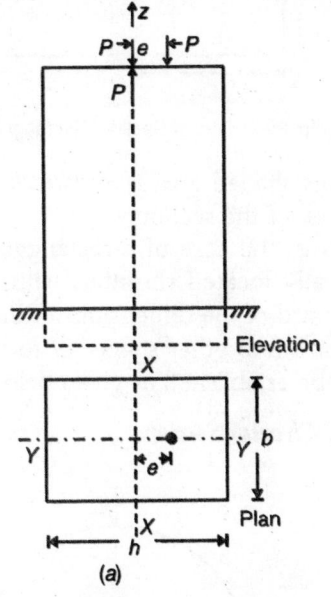

(a)

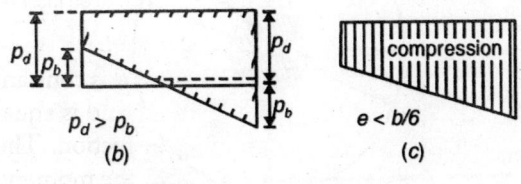

$p_d > p_b$

(b)

$e < b/6$

(c)

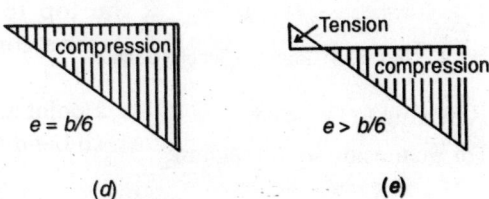

$e = b/6$

(d)

$e > b/6$

(e)

Fig. 29.2: Eccentric Load on a Strut

The load is assumed to act one of the two principal axes, say YY in this case, the eccentricity being about the axis XX. By introducing two forces, P, equal and opposite at the axis, the system reduces to an axial force, P, and a couple $M(= P \cdot e)$, as shown in the figure.

Now, the principle of superposition may be used to determine the resultant stress at any point in any cross-section. This is applicable only in elastic problems where *deformations are small*.

If the point lies at a distance y from the neutral axis, which is XX in this case, the resultant stress, p_z is given by

$$p_z = p_d \pm p_b = \frac{P}{A} \pm \frac{Mc}{I_x} = \frac{P}{A} \pm \frac{P \cdot e \cdot c}{I_x} \text{...(Eq. 29.3)}$$

Since c, with respect to x, is $\dfrac{h}{2}$,

$$p_z = \frac{P}{A} \pm \frac{P \cdot e \cdot h}{2I_x} \qquad \text{...(Eq. 29.4)}$$

For a rectangular section, $b \times h$, substituting for I_x and A,

$$p_z = \frac{P}{bh} \pm \frac{12 P \cdot e \cdot c}{bh^3}$$

$$= \frac{P}{bh} \pm \frac{6P \cdot e}{bh^2} \quad \text{(since } c = h/2)$$

or $\qquad p_z = \dfrac{P}{bh}\left[1 \pm \dfrac{6e}{h}\right] \qquad$ ...(Eq. 29.5)

The distribution of stress across the width h is shown in Fig. 29.2 (c), (d) and (e) for the conditions $e < h/6$, $e = h/6$, and $e > h/6$, respectively.

29.3 CORE OF A SECTION

In Eq. 29.2, direct stress p_d, for the case of the strut under consideration is compressive at all points of the cross-section, while the bending stress, p_b, is compressive or tensile, depending upon the location of the point with respect to the neutral axis in the cross-section.

Thus the resultant stress, p_z, is compressive in the zone in which p_b is compressive, and may be compressive or tensile in the zone in which p_b is tensile. This depends upon the relative values p_d and p_b.

If $p_b \leq p_d$, no tensile stress occurs any where in the cross-section. This condition becomes very significant in the case of materials which are relatively weak in tension, e.g., masonry and plain concrete.

From Eq. 29.3, this condition for no-tension may be written as,

$$\frac{P \cdot e_x \cdot c_x}{I_x} \leq \frac{P}{A},$$

e_x and c_x being, specifically, the values of e and c with respect to the x-axis.

or

$$\frac{P \cdot e_x \cdot c_x}{Ak_x^2} \leq \frac{P}{A}$$

where k_x is the radius of gyration of the section about x-axis.

or

$$e_x \leq \frac{k_x^2}{c_x}$$

In the limiting case, the maximum permissible eccentricity for "no-tension",

$$e_x = \frac{k_x^2}{c_x} \qquad \qquad \text{...(Eq. 29.6)}$$

this being called the "limiting eccentricity"

$$(= I_x / A \cdot c_x)$$

Similarly, when the eccentricity occurs about y-axis, the load being on the x-axis, the limiting eccentricity, e_y, is given by

$$e_y = \frac{k_y^2}{c_y} \left(= \frac{I_y}{A \cdot c_y} \right)$$

Specifically, for a rectangular section, $b \times h$,

$$e_x = \frac{bh^3 \times 2}{12 \times bh \times h} = \frac{h}{6}$$

Since the eccentricity can occur on to any side of the axis, the load must act within the middle-third of the section. This is called the "Middle Third Rule", which is well known for a long time, in the design of masonry structures.

If the logic is extended with respect to eccentricity about y-axis, the limiting eccentricity, $e_y = b/6$, and the load must act only within the middle-third of the dimension 'b'.

In case the load acts at a point which lies neither on the x-axis nor on the y-axis, eccentricity occurs about both the principal axes simultaneously; in this case, the load must act within the "rhombus" (shaded zone) marked (Fig. 29.3).

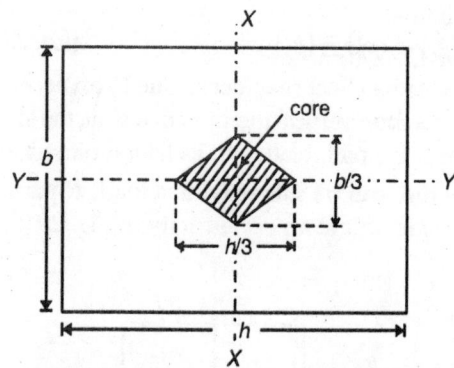

Fig. 29.3: Core of Section (Rectangular Section)

This shaded area is known as the 'core', or 'kernel' of the section.

Thus, the core of a rectangular section is a centrally located rhombus with the diagonals one-third of the dimensions of the section.

For a few other shapes of sections, the core may be established as given below.

Solid Circular Section

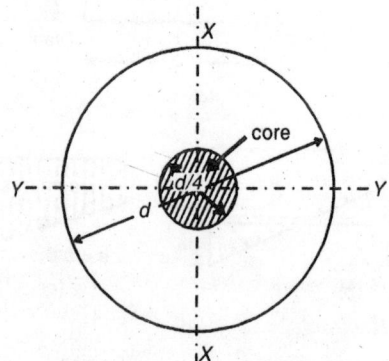

Fig. 29.4: Core of a Circular Section

Referring to Fig. 29.4,

For no tension in the section,

$$\frac{P}{A} = \frac{P \cdot e \cdot c}{I}$$

$$\frac{4P}{\pi d^2} = \frac{P \cdot e(d/2)}{(\pi d^4/64)}$$

whence $e = \dfrac{d}{8}$ (d is the diameter)

The core of the section is obviously a circle with diameter $d/4$.

Hollow Circular Section

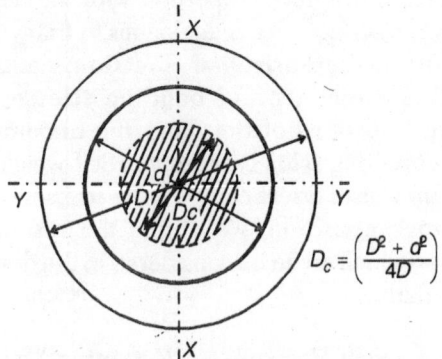

Fig. 29.5: Hollow Circular Section—Core

Referring to Fig. 29.5 for no tension in the section,

$$\frac{P}{A} = \frac{P \cdot e \cdot c}{I}$$

$$\frac{4P}{\pi(D^2 - d^2)} = \frac{P \cdot e(D/2)}{\{\pi(D^4 - d^4)/64\}}$$

whence $e = \dfrac{(D^2 + d^2)}{8D}$

The core is obviously a circle with a diameter

$$\left(\frac{D^2 + d^2}{4D}\right)$$

Note: This is always less than $D/2$.

This is shown shaded in the figure.

Hollow Rectangular Section

Referring to Fig. 29.6,

$$\frac{P}{(BH - bh)} = \frac{P \cdot e_x \cdot (H/2)}{\{(BH^3 - bh^3)/12\}}$$

where $e_x = \dfrac{(BH^3 - bh^3)}{6H(Bh - bh)}$

If the eccentricity is about YY

$$e_y = \frac{HB^3 - hb^3}{6B(BH - bh)}$$

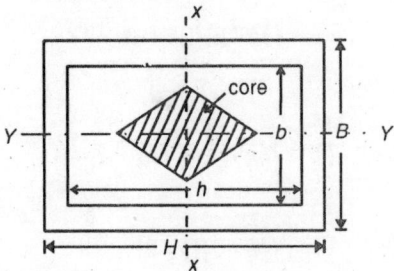

Fig. 29.6: Core of a Hollow Rectangular Section

The core is obviously a rhombus with diagonals of lengths

$$\frac{BH^3 - bh^3}{3H(BH - bh)} \text{ and } \frac{HB^3 - hb^3}{3B(BH - bh)}$$

Triangular Section

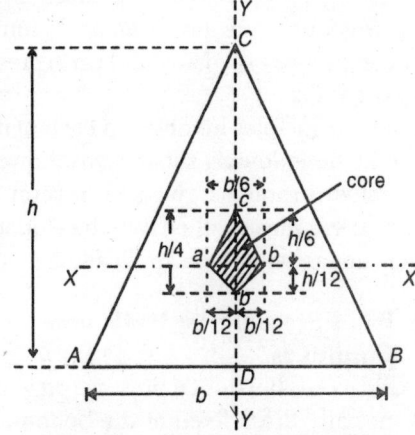

Fig. 29.7: Core of a Triangular Section

Referring to Fig. 29.7,

$$A = \frac{bh}{2}$$

$$I_x = \frac{bh^3}{36}; \ I_y = \frac{hb^3}{48}\left[= \frac{2 \times h(b/2)^3}{12}\right]$$

$$c_x {}_{(\text{max})} = \frac{2h}{3}$$

$$c_x \atop (min) = \frac{h}{3} \; ; \; c_y = \frac{b}{2}$$

e_x (towards D with c_x max.)

$$= \frac{bh^3}{36} \times \frac{1}{(bh/2)(2h/3)} = \frac{h}{12}$$

e_x (towards C with c_x min.)

$$= \frac{bh^3}{36} \times \frac{1}{(bh/2)(h/3)} = \frac{h}{6}$$

$$e_y = \frac{hb^3}{48} \times \frac{1}{(bh/2)(b/2)} = \frac{b}{12}$$

The core is obviously a 'quadrilateral' ($abcd$) with diagonals $\frac{h}{4}$ and $\frac{b}{6}$, which are at right angles to each other.

29.4 APPLICATIONS OF COMBINED FLEXURE AND DIRECT STRESS

There are several practical applications of combined flexure and direct stress in the realm of engineering practice. Some of these are the effect of wind pressure on walls and chimney shafts, water pressure on dams, and earth pressure on retaining walls.

The basic principles involved in each of these are given in the following subsections; however, the illustrative problems given in Section 29.8 relating to these ideas will enable the student to effectively understand these applications.

29.4.1 Wind Pressure on Walls and Chimneys

When wind exerts horizontal pressure on vertical walls, especially those fixed at the bottom, and chimney shafts, it causes bending moment on any section and so bending stresses, while the self-weight causes direct thrust or compressive stress. The resultant stress can be obtained by superimposition; further, the structure will be designed to withstand the maximum compressive stress, and if it is masonry or plain concrete, the effort will be to ensure that no tensile stress occurs in the section, since these are weak in tension.

However, it is necessary for us to know the intensity and the manner of distribution of the wind pressure with respect to the height, which is usually assumed to be uniform for simplicity.

29.4.2 Water Pressure on Dams

Dams are structures built across streams or rivers to impound water in reservoirs behind them. Water, according to Pascal's law, exerts equal pressure in all directions. Thus, water exerts horizontal pressure on the dam which is known to increase in direct proportion with the depth below the surface; this is analogous to triangular loading in the horizontal direction, causing bending moments and bending stresses in different sections of the dam, the maximum values occuring at the base section. Self-weight of the dam causes direct compressive stresses, and on superimposition, we obtain the resultant stresses, which are to be considered in the design of the dam.

29.4.3 Earth Pressure on Retaining Walls

Retaining walls are structures meant to retain earth on one side at a higher elevation than on the other side. The earth exerts lateral pressure on the retaining wall, which may be evaluated using an appropriate earth pressure theory from a knowledge of Geotechnical Engineering. The lateral pressure is known to increase with depth from the top surface of the retained earth. The effect on the wall is similar to water pressure on dams, causing bending moments and bending stresses, which are maximum at the base section. Self-weight causes direct compressive stresses, and the resultant stresses will be got by superimposition as usual, and will be used in the design.

Further consideration of these will be given in Illustrative Problems.

29.5 LONG COLUMNS

In the case of relatively short compression members, the capacity to carry loads depends solely on their strength in compression; but when the length of the members is large in comparison with the transverse dimensions, failure tends to occur by lateral bending or 'buckling', rather than by direct compression. Such compression members are called 'columns' if they are vertical, and 'struts' or 'braces' if they are inclined or diagonally placed. In the case of columns, or 'long' columns as they are some times called, the direct stress is insignificant relative to the flexural stresses due to lateral bending or buckling.

29.5.1 Euler's Theory of Long Columns

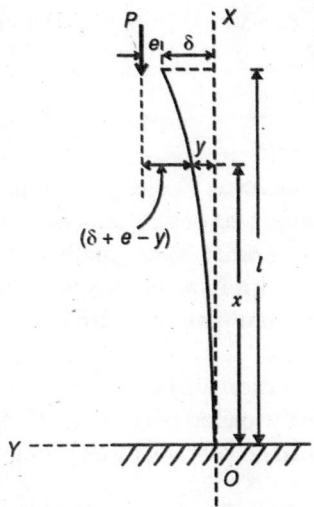

Fig. 29.8: Long Column—Buckling Effect

Referring to Fig. 29.8—a column of length l, fixed at the bottom and free at the top carries a load P parallel to the axis at an eccentricity e. Let the column deflect laterally by δ at the top, the value of lateral deflection being y at a distance x from the bottom, which is taken as the origin. The lever arm of the load from the axis of the column depends, besides the constant value of e, the variable value of the lateral deflection y; thus the bending moment at any section due to the lateral bending as well as the eccentricity is given by

$$M = -P(\delta + e - y)$$

Therefore, the equation to the elastic curve is—

$$EI \cdot \frac{d^2y}{dx^2} = P(\delta + e - y)$$

$$\frac{d^2y}{dx^2} = \left(\frac{P}{EI}\right)(\delta + e - y)$$

If $\dfrac{P}{EI} = k^2$ for convenience,

$$\frac{d^2y}{dx^2} + k^2y = k^2(\delta + e) \qquad \ldots(\text{Eq. 29.7})$$

Multiplying throughout by $\left(2 \cdot \dfrac{dy}{dx}\right)$ to make it exact,

$$2 \cdot \frac{dy}{dx} \cdot \frac{d^2y}{dx^2} + 2k^2 y \cdot \frac{dy}{dx} = 2k^2(\delta + e)\frac{dy}{dx}$$

Integrating,

$$\left(\frac{dy}{dx}\right)^2 + k^2 y^2 = 2k^2(\delta + e)y + C_1$$

C_1 being the constant of integration.

At the bottom end, which is fixed, $x = 0$, $y = 0$ and also,

$$\frac{dy}{dx} = 0$$

Substituting, we get $C_1 = 0$

$$\therefore \quad \left(\frac{dy}{dx}\right)^2 = 2k^2(\delta + e)y - k^2 y^2$$

Rearranging the terms,

$$\frac{dy}{dx} = k\left(\sqrt{2(\delta + e)y - y^2}\right)$$

or

$$\frac{dy}{\sqrt{2(\delta + e)y - y^2}} = k \cdot dx$$

Integrating both sides,

$$\mathrm{Cos}^{-1}\left\{1 - \frac{y}{(\delta + e)}\right\} = kx + C_2$$

C_2 being the constant of integration.

When $x = 0$, $y = 0$ $\therefore C_2 = \cos^{-1} 1 = 0, 2\pi, 4\pi$, and so on.

Taking the least of these values,

$$y = (\delta + e)(1 - \cos kx) \qquad \ldots(\text{Eq. 29.8})$$

For $x = l$, $y = \delta$

$$\therefore \qquad \delta = (\delta + e)(1 - \cos kl)$$

or
$$\delta = \delta + e - \delta \cos kl - e \cos kl$$
$$\delta \cos kl = e(1 - \cos kl)$$

or $\delta = e(\sec kl - 1) = e\left(\sec\sqrt{\dfrac{P}{EI}} \cdot l - 1\right) \ldots(\text{Eq. 29.9})$

If the relation between P and the corresponding δ is plotted for different small, but geometrically decreasing, values of eccentricity, it will be as shown in Fig. 29.9.

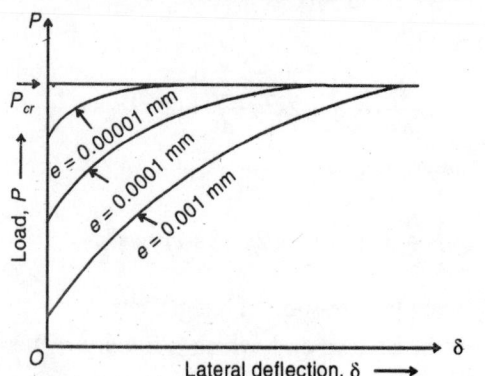

Fig. 29.9: Load Versus Lateral Deflection Relationship for Different Values of Small Eccentricity

It is observed that as the load approaches a certain value, P_{cr}, the curves become asymptotic to a horizontal line through P_{cr}, i.e., the lateral deflection tends to become infinite.

It is interesting to note that, if the eccentricity is reduced to zero, the deflection δ would remain zero until this load P_{cr} is reached; at this stage, the column would be able to retain any small deflection that might be given to it. If the load were to be increased even very slightly beyond P_{cr}, the deflection would continue to increase until failure occurred. Thus P_{cr} has great signi-ficance in that it is the maximum centrally placed load which the column can carry without failure through 'buckling'.

P_{cr} may be derived for this case as follows:

$$\frac{\delta}{e} = \sec\sqrt{\frac{P}{EI}} \cdot l - 1$$

As e tends to zero, $\dfrac{\delta}{e}$ tends to infinity; therefore;

$$\left(\sec\sqrt{\frac{P}{EI}} \cdot l - 1\right) \text{ should also approach infinity.}$$

$$\therefore \quad \sec\sqrt{\frac{P}{EI}} \cdot l \to \infty$$

$$\therefore \quad \sqrt{\frac{P}{EI}} \cdot l = \pi/2,\, 3\pi/2,\, \ldots,\, (2n+1)\pi/2, \text{ and}$$

so on, n being an integer.

The smallast value of P in this series is the values of P_{cr}, since, if failure can occur at a particular load, there is no point in considering any load larger than this load.

$$\sqrt{\frac{P_{cr}}{EI}} \cdot l = \frac{\pi}{2}$$

$$\therefore \quad P_{cr} = \frac{\pi^2 EI}{4l^2} \qquad \ldots\text{(Eq. 29.10)}$$

This is known as "Euler's Column Formula" for the specific case of a column fixed at one end and free at the other, as originally assumed. The load, P_{cr}, is known as the 'Critical load', 'Crippling load', or 'buckling load' for the column.

This expression may also be obtained starting from an axialy loaded column ($e = 0$). Let the top end be displaced laterally by an arbitrary value δ initially.

Then Eq. 29.8 reduces to
$$y = \delta(1 - \cos kx)$$
When $x = l$, $y = \delta$

$$\therefore \quad \delta = \delta(1 - \cos kl) = \delta\left(1 - \cos\sqrt{\frac{P}{EI}} \cdot l\right)$$
$$\ldots\text{(Eq. 29.11)}$$

Then, if $\delta \neq 0$, $\cos\sqrt{\dfrac{P}{EI}} \cdot l = 0$

$$\therefore \quad \sqrt{\frac{P}{EI}} \cdot l = \pi/2,\, 3\pi/2,\, \ldots$$

Taking the least value of $\pi/2$, $P = P_{cr} = \dfrac{\pi^2 \cdot EI}{4l^2}$, as obtained earlier. The peculiarity of the critical load is that any load P less than P_{cr} will make y at the top different from δ, which is contrary to fact, except when $\delta = 0$.

Therefore, from Eq. 29.11, we observe that, for any load P less than P_{cr}, the lateral deflection must be zero, and for $P = P_{cr}$, any deflection δ given to the column may be retained.

It may be easily understood that the critical load depends only on the modulus of elasticity and the length and cross-sectional dimensions of the column; it does not depend upon the *strength*.

(Tubular sections are more economical than solid sections of the same area, since the moment of inertia for the former is greater). Thus it may be seen that the failure of a slender member may be attributed to elastic instability and not to lack of strength on the part of the material.

29.5.2 Critical Load for Different End Conditions

Euler's formula for the critical load of a long column for other end-conditions may be obtained either from the one already derived for the case of one end fixed and the other end free, or more rigorously from fundamentals.

Both Ends Hinged

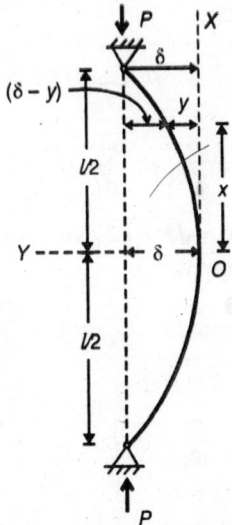

Fig. 29.10: Column with Both Ends Hinged—Euler's Formula

When both ends of the column are pinned or hinged as shown in Fig. 29.10, the lateral buckling under an axial load P, appears to be analogous to that in Fig. 29.8, except that the column appears to be divided into two with one end fixed and the other free.

From symmetry, each half is in the same condition as the column of the previous case with one end fixed and the other free, except that in place of l, $l/2$ is to be substituted.

Thus, setting $l/2$ for l in Eq. 29.10,

we get $P_{cr} = \dfrac{\pi^2 EI}{l^2}$ ...(Eq. 29.10)

This case is encountered more commonly in practice, and is called the "fundamental case" of buckling.

More rigorously, from fundamentals, we may proceed as follows:

$$EI \cdot \frac{d^2 y}{dx^2} = P(\delta - y)$$

$$\frac{d^2 y}{dx^2} = k^2(\delta - y) \text{ where } k^2 = \frac{P}{EI}$$

$$\frac{d^2 y}{dx^2} + k^2 y = k^2 \delta$$

The solution to this differential equation is,

$$y = C_1 \cos kx + C_2 \sin kx + \delta$$

At $x = 0$, $y = 0$

and also $\dfrac{dy}{dx} = 0$

$\therefore \quad C_2 = 0$ and $C_1 = -\delta$

Finally, $y = \delta(1 - \cos kx)$

For $x = l/2$, $y = \delta$

$\therefore \cos(kl/2) = 0$, assuming $\delta \neq 0$

$$kl/2 = \sqrt{\frac{P}{EI}} \cdot \frac{l}{2} = \frac{\pi}{2}, \frac{3\pi}{2}, \ldots$$

Taking the least value of $\pi/2$ and calling P as P_{cr},

$$P_{cr} = \frac{\pi^2 EI}{l^2}, \text{ the same as Eq. 29.12.}$$

Note: For any load less than P_{cr}, y at $x = l/2$ will be different from δ, which is contrary to our assumption; in other words, no lateral deflection will occur in such a case.

Both Ends Fixed

Referring to Fig. 29.11, when both ends are fixed for the column, reactive couples are induced at the ends in order to keep them from rotating during buckling.

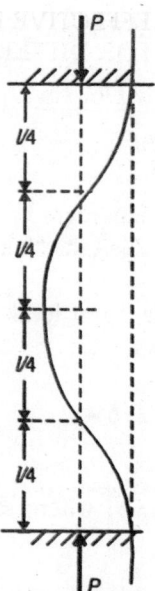

Fig. 29.11: Column with Both Ends Fixed—Euler's Formula

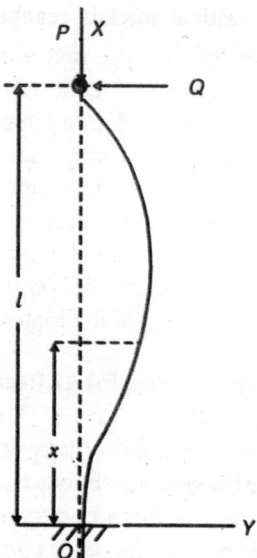

Fig. 29.12: Column with Bottom End Fixed and the Top End Hinged—Euler's Formula

The combination of the axial thrust with these moments is equivalent to the axial thrust applied eccentrically. The points of inflexion occur where the line of action of the thrust P intersects the deflection curve because the bending moments at these points are zero. These points and the mid point divide the span into four equal parts, each of which is in the same condition as the column with one end fixed and the other free. Hence the critical load for the present case may be got by substituting $l/4$ for l in the formula for the critical load for one end fixed and the other free, i.e.,

$$P_{cr} = \frac{\pi^2 EI}{4l^2}$$

or $\qquad P_{cr} = \frac{4\pi^2 EI}{l^2}$ $\qquad$ ...(Eq. 29.13)

Lower End Fixed and the Top End Hinged

For a column with the lower end fixed and the top end hinged (Fig. 29.12), a lateral reaction Q is produced at the hinge during buckling since the position of the hinge is to be retained. The bending moment, M, at a distance x from the bottom end, the origin, is,

$$M = P \cdot y + Q(l - x)$$

(moment causing concavity upward is considered +ve)

The differential equation to the deflection curve, therefore,

$$E \cdot I \frac{d^2 y}{dx^2} = -P \cdot y - Q(l - x)$$

$$\frac{d^2 y}{dx^2} = -\frac{P}{EI} \cdot y - \frac{Q}{EI} \cdot (l - x)$$

Putting $k = \sqrt{P/EI}$, the solution to the differential equation is,

$$y = C_1 \cos kx + C_2 \sin kx - \frac{Q}{P}(l - x)$$

$$\left. \begin{array}{l} \text{At } x = 0, \quad y = 0 \\ \qquad x = 0, \quad \dfrac{dy}{dx} = 0 \\ \text{and} \quad x = l, \ y = 0 \end{array} \right\}$$

Applying these boundary conditions, we get

$$C_1 - \frac{Q}{P} \cdot l = 0; \ C_1 \cos kl + C_2 \sin kl = 0;$$

and $\ kC_1 + \dfrac{Q}{P} = 0$

Substituting the values of the constants C_1 and C_2 from the above into the middle one,

$$\frac{Q}{P}\left[\frac{1}{k} \sin kl - l \cos kl \right] = 0$$

When the critical load is reached, buckling occurs and Q is not zero:

$$\therefore \quad \frac{1}{k}\sin kl - l\cos kl = 0$$

or $\quad \tan kl = kl$

This is a transcendental equation.

The smallest non-zero value of kl which satisfies this equation is

$$kl = 4.49$$

$$\therefore \quad P_{cr} = k^2 EI = \frac{20.2 EI}{l^2} = \frac{\pi^2 EI}{(0.707l)^2}$$

$$= \frac{\pi^2 EI}{\left(l/\sqrt{2}\right)^2}$$

or $\quad P_{cr} = \dfrac{2\pi^2 EI}{l^2} \qquad \text{...(Eq. 29.14)}$

The value of the reaction Q depends upon the maximum value of the deflection δ.

29.5.3 Effective Length

The critical load for the fundamental case of buckling, i.e., when both ends are hinged is found

to be $P_{cr} = \dfrac{\pi^2 \cdot EI}{(l)^2}$. The critical loads for other end conditions may also be put in this form as follows:

$$P_{cr} = \frac{\pi^2 \cdot EI}{(2l)^2} \text{ one end fixed and the other free}$$

$$P_{cr} = \frac{\pi^2 \cdot EI}{(l/2)^2} \text{ both ends fixed.}$$

$$P_{cr} = \frac{\pi^2 \cdot EI}{\left(l/\sqrt{2}\right)^2} \text{ one end fixed and the other hinged}$$

In such a case the term within the brackets in the denominator of each of the above expressions is called the 'equivalent', or the 'effective' length of the column as compared to the fundamental case of buckling.

The effective lengths, l_e, for different end conditions are set out in the table below in terms of the actual length, l, of the column.

TABLE 29.1: EFFECTIVE LENGTH OF COLUMN FOR DIFFERENT END CONDITIONS

S. No.	Case	Effective length, l_e
1.	Fundamental case: Both ends hinged.	$l_e = l$
2.	One end fixed and the other free	$l_e = 2l$
3.	Both ends fixed	$l_e = l/2$
4.	One end fixed and the other hinged	$l_e = l/\sqrt{2}$

The Euler's formula may be put down in terms of l_e:

$$P_{cr} = \frac{\pi^2 \cdot EI}{l_e^2} \qquad \text{...(Eq. 29.15)}$$

29.5.4 Critical Stress and Slenderness Ratio

The critical load being known for the column, the 'critical stress', p_{cr}, may be written as follows:

$$p_{cr} = \frac{P_{cr}}{A} = \frac{\pi^2 \cdot EI}{l_e^2 \cdot A} = \frac{\pi^2 EAr^2}{l_e^2 A}$$

or $\quad P_{cr} = \dfrac{\pi^2 E}{\left(l_e/r\right)^2} \qquad \text{...(Eq. 29.16)}$

(r is the least radius of gyration of the column section, since it can buckle about any axis)

This is known as the generalised Euler-formula.

The ratio (l_e/r) characterises the degree of slenderness of the column, and is known as the "Slenderness Ratio" of the column.

(It must be noted that the radius of gyration, r, must be taken as the least value for the section here).

The relation between P_{cr} and (l_e/r) for mild steel with

$E = 2.05 \times 10^5$ MPa is shown in Fig. 29.13.

The critical stress diminishes indefintely with increase in slenderness of the bar; in other words, a very slender bar fails at a very low critical stress in buckling.

The assumption that the direct component of stress is negligible when compared with the bending stress holds as long as the critical stress is not reached. Since this is the basis of Euler's formula, its validity will be questionable when the slenderness ratio is such that direct stress is not small in relation to the bending stress due to lateral buckling.

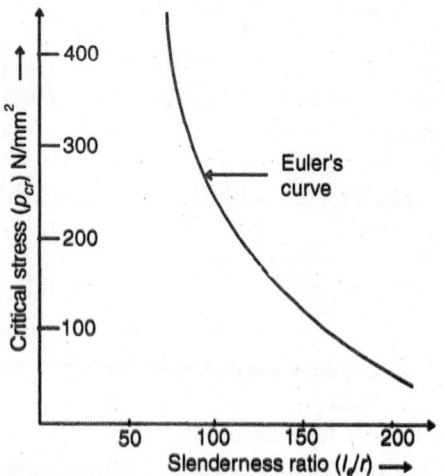

Fig. 29.13: Euler's Curve for Slender Compression Members

Taking the proportional limit of structural steel as 210 N/mm² and the modulus of elasticity as 2.1×10^5 N/mm².

$$P_{cr} < 210$$

or $$\frac{\pi^2 E}{(l_e/r)^2} < 210$$

or $$(l_e/r)^2 > \frac{\pi^2 E}{210}$$

$$> \frac{\pi^2 \times 2.1 \times 10^5}{210}$$

or $$(l_e/r) > 100$$

Thus, it means that Euler's formula can be applied for slender compression members of mild steel with their slenderness ratio greater than 100, if the critical stress is not to exceed the proportional limit of the material.

This limit varies from one material to another since the modulus and proportional limit values vary.

29.5.5 Design of Compression Members

It is seen that Euler's formula cannot be applied for members with slenderness ratio less than 100.

For all practical purposes, the upper limit of critical stress is the yield point of the material. For mild steel with yield point of 300 N/mm², p_{cr} may be taken as p_{yp} at $(l_e/r) = 60$. For slenderness ratio values between 60 and 100, linear variation may be assumed. For $(l_e/r) < 60$, the critical stress may be considered as constant at p_{yp}.

The working stress for the column in compression has to be much below the critical, at which failure is anticipated. The factor of safety depends upon unforeseen increases in the load as also possible errors in the application of the load exactly along the axis, and on the possible initial crookedness of the column. However, a constant factor of safety, say 2.

The "generalised buckling stress diagram" for mild steel will be somewhat as shown in Fig. 29.14.

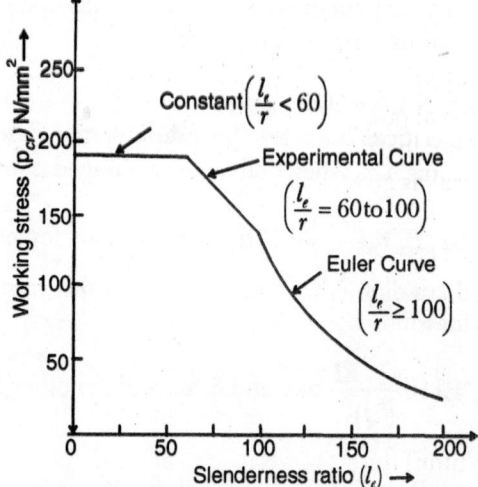

Fig. 29.14: Generalised Buckling Stress Diagram

Experiments show that errors in the application of load and initial deflection increase with increasing values of slenderness ratios of the column. This justifies the use of variable factor of safety which increases with the slenderness ratio of the column.

$l_e/r > 100$ ...factor of safety 3.5

$l_e/r \to 0$ to 100 ...factor of safety varies according to a linear law.

This way, a table or a curve of working stresses for columns of various slenderness ratios can be established, which may be used in the design of compression members.

29.5.6 Empirical Formulae for the Design of Compression Members

Certain empirical formulae were evolved, partly based on experiments, by some authorities, for the critical stress, or more commonly for the working stress, for compression members with known slenderness ratio.

Some of these, in the order of increasing importance, are given below.

(i) Straight Line Formulae

These assume that, beyond the proportional limit, the relation between p_{cr} and (l_e/r) can be represented with sufficient accuracy by the linear law,

$$p_{cr} = a - b\left(\frac{l_e}{r}\right) \qquad \text{...(Eq. 29.17)}$$

a and b being constants, which depend upon the physical properties of the material.

Usually, the working stress, rather than the critical, is given by these.

or $\quad p_w = a - b\left(\frac{l_e}{r}\right)$

For mild steel,

$$p_w = 110 - 0.5\left(\frac{l_e}{r}\right) \text{ (L. V. Tetmajer) ...(Eq. 29.18)}$$

(N/mm^2) (for l_e/r between 30 and 120)

For values of $(l_e/r) < 30$,

$p_w = 100 \text{ N/mm}^2$ is recommended.

(ii) Parabolic Formulae

Prof. J. B. Johnson and Prof. Ostenfeld proposed this concept, according to which,

$$p_{cr} = a - b\,(l_e/r)^2, \qquad \text{...(Eq. 29.19)}$$

a and b being constants which depend upon the material.

These constants are so chosen that the stress versus slenderness ratio curve is tangential to Euler's curve and to make $p_{cr} = p_{yp}$ for short bars.

For mild steel,

$$p_{cr} = 280 - 0.01\left(\frac{l_e}{r}\right)^2 \text{ N/mm}^2 \text{ ...(Eq. 29.20)}$$

for $(l_e/r) \not> 120$

For $(l_e/r) > 120$, Euler's formula is recommended.

With a factor of safety of 2.5,

$$p_w = 110 - 0.004\,(l_e/r)^2 \text{ N/mm}^2 \text{ ...(Eq. 29.21)}$$

for $(l_e/r) \not> 120$

29.5.7 Rankine-Gordon Formula

The effect of the straight-line and parabolic formulae is to reduce the critical compressive stress, applicable for short struts, with increasing values of slenderness ratio.

This can be also effected by using the Rankine-Gordon formula:

$$p_{cr} = \frac{f_c}{1 + b\left(l_e/r\right)^2} \qquad \text{...(Eq. 29.22)}$$

Here f_c is the critical compressive stress for short bars, and b is a constant which depends upon the material. By proper selection of the constants, this may be made to agree satisfactorily with experimental results within certain limits.

This formula may be arrived at as follows:-

Let the ultimate load for a short bar (which fails only in compression) be P_c.

$$P_c = f_c \cdot A$$

For a long bar for which Euler's formula is valid the critical load P_e is

$$P_e = \frac{\pi^2 \cdot EI}{l_e^2}$$

For bars which can neither be classified as very short nor very long, failure will be due to direct stress as well as bending stress due to buckling as well.

If the crippling load be denoted as P_{cr} for such cases, Rankine suggested the relation,

$$\frac{1}{P_{cr}} = \frac{1}{P_c} + \frac{1}{P_e}$$

This is supposed to cover all cases ranging from very long to very short bars.

$$\therefore \quad P_{cr} = \frac{P_c \cdot P_e}{(P_c + P_e)} = \frac{P_c}{\left(1 + \frac{P_c}{P_e}\right)} = \frac{f_c \cdot A}{\left\{1 + \left(\frac{f_c \cdot A}{\pi^2 EI}\right) l_e^2\right\}}$$

$$\text{or} \quad P_{cr} = \frac{f_c \cdot A}{\left\{1 + b\left(\frac{l_e}{r}\right)^2\right\}}, \qquad \text{...(Eq. 29.23)}$$

where $b = \dfrac{f_c}{\pi^2 E}$

$$p_{cr} = \frac{P_{cr}}{A} = \frac{f_c}{\left\{1 + b\left(l_e / r\right)^2\right\}} \qquad \text{...(Eq. 29.24)}$$

Usually, the crippling load, P_{cr}, is found directly from Eq. 29.23. Though the formula is empirical, the value of b is found experimentally by Gordon; so this formula has got practical importance with regard to specific cases.

The average values of f_c and b are given in the table below for certain materials.

TABLE 29.2: RANKINE'S CONSTANTS FOR CERTAIN MATERIALS

S.No.	Material	Rankine's constants	
		f_c (N/mm^2)	b
1.	Mild Steel	320	1/750
2.	Cast Iron	550	1/1600
3.	Wrought Iron	240	1/9000
4.	Timber	45	1/750

The factor of safety may be taken as 3 for mild steel.

29.5.8 I.S. Formula

The I.S. code for the design of columns gives the following equation for the allowable average axial compressive stress for mild steel:

For $\left(\dfrac{l_e}{r}\right) < 160$

$$p_c = \frac{(f_y / m)}{\left[1 + 0.20 \sec\left\{\frac{l_e}{r} \cdot \frac{\sqrt{mp_c}}{4E}\right\}\right]} \qquad \text{...(Eq. 29.25)}$$

and for $\left(\dfrac{l_e}{r}\right) \geq 160$

$$p_c = \left[1.2 - \frac{l_e}{800r}\right] \left[\frac{(f_y / m)}{\left[1 + 0.20 \sec\left\{\frac{l_e}{r} \cdot \frac{\sqrt{mP_c}}{4E}\right\}\right]}\right]$$

Here p_c = allowable average axial compressive stress,

f_y = the guaranteed minimum yield stress,

m = factor of safety (taken as 1.68),

$\left(\dfrac{l_e}{r}\right)$ = slenderness ratio,

E = Modulus of elasticity (taken as 2.010×10^5 N/mm^2)

Some salient values of p_c are shown in the following table:

TABLE 29.3: SALIENT VALUES OF p_c FOR SOME (l_e/r) VALUES FROM I.S. FORMULA

(l_e/r)	p_c (N/mm^2)	(l_e/r)	p_c (N/mm^2)
0	125	160	42.3
10	124.6	180	33.6
50	117.2	190	30
80	100.7	200	27
100	84	250	16.6
120	67.1	300	10.9
140	53.1	350	7.6

This may also be shown diagrammatically as a relation between p_c and l_e/r.

29.5.9 Design Procedure for a Long Column

When a column is to be designed for a given load, it will be a trial and error process. A section is chosen, and the least radius of gyration, and the slenderness ratio for the given length and end conditions.

For the particular value of slenderness ratio, the best and the most appropriate formula is chosen, and the critical compressive stress is determined, and the critical load is got multiplying this with the area of the section. The safe or working load is got by applying a suitable factor of safety. If this compares favourably with the given load (with a little of safety margin), the section is considered satisfactory; otherwise, the procedure is repeated with a different section.

29.6 ECCENTRICALLY LOADED LONG COLUMNS

Precise central loading may be difficult to achieve in the case of long columns. Thus, unintended eccentricity of loading, though small, always may occur for any reason including crookedness along the axis; further, deliberate or intended eccentricity may also be there in certain cases for practical, or structural reasons, such as columns carrying beams on brackets attached to the former.

In such a case Euler's analysis leads one to a some what different expression for the critical stress, which includes both the direct stress component and the bending stress component due to lateral buckling.

29.6.1 Eccentric Loading on Long Columns —Secant formula

Let us consider the fundamental case of buckling of a column with both ends hinged under eccentric loading from first principles (Fig. 29.15)

Let the maximum lateral deflection be δ at mid height.

B.M. at any point distant x from the mid-height of the column is,

$$M = -P(\delta + e - y)$$

(with directions of x and y axes as indicated)

The flexure equation is

$$EI \cdot \frac{d^2y}{dx^2} = P(\delta + e - y)$$

Setting $\frac{P}{EI} = k^2$, for convenience,

$$\frac{d^2y}{dx^2} + k^2y = k^2(\delta + e)$$

Multiplying throughout by $2\left(\dfrac{dy}{dx}\right)$ to make it exact

$$2 \cdot \frac{dy}{dx} \cdot \frac{d^2y}{dx^2} + 2k^2y \cdot \frac{dy}{dx} = 2k^2(\delta + e)\frac{dy}{dx}$$

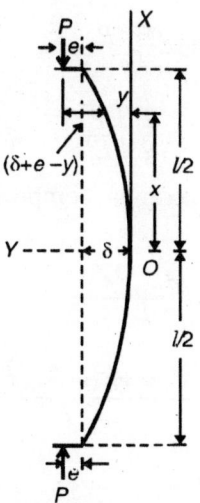

Fig. 29.15: Column with Both Ends Hinged—Eccentric Loading

Integrating, $\left(\dfrac{d^2y}{dx^2}\right)^2 + k^2y^2 = 2k^2(\delta + e)y + C_1$

where C_1 is constant of integration.

When $x = 0$, $y = 0$ and also $\dfrac{dy}{dx} = 0$ (by symmetry);

so, $C_1 = 0$

$\therefore \quad \dfrac{dy}{\sqrt{2(\delta + e)y - y^2}} = k \cdot dx$

Integrating both sides,

$$\cos^{-1}\left[1 - \frac{y}{(\delta + e)}\right] = kx + C_2$$

where C_2 is a constant.

When $x = 0$, $y = 0$, $\therefore C_2 = \cos^{-1}1 = 0, 2\pi, 4\pi, \ldots$

Taking the least of these values for C_2,

$$y = (\delta + e)(1 - \cos kx)$$

At $x = l/2$, $y = \delta$

$\therefore \qquad \delta = (\delta + e)\left(1 - \cos\frac{kl}{2}\right)$

On simplification, $\delta = e\left(\sec\dfrac{kl}{2} - 1\right)$

Maximum B.M. at mid-height,

$$M_{\max} = P(e + \delta)$$

$$= P\left(e + e\sec\frac{kl}{2} - e\right)$$

$$= P \cdot e \sec \frac{kl}{2}$$

or $M_{\max} = P \cdot e \sec \dfrac{l}{2} \sqrt{\dfrac{P}{EI}}$ \qquad ...(Eq. 29.26)

Hence the maximum compressive stress, $p_{\max}$, is.

$$p_{\max} = \frac{P}{A} + \frac{M_{\max} \cdot e}{I}$$

or $p_{\max} = \dfrac{P}{A} + \dfrac{P \cdot ec \sec(l/2)\sqrt{P/EI}}{I}$

...(Eq. 29.27)

$$p_{\max} = \frac{P}{A} + \left[1 + \frac{ec}{r^2} \cdot \sec \frac{l}{2r} \sqrt{\frac{P}{EA}}\right] \quad ...(Eq. 29.28)$$

(since $I = A \cdot r^2$)

This will be the maximum stress provided it does not exceed the elastic limit.

This is called the "Secant Formula".

It is observed that the maximum stress is not proportional to the load, P, but increases more rapidly than the load.

If the factor of safety, η, is defined as the ratio by which one multiplies the working load to obtain the critical load which will produce a stress equal to the yield stress (or in some cases, the ultimate strength), the design equation based on yield stress will be as follows:

$$p_{y.P.} = \frac{\eta P}{A} \left[1 + \frac{ec}{r^2} \cdot \sec\left(\frac{l}{2r} \sqrt{\frac{\eta P}{EI}}\right)\right] \quad ...(Eq. 29.29)$$

If P, e, l, $p_{y.p}$, and η are known, the cross-sectional dimensions may be obtained by trial and error so as to satisfy this equation, which is an intrinsic one.

Instead, curves may be drawn between P/A and l_e/r for various values of ratio, (ec/r^2), which is called the "eccentricity ratio". For hinged ends, with $\eta = 2.5$, these curves are shown in Fig. 29.16 for mild steel. Such curves enable us to calculate readily the value of P or P/A corresponding to a given eccentricity e, column length l, and the properties of the cross-section c and r.

$[r^2/c$ represents the maximum eccentricity of load P on a short post which will cause no tension. Therefore, ec/r^2 is the ratio of the actual eccentricity to the maximum eccentricity which would cause no tension].

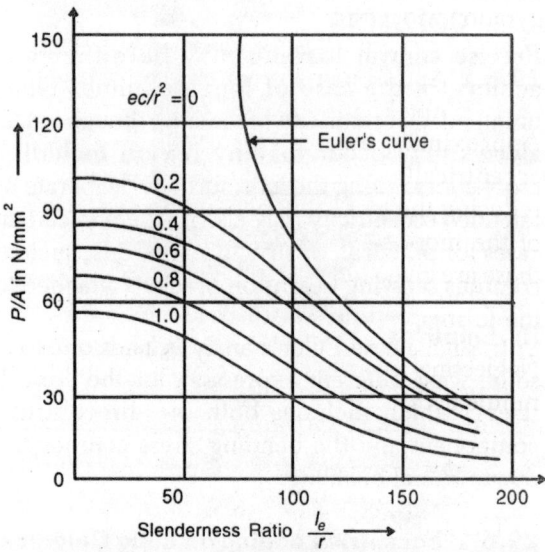

Fig. 29.16: Design Curves for Eccentrically Loaded Long Column

If the column carries more than one load on the same axis of symmetry, the treatment can be with respect to the resultant load and its location. An extra central load also may be treated in a similar way.

Note: The values I and r to be used are those corresponding to the axis m-n about which the bending moment acts (Fig. 29.17).

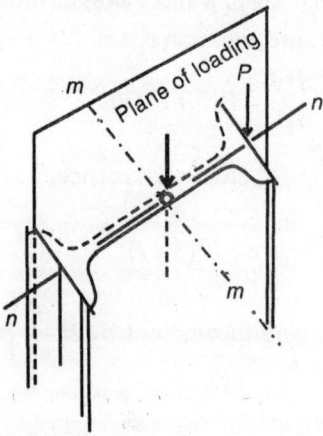

Fig. 29.17: Plane of Loading on the Long Column

If the column is not braced against deflection in a direction parallel to mm and if I and r about the other principal axis nn (lying in the plane of loading) is less than that with respect to the axis mm, then the column should be investigated for strength as for an axial load P, using the radius of gyration with respect to nn].

29.6.2 Approximate Analyses for Eccentrically Loaded Long Columns

Approximate analyses also were evolved for eccentrically loaded columns with a view to reducing the labour involved in the application of the more rigorous Secant formula. Some of these are given below.

(i) Empirical Formulae

Neglecting the effect of the deflection upon the moment, the stress induced is

$$p = \frac{P}{A} + \frac{M \cdot c}{I} = \frac{P}{A} + \frac{P \cdot e \cdot c}{I}$$

The actual stress must not exceed the working stress.

If the column were short, a safe compressive stress like 120 MPa might be satisfactory. For a long column, this is to be reduced according to the slenderness ratio. Such reduction may be effected by using any of the long column formulae for axially applied loads. The least radius of gyration is to be used irrespective of axis of bending, and due regard must be paid to the limiting values of (l_e/r) for which the particular formulae apply.

(ii) Modified Rankine's Formula

$$p_c = \frac{P_{cr}}{A} + \frac{P_{cr} \cdot e \cdot c}{I}$$

$$= \frac{P_{cr}}{A}\left[1 + \frac{ec}{r^2}\right]$$

$$\therefore \quad P_{cr} = \frac{p_c \cdot A}{\left(1 + \dfrac{ec}{r^2}\right)}$$

If the effect of buckling is also considered, we

have seen that in the Rankine formula, the safe axial load is reduced by the factor

$$\left[\frac{p}{1 + b\left(\dfrac{l_e}{r}\right)^2}\right]$$

Thus the modified Rankine's formula for eccentric loading is

$$P_{cr} = \frac{p_c \cdot A}{\left(1 + \dfrac{ec}{r^2}\right)\left(1 + b\dfrac{l_e^2}{r^2}\right)} \qquad \text{...(Eq. 29.30)}$$

This formula is more or less empirical.

A more refined variation of this concept is embodied in the following design equation:

$$\left[\left(\frac{P}{A}\right)/p_c + (M_c/I)/p_b\right] \le 1 \text{ ...(Eq. 29.31)}$$

29.6.3 Prof. Perry's Formula

For the standard case of a column with hinged ends, if p is the permissible stress and e is the eccentricity of loading, the design equation may be written as,

$$p = \frac{P}{A} + \frac{P \cdot e \cdot c \cdot \sec\dfrac{l}{2}\sqrt{P/EI}}{I}$$

(putting p for $p_{\max}$ in Eq. 29.27)

$$= \frac{P}{A}\left[1 + \frac{ec}{r^2} \cdot \sec\frac{l}{2}\sqrt{\frac{P}{EI}}\right]$$

or $\quad p = \dfrac{P}{A}\left[1 + \dfrac{ec}{r^2} \cdot \sec\dfrac{\pi}{2}\sqrt{\dfrac{P}{P_e}}\right]$ ...(Eq. 29.32)

where $P_e = \dfrac{\pi^2 EI}{l^2}$, the Eulerean crippling load.

Prof. Perry found that the expression $\sec \cdot \dfrac{\pi}{2}\sqrt{\dfrac{P}{P_e}}$ approximated closely to $\dfrac{1.2 P_e}{(P_e - P)}$ or

$\dfrac{1.2 p_e}{(p_e - p_o)}$, where $p_o = \dfrac{P}{A}$ and $p_e = \dfrac{P_e}{A}$.

Substituting in Eq. 29.32,

$$p = p_o\left[1 + \frac{ec}{r^2}\cdot\frac{1.2p_e}{(p_e - p_o)}\right]$$

or $\left(\dfrac{p}{p_o}-1\right)\left(1-\dfrac{p_o}{p_e}\right) = \dfrac{1.2ec}{r^2}$ ...(Eq. 29.33)

This is Prof. Perry's approximate formula.

It is easy to calculate p_o for given values of p and e, and hence the value of p, because p_e, c, and r are known for the assumed section.

29.6.4 Effect of Initial Curvature of Column with Axial Loading

Referring to Fig. 29.18, a column *AB* with hinged ends has, say an initial curvature such that the central lateral deflection is e', and loaded axially by *P*. Assuming the initial shape of the column as a line curve, for the sake of mathematical convenience, $y' = e'\cdot\sin\dfrac{\pi x}{l}$, which satisfies the condition at the ends and at the mid-height.

$$\frac{dy'}{dx} = \frac{\pi e'}{l^2}\cos\frac{\pi x}{l}$$

$$\frac{d^2y'}{dx^2} = -\frac{\pi^2 e'}{l^2}\cdot\sin\frac{\pi x}{l}$$

At the critical load *P*, let the column buckle as shown dotted.

Then $EI\cdot\dfrac{d^2(y-y')}{dx^2} - P\cdot y$

Rearranging the terms and substituting for $\dfrac{d^2y'}{dx^2}$,

$$\frac{d^2y}{dx^2} + \left(\frac{P}{EI}\right)\cdot y = -\frac{\pi^2 e'}{l^2}\cdot\sin\frac{\pi x}{l}$$

If the solution is taken in the form

$$y = \lambda e'\cdot\sin\frac{\pi x}{l}$$

we obtain $\lambda\left(\dfrac{\pi^2}{l^2} - \dfrac{P}{EI}\right) = \dfrac{\pi^2}{l^2}$

or $\lambda\left(1 - \dfrac{P}{P_e}\right) = 1$

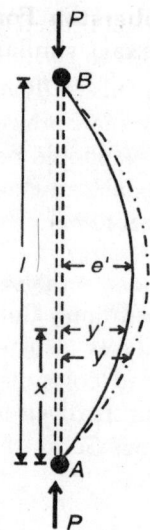

Fig. 29.18: Effect of Initial Curvature on an Axially Loaded Long Column

or $\lambda = \dfrac{P_e}{(P_e - P)}$

$\therefore$ $y = \dfrac{P_e}{(P_e - P)}e'\sin\dfrac{\pi x}{l}$

At $x = \dfrac{l}{2}$, $y_c = \dfrac{P_e\cdot e'}{(P_e - P)}$

Maximum bending moment, $M_{max} = \dfrac{P\cdot P_e\cdot e'}{(P_e - P)}$

The maximum compressive stress, p, is,

$$p = \frac{P}{A} + \frac{M_c}{I} = \frac{P}{A} + \frac{P\cdot P_e}{(P_e - P)}\cdot\frac{e'c}{Ar^2}$$

$$= \frac{P}{A}\left[1 + \frac{P_e}{(P_e - P)}\cdot\frac{e'c}{r^2}\right]$$

$$= p_o\left[1 + \frac{p_e}{(p_e - p_o)}\cdot\frac{e'c}{r^2}\right]$$

or $\left(\dfrac{p}{p_o}-1\right)\left(1-\dfrac{p_o}{p_e}\right) = \dfrac{e'c}{r^2}$ ...(Eq. 29.34)

It is observed that this is of the same form as Prof. Perry's formula (Eq. 29.33) for eccentric loading on long column.

29.6.5 Perry-Roberston Formula

In view of the exact similarity between the formulae for the stress in an axially loaded column with initial curvature and eccentrically loaded column (Prof. Perry's), a column with initial curvature and also eccentrically loaded may be trated as equivalent to an axially loaded column with augmented initial curvature, given by $e_1 = (1.2e + e')$

The maximum compressive stress for such a case is given by

$$p = \left(\frac{p}{p_o} - 1\right)\left(1 - \frac{p_o}{p_e}\right) = \frac{e_1 c}{r^2} \quad \text{...(Eq. 29.35)}$$

where $e_1 = 1.2e + e'$

If we call $\frac{e_1 c}{r^2}$ as β and p as the permissible stress,

$$\left(\frac{p}{p_o} - 1\right)\left(1 - \frac{p_o}{p_e}\right) = \beta$$

$$(p - p_o)(p_e - p_o) = \beta p_o p_e$$

$$p_o^2 - p_o\left(p + p_e \overline{1 + \beta}\right) + p \cdot p_e = 0$$

Solving for p_o,

$$p_o = \frac{p + p_e(1 + \beta)}{2} - \sqrt{\left\{\frac{p + p_e(1 + \beta)}{2}\right\}^2 - pp_e}$$

...(Eq. 29.36)

where $\beta = \frac{e_1 c}{r^2}$ and $p_e = \frac{\pi^2 EI}{Al^2}$.

This is Prof. Perry's formula, modified to take into account imperfections like initial curvature and eccentrcity of loading.

Prof. Robertson, based on experimental evidence, concluded that $\beta = 0.003\frac{l_e}{r}$ is reasonable.

For mild steel, the usual values are: $p = 100$ N/mm^2;

$E = 2 \times 10^5$ N/mm^2 and $\eta = 2.5$.

Another approach (recommended by the I.S.) is to take all imperfections equivalent to an initial eccentricity ratio of 0.20 and apply the Secant formula *even for axially loaded columns*.

29.7 LATERALLY LOADED COLUMNS

When a prismatic bar of material is subjected to axial as well as lateral forces, it may be loooked upon a beam with an axial thrust or pull, or as a strut or tie-rod with lateral bending forces. The stress at any cros-section is given by the algebraic sum of the bending stress, and the direct stress which the axial thrust would cause if there were no lateral loads.

Thus the bending moment due to transverse loading will be generated by that caused due to the lateral buckling of a long column; the resutling bending stress has to be algebraically added to the direct component of stress due to the axial load to obtain the final resultant stress for design. The principle is illustrated in one of the relevant problems in the next section.

ILLUSTRATIVE PROBLEMS

PROBLEM 29.1: A square bar of side a, fixed at the top and free at the bottom, has a notch of width $a/2$ for a small length as shown in Fig. 29.19. It carries an axial pull P. Determine the maximum tensile stress.

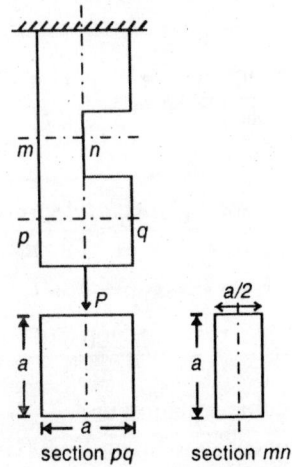

Fig. 29.19: Square Bar with Notch under Pull
(Illustrative Problem 29.1)

Solution:
Section mn at the notch:

Eccentricity of pull P is $\frac{a}{4}$.

∴   Bending moment, $M = P \cdot \dfrac{a}{4}$

∴   Bending stress,

$$p_b = \frac{Mc}{I} = \frac{P \cdot a}{4} \times \frac{a}{4} \times \frac{1 \times 12}{a \times (a/2)^3}$$

$$= 6 \cdot \frac{P}{a^2}$$

Direct stress, $p_d = \dfrac{P}{(a \times a/2)} = 2 \cdot \dfrac{P}{a^2}$

Maximum tensile stress, $p = 6\dfrac{P}{a^2} + 2\dfrac{P}{a^2}$

$$= 8 \cdot \frac{P}{a^2}$$

PROBLEM 29.2: A drum shaft used as an electric hoist is subjected to an axial pull of 33 kN and a bending moment of 3,600 N.m. Calculate the diameter of the shaft required based on a working stress in tension of 40 MPa.

Solution:

Let the diameter of the shaft required be d mm

Pull, $P = 33$ kN $= 33,000$ N.; B.M., $M = 3,600$ N.m.

Direct stress,

$$p_d = \frac{3300 \times 4}{\pi d^2} \text{ MPa} = \frac{132 \times 10^3}{\pi d^2} \text{ MPa}$$

Bending stress

$$p_b = \frac{3600 \times 1000 \times d \times 64}{2 \times \pi d^4} = \frac{115200 \times 10^3}{\pi d^3} \text{ MPa}$$

Maximum tensile stress, $p = p_d + p_b$

∴ $40 = \dfrac{132 \times 10^3}{\pi d^2} + \dfrac{115200 \times 10^3}{\pi d^3}$

This is a cubic equation in d.

Solving by trial and error, $d \approx 100$ mm.

So, a shaft of 100 mm diameter is required.

PROBLEM 29.3: A hollow circular column, 240 mm external diamter and 200 mm internal diameter, is subjected to a load of 10 kN at an eccentricity of 500 mm through a bracket. What are the maximum and minimum stresses induced? What is the maximum permissible

eccentricity for no tensile stress to occur? What is the minimum weight of the column required to avoid tension for the above load acting at an eccentricity of 500 mm? The column section and the load position are shown in plan (Fig. 29.20).

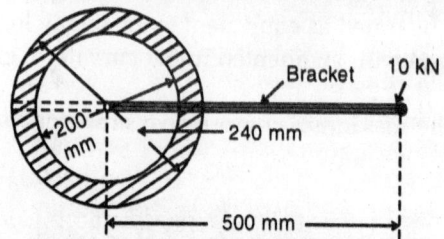

Fig. 29.20: Plan of Column Section and Load through Bracket (Illustrative Problem 29.3)

Solution:

$$M = P \cdot e = 10000 \times 500 \text{ N.mm}$$

$$= 5 \times 10^6 \text{ N.mm}$$

$$A = \frac{\pi}{4}(240^2 - 200^2)$$

$$= 4400\pi \text{ mim}^2$$

$$I = \frac{\pi}{4}(120^2 - 100^2)$$

$$= 26.84\pi \times 10^6 \text{ mm}^4$$

Direct stress,

$$p_d = \frac{P}{A} = \frac{10000}{4400\pi}$$

$$= 0.725 \text{ N/mm}^2 \text{ } (C)$$

Bending stress,

$$p_b = \frac{Mc}{I} = \frac{5 \times 10^6 \times 120}{26.84\,\pi \times 10^6}$$

$$= 7.140 \text{ N/mm}^2 \text{ } (C \text{ or } T)$$

∴ Maximum stress,

$$p_{max} = 0.725 + 7.140$$

$$= 7.865 \text{ MPa } (C)$$

Minimum stress,

$$p_{min} = 0.725 - 7.140$$

$$= 6.415 \text{ MPa } (T)$$

Extra direct stress required to eliminate tension

$$= 6.415 \text{ MPa}$$

Minimum weight of column required

$$= 6.415 \times 4400\pi \text{ N} = 88.7 \text{ kN}$$

For no tension to occur, if the maximum permissible eccentricity is e mm, with the same load of 10 kN, and without considering the weight of the column.

$$p_b = p_d$$

or $\quad 0.725 = \dfrac{10000 \times e \times 120}{26.84 \times \pi \times 10^6}$

$\therefore \quad e = \dfrac{0.725 \times 26.84 \times \pi \times 10^6}{120 \times 10000} \text{ mm}$

$$= 50.94 \text{ mm}$$

So, the maximum permissible eccentricity may be limited to 50 mm. (However, since the direct stress gets enhanced by the self-weight of the column, this value gets increased to some extent accordingly).

PROBLEM 29.4: A short hollow cast-iron column has an external diamter 200 mm and an internal diameter 100 mm, and is subjected to a vertical force at an eccentricity of 80 mm from the axis of the column. Determine the maximum permissible load if the allowable stresses are 100 MPa and 20 MPa in compression and tension, respectively. The section and the load position are shown in plan in Fig. 29.21.

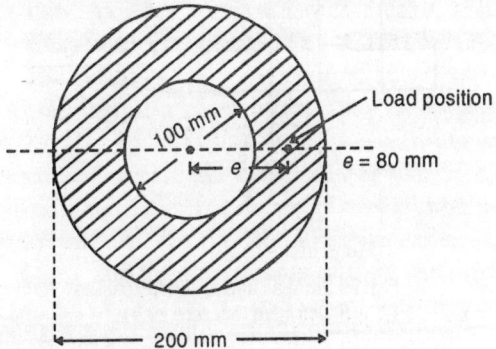

Fig. 29.21: Column with Eccentric Load
(Illustrative Problem 29.4)

Solution: Let the load be P kN.

$$A = \frac{\pi}{4}\left(200^2 - 100^2\right) = 7500\pi \text{ mm}^2$$

$$I = \frac{\pi}{4}\left(100^4 - 50^4\right) = \frac{9375\pi}{4} \times 10^4 \text{ mm}^4$$

Direct stress,

$$p_d = \frac{P}{A} = \frac{1000P}{7500\pi} = \frac{2P}{15\pi} \text{ MPa }(C)$$

$$= 0.04244P\,(C)$$

Bending stress,

$$p_b = \frac{P \cdot e \cdot c}{I} = \frac{1000P \times 80 \times 100 \times 4}{9375\pi \times 10^4}$$

$$= 0.10865P\,(C \text{ or } T)$$

Maximum compressive stress,

$$p_c = 0.15109\,P = 100$$

$\therefore \quad P = \dfrac{100}{0.15109} \text{ kN} = 661.86 \text{ kN}$

Maximum tensile stress,

$$p_t = (0.10865 - 0.04244)P$$

$$= 0.06621P = 20$$

$\therefore \quad P = \dfrac{20}{0.06621} = 302.07 \text{ kN}$

The maximum permissible load is the smaller of these two values, i.e., 300 kN (say).

PROBLEM 29.5: A tower of circular cross-section, 15 m high, inside diameter 2 m and outside diameter 3 m, begins to lean slightly owing to settlement. Determine the critical angle of inclination from the vertical, so that no tension is produced in the tower base. Consider only the self-weight of the tower. The plan and elevation of the tower are shown in Fig. 29.22.

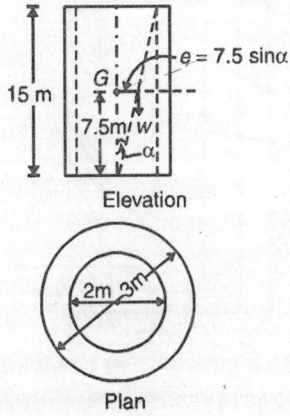

Fig. 29.22 Tower (Illustrative Problem 29.5)

Solution: Eccentricity of the load, W, the self-weight, is $e = 7.5 \sin \alpha$ m, if α is the inclination of the tower with vertical.

$W = \gamma \cdot A \cdot h$ kN if γ is in kN/m³ and A in m²,
h in m.

$$p_d = \frac{\gamma Ah}{A} = \gamma h \ \text{kPa}$$

$$M = W \cdot e = \gamma Ah\,(7.5\sin\alpha)$$

$$p_b = \frac{Mc}{I} = \frac{\gamma Ah \times 7.5\sin\alpha \times 1.5 \times 64}{\pi\left(3^4 - 2^4\right)} \ \text{kPa}$$

For no tension any where,

$$p_d = p_b$$

$$\gamma h = \frac{\gamma Ah \times 7.5\sin\alpha \times 1.5 \times 64}{\pi(3^4 - 2^4)}$$

$$\sin\alpha = 0.0722$$

$$\alpha = 4°08'$$

This is the maximum inclination with vertical for no tension.

PROBLEM 29.6: Compute the maximum and minimum stresses in an eccentricially loaded strut of hollow rectangular section, 180 mm × 120 mm outside and 15 mm thick, if the eccentric load is 45 kN and the eccentricity is 25 mm parallel to the longer side and 15 mm parallel to the shorter side.

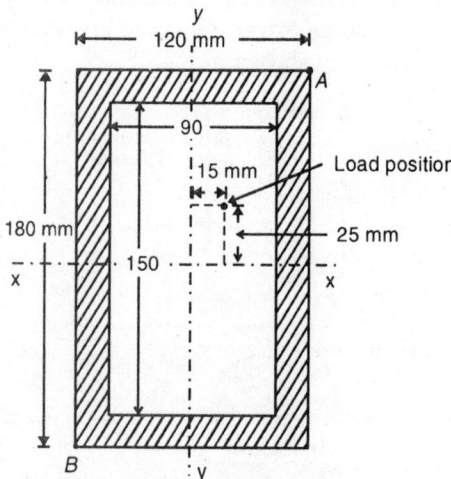

Fig. 29.23: Hollow Rectangular Section
(Illustrative Problem 29.6)

Solution: The section and the load position are shown in Fig. 29.23.

$$A = (120 \times 180 - 90 \times 150) = 8100 \ \text{mm}^2$$

$e_x = 25$ mm ; $e_y = 15$ mm

$$I_x = \frac{120 \times 180^3}{12} - \frac{90 \times 150^3}{12}$$
$$= 33.0075 \times 10^6 \ \text{mm}^4$$

$$I_y = \frac{180 \times 120^3}{12} - \frac{150 \times 90^3}{12}$$
$$= 16.8075 \times 10^6 \ \text{mm}^4$$

Maximum compressive stress at the corner A

$$= \frac{45 \times 10^3}{8100} + \frac{45 \times 10^3 \times 25 \times 90}{33.0075 \times 10^6}$$
$$+ \frac{45 \times 10^3 \times 15 \times 60}{16.8075 \times 10^6}$$
$$= 5.56 + 3.07 + 2.41 = 11.04 \ \text{N/mm}^2$$

Minimum compressive stress at the corner B
$$= 5.56 - 3.07 - 2.41 = 0.08 \ \text{N/mm}^2$$

PROBLEM 29.7: A brickwall, 1.8 m thick, 4.5 m high, is subjected to a horizontal wind pressure of 100 kPa, which may be assumed to act at mid-height of the wall. Determine the maximum and minimum intensities of stress at the base of the wall. Unit weight of masonry is 22.5 kN/m³.

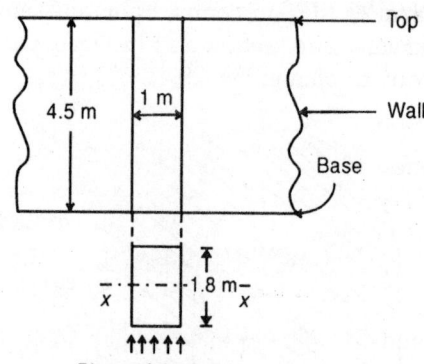

Fig. 29.24: Wall under Wind Pressure
(Illustrative Problem 29.7)

Solution: Let us consider 1 m length of the wall and consider the stresses which occur on the rectangular cross-section, 1 m × 1.8 m (thickness).

Wind pressure, $w = 100$ kN/m²

γ, the unit weight of masonry = 22.5 kN/m³

Weight, W, per metre run = $1 \times 1.8 \times 4.5 \times 22.5$ kN
$$= 182.25 \ \text{kN}$$

Direct stress, $p_b = \dfrac{W}{A} = \dfrac{182.25}{1.8}$

$= 101.25 \text{ kN/m}^2 \,(C)$

Bending moment, M, due to wind pressure

$= 1 \times 4.5 \times 100 \times \dfrac{4.5}{2} \text{ kN.m}$

$= 1012.5 \text{ kN.m}$

Bending stress, $p_b = \dfrac{Mc}{I} = \dfrac{1012.5 \times 0.9}{\left(1 \times 1.8^3 / 12\right)} \text{ kN/m}^2$

$= 1875 \text{ kN/m}^2 \,(\text{or } T)$

(Since the bending is about an axis parallel to the dimension 1 m)–Fig. 29.24

$\therefore$ Maximum stress

$= 101.25 + 1875 = 1976.25 \text{ kPa } (C)$

Minimum stress

$= 101.25 - 1875 = 1773.75 \text{ kPa } (T)$

PROBLEM 29.8: A square chimney, 30 m high, is of uniform hollow square section, 3 m outside and 1 m inside. It is subjected to a uniform wind pressure of 1.5 kPa. Unit weight of brickwork is 18 kN/m^3. If the permissible compressive stress is 800 kPa and no tension is allowed, examine the safety of the chimney.

Solution:

Referring to Fig. 29.25,

$A = (3 \times 3 - 1 \times 1) = 8 \text{ m}^2$

$I = \dfrac{3^4}{12} - \dfrac{1^4}{12} = 6.67 \text{ m}^4$

Wind pressure $= 1.5 \text{ kN/m}^2$

Wind moment at the base,

$M = 1.5 \times 30 \times 3 \times \dfrac{30}{2} \text{ kN.m}$

$= 2025 \text{ kN.m}$

Self weight of chimney,

$W = 8 \times 30 \times 18 = 4320 \text{ kN}$

$\therefore$ Direct stress,

$p_d = \dfrac{4320}{8} = 540 \text{ kPa } (C)$

Bending stress,

$p_b = \dfrac{Mc}{I} = \dfrac{2025 \times 1.5 \times 3}{20}$

$= 455.63 \text{ kPa } (C \text{ or } T)$

Maximum compressive stress,

$p_c = (540 + 455.63) = 995.63 \text{ kPa}$

Minimum pressure $= (540 - 455.63)$

$= 84.37 \text{ kPa } (C)$

Although no tension occurs, the maximum compressive stress *exceeds the permissible value of* 800 kPa. The section *is not adequate*. Hence the section is to be increased and the analysis repeated.

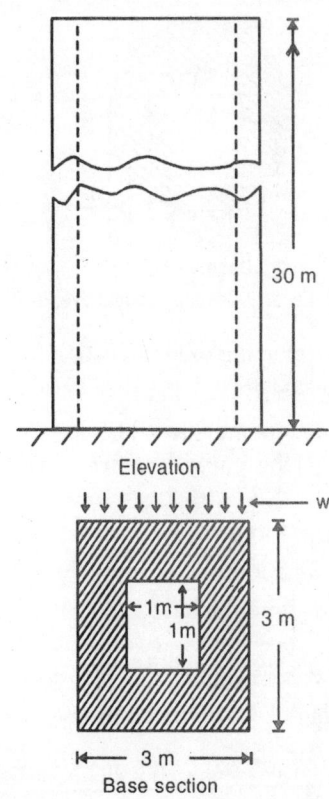

Elevation

Base section

Fig. 29.25: Chimney (Illustrative Problem 29.8)

PROBLEM 29.9: A factory chimney of hollow circular section, 3 m external diamter and 2 m internal diameter, is 20 m high. Wind pressure of 1.5 kPa acts horizontally and a reduction co-efficient of 0.65 is recommended. Unit weight of masonry is 22 kN/m^3. Determine extreme intensities of pressure developed at the base. What will be the limiting wind pressure to cause no tensile stress in the masonry?

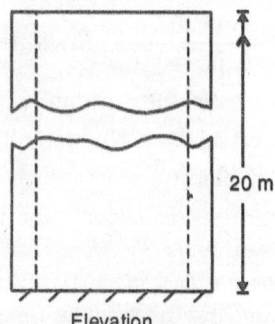

20 m

Elevation

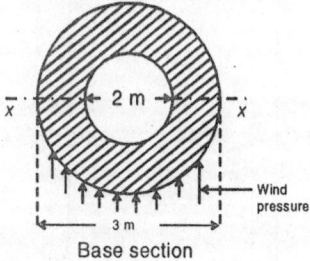

2 m

3 m

Wind pressure

Base section

Fig. 29.26: Factory Chimney (Illustrative Problem 29.9)

Solution: Referring to Fig. 29.26,

Wind pressure = 1.5 kPa

Reduction coefficient = 0.65

(owing to the curved nature of the exposed surface)

Total wind force = $0.65 \times 20 \times 3 \times 1.5$ kN

= 58.5 kN

Moment about the base,

$$M = 58.5 \times \frac{20}{2} = 585 \text{ kN.m}$$

Total weight, W, of the chimney

$$= \frac{\pi}{4}(3^2 - 2^2) \times 20 \times 22 \text{ kN} = 1728 \text{ kN}$$

$$A = \frac{\pi}{4}(3^2 - 2^2) = 1.25\pi \text{ m}^2$$

$$I = \frac{\pi}{64}(3^4 - 2^4) = \frac{65}{64}\pi \text{ m}^4$$

Direct stress,

$$p_d = \frac{W}{A} = \frac{1728}{1.25\pi} = 440 \text{ kPa } (C)$$

Bending stress,

$$p_b = \frac{Mc}{I} = \frac{585 \times 1.5 \times 64}{65 \times \pi} \text{ kPa} = 275 \text{ kPa } (C \text{ or } T)$$

Maximum pressure,

$$p_{max} = 440 + 275 = 715 \text{ kPa } (C)$$

Minimum pressure,

$$p_{min} = 440 - 275 = 165 \text{ kPa } (C)$$

If w kPa is the limiting wind pressure for no tension, $p_b = p_d$

$$M = 0.65 \times 20 \times 3w \times \frac{20}{2} = 390w \text{ kN.m}$$

$$\therefore \quad \frac{390w \times 1.5 \times 64}{65\pi} = 440$$

or $$w = \frac{65\pi \times 440}{64 \times 1.5 \times 390} \text{ kPa} = 2.4 \text{ kPa}$$

PROBLEM 29.10: A tapering chimney of circular section, 50 m high, 4 m external diameter at the base and 3 m external diameter at the top, is subjected to a uniform wind pressure of 1 kPa. Calculate the overturning moment at the base. If the weight of the chimney is 6000 kN and the internal diameter at the base is 2 m, calculate the extreme intensities of stress at the base section.

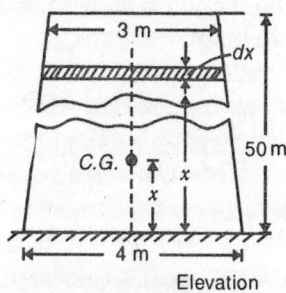

3 m

dx

C.G.

50 m

x

x

4 m

Elevation

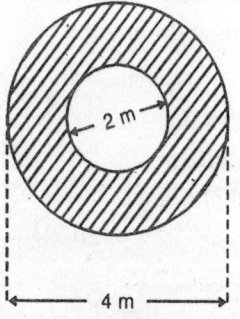

2 m

4 m

Base section

Fig. 29.27: Tapering Chimney (Illustrative Problem 29.10)

Solution: Referring to Fig. 29.27,

At a height of x m above the base, width of strip

$$= \left(4 - \frac{x}{50}\right) m$$

Wind force on the strip $= \left(4 - \frac{x}{50}\right) \times 1 \times dx$ kN

Moment of this force about the base, $dM =$

$\left(4 - \frac{x}{50}\right) x.\, dx$ kNm. Total overturing moment about the base,

$$M = \int_0^{50} \left(4x - \frac{x^2}{50}\right) \cdot dx$$

$$= 4,167 \text{ kN.m}$$

Direct stress at the base section,

$$p_d = \frac{6000}{\frac{\pi}{4}(4^2 - 2^2)} = 636 \text{ kPa } (C)$$

Bending stress at the base section,

$$p_b = \frac{Mc}{I} = \frac{4,167 \times 2}{\frac{\pi}{64}(4^4 - 2^4)} = 700 \text{ kPa } (C \text{ or } T)$$

Maximum stress,

$$p_{max} = (636 + 700) = 1336 \text{ kPa } (C)$$

Minimum stress,

$$p_{min} = (700 - 636) = 64 \text{ kPa } (T)$$

Note: M may also be obtained as follows:

The height above the base of the centroid of the trapezoidal surface exposed to wind pressure.

$$\bar{x} = \left(\frac{4 + 2 \times 3}{4 + 3}\right) \times \frac{50}{3} = 23.81 \text{ m}$$

Total wind force

$$= \left(\frac{4 + 3}{2}\right) \times 50 \times 1 = 175 \text{ kN}$$

Total wind moment about the base

$$= 175 \times 23.81$$

$$= 4,167 \text{ kN.m}$$

PROBLEM 29.11: A mosonry dam, 6 m high, is 1 m wide at the top and 3.5 m wide at the base, with its water face vertical. Maximum water level is 0.5 m below the top. Calculate the stresses at the base section. Unit weight of masonry is 22 kN/m³.

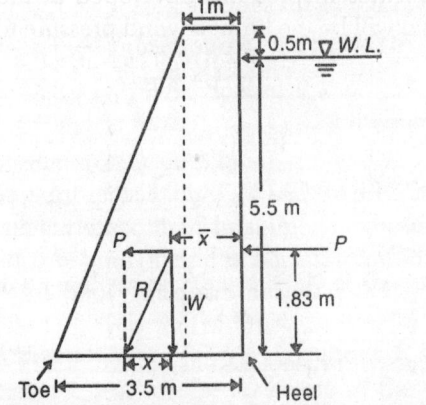

Fig. 29.28: Dam Section (Illustrative Problem 29.11)

Solution:

Per metre run

Wt. of dam, $W = \frac{(1 + 3.5)}{2} \times 6 \times 22 = 297 \text{ kN}$

Centroidal distance from the vertical face,

$$\bar{x} = \frac{(1^2 + 1 \times 3.5 + 3.5^2)}{3(1 + 3.5)} = 1.24 \text{ m}$$

Water force, $P = \frac{1}{2}\gamma_w H^2 = \frac{1}{2} \times 9.81 \times 5.5^2$

$$= 148.23 \text{ kN/m}$$

This acts at $H/3$ or 1.83 m above the base.

The resultant, R, of P and W is shown in the figure.

If x m is the distance from the vertical through C.G. to the point where the resultant, R, strikes the base,

$$\frac{P}{W} = \frac{x}{1.83}$$

or $\quad x = 1.83 \times \frac{P}{W} = \frac{1.83 \times 148.23}{297}$ m

$$= 0.91 \text{ m}$$

∴ Eccentricity, e, of the resultant with respect to the middle of the base

$$= (1.24 + 0.91 - 1.75) = 0.40 \text{ m}$$

Maximum stress at the toe

$$= \frac{W}{b}\left(1 + \frac{6e}{b}\right) = \frac{297}{3.5}\left(1 + \frac{6 \times 0.4}{3.5}\right)$$

$$= 145 \text{ kPa } (C)$$

Minimum stress at the heel

$$= \frac{W}{b}\left(1 - \frac{6e}{b}\right) = \frac{297}{3.5}\left(1 - \frac{6 \times 0.4}{3.5}\right)$$

$$= 26.7 \text{ kPa } (C)$$

PROBLEM 29.12: A mild steel tube, 25 mm external diameter, and 20 mm internal diameter, 2.4 m long, is used as a strut, hinged at both the ends. What is the crippling load by Euler's formula? $E = 2 \times 10^5$ MPa.

Solution: I for the tube section

$$= \frac{\pi}{64}\left(25^4 - 20^4\right) = 11,250 \text{ mm}^4$$

Crippling load, P_{cr}, by Euler's formula,

$$P_{cr} = \frac{\pi^2 EI}{l^2} = \frac{\pi^2 \times 2 \times 10^5 \times 11250}{(2400)^2 \times 1000} \text{ kN}$$

$$= 3.85 \text{ kN}$$

PROBLEM 29.13: A steel bar of rectangular section 25 mm × 50 mm with hinged ends is axially compressed. Determine the minimum length at which Euler's equation can be applied if $E = 2.1 \times 10^5$ MPa and the proportional limit is 210 MPa. If the length is 1.5 m, determine the magnitude of the critical stress.

Solution:

$$I_y \text{ (least value)} = \frac{50 \times 25^3}{12} \text{ mm}^4$$

$$r_y = \sqrt{\frac{50 \times 25^3}{13 \times 1250}} = 7.24 \text{ mm}$$

$$p_{cr} = \frac{\pi^2 E}{(l_e / r)^2}$$

$$\therefore \quad \left(\frac{l_e}{r}\right) = \pi\sqrt{\frac{E}{p_{cr}}} = \pi\sqrt{\frac{2.1 \times 10^5}{210}} = 100$$

$$\therefore \quad l_e = l = 100 \times 7.24 = 724 \text{ mm} = 0.724 \text{ m}$$

If $l = 1.5$ m, $l_e/r = l/r = \dfrac{1500}{7.24} = 207$

$$p_{cr} = \frac{\pi^2 \times 2.1 \times 10^5}{(207)^2} \text{ MPa} = 48.7 \text{ MPa}$$

PROBLEM 29.14: A built-up section is used as a strut, 2.5 m long, with fixed ends. The working stress is given by,

$$p_w = \left[112 - \frac{1}{450}\left(\frac{l_e}{r}\right)^2\right].$$

Area of section = 4000 mm²;
Minimum moment of inertia = 1×10^7 mm⁴. Calculate the safe axial load.

Solution:

$$r = \sqrt{\frac{1 \times 10^7}{4000}} \text{ mm} = 50 \text{ mm}$$

$$l_e = 1.25 \text{ m} = 1250 \text{ mm}$$

$$(l_e / r) = \frac{1250}{50} = 25$$

$$p_w = \left[112 - \frac{1}{450} \times 25^2\right] \text{ N/mm}^2$$

$$= 110.6 \text{ MPa}$$

Safe load, $P = \dfrac{110.6 \times 4000}{1000} \text{ kN} = 442.4 \text{ kN}$

PROBLEM 29.15: Compare the critical loads given by Euler and Rankine formulae for a tubular strut, 2.25 m long, with hinged ends. The external and internal diameters are 75 mm and 65 mm. $E = 2 \times 10^5$ MPa; $f_c = 400$ MPa; and,

Rankine's $b = \dfrac{1}{7500}$.

Solution: $D = 75$ mm; $d = 65$ mm; Hinged ends.

$$\therefore \quad A = \frac{\pi}{4}\left(75^2 - 65^2\right) = 350\pi \text{ mm}^2$$

$$I = \frac{\pi}{64}\left(75^4 - 65^4\right) = \frac{1379}{64} \times 10^4 \pi \text{ mm}^4$$

Euler crippling load,

$$P_e = \frac{\pi^2 EI}{l^2} = \frac{\pi^2 \times 2 \times 10^5 \times 1379 \times 10^4}{64 \times (2250)^2 \times 10} \text{ kN}$$

$$= 263.94 \text{ kN}$$

Rankine crippling load, $P_R = \dfrac{f_c \cdot A}{1 + b\left(\dfrac{l_e}{r}\right)^2}$

$$= \frac{400 \times 350\pi}{10^3 \left[1 + \frac{1}{7500} \times \frac{(2250)^2 \times 64 \times 350\pi}{1379 \times 10^4 \times \pi}\right]}$$

$$= 209.8 \text{ kN.}$$

PROBLEM 29.16: A hollow cylindrical steel strut has to be designed for the following conditions: Length $= 2$ m; Axial load $= 240$ kN; Load factor $= 5$; Ratio of internal to external diameter $= 0.8$. Determine the necessary external diameter and the thickness of metal, if the ends of the strut are fixed. Assume Rankine's constants as $f_c = 320$ N/mm^2 and $b = \frac{1}{7500}$.

Solution: Let the external diameter be D mm; the internal diameter is therefore $0.8\,D$ mm.

Area of section,

$$A = \frac{\pi D^2}{4}\left(1 - 0.8^2\right) = 0.09\pi D^2 \text{ mm}^2$$

$$I = \frac{\pi D^2}{64}(1 - 0.8^4) = \frac{0.59\pi D^4}{64} \text{ mm}^4$$

$$\frac{A}{I} = \frac{0.09\pi D^2 \times 64}{0.59\pi D^4} = \frac{9.763}{D^2}$$

$$l_e = \frac{l}{2} = \frac{2}{2} = 1 \text{ m} = 1000 \text{ mm}$$

$$P_w = 240 \text{ kN}; \quad \text{Load factor} = 5$$

$$\therefore \quad P_{cr} = 5 \times 240 = 1200 \text{ kN,}$$

$$P_{cr} = \frac{f_c \cdot A}{\left\{1 + b\left(\frac{l_e}{r}\right)^2\right\}} \qquad \left(\text{Since } r^2 = \frac{I}{A}\right)$$

$$1200 \times 10^3 = \frac{320 \times A}{\left(1 + \frac{1}{7500} \times \frac{1000 \times 1000 \times A}{I}\right)}$$

$$\left(\text{Since } r^2 = \frac{I}{A}\right)$$

$$120000 + \frac{120000}{7500} \times \frac{10^6 \times A}{I} = 32 \times A$$

$$= 32 \times 0.09\pi D^2 = 9.05 D^2$$

$$120000 + \frac{12 \times 10^2}{75} \times 10^6 \times \frac{9.763}{D^2} = 9.05 D^2$$

$$12 + \frac{15621}{D^2} = 9.05 \times 10^{-4} D^2$$

$$9.05 \times 10^{-4} D^4 - 12 D^2 - 15621 = 0$$

$$D^4 - 1.326 \times 10^4 - 1726 \times 10^4 = 0$$

$$D^2 = \frac{13260 \pm \sqrt{(13260)^2 + 4 \times 1726 \times 10^4}}{2}$$

$$= 14454$$

$$D = \sqrt{14454} = 120 \text{ mm}$$

$\therefore$ External diameter $= 120$ mm
　Internal diameter $= 96$ mm

$$\text{Thickness of metal} = \frac{(120 - 96)}{2} = 12 \text{ mm}$$

Note: If, instead of the ratio between the internal and external diameter, the thickness of metal is given, a cubic equation is to be solved.

PROBLEM 29.17: A mild steel tube is 80 mm external diameter and 75 mm internal diameter. A short length of this tube when tested in compression is found to yield at 400 kN. A 2 m length of the above tube tested as a strut with hinged ends failed at a load of 200 kN. Calculate the Rankine's constant for mild steel, taking yield stress as obtained from the above test. Find the crushing load by Euler's formula. $E = 220$ MPa.

Solution:

$$\text{Area, } A = \frac{\pi}{4}\left(80^2 - 75^2\right) = 608.7 \text{ mm}^2$$

$$I = \frac{\pi}{64}\left(80^4 - 75^4\right) = 4.5746 \times 10^5 \text{ mm}^4$$

Square of radius of gyration

$$r^2 = \frac{I}{A} = \frac{4.5746 \times 10^5}{608.7}$$

$$= 751.536 \text{ mm}^2$$

$$l = l_e = 2 \text{ m} = 2000 \text{ mm}$$

$$f_c = \frac{P_c}{A} = \frac{400 \times 1000}{608.7} \text{ MPa} = 657 \text{ MPa}$$

Critical load $= 200$ kN.

$$200 \times 10^3 = \frac{657 \times 608.7}{1 + b\left(\dfrac{2000 \times 2000}{751.536}\right)}$$

$$1 + b(5322.433) = \frac{400 \times 1000}{200 \times 1000} = 2$$

$$b = \frac{1}{5322.433}, \text{ say } \frac{1}{5320}$$

PROBLEM 29.18: A hollow circular cast-iron column of 250 mm external diameter and 25 mm wall thickness is 5 m long, and is fixed at both ends. It carries a load of 250 kN at an eccentricity of 25 mm from the column axis. Determine the maximum compressive stress. Determine also the maximum eccentricity without allowing tension in the section. $E = 25000 \text{ N/mm}^2$.

Solution:

$$D = 250 \text{ mm}$$

$$d = (250 - 2 \times 25) = 200 \text{ mm}$$

$$A = \frac{\pi}{4}\left(250^2 - 200^2\right) = 225 \times 25\pi \text{ mm}^2$$

$$I = \frac{\pi}{64}\left(250^4 - 200^4\right) = 1.132 \times 10^8 \text{ mm}^4$$

Eccentricity of the load, $e = 25$ mm

$$P = 250000 \text{ N}$$

$$l_e = \frac{l}{2} = \frac{5}{2} = 2.5 \text{ m} = 2500 \text{ mm}$$

$$E = 25000 \text{ N/mm}^2$$

$$p_c = \frac{P}{A} + \frac{P \cdot e \cdot c \cdot \sec\left[(l_e/2)\sqrt{P/EI}\right]}{I},$$

by the secant formula.

$$= \frac{250000}{225 \times 25\pi} + \frac{250000 \times 25 \times 125}{1.132 \times 10^8}$$

$$\sec\left(1250\sqrt{\frac{250000}{25000 \times 1.132 \times 10^8}}\right)$$

$$= 14.147 + 7.405$$

or $\quad p_c = 21.552 \text{ MPa}$

Let the maximum eccentricity for no tension be e_{max} mm.

For this—

$$14.147 = \frac{P}{A} = \frac{P \cdot e_{max} \cdot c \cdot \sec\left[(l_e/2)\sqrt{P/EI}\right]}{I}$$

$$\therefore \quad e_{max} = \frac{14.147 \times I}{P \cdot c \cdot \sec\left(\dfrac{l_e}{2}\sqrt{\dfrac{P}{EI}}\right)}$$

$$= \frac{14.147 \times 1.132 \times 10^8}{250 \times 1000 \times 125 \times 1.073}$$

$$= 47.76 \text{ mm} \approx 50 \text{ mm}$$

PROBLEM 29.19: A tubular steel strut has the following dimensions: length $= 1$ m; outer diameter $= 60$ mm; inner diameter $= 50$ mm. The ends are hinged. Determine the maximum load that may be applied at an eccentricity of 50 mm. $E = 2 \times 10^5$ MPa and $p_w = 225$ MPa

Solution: In this case, modified Rankine formula is easier to apply since the secant formula requires trial and error to solve for P.

$$A = \frac{\pi}{4}\left(60^2 - 50^2\right) = 275\pi \text{ mm}^2;$$

$$l = l_e = 1000 \text{ mm}$$

$$I = \frac{\pi}{4}\left(30^4 - 25^4\right) = 104850\pi \text{ mm}^4$$

$$r^2 = \frac{I}{A} = \frac{104850}{275} = 381.27 \text{ mm}^2$$

Modified-Rankine formula:

$$P_w = \frac{225 \times 275\pi}{10^3\left(1 + \dfrac{50 \times 30}{381.27}\right)\left(1 + \dfrac{1}{7500} \times \dfrac{10^6}{381.27}\right)} \text{ kN}$$

$$= \frac{0.225 \times 275\pi}{6.659756} \text{ kN}$$

$$= 29.2 \text{ kN}$$

Hence the maximum working load $= 29$ kN, Say

PROBLEM 29.20: A column of length l with hinged ends carries a uniformly distributed load w per unit length transversely besides an axial thrust P. Determine the maximum bending moment and the maximum compressive stress.

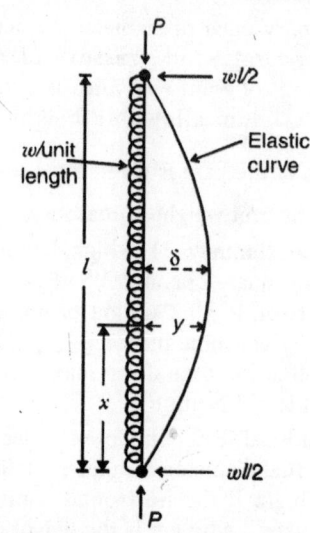

Fig. 29.29: Long Column with Lateral Load
(Illustrative Problem 29.20)

Solution: Referring to Fig. 29.29, B.M. at a section distant x from the lower end A,

$$M = \frac{wlx}{2} - \frac{wx^2}{2} + P \cdot y$$

$$\therefore \quad EI\frac{d^2y}{dx^2} = -\frac{wlx}{2} + \frac{wx^2}{2} - P \cdot y$$

$$\frac{d^2y}{dx^2} + \left(\frac{P}{EI}\right) \cdot y = -\frac{wx(l-x)}{2EI}$$

The solution to this differential equation is,

$$y = C_1 \cos x\sqrt{\frac{P}{EI}} + C_2 \sin x\sqrt{\frac{P}{EI}} - \frac{wx(l-x)}{2P} - \frac{wEI}{P^2}$$

At $x = 0$, $y = 0$ $\therefore C_1 = \frac{wEI}{P^2}$.

At $x = l/2$, $\frac{dy}{dx} = 0$

$$0 = -C_1\sqrt{\frac{P}{EI}} \cdot \sin\frac{l}{2}\sqrt{\frac{P}{EI}} + C_2\sqrt{\frac{P}{EI}} \cdot \cos\frac{l}{2}\sqrt{\frac{P}{EI}}$$

$$\therefore \quad C_2 = C_1 \tan\frac{l}{2}\sqrt{\frac{P}{EI}} = \frac{wEI}{P^2}\tan\frac{l}{2}\sqrt{\frac{P}{EI}}$$

Substituting,

$$y = \frac{wEI}{P^2}\left[\cos x\sqrt{\frac{P}{EI}} + \tan\frac{l}{2}\sqrt{\frac{P}{EI}} \cdot \sin x\sqrt{\frac{P}{EI}}\right]$$

$$- \frac{wx(l-x)}{2P} - \frac{wEI}{P^2}$$

When $x = \dfrac{l}{2}$, $y = \delta$

$$\therefore \delta = \frac{wEI}{P^2}\left[\cos\frac{l}{2}\sqrt{\frac{P}{EI}} + \tan\frac{l}{2}\sqrt{\frac{P}{EI}} \cdot \sin\frac{l}{2}\sqrt{\frac{P}{EI}}\right]$$

$$- \frac{wl^2}{8P} - \frac{wEI}{P^2}$$

or $\quad \delta = \dfrac{wEI}{P^2}\left[\sec\dfrac{l}{2}\sqrt{\dfrac{P}{EI}} - 1\right] - \dfrac{wl^2}{8P}$

The maximum value of the bending moment occurs at the centre of the height.

$$M_{max} = \frac{wl^2}{8} + P \cdot \delta$$

$$= \frac{wl^2}{8} + P\left[\left(\frac{wEI}{P^2}\sec\frac{l}{2}\sqrt{\frac{P}{EI}} - 1\right) - \frac{wl^2}{8P}\right]$$

or $M_{max} = \dfrac{wEI}{P}\left(\sec\dfrac{l}{2}\sqrt{\dfrac{P}{EI}} - 1\right)$ ...(Eq. 29.37)

Using the expansion for $\sec\theta$,

$$\sec\theta = 1 + \frac{\theta^2}{2!} + \frac{50^4}{4!} + \dots$$

$$M_{max} = \left[\frac{wl^2}{8} + P \cdot \left(\frac{5wl^2}{384EI}\right)\right] \text{(approx.)}$$

...(Eq. 29.38)

(neglecting higher powers in the series expansion for $\sec\theta$)

The first term represents the B.M. due to transvers load, and the second term, the B.M. due to axial load to the extent of the maximum lateral deflection under the transverse load.

The maximum compressive stress, p_{max}, is the sum of the values due to direct axial thrust and due to the maximum bending moment:

$$p_{max} = \left(\frac{P}{A} + \frac{M_{max} \cdot c}{I} \right) \qquad \text{...(Eq. 29.39)}$$

PRACTICE PROBLEMS

29.1 A cylindrical bar of diameter d is fixed at the upper end and free at the lower. It has a notch for a small length at the middle such that the notch portion is also cylindrical with diameter $d/2$. The bar carries a pull P along the axis of the bar (Fig. 29.30). Determine the maximum tensile stress.

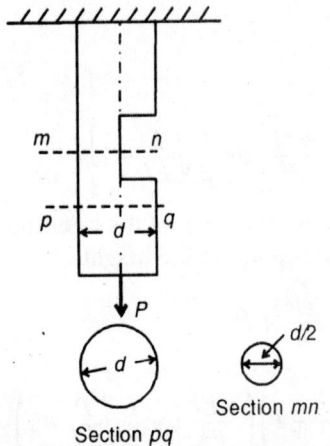

Fig. 29.30: Bar with Notch (Practice Problem 29.1)

29.2 A short concrete pillar of rectangular section 400 mm × 800 mm carries an eccentric load P, acting at an eccentricity of 100 mm with respect to an axis parallel to the 400 mm side and lying on the axis of symmetry parallel to the 800 mm side. Determine the load P if the maximum stress is not to exceed 4 MPa.

29.3 A short hollow column, 240 mm external diameter and 200 mm internal diameter, carries a load of 10 kN through a bracket at an eccentricity of 0.5 mm. Determine the extreme stresses. What is the limiting eccentricity for no tension.

29.4 A short column of 180 mm × 300 mm section carries an axial thrust of 30 kN placed at a point, 90 mm, from either axis. Determine the maximum values of compressive and tensile stresses.

29.5 A short hollow C.I. column with 200 mm external diameter and 120 mm internal diameter is subjected to a compressive load at an eccentricity of 60 mm. Find the allowable load for a maximum compressive stress of 80 MPa.

29.6 A masonry pillar of diameter d is acted on by a horizontal wind pressure of w. If the coefficient of wind resistance is c, prove that the maximum allowable height h for no tension at the base is given by $h = \dfrac{\pi d^2 \gamma}{16wc}$, γ being the unit weight of masonry.

29.7 A square chimney, 24 m high, has an opening of 1.2 m square opening. Wind pressure of 14 kPa acts on it. Unit weight of masonry is 18 kN/m³. Determine the necessary thickness of the wall at the base if the maximum stress is limited to 0.8 N/mm².

29.8 A cylindrical shaft of hollow circular section, 2 m external diameter, 1 m internal diameter, is 30 m high. If the horizontal wind pressure varies as $x^{2/3}$ where x is the height above the ground, calculate the overturning moment at the base, the coefficient of wind pressure being as 0.6. Given that the horizontal wind pressure at a height of 20 m is 1 kPa. Calculate the extreme stresses at the base, assuming the unit weight of masonry as 20 kN/m³.

29.9 A steel bar of rectangular section, 120 mm × 240 mm, with hinged ends, is axially compressed. Determine the minimum length at which Euler's equation can be applied if $E = 2 \times 10^5$ MPa and the proportional limit is 200 MPa. Determine also the critical stress if the length is 5 m.

29.10 A hollow C.I. column with fixed ends is 6 m long. Internal and external diameters are 180 mm and 200 mm, respectively. Calculate the maximum axial load this column can carry, with a factor of safety $2 \cdot f_c = 560$ MPa and Rankine's constant $b = \dfrac{1}{1600}$.

29.11 A stanchion, 7 m long, is made up of an I-section with a 250 mm × 25 mm plate riveted to each of the flanges. It is hinged at the both ends. Calculate the safe load using Rankine's formula with $f_c = 350$ MPa and $b = \dfrac{1}{7500}$.

Factor of safety = 4. For the I–section, $I_x = 2.0458 \times 10^8$ mm^4; $I_y = 6.22 \times 10^6$ mm^4; $A = 7846$ mm^2.

29.12 A hollow C.I. column, 6 m long, with an external diameter of 300 mm has to support a load of 600 kN with hinged ends. Determine the thickness of the metal required if $f_c = 600$ MPa, $b = \dfrac{1}{1600}$, and factor of safety desired is 8.

29.13 A 250 mm × 150 mm RSJ is used as a strut, 6 m long, has one end fixed and the other hinged. Calculate Rankine's crippling load.
$f_c = 330$ MPa and $b = \dfrac{1}{7500}$. $E = 2 \times 10^5$ MPa. Area of section = 7600 mm^2, $I_x = 8.5 \times 10^7$ mm^4, $I_y = 9 \times 10^6$ mm^4. Compare the Rankine value with the Euler-load. For what length of this column will the two formulae yield the same result?

29.14 By means of Rankine's formula, the buckling load for hollow cylindrical pillar, 6 m long, 200 mm external diameter and 150 internal diameter, is found to be 3300 kN. The direct compressive strength of the same material is 570 MPa and the ends are considered fixed. What would be the buckling load if the length were to be 4.5 m, under the same conditions.

29.15 A steel bar of section 30 mm × 60 mm, with hinged ends, is axially compressed. Determine the length of the bar such that the Euler critical stress equals 1.2×10^5 kPa. $E = 2 \times 10^8$ kPa.

29.16 A straight steel pin-jointed strut is 50 mm in diameter and 1.25 m long. Calculate (*a*) the Euler crippling load for axial loading; and (*b*) the eccentricity which will cause failure at 50 % of this load. Yield stress = 285 MPa, $E = 2 \times 10^5$ MPa. Use secant formula and also modified Rankine formula.

29.17 A vertical stanchion, 6 m long, consists of a R.S.J. ($A = 6600$ mm^2; $I_x = 4.72 \times 10^7$ mm^4). An eccentric vertical load P is applied on *y*-axis and at an eccentricity of 150 mm with respect to *x*-axis. Ends are hinged. Find the value of P if the maximum stress is limited 115 MPa (*a*) neglecting deflection, and (*b*) considering deflection. $E = 2 \times 10^5$ MPa. Depth = 200 mm.

29.18 A R.S.J., 250 mm × 150 mm × 6 m long, is hinged at the ends, and carries an axial thrust of 300 kN besides a uniformly distributed load of 6.4 kN/m of a 150 mm side. If the beam is prevented from a lateral buckling, find the maximum deflection, the maximum value of bending moment and the maximum stress in the material. $A = 7280$ mm^2; I about the axis of bending $= 6.78 \times 10^7$ mm^4; $E = 2 \times 10^5$ MPa.

29.19 The coupling rod of a locomotive is 2.5 m long and has an *I*-section, 120 mm deep. $A = 2840$ mm^2 and $I_x = 6 \times 10^6$ mm^4. Internal load due to inertia at full speed is 4 kN/m and the thrust on the rod is 180 kN. Estimate the maximum compressive stress in the rod. $E = 2 \times 10^5$ MPa.

29.20 A column of height l carries an axial thrust P and a horizontal load W at mid-height. Determine the maximum bending moment.

29.21 A vertical strut of length l is fixed at the bottom and free at the top. A horizontal force W and a vertical thrust P act through the centroid of the section. Determine the horizontal deflection at the top in terms of the flexural rigidity, $E I$.

Chapter 30

Stresses in Pressure Vessels

30.1 INTRODUCTION

When vessels such as steam boilers and large pipes are subjected to uniform internal fluid pressure, the material is subjected to principal stresses–one in the direction tangential to the circumference (called the "hoop stress"), another in the direction of the longitudinal axis (called the "longitudinal" or "axial" stress), and the third in a radial direction (called the "radial compression" in this case). Such vessels have large cross-dimensional dimensions in relation to the wall thickness, and are therefore said to be "thin". In such a case, the first principal stress or the hoop stress may be taken to be uniform along the thickness from the inner surface to the outer; further, the third or the radial pressure may be considered to be negligible.

Instead of cylinders, spheres also may be used as pressure vessels. The stresses are similar but the expressions may be a little different.

Thin pressure vessels are not suited to the application of external pressure as they tend to form "lobes", and the assumptions with regard to uniform strains made under internal pressure will no longer be valid.

Thin vessels may be some times fortified by winding wire under pressure on the exterior surface, which induces hoop compression in the vessel; this has to be first neutralised by the hoop tension produced by internal pressure. Thus, the strength of the material is utilised more economically than otherwise. (However, the wire will be under high tensile stresses, which it must be capable of withstanding without failure).

Pressure vessels with a thickness which is not small compared to the cross–sectional dimensions are considered to be "thick", the terms "thick" and "thin" should be based on the non–uniform or uniform distribution of the stresses along the wall thickness, and not on the arbitrarily imposed ratios of diameter to thickness. Stresses in such vessels under the action of internal pressure, external pressure, or both, may be determined at any point in the material, by a theory due to Frenchman, Lame. Behaviour under internal and external pressures will be similar in such vessels.

Thick spherical shells may also be treated in a similar way.

In a manner similar to fortifying thin cylinders by winding a high tensile wire, two thick cylinders can be slipped one over the other to make a "compound" or a "composite" cylinder with a view to economically utilising the strength of the material.

All these concepts are set out in the following sections.

30.2 THIN-WALLED PRESSURE VESSELS

As already stated, a pressure vessel is considered "thin" if the stress distribution across the wall thickness can be treated to be uniform; this is justified when the thickness is small in cmparison with the diameter. Let us now see the analysis of these vessels under the action of internal pressure.

30.2.1 Thin Cylinders

Let a thin cylinder of diameter d and wall thickness t be subjected to uniform internal radial pressure p (Fig. 30.1).

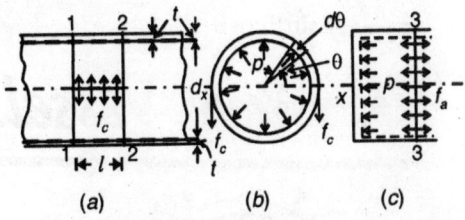

(a)	(b)	(c)

Fig. 30.1: Stresses In a Thin Cylinder under Internal Pressure

Let us consider the equilibrium of a length l between the sections 1–1 and 2–2 of the cylinder [(a)].

The radial force acting on a small area subtending a small angle,

$$d\theta = p\left(l \cdot \frac{d}{2} d\theta\right) \quad [(b)].$$

The component of this force perpendicular to the diameter XX

$$= p \cdot l\left(\frac{d}{2} d\theta\right) \sin\theta$$

$$= pl\frac{d}{2} \sin\theta \cdot d\theta$$

The total force perpendicular to the diameter

$$= \int_0^\pi pl\frac{d}{2} \sin\theta\, d\theta$$

$$= p \cdot l\, d$$

This could have been obtained from the fluid statics principle that the total force in any direction due to uniform radial pressure on a curved surface equals that on the projected area normal to that direction under the same pressure (Of course, the proof for this also is given in precisely the same way as set out here).

If f_c is the circumferential or hoop stress,

resisting force $= 2 \cdot f_c \cdot l \cdot t$

Hence, for equilibrium, $2 f_c \cdot l \cdot t = p \cdot l \cdot d$

$$\therefore \quad f_c = \frac{pd}{2t} \qquad \qquad \text{...(Eq. 30.1)}$$

If the ends are closed by the material, there will be a longitudinal or axial stress f_a in the material [(c)].

The total axial force on the ends (whether plane or curved), due to the fluid pressure

$$= \frac{\pi d^2}{4} \cdot p$$

Resisting force $= \pi d \cdot t \cdot f_a$

Hence, for equilibrium,

$$\pi dt f_a = \frac{\pi d^2}{4} \cdot p$$

$$f_a = \frac{pd}{4t} \qquad \qquad \text{...(Eq. 30.2)}$$

Thus the axial or longitudinal stress is only half the intensity of hoop stress and these are the principal stresses in the material of the cylinder; obviously, both are tensile for an internal radial pressure.

Seamed Cylinders

Large cylinders are usually built-up of plates with riveted joints. The resisting force will be reduced depending upon the efficiency of the riveted joint. The efficiency of longitudinal joint affects the circumferential stress and that of circumferential joint affects the longitudinal or axial stress; these stresses tend to get increased compared to those in a seamless cylinder, since the efficiency of a riveted joint is always less than 100 % or unity.

Thus equations 30.1 and 30.2 will get modified as follows:

$$f_c = \frac{p \cdot d}{2t \cdot \eta_a} \qquad \qquad \text{...(Eq. 30.3)}$$

(η_a: efficiency of longitudinal joint)

$$f_a = \frac{p \cdot d}{4t \cdot \eta_c} \qquad \qquad \text{...(Eq. 30.4)}$$

(η_c: efficiency of circumferential joint)

Strains

Since the principal stresses f_c and f_a are known, principal strains—circumferential or hoop strain, e_c, and axial or longitudinal strain, e_a—can be easily determined.

$$e_c = \frac{f_c}{E} - \frac{f_a}{mE}$$

or $$e_c = \frac{pd}{2tE}\left(1 - \frac{1}{2m}\right) \qquad \text{...(Eq. 30.5)}$$

$$e_a = \frac{f_a}{E} - \frac{f_c}{mE}$$

or $$e_a = \frac{pd}{2tE}\left(\frac{1}{2} - \frac{1}{m}\right) \qquad \text{...(Eq. 30.6)}$$

Since, volume V is given by

$$V = \frac{\pi d^2}{4} \cdot l$$

Differentiating,

$$dV = \frac{\pi}{4}\left[d^2 \cdot dl + 2dl\, d(d)\right]$$

$$\frac{dV}{V} = \frac{dl}{l} + 2\frac{d(d)}{d}$$

$\therefore \qquad e_v = (2e_c + e_a) \qquad \text{...(Eq. 30.7)}$

The changes in diameter, length and volume are, therefore, given by

$$\delta d = e_c \cdot d$$
$$\delta l = e_a \cdot l$$
$$\delta V = e_v \cdot V$$

30.2.2 Thin Spherical Shells

Some times spherical shells also may be employed as pressure vessels for certain special purposes. If the wall thickness of the shell is small in relation to the diameter of the sphere, the variation of srtresses across the thickness may be neglected; in such a case, the shell is considered "thin".

Let us now analyse a thin spherical shell of diameter d under uniform internal fluid pressure p, the wall thickness being t (Fig. 30.2)

The shell will tend to fail along a diameter plane like XX,

Let a thin ring of internal surface $AC-DB$ subtending a small angle $d\theta$ at the centre and situated at an angle θ from the axis XX.

The force acting on this thin ring,

$$= \left(\pi d\cos\theta\right)\left(\frac{d}{2}d\theta\right)p$$

The vertical component of this force

$$= \left(\pi d\cos\theta\right)\left(\frac{d}{2}d\theta\right)p\sin\theta$$

$$= \frac{\pi d^2}{2}\,p\sin\theta\cos\theta \cdot d\theta$$

$$= \frac{\pi d^2}{4} \cdot p\sin 2\theta \cdot d\theta$$

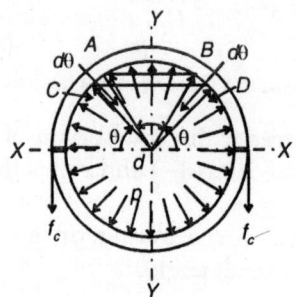

Fig. 30.2: Thin Spherical Shell under Internal Pressure

The bursting force along XX

$$= \frac{\pi d^2}{4}\,p\int_0^{\pi/2}\sin 2\theta \cdot d\theta$$

$$= \frac{\pi d^2}{8} \cdot p[\cos 2\theta]_{\pi/2}^{0} = \frac{\pi d^2}{4} \cdot p$$

(This may also be obtained from the fluid statics principle given in subsection 30.2.1).

Resisting force $= \pi d \cdot t \cdot f_c$

For equilibrium of the shell,

$$\pi d t \cdot f_c = \frac{\pi d^2}{4} \cdot p$$

$\therefore \qquad f_c = \frac{pd}{4t} \qquad \text{...(Eq. 30.8)}$

This is the same as the axial or longitudinal stress in a thin cylindrical shell. Owing to summetry, the principal stresses are $\frac{pd}{4t}$, $\frac{pd}{4t}$, and radial pressure, which may be neglected.

It is obvious that the nature of the stress is tensile for internal pressure. The shell is completely free of shear as the principal stresses are like and equal.

Seamed Shells

Large spherical shells are built-up of plates with riveted joints, the efficiency of which reduce the resisting force as in the case of cylindrical shells. If the joint-efficiency is η, equation 30.8 gets modified as

$$f_c = \frac{pd}{4t\,\eta} \qquad \ldots\text{(Eq. 30.9)}$$

as in the case of cylindrical shells.

The hoop strain, e, is given by

$$e = \frac{f_c}{E} - \frac{f_c}{mE}$$

$$= \frac{f_c}{E}\left(1 - \frac{1}{m}\right) = \frac{pd}{4t\,E}\left(1 - \frac{1}{m}\right) \ldots\text{(Eq. 30.10)}$$

Owing to symmetry, the volumetric strain, e_v, is given by

$$e_v = 3 \cdot e = \frac{3pd}{4t\,E}\left(1 - \frac{1}{m}\right) \qquad \ldots\text{(Eq. 30.11)}$$

The changes in diameter and volume are
$$\delta d = e \cdot d$$
$$\delta V = e_v \cdot V$$

30.2.3 Wire-wound Thin Cylinders

If a wire under tension is closely wound round a pipe, the pipe section will be initially under compression. If now a fluid under pressure is allowed into the pipe, the bursting force will be resisted by the pipe and the wire sections jointly, each offering the necessary tensile resistance. In view of this, a wire–wound cylinder can withstand a greater intensity of internal pressure for a given working stress than an ordinary one.

The principle will be clear from the illustrative problem involving a wire-would cylinder in Section 30.5.

30.3 THICK-WALLED PRESSURE VESSELS

As already observed pressure vessels like water mains, gas mains, and the like, the wall-thickness of which is not small in relation to the cross-sectional dimensions, are called "thick-walled pressure vessels"; when such vessels are subjected to radial pressures, the resulting stresses in the material can be no longer considered to be constant or uniform across the thickness—from the inner surface to the outer. In all these cases, of course, the material is taken to be homogeneous and isotropic.

It was pointed out that a thin cylinder or spherical shell behaves differently under internal and external pressures (forming 'lobes' just before failure in the case of the latter), in much the same way as a slender bar used as a tie behaves differently when used as a strut. Under internal pressure, the cylindrical shape is maintained until failure, while it is not so under external pressure. But in the case of a thick cylinder, the shape is maintained until failure in the case of both internal and external pressures; consequently, the treatment for internal and external pressures is similar, except for the reversal of signs for the stresses. (Also thin tubes with length less than six diameters may behave in a similar way under internal and external pressures, due to the lateral support received from the ends).

It can be easily understood that the circumferential or hoop stress is tensile, for purely internal pressure, and it is compressive for purely external pressure. However, for the simultaneous application of internal and external pressures, the hoop stress depends upon their relative values, and the position of the point considered within the wall thickness.

30.3.1 Thick Cylinders under Internal and External Pressures—Lame's Formula

Now let us consider a thick cylinder under internal and external pressures (radial compressive stresses). The following derivation is after Frenchman Lame.

Let us consider a unit length of the cylinder, with internal and external radii r_i and r_o respectively (Fig. 30.3).

Let p_i and p_o be the internal and external fluid pressures (radial compressive stresses),

$\quad p$ be the internal pressure (compressive) on the ring of thickness δr at a radial distance r,

$(p + \delta p)$ be the external pressure (compressive) on the outside of the ring and f be the circumferential stress (tensile) on the elemental ring at a radius r.

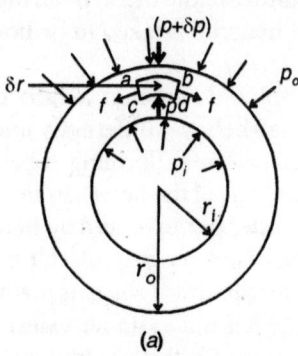

(a)

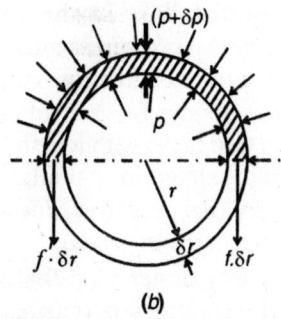

(b)

Fig. 30.3: Thick Cylinder under Radial Pressures—Lame's Analysis

The third principal stress is parallel to the axis of the cylinder, or the longitudional or axial stress.

Considering the equilibrium of half of any thin cylindrical element of radius r, thickness δr, and of unit length, one can write

$$2r \cdot p - 2(r + \delta r)(p + \delta p) = 2f \cdot \delta r$$

$$f \cdot \delta r = -(p \cdot \delta r + r \cdot \delta p), \text{ neglecting the } \delta p$$
$\times \delta r$ as the product of two infinitesimals.

or $\quad f = -\left(p + r \cdot \dfrac{\delta p}{\delta r}\right)$

In the limit when δr is reduced indefinitely,

$$f = -\left(p + r \cdot \frac{dp}{dr}\right)$$

$$= -\frac{d}{dr}(p \cdot r) \qquad \qquad \text{... (Eq. 30.12)}$$

Another relation between p and f may be derived on the basis of longitudinal strains.

It is assumed that plane transverse sections remain plane under pressure—this is almost true at considerable distance from the ends of the cylinder, irrespective of the end conditions. This implies that the longitudinal strain is independent of r, or is a constant. Also the longitudinal stress, f_z, (say tensile), is taken to be uniformly distributed. This can be shown to follow the assumption of uniform longitudinal strain, although it appears to be less obvious.

The longitudinal strain, e_z, is given by

$$e_z = \frac{f_z}{E} - \frac{f}{mE} + \frac{p}{mE}$$

$$= \frac{1}{E}\left[f_z - \frac{(f - p)}{m}\right]$$

Since, according to the assumptions, e_z, f_z and $1/m$ and E are all constants, or independent of r, $(f - p)$ must be constant.

So, let us take

$$(f - p) = 2a \qquad \qquad \text{...(Eq. 30.13)}$$

Substituting this in Eq. 30.12

$$-\left(2p + r \cdot \frac{dp}{dr}\right) = 2a$$

$$\frac{dp}{(p + a)} = -2 \cdot \frac{dr}{r}$$

Integrating both sides,

$$\log(p + a) = -\log r^2 + a \text{ constant}$$

or $\quad (p + a) = \dfrac{b}{r^2}$, b being another constant.

$\therefore \quad p = \dfrac{b}{r^2} - a \qquad \qquad \text{... (Eq. 30.14)}$

Also, $f = \dfrac{b}{r^2} + a \qquad \qquad \text{...(Eq. 30.15)}$

This is the set of equations known as Lame's equations, used in the analysis of thick cylinders under fluid pressure.

The constants a and b may be solved from any two known boundary conditions for p, f, or both.

Applying the boundary conditions—$p = p_i$ for $r = r_i$ and $p = p_o$ for $r = r_o$, the constants b and a can be solved.

$$b = \frac{r_i^2 r_o^2}{(r_o^2 - r_i^2)}(p_i - p_o)$$

and $$a = \frac{(p_i r_i^2 - p_o r_o^2)}{(r_o^2 - r_i^2)}$$

If these are substituted in Eqs. 30.14 and 30.15, we obtain the general expressions for p and f. It is important to note that *positive* values from these equations will indicate *compressive for p* and *tensile for f.*

The formulae appear slightly different with a different, consistent sign convention—positive values indicate compressive stress and negative values tensile stress for both p and f. This is demonstrated in an *alternative proof* given below.

Referring to Fig. 30.4, let the hoop stress f also be assumed to be compressive. The equilibrium of a small element abcd of the ring subtending a small angle $\delta\theta$ at the centre—with a unit length perpendicular to the plane of the figure, or in the axial or longitudinal direction.

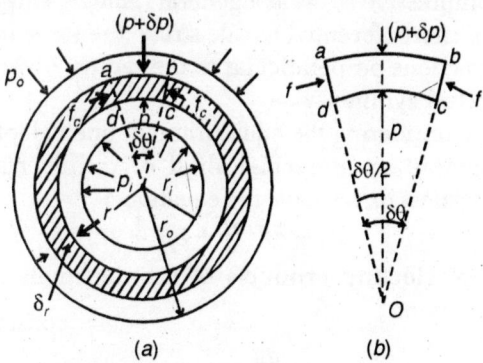

Fig. 30.4: Alternative Proof for Lame's Formula

$$(p + \delta p)(r + \delta r)\delta\theta - p \cdot r \cdot \delta\theta$$

$$= 2f \cdot \delta r \cdot \sin\frac{\delta\theta}{2}$$

Since $\delta r \cdot \delta\theta$ can be neglected and $\sin\dfrac{\delta\theta}{2} = \dfrac{\delta\theta}{2}$

(in radians) for small values,

$$(p \cdot r + r \cdot \delta p + p \cdot \delta r) - p \cdot r \cdot \delta\theta$$

$$= f \cdot \delta r \cdot \delta\theta$$

$$r \cdot \delta p + p \cdot \delta r = f \cdot \delta r$$

$$f = p + r \cdot \frac{\delta p}{\delta r}$$

In the limit when δr is infinitesimally small,

$$f = p + r \cdot \frac{dp}{dr} \qquad \text{...(Eq. 30.16)}$$

If f_z is the longitudinal tensile stress in the ring,

$$e_z = \frac{f_z}{E} + \frac{f}{mE} + \frac{p}{mE}$$

(since p and f are compressive)

$$= \frac{1}{E}\left\{f_z + \frac{(f + p)}{m}\right\}$$

If plane transverse sections remain plane under pressure, e_z is constant. Also if f_z is assumed to be uniform, $(f + p)$ must be independent of r and so a constant.

Let $(f + p) = 2a$ $\qquad$ ...(Eq. 30.17)

From Eqs. 30.16 and 30.17,

$$2p + r \cdot \frac{dp}{dr} = 2a$$

$$2pr + r^2 \cdot \frac{dp}{dr} = 2ar$$

$$\frac{d}{dr}(pr^2) = \frac{d}{dr}(ar^2)$$

$$pr^2 = ar^2 + b$$

or $$p = a + \frac{b}{r^2} \qquad \text{...(Eq. 30.18)}$$

and $$f = a - \frac{b}{r^2} \qquad \text{...(Eq. 30.19)}$$

If positive values are obtained from these equations, compressive stresses are indicated both for p and f, and negative values tensile stresses.

However, the former notation will be employed as embodied in Eqs. 30.14 and 30.15 throughout this book in solving the problems, in order to avoid possible confusion between the two sets of conventions and formulae.

Internal Pressure

For purely internal pressure p_i and no external pressure (*ex*: hydraulic mains above ground), we obtain, from Eqs.30.14 and 30.15,

$$p = p_i \frac{r_i^2}{\left(r_0^2 - r_i^2\right)} \left(\frac{r_0^2}{r^2} - 1\right)$$

and $\quad f = p_i \frac{r_i^2}{\left(r_0^2 - r_i^2\right)} \left(\frac{r_0^2}{r^2} + 1\right)$

The maximum hoop tension occurs at the inner surface (when $r = r_i$) and it is equal to

$$f_{max} = p_i \left(\frac{r_0^2 + r_i^2}{r_0^2 - r_i^2}\right) = p_i \left(\frac{k^2 + 1}{k^2 - 1}\right) \quad \ldots\text{(Eq. 30.20)}$$

where $\quad k = \dfrac{r_0}{r_i}$

This is to be limited to the allowable tensile stress in the material.

The greatest radial compressive stress also occurs at the inner surface ($p = p_i$), which is obvious.

External Pressure

When the cylinder is subjected to external pressure p_o and no internal pressure, we obtain, similarly,

$$p = p_o \frac{r_0^2}{\left(r_0^2 - r_i^2\right)} \left(1 - \frac{r_i^2}{r^2}\right)$$

and $\quad f = -p_o \frac{r_0^2}{\left(r_0^2 - r_i^2\right)} \left(1 + \frac{r_i^2}{r^2}\right)$

(i.e., hoop compression)

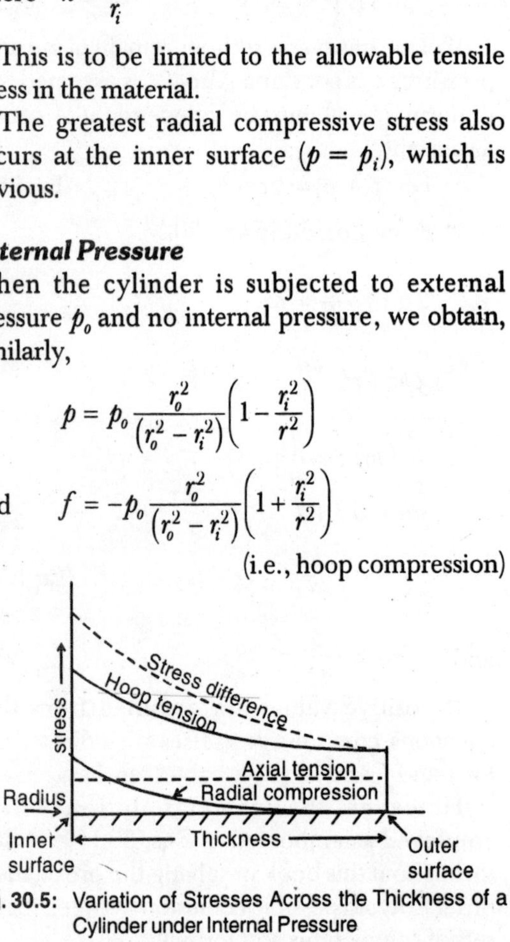

Fig. 30.5: Variation of Stresses Across the Thickness of a Cylinder under Internal Pressure

The maximum hoop compression occurs at the inner surface and is given by

$$f = -\frac{2 p_o \cdot r_0^2}{\left(r_0^2 - r_i^2\right)} \quad \ldots\text{(Eq. 30.21)}$$

In the case of pure internal pressure, the variation of stresses across the thickness is shown qualitatively in Fig. 30.5.

30.3.2 Thick Spherical Shells

Spherical shells subjected to uniform radial pressures can also be dealt with in a similar way as can cylinders be.

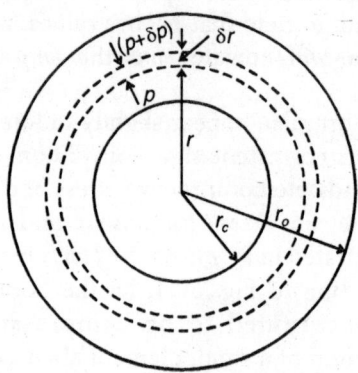

Fig. 30.6: Thick Spherical Shell under Uniform Radial Pressure

Referring to Fig. 30.6, let p be the radial compressive stress at a general radius r, and f be the circumferential tensile stress (the same in all directions perpendicular to the radius, owing to perfect symmetry).

Considering the equilibrium of one half of an elementary spherical shell of radius r and thickness δr, we have the equation

$$\pi r^2 \cdot p - \pi (r + \delta r)^2 (p + \delta p) = 2\pi r \cdot \delta r \cdot f$$

Neglecting products of infinitesimals and simlifying;

$$2f = -2p - r \cdot \frac{dp}{dr}, \text{ in the limit when } \delta r \text{ is}$$

made indefinitely small.

or $\quad 2f = -\dfrac{1}{r} \cdot \dfrac{d}{dr} (r^2 \cdot p) \qquad \ldots\text{(Eq. 30.22)}$

Another equation involving p and f may be obtained from the consideration of strains.

If the displacement of a point in the shell at a radial distance r from the centre is u, the circumferential and radial strains are

$$e_c = \frac{u}{r} \text{ and } e_r = \frac{du}{dr}, \text{ respectively.}$$

[If the radius r increases to $(r + u)$,

circumferential strain, $e_c = \dfrac{2\pi(r + u) - 2\pi r}{2\pi r} = \dfrac{u}{r}$]

[The radial width of the element after the strain will be $r + \delta r + u + \delta u - (r + u) = \delta r + \delta u$

The radial strain, e_r, is the limiting value of

$$\frac{\delta r + \delta u - \delta r}{\delta r} = \frac{\delta u}{\delta r} \text{ or } \frac{du}{dr} \Big]$$

The circumferential strain is the same in all tangential directions, owing to perfect symmetry. But

$$e_c = \frac{1}{E}\left(f - \frac{f}{m} + \frac{p}{m}\right) = \frac{u}{r}$$

$$= \frac{1}{E}\left\{f\left(1 - \frac{1}{m}\right) + \frac{p}{m}\right\} \qquad \text{...(Eq. 30.23)}$$

and

$$e_r = \frac{1}{E}\left(-p - \frac{2f}{m}\right) = \frac{du}{dr}$$

$$= -\frac{1}{mE}(2f + mp) \qquad \text{...(Eq. 30.24)}$$

We have to eliminate u from these two equations.

From Eq. 30.23,

$$u = \frac{r}{mE}\{f(m - 1) + p\}$$

Differentiating with respect to r,

$$\frac{du}{dr} = \frac{r}{mE}\left[(m - 1)\frac{df}{dr} + \frac{dp}{dr}\right] + \frac{1}{mE}[f(m - 1) + p]$$

$$= \frac{1}{mE}\left[r(m - 1) + \frac{df}{dr} \times \frac{r \cdot dp}{dr} + f(m - 1) + p\right]$$

But $\dfrac{uu}{dr} = e_r = -\dfrac{1}{mE}(2f + mp)$ from Eq. 30.24.

Equating these two expressions for $\dfrac{du}{dr}$, and simplifying,

$$r(m - 1)\frac{df}{dr} + r\frac{dp}{dr} + (f + p)(m + 1) = 0$$

$$\text{...(Eq. 30.25)}$$

From Eq. 30.22, $2(f + p) = -r\dfrac{dp}{dr}$

Also differentiating Eq. 30.22 with respect to r,

$$2\frac{df}{dr} = -2\frac{dp}{dr} - r \cdot \frac{d^2p}{dr^2} - \frac{dp}{dr} = -3\frac{dp}{dr} - r\frac{d^2p}{dr^2}$$

Substituting for $(f + p)$ and $\dfrac{df}{dr}$ in Eq. 30.25, and simplifying, we obtain,

$$\frac{d^2p}{dr^2} + \frac{4}{r} \cdot \frac{dp}{dr} = 0 \qquad \text{...(Eq. 30.26)}$$

Solving this as an equation in

$$\frac{dp}{dr}, r^4 \cdot \frac{dp}{dr} = \text{a constant}$$

$$= -6b, \text{ say.}$$

or $$\frac{dp}{dr} = -\frac{6b}{r^4}$$

Integrating, $p = \dfrac{2b}{r^3} + a$ ...(Eq. 30.27)

and $f = \dfrac{b}{r^3} - a$ ...(Eq. 30.28)

(a being the constant of integration)

(According to our original assumption, positive values indicate compressive stress for p and tensile stress for f, in these equations).

As in the case of cylinders, the equations will apear some what similar but slightly different if a consistent sign convention—say, compressive stresses being positive and tensile stresses being negative is used both for p and f. This is demonstrated in the alternative proof given below.

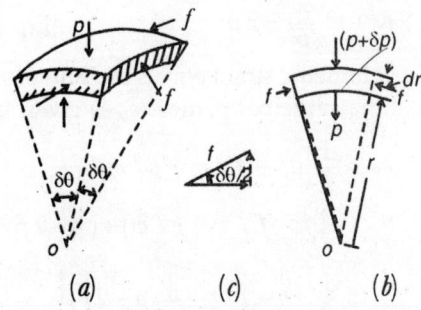

Fig. 30.7: Alternative Proof for Stresses in Spherical Shell

Referring to Fig. 30.7, consider a small portion of the shell, nearly square, with each side subtending an angle $\delta\theta$ at the centre. Let f be the circumferential compressive stress on an elemental ring of radius r and thickness δr, p being internal radial pressure, and $(p + \delta p)$ the external radial pressure on the ring. As usual, let p_i and p_o be the internal and external pressures, the internal and external radii being r_i and r_o respectively.

Net downward radial force

$$= (p + \delta p)\,(r + \delta r)^2\,(\delta\theta)^2 - pr^2(\delta\theta)^2$$

$$= (p + \delta p)\,[r^2 + 2r \cdot \delta r + (\delta r)^2]\,(\delta\theta)^2 - pr^2(\delta\theta)^2$$

Neglecting $(\delta r)^2$ and $\delta p \cdot \delta r$ since they are small, this comes to,

$$pr^2(\delta\theta)^2 + 2pr\,\delta r\,(\delta\theta)^2 + \delta p\,r^2(\delta\theta)^2 - pr^2(\delta\theta)^2$$

Net force $\downarrow V_d = 2pr\,\delta r\,(\delta\theta)^2 + \delta p \cdot r^2 \cdot (\delta\theta)^2$

$$\qquad\qquad\qquad\qquad \ldots(\text{Eq. 30.29})$$

This will be balanced by the vertical components of hoop stress.

Vertical component of hoop stress

$$= f \cdot \left(r + \frac{\delta r}{2}\right) \cdot \delta\theta \cdot \delta r \frac{\delta\theta}{2}, \text{ nearly.}$$

Considering the four faces, this force comes to

$$V_u = 4f \cdot \left(r + \frac{\delta r}{2}\right) \cdot \delta\theta \cdot \delta r \frac{\delta\theta}{2}$$

or $\quad V_u = 2\,fr \cdot \delta r \cdot (\delta\theta)^2 \qquad \ldots(\text{Eq. 30.30})$

Equating V_d and V_u from Eqn. 30.29 and 30.30 for equilibrium, and dividing by $(\delta\theta)^2$, we get

$$2pr \cdot \delta r + r^2 \cdot \delta p = 2\,fr \cdot \delta r$$

or $\quad 2\,pr + r^2 \dfrac{\delta p}{\delta r} = 2\,fr$

In the limit when $\delta r \to 0$, this becomes,

$$2\,pr + r^2 \frac{dp}{dr} = 2\,fr \qquad \ldots(\text{Eq. 30.31})$$

The volumetric strain equals the sum of strains in the three principal planes. e_V is given by

$$e_V = \frac{1}{E}\big[(f - f/m - p/m)$$

$$+ (f - f/m - p/m) + (p - 2f/m)\big]$$

$$= \frac{1}{E}\left[(p + 2f) - \frac{2}{m}(p + 2f)\right]$$

or $\quad e_V = \dfrac{(p + 2f)}{E}\left(1 - \dfrac{2}{m}\right) \qquad \ldots(\text{Eq. 30.32})$

This is a constant since the strain in all the elements will be the same, by symmetry.

$\therefore \ (p + 2f) = \text{a constant} = 3a, \text{ say} \ \ldots(\text{Eq. 30.32})$

Substituting in Eq. 30.31 for $2f$ from this equation,

$$2pr + r^2 \cdot \frac{dp}{dr} = (3a - p)r$$

or $\quad 3pr + r^2 \dfrac{dp}{dr} - 3ar = 0$

or $\quad 3pr^2 + r^3 \dfrac{dp}{dr} - 3ar^2 = 0$

Integrating, $pr^3 - ar^3 = \text{a constant} = b, \text{ say}$

$$\qquad\qquad\qquad\qquad\qquad \ldots(\text{Eq. 30.33})$$

or $\quad p = a + \dfrac{b}{r^3} \qquad\qquad \ldots(\text{Eq. 30.34})$

and $\quad f = a - \dfrac{b}{2r^3} \qquad\qquad \ldots(\text{Eq. 30.35})$

Positive values of p and f indicate compression and negative values tension, when these equations are used.

Irrespective of the set of equations chosen for p and f, the constants a and b can be solved if any two boundary conditions for p, for f, or for both, are known.

The general rexpressions may be written down for p and f, and the salient or maximum values of stresses may be solved for use in design.

Seamed shells may be dealt with as usual by increasing the stresses by dividing them by the joint efficiency.

Volumetric strain is given by Eq. 30.32 and linear, e, strain is one-third of this.

Change in diameter, $\delta d = e \cdot d$
and Change in volume, $\delta V = e_V \cdot V$ $\Big\}$ as usual.

30.4 COMPOUND CYLINDERS

It is seen that when a thick cylinder is subjected to internal pressure, the material near the inner surface carries a much greater hoop stress than that near the outer surface. A more uniform distribution of hoop stress may be achieved by inducing an initial hoop compression near the inner surface. One method to achieve this is to shrink cylinders onto smaller tubes, to produce a compound cylinder; in this, the initial hoop stress in the outer portion will be tensile, that in the inner part being compressive.

When the compound tube sustains an internal fluid pressure, the stresses will be the algebraic sum of initial stresses, and those resulting from this internal pressure as calculated for a single cylinder. The initial stresses may also be calculated as uaual from Lame's equations.

Let us a consider a compound tube with the inner, junction, and outer radii r_i, r_j, and r_o, respectively (Fig. 30.8).

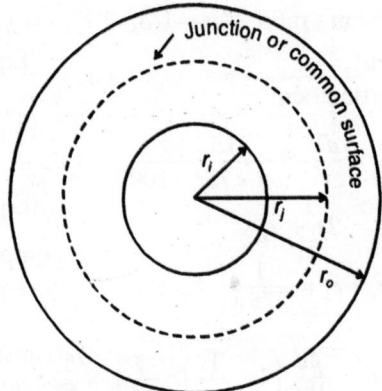

Fig. 30.8: Compound Cylinder

For the inner cylinder

$$p = \frac{b}{r^2} - a \left.\vphantom{\frac{b}{r^2}}\right\}$$

$$f = \frac{b}{r^2} + a$$

p being compressive and
f being tensile, when positive.

For the outer cylinder

$$p = \frac{b_1}{r_o^2} - a_1 \left.\vphantom{\frac{b_1}{r_o^2}}\right\}$$
$$, \text{ similarly.}$$
$$f = \frac{b_1}{r_o^2} + a_1$$

The four conditions necessary to solve for the constants a, b, a_1 and b_1,

(i) $p = 0$ for $r = r_i$;
(ii) $p = 0$ for $r = r_o$;
(iii) p is the same for each cylinder at $r = r_j$;
(iv) The algeberaic difference of the hoop stresses for the cylinders at $r = r_j$, divided by E, equals the original difference of radii at the junction radius r_j before shrinking, divided by r_j, or equals the algebraic sum of the hoop strains.

If a value, p_j, is assigned for $r = r_j$, the conditions (iii) and (iv) will be replaced by—
(iii) $p = p_j$ for $r = r_j$ for the inner cylinder; and
(iv) $p = p_j$ for $r = r_j$ for the outer cylinder.

To explain the condition (iv) on hoop strains more completely—at the junction of the cylinders, the circumferencential strain of the outer cylinder is $\dfrac{f}{E} + \dfrac{p}{mE}$, and so the original dius is increased after the shrinking-on process an amount,

$$r_j\left(\frac{f}{E} + \frac{p}{mE}\right) = r_j\left\{\frac{1}{E}\left(\frac{b_1}{r_j^2} + a_1\right) + \frac{1}{mE}\left(\frac{b_1}{r_j^2} - a_1\right)\right\}$$

The decrease in radius of the inner cylinder at radius r_j is similarly,

$$-r_j\left\{\frac{1}{E}\left(\frac{b}{r_j^2} + a\right) - \frac{1}{mE}\left(\frac{b}{r_j^2} - a\right)\right\}$$

The difference of original radii at the junction surface is

$$\frac{r_j}{E}\left\{\left(\frac{b_1}{r_j^2} + a_1\right) - \left(\frac{b}{r_j^2} + a\right)\right\}$$

Since by condition (iii), $\left(\dfrac{b}{r_j^2} - a\right) = \left(\dfrac{b_1}{r_j^2} - a_1\right)$

Hence we obtain

$$\frac{1}{E}\left\{\frac{(b_1 - b)}{r_j^2} + (a_1 - a)\right\}$$
$$= \frac{\text{Original diffrerence of radial jnction}}{r_j}$$

Instead of finding the unknown constants,

since the two quantities $\left(\dfrac{b_1}{r_j^2}+a_1\right)$ and $\left(\dfrac{b}{r_j^2}+a\right)$ are the values of f in the outer and inner cylinders respectively at the junction surface, their values in terms of p_j may be written with the necessary change of notation for the outer cylinder. Then the above equation reduces to

$$\frac{p_j}{E}\left\{\left(\frac{r_o^2+r_j^2}{r_o^2-r_j^2}\right)+\left(\frac{r_j^2+r_i^2}{r_j^2-r_i^2}\right)\right\}=\frac{\delta}{2r_j}$$

where δ is the original difference in diameter at the junction.

If δ is known, p_j can be found, and vice versa.

All these principles are illustrated by means of the illustrative problems in the next section.

Note: The student need not remember all these complex equations as it is easy to proceed from fundamentals by applying Lame's equations from the appropriate boundary conditions known, as will be seen from the problems in the next section.

ILLUSTRATIVE PROBLEMS

PROBLEM 30.1: A cylindrical shell, 400 mm in diameter, wall thickness 10 mm, and 2 m long, is subjected to an internal pressure of 4 MPa. Calculate the changes in diameter, length and volume of the shell under pressure. $E=2\times10^5$ MPa; Poisson's ratio $=0.3$

Solution:

Hoop stress, $f_c=\dfrac{pd}{2t}=\dfrac{4\times400}{2\times10}=80$ MPa

Axial stress, $f_a=\dfrac{pd}{4t}=40$ MPa

Hoop strain,

$e_c=\dfrac{f_c}{E}-\dfrac{f_a}{mE}=\dfrac{1}{E}(80-0.3\times40)=\dfrac{68}{2\times10^5}$

$=34\times10^{-5}$

Axial strain,

$e_a=\dfrac{f_a}{E}-\dfrac{f_c}{mE}=\dfrac{1}{E}(40-0.3\times80)=\dfrac{16}{2\times10^5}$

$=8\times10^{-5}$

Volumetric strain, $e_v=(2e_c+e_a)=76\times10^{-5}$

Change in diameter $=e_c\cdot d=34\times10^{-5}\times400$

$=0.136$ mm

Change in length $=e_a\cdot l=8\times10^{-5}\times2000$

$=0.160$ mm

Change in volume $=e_V\cdot V=76\times10^{-5}\times\pi\times(200)^2\times2000$

$=1,91,009$ mm$^3\approx191$ ml

PROBLEM 30.2: A cylindrical shell, 150 mm internal diameter, wall thickness 5 mm, and 1 m long, is filled with a fluid at atmospheric pressure. If an additional fluid of 15 ml is pumped into the cylinder, calculate the pressure exerted by the fluid on the wall of the cylinder, and the hoop stress induced in the section. $E=2\times10^5$ MPa and Poisson's ratio, $1/m=1/3$.

Solution:

Volumetric strain,

$e_V=\dfrac{dV}{V}=\dfrac{15\times10^3}{\pi\times75^2\times1000}=\dfrac{1}{375\pi}$

$e_V=2e_c+e_a$

$\therefore\ 2e_c+e_a=\dfrac{1}{375\pi}$ ...(i)

But $e_c=\dfrac{pd}{2tE}\left(1-\dfrac{1}{2m}\right)=\dfrac{pd}{2tE}\times\dfrac{5}{6}$

$e_a=\dfrac{pd}{2tE}\left(\dfrac{1}{2}-\dfrac{1}{m}\right)=\dfrac{pd}{2tE}\times\dfrac{1}{6}$

$\therefore\ \dfrac{e_c}{e_a}=5$ ...(ii)

Solving (i) and (ii), $e_a=\dfrac{1}{375\pi\times11}$ and

$e_c=\dfrac{5}{375\pi\times11}$

Also $\dfrac{pd}{2tE}=e_c\times\dfrac{6}{5}=\dfrac{5}{375\pi\times11}\times\dfrac{6}{5}=\dfrac{6}{375\pi\times11}$

$\therefore\ f_c=\dfrac{pd}{2t}=\dfrac{6}{375\pi\times11}\times2\times10^5=92.6$ MPa

Also $p=\dfrac{2t}{d}\cdot f_c=\dfrac{2\times5}{150}\times92.6=6.2$ MPa

PROBLEM 30.3: A thin copper spherical shell, 500 mm in diameter and 5 mm thick, is full of water at atmospheric pressure. If 25 ml of water is further pumped inside the shell, find the increase in the internal pressure of the water. $E = 2.05 \times 10^5$ N/mm². Poisson's ratio $= 0.3$.

Solution:

Volumetric strain, $e_V = 3 \cdot e$

$$e = \frac{e_V}{3} = \frac{25 \times 10^3 \times 3}{4 \times \pi \times (250)^3 \times 3} = \frac{1}{2500\pi}$$

$$e = \frac{f}{E}\left(1 - \frac{1}{m}\right) = 0.7(f/E)$$

$$\therefore \quad f = \frac{E \cdot e}{0.7} = \frac{2.05 \times 10^5 \times 1}{2500\pi \times 0.7} = 37.3 \text{ MPa}$$

But $\quad f = \dfrac{pd}{4t}$

$$\therefore \quad p = \frac{4t}{d} \times f = \frac{4 \times 5}{500} \times 37.3 \approx 1.5 \text{ MPa}$$

PROBLEM 30.4: A thin spherical shell is used to store gas under a pressure of 20 N/mm². The compressed volume of the gas is 2×10^9 mm³. Find the diameter of the spherical vessel when empty, and the thickness of the wall. Allowable stress in the material is 150 N/mm². $E = 2 \times 10^5$ N/mm². Poisson's ratio $= 0.3$.

Solution:

$$f = \frac{pd}{4t} \quad \therefore 150 = \frac{20}{4} \times \frac{d}{t} = \frac{5 \times d}{t}$$

$$\therefore \quad d = 30t \qquad \qquad \dots(i)$$

$$e_V = 3 \cdot e = 3\left[\frac{f}{E}\left(1 - \frac{1}{m}\right)\right] = \frac{3 \times 150}{2 \times 10^5} \times 0.7 = \frac{\delta V}{V}$$

$$= 1.575 \times 10^{-3}$$

$$V + \delta V = 2 \times 10^9$$

$$1 + \frac{\delta V}{V} = \frac{2 \times 10^9}{V}$$

$$\therefore \quad \frac{2 \times 10^9}{V} = 1.001575$$

$$\therefore \quad V = \frac{2 \times 10^9}{1.001575} \text{ mm}^3 = \frac{\pi d^3}{6}$$

$$\therefore \quad d = \sqrt[3]{\frac{6 \times 2 \times 10^9}{1.001575 \times \pi}} \text{ mm} \approx 1560 \text{ mm}$$

Using (i), $t = \dfrac{d}{30} = \dfrac{1560}{30} = 52$ mm

PROBLEM 30.5: A copper tube of inside and outside diameters 50 mm and 52.5 mm, is wound closely with a 0.7 mm diameter steel wire under tension. When an internal pressure of 14 bar is applied to the copper tube, the net tensile stress in the tube is to be 7 MN/m². Find the winding tension. $E_c = 0.6\, E_s$. Poisson's ratio of copper $= 0.3$. 1 bar $= 10^5$ N/m².

Solution:

Before application of internal pressure

Consider 7 mm length of tube for convenience in mathematical manipulation.

Number of turns of steel wire $= \dfrac{7}{0.7} = 10$

Number of sections of steel wire $= 10 \times 2 = 20$

Crushing force, P_c, on the pipe due to 20 sections of steel wire under a tensile stress of f N/m² is

$$P_c = 20 \times \frac{\pi}{4} \times (0.7)^2 \times f \text{ newtons.}$$

If f_c is the compressive stress induced in the pipe, the resisting force is $f_c \cdot 2t \times 7$

$$\therefore \quad f_c = \frac{20 \times \pi \times 0.49 \times f}{4 \times 2 \times 1.25 \times 7} = 0.14\,\pi f \text{ N/mm}^2$$

(compressive)

Fluid pressure $= 14 \times 10^5$ N/m² $= 1.4$ N/mm²

After application of fluid pressure

Let the pipe section resist it with a stress f_p and the wire with a stress f_w.

Bursting force for a length of 7 mm

$$= 1.4 \times 50 \times 7 = 490 \text{ N}$$

Resisting force,

$$f_p \times 2.5 \times 7 + f_w \times 20 \times \frac{\pi}{4} \times 0.49 = 490$$

$$\therefore \quad f_p + 0.14\pi f_w = 28 \qquad \dots(i)$$

The circumferential strain in the pipe and wire must be the same.

$$e_{c_p} = \left(\frac{f_p}{E_c} - \frac{1}{m} \cdot \frac{pd}{4t} \cdot \frac{1}{E_c}\right)$$

$$= \frac{1}{E_c}\left[f_p - \frac{0.3 \times 1.4 \times 50}{4 \times 1.25}\right]$$

$$= \frac{1}{E_c}\left[f_p - 4.2\right]$$

$$e_{c_w} = \frac{f_w}{E_s}$$

$$\therefore \quad \frac{f_w}{E_s} = \frac{1}{E_c}(f_p - 4.2) = \frac{(f_p - 4.2)}{0.6\,E_s}$$

$$f_p - 0.60\,f_w = 4.2 \qquad \qquad ...(ii)$$

Solving (*i*) and (*ii*), $f_w = 31$ N/mm^2 (Tensile)

$$f_p = 22.8 \text{ N/mm}^2 \text{ (Tensile)}$$

Resultant tensile stress in the tube

$$= 22.8 - 0.44\,f = 7$$

$$\therefore \quad f = 35.91 \text{ N/mm}^2$$

Hence, the winding tension is 36 N/mm^2, nearly.

PROBLEM 30.6: Determine the circumferential stresses at the inside and outside of a pipe, 100 mm internal diameter and 180 mm external diameter when an internal fluid pressure of 7 MPa acts.

Solution:

$p = 7$ MPa at $r = 50$ ⎱ are the known boundary
$p = 0$ at $r = 90$ ⎰ conditions.

Applying Lame's equation for p,

$$7 = \frac{b}{2500} - a \quad \Big\} \quad \text{solving for } a \text{ and } b,$$

$$0 = \frac{b}{8100} - a \quad \Big\} \quad a = \frac{625}{2} \text{ and } b = \frac{81 \times 625}{2}$$

Now substituting in Lame's equation for f, and simplifying,

Circumferential stress at the inside = 13.25 MPa

Circumferential stress at the outside = 6.25 MPa

PROBLEM 30.7: A hydraulic main is 100 mm internal diamter and 150 mm external diameter. What is the allowable internal pressure with a permissible tensile stress of 28 MPa?

Solution:

$p = 0$ at $r = 75$ ⎱ are the boundary
$f = 28$ MPa   at $r = 50$ ⎰ conditions known.

(since f_{max} occurs at the inside of the main)

Applying Lame's equations for p and f respectively,

$$0 = \frac{b}{75^2} - a \quad \Big\} \text{ Solving,}$$

$$28 = \frac{b}{50^2} + a \quad \Big\} \, a = 12.44 \text{ and } b = 5625 \times 12.44$$

The required internal pressure p_i is got from

$$p_i = \frac{b}{50^2} - a = \left(\frac{5625 \times 12.44}{50^2} - 12.44\right)$$

$$= 15.56 \text{ MPa}$$

PROBLEM 30.8: Determine the necessary thickness of a 120 mm hydraulic main to resist a pressure of 7 MPa if the allowable tensile stress is 10.5 MPa. What is the intensity of stress at the outer surface?

Solution:

$p = 7$ MPa   at $r = 60$ mm
$f = 10.5$ MPa   at $r = 60$ mm

$$7 = \frac{b}{60^2} - a \quad \Big\} \text{ Solving,}$$

$$10.5 = \frac{b}{60^2} + a \quad \Big\} \, a = 1.75 \text{ and } b = 3600 \times 8.75$$

We know $p = 0$ at $r = r_o$

$$\therefore \quad 0 = \frac{3600 \times 8.75}{r_o^2} - 1.75$$

or $r_o = \sqrt{\dfrac{3600 \times 8.75}{1.75}} = 134.2$ mm

The thickness,

$$t = (r_o - r_i) = 134.2 - 60$$

$$= 74.2 \text{ mm, or say 75 mm}$$

Stress at the outer surface $\left\{\dfrac{3600 \times 8.75}{(134.2)^2} + 1.75\right\}$

$$= 3.5 \text{ MPa}$$

PROBLEM 30.9: A cylinder, 150 mm internal diameter and 25 mm thick, is subjected to internal fluid pressure. A thrust of 500 kN is applied at the ends. Find the greatest fluid pressure such that the maximum stress in the material does not exceed 60 N/mm^2.

Solution: The axial thrust merely tends to decrease the longitudinal tensile stress due to fluid pressure, which itself is less than the maximum hoop tensile stress. Thus, the greatest principal stress is nothing but the maximum hoop stress. Therefore, we can proceed as usual:

$$\left.\begin{array}{l} p=0 \ \text{ at } r=100 \text{ mm} \\ f=60 \ \text{ at } r=75 \text{ mm} \end{array}\right\} \begin{array}{l} \text{are the known} \\ \text{conditions.} \end{array}$$

$$\left.\begin{array}{l} 0=\dfrac{b}{(100^2)}-a \\ \\ 60=\dfrac{b}{(75^2)}+a \end{array}\right\} \begin{array}{l} \text{solving for } a \text{ and } b, \\ \\ a=21.6 \ b=21.6\times10^4 \end{array}$$

Internal fluid pressure $p_i = \dfrac{b}{75^2}-a$

$$= 16.8 \text{ N/mm}^2$$

PROBLEM 30.10: A cylinder, 150 mm internal diameter and 250 mm extenal diameter, is subjected to an internal pressure of 25 MPa and an external pressure of 5 MPa. Find the maximum and minimum hoop stresses, and the radial pressure at 25 mm from the surface.

Solution:

$$p_i = 25 \text{ MPa at } r = 75 \text{ mm}$$

$$p_o = 5 \text{ MPa at } r = 125 \text{ mm}$$

$$\therefore \quad \left.\begin{array}{l} 25=\dfrac{b}{75^2}-a \\ \\ 5=\dfrac{b}{125^2}-a \end{array}\right\} \begin{array}{l} \text{Solving, } b=175781 \\ \\ \text{and } a=6.25 \end{array}$$

Maximum hoop stress at $r = 75$ mm

$$f_{max} = \frac{175781}{5625}+6.25 = 37.5 \text{ N/mm}^2$$

Maximum hoop stress at $r = 125$ mm

$$f_{min} = \frac{175781}{125^2}+6.25 = 17.5 \text{ N/mm}^2$$

Radial pressure p at 25 mm from the surface,

$$p = \frac{175781}{(100)^2}-6.25 = 11.35 \text{ N/mm}^2$$

PROBLEM 30.11: Determine the ratio of thickness to internal diameter for a tube subjected to internal pressure when the ratio of internal pressure to the greatest circumferential stress is 0.5. Find also the change in thickness for a tube, 250 mm in diameter, when the internal pressure is 75 N/mm^2. Take Poisson's ratio = 0.3 and $E = 2 \times 10^5$ N/mm^2.

Solution: Let the internal radius be r_i external radius be r_o, and thickness t.

Let the internal pressure p_i.

Applying Lame's equations,

$$\left.\begin{array}{l} 0=\dfrac{b}{r_o^2}-a \\ \\ p_i=\dfrac{b}{r_i^2}-a \end{array}\right\} \begin{array}{l} \text{Solving these equations} \\ \text{for } b \text{ and } a, \text{ we get the} \\ \text{general equations for } p \\ \text{and } f. \end{array}$$

It has been shown that the maximum hoop stress, f_{max}, at the inner surface is given by,

$$f_{max} = p_i\left(\frac{k^2+1}{k^2-1}\right), \text{ where } k = \frac{r_o}{r_i}$$

But $f_{max} = 2p_i$

$$\therefore \quad 2p_i = p_i\left(\frac{k^2+1}{k^2-1}\right)$$

$$\therefore \quad \left(\frac{k^2+1}{k^2-1}\right)=2$$

or $\quad k^2 = 3$

$$k = \sqrt{3} \quad \therefore \ \frac{r_o}{r_i}=\sqrt{3}$$

$r_o = \sqrt{3} \ r_i$ Thickness, $t = (r_o - r_i) = r_i(\sqrt{3}-1)$

$$\frac{t}{r_i}=(\sqrt{3}-1) \ \therefore \ \frac{t}{2r_i}=\frac{(\sqrt{3}-1)}{2}=0.366$$

If $r_i = 125$ mm, $\quad r_o = 125\sqrt{3}$ mm

$$f_{max} = 2p_i = 2\times75 = 150 \text{ N/mm}^2 \text{ (Tensile)}$$

Hoop strain,

$$e_c = \frac{150}{E}+\frac{0.3\times75}{E}=\frac{172.5}{E} \text{ (Tensile)}$$

Change of thickness

$$e_c(r_o - r_i) = e_c \cdot t$$

$$= \frac{172.5}{2 \times 10^5} \times 125(\sqrt{3}-1) \text{ mm} \approx 0.08 \text{ mm}$$

PROBLEM 30.12: Two thick steel cylinders A and B, closed at the ends, have the same conditions, the outer diamter of each being 1.6 times the inner. The cylinder A is subjected to internal fluid pressure only and cylinder B to external fluid pressure only. Find the ratio of the pressures on these cylinders:

(*i*) When the greatest noop strain is numerically the same for both the cylinders; and

(*ii*) When the greatest hoop stress is numerically the same for both the cylinders. Poisson's ratio = 0.3.

Solution: Referring to Fig. 30.9,

(*i*) Cylinder A

$$f_{max} = p_i \left(\frac{k^2 + 1}{k^2 - 1} \right)$$

$$k = \frac{r_e}{r_i} = 1.6$$

$$\therefore \quad f_{max} = p_i \left(\frac{1.6^2 + 1}{1.6^2 - 1} \right) = 2.282 \, p_i$$

Maximum hoop strain,

$$e_{max} = \frac{f_{max}}{E} + \frac{0.3 \, p_i}{E} = \frac{2.282 \, p_i}{E} + \frac{0.3 \, p_i}{E}$$

$$= \frac{2.582 \, p_i}{E}$$

Cylinder B

$$f_{max} = \frac{-2 p_o \cdot r_o^2}{(r_o^2 - r_i^2)} = -\frac{2 p_o}{\left(1 - \frac{1}{k^2}\right)}$$

$$= -\frac{2 p_o}{\left(1 - \frac{1}{1.6^2}\right)}$$

$$= -3.282 \, p_o$$

$$e_{max} = \frac{f_{max}}{E} = \frac{-3.282 \, p_o}{E}$$

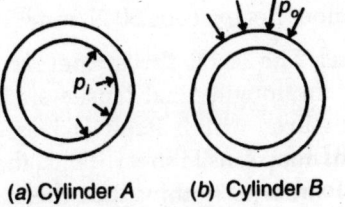

(a) Cylinder A   **(b)** Cylinder B

Fig. 30.9: Cylinders under Internal and External Pressures (Illustrative Problem 30.12)

Equating the numerical values of e_{max} for cylinders A and B,

$$2.582 \, p_i = 3.282 \, p_o$$

$$\therefore \quad \frac{p_i}{p_o} = \frac{3.282}{2.582} = 1.27$$

(*ii*) Cylinder A

$$f_{max} = 2.282 \, p_i$$

Cylinder B

$$f_{max} = -3.282 \, p_o$$

$\therefore$ Equating f_{max} values numerically,

$$2.282 \, p_i = 3.282 \, p_o$$

$$\therefore \quad \frac{p_i}{p_o} = \frac{3.282}{2.282} = 1.44$$

PROBLEM 30.13: A spherical shell is 300 mm external diameter and 50 mm thick. What is the tensile stress at the inner and outer surfaces when the internal pressure is 7 MPa?

Soution:

$$r_i = 100 \text{ mm} \quad r_o = 150 \text{ mm}$$

$$\left. \begin{array}{l} p = \dfrac{2b}{r^3} + a \\[2mm] f = \dfrac{b}{r^3} - a \end{array} \right\} \begin{array}{l} \text{When } r=150, \, p=0 \; \therefore 0 = \dfrac{2b}{150^3} + a \\[2mm] \quad r=100, \, p=7 \\[2mm] \qquad\qquad 7 = \dfrac{2b}{100^3} + a \end{array}$$

Solving for a and b, $b = 4.9737 \times 10^6$ and $a = -2.9474$

At $r = 100$ mm,

$$f = \frac{4.9737 \times 10^6}{100 \times 100 \times 100} + 2.9474$$

$$= 7.92 \text{ MPa (Tensile)}$$

At $r = 150$ mm,

$$f = \frac{4.9737 \times 10^6}{15^3 \times 10^3} + 2.9474$$

$$= 4.42 \text{ MPa (Tensile)}$$

PROBLEM 30.14: What should be the thickness of a metal in a spherical shell, 250 mm radius, resisting an internal pressure of 1.4 MPa if the maximum tensile stress is limited to 3.5 MPa?

Solution:

$$\left. \begin{array}{l} p = \dfrac{2b}{r^3} + a \\[3mm] f = \dfrac{b}{r^3} - a \end{array} \right\} \begin{array}{l} \text{When } r = 250, \ p_i = 1.4 \text{ MPa} \\ r = 250, f_i = 3.5 \text{ MPa} \quad \text{Let } r_o \text{ be the} \\ \text{outer radius} \\ \text{in mm} \end{array}$$

$$\left. \begin{array}{l} 1.4 = \dfrac{2b}{250^3} + a \\[3mm] 3.5 = \dfrac{b}{250^3} - a \end{array} \right\} \begin{array}{l} \text{Solving,} \\[2mm] b = \dfrac{4.9 \times 250^3}{3} \quad a = -1.867 \end{array}$$

But the external pressure is known to be zero.

$$\therefore \qquad 0 = \frac{2b}{r_o^3} + a$$

or $\qquad \dfrac{2 \times 4.9 \times 250^3}{3 r_o^3} - 1.867 = 0$

or $\qquad r_o = \sqrt[3]{\dfrac{2 \times 4.9 \times 250^3}{3 \times 1.867}}$ mm

$$= 250 \sqrt[3]{\frac{9.800}{5.600}} \text{ mm} = 300 \text{ mm}$$

$\therefore$ Thickness $= (r_o - r_i) = (300 - 250) = 50$ mm

PROBLEM 30.15: A compound cylinder is made by shrinking one tube on to another, the final dimensions being—internal diameter 100 mm; external diamter 200 mm; junction diameter = 150 mm. If the radial pressure at the junction due to shrinking on process is 17.5 MPa, find the maximum hoop stress in the compound cylinder. What should be the difference between the external diameter of inner tube and the internal diameter of the outer tube before shrinking-on? What should be the least difference

of temperature necessary to allow the outer tube to pass over the inner?

If the compound cylinder is subjected to an internal pressure of 100 MPa, find the hoop stress at salient points.

If a single cylinder is designed to withstand the same pressure with the same maximum hoop tension, compare its weight with that of the compound cylinder. $E = 2 \times 10^5$ MPa and coefficient of linear expansion $= 11 \times 10^{-6}/^\circ$C

Solution:

Shrinking-on stresses

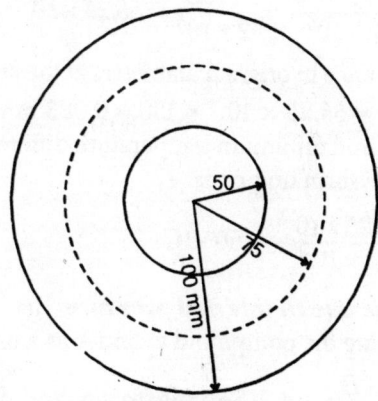

Fig. 30.10: Compound Cylinder (Illustrative Problem 30.15)

Inner tube:

$$0 = \frac{b}{50^2} - a$$

$$17.5 = \frac{b}{75^2} - a$$

Solving for a and b, and substituting in Lame's equation for f,

$$f_{50} = -63 \text{ MPa (i.e., compressive)}$$

$$f_{75} = -45.5 \text{ MPa (i.e., compressive)}$$

Outer tube:

$$0 = \frac{b_1}{100^2} - a_1$$

$$17.5 = \frac{b_1}{75^2} - a_1$$

Solving for a_1 and b_1 and substituting in Lame's equation for f,

$$f_{75} = 65 \text{ MPa (tensile)}$$

$f_{100} = 45$ MPa (tensile)

Circumferential tensile strain at the junction, in the outer tube is,

$$\frac{65}{2 \times 10^5} + \frac{1}{m} \cdot \frac{(17.5)}{2 \times 10^5}$$

Circumferential compressive strain at the junction, in the inner tube is,

$$\frac{45.5}{2 \times 10^5} - \frac{1}{m} \cdot \frac{(17.5)}{2 \times 10^5}$$

Net strain in junction is

$$\frac{(65 + 45.5)}{2 \times 10^5} = \frac{110.5}{2 \times 10^5} = 55.25 \times 10^{-5}$$

∴ Difference in original diameters at the junction
$$= 55.25 \times 10^{-5} \times 150 = 0.083 \text{ mm}$$

Required minimum temperature difference to allow the shrinking-on is

$$\frac{55.25 \times 10^{-5}}{11 \times 10^{-6}} = 50°\cdot 3 \text{ C}$$

Stresses due to internal pressure

Treating the compound cylinder as a unit,

$$\left. \begin{array}{l} 0 = \dfrac{B}{100^2} - A \\[3mm] 100 = \dfrac{B}{50^2} - A \end{array} \right\}$$
Solving for A and B, and substituting in the Lame's formula for f, we can get the hoop stresses.

$f_{50} = 166.7$ MPa (tensile)

$f_{75} = 92.7$ MPa (tensile)

$f_{100} = 66.7$ MPa (tensile)

Resultant hoop stresses

On combining algebraically,

$f_{50} = 103.7$ MPa (tensile);

$f_{75} = 47.2$ MPa (tensile) (inner tube)

$f_{75} = 157.7$ MPa (tensile) (outer tube);

$f_{100} = 111.7$ MPa (tensile)

Single tube

$$\left. \begin{array}{l} 100 = \dfrac{B_1}{50^2} - A_1 \\[3mm] 157.7 = \dfrac{B_1}{50^2} + A_1 \end{array} \right\}$$
Solving for A_1 and B_1, and substituting in the Lame's formula for p at the outer surface.

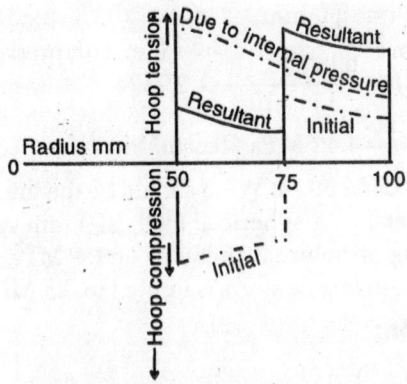

Fig. 30.11: Variation of Stresses in Compound Cylinder (Illustrative Problem 30.15)

$$0 = \frac{B_1}{r_o^2} - A_1$$

and solving for r_o^2, we get $r_o^2 = 11,200$

Area of the single tube is proportional to $(11200 - 2500)$; and area of the compounding tube is proportional to $(10000 - 2500)$.

Ratio of the weights of single tube and compound tube is given by

$$\frac{11200 - 2500}{7500} \text{ or } 1.16$$

The variation of the hoop stresses across the thickness is shown in Fig. 30.11.

PROBLEM 30.16: A compound cylinder is formed by shrinking a tube of 160 mm diameter external and 120 mm diameter internal, on to another tube with an internal diameter of 80 mm. The initial difference of the diameters of the tubes at the junction was 0.1 mm. Determine the radial pressure at the junction due to shrinking-on. $E = 2 \times 10^5$ MPa.

Solution: Let the junction pressure be p N/mm².

Inner tube

$$\left. \begin{array}{l} 0 = \dfrac{b}{40^2} - a \\[3mm] p = \dfrac{b}{60^2} - a \end{array} \right\}$$
Solving for a and b,
$a = -1.8p$
$b = -2880p$

∴ $f_{40} = \dfrac{b}{40^2} + a = -3.6\,p$ (i.e., compressive)

$$f_{60} = \frac{b}{60^2} + a = -2.6\,p \text{ (i.e., compressive)}$$

Outer tube

$$\left. \begin{array}{l} 0 = \dfrac{b_1}{80^2} - a_1 \\[2mm] p = \dfrac{b_1}{60^2} - a_1 \end{array} \right\} \quad \begin{array}{l} \text{Solving for } a_1 \text{ and } b_1 \text{ and} \\ \text{substituting in Lame's} \\ \text{formula for } f, \text{ we have} \end{array}$$

$$f_{60} = 3.57p \text{ (tensile)}$$

$$f_{80} = 2.57p \text{ (tensile)}$$

Hoop tensile strain at the junction in the outer tube

$$\frac{3.57p}{E} + \frac{p}{mE}$$

Hoop compressive strain at the junction in the inner tube

$$\frac{2.6p}{E} - \frac{p}{mE}$$

∴ Resultant strain at the junction (algebraic difference)

$$= \frac{(3.57 + 2.60)}{E}\,p = \frac{6.17\,p}{E}$$

∴ The difference in junction diameters of the two tubes is

$$\frac{120 \times 6.17\,p}{E}$$

Equating this to the given value of 0.1 mm,

$$\frac{120 \times 6.17\,p}{2 \times 10^5} = 0.1$$

or $$p = \frac{2 \times 10^5 \times 0.1}{120 \times 6.17} \text{ MPa} = 27 \text{ MPa}$$

∴ The radial pressure at the junction due to shrinking-on = 27 MPa

PROBELM 30.17: A thick cylinder of inner radius a is made by shrinking, an outer cylinder of inner radius $2a$ and outer radius $4a$ on an inner cylinder of inner radius a and outer radius, $2a$, the material being the same for both. Show that an interference δ gives rise to a shrinkage pressure,

$$p_s = \frac{3E\delta}{20a} \text{ at the common surface, } E \text{ being}$$

the Youngs modulus

Solution:

Compound Cylinder

$$r_i = a; \; r_j = 2a; \text{ and } r_o = 4a$$

Interference (shrinkage allowance in radius) $= \delta$

Outer tube

$$\left. \begin{array}{l} \dfrac{b_o}{16a^2} - a_o = 0 \\[2mm] \dfrac{b_o}{4a^2} - a_o = p_s \end{array} \right\}$$

Solving for a_o and b_o, and substituting in the Lame's formula for f, we get,

$$f_{2a} = \frac{5}{3}\,p_s \text{ (Tensile)}$$

$$f_{4a} = \frac{2}{3}\,p_s \text{ (Tensile)}$$

Inner tube

$$\left. \begin{array}{l} \dfrac{b_i}{a^2} - a_i = 0 \\[2mm] \dfrac{b_i}{4a^2} - a_i = p_s \end{array} \right\}$$

Solving for a_i and b_i, and substituting in the Lame's formula for f, we obtain,

$$f_a = -\frac{8}{3}\,p_s \text{ (i.e., compressive)}$$

$$f_{2a} = -\frac{5}{3}\,p_s \text{ (i.e., compressive)}$$

Hoop tensile strain at the junction in the outer tube $= \dfrac{5p_s}{3E} + \dfrac{p_s}{mE}$

Hoop compressive strain at the junction in the inner tube

$$= \frac{5\,p_s}{3E} - \frac{p_s}{mE}$$

Resultant tensile strain at the junction $= \dfrac{10p_s}{3E}$

∴ Shrinkage allowance in junction radius

$$= \frac{10\,p_s}{3E} \times 2a$$

or, $$\delta = \frac{20\,p_s\,a}{3E}$$

or $$p_s = \frac{3E\delta}{20a}, \text{ as given.}$$

PRACTICE PROBLEMS

30.1 A steam boiler, 3 m internal diameter, wall thickness 20 mm, sustains an internal pressure 2 MPa. Calculate the maximum values of hoop stress, axial stress, and maximum shear stress in the material.

30.2 A marine boiler 4.5 m in diameter and carries a working pressure of 1.2 MPa. The maximum allowable stress in the material is 80 MPa and the efficiency of the riveted joints is to be taken at 80 %. What is the plate thickness necessary?

30.3 A thin spherical shell, 1.2. m in diameter, is subjected to an internal pressure of 2 MPa. Determine the thickness of plate required if the permissible tensile stress is 120 MPa. The joint efficiency may be taken as 80 %.

30.4 A thin spherical shell of 0.45 m internal diameter and thickness of metal 3 mm, is full of fluid at atmospheric pressure. Find the radial pressure exerted on the wall of the shell if 20 ml of the fluid are pumped in. Determine the resulting hoop stress and change in diameter. $E = 2 \times 10^5$ MPa; $1/m = 1/3$.

30.5 A cast-iron pipe, 270 mm in diameter, thickness of metal 9 mm, is closely wound with a layer of 5 mm diameter steel wire under a tensile stress of 50 MPa. If water under a pressure of 3 MPa is admitted into the pipe, calculate the stresses induced in the pipe and the steel wire. E for cast-iron $= 9 \times 10^4$ MPa; E for steel $= 1.8 \times 10^5$ MPa; and Poisson's ratio for cast-iron $= 0.28$.

30.6 A hydraulic main 150 mm internal diameter and 50 mm thickness carries water under a pressure of 75 MPa. Determine the maximum and minimum values of hoop stress, and the hoop stress and radial pressure at the middle of the thickness.

30.7 A hydraulic main is 100 mm diameter and 25 mm thick. What is the allowable internal pressure, if the stress in the material is not to exceed 30 MPa?

30.8 The internal diameter of a hydraulic main is 200 mm and the working stress in the material is 24 MPa. Calculate the thickness required to carry an internal pressure of 15 MPa.

30.9 A thick cylinder of internal diameter 100 mm and external diameter 200 mm is subjected to internal fluid pressure only. If the maximum stress is not to exceed 120 MPa and the maximum principal strain is not to exceed 6×10^{-4}, determine the safe intensity of fluid pressure. $E = 2 \times 10^4$ MPa and $\mu = 0.25$.

30.10 A cylinder, 150 mm internal diameter and 50 mm thick, is subjected to an internal pressure of 30 MPa and an external pressure of 10 MPa. Determine the maximum and minimum hoop stresses, and the radial pressure at mid-thickness.

30.11 A cylinder of internal and external diameters of 400 mm and 500 mm respectively is subjected to an internal pressure of 2 N/mm². Find the maximum circumferential stress, and the percentage error if this stress is computed from the thin cylinder formula.

30.12 A spherical shell of 200 mm external diameter is 50 mm thick. If an internal pressure of 10 MPa acts determine the maximum stress.

30.13 What is the thickness of the metal required for a spherical shell, 200 mm internal diameter, to resist a pressure of 1 MPa with a maximum tensile stress of 2 MPa?

30.14 A steel cylinder of 300 mm extenal diameter is shrunk onto another steel cylinder of 150 mm internal diameter. After shrinking-on, the radial pressure at the junction surface of 250 mm diameter is 28 MPa. Determine the original difference in the radii at the junction. $E = 2 \times 10^5$ MPa.

30.15 A compound cylinder has an external diameter of 200 mm and an internal diameter of 100 mm. The junction diameter is 150 mm. If the shrinkage allowance in diameter is 0.15 mm, find the radial pressure induced. What is the least difference in temperature necessary for shrinking-on. $E = 2 \times 10^5$ MPa, $\alpha = 12 \times 10^{-6}/°C$.

30.16 A compound cylinder is 120 mm internal diameter and 240 mm external diameter, the junction diameter being 180 mm. The junction pressure is 8 MPa. If a fluid under a pressure of 60 MPa is allowed, determine the final hoop stresses in the tube.

30.17 A compound cylinder has inner, junction, and outer diameters of 150, 200 and 250 mm. If the radial pressure at the junction on shrinking-on is 40 MPa, and if a fluid under a pressure 80 MPa is admitted, calculate the final stresses.

30.18 A compound tube is made by shrinking one tube on to another, the final dimensions being: internal diameter = 100 mm; external diameter = 200 mm; junctin diameter = 150 mm. If the junction pressure is 20 MPa, what is the maximum hoop tension? What is the shrinkage allowance? What is the least difference in temperature needed to allow the shrinking-on?

If the compound tube is subjected to an internal pressure of 100 MPa, find the greatest hoop stress?

How much percent heavier will a single tube be in order to withstand this pressure with the same maximum hoop tension? $E = 2 \times 10^5$ MPa and $\alpha = 11 \times 10^{-6}/°C$.

APPENDICES

Appendix A

Mechanical Properties of Engineering Materials

MECHANICAL PROPERTIES OF CERTAIN COMMON ENGINEERING MATERIALS

Sl. No.	Material	Ultimate Strength (MN/m²)			E (MN/m²)	N (MN/m²)	μ
		Tension	Compression	Shear			
1.	Mild steel	450	450	350	2.1×10^5	0.84×10^5	0.25 to 0.30
2.	Tool steel (Carbon, hardened)	1,050	1,000	675	–	–	For most Metals, μ ranges between 0.25 to 0.33
3.	Cast iron	120	600	120	1.0×10^5	0.42×10^5	
4.	Wrought iron	300	280	280	1.8×10^5	0.70×10^5	
5.	Aluminium	120	–	90	0.8×10^5	0.28×10^5	
6.	Copper (cast)	180	250	180	0.7×10^5	0.28×10^5	
7.	Copper (hard drawn)	320	350	210	1.2×10^5	0.42×10^5	
8.	Brass	175	75	110	0.8×10^5	0.30×10^5	
9.	Timber	30 to 90	40 to 100	02 to 10	0.15×10^5	–	–

*These are average values and not unique; they may be determined experimentally wherever possible in all important situations.

Appendix B

Notation

A : Area (Ch. 5)
Area of Section (Ch. 16)
Amplitude (Ch. 20)

A' : Final sectional area of a bar at failure (Ch. 21)

A_m : Mechanical advantage (Ch. 9)

A_1, A_2: Area (Ch. 5)
Modulus of figure (Ch. 7)

a : x-coordinate of the vertex of a parabola (Ch. 5)
Instantaneous accelera-tion (Ch. 11)
Cross-sectional area of belt (Ch. 19)

a_{av} : Average acceleration (Ch. 11)

a_n, a_t: Normal and tangential components of acceleration (Ch. 13)

a_x, a_y: x-and y-components of acceleration (Ch. 13)

a_r, a_θ: Radial and transverse components of acceleration (Ch. 13)

b : Base width of a triangle (Ch. 5 & 6)
Breadth of a rod (Ch. 6)
Breadth of a plate in a leaf spring (Ch. 27)
Rankine's constant (Ch. 29)

b' : Width of a strip (Ch. 5)

C : Constant in the law of machine (Ch. 9)
Instantaneous centre of rotation (Ch. 14)
Coefficient of damping (Ch. 20)
Shear modulus (Ch. 21)

C_c : Coefficient of critical damping (Ch. 20)

C_1, C_2: Constants of integration (Chs. 11 & 20)

c : Extreme fibre distance (Ch. 25)

c_x, c_y : Extreme fibre distances w.r.t. x-and y-axes (Ch. 29)

D : Diameter of circle, plate and sphere (Ch. 6)
Number in Routh's rule (Ch. 6)
Distance moved by effort in a given time (Ch. 9)
Diameter of wheel (Ch. 9)
Diameter of bigger pulley in Weston block (Ch. 9)

d : Depth of a rectangle (Ch. 6)
Diameter of a circle
Lever arm of force (Ch. 3)
Distance moved by load in a given time (Ch. 9)
Diameter of axle (Ch. 9)
Diameter of smaller pulley of a Weston block (Ch. 9)
Mean diameter of screw (Ch. 9)
Distance centre to centre of pulleys (Ch. 19)
Damping factor (Ch. 20)
Diameter of spring wire (Ch. 28)
Diameter of Cylinder (Ch. 30)

d_1, d_2: Diameter of bigger and smaller axles (Ch. 9)

$\left. \begin{array}{l} d_1, d_2 : \\ d'_1 \ d'_2 : \end{array} \right\}$ Lever arms of forces (Ch. 3)

dm	:	Elemental mass (Ch. 6)
dt	:	Elemental time (Ch. 12)
dr	:	Elemental radial thickness (Chs. 5 & 6)
$d\theta$	:	Elemental angle (Ch. 5)
E	:	Modulus of Elasticity (Chs. 21 & 25)
E_k	:	Kinetic energy ⎫ (Ch.16)
E_p	:	Potential energy ⎭
e	:	Coefficient of restitution (Ch. 18)
		Eccentricity of loading (Ch. 29)
e_x, e_y	:	Ecentricities w.r.t. x-and y-axes (Ch. 29)
e_1, e_2	:	Constants in Eq. 21.23 (Ch. 21)
F	:	Frictional force (Ch. 8)
		Frction loss in a machine (Ch. 9)
		Shearing force (Ch. 24)
F_c	:	Centrifugal force (Ch.19)
F_k	:	Kinetic friction force (Ch. 8)
F_s	:	Static friction force (Ch. 8)
f	:	Frequency or number of cycles per second (Ch. 20)
		Normal stress in bending (Ch. 25)
		Hoop stress (Ch. 30)
f_a	:	Axial stress in a cylinder (Ch.30)
f_c, f_t	:	Compressive and tensile stresses in bending (Ch. 25)
f_c	:	Critical compressive stress for short bars (Ch. 29)
		Circumferential stress in a cylinder (Ch. 30)
f_p	:	Principal stress in a shaft under tension and bending (Ch. 28)
f_s	:	Shear stress in bending (Ch. 26)
$f_{s_{max}}$	:	Maximum shear stress in bending (Ch. 26)
$f_{s_{mean}}$	:	Mean shear stress in bending (Ch. 26)
$f_y, f_{y,p}$	:	Yield stress (Ch. 29)
G	:	Universal constant of gravitation (Ch. 15)
		Shear modulus (Ch. 21)
g	:	Acceleration due to gravity (Ch. 11)
H	:	Maximum height reached by a projectile (Ch. 12)
h	:	Altitude of a triangle (Ch. 5)
		Height of cylinder (Ch. 6)

		Difference in elevation (Ch. 16)
		Height of fall of hammer (Ch. 17)
		Perpendicular distance between parallel axes (Ch. 6)
I	:	Second moment of area/Moment of inertia (Ch. 6)
		Input (Ch. 9)
		Path of motion (Ch. 16)
		Moment of inertia of a beam section about neutral axis (Ch. 27)
I_g	:	Moment of inertia about centroidal axis (Ch. 6)
I_x, I_y	:	Second moment of area about x-, y -axis (Ch. 6)
I_z	:	Polar moment of inertia (Ch. 6)
i	:	Impulse (Ch. 17)
J	:	Polar moment of inertia of a shaft section (Ch. 28)
j	:	Number of joints of a truss (Ch. 4)
K	:	Bulk modulus (Ch. 21)
k	:	Constant of proportionality (Chs. 5 & 11)
		Radius of gyration (Ch. 6)
		Spring constant or a stiffness of a spring (Chs. 16, 17, & 20)
k_{eq}	:	Equivalent spring constant for compound springs (Ch. 20)
k_g, k_x, k_y	:	Radius of gyration about g-, x-, y-axis (Ch. 6)
k_t	:	Torsional stiffness (Ch. 28)
L	:	Couple (Ch. 3)
		Length of belt (Ch. 19)
$L_1, L_2\ldots$	:	Couples (Ch. 3)
l	:	Length of line segment/arc (Ch. 5)
		Length of inclined plane (Ch. 9)
		Initial length of a bar (Ch. 21)
		Length of spring wire (Ch. 28)
l'	:	Final length of a bar (Ch. 21)
l_e	:	Effective length of a column (Ch. 29)
M	:	Moment of a couple (Ch. 3)
		Magnification factor (Ch. 20)
		Bending moment/couple (Ch. 24)
M_{am}	:	Angular momentum (Ch. 17)
M_m	:	Momentum (Ch. 17)
M_E	:	Equivalent bending moment (Ch. 28)
M_o, M_1, M_2	:	Moments of Couples (Ch. 3)

m : Number of members of a truss (Ch. 4)
Constant of proportionality (Ch. 9)
Mass of a body (Chs. 6 and 15)
Reciprocal of Poisson's ratio (Ch. 21)
Factor of safety (Ch. 29)

m_g : Mass of gun ⎫
⎬ (Ch. 17)
m_s : Mass of shell ⎭

N : Newton (Ch. 1)
Normal reaction (Ch. 8)
Modulus of rigidity (Ch. 21)

n : Number of movable pulleys (Ch. 21)
Revolutions per minute (rpm) (Ch. 14)
Frequency ratio (Ch. 20)
Number of leaves or plates in a leaf spring (Ch. 27)
Number of coils in a helical spring (Ch. 28)

O : Moment centre (Ch. 3)
Output (Ch. 9)

O_1, O_2 : Moment centres (Ch. 3)

P : Force (Ch. 1)
Effort (Ch. 9)
Power transmitted (Ch. 19)
Particle (Ch. 20)

P_{cr} : Critical load on a column (Ch. 29)

P_e : Euler's crippling load (Ch. 29)

P_i : Ideal effort (Ch. 9)

P_{max} : Maximum Power (Ch. 19)

P_x, P_y : x-, y-components of force P (Ch. 2)

P_1, P_2 : Forces (Ch. 2)

P'_1, P'_2 : Forces (Ch. 3)

p : Pitch of screw thread (Ch. 8)
Pressure
Radial pressure (Ch. 30)

p_b : Bending stress ⎫
p_{cr} : Critical stress ⎬ (Ch. 29)
p_d : Direct stress ⎭

p_i : Internal radial pressure (Ch. 30)

p_e : External radial pressure (Ch. 30)

p_j : Radial pressure at junction of compound cylinder (Ch. 30)

p_w : Working stress (Ch. 29)

p_z : Resultant vertical stress (Ch. 29)

Q : Periodic external exciting force (Ch. 20)
First moment of the area of section above a fibre about N.A. (Ch. 26)

Q_o : Maximum exciting force (Ch. 20)

q : Direct shear stress (Ch. 21)

q_{max} : Maximum shear stress (Ch. 22)

R : Resultant force (Ch. 2)
Radius of a circle, sphere, hemisphere, base of cone (Ch. 5)
Radius of crank (Ch. 9)
Range (Ch. 12)
Mean radius of earth (Ch. 15)
Average resistance of soil (Ch. 17)
Internal resistance to an external force (Ch. 21)
Radius of curvature in bending (Ch. 25)
Radius of shaft (Ch. 28)
Mean radius of coils (Ch. 28)

R_x, R_y : x- and y-components of force R (Ch. 2)

R_A, R_B : Reactions at supports A&B (Ch. 24)

R_L, R_R : Reactions at left and right supports (Ch. 7)

R_{max} : Maximum range (Ch. 12)

R_1, R_2 : Forces exerted by a person on a lift (Ch. 15)

r : Radial distance (Ch. 6)
Radius of drum (Ch. 9)
Distance between bodies (Ch. 15)
Junction radius of a compound cylinder (Ch. 30)

S : Force (Ch. 3)

s : Distance travelled (Ch. 11)
Pile set (Ch. 17)
Stiffness of a spring (Ch. 28)

T : Tension in a belt (Ch. 8)
Time of travel of a projectle (Ch. 12)
Period (Chs. 14 & 20)
Torque or twisting moment (Ch. 28)

T_E : Equivalent twisting moment (Ch. 28)

T : Equivalent twisting moment for shear (Ch. 28)

T_i : Initial tension of the belt (Ch. 19)

T_o : Torque (Ch. 20)

T_1, T_2: Tensions on either side of belt (Ch. 8)

T_1 : Number of teeth on pinions (Ch. 9)

T_2 : Number of teeth on spur wheels (Ch. 9)

t : Thickness of a plate (Ch. 6)
Time of travel (Ch. 11)
Time of flight (Ch. 12)
Thickness of a plate in a leaf spring (Ch. 27)
Thickness of cylinder (Ch. 30)

$U^\cdot$: Work done by a force (Ch. 16)
Internal strain energy (Ch. 23)

U_b : Strain energy in bending (Ch. 25)

u : Initial velocity (Ch. 11)
Velocity of projection (Ch. 12)
Strain energy per unit volume (Ch. 23)

u_1, u_2: Velocities of masses m_1 and m_2 before impact (Ch. 18)

V : Velocity Ratio (Ch. 9)
Velocity of hammer and pile immediately after impact (Ch. 17)
Volume of a body (Ch. 21)
Vertical shear force (Chs. 26 & 27)

v : Instantaneous velocity (Ch. 11)
Final velocity (Ch. 11)
Hammer velocity at the instant of striking the pile (Ch. 17)

v_{av} : Average velocity (Ch. 11)

v_A, v_B: Velocities at A, B (Ch. 11)

v_{BA} : Relative velocity of B w.r.t. A (Ch. 11)

$v_{b/a}$: Relative velocity of B w.r.t. A associated with rotation (Ch. 14)

v_g : Velocity of gun ⎫
v_s : Velocity of shell ⎬ (Ch. 17)

v_h, v_v: Horizontal, vertical components of velocity (Ch. 12)

v_u, v_t: Normal, tangential components of velocity (Ch. 13)

v_r, v_θ: Radial, transverse components of velocity (Ch. 13)

v_x, v_y: x-, y-components of velocity (Ch. 13)

W : Weight of a body (Chs. 8 & 16)
Load lifted or Resistance (Ch. 9)
Work done by a force (Ch. 10)

Oscillating weight (Ch. 20)
Point load or concentrated load (Ch. 24)

W_h : Weight of hammer (Ch. 17)

W_i : Ideal load (Ch. 9)

W_p : Weight of pile (Ch. 17)

W_1, W_2 : Point loads (Ch. 24)

w : Weight per unit length of belt (Ch. 19)
Intensity of uniform load (Chs. 3 & 24)

x : x-coordinate (Ch. 5)
Distance of a general section from the origin (Ch. 24)

$\bar{x}$: x-coordinate of centroid (Ch. 5)

$\dot{x}$: Velocity in the x-direction (Ch. 20)

$\ddot{x}$: Acceleration in the x-direction (Ch. 20)

y : y-coordinate (Ch. 5)
Vertical deflection in a beam section (Ch. 27)

$\bar{y}$: y-coordinate of centroid (Ch. 5)

y_{max} : Maximum deflection in a beam (Ch. 27)

Z : Modulus of section (Ch. 25)

Z_c, Z_t: Modulii of section in compression and intension, respectively (Ch. 25)

z : z-coordinate (Ch. 5)

Z : z-coordinate of centroid (Ch. 5)

GREEK SYMBOLS

α : Angle between forces (Ch. 2)
Half-angle of an arc (Ch. 5)
Angle of inclined plane (Ch. 8)
Angle of projection (Ch. 12)
Angular acceleration (Ch. 14)
Coefficient of linear expansion (Ch. 21)
Angle made by plane of coils of a helical spring with horizontal (Ch. 28)

β : Angle of contact of belt with pulley (Ch. 8)
Angle of inclined plane (Ch. 12)

γ : Angle between forces (Ch. 2)
Half-angle of groove (Ch. 19)
Unit weight of the material of a bar (Ch. 21)

Δ	:	Area of a triangle
ΔV	:	Change in volume (Ch. 21)
δ	:	Maximum recoil of gun (Ch. 17) Yield of grips (Ch. 21) Axial deflection of spring Ch. 28 Maximum lateral deflection (Ch. 29)
δl	:	Small change in length (Ch. 21)
δ_{st}	:	Static displacement of spring under the load (Ch. 28)
$\varepsilon_c, \varepsilon_t$	:	Compressive, tensile strain (Ch. 21)
ε_v	:	Volumetric strain (Ch. 21)
$\varepsilon_x, \varepsilon_y, \varepsilon_v$	:	Strain in x-, y-, z-direction (Ch. 21)
η	:	Efficiency of a machine (Ch. 9) Factor of safety (Ch. 21)
θ	:	Angle between forces (Ch. 2) Angle of inclined plane (Ch. 9) Angle of slope of a straight line (Ch. 11) Angular distance moved (Ch. 14) Angle of lap of pulley (Ch. 19) Slope or rotation at a beam section (Ch. 27) Angle of twist (Ch. 28)
θ_{max}	:	Maximum slope (Ch. 27)
μ	:	Coefficient of friction (Ch. 8) Poisson's Ratio (Ch. 21)
ω	:	Angular velocity (Ch. 14) Circular frequency of natural vibrations (Ch. 20)
ω_0	:	Initial angular velocity (Ch. 14)
π	:	Rate of circumference to diameter of a circle (Ch. 21)
σ	:	Stress (Ch. 21)
σ_c/σ_t	:	Compressive/tensile stres (Ch. 21)
σ_e	:	Proof stress or elastic limit (Ch. 23)
σ_n	:	Normal component of stress (Ch. 22)
σ_r	:	Resultant stress on plane (Ch. 22)
σ_x, σ_y	:	Principal stresses (Ch. 22)
τ	:	Shear component of stress (Ch. 22)
ϕ	:	Angle between forces (Ch. 2) Angle of friction (Ch. 8) Shear strain (Ch. 21) Angle of obliquity of resultant stress on a plane (Ch. 22)

Appendix C

Objective Questions

1. "When a number of forces act simultaneously on a body, their combined effect may be obtained as the sum of their individual effects when each acts alone". This is called the

    (a) principle of resolution
    (b) principle of transmissibilty
    (c) principle of superposition
    (d) none of the above.

2. The polygon of forces and the funicular polygon are one and the same.

    True or False?

3. The resultant of two forces P_1 and P_2 acting at an angle α with each other is given by

    (a) $\sqrt{P_1^2 + P_2^2 - 2P_1P_2 \sin \alpha}$

    (b) $\sqrt{P_1^2 + P_2^2 - 2P_1P_2 \cos \alpha}$

    (c) $\sqrt{P_1^2 + P_2^2 + 2P_1P_2 \sin \alpha}$

    (d) $\sqrt{P_1^2 + P_2^2 + 2P_1P_2 \cos \alpha}$.

4. "The algebraic sum of the moments of all the forces about any point is equal to the moment of their resultant about the same point". This principle is known as

    (a) principle of complementary shear
    (b) Varignon's principle of moments
    (c) principle of superposition
    (d) none of the above.

5. Fill in the blanks:

    "The moment of a force about any point is geometrically equal to _____ the area of the triangle with base equal to the force vector and vertex the moment centre.

6. A couple can be balanced by a single force.

    True or False?

7. If the lever arm of a couple is halved, its moment will

    (a) remain unaltered
    (b) be doubled·
    (c) be halved
    (d) none of these.

8. A couple consists of

    (a) two like parallel forces of different magnitudes
    (b) two unlike parallel forces of different magnitudes
    (c) two like parallel forces of the same magnitude
    (d) two unlike parallel forces of the same magnitude.

9. Lami's theorem is not applicable to concurrent forces.　　　　True/False

10. A body is said to be in equilibrium if it has no translatory motion.　　　True/False

11. If the resultant of all the forces acting on a body is zero, then the body may be in equilibrium if the forces are

    (a) parallel

(*b*) concurrent

(*c*) unlike and parallel

(*d*) like and parallel.

12. If a body is in equilibrium, the conclusion is that

(*a*) no force is acting on it

(*b*) the resultant of the forces is zero

(*c*) the algebraic sum of the moments of all the forces about any point is zero

(*d*) both (*b*) and (*c*) must be satisfied.

13. The three basic kinds of stress are:

(*i*) _____ (*ii*) _____ (*iii*) _____

14. The equilibrant and resultant of a force system are _____.

15. The resultant of a force system as shown in the Figure below is _____ (IES-92)

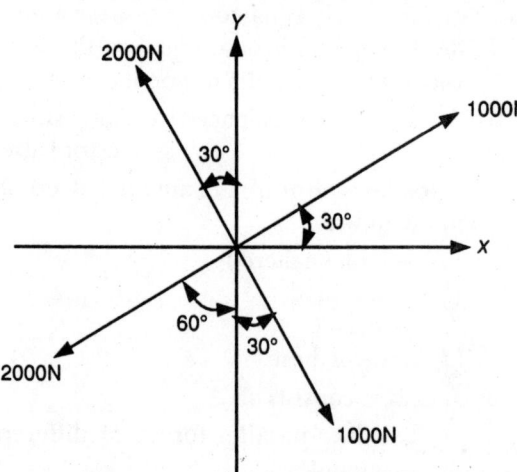

16. For a non-concurrent force system to be in equilibrium   (IES-94)

(*a*) closure of force polygon alone is sufficient

(*b*) closure of funicular polygon alone is sufficient

(*c*) closure of both force polygon and funicular polygon is necessary

(*d*) the above conditions in as (*b*) and (*c*) are not relevant.

17. For a given system of coplanar forces if the pole *O* of the force polygon moves along a straight line *OP*, then the sides of the funicular polygon would rotate about fixed points all of which lie on   (IES-94)

(*a*) a straight line parallel to *OP*

(*b*) a circle with centre *O*

(*c*) an ellipse with *OP* as the major axis

(*d*) a circle with centre at *P*.

18. **I** The intersection of the first and last ray in a funicular polygon lies on the line of action of the resultant.

II The line joining the ends of the first and last in the polar diagram represents the magnitude of the resultant.   (IES-94)

(*a*) both I and II are true

(*b*) I is false but II is true

(*c*) I is true but II is false

(*d*) both I and II are false.

19. Match List I with List II: (IES-94)

List I	*List* II
A. Segments of force polygon	1. Parallel to the lines joining the ends of the vector and the pole.
B. Segments of funicular polygon	2. Parallel to the line of action of forces.
C. Direction of resultant	3. Line joining the first and the last points of the force diagram.
D. Direction of equilibrant	4. Line joining the last point and first point of the force diagram.

20. For the coplanar concurrent system of forces shown here, the system   (IES-95)

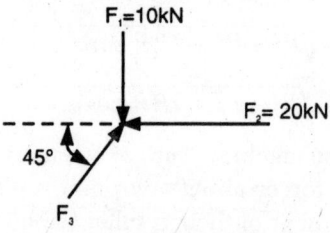

(*a*) will be in equilibrium if $F_3 = 10$ kN

(*b*) will be in equilibrium if $F_3 = 10\sqrt{2}$ kN

(*c*) will be in equilibrium if $F_3 = 20$ kN

(d) will not be in equilibrium whatever be the magnitude of F_3.

21. The coplanar force $P_1 = 20$ kN and $P_2 = 20$ kN meeting at D act on a lamina at 45° as shown in the figure. From the force diagram the equilibrant R is given by

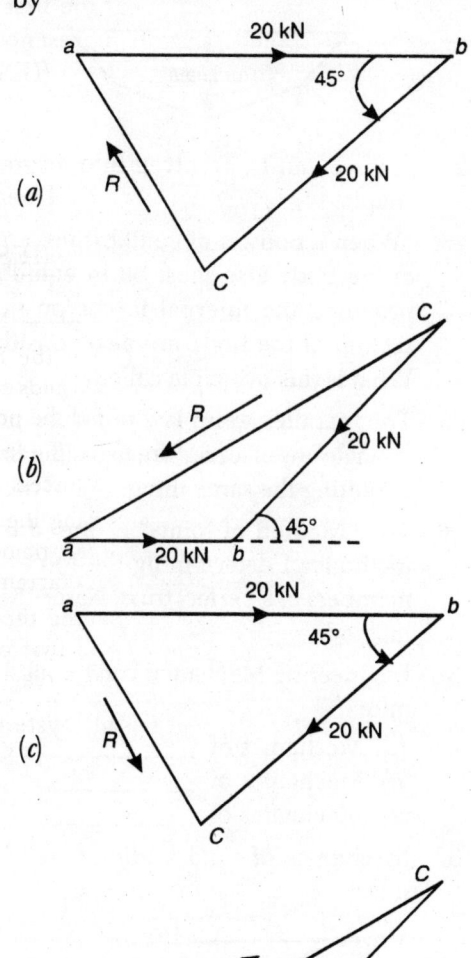

22. According to the parallelograin law of forces, if two forces acting simultaneously

at a point be represented in magnitude and direction by the adjacent side of a parallelogram, the resultant is represented by

(a) longer side of the other two sides

(b) shorter side of the other two sides

(c) diagonal of the parallelogram which does not pass through their point of inter-section

(d) diagonal of the parallelogram which passes through their point of inter-section. (IES-95)

23. In order to find the resultant of a system of coplanar parallel system of forces, the correct sequence of the graphical procedure to be followed is (IES-96)

(a) force diagram, space diagram, funicular polygon, and polar diagram

(b) funicular polygon, force diagram, space diagram and polar diagram

(c) space diagram, force diagram, polar diagram, and funicular polygon

(d) space diagram, funicular polygon, force diagram, and polar diagram.

24. Three coplanar forces $P_1 = P_2 = P_2 = 20$ kN act at a joint O as shown in the figure below:

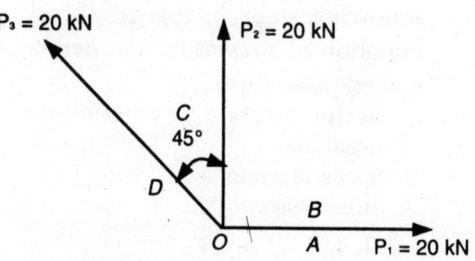

From the force diagram, the force R to be applied at O to keep the joint O in equilibrium is given by

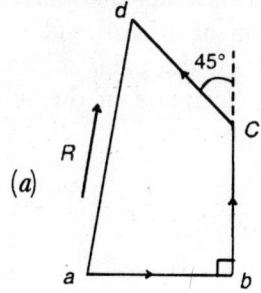

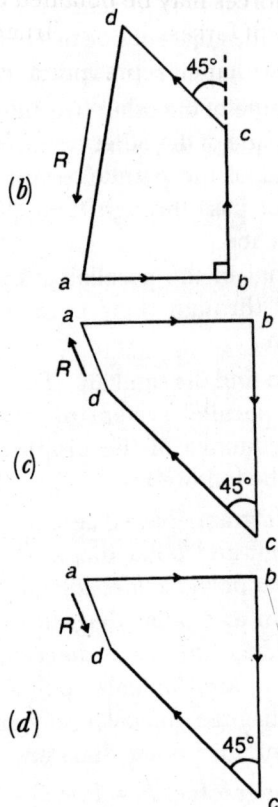

(b)

(c)

(d)

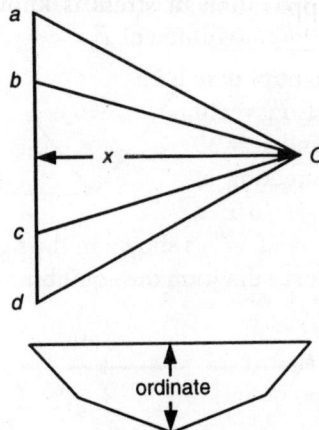

ordinate

25. What is the correct sequence of the following steps in the graphical determination of stresses in the members of a loaded plane truss:
    1. Vector diagram to determine the end reactions
    2. Space diagram
    3. Stress diagram   (IES-97)
    (a) 1, 2, 3
    (b) 1, 3, 2
    (c) 2, 1, 3
    (d) 2, 3, 1

26. The figure shows a funicular polygon and polar diagram of a beam subjected to vertical loads. If the polar distance x is doubled, the ordinate of the funicular polygon will be
    (a) doubled
    (b) halved
    (c) unaffected
    (d) tripled   (IES-97)

27. A force can be resolved into a force and couple.   True/False

28. "When a body is in equilibrium, any part of the body also must be in equilibrium, provided the internal forces on the cut-section of the body are also considered". What is this principle called?

29. The 'parallelogram law of forces' and the 'triangle law of forces' are two different ways of putting the same thing.   True/False

30. The "Method of Joints" is one of the two methods of determining the forces in the members of a perfect truss. Name the other method.

31. Engineering Mechanics can be subdivided into:
    (i) Mechanics of _____
    (ii) Mechanics of _____
    (iii) Mechanics of _____

32. Mechanics of rigid bodies is subdivided into
    (i) _____ (ii) _____

33. Kinematics and kinetics are the branches of study which fall under _____.

34. The study of motion of a body without reference to the causes of motion is called _____.

35. _____ deals with the motion as well as the forces responsible for it.

36. Change in optical properties of a material on application of stress is known as the _____ effect.

37. A quantity which has not only magnitude but also direction is called a _____.

38. _____ is a measure of its inertia, or its resistance to change of velocity.

39. The force with which any body is attracted towards the centre of the earth is called its _____.

40. "The condition of equilibrium or of motion of a rigid body will remain unchanged if the point of application of a force acting on it is transmitted to act at any other point along its line of action". This is the principle of _____.

41. When a body is stationary, or at rest, it is said to be in _____.

42. A force, equal and opposite to the resultant of a force system, is called the _____.

43. The process of finding the resultant of a force system is called _____ of forces.

44. "If a body is in equilibrium under the action of three coplanar concurrent forces, each force is proportional to the sine of the angle between the other two". This statement is called _____.

45. "If only three forces keep a body in equilibrium, they shall be concurrent".
     True/False

46. "Theoretically speaking, a given force may be resolved into any number of sets of components". True/False

47. "The parallelogram law of forces cannot be extended to forces in space, or those which do not all act in one plane". True/False

48. "The component of the resultant of a coplanar concurrent force system in a given direction is equal to the algebraic sum of the components of all the forces of the system in that direction". This is called the theorem of _____.

49. The polygon law of forces is an extension of the triangle law of forces. True/False

50. "The equilibrant of a system of coplanar concurrent forces may be obtained by the polygon law of forces. True/False

51. A parallel force system is a special case of non-concurrent force systems. True/False

52. Moment of a force about a point is a measure of its_____ effect about the point.

53. The perpendicular distance of the point about which the moment of a force is required, from the line of action of the force, is called _____.

54. If the moment of a force about a point is zero, the point lies on _____.

55. The resultant of two unlike parallel forces lies in between these two. True/False

56. A couple may be rotated through any angle without changing its effect. True/False

57. Two equal and opposite couples acting on a rigid body will keep it in equilibrium.
     True/False

58. "A force acting at a point on a body is equivalent to a force acting at another point and a couple. True/False

59. The conditions of static equilibrium for a general coplanar force system are:
     (*i*)_____. (*ii*)_____. (*iii*)_____.

60. The proposition that three forces keeping a body in equilibrium have to be necessarily concurrent can be proved from the equilibrium conditions of a general coplanar force system. True/False

61. The simplest perfect frame is a _____ frame.

62. A redundant frame has _____ number of members than that required for a perfect frame with the same number of joints.

63. A _____ frame cannot retain its shape when loaded.

64. The joints in trusses cannot resist moments.
     True/False

65. All the loads on a truss are assumed to be transmitted through the nodes or joints, irrespective of their location. True/False

66. A truss is completely analysed if the internal forces in all the members are solved.
                                                                    True/False

67. The indeterminacy of a truss with respect to reactions at the supports is called _____ indeterminacy.

68. The total indeterminacy of a truss is the sum of_____ and _____ indeterminacy.

69. The analytical methods for the determination of forces in the members of a truss are:

    (i)_____.   (ii)_____.

70. The force in any member can be directly determined by the 'method of joints'.
                                                                    True/False

71. The method of _____ is particularly useful in the analysis of forces in the members of complex trusses.

72. Graphical method is also available for the analysis of forces in the members of perfect frames.   True/False

73. Centroid applies only to line segments and plane areas.   True/False

74. The centre of mass and centre of gravity of a body are not identical, strictly speaking; they are identical only for uniform gravitational field.   True/False

75. The mathematical technique of _____ will be useful in the location of centroid and centre of gravity.

76. The centroid is the point of intersection of the axes of symmetry.   True/False

77. The height of the centroid of a semi-circular lamina above the base (R being the radius) is

    (a) $\dfrac{R}{3\pi}$ (b) $\dfrac{3R}{\pi}$

    (c) $\dfrac{2R}{3\pi}$ (d) $\dfrac{4R}{3\pi}$.

78. The height of the centre of gravity of a right circular cone above its base is (height h)

    (a) $h/3$ (b) $h/4$

79. The height of the centre of gravity of a hemisphere above the base is

    (a) $\dfrac{R}{4}$ (b) $\dfrac{R}{8}$

    (c) $\dfrac{3R}{8}$ (d) $\dfrac{5R}{8}$

80. The moment of inertia of an area should be called, strictly speaking, the 'Second Moment of Area' or the 'Area Moment of Inertia'.   True/False

81. The polar moment of inertia of an area about an axis is the sum of _____.

82. A circular hole of radius r is cut from a circular disc of radius $2r$ such that the diameter of the hole is the radius of the disc. The $C.G.$ of the section lies at

    (a) center of the disc
    (b) centre of the hole
    (c) in the hole
    (d) in the disc.

83. The radius of gyration is not related to the polar moment of inertia.   True/False

84. Among the following, the units of radius of gyration are:

    (a) mm   (b) $(mm)^2$
    (c) $(mm)^3$ (d) $(mm)^4$

85. The parallel axis theorem relates to the shift of the axis in the plane of the area.
                                                                    True/False

86. The perpendicular axis theorem is not useful in the determination of the polar moment of inertia of an area.   True/False

87. The moment of inertia of a triangle of base 'b' and height 'h' about its base is given by

    (a) $\dfrac{bh^3}{3}$ (b) $\dfrac{bh^3}{36}$

    (c) $\dfrac{bh^3}{12}$ (d) $\dfrac{bh^3}{6}$

88. For a circular area, the moment of inertia about a diametral axis may be derived by using the perpendicular axis theorem.
                                                                    True/False

89. The moment of inertia of a quadrant of a circle (diameter d) is given by

(a) $\dfrac{\pi d^4}{64}$ (b) $\dfrac{\pi d^4}{32}$

(c) $\dfrac{\pi d^4}{128}$ (d) $\dfrac{\pi d^4}{256}$.

90. The polar moment of inertia of an annular area (external diameter D and internal diameter d) is given by _____.

91. The product of inertia of an area about a set of axes, one of them being an axis of symmetry for the area, is _____.

92. The product of inertia of a rectangle (breadth b and depth d) about axes coinciding with the adjacent edges is given by _____.

93. A·parallel axis theorem may be enunciated for product of inertia similar to that for moment of inertia. True/False

94. The principal axes of an area are those about one of which the moment of inertia is a _____ and about the other it is a _____.

95. The product of inertia about the principal axes of an area is _____.

96. The principal axes of an area passing through its _____ constitute the most important set.

97. The principal moments of inertia may be determined by using the Mohr's circle of inertia. True/False

98. Mass moment of inertia refers to that of a body while second moment of area refers to that of a lamina. True/False

99. Parallel axis theorem and perpendicular axes theorem are not applicable to mass moments of inertia of a body. True/False

100. The concept of principal moments of inertia is not applicable to mass moments of inertia of a body. True/False

101. The mass polar moment of inertia of a thin circular plate of mass m and diameter d is given by _____.

102. The centroid of an area can be located using a graphical approach. True/False

103. Graphical approach is not available for the determination of support reactions in beams. True/False

104. The second moment of area of a lamina about an axis in its plane may be obtained by graphical method. True/False

105. The force diagram for a truss obtained graphically is known as a _____.

106. The first step in the determination of the forces in the members of a cantilever truss is the determination of the reactions. True/False

107. The graphical solution of a compound truss requires the use of the method of_____.

108. Frictional force is

(a) perpendicular to the surface of contact
(b) tangential to the surface of contact
(c) inclined to the surface of contact
(d) normal to the surface of contact.

109. Friction has desirable as well as undesirable effects. True/False

110. Broadly speaking, friction is categorised as:
(i)_____. (ii)_____. (iii)_____.

111. The maximum frictional force developed is called the _____.

112. The laws of friction were first enunciated by _____.

113. The frictional force is independent of the area of contact between the two bodies under consideration. True/False

114. Under the condition of impending motion, the magnitude of the limiting friction is proportional to the normal force across the surface contact. True/False

115. The ratio of the frictional force to the normal force at the surface of contact is called the _____.

116. The semi–apex angle of the cone of friction is the _____.

117. The coefficient of friction (ϕ: angle of friction) equals
    - (a) $\tan\phi$
    - (b) $\sin\phi$
    - (c) $\cot\phi$
    - (d) $\sec\phi$.

118. A body lying on a rough inclined plane can be kept in equilibrium by applying an external force
    - (a) in a horizontal direction
    - (b) in a direction parallel to the inclined plane
    - (c) in an appropriate direction other than (a) and (b)
    - (d) in a direction given in (a), (b) or (c).

119. An important application of friction on an inclined plane is a _____.

120. Dynamic friction is classified as
    - (i) _____.
    - (ii) _____.

121. The efficiency of a machine is the ratio of the _____ to the _____.

122. In an ideal machine, the mechanical advantage is _____ to the velocity ratio.
    - (a) less than
    - (b) greater than
    - (c) equal to
    - (d) not related.

123. The law of machine is
    - (a) $P = mW + C$
    - (b) $P = \dfrac{m}{W} + C$
    - (c) $P = mW - C$
    - (d) $P = \dfrac{m}{W} - C$.

    [P = effort; W = load; m and C are constants].

124. The maximum possible mechanical advantage of a lifting machine is
    - (a) $1/m$
    - (b) m
    - (c) $(1 + m)$
    - (d) $(1 - m)$.

    (m is the multiplying constant in the law of the machine)

125. The maximum efficiency of a lifting machine is
    - (a) $\dfrac{1}{m}$
    - (b) $\dfrac{1}{V_r}$
    - (c) $\dfrac{1}{m \cdot V_r}$
    - (d) $\dfrac{V_r}{m}$.

    [V_r is the velocity of ratio and m is the multiplying constant in the law of the machine].

126. A common example of _____ of a machine is the manual lifting of water from a well by a bucket with a rope passing over a pulley.

127. A common example of _____ of a machine is the lifting of a car by means of a screw-jack.

128. A lifting machine with an efficiency less than 50% is said to be
    - (a) reversible
    - (b) self locking
    - (c) ideal
    - (d) practical.

129. A lifting machine with an efficiency more than 50% is said to be
    - (a) self-locking
    - (b) ideal
    - (c) practical
    - (d) reversible.

130. If n is the number of movable pulleys in a first-order system of pulleys, the mechanical advantage is
    - (a) n^2
    - (b) 2^n
    - (c) $2^{(n-1)}$
    - (d) $2^{(n+1)}$

131. If n is the number of the parts of the string supporting the lower block in a second order system of pulleys, the mechanical advantage is
    - (a) $2^{(n-1)}$
    - (b) 2^n
    - (c) n
    - (d) n^2

132. If n is the number of pulleys in a third order system, the mechanical advantage is
    - (a) $2^n - 1$
    - (b) 2^n
    - (c) n
    - (d) n^2

133. The velocity ratio of a wheel-and-axle system is
    - (a) $\dfrac{d}{D}$
    - (b) $\dfrac{d^2}{D^2}$
    - (c) $\dfrac{D^2}{d^2}$
    - (d) $\dfrac{D}{d}$

    [D is the diameter of the wheel and d that of the axle]

134. The velocity of ratio of a differential wheel and axle with wheel diameter D and axle diameter d_1 and d_2 ($d_2 > d_1$) is _____.

135. In a Weston differential pulley block with the diameter of the co-axial pulleys D and d ($D > d$) the velocity ratio is _____.

136. The velocity ratio of an inclined plane of angle θ with horizontal is

    (a) $\dfrac{1}{\sin\theta}$ (b) $\dfrac{1}{\cos\theta}$

    (c) $\dfrac{1}{\tan\theta}$ (d) $\dfrac{1}{\sin^2\theta}$.

137. For an ideal machine with negligible friction, the mechanical advantage is the same as the velocity ratio.   True/False

138. In a screw jack with radius of handle R and pitch of screw p, the velocity ratio is _____.

139. In a differential screw jack with radius of handle R and pitches of screw p_A and p_B ($p_A > p_B$), the velocity ratio is _____.

140. For a worm and worm-wheel (with the diameter of effort wheel and load drum D and d respectively, and the number of teeth on the worm wheel T), the velocity ratio is _____.

141. In a worm and worm wheel, if the number of teeth on the worm wheel is halved, the velocity ratio is
    (a) halved
    (b) doubled
    (c) decreased to one-fourth
    (d) unaltered.

142. In a single-purchase winch crab, with the radius of the crank R and that of the load drum r, the velocity ratio is unaltered if R is doubled and r is halved.   True/False

143. In a double-purchase winch crab, with the radius of the crank R and that of the load drum r, the velocity ratio is unaltered if both R and r are doubled.   True/False

144. 'Work' done by a force is a scalar quantity.   True/False

145. The unit of newton-metre is also called _____.

146. Work done by a force is zero if either the displacement is zero, or the force is in a _____ direction to that of the displacement.

147. Work done may sometimes be negative.   True/False

148. Virtual work refers to
    (a) virtual work done by virtual forces
    (b) virtual work done by actual forces
    (c) actual work done by virtual forces
    (d) actual work done by actual forces.

149. The external forces which do work are known as _____ forces.

150. The principle of virtual work is a necessary and sufficient condition for equilibrium.   True/False

151. The number of unique virtual displacements that may be given to a system equals its _____.

152. The principle of virtual work has very wide and versatile applications.   True/False

153. The internal force in any member of a truss under given load system may be determined by using the principle of virtual work.   True/False

154. The two basic types of motion are:
    (i) _____. (ii) _____.

155. Translation along a straight line is called _____.

156. A general plane motion consists of a combination of both _____ and _____.

157. Velocity is rate of change of _____.

158. If the direction is of no interest, velocity and speed are one and the same. True/False

159. Rate of change of velocity with respect to time is called _____ acceleration.

160. The acceleration at any instant of time is called _____ acceleration.

161. Negative acceleration is called _____.

162. The motion of a body may be represented by means of a graph.   True/False

163. The slope of the time-displacement curve at any instant of time is the _____.

164. The slope of time-velocity curve at any instant of time is the _____.

165. The simplest motion is _____ motion in a straight line.

166. The resultant velocity in a given case may be got by procedures similar to those for forces.   True/False

167. The _____ velocity of *B* with respect to *A* is the vector difference of the velocity of *B* and *A*.

168. Two bodies are involved in the concept of relative velocity while only one is involved in that of resultant velocity.   True/False

169. The most common case of motion under uniform acceleration is that under the influence of _____.

170. When a body is projected vertically upwards the time of ascent is less than that of descent.   True/False

171. Motion of a particle in a viscous medium like oil is a special case of motion with _____ acceleration.

172. The motion of a projectile has both vertical and horizontal components, and hence traces a _____ path.

173. In a projectile motion the _____ component of motion remains constant.

174. Two common examples of projectile motion are:
     (*i*) _____. (*ii*)_____.

175. The angle of projection is defined as the angle between the direction of projection and the vertical.   True/False

176. The path traced by the projectile is called its _____.

177. In a projectile motion, the time required to reach the maximum height is half the time of flight when the point of striking and the point of projection are at the same level.   True/False

178. The trajectory of a projectile is a _____.

179. The maximum height reached by a projectile depends upon the veloctiy of projection and the angle of projection.   True/False

180. For the range to be a maximum, the angle of projection should be _____.

181. The angle of projection for a specified range is unique.   True/False

182. The range of a projectile is directly proportional to the velocity of projection.
                                    True/False

183. The motion of a coupling-rod of a locomotive moving on a straight level railway track is an example of planar _____.

184. The normal component of acceleration is equal to the square of speed divided by radius of curvature of the path at the point.   True/False

185. The word 'centripetal' means away from the centre.   True/False

186. The vector sum of the radial and transverse components of the velocity gives the total velocity at the point.   True/False

187. Just as translation involves linear motion, rotation involves _____.

188. When all the particles of a rigid body describe concentric circles about a fixed axis, the motion of the rigid body is _____.

189. A combination of translation and rotation is called _____.

190. Two common examples of general plane motion of a body are:
     (*i*) _____. (*ii*)_____.

191. The units of angular velocity are metres/second.   True/False

192. The units of angular acceleration are _____.

193. The relation between linear velocity (*v*) and angular velocity (ω) in the circular path (*r*) is _____.

194. The relation between angular velocity (ω) and revolutions per minute (*n*) is _____.

195. The instantaneous centre of rotation is a fixed point.   True/False

196. The units of momentum in S.I. units are _____.

197. The Newton's second law of motion states that _____.

198. "The forces of action and reaction between any two bodies are equal and opposite". This is
    (*a*) Newton's first law of motion
    (*b*) *d*'Alembert's principle
    (*c*) condition of static equilibrium
    (*d*) Newton's third law of motion.

199. The value of the Universal constant of gravitation is _____.

200. The fundamental quantities in gravitational units are length (L) _____, and time (T).

201. A newton is the amount of force which causes a mass of one gram to move with an acceleration of 1 cm/s². True/False

202. A force which causes a mass of one kilogram to move with an acceleration equal to _____ is called one kilogram-weight.

203. Engineers prefer absolute units of force to gravitational units. True/False

204. *d'* Alembert's principle may be looked upon as the equation of _____ equilibrium.

205. The force exerted on the floor of an elevator is more for downward motion than for upward motion. True/False

206. The principle which enables one to convert a dynamics problem to one of static equilibrium is
    (*a*) principle of conservation of energy
    (*b*) *d*'Alembert's principle
    (*c*) principle of conservation of momentum
    (*d*) work-energy principle.

207. Work and energy are vector quantities. True/False

208. In the work-energy method, the velocity can be obtained directly without the need to obtain acceleration. True/False

209. One newton-metre of work is also called one _____.

210. Work done by a force is zero if the force acts in a direction _____ to that of motion.

211. Work done by self-weight is negative if the body moves upward. True/False

212. Work done by a force does not depend upon the path between specified points if the nature of the force is _____.

213. Two examples of conservative forces are
    (*i*) _____ (*ii*) _____

214. One example of non–conservative force is _____.

215. Energy is defined as the _____ to do work.

216. Energy has the same units as work. True/False

217. Mechanical energy is of two forms:
    (*i*) _____.
    (*ii*) _____.

218. Work done against gravity is stored as increase in _____.

219. "The work done by a system of forces on a body during a certain displacement equals the change in kinetic energy during the same displacement". This principle is called the
    (*a*) principle of conservation of energy
    (*b*) *d*'Alembert's principle
    (*c*) principle of conservation of momentum
    (*d*) work-energy principle.

220. "The sum of the potential energy and kinetic energy of a body in motion under the action of conservative forces remains constant". This law is called the_____.

221. For a system in equilibrium the derivative of its potential energy is zero. True/False

222. The following are the three kinds of equilibrium possible for a system:
    (*i*) _____;
    (*ii*) _____;
    (*iii*) _____.

223. When the potential energy of a system remains constant, the system will be in _____ equilibrium.

224. The force required to produce unit displacement parallel to the axis of a spring is the called the _____.

225. The time-rate of doing work is called_____.

226. One Joule per second is called one_____.

227. A large force acting over a short period of time is called
    (a) moment   (b) momentum
    (c) impulse   (d) power.

228. The units of momentum are the same as those of impulse,   True/False

229. "When the resultant force of a system is zero, the vector sum of the impulses of all external forces is zero, and the momentum of the system remains constant". This principle is called the
    (a) principle of conservation of momentum
    (b) principle of conservation of energy
    (c) d'Alembert's principle
    (d) work-energy principle.

230. Angular momentum is also known as moment of momentum.   True/False

231. Two examples of problems that may be solved by the impulse–momentum principle are:
    (i) _____.   (ii)_____.

232. The collision between two elastic bodies for a very short interval of time, exerting a large force on each other, is called_____.

233. If the motion of the two bodies before collision is along the line of impact, it is called _____.

234. If the centres of mass of the colliding bodies lie on the line of impact, it is called _____.

235. The sum of the periods of _____ and _____ is the total period of collision or impact.

236. The coefficient of restitution is a measure of the energy loss during impact.
                                        True/False

237. The value of the coefficient of restitution always lies between 1/2 and 1.   True/False

238. In a perfectly elastic impact of two equal masses, the masses exchange their velocities of approach.   True/False

239. A body striking a rigid plane vertically will rebound with the same velocity with which it strikes the rigid plane.   True/False

240. The loss of kinetic energy in an impact may be expressed in terms of the masses of the colliding bodies, their velocities of approach, and _____.

241. Based on the cross-section, belts are classified as
    (i) _____
    (ii) _____
    (iii) _____

242. The rotational speeds of the pulleys are directly proportional to their radii.
                                        True/False

243. The two different arrangements of a belt over two pulleys are
    (i) _____
    (ii) _____

244. The angle of lap for the smaller pulleys is more than that for the bigger pulley.
                                        True/False

245. In cross-belt drive, the pulleys rotate in opposite directions.   True/False

246. Angle of lap for both pulleys is the same in cross-belt drive.   True/False

247. The length of belt required is a little more in open belt drive than in cross-belt drive.
                                        True/False

248. The ratio of tight side to slack side tensions, T_1 and T_2, on either side of the pulley in a belt drive with angle of lap θ, and coefficient of friction being μ, is given by

    (a) $\dfrac{T_1}{T_2} = \mu\theta$ (b) $\dfrac{T_2}{T_1} = e^{\mu\theta}$

(c) $\dfrac{T_1}{T_2} = e^{\mu\theta}$ (d) $\dfrac{T_2}{T_1} = \mu\theta$

249. The centrifugal tension in belts is an additional one which gets added to the tensions on either side in power transmission.   True/False

250. The initial tension in a belt is merely the average of the tensions on either side during the transmission of power.   True/False

251. The maximum tension in the belt for design purpose is tight-side tension plus centrifugal tension.   True/False

252. Power, P, transmitted by a belt with tensions, T_1 and T_2, and velocity of rotation v, is given by

    (a) $P = (T_1 - T_2)v$ (b) $P = (T_1 + T_2)v$

    (c) $P = \dfrac{T_1}{T_2}v$ (d) $P = T_1 T_2 v$.

253. The centrifugal tension in a belt is proportional to the velocity of rotation.   True/False

254. The power transmitted is maximum when the centrifugal tension is one-third the maximum tension in the belt.   True/False

255. The maximum displacement from the mean position in an oscillation is called the _____.

256. The time required to complete one cycle is called the _____.

257. The number of cycles per second is called the _____.

258. Frequency and period of an oscilaltion are directly related.   True/False

259. The frequency of undamped free vibrations of a system is known as its _____.

260. The equivalent spring constant for a system of two springs in parallel is the_____ of their spring constants.

261. The units of viscous damping coefficient are

    (a) $\dfrac{N}{m \cdot s}$ (b) $\dfrac{N \cdot s}{m}$

(c) $\dfrac{N \cdot m}{s}$ (d) $\dfrac{m}{N \cdot s}$

262. The minimum damping coefficient required to make the motion of a system aperiodic or non-vibrating is called_____ damping coefficient.

263. The ratio of actual damping coefficient to the critical damping coefficient is the _____.

264. Damping is classified by its nature as

    (i) _____ ;

    (ii) _____ ; and

    (iii) _____ .

265. The ratio of exciting frequency in forced vibrations to the natural frequency is called the _____.

266. The magnification factor tends to become infinite (when the damping is insignificant) at a frequency ratio of _____.

267. When a rigid body oscillates about a fixed point in it other than the centre of mass, it is known as _____.

268. When the equivalent length of a compound pendulum is stated with respect to a simple pendulum, the criterion of equivalence is _____.

269. The torsional stiffness k_t of a torsional pendulum is given in terms of the length l, polar moment of inertia J, and the modulus of rigidity of the material N, is given as

    (a) $k_t = \dfrac{N}{lJ}$ (b) $k_t = \dfrac{J}{Nl}$

    (c) $k_t = \dfrac{l}{NJ}$ (d) $k_t = \dfrac{NJ}{l}$.

270. When a disc is suspended from three wires of actual length, attached to it at equidistant points along its circumference, it is called a _____ suspension.

271. The three major aspects in the design of any structure are:

    (i) _____ ;

    (ii) _____ ; and

    (iii) _____ .

272. Shear stress is also called _____ stress.

273. Shear strain is measured in _____.

274. The residual strain on removal of load beyond elastic limit is called the _____.

275. For homogeneous materials, the modulus of elasticity is practically the same in compression and in tension.   True/False

276. In the case of a cube, the ratio of volumetric strain to linear strain is _____.

277. In the case of a rectangular parallelopiped, the volumetric strain equals the sum of the linear strains in the three principal directions.   True/False

278. "A shear stress in a given plane will be automatically accompanied by a balancing shear stress of equal magnitude in a perpendicular plane for equilibrium". This principle is called the principle of _____.

279. When an element is in equilibrium under a state of simple shear, the diagonal planes of the element will be subjected to _____ and _____ stresses of equal magnitude.

280. In a state of pure shear, the 45° planes will be free of shear.   True/False

281. The lateral strain is always less than the longitudinal strain in the case of a uniaxial stress.   True/False

282. The ratio of shear strain to the linear strain in a diagonal in the case of pure shear is _____.

283. The ratio of the modulus of elasticity (E) and the rigidity modulus (N) (in terms of the Poisson's ratio) is

(a) $2\left(1-\dfrac{1}{m}\right)$ (b) $3\left(1-\dfrac{2}{m}\right)$

(c) $2\left(1+\dfrac{1}{m}\right)$ (d) $3\left(1+\dfrac{2}{m}\right)$

284. The ratio of the modulus of elasticity (E) and the bulk modulus (K) (in terms of the Poisson's ratio) is

(a) $2\left(1-\dfrac{1}{m}\right)$ (b) $3\left(1-\dfrac{2}{m}\right)$

(c) $2\left(1+\dfrac{1}{m}\right)$ (d) $3\left(1+\dfrac{2}{m}\right)$

285. The relation between the three moduli of elasticity is

(a) $E=\dfrac{3KN}{(9K+N)}$ (b) $E=\dfrac{9KN}{(3K-N)}$

(c) $E=\dfrac{3KN}{(9K-N)}$ (d) $E=\dfrac{9KN}{(3K+N)}$.

286. "Brittleness" is lack of _____.

287. Cast iron is an example of a _____ material.

288. The limit of proportionality and the elastic limit are far different in the case of wrought iron.   True/False

289. Percentage elongation and percentage reduction in cross-sectional area of a bar under uniaxial tension are indices of the property of _____ of the material.

290. The ratio of ultimate strength to working stress for a material is called the _____ in design.

291. The principle of superposition is applicable for elastic materials irrespective of the magnitude of the stress.   True/False

292. The ratio of the moduli of elasticity of a composite bar consisting of two materials is called the _____.

293. The strain due to a temperature change, t, is

(a) $\alpha \cdot t$ (b) α/t
(c) t/α (d) $\alpha + t$.
(α is the coefficient of linear expansion of the material)

294. One of the principal stresses is the largest and the other will be the smallest; the former is called the _____ principal stress and the latter the _____ principal stress.

295. Principal planes will be free of shear.
                                                                 True/False

296. Shear stress will be _____ the on 45°–planes with respect to the principal

planes and its magnitude is _____ the difference of the principal stresses.

297. The plane of maximum shear is the same as the plane of maximum obliquity.
                                    True/False

298. On two perpendicular planes there are normal stress, σ_1 and σ_2 and shear stress q. If $q^2 = \sigma_1\sigma_2$, the major principal stress is _____ and the minor principal stress is _____ .

299. In the Mohr-circle of stress with the principal stresses σ_1 and σ_2 being known, the radius of the Mohr's circle is _____ .

300. Half the difference of principal strains equals the maximum shear strain.
                                    True/False

301. Work done on a body by an externally applied load is stored as _____ in the material of the body.

302. The concept of strain energy is applicable to any kind of straining action.
                                    True/False

303. Graphically speaking, the area under the load-deformation curve represents the _____ by the load. This also represents the _____ in the body.

304. If σ is the uniform stress and E is the modulus of elasticity of the material, the strain energy per unit volume, u, is given by _____ .

305. The maximum strain energy which can be stored in a piece of material without permanent set is called its _____ .

306. The proof resilience per unit volume of the material is called the _____ .

307. The modulus of resilience is a property of the material.
                                    True/False

308. The expression for strain energy in shear is of similar form for that in uniaxial stress.
                                    True/False

309. The stress due to a suddenly applied load is _____ that due to the same load gradually applied.

310. The nature of the straining action caused in beams is primarily _____ .

311. A beam is fully analysed if the S.F. and B.M. diagrams under the given transverse loads are determined.   True/False

312. The rate of change of shear force with respect to the span is the _____ at the section in a loaded beam.

313. The rate of change of bending moment with respect to the span is the _____ at the section in a loaded beam.

314. Maximum bending moment always occurs at a section where the shearing force is zero.
                                    True/False

315. The support reactions shall first be determined before any beam can be analysed.
                                    True/False

316. The bending moment at point of contra-flexure is _____ or _____ .

317. In a simple beam under triangular loading, the nature of the variation of bending moment along the span is
    (*a*) parabolic
    (*b*) linear
    (*c*) cubic parabolic
    (*d*) a straight line parallel to the span.

318. An overhang portion of a beam acts virtually as a cantilever.   True/False

319. In an overhanging beam with equal overhangs on either side with equal point loads at the free ends, the portion of the beam between the supports is in a state of _____ .

320. In an overhanging beam with equal overhangs on either side loaded uniformly throughout the entire length, the ratio of the length of overhang to the span for the bending moment at mid-span to be zero is _____ .

321. The load diagram and the beam may be determined if the S.F. and B.M. diagrams are known.   True/False

322. A beam which can be fully solved purely with the aid of the laws of static equilibrium is called a _____ beam.

323. The number of additional conditions over and above the useful laws of static equilibrium required to solve an indeterminate beam is known as _____.

324. For determining the degree of indeterminacy both the nature of the beam and the nature of loading must be known.
                                        True/False

325. Indeterminate beams are preferred in practice in view of the _____ in their design.

326. The nature of stresses caused by bending moment on a cross-section is _____ or _____.

327. The surface on which no stress occurs when a bending moment acts on a cross-section of a beam is known as the _____ surface; the trace of this surface on the cross-section is known as the _____.

328. The bending formula is, strictly speaking, applicable only for _____ or _____ bending.

329. "Plane transverse sections of the beam remain plane and normal after bending". This is known as _____ assumption in the theory of simple bending.

330. The modulus of elasticity is assumed to be the same value both in compression and in tension.   True/False

331. The radius of curvature of the beam after bending is assumed to be very large in comparison with the dimensions of the cross-section.   True/False

332. The bending stress is inversely proportional to the distance of the fibre from the neutral axis.   True/False

333. The radius of curvature is constant for a constant value of bending.   True/False

334. The maximum value of the bending stress is the same on the tension side and the compression side when the cross-section is _____ with respect to the neutral axis.

335. In simple bending, the neutral axis passes through the _____ of the cross–section.

336. The application of simple bending formula to practical cases of bending can be justified primarily because the shear force at the section of maximum bending moment is zero or changes sign, thus its effect being negligible.   True/False

337. The ratio of the moment of inertia of the cross-section and the extreme fibre distance from the neutral axis is called the _____.

338. A rectangular section is a better shape than an I- section in bending.   True/False

339. If a beam section is proportioned such that the modulus of section varies in direct proportion to the bending moment along the span, the beam is said to be one of_____.

340. The diameter of a cylindrical beam should be proportional to the square roof of the bending moment at every section along the span if the bending stress has to be uniform.
                                        True/False

341. The strain energy in bending is directly proportional to the square of the bending stress.   True/False

342. The principle of complentary shear is utilised in deriving an expression for the vertical shear stress in a beam section acted on by a vertical shearing force.
                                        True/False

343. The shear stress distribution with depth in a rectangular section and in a circular section is _____.

344. The ratio of the maximum to mean shear stress in a rectangular section is $1\frac{1}{3}$.
                                        True/False

345. The ratio of the maximum to mean shear stress in a circular section is $1\frac{1}{2}$.
                                        True/False

346. The shear stress distribution diagram integrated over the width and depth of the

cross-section equals the _____ at the section.

347. The shear stress distribution diagram for a diamond section is in the shape of the letter _____.

348. The ratio of the maximum to the mean shearing stress in the case of a diamond section is _____.

349. The maximum shear stress for a diamond section occurs at the neutral axis. True/False

350. The maximum shear stress for a triangular section with its vertex at the top occurs at the neutral axis. True/False

351. The maximum shear stress for a rectangular section with its vertex at the top occurs at _____ of the section.

352. The ratio of maximum to mean shear stress for a rectangular section with vertex at the top is _____.

353. Most of the shear is carried by the _____ in the case of an I-section.

354. The shear stress distribution for a channel section is similar to that for an I-section. True/False

355. A moment acting in the plane of a cross-section causes _____ of the section.

356. The shear stress caused by a torque acting on a circular cross-section varies directly with the radial distance of the fibre from the centre. True/False

357. The shear strain and the angle of twist are one and the same when a cylindrical shaft is subjected to torsion. True/False

358. The torsion formula is a dimensionally correct formula. True/False

359. The polar moment of inertia occurs in the torsion formula. True/False

360. The modulus of elasticity occurs in the torsion formula. True/False

361. The torsional stiffness is dependent purely upon the geometrical and material properties of shaft. True/False

362. The torsional stiffness for unit length of the shaft is called the _____.

363. A hollow cylindrical section is more efficient in torsion than a solid circular section of the same area. True/False

364. Maximum torque is considered in the computation of the power that may be transmitted by a shaft. True/False

365. "Equivalent twisting moment" is just a concept used in combined bending and torsion of a cylindrical shaft, and not necessarily a twisting moment.

366. The resilience per unit volume of a cylindrical shaft in pure torsion is given by

   (a) $\dfrac{1}{4} \cdot \dfrac{f_s^2}{N}$ (b) $\dfrac{1}{2} \cdot \dfrac{f_s^2}{N}$

   (c) $\dfrac{1}{8} \cdot \dfrac{f_s^2}{N}$ (d) $\dfrac{1}{6} \cdot \dfrac{f_s^2}{N}$.

   (f_s = maximum torsional shear stress and N = modulus of rigidity of the shaft material).

367. A rectangular beam is to be cut from a circular log of wood of diameter D. The ratio of the breadth to depth for strongest section in bending should be

   (a) $\sqrt{\dfrac{2}{3}}$ (b) $\dfrac{3}{2}$

   (c) $\dfrac{1}{\sqrt{2}}$ (d) $\dfrac{3}{4}$ (IES-92)

368. For engineering materials, Poisson's ratio lies between
   (a) 0 and 1   (b) –1 and +1
   (c) –1/2 and +1/2   (d) 0 and 1/2
   (IES-92)

369. Under torsion, brittle materials generally fail
   (a) along a plane perpendicular to the longitudinal axis
   (b) in the direction of minimum tension
   (c) along surfaces forming a 45° angle with the longitudinal axis
   (d) not in any specific manner.   (IES-92)

370. In a compression test on mild steel
     (a) necking does not occur.
     (b) Hooke's law is not valid.
     (c) Hooke's law is valid beyond yield point.
     (d) strength in compression is much greater than that in tension.   (IES-92)

371. For most brittle materials generally ultimate strength in compression is much larger than that in tension because
     (a) of flows such as microscopic cracks or cavities.
     (b) compression failure is due to normal stress and failure in tension is due to shear stress.
     (c) yield point does not occur in compression.
     (d) of inherent properties of materials.
                                              (IES-92)

372. The SFD and BMD are shown below:

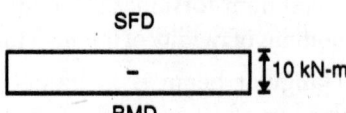

     The corresponding loading diagram will be

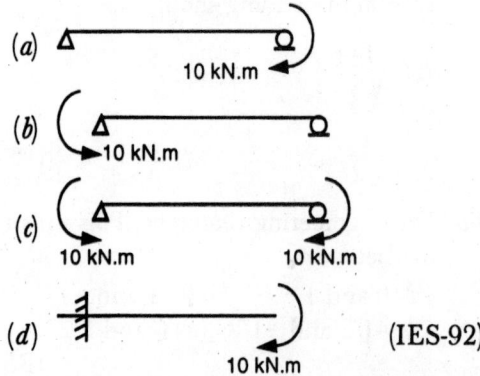
                                              (IES-92)

373. A simple beam *A* carries a point load at mid-span. Another identical beam *B* carries the same load but as uniformly distributed load over the entire span. The ratio of the maximum B.M. in beam *A* to that in beam *B* will be

     (a) $\frac{1}{2}$ (b) 2

(c) 4   (d) $\frac{1}{4}$

374. A free body diagram on the left side of the section *A–A* of the truss loaded as shown is

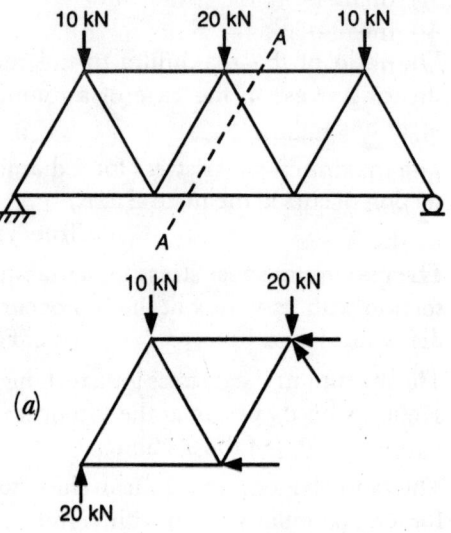

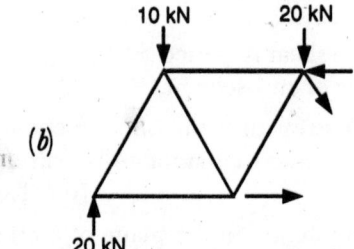

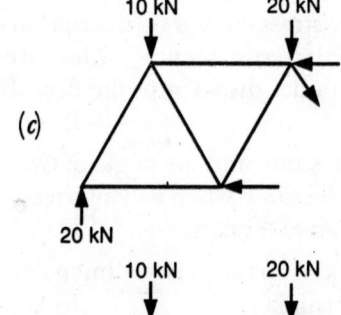

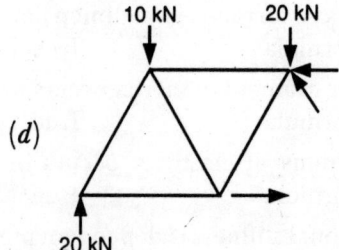

375. When a close-coiled helical spring is subjected to an axial load, the deflection of

the spring is directly proportional to the

(*a*) modulus of rigidity of the spring material

(*b*) diameter of the spring wire

(*c*) mean diameter of the spring

(*d*) number of turns of the spring

(IES-92)

376. Two circular bars *A* and *B* of the same material and same length are of diameters d_A and d_B respectively; they are subjected to the same torque. The ratio of strain energies stored in the bars *A* and *B* is related to (IES-92)

(*a*) d_A/d_B (*b*) d_B/d_A

(*c*) $(d_A/d_B)^2$ (*d*) $(d_B/d_A)^2$.

377. For a cantilever loaded with a uniformly varying load, zero at the free end and *w/* unit run at the fixed end, then the shape of the bending moment diagram is

(*a*) cubic parabola (*b*) parabola

(*c*) straight line (*d*) triangle.

(IES-92)

378. The fixed-end B.M. of a cantilever of length *l* with a couple *M* acting at its free end is

(*a*) *M* (*b*) *M*/2

(*c*) *M/l* (*d*) *Ml*.

379. The maximum negative B.M. in a fixed beam carrying a uniform load is at

(*a*) mid-span (*b*) $\frac{1}{3}$ of the span

(*c*) quarter span (*d*) supports.

(IES-92)

380. A thin cylindrical shall of diameter *d*, length *l*, thickness *t*, is subjected to an internal pressure *p*. The ratio of longitudinal strain to hoop strain is

(*a*) $\frac{pd}{2t}$ (*b*) $\frac{pd}{2t}\left(1-\frac{1}{m}\right)$

(*c*) $\left(\frac{m-2}{2m-1}\right)$ (*d*) $\left(\frac{2m-1}{m-2}\right)$

(IES-92)

381. In the case of cantilever with uniformly varying load, the ratio of the maximum

B.M. when the intensity increases from zero at the fixed end to *w* at the free end to that when it increases from zero at the free end to *w* at the fixed end is

(*a*) 1 (*b*) 2

(*c*) 1/2 (*d*) 1/4. (IES-92)

382. If the normal cross-section *A* of a member is subjected to a tensile force *P*, the resulting normal stress in a oblique plane inclined at an angle θ.to the cross-sectional plane will be

(*a*) $\frac{P}{A}\cdot\sin^2\theta$ (*b*) $\frac{P}{A}\cdot\cos^2\theta$

(*c*) $\frac{P}{2A}\cdot\sin 2\theta$ (*d*) $\frac{P}{2A}\cdot\cos 2\theta$

(IES-92)

383. The shear stress developed at a radial distance *r* is *q*. The shear stress developed at a radial stress *r*/2 is

(*a*) *q* (*b*) 0.25*q*

(*c*) 0.50*q* (*d*) 0.75*q*. (IES-93)

384. In a uni-dimensional stress system, the principal plane is defined as one on which the

(1) shear stress is zero

(2) normal stress is zero

(3) shear stress is maximum

(4) normal stress is maximum (IES-93)

(*a*) 1 and 2 are correct

(*b*) 2 and 3 are correct

(*c*) 1 and 4 are correct

(*d*) 3 and 4 are correct.

385. If an element is subjected to pure shear stress q_{xy}, then the maximum principal stress is equal to

(*a*) $2q_{xy}$ (*b*) $q_{xy/2}$

(*c*) q_{xy} (*d*) $\sqrt{1-\left(q_{xy}\right)^3}$.(IES-93)

386. A steel cable of 20 mm diameter is used to lift a load of 500π N. Given that $E = 2 \times 10^5$ N/mm^2 and the length of the cable is 10 m, elongation of the cable due to the load will be

(*a*) 5 mm (*b*) 2.5 mm

(c) 10 mm   (d) $\dfrac{10}{\pi}$ mm.   (IES-93)

387. The ratio of the greatest to the least of the elongations of the three parts of the following steel bar ($E = 2 \times 10^5$ MPa) is

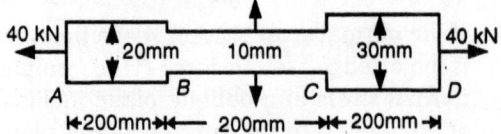

   (a) 9   (b) 4
   (c) 3   (d) 2.   (IES-93)

388. A steel bar of 2 m length is fixed at both ends at $20°C$. The coefficient of thermal expansion is $11 \times 10^{-6}/°C$. The modulus of elasticity is 2×10^5 N/mm². If the temperature is reduced to $18°C$, the bar will experience a stress of

   (a) 2.2 MPa (tensile)
   (b) 2.2 MPa (compressive)
   (c) 4.4 MPa (compressive)
   (d) 4.4 MPa (tensile).   (IES-93)

389. A cast-iron block 500 mm² cross-section carries a tensile load of 100 kN. The maximum shear stress in the block is given by
   (a) 200 MPa   (b) 100 MPa
   (c) 50 MPa   (d) 25 MPa.   (IES-93)

390. The principle of superposition is applied to (1) linear elastic bodies, (2) bodies subjected to small deformations. (IES-93)
   (a) 1 alone is correct
   (b) 1 and 2 are correct
   (c) 2 alone is correct
   (d) neither 1 nor 2 is correct.

391. A prismatic bar of volume V is subjected to a compressive force in the longitudinal direction. If the Poisson's ratio is μ and the longitudinal strain is e, the reduction in volume is   (IES-93)
   (a) $eV(1 - 2\mu)$ (b) $eV(1 + 2\mu)$
   (c) $eV(1 - 2\mu)^2$ (d) $eV(1 + 2\mu)^2$.

392. A bar of length l, cross-sectinal area A, and self-weight W hangs vertically, and supports a load P at the bottom end. If the Young's modulus is E, the total elongation of the bar is   (IES-93)

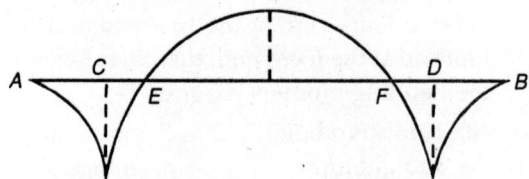

393. In the graphical method of determining reactions in a simple beam with inclined loads, the funicular polygon must be started through the roller end of the beam.

                             True/False

394. The bending moment diagram for an overhanging beam is shown below:

The points of contraflexure are
   (a) A and B (b) C and D
   (c) A and C (d) E and F. (IES-93)

395. A Mohr's circle reduces to a point when the body is subjected to   (IES-93)
   (a) pure shear
   (b) uniaxial stress only
   (c) equal and opposite normal stresses on two mutually perpendicular planes being free of shear.
   (d) equal normal stresses on two mutually perpendicular planes, the planes being free of shear.

396. A T-section is used as a simple beam with uniform loading. The maximum bending stress for a given load will occur at the
   (a) top of the section
   (b) C.G. of the section
   (c) bottom of the section
   (d) mid-depth of the section   (IES-93)

397. An I-section beam has an overall depth of 300 mm. If the flange stresses due to bend-

ing at the top and bottom are 120 MPa and 30 MPa respectively, the depth of the N.A. from the top is (IES-93)

(a) 250 mm   (b) 240 mm
(c) 200 mm   (d) 180 mm.

398. The shear stress distribution with depth in a rectangular section is (IES-93)

(a) parabolic   (b) linear
(c) elliptical   (d) circular.

399. The ratio of the shear stress at N.A. in a square with side vertical and with diagonal vertical is (IES-93)

(a) 0.5   (b) 1.0
(c) 1.5   (d) 2.0.

400. The force in the member marked ϕ is

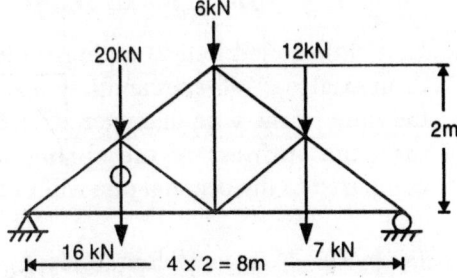

(a) 4 kN (compressive)
(b) 36 kN (tensile)
(c) 16 kN (tensile)
(d) 20 kN (tensile).   (IES-93)

401. The maximum energy stored at elastic-limit of a material is called

(a) resilience
(b) proof resilience
(c) modulus of resilience
(d) none of these.   (IES-93)

402. In order to produce bending and shear stresses of equal magnitudes at the extreme fibres of a circular section under the action of combined bending and torsion, the ratio of bending and twisting moments must be

(a) 1/4   (b) 1/2
(c) 1   (d) 2.   (IES-93)

403. A thin circular plate is subjected to a radial stress σ throughout the circumference. The unit volume change of the plate is

(a) $\frac{\sigma}{E}(1-\mu)$ (b) $\frac{2\sigma}{E}(1-\mu)$

(c) $\frac{\sigma}{E}(1-2\mu)$ (d) $\frac{2\sigma}{E}(1-2\mu)$.

(IES-94)

404. If all the dimensions of a prismatic bar elongating under its own weight are increased m times, then the total elongation will increase in the ratio (IES-94)

(a) $1 : m$ (b) $1 : m^2$
(c) $1 : m^3$ (d) $1 : m^4$.

405. A bar of 4 cm diameter is subjected to an axial load of 4 kN. The extension of the bar over a gauge length of 20 cm is 0.03 cm. The decrease in diameter is 0.0018 cm. The Poisson's ratio is

(a) 0.25   (b) 0.30
(c) 0.33   (d) 0.35.   (IES-94)

406. A compound bar consisting of materials A and B is tightly secured at the ends. The coefficient of thermal expansion of A is more than that of B. When the temperature is increased, the stresses induced will be

(a) tensile in both materials
(b) tensile in A and compressive in B
(c) compressive in A and tensile in B
(d) compressive in both materials.

(IES-94)

407. A single direct stress p acts along the longitudinal (x-axis) axis of a bar. Match List I with List II and select the correct answer using the codes given: (IES-94)

List I	List II
A. Strain along x-axis	1. p/mE
B. Strain along y-axis	2. p/E
C. Strain energy per unit volume	3. $(p/E)(1-2/m)$
D. Volumetric strain	4. $p^2/2E$

Codes:

(a) $A-1, B-2, C-3, D-4$
(b) $A-1, B-2, C-4, D-3$
(c) $A-2, B-1, C-3, D-4$
(d) $A-2, B-1, C-4, D-3$.

408. If the principal stresses at a point are p_1 and p_2. The plane on which the resultant stress is the maximum is inclined to the major principal plane at an angle
    (a) 30°   (b) 45°
    (c) 60°   (d) 75°   (IES-94)

409. If two planes at right angles carry only shear stress of magnitude q, then the   (IES-94)
    (1) diameter of the Mohr's circle would equal $2q$
    (2) centre of the Mohr's circle would lie at the origin
    (3) principal stresses are unlike and have magnitude q
    (4) angle between the principal plane and the plane of maximum shear would be equal to 45°

    Of these statements
    (a) 1 and 2 are correct
    (b) 2 and 4 are correct
    (c) 3 and 4 are correct
    (d) 1, 2, 3 and 4 are correct.

410. A bar of square section is subjected to a pull of 100 kN. If the maximum allowable shear stress is 50 N/mm², then the side of the square section will be   (IES-94)

    (a) $\sqrt{5}$ cm   (b) $\sqrt{10}$ cm

    (c) $\sqrt{15}$ cm   (d) $\sqrt{20}$ cm.

411. A simply supported beam of span L carries a concentrated load W at its mid-span. If the width b of the beam is constant throughout the span, then the depth at mid-span (f being the permissible bending stress) will be

    (a) $\dfrac{3WL}{2bf}$ (b) $\sqrt{\dfrac{3WL}{2bf}}$

    (c) $\dfrac{6WL}{bf}$ (d) $\sqrt{\dfrac{6WL}{2bf}}$. (IES-94)

412. Which of the following pairs are correctly matched?
    1. Shear stress distribution diagram ––varies for different shapes of sections.

2. Bending stress distribution diagram –– similar for different shapes of sections.
3. Ratio of bending stress to shear stress for a shaft subjected to a torque T and moment M_____$M/2T$
4. Polar moment of inertia _____ sum of MI about the XX and YY-axes.
    Select the correct answer using the codes given below: (IES-94)
    Codes:
    (a) 1, 2 and 3   (b) 1, 3 and 4
    (c) 1, 2 and 4   (d) 2, 3 and 4.

413. In a plane truss if M is the number of members, J the number of joints, and R the number of support reactions, then for the truss to be determinate,   (IES-94)
    (a) $J = M + R$ (b) $J = 2M + R$
    (c) $3J = M + 2R$ (d) $2J = M + R$.

414. In a close-coiled helical spring subjected to an axial load, other quantities remaining the same, if the wire diameter is doubled, then the stiffness of the spring when compared to the original one will become   (IES-94)
    (a) twice   (b) four times
    (c) eight times,   (d) sixteen times.

415. A steel wire of 20 mm diameter is bent into a circular shape of 10 m radius. If $E = 2 \times 10^5$ MPa, then the maximum stress induced is   (IES-95)
    (a) 10^2 MPa   (b) 2×10^2 MPa
    (c) 4×10^2 MPa   (d) 6×10^2 MPa.

416. A solid metal bar of uniform diameter D and length L is hung vertically from a ceiling. If the density of the material of the bar is ρ and the modulus of elasticity is E. Then the total elongation due to its own weight is

    (a) $\dfrac{\rho L}{2E}$ (b) $\dfrac{\rho L^2}{2E}$

    (c) $\dfrac{\rho E}{2L}$ (d) $\dfrac{\rho E}{2L^2}$. (IES-95)

417. A steel cube of volume 8×10^6 mm³ is subjected to all round stress of 133 N/mm².

The bulk modulus of the material is 1.33×10^5 N/mm^2. The volumetric change is

(a) 8000 mm^3　　(b) 6000 mm^3
(c) 800 mm^3　　(d) 1 mm^3.　(IES-95)

418. In terms of bulk modulus K, and modulus of rigidity, N, the Poisson's ratio can be expressed as　　(IES-95)

(a) $\dfrac{3k - 4N}{6k + 4N}$　　(b) $\dfrac{3k + 4N}{6k - 4N}$

(c) $\dfrac{3k - 2N}{6k + 2N}$　　(d) $\dfrac{3k + 3N}{6k - 2N}$

419. The cross-section of a bar is subjected to a uniaxial tensile stress p. The tangential stress on a plane inclined at θ to the cross-section would be

(a) $\dfrac{p}{2}\sin2\theta$　　(b) $p\sin2\theta$

(c) $\dfrac{p}{2}\cos2\theta$　　(d) $p\cos2\theta$.　(IES-95)

420. Consider the following statements:
　　(IES-95)
(1) On planes having principal stresses, there will be no tangential stress.
(2) Shear stress on mutually perpendicular planes are numerically equal.
(3) Maximum shear stress is equal to half the sum of the maximum and minimum principal stresses.
Of these statements
(a) 1, 2 and 3 are correct
(b) 1 and 2 are correct
(c) 2 and 3 are correct
(d) 1 and 3 are correct.

421. In a stressed body, tensile stresses of 15 and 9 kN/cm^2 are observed on two perpendicular planes along with shear stresses of 4 kN/cm^2.　　(IES-95)
The principal stresses at the point are
(a) 12 kN/cm^2 (ten) and 3 kN/cm^2 (ten)
(b) 17 kN/cm^2 (ten) and 7 kN/cm^2 (ten)
(c) 9.5 kN/cm^2 (com) and 6.5 kN/cm^2 (com)

(d) 19 kN/cm^2 (ten) and 13 kN/cm^2 (ten)

422. Consider the following statements:
(1) If a beam has two axes of symmetry, even then the shear centre does not coincide with the centroid.
(2) For a section with one axis of symmetry, the shear centre does not coincide with the centroid, but lies on the axis of symmetry.
(3) If a load passes through the shear centre then there will be only bending in the section and no twisting.　(IES-95)
(a) 1, 2 and 3 are correct
(b) 1 and 2 are correct
(c) 2 and 3 are correct
(d) 1 and 3 are correct.

423. Match List I with List II and select the correct answer using the codes given below:

List I *Type and position of load on cantilever*	List II *Shape of moment diagram*
A. Triangular load with zero at the free end to maximum at fixed end.	1. Parabola
B. Uniformly distributed load	2. Rectangle
C. Concentrated load at free end.	3. Cubic parabola
D. Bending couple at free end	4. Triangle

Codes:
(a) $A - 3$, $B - 2$, $C - 1$, $D - 4$
(b) $A - 4$, $B - 3$, $C - 2$, $D - 1$
(b) $A - 3$, $B - 1$, $C - 4$, $D - 2$
(d) $A - 2$, $B - 4$, $C - 1$, $D - 3$　(IES-95)

424. A simply supported beam is loaded shown in the figure. The bending moment at E would be

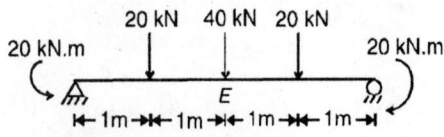

(a) 60 kN.m (sagging)
(b) 40 kN.m (hogging)

(c) 60 kN.m (hogging)

(d) 40 kN.m (sagging).   (IES-95)

425. Tick the correct one of the following statements:

(a) Shear force is the first derivative of the bending moment.

(b) Shear force is the first derivative of the intensity of the load.

(c) Load intensity on a beam is the first derivative of the bending moment.

(d) Bending moment is the first derivative of the shear force.   (IES-95)

426. The ratio of the moment-carrying capacity of a beam of diameter D and square beam of size D is

(a) $\pi/4$ (b) $3\pi/8$

(c) $\pi/3$ (d) $3\pi/16$.   (IES-95)

427. A solid shaft of circular cross–section is subjected to a torque T and a bending moment M; the equivalent bending moment is given by

(a) $\dfrac{M + \sqrt{M^2 + T^2}}{4}$

(b) $\dfrac{M^2 + \sqrt{M + T}}{2}$

(c) $\dfrac{M - \sqrt{M^2 + T^2}}{2}$

(d) $\dfrac{M + \sqrt{M^2 + T^2}}{2}$ (IES-95)

428. A beam has triangular cross-section, having base b and altitude h; if a section of the beam is subjected to a shear force F, the shear stress at the level of neutral axis in the cross–section is given by

(a) $\dfrac{4F}{3\,bh}$ (b) $\dfrac{3F}{4\,bh}$

(c) $\dfrac{8F}{3\,bh}$ (d) $\dfrac{3F}{8\,bh}$.   (IES-95)

429. A solid shaft of circular cross–section is subjected to a torque T, which produces a

maximum shear stress f_s in the shaft. The diameter of the shaft is given by

(a) $\sqrt{\dfrac{\pi\,f_s}{16\,T}}$ (b) $\sqrt[3]{\dfrac{\pi\,f_s}{16\,T}}$

(c) $\sqrt{\dfrac{16\,T}{\pi\,f_s}}$ (d) $\sqrt[3]{\dfrac{16\,T}{\pi\,f_s}}$ (IES-95)

430. A rectangular block of size 200×50 is subjected to a shear stress of 50 MPa. If the rigidity modulus of the material is 1×10^5 MPa, the strain energy stored will be

(a) 10^5 N.mm   (b) 5×10^4 N.mm

(c) 12,500 N.mm   (d) 10,000 N.mm.

(IES-95)

431. If $E = 2.06 \times 10^5$ MPa, an axial pull of 60 kN suddenly applied to a steel rod 50 mm in diameter and 4 m long causes an instantaneous elongation of the order of

(IES-95)

(a) 1.19 mm   (b) 2.19 mm

(c) 3.19 mm   (d) 11.9 mm.

432. A cantilever of constant depth carries a uniform load on the whole span. To make the maximum stress at all sections the same, the breadth of the section at a distance x from the free end should be proportional to

(a) x (b) $\sqrt{x}$ (c) x^2 (d) x^3.

(IES-96)

433. A section of the solid circular shaft with diameter D is subjected to a bending moment M and a torque T. The expression for maximum principal stress is

(a) $\dfrac{(2M + T)}{\pi D^3}$

(b) $\dfrac{16\pi}{D^3}\left(M + \sqrt{M^2 + T^2}\right)$

(c) $\dfrac{16}{\pi D^3}\sqrt{M^2 + T^2}$

(d) $\dfrac{16}{\pi D^3}\left(M + \sqrt{M^2 + T^2}\right)$ (IES-96)

434. The stress at which a material fractures under large number of reversals of stress is called
    (*a*) endurance limit
    (*b*) creep
    (*c*) ultimate strength
    (*d*) plastic limit. (IES-96)

435. The length, coefficient of thermal expansion, and young's modulus of material of a bar *A* are twice that of bar *B*. If the temperature of both bars is increased by the same amount while preventing any expansion, then the ratio of stress developed in bar *A* to that in bar *B* will be
    (*a*) 2   (*b*) 4   (*c*) 8   (*d*) 16.
    (IES-96)

436. The linear strain in the diagonal of a square block under pure shear, with a shear strain φ, is given by
    (*a*) $\phi/2$
    (*b*) $\phi/\sqrt{2}$
    (*c*) $\phi\sqrt{2}$
    (*d*) ϕ. (IES-96)

437. The lists given refer to a bar of length *l*, cross-sectional area *A*, Young's modulus *E*, poisson's ratio μ, and is subjected to axial stress *p*. Match List I and with List II using the codes given below:

List I	*List* II
A. Volumetric strain	1. $2(1+\mu)$
B. Strain energy per unit volume	2. $3(1-2\mu)$
C. Ratio of young's modulus to bulk modulus	3. $(p/E)(1-2\mu)$
D. Ratio of young's modulus to rigidity modulus	4. $p^2/2E$
	5. $2(1-\mu)$
	(IES-96)

   Codes:
   (*a*) $A-3, B-4, C-2, D-1$
   (*b*) $A-5, B-4, C-1, D-2$
   (*c*) $A-5, B-4, C-2, D-1$
   (*d*) $A-2, B-3, C-1, D-5$.

438. If all dimensions of a prismatic bar of square cross-section suspended freely from the ceiling of a roof are doubled then the total elongation produced by its own weight will increase.
    (*a*) eight times
    (*b*) four times
    (*c*) three times
    (*d*) two times. (IES-96)

439. For the beam shown in the figure, the maximum positive B.M. equals the maximum negative B.M. The value of L_1 is

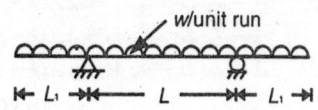

   (*a*) $L/\sqrt{2}$
   (*b*) $L/\sqrt{3}$
   (*c*) $L/2$
   (*d*) $L/(2\sqrt{2})$.
    (IES-96)

440. A simply supported beam of span *L* carries a concentrated load *W* at mid span. If the maximum permissible stress is *f* and the width of the beam section *b*, the depth *d* is
    (*a*) $\sqrt{\dfrac{6WL}{bf}}$
    (*b*) $\dfrac{6WL}{bf}$
    (*c*) $\sqrt{\dfrac{3WL}{2b\cdot f}}$
    (*d*) $\dfrac{3WL}{2bf}$.
    (IES-96)

441. Match List I with List II and select the correct answer using the codes given below:
    (IES-96)

List I *Type of beam with loading*	*List* II *Maximum bending moment*
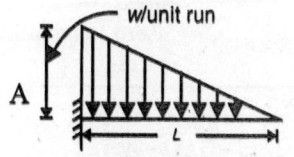 A	1. $wL^2/12$
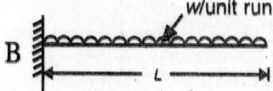 B	2. $wL^2/6$

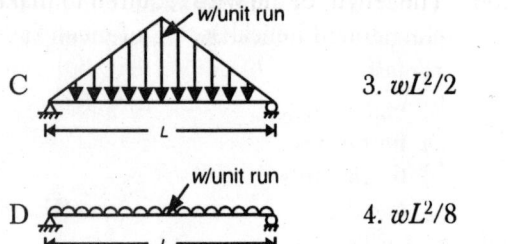

C   3. $wL^2/2$

D   4. $wL^2/8$

Codes:

(a) $A-4$, $B-3$, $C-2$, $D-1$
(b) $A-1$, $B-3$, $C-2$, $D-4$
(c) $A-2$, $B-3$, $C-1$, $D-4$
(d) $A-2$, $B-4$, $C-1$, $D-3$.

442. A round steel bar of length 400 mm consists of 100 mm ϕ for 200 mm and 80 mm ϕ for the remaining 200 mm. If the rod is subjected to a tensile load of 100 kN, the elongation will be ($E = 2 \times 10^5$ MPa).

(a) $\dfrac{1}{\pi}\left(\dfrac{1}{25} + \dfrac{1}{16}\right)$ mm

(c) $\dfrac{2}{\pi}\left(\dfrac{1}{25} + \dfrac{1}{16}\right)$ mm

(c) $\dfrac{3}{\pi}\left(\dfrac{1}{25} + \dfrac{1}{16}\right)$ mm

(d) $\dfrac{4}{\pi}\left(\dfrac{1}{25} + \dfrac{1}{16}\right)$ mm.   (IES-97)

443. Match List I with List II using the codes given: (IES-97)

List I	*List* II
A. Moment of inertia	1. Tensile stress
B. Elongation	2. Modulus of rupture
C. Neutral axis	3. Zero shear stress
D. Top fibre	4. Zero longitudinal stress

Codes:

(a) $A-2$, $B-1$, $C-3$, $D-4$
(b) $A-1$, $B-2$, $C-4$, $D-3$
(c) $A-3$, $B-4$, $C-1$, $D-2$
(d) $A-2$, $B-1$, $C-4$, $D-3$.

444. A given material has a young's modulus E, modulus of rigidity N, and Poisson's ratio 0.25. The ratio of E to N of this material is

(a) 0.75   (b) 3.00
(c) 2.50   (d) 1.50.   (IES-97)

445. A prismatic bar of uniform cross–sectional area 500 mm² is subjected to axial loads as shown in the figure:

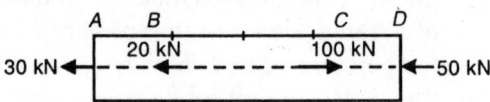

Portion $B\,C$ is subjected to an axial stress of

(a) 40 N/mm² (ten)
(b) 200 N/mm²(com)
(c) 100 N/mm² (ten)
(d) 60 N/mm² (ten).   (IES-97)

446. Two wires, one of copper and the other of steel ($E_{cu} = 1 \times 10^5$ MPa, $E_{st} = 2 \times 10^5$ MPa) are hung vertically from a ceiling. A load of 10 kN is suspended symmetrically to a rigid horizontal bar to which the wires are fixed. The areas of section are 400 mm² for the copper wire and 200 mm for steel wire. The ratio of the elongations of copper and steel wire is

(a) 0.25   (b) 0.50
(c) 2.00   (d) 1.00.   (IES-97)

447. At a point in a strained material, if two tensile stresses 200 MPa and 100 MPa act in two perpendicular directions, then the tangential stress on a plane inclined at 15° to the axis of the minor stress will be

                             (IES-97)

(a) 12.5 MPa   (b) 25 MPa
(c) 50 MPa   (d) 100 MPa.

448. At a point in a strained material, if the normal stresses on two perpendicular planes are 100 MPa (tensile) and 50 MPa (compressive); the normal stress acting on the plane of maximum shear is

(a) 25 MPa (compr)
(b) 75 MPa (compr)

(c) 25 MPa (ten)

(d) 75 MPa (ten).   (IES-97)

449. The shear centre of a section is defined as that point

(a) through which the load must pass to produce zero twisting moment on the section.

(b) at which the shear force is zero.

(c) at which the shear force is a maximum.

(d) at which the shear force is a minimum.   (IES-97)

450. The ratio of the flexural strengths of two beams of square cross-section, the first beam placed with its side horizontal, and the second placed with a diagonal horizontal, is

(a) $\sqrt{3}$

(b) $1/\sqrt{3}$

(c) $1/\sqrt{2}$

(d) $\sqrt{2}$.   (IES-97)

451. A device in which the material is so arranged that it can suffer considerable changes in form without being permanently damaged or distorted is called a _____.

452. A spring which has coils wound into the shape of a helix is called a _____.

453. The purpose of a _____ is to absorb energy and restore it as required.

454. Springs with a very small helix angle are called a _____ helical springs.

455. In a spring loaded axially, the direct shear stress is considered negligible when compared with torsional shear stress.
     True/False

456. The direct tensile stress is negligible when compared with the bending stresses when an open-coiled spring is axially loaded.
     True/False

457. In axially loaded close-coiled helical spring, the wire section is subjected only to normal stresses.   True/False

458. The length of the wire required to make a close-coiled helical spring of mean radius R and number of coils n is a little more than

(a) $2\pi Rn$ (b) πRn

(c) $4\pi Rn$ (d) $\pi R^2 n$.

459. The _____ of a spring is defined as the axial force required to cause unit axial deflection.

460. An axial torque applied to a close-coiled helical spring causes a constant _____ on the coils, of magnitude equal to _____.

461. The absolute maximum shear stress in the wire section of a close-coiled helical spring subjected to an axial torque is numerically the same as that when it is subjected to an axial load which causes the same torque.
     True/False

462. The absolute maximum shear stress in the wire section of an open-coiled helical spring subjected to axial load is the same as that in a corresponding close-coiled spring.
True/False

463. In the case of parallel helical springs, one inside the other, the axial load is the same in both.   True/False

464. In the case helical springs, arranged in series, the axial deflection under an axial load is the same in both springs.
     True/False

465. The nature of the stress caused in a leaf spring is shear.   True/False

466. The nature of the stress caused in a leaf spring is bending stress.   True/False

467. In the case of a close-coiled spring carrying an axial load, the spring is subjected to

(1) torsion,

(2) axial deflection

(3) bending moment, the effect of which may be neglected.   (IES-93)

Of these statements,

(a) 1, 2 and 3 are correct

(b) 1 and 2 are correct

(c) 1 and 3 are correct

(d) 2 and 3 are correct.

468. If two springs of stiffness k_1 and k_2 are connected in series, the stiffness of the combined spring is

(a) $\dfrac{k_1}{(k_1 + k_2)}$ (b) $\dfrac{(k_1 + k_2)}{k_1 k_2}$

(c) $(k_1 + k_2)$ (d) $k_1 k_2$.   (IES-95)

469. The load which causes the initially curved plates of a laminated spring to straigthen is known as the_____.

470. The length of the plates of a carriage spring is gradually curtailed from the top to the bottom because the bending moment caused by the central load is maximum at mid-span and decreases to zero value at the end supports.   True/False

471. Each plate in a leaf spring is bent into the shape of the arc of a circle.   True/False

472. The deflection of a leaf spring is inversely proportional to the _____ to which the leaves are bent.

473. Knowledge of deflections helps in solving statically indeterminate structures.
                                                                          True/False

474. Macaulay's approach of determining deflections of beams is a modification of moment-area method.   True/False

475. The first derivative of deflection of a beam with respect to its span is the _____ of the beam.

476. The second derivative of deflection of a beam with respect to its span is numerically the _____ at the particular section of the beam, divided by the flexural rigidity.

477. The_____derivative of the deflection of a beam is related to the rate of loading.

478. The maximum slope of a symmetrically loaded simply supported beam occurs at the mid-span section.   True/False

479. For a simple beam of span l, loaded uniformly with w per unit length, the maximum deflection is

(a) $\dfrac{7}{384} \cdot \dfrac{wl^4}{EI}$ (b) $\dfrac{11}{384} \cdot \dfrac{wl^4}{EI}$

(c) $\dfrac{5}{384} \cdot \dfrac{wl^4}{EI}$ (d) $\dfrac{1}{48} \cdot \dfrac{wl^4}{EI}$.

480. Young's modulus of a material is invariably greater than the rigidity modulus.
                                                                          True/False

481. If the depth of a beam carrying an isolated load is doubled, the deflection of the beam at mid-span will change by a factor of

(a) $\dfrac{1}{6}$ (b) $\dfrac{1}{4}$

(c) $\dfrac{1}{2}$ (d) $\dfrac{1}{8}$ (IES-92)

482. The maximum values of slope and deflection occur at the free end of a cantilever.   True/False

483. The maximum deflection of a cantilever of length l with a uniform load of w per unit run is

(a) $\dfrac{wl^4}{8EI}$ (b) $\dfrac{wl^3}{6EI}$

(c) $\dfrac{wl^4}{6EI}$ (d) $\dfrac{wl^3}{48EI}$.

484. The maximum deflection in a simple beam with a non-central concentrated load occurs in the longer portion of the span, between the load and the mid-span section.
                                                                          True/False

485. A simple beam "A" carries a point load at mid-span. Another identical beam "B" carries the same load but uniformly distributed over the entire span. The ratio of the maximum deflection of beam "A" to that of beam "B" is

                                                                          (IES-92)

(a) $\dfrac{5}{3}$ (b) $\dfrac{8}{5}$

(c) $\dfrac{5}{8}$ (d) $\dfrac{3}{5}$.

486. The free-end deflection of a cantilever of length l with a couple M acting at its free end is (IES-92)

(a) $\dfrac{Ml^2}{2EI}$ (b) $\dfrac{Ml^2}{8EI}$

(c) $\dfrac{2Ml^2}{EI}$ (d) $\dfrac{Ml}{4EI}$.

487. In a simple beam with a non-central concentrated load, the deflection at mid-span is a close-approximation to the maximum deflection. True/False

488. In a simple beam of span l and flexural rigidity EI, acted on by a non-central load W at a distance a from one end and b from the other, the deflection under the load is

(a) $\dfrac{W\,a^2b^2}{3EI\,l}$ (b) $\dfrac{Wa^3b^3}{3EIl}$

(c) $\dfrac{Wab}{3EIl}$ (d) $\dfrac{Wa^2b^2}{6EIl}$

489. A smooth sphere of weight W is supported in contact with a smooth vertical wall by a string fastened to a point on its surface, the other end being attached to a point in the wall as shown in the figure. If the length of the string is equal to the radius of the sphere, then the tension in the string and the reaction of the wall will be respectively.

(IES-93)

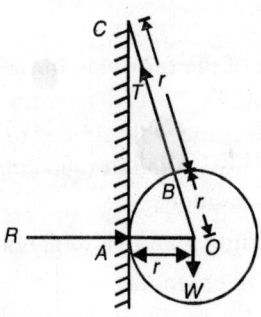

(a) $\dfrac{4W}{\sqrt{3}},\dfrac{2W}{\sqrt{3}}$ (b) $\dfrac{2W}{\sqrt{3}},\dfrac{W}{\sqrt{3}}$

(c) $\dfrac{5W}{\sqrt{3}},\dfrac{W}{\sqrt{3}}$ (d) $\dfrac{3W}{\sqrt{3}},\dfrac{W}{\sqrt{3}}$.

490. A simple beam of span l and flexural ridigity EI carries a central point load W and a total uniform load W spread throughout the span. The maximum deflection is given by

(a) $\dfrac{13Wl^3}{96EI}$ (b) $\dfrac{5Wl^3}{384EI}$

(c) $\dfrac{5Wl^3}{96EI}$ (d) $\dfrac{13Wl^3}{384EI}$. (IES-93)

491. A cantilever beam of uniform EI has a span equal to l. An upward force P acts at mid-section and a downward load W acts at the free end. If the free-end deflection is zero, the relation between P and W is

(IES-93)

(a) $P = \dfrac{2W}{3}$ (b) $P = \dfrac{W}{5}$

(c) $P = \dfrac{5W}{16}$ (d) $P = \dfrac{W}{2}$.

492. A simple beam of span l carries two equal unlike couples M at the two ends. The central deflection of the beam, EI being constant, is given by (IES-95)

(a) $\dfrac{Ml^2}{4EI}$ (b) $\dfrac{Ml^2}{16EI}$

(c) $\dfrac{Ml^2}{64EI}$ (d) $\dfrac{Ml^2}{8EI}$.

493. A cantilever of length L carries a load P at C at a distance l from the fixed-end A. The deflection at the free end B is

(a) $\dfrac{Pl^2}{2EI}(L-l)$ (b) $\dfrac{Pl^2}{3EI}(L-l)$

(c) $\dfrac{Pl^2}{2EI}(L+l/3)$ (d) $\dfrac{Pl^2}{2EI}(L-l/3)$.

(IES-96)

494. When a simple beam of span l is acted on by a bending couple M at one end, the maximum deflection is given by

(a) $\dfrac{Ml^2}{9EI}$ (b) $\dfrac{Ml^2}{\sqrt{3}EI}$

(c) $\dfrac{Ml}{9\sqrt{3}EI}$ (d) $\dfrac{Ml^2}{9\sqrt{3}EI}$

495. Macaulay's approach is especially useful in the determination of deflections when a simple beam is acted on by a number of concentrated loads or by a uniform load on part of the span.   True/False

496. The effect of a load inclined to the longitudinal axis of a short strut, fixed at the bottom and free at the top, is to cause a _____ stress and a _____ stress in the cross-section of the strut.

497. The effect of an eccentric vertical load on the top of a short vertical strut is precisely similar to that of a load inclined to the longitudinal axis of the strut.   True/False

498. The limiting eccentricity for "no tension" when a vertical load acts on a short vertical strut of rectangular cross section of width b is

(a) $\dfrac{b}{4}$ (b) $\dfrac{b}{6}$

(c) $\dfrac{b}{3}$ (d) $\dfrac{b}{2}$.

499. The safe zone in which a vertical compressive load can act on the cross-section of a short strut so that no tensile stresses occur any where in the cross-section is called the _____ of section.

500. The shape of the core of a rectangular section is
   (a) square   (b) rhombus
   (c) circle   (d) rectangle.

501. When a vertical load acts at the limiting eccentricity on the top of a short vertical strut, the ratio of the maximum compressive stress to the average value is _____.

502. Determination of the effect of wind pressure on tall chimney shafts involves the use of the principle of combined flexure and direct stress.   True/False

503. The maximum axial load which a long column can carry without buckling is called the _____.

504. The ratio of the effective length to the least radius of gyration of a long column is known as its _____.

505. The "effective length" of a long column of length l, fixed at the bottom and free at the top, is
   (a) $2l$ (b) l
   (c) $l/2$ (d) $l/\sqrt{2}$.

506. The critical load of a long column does not depend on the strength of the material.   True/False

507. The minimum slenderness ratio of a mild steel column for the Euler–formula to be applicable is _____ approximately.

508. The design procedure of slender compression members is one of trial and error.   True/False

509. The critical stress for slender compression members is directly proportional to the slenderness ratio.   True/False

510. The factor of safety in the design of slender compression members can be varied with the range of slenderness ratio.
     True/False

511. "Secant formula" is the rigorous formula for determining the critical stress for an eccentrically loaded long column.
     True/False

512. The form of the formulae for critical stress of an initially curved column with axial loading is some what similar to that for an eccentrically loaded straight column under certain conditions.   True/False

513. The buckling load in a steel column is
   (a) related to length
   (b) directly proportional to slenderness ratio.
   (c) inversely proportional to the slenderness ratio.

(d) non-linearly related to the slenderness ratio. (IES-92)

514. If K is the ratio of young's modulus to the permissible compressive stress in the material of a column, then the Rankine's constant is proportional to (IES-93)

(a) $\dfrac{1}{K}$

(b) K

(c) $\sqrt{K}$

(d) $\dfrac{1}{\sqrt{K}}$

515. A load P acts on the diagonal of a square column of size D. The maximum distance of the load from the centre for no tension is

(IES-93)

(a) $\dfrac{D}{8}$

(b) $\dfrac{D}{6}$

(c) $\dfrac{\sqrt{2}}{6} \cdot D$

(d) $\dfrac{\sqrt{2}\,D}{12}$.

516. A circular shaft subjected to torsion undergoes a twist of $1°$ in a length of 1.2 m. If the maximum shear stress induced is 100 MPa and the rigidity modulus is 0.8×10^5 MPa, the radius of the shaft in mm should be (IES-93)

(a) $\dfrac{\pi}{180}$

(b) $\dfrac{\pi}{270}$

(c) $\dfrac{180}{\pi}$

(d) $\dfrac{270}{\pi}$.

517. A short column of external diameter D and the internal diameter d, is subjected to a load W at an eccentricity e, causing zero stress at an extreme fibre. Then the value of e must be (IES-94)

(a) $\dfrac{\left(D^2 + d^2\right)}{8\pi D}$

(b) $\dfrac{\left(D^2 + d^2\right)}{8 D}$

(c) $\dfrac{\left(D^2 - d^2\right)}{8 D}$.

(d) $\dfrac{\left(D^3 + d^3\right)}{8 D^2}$

518. Four vertical columns of the same material, height and weight have the same end conditions. The buckling load will be the largest for a column having the cross-section of a

(a) solid square

(b) thin hollow circle

(c) solid circle

(d) I-section. (IES-94)

519. If the crushing stress in the material of a mild steel column is 330 N/mm^2, Euler's formula for critical load is applicable for slenderness ratio equal to/greater than

(a) 40

(b) 50

(c) 60

(d) 80. (IES-95)

520. The resultant cuts the base of a circular column of diameter d with an eccentricity $d/4$. The ratio between the maximum values of compressive and tensile stress is

(IES-95)

(a) 3

(b) 4

(c) 5

(d) infinity

521. For a circular column having its ends hinged, the slenderness ratio is 160. The l/d ratio of the column is

(a) 80

(b) 57

(c) 40

(d) 20. (IES-96)

522. A hollow circular column of internal diameter d and external diameter $(1.5\,d)$ is subjected to compressive load. The maximum eccentricity for no tension is

(IES-96)

(a) $\dfrac{d}{8}$

(b) $\dfrac{13\,d}{48}$

(c) $\dfrac{d}{4}$

(d) $\dfrac{13\,d}{96}$.

523. Which one of the following pairs is not correctly matched?

Boundary conditions of the column	Euler's buckling load
	(IES-95)
(a) Pin–Pin	$\pi^2\ EI/l^2$
(b) Fixed–Fixed	$4\,\pi^2\ EI/l^2$
(c) Fixed–Free	$0.25\,\pi^2\ EI/l^2$
(d) Fixed–Pin	$\sqrt{2}\ \pi EI/l^2$

524. Match List I with List II to ensure 'no tension' condition and select the correct answer using the codes given below:

List I *(Sections)*	List II *(Cores)*
A. Rectangular	1. Circle
B. I-Section	2. Annulus
C. Hollow Circular	3. Rhombus
D. Square	4. I-section
	5. Square
	6. Rectangular

   Codes:
   (a) $A-4, B-6, C-1, D-2$
   (b) $A-3, B-3, C-1, D-5$
   (c) $A-6, B-4, C-2, D-5$
   (d) $A-3, B-3, C-5, D-1$.   (IES-96)

525. Two bars, one of material A and the other of material B of the same length are tightly secured between two unyielding walls. The coefficient of thermal expansion of bar A is more than that of B. When temperature rises, the stresses induced are
   (a) tension in both materials.
   (b) tension in material A and compression in material B.
   (c) compression in material A and tension in material B.
   (d) compression in both materials.
                             (IES-95)

526. The deflection of a leaf spring is proportional to_____the span.

527. Two long columns A and B are of the same material, the same length, and the same cross-section. Column A is hinged at both ends, while column B is fixed at one end and hinged at the other. The ratio of the Euler-crippling loads for A and B is

   (a) $\dfrac{1}{2}$ (b) 2

   (c) $\dfrac{1}{\sqrt{2}}$ (d) $\sqrt{2}$.

528. In a thin long closed cylindrical container, the fluid pressure induces   (IES-94)
   (a) only hoop stress.
   (b) only longitudional stress.
   (c) longitudinal stress equal to twice the hoop stress.
   (d) longitudinal stress equal to half the hoop stress.

529. The ratio of the hoop stress to the longitudinal stress in a thin cylinder under internal pressure is
   (a) 1/2   (b) 2
   (c) 4   (d) 1/4.

530. Longitudinal and hoop stresses in a thin cylinder under internal pressure are "principal stresses".   True/False

531. Riveted seams in thin cylinders tend to decrease the hoop stresses in pressure vessels.   True/False

532. A thin cylindrical steel pressure vessel of diameter 60 mm and wall thickness 3 mm is subjected to an internal fluid pressure of intensity p. If the ultimate strength of steel is 360 N/mm^2, the bursting pressure in N/mm^2 will be
   (a) 1.8   (b) 3.6
   (c) 18   (d) 36   (IES-95)

533. In a thin spherical shell under internal pressure the stress in the material is the same in all directions.   True/False

534. The hoop stress in a thin spherical shell of diameter d and wall thickness t under internal pressure is

   (a) $\dfrac{pd}{2t}$ (b) $\dfrac{pd}{3t}$

   (c) $\dfrac{pd}{4t}$ (d) $\dfrac{pd}{t}$.

535. A thin cylinder wound by a wire under tension is capable of withstanding greater internal pressure than otherwise.
                                 True/False

536. A cylinder is said to be "thick", when the stress in the material across the wall thickness are not uniform but vary from point to point.   True/False

537. A thin cylinder maintains its cylindrical shape under uniform external radial pressure until failure.　　　True/False

538. Under increasing uniform external radial pressure, a thin cylinder tends to fail by forming_____.

539. For internal pressure, the hoop stresses in a cylinder are compressive.　　True/False

540. In a thick cylinder under simultaneous internal and external pressure, the nature of the hoop stress depends upon the relative values of these pressures and the position of the point considered along the radius.
　　　　　　　　　　　True/False

541. _____ equations are useful for determining the stresses in thick cylinders.

542. The maximum hoop stress in a thick cylinder of internal and external radii r_1 and r_2 respectively when subjected to an internal pressure p_i is given by (k : ratio of external to internal radius).

　(a) $p_i\left(\dfrac{k^2-1}{k^2+1}\right)$　　(b) $p_i\left(\dfrac{1-k^2}{1+k^2}\right)$

　(c) $p_i\left(\dfrac{k^2+1}{k^2-1}\right)$　　(d) $p_i\left(\dfrac{1+k^2}{1-k^2}\right)$

543. Equations similar to Lame's equations can be derived for the stresses in a thick spherical shell under uniform internal radial pressure.
　　　　　　　　　　　True/False

544. The form of Lame's equations will be some what different depending upon whether the sign convention for the radial pressure and hoop stress is consistent or not.　True/False

545. Axial stress in a thick cylinder is considered to be negligible in comparison with the radial pressure and hoop stress.　True/False

546. The greatest stress in the material of a thick spherical shell under uniform internal pressure occurs at the outer surface.
　　　　　　　　　　　True/False

547. The effect of shrinking one tube over another is to cause initial hoop compression in the inner tube and hoop tension in the outer tube.　　　　　True/False

548. There is relationship between the shrinkage allowance in the junction diameters of the two tubes of a compound cylinder and the radial pressure induced by the shrinking-on process at the common surface.
　　　　　　　　　　　True/False

549. In cylindrical shells with hemispherical ends, the cylindrical section is thinner than the hemispherical ends.　　True/False
　　　　　　　　　　　(IES-92)

550. Which one of the following pairs is not correctly matched?
　(a) Lame's constants : Thick cylinders
　(b) Macaulay's method : Deflection of beams
　(c) Euler's method : Theory of columns
　(d) Eddy's theorem : Torsion of shafts .
　　　　　　　　　　　(IES-96)

551. Two closed thin vessels, one cylindrical and the other spherical with equal internal diameter and wall thickness are subjected to equal internal fluid pressure. The ratio of hoop stresses in the cylindrical vessel to that in the spherical one is
　(a) 4　　　　　　　(b) 2
　(c) 1　　　　　　　(d) 0.5.　　(IES-96)

552. A thin cylindrical shell of internal diameter D and thickness t is subjected to internal pressure p. The change in diameter is given by

　(a) $\dfrac{pD^2}{4tE}(2-\mu)$　　(b) $\dfrac{pD^2}{4tE}(1-2\mu)$

　(c) $\dfrac{pD^2}{2tE}(1-2\mu)$　　(d) $\dfrac{pD^2}{2tE}(2-\mu)$.
　　　　　　　　　　　(IES-97)

553. The indeterminacy of a truss with respect to the forces in the members is called _____indeterminacy.

Answers to Practice Problems

PART-I: STATICS

Chapter-1

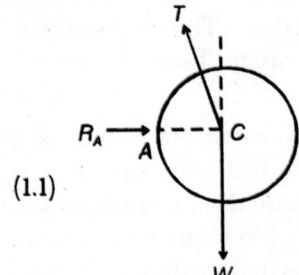

(1.1)

Free Body Diagram for the Ball.

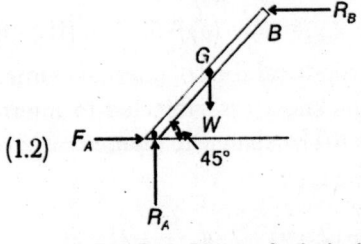

(1.2)

Free Body Diagram for the Ladder

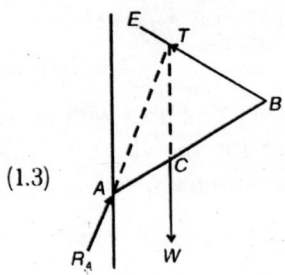

(1.3)

Free Body Diagram for the Bar

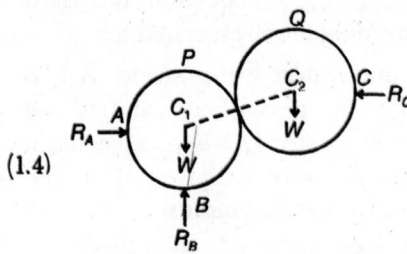

(1.4)

Free Body Diagram of Spheres P and Q

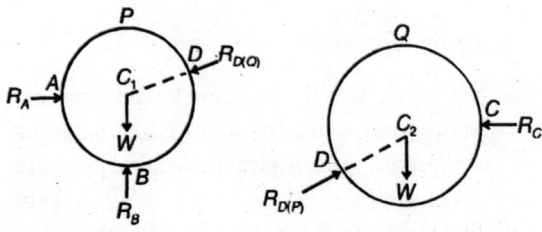

Free Body Diagram of   Free Body Diagram of Sphere Q
Sphere P

Chapter-2

(2.1) 250 N

(2.3) P; 90°

(2.4) 24.18 N

(2.5) 139.9 N; 45°21′40″ with the x-axis

(2.6) 222.67 N; 118.48 N

(2.7) 75 N & 129.9 N

(2.8) $Ox = 61.28$ units;
$Oy = 61.28$ units;
$Oz = 50$ units; 52°.20
with x-axis; 52°.20 with y-axis;
60° with z-axis.

(2.9) 374.99 N & 114°24′

(2.10) 107.12 N & 141°44′

(2.11) 39.6 kN & 169°12′

(2.12) 25.16 N & 349°

(2.13) Resultant = 100 N acting to the right;
Equilibrant = 100 N but acting to the left.

(2.14) Force in the Jib = 5 kN;
Force in the Tie Rod = 4 kN.

(2.15) Tension in chain No. 1 = 450 N;
Tension in chain No. 2 = 779.42 N

(2.16) Force in the string AC = 14.64 N;
Force in the string BC = 10.35 N

(2.17) $R_A = W \operatorname{cosec} \theta$ & $R_B = W \cot \theta$

(2.18) T = 127.70 N; R = 43.68 N

(2.19) R = 24.36 kN (Reaction from the floor)
S = 1.749 kN (Tensile force in the bar)

(2.20) 11.25 kN

(2.21) The tensions in the ropes AB and CD are
866 N and 500 N respectively.

Chapter-3

(3.1) 13.856 N.m

(3.2) R_A = 0.96 kN; R_B = 0.96 kN; R_C = 0.68 kN

(3.3) 40 N

(3.4) P = 320 N; Q = 480 N

(3.5) R = 1882.49N;
θ = 59°52′; x = 3.19 m
where x = the distance between the point P
and the line of action of the resultant force.

(3.6) 28.993 kN; 193°42′, 42.43 mm
(Perpendicular distance from A to the line
of action of the resultant).

(3.7) 42.44 N; 124°44; 6.13 m
(Perpendicular distance from 0 to the line
of action of the resultant).

(3.8) (i) R = 91.98 kN; θ = 301°40′
(ii) 13.63 kN; 80.71 kN

(3.9) (a) R_y = 55N (downwards);
x = 0.9 m from A
(b) 55 N; M_A = 49.5 N-m (clockwise)
(c) 55 N; M_B = 16.5 N-m (Anticlockwise)

(3.10) R = 300 N; x = 0.63 m from A

(3.11) Resultant will be 175 N upwards and is
acting at a distance of 4.07 m to the right of
point A.

(3.12) The system is in equilibrium.

(3.13) R_A = 57.66 kN; θ = 72°31′;
M_A = 160 kN.m (Anticlockwise)

(3.14) R_A = 21.67 kN; R_B = 28.33 kN

(3.15) 4.167 kN acting at a distance of 3.75 m from
A;

R_A = 1.04175 kN; R_B = 3.12525 kN

(3.16) R_A = 400 N (acting vertically down);
R_E = 1000 N (acting vertically up);
R_F = 4200 N (acting vertically up);
R_D = 1200 N (acting vertically down)

(3.17) R_A = 76.869 kN; H_B = 21.213 kN;
V_B = 64.344 kN; R_B = 67.75 kN,
θ = 71°45′

(3.18) R_A = 5 kN & R_B = 6 kN

(3.19) R_A = 70 kN; R_B = 20 kN

(3.20) P_{min} = 13.0 kN; α = 60°;
R = 7.5 kN; θ = 90°

(3.21) P_{min} = 9.478 kN; α = 90°

(3.22) b = 1.865 m

Chapter-4

(4.1) F_{AB} = 3300 N(C); F_{BC} = 3960 N(C);
F_{AC} = 3510 N(T)

(4.2) F_{AB} = 4.0415 kN (C);
F_{BC} = 2.309 kN (C);
F_{CD} = 5.196 kN (C);
F_{DE} = 2.598 kN (T);
F_{EA} = 2.0207 kN (T);
F_{EB} = 0.5775 kN (T);
F_{EC} = 0.5775 kN (C);

(4.3) F_{AB} = $8\sqrt{3}$ kN (C); F_{BC} = 14 kN (C);
F_{CD} = 20 kN (C);
F_{DE} = $10\sqrt{3}$ kN (T)
F_{EA} = $4\sqrt{3}$ kN (T);
F_{EB} = $6\sqrt{3}$ kN (T);
F_{EC} = $6\sqrt{3}$ kN(C)

(4.4) F_{AB} = 12 kN (C); F_{BC} = 8 kN (C);
F_{CD} = 8 kN (C); F_{DE} = 12 kN (C);
F_{EF} = 10.392 kN (T);
F_{FG} = 10.392 kN (T);
F_{GH} = 10.392 kN (T);
F_{HA} = 10.392 kN (T);
F_{GB} = 4 kN (C); F_{GC} = 4 kN (T);
F_{GD} = 4 kN (C); F_{BH} = 0; F_{DF} = 0;

(4.5) F_{AB} = 7.5 kN (C); F_{BC} = 5.0 kN (C);
F_{CD} = 4.0 kN (T); F_{DE} = 4.0 kN (T);
F_{BD} = 3.0 kN (C); F_{BE} = 2.5 kN (T)

(4.6) F_{AB} = $120/\sqrt{3}$ (C);
F_{BC} = $20/\sqrt{3}$ (C);
F_{CD} = $40/\sqrt{3}$ (T);
F_{DE} = $60/\sqrt{3}$ (T);

$F_{EF} = 200/\sqrt{3}$ (T);

$F_{AE} = 160/\sqrt{3}$ (C);

$F_{EB} = 120/\sqrt{3}$ (T);

$F_{BD} = 80/\sqrt{3}$ (C);

Chapter-5

(5.1) $\cdot$ (4, 2)

(5.2) $\dfrac{l+a}{3}$ from the left end; $\dfrac{l+b}{3}$ from the right end.

(5.3) $\dfrac{b^3 - a^3}{3(b^2 - a^2)}$; (5.4) $\bar{x} = \dfrac{3}{4}a$; $\bar{y} = \dfrac{3}{10}b$

(5.5) (30 mm; 40 mm)

(5.6) (33.421 mm; 43.421 mm)

(5.7) 52.5 mm

(5.8) 39.815 mm

(5.9) 43.33 mm

(5.10) (282.83 mm; 178.61 mm)

(5.11) (131.15 mm; 92.77 mm)

(5.12) (1.509 m; 2.885 m)

(5.13) (54.73 mm; 20.15 mm)

(5.14) (91.57 mm; –42 mm)

(5.15) (97.69 mm; 72.43 mm)

(5.16) (4.355 m; 2.85 m)

(5.17) 36.923 mm

(5.18) $\left(\dfrac{5a}{12} ; \dfrac{5a}{12} \right)$

(5.19) 50.67 mm

(5.20) (194.05 mm; 120.71 mm)

(5.21) (202.57 mm; 0)

(5.22) 86.3 mm

(5.23) 46.635 mm

(5.24) (55.808 mm; 101.856 mm)

(5.25) 98.60 mm

(5.26) (22 mm; 60 mm)

(5.27) $\dfrac{3}{8}r$ above the base

(5.28) $\dfrac{r}{2}$ above the base (5.29) 60 mm

(5.30) $\bar{y} = 20.357$ mm from the base

(5.31) $\bar{x} = 0.4560$m; $\bar{y} = 0.3648$ m;

$\bar{z} = 0.2952$ m

Chapter-6

(6.1) 3,41,333 mm^4

(6.2) 0.5333 m^4

(6.3) $I_{xx} = 72 \times 10^4$ mm^4;

$I_{yy} = 32 \times 10^4$ mm^4

(6.4) $I_{xx} = 15.1875 \times 10^4$ mm^4;

$I_{AB} = 45.5625 \times 10^4$ mm^4

(6.5) $I_{AA} = 3618 \times 10^4$ mm^4;

$I_{BB} = 4266 \times 10^4$ mm^4

(6.6) $I_{yy} = 424.66 \times 10^4$ mm^4

(6.7) 0.2036 m^4

(6.8) 2166.75 $\times 10^4$ mm^4

(6.9) 18,816 $\times 10^4$ mm^4

(6.10) 5184 $\times 10^4$ mm^4

(6.11) 474.992 $\times 10^4$ mm^4

(6.12) 512.29 $\times 10^4$ mm^4

(6.13) 81.68 $\times 10^4$ mm^4

(6.14) 300.78125 $\times 10^4$ mm^4

(6.15) 3889.66 $\times 10^4$ mm^4

(6.16) 2265.79 $\times 10^4$ mm^4

(6.17) $I_{xy} = \dfrac{b^2 h^2}{24}$; $I_{x'y'} = (-)\dfrac{b^2 h^2}{72}$

(6.18) $I_{xy} = \dfrac{bb_1}{8}(h^2 - h_1^2)$

(6.19) $I_{xy} = \dfrac{r^4}{8}$

(6.20) $\theta_m = 149°.53$ or $59°53$;

$I_{max} = 23.29$ m^4; $I_{min} = 2.7$ m^4

(6.21) $I_{xy} = 39.23 \times 10^4$ mm^4

(6.22) $\theta_m = 150°.13$ & $60°.13$

(6.23) (i) 90 x 10^4 mm^4 ; (ii) -6×10^4 mm^4

(6.24) $\theta_m = 154°.33$ & $64°.33$;

$I_{max} = 227.867 \times 10^4$ mm^4;

$I_{min} = 22.967 \times 10^4$ mm^4

(6.25) $I_{LL} = 28$ m^4 (6.26) 97.8853 kg.m^2

Chapter-7

(7.1) $R = 72.5$ N; $\theta = 35°$

(7.2) $R = 4.5$ N; $\theta = 342°$

(7.3) $R = 50$ N $\downarrow$ & $\bar{x} = 8.6$ m from P

(7.4) $R = 75$ N; $\theta = 59°.5$; $\bar{x} = 4.30$ m

(7.5) $F_2 = 3.7$ N; $F_3 = 2.6$ N

(7.6) $F_5 = 21$ N & $\theta = 56°.5$

(7.7) 96 mm from bottom

(7.8) $\bar{x} = 50$ mm & $\bar{y} = 19.5$ mm from bottom

(7.9) $\bar{x} = 69$ mm; $\bar{y} = 36$ mm

(7.10) 710×10^4 mm^4

(7.11) 790×10^4 mm^4

(7.12)   40 kN; 30 kN

(7.13)   47.5 kN; 42.5 kN

(7.14)   55 kN; 40 kN

(7.15) $F_{AB} = 4.0$ kN (C); $F_{BC} = 2.3$ kN(C);
$F_{CD} = 5.2$ kN (C); $F_{DE} = 2.6$ kN(T);
$F_{EA} = 2.0$ kN (T); $F_{EB} \approx 0.6$ kN(T);
$F_{EC} \approx 0.6$ kN (T);

(7.16) $F_{AB} = 13.9$ kN (C); $F_{BC} = 14$ kN (C);
$F_{CD} = 20$ kN (C);
$F_{DE} = 17.3$ kN (T); $F_{EA} = 6.9$ kN (T);
$F_{EB} = 10.4$ kN (T);
$F_{EC} = 10.4$ kN (C);

(7.17) $F_{AB} = 12$ kN (C); $F_{BC} = 8$ kN (C);
$F_{CD} = 8$ kN (C); $F_{DE} = 12$ kN (C);
$F_{EF} = 10.4$ kN (T);
$F_{FG} = 10.4$ kN (T);
$F_{GH} = 10.4$ kN (T);
$F_{HA} = 10.4$ kN (T);
$F_{GB} = 4$ kN (C); $F_{GC} = 4$ kN (T);
$F_{GD} = 4$ kN (C);
$F_{BH} = 0$; $F_{DF} = 0$.

(7.18) $F_{AB} = 11.5$ kN (C); $F_{BC} = 23$ kN (T);
$F_{CD} = 23$ kN (T); $F_{CA} = 23$ kN (T);

(7.19) $F_{AB} = 69$ kN (C);

$F_{BC} = 11.5$ kN (C);
$F_{CD} = 23$ kN (T);
$F_{DE} = 34.5$ kN (T);
$F_{EF} = 115$ kN (T);
$F_{AE} = 92$ kN (C);
$F_{EB} = 69$ kN (T); $F_{DB} = 46$ kN (C);

Chpater-8

(8.1)   202.43 N

(8.2)   845.15 N; 0.2383

(8.3)   0.4142

(8.4)   0.1957

(8.5)   376.62 N

(8.6)   29.18 N

(8.7)   (*a*) 19.82 N; (*b*) 0.286; (*c*) 5.67 N

(8.8) $P = 348.66$ N

(8.9) $P = \dfrac{W}{2} \tan \theta$

(8.10)   51.13 N

(8.11)   3.33 m

(8.12)   0.259

(8.13)   62.5 N

(8.14)   21°46″

(8.15) $M = \dfrac{W \cdot r \cdot \mu (1 + \mu)}{(1 + \mu^2)}$

(8.16)   4.167 N

(8.17)   55.84 %; No (not a self-locking machine)

(8.18)   10.405 N

(8.19)   2 turns

(8.20)   648.23 N

(8.21)   56.25 N.m; 589.05 watts

(8.22)   289.78 N.m (Assuming uniform intensity of pressure);
217.33 N.m (Assuming uniform rate of power)

Chapter-9

(9.1)   400 N

(9.2)   75 %

(9.3) $\eta = 75\%$ 750% ∴ The machine can work in the reverse direction

(9.4) $\eta = 41.67\% < 50\%$ ∴ Self locking machine;
Frictional resistance = 7000 N

(9.5)   4; 80%; 20 N; 5 N; 125 N; 25 N

(9.6)   60 N; 16.67; 66.67 %

(9.7)   13.5 N

(9.8)   90 %; 10 N; 80 N

(9.9)   500 N

(9.10)   75 %; 50 N; 500 N

(9.11)   50 N

(9.12)   250 mm

(9.13)   70 %

(9.14)   119 N

(9.15)   60%; 8 N; 80 N

(9.16)   24; 15; 62.5 %

(9.17)   39.55 N

(9.18)   36.2 N; 44%

(9.19)   29.71 %

(9.20)   25 %

(9.21) 50 %; 80 %; 1200 N; 20 N

(9.22) 2700 N

Chapter-10

(10.1) $R_A = 24$ kN; $R_B = 36$ kN

(10.2) $R_A = 23.75$ kN; $R_B = 26.25$ kN

(10.3) $R_A = 20$ kN; $R_B = 35$ kN

(10.4) 85 kN

(10.5) $R_A = 21$ kN; $R_B = 24$ kN

(10.6) $R_A = 82$ kN; $R_B = 117$ kN

(10.7) $P = \dfrac{W}{2}\tan\theta$

(10.8) Tension in the rope = 225 N

(10.9) $P = 2000$ N (10.10) $P = 3.5$ kN

(10.11) $P = 990$ N

(10.12) $\theta = \tan^{-1}(Q\sqrt{3}\,/\,P)$

(10.13) $P = 500$ N

(10.14) Force in the member $CD = 8$ kN

(10.15) Tension in the string = 3 W

(10.16) $H_A = 400$ N; $V_A = 1600$ N;
$H_B = 400$ N; $V_B = 800$ N

PART-II: DYNAMICS

Chapter-11

(11.1) 88 m; 92 m/sec; 66 m/se²; 63 m/sec

(11.4) 15.65 m

(11.5) 120.8328 sec; 833.33 m

(11.6) $-25 \sin 4x$

(11.7) 60° to the horizontal or 30° to the vertical

(11.8) 19°06′ North of East; 21.166 kmph

(11.9) 20 kmph at 36°52′ west of south

(11.10) 2 sec; 40.38 m; 30 m/sec

(11.11) 5.855 m/sec

(11.12) 29.36 m; 4.893 sec; 1.957 sec; 1.92 sec and 2.973 sec; 5.866 sec.

(11.13) 341.82 m/sec

(11.14) 1 m/sec; −4 m/sec²; 56 m/sec; 26 m/sec²; −0.33 m/sec (minimum velocity); 1 and 0.33 sec

(11.15) $-1.929\ s$ where s is distance

(11.16) 13.33 m/sec; 397.5 sec

Chapter-12

(12.1) 17.487 sec; 3.50 km

(12.2) 4.521 m/sec

(12.3) 53°07′48″

(12.4) 107.94 m

(12.5) 8.93 m/sec

(12.6) 30°16′30″

(12.7) 102.47 m

(12.8) 286.70 m; 2.176 km

(12.9) 9°52′ & 53°34′21″; 45.57 m/sec; 49°34′ to the horizontal and 45.57 m/sec; 67° to the horizontal

(12.10) 442.34 m; 1620.98 m

(12.12) 121.53 m; 4.978 sec

Chapter-13

(13.1) 8.485 m/sec; −45° with x-axis, 26.83 m/sec²; 63°26′ with x-axis.

(13.2) $v_x = -2.7264$ m/sec;
$v_y = 2.9268$ m/sec;
$a_x = -2.8266$ m/sec²;
$a_y = 0.1002$ m/sec²

(13.3) $\pm s\sqrt{g\,/\,l}$; $\pm S(g/l)$; $S^2\,(g/l^2)$

(13.4) 2.25 m/sec² & 0; 2.25 m/sec² & −0.75 m/sec² (−ve sign shows there is deceleration)

(13.5) 0.075 m/sec² & 0.0253 m/sec²

(13.6) 11.18 sec; 37.50 m

(13.7) 9.335 m/sec & 57°.52; 35.14 m/sec² & −63°.39

(13.8) 10 m/sec & 0°; 125.66 m/sec² & 90°.

(13.9) 0.3619 m/sec & 46°.32; 2.823 m/sec² & 10°.69; −2.50 m/sec² (−ve sign indicates that it is directed towards 0)

Chapter-14

(14.1) 83 rad; 81 rad/sec; 54 rad/sec²

(14.2) 31.416 rad/sec; 31.416 m/sec

(14.3) 15 rad.; 18 rad/sec; 20 rad/sec²

(14.4) 12 m/sec; 4 m/sec²,
(2/3) sec; 13.33 m/sec

(14.5) 4 rad/sec²

(14.6) 100 rad

(14.7)   5.2359 rad/sec; 6 min 45 sec

(14.8)   1031.324 rpm; 6 min

(14.9) 0.2793×10^{-3} rad/sec^2; 13.965×10^{-2} rad/sec; 0.5586×10^{-3} rad/sec^2

(14.10) 1.75 m/sec

Chapter-15

(15.1)   2.4 kN

(15.2)   61.162 N

(15.3)   37.5 kg

(15.4)   539.55 N

(15.5)   0.278 m/sec^2

(15.6)   80 m/sec & –20 m/sec (–ve sign indicates that the vehicle is moving in the reverse direction).

(15.7)   152.905 sec

(15.8) $g/5$ & 20 N

(15.9)   3.27 m/sec^2; 6.54 m/sec^2; 0

(15.10) 503.87 N

(15.11) 50 N; 1.962 m/sec^2

(15.12) 1.962 m/sec^2; 16N

(15.13) 1.401 m/sec^2; 64.29 N

(15.14) 1.8075 m/sec^2; 15.7875 N

(15.15) 2.69 m/sec^2

(15.16) 661.162 N; 538.838 N; 2 m/sec^2

(15.17) 5764.526 N; 508.26 N

(15.18) 722.3242 N; 4333.945 N

Chapter-16

(16.1)   100 kJ

(16.2)   14.4 kJ (–ve)

(16.3)   19.5 kJ

(16.4)   2752.2 J

(16.5)   6371 N

(16.6)   150 m/sec; 45 kN

(16.7)   19.11 m

(16.8)   349 N

(16.9)   375 N

(16.10) 7.17 m

(16.11) 437.19 mm

(16.12) 333.33 kW

(16.13) 100 kW; 200 kW; 50 kW

(16.14) 1529 N; 16.49 kW

(16.15) 113.1 N.m

Chapter-17

(17.1)   293.27 N; 23°30' to horizontal

(17.2) $0.8756 \dfrac{kg-m}{sec}$; 17.512 N

(17.3)   12.82 m/sec; 12.783 m/sec

(17.4)   1.308 m/sec^2; 85.02 N

(17.5)   35.9 sec; 1207.11 N

(17.6)   500N; 3.679 m/sec

(17.7)   738.05N; 72°30' with the x-axis in fourth quadrant

(17.8)   81.67 kN; 0.0479

(17.9)   7.5 m/sec in the direction opposite to that of shell; 0.643 m; 0.171 sec

Chapter-18

(18.1)   4 m/sec

(18.2)   –3 m/sec; –0.5 m/sec

(18.3)   4 m/sec; 8 m/sec; 0.5 m/sec; 6.5 m/sec

(18.4)   0.8409

(18.5) $\dfrac{1}{4}$; 180 J

Chapter-19

(19.1)   540 rpm

(19.2)   535.61 rpm

(19.3)   1440 rpm; 1310.4 rpm

(19.4)   27.477 m; 27.557 m

(19.5)   18 kW

(19.6)   26 kW considering T_c; 32.51 kW neglecting T_c

(19.7)   145.1 mm say 150 mm

(19.8)   23.63 kW; 25.82 m/sec; 25.42 kW

Chapter-20

(20.1)   2.533 mm

(20.2)   5.269 m/sec; 44.41 m/sec

(20.3)   178.77 rad/sec; 0.1025 m; 2396.9 m/sec^2

(20.4)   9.19 Hz

(20.5)   3.06 kg; 22.38 kN/m

(20.6) 5.43 N-sec/m

(20.7) +0.11574 mm

(20.8) 0.0353 mm

(20.9) 71.88 seconds

(20.10) 1000 m

(20.11) 0.825 Hz

(20.12) 8.842 mm

(20.13) 0.3411 sec

(20.14) 1.4185 sec

PART-III: ELEMENTS OF SOLID MACHANICS AND STRUCTURAL ANALYSIS

Chapter-21

(21.1) 0.15 mm

(21.2) 2.04×10^5 MPa

(21.3) 76.4 MPa; 3.82×10^{-4}; 0.573 mm

(21.4) 0.20 mm decrease

(21.5) 42 mm; 410 mm

(21.7) 114.6 MPa; 509.3 MPa; 183.3 MPa; 0.188 mm

(21.8) 70 kN; 30 kN; 750 MPa; 150 MPa

(21.9) 37 mm, 188.5 mm

(21.10) 0.016 mm

(21.12) 0.611 mm

(21.13) $\dfrac{rl^2}{6E}$ or $\dfrac{Wl}{2AE}$

(21.14) 83.7 mm

(21.15) 14 mm

(21.16) 0.3; 2.04×10^5 MPa; 0.785×10^5 MPa; 1.7×10^5 MPa

(21.17) 0.25; 0.67×10^5 MPa; 0.01875 mm

(21.18) Tensile stress in the bolt = 222 MPa; and compressive
stress in the tube = 133 MPa

(21.19) 0.003 mm; 67.86 mm^3

(21.20) 8.13 MPa; 126.65 MPa

(21.21) 234 kN

(21.22) 1.999 m, 45°.5

(21.23) 93.3 kN; 50.9 kN

(21.24) $\sigma_s = 195.93$ MPa (tensile),
$\sigma_c = 110.36$ MPa (tensile)

Chapter-22

(22.1) 75 MPa (tensile); 25 MPa;
79 MPa (tensile) $\phi = 18°26'$

(22.2) 90.2 MPa (tensile); $\phi = 46°07'$;
$\tau_{max} = 75$ MPa

(22.3) 81 MPa (tensile); 21 MPa (compressive);
$\theta_1 = 39°28'$,
$\tau_{max} = 51$ MPa; $p_n = 30$ MPa (tensile)

(22.4) 35 MPa (tensile); 8.66 MPa; 36 MPa (tensile);
10 MPa; 19°30'; 54°45'

(22.5) 33.5 MPa; 90 MPa (tensile)

(22.6) 121 MPa (tensile); 71 MPa (compressive);
$\theta_1 = 19°20'$;
96 MPa; 110.6 MPa (tensile)

(22.7) 79.5 MPa

(22.8) 1.98

(22.9) 30°; 109.3 MPa (tensile); 29.3 MPa (tensile)

(22.10) 90 MPa (tensile); 17.32 MPa; 4.25×10^{-4}

(22.11) 5.25×10^{-4} (tensile); 4.75×10^{-4} (compressive);
1.5×10^{-4} (tensile); 2×10^{-4} (+ve)

(22.12) 150.056175 mm; 149.987925 mm

(22.13) 80.24 mm; 79.8134 mm

(22.14) 144 MPa (tensile); 64 MPa (compressive);
104 MPa

Chapter-23

(23.1) 15,470 N.mm; 5.25×10^{-3} N.mm/mm^3

(23.2) 20,000 N.mm

(23.3) 38,197 N.mm

(23.4) 66.7 MPa; 1 mm; 30,000 N.mm

(23.5) 25,000 N.mm

(23.6) 224.2 MPa; 2.8025 mm

(23.7) 200 MPa; 1935.5 N

(23.8) 100.62 MPa

(23.9) 163.3 MPa; 2/3

(23.10) 5906.2 N.mm

(23.11) 2.125

(23.12) 50,000 N.mm

Chapter-24

(24.1) $F_{max} = 60$ kN; $M_{max} = 210$ kN.m at fixed end

(24.2) $F_{max} = 50$ kN; $M_{max} = 180$ kN.m at fixed end

(24.3) $F_{max} = 50$ kN; $M_{max} = 67.7$ kN.m at 3.15 m from left end

(24.4) $F_{max} = 10$ kN; $M_{max} = 15$ kN.m

(24.5) $F_{max} = 6$ kN; $M_{max} = 3$ kN.m

(24.6) $F_{max} = -4$ kN; $M_{max} = \pm 8$ kN.m at mid-span

(24.7) $W = 11.43$ kN; $F_{max} = 45.72$ kN; $M_{max} = 53$ kN.m at 2.3 m from the hinge

(24.8) $M_{max} = -240$ kN.m at the roller end

(24.9) $F_{max} = \pm 20$ kN; $M_{max} = -15$ kN.m and $+5$ kN.m

(24.10) $F_{max} = +110$ kN & -100 kN; $M_{max} = 240$ kN.m (–) and 10 kN.m (+)

(24.11)

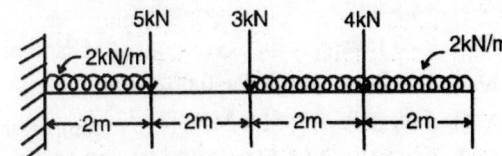

(24.12) $F_{max} = 5$ kN; $M_{max} = 15$ kN.m
$P_{max} = 15$ kN

Chapter-25

(25.1) 100 MPa

(25.2) 0.4 MPa

(25.3) 2.80 m

(25.4) 16 kN/m

(25.5) 180 kN

(25.6) 28.32 MPa (Compressive) and 70.44 MPa (tensile)

(25.7) 200 mm × 300 mm

(25.8) $W_s/W_h = 1.28$

(25.9) $\pi : 2.50$ or $1.257 : 1$

(25.10) $4\sqrt{3}\pi : 15$ or $1.4434 : 1$

(25.11) $6.38 : 1.67 : 1$

(25.12) 16.66 kN/m

Chapter-26

(26.1) 1.2 MPa

(26.2) 8.6 MPa; 3.3

(26.3) 2.07 MPa; 2.25

(26.4) 1.5

(26.5) 6.25 MPa

(26.6) 6.35 MPa

(26.7) 10 MPa

(26.8) 10.1 MPa

(26.9) 7.92 MPa

(26.10) Shape of shear stress distribution is as for diamond section; $f_{s_{max}} = 7.75$ MPa

(26.11) 7.42 MPa; 2.82

(26.12) 315 MPa (compressive) and 0.30 MPa (tensile)

Chapter-27

(27.1) 0°07′; 5.4 mm

(27.2) $\dfrac{Wl^2}{8EI} ; \dfrac{5Wl^3}{48EI}$

(27.3) 19.4 mm; 19.2 mm; 17.8 mm

(27.7) 3.42 mm; 4.44 mm; 4.76 mm at 4.033 from the left end

(27.8) $\dfrac{2Wl^2}{9EI} ; \dfrac{13Wl^3}{108EI}$

(27.9) $\dfrac{280}{EI} ; \dfrac{200}{EI} ; \dfrac{480}{EI}$

(27.10) $a/l = 0.453$

(27.12) 5,23.5 mm

Chapter-28

(28.1) 82.5 kW

(28.2) 2.1 m

(28.3) 1964 kW

(28.4) $T_{solid} : T_{hollow} = 10 : 17$ or $0.588 : 1$

(28.5) 100 mm

(28.6) 115 mm; 130 mm and 104 mm; 53%

(28.7) $T_{hollow}/T_{solid} = (k^2+1)/(k\sqrt{k^2-1})$; 2.733; 1.93

(28.8) $W_h/W_s = [(k^2(k^2-1))/(k^2+1)^2]^{1/3}$; 0.65; 0.79

(28.9) 162 mm

(28.10) 150 mm

(28.11) 2380 N.m; 1940 N.m

(28.12) 15 MPa (compressive)

(28.13) 270 mm

(28.14) 354 kN

(28.15) 673.2 N-m

(28.16) 76.4 MPa; 120 mm

(28.17) 409N; 141 mm

(28.18) 1.9 m

(28.19) 13 mm; 79010 N.mm/rad.

(28.20) 26

(28.21) 580 MPa; 2.1 rad.

(28.22) 0.316

(28.23) 2.4 mm; 36 N

(28.24) $W_o = 450$ N; $W_i = 300$ N; $\delta_o = 90$ mm; $\delta_i = 60$ mm

(28.25) 60 mm; 127 MPa; 110 MPa

Chapter-29

(29.1) $\dfrac{80P}{\pi d^2}$

(29.2) 731.4 kN

(29.3) 7.87 MPa (Compressive); 6.42 MPa (tensile); 50 mm

(29.4) 3.22 MPa (compressive); 2.11 MPa (tensile)

(29.5) 581.8 kN

(29.7) 3.46 m

(29.8) 531.8 kN.m; 1317 kPa (compressive); 117 kPa (tensile)

(29.9) 3.464 m; 94.1 MPa (29.10) 745 kN

(29.11) 620 kN

(29.12) 33 mm

(29.13) 788 kN; 990 kN; 11 m

(29.14) $b = \dfrac{1}{1665}$; 4130 kN

(29.15) 1.11 m

(29.16) 387.6 kN; 1.485 mm; 5.24 mm

(29.17) 245 kN; 146 kN by Modified Rankine's formula

(29.18) 9.105 mm; 31.53 kN.m; 99.34 MPa

(29.19) 100 MPa

(29.20) $M_{\max} = \dfrac{Wl}{4} + \left(\dfrac{Wl^3}{48\,EI} \right) P$

(29.21) $\delta = \dfrac{W}{P} \left(\dfrac{\tan kl}{k} - l \right)$

where $k^2 = \dfrac{P}{EI}$

Chapter-30

(30.1) 150 MPa; 75 MPa; 37.5 MPa

(30.2) 42.2 mm

(30.3) 6.25 mm

(30.4) 1.04 MPa; 41.6 MPa; 0.0625 mm

(30.5) 5.08 MPa (T); 91.2 MPa (T)

(30.6) 160 MPa; 84.5 MPa; 105 MPa; 18 MPa

(30.7) 11.5 MPa

(30.8) 110 mm

(30.9) 72; 62.61; 45; 55.38; 51.96 MPa; Safe intensity of fluid pressure is 45 MPa

(30.10) 32.5; 12.5; 16.4 MPa

(30.11) 9.11 MPa (T); 12.2%

(30.12) 7.14 MPa

(30.13) 26 mm

(30.14) 0.1343 mm

(30.15) 27.9 MPa; 83°.4 C

(30.16) 71.2 (T); 348 (T); 84.2 (T); 60.6 MPa (T)

(30.17) 13 (C); 28 (C); 297 (T); 232 MPa (T)

(30.18) 71.8 MPa; 0.093 mm; 57°C; 164.5 MPa; 2.7 %

Answers to Objective Questions

1. (c)
2. False
3. (d)
4. (b)
5. Double
6. False
7. (c)
8. (d)
9. False
10. False
11. (b)
12. (d)
13. Tensile, compressive; and shear
14. Equal and opposite
15. $100\sqrt{2}$ N
16. (c)
17. (a)
18. (a)
19. A-2, B-1, C-3, D-4
20. (d)
21. (a)
22. (d)
23. (c)
24. (b)
25. (c)
26. (b)
27. True
28. Free body concept
29. True
30. Method of sections
31. Rigid bodies; deformable bodies, and fluids
32. Statics, Dynamics
33. Dynamics
34. Kinematics
35. Kinetics
36. Photoelastic effect
37. Vector
38. Mass of a body
39. Weight
40. Transmissibility
41. Static equilibrium
42. Equilibrant
43. Composition
44. Lami's theorem
45. True
46. True
47. False
48. Resolved parts
49. True
50. True
51. True
52. Rotational
53. Lever arm or moment arm
54. The line of action of the force
55. False
56. True
57. True
58. True
59. (i) $\Sigma F_x = 0$,
    (ii) $\Sigma F_y = 0$; and
    $\Sigma M = 0$
60. True
61. Triangular
62. More
63. Deficient
64. True
65. True
66. True
67. External
68. External and internal
69. (i) Method of joints
    (ii) Method of sections
70. False
71. Sections
72. True
73. True
74. True
75. Integration
76. True
77. (d)
78. (b)
79. (c)
80. True
81. The moments of inertia about a set of rectangular axes in a plane normal to the area.
82. (d)
83. False
84. (a)

85. True
86. False
87. (c)
88. True

89. (d)
90. $\frac{\pi}{32}(D^4 - d^4)$

91. Zero
92. $\frac{1}{4}b^2d^2$

93. True
94. Maximum; Minimum

95. Zero
96. Centroid
97. True
98. True
99. False
100. False

101. $\frac{1}{8}md'^2$
102. True

103. False
104. True
105. Maxwell diagram
106. False
107. Substitution
108. (b)
109. True
110. (i) Solid friction;
     (ii) Fluid or viscous friction
     (iii) internal friction
111. Limiting friction
112. Coulomb
113. True
114. Coulomb
115. Coefficient of friction
116. Friction angle
117. (a)
118. (d)
119. Screw jack
120. (i) Sliding friction;
     (ii) Rolling friction
121. Output; input
122. (c)
123. (a)
124. (a)
125. (c)
126. Reversibility
127. Self-locking
128. (b)
129. (d)
130. (b)
131. (c)
132. (a)
133. (d)
134. $2D/(d_1 - d_2)$
135. $2D/(D - d)$
136. (a)
137. True
138. $2\pi R/\phi$
139. $2\pi R/(p_1 - p_2)$
140. $T \cdot D/d$
141. (a)
142. False
143. True
144. True
145. Joule
146. Perpendicular

147. True
148. (b)
149. Active
150. True
151. Degree of freedom
152. True
153. True
154. (i) Translation;
     (ii) Rotation
155. Rectilinear motion
156. Translation; rotation
157. Displacement
158. True
159. Acceleration
160. Instantaneous
161. Retardation
162. True
163. Velocity
164. Acceleration
165. Uniform
166. True
167. Relative
168. True
169. Gravity
170. False
171. Varying
172. Curved
173. Horizontal
174. (i) a bar thrown into the air.
     (ii) a bullet fired from a gun.
175. False
176. Trajectory
177. True
178. Parabola
179. True
180. 45°
181. False
182. False
183. Curvilinear translation
184. True
185. False
186. True
187. Angular motion
188. Rotation about the fixed axis
189. General plane motion
190. (i) Rotating wheel
     (ii) Ladder against a wall
191. False
192. Radians/second^2
193. $v = r \cdot \omega$
194. $\omega = 2\pi n$
195. False
196. Newton-seconds
197. The rate of change of momentum of a body is directly proportional to the force acting on it

198. (*d*)

199. $6.673 \times 10^{-11} \, \text{Nm}^2/\text{kg}^2$ 200. Force (*F*)

201. False

202. That due to gravity

203. False   204. Dynamic

205. False   206. (*b*)

207. False   208. True

209. Joule   210. Perpendicular

211. True   212. Conservative

213. (*i*) Gravity force   214. Frictional force
     (*ii*) Spring force

215. Capacity   216. True

217. (*i*) Potential energy
     (*ii*) Kinetic energy

218. Potential energy   219. (*d*)

220. Law of Conservation   221. True
     of energy

222. (*i*) Stable;   223. Neutral
     (*ii*) Unstable
     (*iii*) Neutral

224. Spring constant   225. Power

226. Watt   227. (*c*)

228. True   229. (*a*)

230. True

231. (*i*) Force of jet on a vane
     (*ii*) Recoil of a gun

232. Impact   233. Direct impact

234. Central impact

235. Deformation; restitution

236. True   237. False

238. True   239. True

240. The coefficient of restitution

241. (*i*) Flat belt;
     (*ii*) V-belt
     (*iii*) Circular belt or rope

242. False

243. (*i*) Open-belt drive;
     (*ii*) Cross-belt drive

244. False   245. True

246. True   247. False

248. (*c*)   249. True

250. True   251. True

252. (*a*)   253. False

254. True   255. Amplitude

256. Period   257. Frequency

258. False

259. Natural frequency   260. Sum

261. (*b*)   262. Critical

263. Damping factor

264. (*i*) viscous damping;
     (*ii*) friction damping;
     (*iii*) structural damping

265. Frequency ratio   266. Unity

267. Compound pendulum

268. The period   269. (*d*)

270. Trifilar

271. (*i*) Strength;
     (*ii*) Stiffness; and
     (*iii*) Stability

272. Tangential   273. Radians

274. Permanent set   275. True

276. Three   277. True

278. Complementary shear

279. Tensile; Compressive

280. True   281. True

282. Two   283. (*c*)

284. (*b*)   285. (*d*)

286. Ductility   287. Brittle

288. False   289. Ductility

290. Factor of safety   291. False

292. Modular ratio   293. (*a*)

294. Major; Minor   295. True

296. Maximum; half   297. False

298. $(\sigma_1 + \sigma_2)$; zero   299. $(1/2)\,(\sigma_1 - \sigma_2)$

300. False

301. Internal strain energy

302. True

303. Work done; internal
     strain energy stored.

304. $\dfrac{1}{2} \cdot \dfrac{\sigma^2}{E}$

305. Proof resilience

306. Modulus of resilience

307. True

308. True	309. Twice	376. (*d*)	377. (*a*)
310. Bending	311. True	378. (*a*)	379. (*d*)
312. Intensity of loading	313. Shear force	380. (*c*)	381. (*b*)
314. False	315. False	382. (*b*)	383. (*c*)
316. Zero; changes sign	317. (*c*)	384. (*c*)	385. (*c*)
318. True	319. Pure bending	386. (*b*)	387. (*a*)
320.	1/2	388. (*d*)	389. (*b*)
321. True		390. (*b*)	391. (*a*)
322. Statically determinate		392. (*b*)	393. False
323. The degree of static indeterminacy		394. (*d*)	395. (*d*)
324. True	325. Economy	396. (*c*)	397. (*b*)
326. Tensile; compressive		398. (*a*)	399. (*c*)
327. Neutral; neutral axis	328. Pure; simple	400. (*c*)	401. (*b*)
329. Bernoulli's	330. True	402. (*b*)	403. (*d*)
331. True	332. False	404. (*c*)	405. (*b*)
333. True	334. Symmetrical	406. (*c*)	407. (*d*)
335. Centroid	336. True	408. (*b*)	409. (*d*)
337. Modulus of section	338. False	410. (*b*)	411. (*b*)
339. Uniform stress	340. False	412. (*c*)	413. (*d*)
341. True	342. True	414. (*d*)	415. (*b*)
343. Parabolic	344. False	416. (*b*)	417. (*a*)
345. False	346. Shear force	418. (*c*)	419. (*a*)
347. "*B*"		420. (*b*)	421. (*b*)
		422. (*c*)	423. (*c*)
348. $1\frac{1}{8}$	349. False	424. (*d*)	425. (*a*)
350. False	351. Mid-depth	426. (*d*)	427. (*d*)
		428. (*c*)	429. (*d*)
352. $1\frac{1}{2}$	353. Web	430. (*c*)	431. (*a*)
354. True		432. (*c*)	433. (*d*)
355. Twisting or torsion	356. True	434. (*a*)	435. (*b*)
357. False	358. True	436. (*a*)	437. (*a*)
359. True	360. False	438. (*b*)	439. (*d*)
361. True		440. (*c*)	441. (*c*)
362. Torsional rigidity	363. True	442. (*a*)	443. (*d*)
364. False	365. True	444. (*c*)	445. (*c*)
366. (*a*)	367. (*a*)	446. (*d*)	447. (*b*)
368. (*d*)	369. (*c*)	448. (*c*)	449. (*a*)
370. (*a*)	371. (*a*)	450. (*d*)	451. Spring
372. (*c*)	373. (*b*)	452. Helical spring	453. Spring
374. (*d*)	375. (*d*)	454. Close-coiled	455. True

456. True
457. False
458. (*a*)
459. Stiffness
460. Bending moment; that of the torque
461. True
462. True
463. False
464. False
465. False
466. True
467. (*a*)
468. (*b*)
469. Proof load
470. True
471. True
472. Radius
473. True
474. False
475. Slope
476. Bending moment
477. Fourth
478. False
479. (*c*)
480. True
481. (*d*)
482. True
483. (*a*)
484. True
485. (*b*)
486. (*a*)
487. True
488. (*a*)
489. (*b*)
490. (*d*)
491. (*c*)
492. (*d*)
493. (*d*)
494. (d)
495. True
496. Direct compressive; bending
497. True
498. (*b*)
499. "Core"
500. (*b*)
501. 2
502. True

503. Critical or crippling load
504. Slenderness ratio
505. (*a*)
506. True
507. 100
508. True
509. False
510. True
511. True
512. True
513. (*a*)
514. (*a*)
515. (*c*)
516. (*d*)
517. (*b*)
518. (*b*)
519. (*d*)
520. (*a*)
521. (*c*)
522. (*b*)
523. (*d*)
524. (*b*)
525. (*c*)
526. The square of
527. (*a*)
528. (*d*)
529. (*b*)
530. True
531. True
532. (*b*)
533. True
534. (*c*)
535. True
536. True
537. True
538. Lobes
539. False
540. True
541. Lame's
542. (*c*)
543. True
544. True
545. True
546. False
547. True
548. True
549. False
550. (*d*)
551. (*b*)
552. (*c*)
553. Internal

References

1. Aggarwal, S. K. and Gupta, P. K.: *Strength of Materials*, First Edition, Metropolitan Book Company Limited, New Delhi, 1994.

2. Andrews, E. S.: *Further Problems in the Theory and Design of Structures*, Second Edition, Chapman and Hall, London, 1949.

3. Arthur Morley: *Strength of Materials*, Eleventh Edition, Longmans, Green and Company, London, 1960.

4. Bansal, R. K.: *Engineering Mechanics*, Third Edition Fourth Reprint, Laxmi Publications, New Delhi, 2000.

5. Bari, S. A.: *Elements of Structural Analysis*, First Edition, S. Chand and Company Limited, New Delhi, 1997.

6. Beer, F. P., and Johnston, E. R.: *Vector Mechanics for Engineers* (Statics), Third SI Metric Edition, Tata-McGraw Hill Book Company Limited, 2000.

7. Beer, F. P. and Johnston, E. R.: *Vector Mechanics for Engineers* (Dynamics), Third SI Metric Edition, Tata McGraw-Hill Book Company Limited, 2000.

8. Beer, F. P. and Johnston E. R.: *Mechanics of Materials*, Second Printing, McGraw-Hill Book Company, Inc., 1985.

9. Bhavikatti, S. S.: *Strength of Materials*, Reprint, Vikas Publishing House Private Limited, New Delhi, 1999.

10. Bhavikatti, S. S. and Rajashekarappa, K. G.: *Engineering Mechanics*, Second Reprint, New Age International Private Limited, New Delhi, 1996.

11. Clark, D. A. R.: *Materials and Structures*, First Edition-Seventh Reprint, Blackie and Son Limited, London, 1959.

12. Clark, D. A. R.: *Advanced Strength of Materials*, First Edition, Blackie and Son Limited, London, 1951.

13. Gere, J. M., and Timoshenko, S. P.: *Mechanics of Materials*, First Indian Edition, C.B.S. Publishers and Distributors, Delhi, 1986.

14. Ghose, D. N.: *Applied Mechanics and Strength of Materials*, Third Edition, C.B.S. Publishers and Distributors, Delhi, 1985.

15. Hibbeler, R. C.: *Engineering Mechanics* (Statics), First Indian Reprint, Addison Wesley Longman (Singapore) Private Limited, 1997.

16. Hibbeler, R. C.: *Engineering Mechanics* (Dynamics), First Indian Reprint, Addison Wesley Longman (Singapore) Private Limited, 1997.

17. Higdon, A. and Stiles, W. B.: *Engineering Mechanics*, Prentice Hall, New York, 1950.

18. Higdon, A., Ohlsen, E. H., Stiles, W. B., Weese, J. A. and Riley, W. F.: *Mechanics of Materials*, Third Edition, John Wiley and Sons, Inc., New York, 1986.

19. Junnarkar, S. B. and Shah, H. J.: *Applied Mechanics*, Fifteenth Revised Edition, Charotar Publishing House, Anand, 2000.

20. Junnarkar, S. B. and Shah, H. J.: *Mechanics of Structures*, Volume I, Twenty-Second Revised Edition, Charotar Publishing House, Anand, 1997.

21. Junnarkar, S. B. and Shah, H. J.: *Mechanics of Structures*, Volume II, Sixteenth Edition, Charotar Publishing House, Anand, 1999.

22. Kumar, K. L.: *Engineering Mechanics*, Third Edition -First Reprint, Tata McGraw-Hill Publishing Company Limited, New Delhi, 1998.

23. Khurmi, R. S.: *Strength of Materials*, Eighteenth Edition-Reprint, S. Chand and Company Limited, New Delhi, 1987.

24. Khurmi, R. S.: *Engineering Mechanics*, Nineteenth Edition - Reprint, S Chand and Company Limited, New Delhi, 1999.

25. Low, B. B.: *Strength of Materials*, E.L.B.S. Edition First Published, and Longmans, Green and Company Limited, London, 1965.

26. McLean and Nelson: *Engineering Mechanics* (Statics and Dynamics), Third (SI Metric) Edition, McGraw-Hill Book Company, Inc., 1980.

27. Mokashi, V. S.: *Engineering Mechanics-I* (Statics), First Edition, Tata McGraw-Hill Publishing Company Limited, New Delhi, 1995.

28. Mokashi, V. S.: *Engineering Mechanics-II* (Dynamics), First Edition, Tata McGraw-Hill Publishing Company Limited, New Delhi, 1995.

29. Nash, W. A.: *Strength of Materials*, Asian Student Edition-First Impression, McGraw-Hill International Book Company, Singapore, 1983.

30. Paradise, R. S. and Church, G. A.: *Strength of Materials*, First Edition, Blackie and Son Limited, London, 1959.

31. Pytel, A. and Singer, F. L.: *Strength of Materials*, Third Edition, Harper and Row Publishers, New York, 1980.

32. Popov, E. P.: *Engineering Mechanics of Solids*, Second Edition, Pearson Education, Asia, 2001.

33. Prasad, I. B.: *Applied Mechanics*, Sixteenth Edition, Khanna Publishers, New Delhi, 1994.

34. Prasad, I. B.: *Strength of Materials*, Eleventh Edition, Khanna Publishers, New Delhi, 1998.

35. Punmia, B. C., Ashok Kumar Jain, and Arun Kumar Jain: *Mechanics of Materials*, First Edition, Laxmi Publica-tions Private Limited, New Delhi, 2002.

36. Rajput, R. K.: *Strength of Materials*, First Edition, S. Chand and Company Limited, New Delhi, 1996.

37. Riley, W. F., Sturges, L. D. and Morris, D. N.: *Mechanics of Materials*, Fifth Edition, John Wiley, Newyork, 2001.

38. Ramamrutham, S.: *Strength of Materials*, Seventh Edition Reprint, Dhanpat Rai and Sons, Delhi, 1984.

39. Ramamrutham, S.: *Applied Mechanics*, Fifth Revised and Enlarged Edition, Dhanpat Rai and Sons, Delhi, 1994.

40. Sadhu Singh: *Strength of Materials*, Fifth Edition, Khanna Publishers, Delhi, 1992.

41. Seely, F. B., and Smith, J. O.: *Advanced Mechanics of Materials*, Second Edition, John Wiley and Sons, New York, 1952.

42. Shah, V. L.: *Strength of Materials*, First Edition, Structures Publishers, Pune, 1998.

43. Shames, I. H.: *Engineering Mechanics* (Statics and Dynamics), Second Edition, Prentice-Hall of India Private Limited, New Delhi, 1986.

44. Surendra Singh: *Strength of Materials*, Third Revised Edition, Konark Publishers Private Limited, Delhi, 1991.

45. Tayal, A. K.: *Engineering Mechanics* (Statics and Dynamics), Eleventh Edition, Umesh Publications, Delhi, 2001.

46. Timoshenko, S. P. and MacCullough, G. H.: *Elements of Strength of Materials*, Third Edition, D. Van Nostrand Company, Inc., Princeton, 1960.

47. Timoshenko, S. P. and Young, D. H.: *Elements of Strength of Materials*, Fifth Edition, Affiliated East-West Press Private Limited, New Delhi, 1968.

48. Timoshenko, S. P. and Young, D. H.: *Engineering Mechanics*, Twenty-Ninth Printing, McGraw-Hill Book Company, Inc., 1983.

49. Urry, S. A.: *Solution of Problems in Strength of Materials*, Second Edition, Sir Isaac Pitman and Sons Limited, London, 1959.

50. Urry, S. A. and Turney, P. J.: *Solution of Problems in Strength of Materials and Mechanics of Solids*, Fourth Edition, Sir Isaac Pitman and Sons Limited, London, 1981.

51. Vazirani, V. N. and Ratwani, M. M.: *Analysis of Structures-Volume I*, Sixteenth Edition-Third Reprint, Khanna Publishers, Delhi, 1999.

52. Vazirani, V. N. and Ratwani, M. M.: *Analysis of Structures-Volume II*, Fifteenth Edition- Second Reprint, Khanna Publishers, Delhi, 1998.

53. Warnock, F. V.: *Strength of Materials*, Eighth Edition, Sir Isaac Pitman and Sons Limited, London, 1957.

54. Warnock, F. V. and Benham, P. P.: *Mechanics of Solids and Strength of Materials*, Sir Isaac Pitman and Sons Limited, London, 1965.

Index